Lecture Notes in Electrical Engineering 1135

The book series *Lecture Notes in Electrical Engineering* (LNEE) publishes the latest developments in Electrical Engineering—quickly, informally and in high quality. While original research reported in proceedings and monographs has traditionally formed the core of LNEE, we also encourage authors to submit books devoted to supporting student education and professional training in the various fields and applications areas of electrical engineering. The series cover classical and emerging topics concerning:

- Communication Engineering, Information Theory and Networks
- Electronics Engineering and Microelectronics
- Signal, Image and Speech Processing
- Wireless and Mobile Communication
- Circuits and Systems
- Energy Systems, Power Electronics and Electrical Machines
- Electro-optical Engineering
- Instrumentation Engineering
- Avionics Engineering
- Control Systems
- Internet-of-Things and Cybersecurity
- Biomedical Devices, MEMS and NEMS

For general information about this book series, comments or suggestions, please contact leontina.dicecco@springer.com.

To submit a proposal or request further information, please contact the Publishing Editor in your country:

China

Jasmine Dou, Editor (jasmine.dou@springer.com)

India, Japan, Rest of Asia

Swati Meherishi, Editorial Director (Swati.Meherishi@springer.com)

Southeast Asia, Australia, New Zealand

Ramesh Nath Premnath, Editor (ramesh.premnath@springernature.com)

USA, Canada

Michael Luby, Senior Editor (michael.luby@springer.com)

All other Countries

Leontina Di Cecco, Senior Editor (leontina.dicecco@springer.com)

**** This series is indexed by EI Compendex and Scopus databases. ****

Limin Jia · Yong Qin · Jianwei Yang ·
Zhigang Liu · Lijun Diao · Min An
Editors

Proceedings of the 6th International Conference on Electrical Engineering and Information Technologies for Rail Transportation (EITRT) 2023

Energy Traction Technology of Rail Transportation

Springer

Editors
Limin Jia
Beijing Jiaotong University
Beijing, China

Jianwei Yang
Beijing University of Civil Engineering and Architecture
Beijing, China

Lijun Diao
Beijing Jiaotong University
Beijing, China

Yong Qin
Beijing Jiaotong University
Beijing, China

Zhigang Liu
Beijing Jiaotong University
Beijing, China

Min An
University of Salford
Salford, UK

ISSN 1876-1100 ISSN 1876-1119 (electronic)
Lecture Notes in Electrical Engineering
ISBN 978-981-99-9306-2 ISBN 978-981-99-9307-9 (eBook)
https://doi.org/10.1007/978-981-99-9307-9

This Springer imprint is published by the registered company Springer Nature Singapore Pte Ltd.
The registered company address is: 152 Beach Road, #21-01/04 Gateway East, Singapore 189721, Singapore

Contents

Wide-Band Harmonic Interaction and Characteristic Analysis of FCTPSS 1
Yan Zhang, Yonggang Wei, and Gaotao Cao

An Energy Management Strategy for Traction Substation with ESS Based on Reinforcement Learning 11
Jiaming Luo, Xiaoguang Wei, Shibin Gao, Xiaoran Guo, and Tianbao Zhong

A Global Optimized Energy Management Strategy for Reducing Fuel Consumption in Diesel-Electric Hybrid Power Supply System 20
Shuo Zhang, Wanrui Ren, Cunguo Liu, and Xuhui Dai

A Dynamic Tuning Method of Wireless Power Transfer System Based on Phase-Controlled Inductor Circuit 28
Shan He, Xiaona Li, Yuhao Peng, Weige Zhang, Guohong Zeng, and Xiaogong Du

Energy-Efficient Train Control Incorporating Realistic Traction Characteristics Using Convex Optimization 36
Yang Peng, Fuwang Chen, and Shaofeng Lu

Design and Modeling of Subway Train Traction System Based on MBSE 46
Jiahao Zhao, Baomin Wang, Lujie Zhou, Tingli Huang, and Qingyong Wang

Research on Electromagnetic Characteristics of Linear Synchronous Motor of Superconducting High-Speed Maglev Train 56
Yuhao Zheng, Jingyu Huang, and Xiaonong Wang

NSGA-II Genetic Algorithm Based Optimization of Parameters of Multiphase Interleaving Buck/Boost DC-DC Converter 68
Haotian Yang, Qing Lv, and Keling Song

Research on Optimal Power Flow Control Strategy for Urban Rail Flexible Traction Power Supply System 80
Junjie Zhang, Shuai Jiang, Zhigang Liu, and Jie Chen

Research on Harmonic Noise Characteristics of Train Four-Quadrant Converter Using SiC Devices 89
Yongxiang Wang, Kan Liu, Fanyang Meng, and Jianqiang Liu

Research on the Common Mode EMI Suppression Method for SiC Traction Converters 99
Kan Liu, Donghui Liu, Yongxiang Wang, Fanyang Meng, and Jianqiang Liu

Inductor Current Ripple Modeling and Mitigation Method of Cascaded Half-Bridge DC/DC Converter Under Unbalanced Conditions 108
Tianmu Chen, Guohong Zeng, Fen Tang, and Luoqi Wu

Weak Magnetic Control of Automotive Permanent Magnet Synchronous Motor Based on Sliding Mode Control 117
Huchuan Yang, Jing Chen, Meng Zhang, Zhuo Wang, and Wei Wu

Fuzzy Sliding Mode Control of DC/DC Converter for Hydrogen Vehicles 125
Zhuo Wang, Jing Chen, Wei Wu, Huchuan Yang, and Meng Zhang

Research on Observer-Based Vector Control Strategy for Battery Vehicle Drive System 134
Congding Yang, Jing Chen, Chengbo Xu, and Junjie Zhao

Simulation Study of Bidirectional DC/DC Converter for New Energy Vehicles Based on RBF Neural Variable Structure 141
Junjie Zhao, Jing Chen, Congding Yang, Jiawei Gao, and Mu Yang

Rotor Position Estimation Method for PMSM Based on Improved Sliding Mode Observer 149
Jing Wang, Jiale Chen, Guohong Zeng, and Jingdou Liu

Application Research of Photovoltaic Power Generation Technology in Railway System 157
Yajing Han, Jie Jin, Feng Lin, Dawei Li, and Jianfan Wu

Improvement on Command Current Extraction of Railway Static Power Conditioner 170
Yunpeng Zhang, Ying Wang, Xiaoqiang Chen, and Chenyu Song

Research on High Frequency Current Measurement Method for the SiC MOSFET 178
Shuai Liang, Yuming Liu, Chunhui Miao, and Ke Sun

Bandwidth Compensation Method for Rogowski Coil for the SIC MOSFET ... 188
Shuai Liang, Huiqing Du, Yuming Liu, and Xiran Sun

An Interleaved Control Strategy for Input-Series Output-Parallel System with Three-Level Boost Converter+Full-Bridge LLC Converter 198
Ruiqi Ma, Xinbo Liu, Shuiyuan He, Jiepin Zhang, Yingtao Ma, Zheming Jin, and Lijun Diao

Design and Simulation Analysis of Hybrid Pulse Width Modulation for Urban Rail Traction System 207
Jianyu Yang, Xiaoxi Lu, and Yilin Wang

A Review of Hydrogen Fuel Cell Applications in Rail Transit 217
Suyao Liu, Yifei Zhang, Chunmei Xu, and Lijun Diao

Power Integration and Energy Management Strategy of CJ5 Hybrid EMU 225
Xuefei Li, Wei Han, Suyao Liu, Lijun Diao, and Chengtao Wang

Design and Application of Complex Vector Current Controller for IM Based on Ripple-Free Minimum Beat System 234
Xuelei Yue, Lei Wang, and Di Tian

Design and Analysis of Energy Storage Converters for Regenerative Braking Energy Absorption in Urban Rail Transit 247
Yong Jin, Weijie Pu, Xianjin Huang, Zhongping Yang, and Seiki Igarshi

Network-Side Power Optimization Control Considering Photovoltaic and Energy Storage Connection to RPC-Type Traction Power Supply System 255
Xiuqing Mu, Xiaoqiang Chen, Ying Wang, Qiaoli Yang, and Yunpeng Zhang

Development of Double Suspension Frame Electromagnetic Permanent Magnet Hybrid Maglev Controller 264
Xiaolei Shi, He Xi, Yuanzhi Li, Yanchao Wang, and Deyu Jiao

Current Control Calculation and Simulation Analysis of Urban Railway Traction System 276
Yilin Wang

Research on Wireless Power Transfer Systems Based on Semiactive Rectifier 287
Kangli Luo, Weiyao Mei, Jiangjun Yuan, Tongzhao Li, and Lijun Diao

Resolver-To-Digital Conversion and Sensor Fault Identification of PMSM 297
Hao Zhu, Chunmei Xu, Qiya Wu, Zichen Gao, and Jia Zhang

Modeling and Control of Long-Stator Linear Synchronous Motors for High-Speed Maglev 305
Shanhui Zhang, Chunmei Xu, Lei Wang, Hao Zhu, Zichen Gao, Qiya Wu, and Lijun Diao

Design and Research of Variable Frequency Direct Feed Power Supply for Urban Rail Transit 313
Yifei Miao and Xueyan Dong

Research on Power Calculation Method for Auxiliary Converters of Multiple Unit 322
Bo Zhang, Qingwen Sun, Junbo Zhao, Xuan Su, Xinyu He, Haining Hu, Qiang Li, and Zheng Xie

Demagnetization Characteristics Diagnosis of Permanent Magnet Synchronous Motors 331
Zichen Gao, Haiting Wang, Qiya Wu, Yaoheng Li, and Lijun Diao

Full Speed Range Field Oriented Control Method for PMSM 339
Donghui Liu, Leiting Zhao, Liu Kan, Liran Li, Chi Ma, Zheng Yue, Zheming Jin, and Lijun Diao

Multi-device Collaborative Anomaly Diagnosis Method for Electric Mining Dump Trucks 347
Yichao Ma, Lei Wang, Benchao Zhu, and Qubo Xie

A New DC Auxiliary Power Supply System Architecture Designed for Electric Multiple Units 356
Chao Wang, Yingtao Ma, Zheming Jin, and Lijun Diao

Topology Design of High-Frequency DC Auxiliary Converter for Electric Multiple Units 364
Kan Dong, Yingtao Ma, Long Cheng, Chao Wang, Zheming Jin, and Lijun Diao

Residual Energy Estimation of Battery Packs for Energy Storage Based on Working Condition Prediction and the Representative Cell 371
Yong Chen, Tao Wen, Lintao Hou, Chunhua Hou, and Xizheng He

Research on the Design Scheme of Routine Conductive High-Speed Maglev Integrated Grounding System 379
Guosheng Li, Di Su, and Ruichao Chen

Design and Research of Propulsion Substation for High-Speed Maglev Railway 387
Di Su and Guosheng Li

Design of IGBT Drive and Protection Based on M57962AL 396
Jiaxi Gou, Zhankai Fang, Guo Chen, Qiang Lai, and Shuang Liu

Speed Balance Control of the Dual-Motor System Based on Three-Vector Model Prediction 404
Yuqi Liu, Yuanfeng Huang, Bin Xiong, and Haifeng Wang

Initial Rotor Position Detection Based on Peak Dc-Link Current Comparison for SPMSM 413
Peng Chen and Ruiqing Ma

Research on Electromagnetic Force Characteristics Based on Asymmetric 120° Pitch Figure-8 Coils in EDS System for Maglev Train 422
Zhigang Xue, Rong Wei, Yukang Gao, and Qiongxuan Ge

Performance Evaluation on Levitation Control System of High-Speed Maglev Train 431
Mingda Zhai, Pinglang Yi, Xiaolong Li, and Zhiqiang Long

Design of a Three-Level APF Controller Based on Dual DSP 439
Fang Zhankai, Gou Jiaxi, Liu Shuang, Huang Yunbiao, Chen Guo, and Yang Ziqing

Study on Dynamic Characteristics and Electromagnetic Thrust of Unilateral Linear Induction Motor for Maglev Train 448
Shuncai Yao, Zhiping Cheng, Huixing Chen, and Bailin Li

Research on Magnetic Levitation Bearing Three-Level Switching Power Amplifier 456
Yong Yu and Lei Mei

Research on Damping System of High Speed Superconducting Maglev Train 463
Haoyun Wang and Ruikun Mai

A High and Low Voltage DC Distribution System for Special Vehicles 477
Fengpu Qiao, Yanming Li, Zhiyuan Wang, and Lu Xu

IGBT Open-Circuit Fault Diagnosis for MMC Based on OCDWPT 489
Qinyue Zhu, Yichen Yu, Yanwen Zhan, Jie Li, and Yanbiao Yang

Sensorless Control with Sliding-Mode Observer for Superconducting Linear Motors 498
Shaobo Yin, Xin Jiang, and Kekang Wei

Review on Energy Management Strategies of On-Board Hybrid Energy Storage Systems for Urban Rail Vehicles 506
Xinni Peng, Haiquan Liang, and Jingtai Hu

Energy-Efficient Train Control Based on Improved Dynamic Programming Algorithm for Online Applications 518
Fuwang Chen, Yang Peng, and Shaofeng Lu

Long Distance Two Terminal Power Supply Simulation Technology Based on RTLAB Hardware-in-The-Loop Simulation Platform 530
Xin Jiang, Shaobo Yin, Yongyong Guo, and Kekang Wei

Research on Maglev Pipeline Trains Speed and Positioning Based on Improved Kalman Filtering Algorithm 538
Jusong Jiang, Jie Yang, Chuanshu Meng, and Zhixin Li

Dynamic Trajectory Planning Method on ARM Platform for the Vacuum Pipelines High-Speed Maglev Train 548
Yongyong Guo, Zhihong Yang, Kekang Wei, and Zhiyong Zhao

The Top-Level Scheme Analysis of Medium Speed Maglev Operation Control System 556
Kuanxin Li, Xinghu Chen, Yuefeng Liu, and Yonggang Wang

Research on Key Technologies of Hydrogen Energy Powered Commuter Trains 564
Hang Yin, Caihui Zheng, Zhonghai Wang, and Xuekun Li

Optimal Sizing of Hybrid Power System Based on Typical Driving Cycle of the Shunting Locomotive 572
Mingming Kang, Mingjie Jiang, Chenguang Fan, Bo Yao, Xin Bi, and Xiaoxu Wang

Application Research of New Contactless Power Supply System for Medium Speed Maglev Train 587
Kuanxin Li, Yonggang Wang, Xinghu Chen, Zhenggang Yin, and Liming Shi

An Enhanced Feedforward Compensation Speed Closed Loop Control Strategy for the Vacuum Pipelines High-Speed Maglev Train 596
Zhihong Yang, Yongyong Guo, and Kekang Wei

Study on Variable Frequency Power Supply Cable and Leakage Current Characteristics of High-Speed Maglev Train 605
Qinglai Li, Pengcheng Du, Yongyong Guo, and Kekang Wei

Superconducting Electrodynamic Suspension Traction Numerical Simulation 613
Xu Ji, Jiansheng Hao, Ying Kang, and Kekang Wei

A Low Switching Frequency Model Predictive Control for Surface-Mounted Permanent Magnet Synchronous Motor Without Weight Factor 628
Jiangming Deng, Han Xiao, Wei Xu, and Yirong Tang

Optimization Control of Energy Management for Diesel-Electric Hybrid Trains Based on MPC-PMP 638
Zhe Chen, Xiaoxu Wang, Xiongwen Zhu, Chi Zhang, and Xinjie Zheng

A High Pulse Power and Efficiency Integrated Supercapacitor Energy Package 652
Chuanwei Guo, Tongzhao Li, Xinchen Cui, Wenli Lin, Weiyao Mei, and Lijun Diao

Research on Improving the Elman Neural Network for the INS/GNSS Integrated Velocity and Positioning Algorithm 660
Junteng Wu, Jie Yang, Chuanshu Meng, Jusong Jiang, and Yueyun Zhu

Calculation of Air Gap Electromagnetic Field of High-Speed Magnetic Levitation Linear Motor Based on Equivalent Current Layer 669
Jiahui Zhang, Guodong Wang, and Dan Zhang

Author Index 679

Wide-Band Harmonic Interaction and Characteristic Analysis of FCTPSS

Yan Zhang[1], Yonggang Wei[1], and Gaotao Cao[2(✉)]

[1] Shuohuang Railway Development Limited Liability Company, Beijing 100080, China
58501853@qq.com, 11094503@ceic.com

[2] School of Electrical Engineering, Beijing Jiaotong University, Beijing 100044, China
22110448@bjtu.edu.cn

Abstract. Connecting renewable energy system (RES) and hybrid energy storage system (HESS) to railway traction power supply system is the future development trend, which will cause new harmonic resonance problems.In order to analyze the harmonic resonance problems, a flexible cooperative traction power supply system (FCTPSS) connected with RES and HESS is studied in this paper. Based on the harmonic transmission theory, a mathematical models of harmonic transmission of traction power supply system (TPSS) and FCTPSS are established. The simulation models of the TPSS and the FCTPSS are built in MATLAB/Simulink, and the harmonic transmission characteristics of TPSS and FCTPSS are compared and analyzed. The results show that the location of the locomotive has an influence on the harmonic transmission characteristics, and the access of RES and HESS can make the resonant frequency to shift to the right.

Keywords: Traction power supply system · Renewable energy system · Energy storage system · Harmonic transmission characteristics

1 Introduction

At present, there have been many studies on connecting RES and energy storage system (ESS) to the grid [1]. RES can generate electricity green and clean, reduce carbon emissions, ESS can smooth the output volatility of renewable energy, improve power quality. Similarly, some studies have conducted on the integration of RES and ESS into TPSS. East Japan Railway Company has connected photovoltaic and energy storage into the traction power supply system [2]. Connecting RES and HESS into the railway is not only conducive to the green development of the railway, but also can promote the consumption of renewable energy.

The integration of RES and HESS into TPSS is a new research direction, and most of the researchers are just starting at present. Reference [3] proposed a photovoltaic (PV) DC microgrid topology with HESS, which connected to TPSS through railway power conditioner (RPC). And this paper proposed the control strategy of the system, and simulated the operation scheduling and energy flow of the system. Wu put forward a back-to-back PV power generation system for electrified railway, designed its control

L. Jia et al. (Eds.): EITRT 2023, LNEE 1135, pp. 1–10, 2024.
https://doi.org/10.1007/978-981-99-9307-9_1

strategy, established a simulation model and verified the feasibility of the system in [4]. In [5], Deng proposed a hybrid traction power supply system integrated with photovoltaic sources (HTPSS-PVSs), which can realize on-site access and consumption of PV power along the line without affecting the original structure of TPSS. In [6], the scheme of integrating RES and HESS into the TPSS was put forward, which reduced the demand for power grid energy and carbon emissions. In [7], Ying proposed a flexible intelligent traction power supply system that consisted of AC-DC-AC traction substation and the microgrid along the railway, which can effectively alleviate the problem of negative sequence current and improve the utilization rate of regenerative braking energy and RES.

Although the above references have done a lot of research on the topology and energy control strategy of the TPSS system accessed by res and HESS, the research on the harmonic transmission characteristics of the new system is insufficient.Currently, AC-DC-AC electric locomotive used in electrified railway has the characteristics of high power factor and low harmonic content. Its harmonic spectrum is relatively wide, coupled with the spatiotemporal nature of locomotive movement mode, the introduced harmonics may appear harmonic resonance and amplification in the propagation process of traction network with complex structure [8]. In addition, the access of RES and HESS will change the topology of TPSS, and the harmonic interaction characteristics may appear new changes. Therefore, this paper has carried out research on this problem.

This paper is organized as follows: Sect. 2 introduces the topology of FCTPSS and the mathematical models of harmonic transmission of TPSS and FCTPSS. Section 3 introduces the simulation models of TPSS and FCTPSS established in MATLAB/Simulink. In Sect. 4, the harmonic transmission characteristics are simulated and verified and the result are given, followed by the conclusion in Sect. 5.

2 Topological Structure and Theoretical Analysis of Harmonic Characteristics of FCTPSS

2.1 Topology of FCTPSS

Co-phase power supply provides an idea to solve the power quality problem and the power phase separation problem of high-speed railway system [9], and also provides a good integration method for the access of RES and HESS [7, 10]. Based on the scheme of co-phase power supply, this paper constructs a FCTPSS including wind power generation, photovoltaic power generation and HESS. The scheme proposed in this paper is to summarize the wind power generation, photovoltaic power generation and HESS by connecting the converter to the DC bus, and then connect to TPSS through the DC/AC inverter connected to the step-down transformer, so as to supply power together with the power grid, as shown in Fig. 1.

2.2 Theoretical Analysis of Harmonic Transmission in FCTPSS

Both RES and locomotive generate harmonic during operation. When the harmonic frequency matches the system parameters, harmonic resonance phenomenon will occur

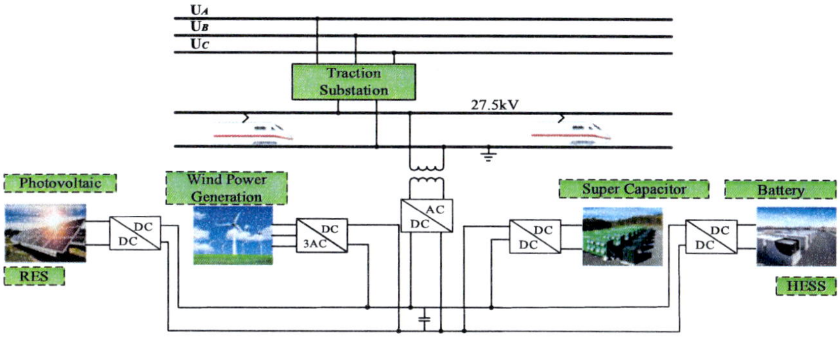

Fig. 1. Topology of FCTPSS.

[11, 13]. In this section, the mathematical models of harmonic characteristics of TPSS and FCTPSS are established respectively to analyze their harmonic transmission characteristics.

The circuit network of traction power supply system is equivalent to a Π equivalent circuit model, as shown in Fig. 2. $\boldsymbol{Z_{ss}}$ is the system impedance of traction substation, including power reactance and traction transformer reactance; $\boldsymbol{I_1}$ is locomotive current flowing to traction substation; $\boldsymbol{I_2}$ is the locomotive current flowing to section post; $\boldsymbol{I_{load}}$ is locomotive current; $\boldsymbol{l_1}$ is the distance from locomotive to substation, km; $\boldsymbol{l_2}$ is the distance from the locomotive to the section post, km. $\boldsymbol{Z_L}$ and $\boldsymbol{Y_L}$ are the equivalent impedance between train and substation; $\boldsymbol{Z_R}$ and $\boldsymbol{Y_R}$ are the equivalent impedance between the train and the section post.

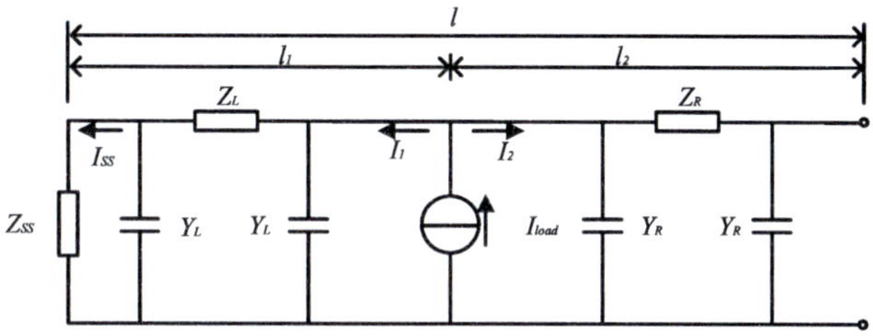

Fig. 2. Equivalent circuit model of Π type of TPSS..

The equivalent impedance of traction network is:

$$Z_L = Z_0 sh(\gamma l_1),\ Y_L = \frac{(ch(\gamma l_1)-1)}{Z_0 sh(\gamma l_1)},\ Z_R = Z_0 sh(\gamma l_2),\ Y_R = \frac{(ch(\gamma l_2)-1)}{Z_0 sh(\gamma l_2)} \quad (1)$$

where, $Z_0 = \sqrt{Z/Y}$ is the characteristic impedance of traction network line, Z is the equivalent impedance per unit length of traction network line, and Y is the equivalent admittance per unit length. $\gamma = \sqrt{ZY}$ is the propagation coefficient of traction network line.From this point of the train, the equivalent input impedance $\boldsymbol{Z_q}$ of the traction network is:

$$Z_q = \frac{Z_0 ch\gamma l_2 (Z_{ss} ch\gamma l_1 + Z_0 sh\gamma l_1)}{Z_{ss} sh\gamma l + Z_0 ch\gamma l} \quad (2)$$

The locomotive harmonic current will change through the transmission of traction network. Assume that the train current at x1 from the substation is $\boldsymbol{I_{x1}}$, as shown in Fig. 3.

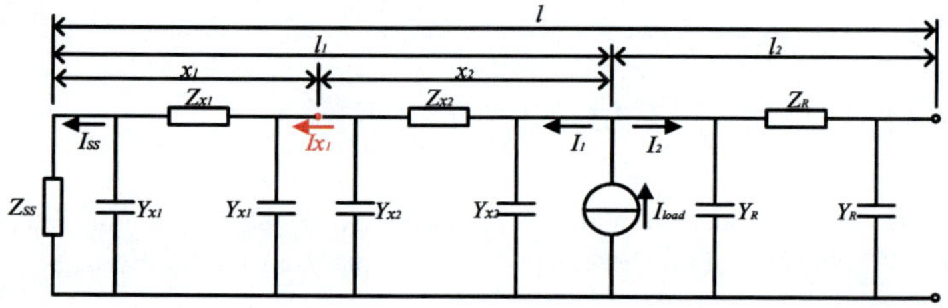

Fig. 3. Equivalent circuit model of Π type of TPSS at distance x1 from substation.

Then $\boldsymbol{I_1}$ and $\boldsymbol{I_{x1}}$ are respectively:

$$\begin{cases} I_1 = \dfrac{ch\gamma l_2 (Z_{ss} sh\gamma l_1 + Z_0 ch\gamma l_1)}{Z_{ss} sh\gamma l + Z_0 ch\gamma l} I_{load} \\ I_{x1} = \dfrac{(Z_{ss} sh\gamma x_1 + Z_0 ch\gamma x_1)}{Z_{ss} sh\gamma l_1 + Z_0 ch\gamma l_1} I_1 = \dfrac{ch\gamma l_2 (Z_{ss} sh\gamma x_1 + Z_0 ch\gamma x_1)}{Z_{ss} sh\gamma l + Z_0 ch\gamma l} I_{load} \end{cases} \quad (3)$$

$\boldsymbol{Q_{am}}$ is defined as the amplification factor of locomotive current amplified by traction network, and the $\boldsymbol{Q_{am}}$ of current $\boldsymbol{I_{x1}}$ at distance $\boldsymbol{x_1}$ from substation is:

$$Q_{am} = \frac{I_{x1}}{I_{load}} = \frac{ch\gamma l_2 (Z_{ss} sh\gamma x_1 + Z_0 ch\gamma x_1)}{Z_{ss} sh\gamma l + Z_0 ch\gamma l} \quad (4)$$

According to formula (4), when the denominator in the equation is 0 or close to 0, $\boldsymbol{Q_{am}}$ reaches the maximum, that is, harmonic resonance occurs. The resonant condition is: $Z_{SS} sh\gamma l + Z_0 ch\gamma l = 0$, because $\gamma l << 1$, $th\gamma l \approx \gamma l$, $Z_{SS} = -Z_0/(\gamma l)$. Z_{SS} is expressed by traction transformer inductance L_{SS} as $Z_{SS} = \omega L_{SS}$, and the resonant frequency $f_0 = 1/2\pi\sqrt{L_{SS} C}$.

When $x_1 = 0$, the current I_{ss} flowing into the traction substation can be obtained, and the $\boldsymbol{Q_{am}}$ of the current $\boldsymbol{I_{ss}}$ at the substation is:

$$Q_{am} = \frac{I_{ss}}{I_{load}} = \frac{Z_0 ch\gamma l_2}{Z_{ss} sh\gamma l + Z_0 ch\gamma l} \quad (5)$$

According to Formula (5), the $\boldsymbol{Q_{am}}$ of Iss is shown in Fig. 4a. It is known that the peak value of $\boldsymbol{Q_{am}}$ appears around the 25th and 81st harmonics.

(2) It is assumed that RES and HESS are connected to the traction power supply system at the traction substation. RES and HESS are equivalent to the equivalent impedance,

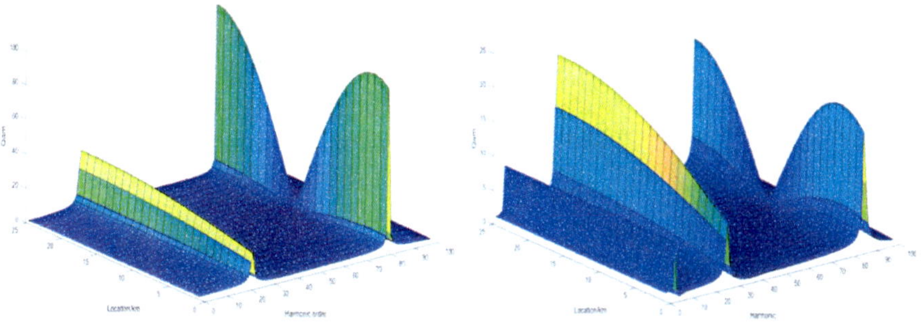

(a) (b)

Fig. 4. **a** Image of Q_{am} varying with locomotive location and harmonic frequency; **b** Image of Q_{am}' changing with locomotive location and harmonic frequency.

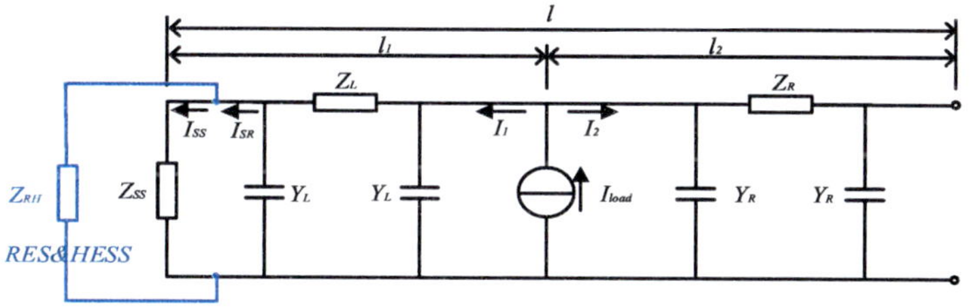

Fig. 5. Equivalent circuit model of Π type of FCTPSS.

and the equivalent circuit of FCTPSS can be simplified into the circuit structure shown in Fig. 5.

Make $Z_{SR} = Z_{SS}//Z_{RH}$, from this point of the train, the equivalent impedance of the traction network is Z_q'.

$$Z_q' = \frac{Z_0 ch\gamma l_2 (Z_{SR} ch\gamma l_1 + Z_0 sh\gamma l_1)}{Z_{SR} sh\gamma l + Z_0 ch\gamma l} \tag{6}$$

At this point, Formula (7) can be derived.

$$\begin{cases} I_1 = \dfrac{ch\gamma l_2 (Z_{SR} sh\gamma l_1 + Z_0 ch\gamma l_1)}{Z_{SR} sh\gamma l + Z_0 ch\gamma l} I_{load} \\ I_{SR} = \dfrac{Z_0}{Z_{SR} sh\gamma l_1 + Z_0 ch\gamma l_1} I_1 = \dfrac{Z_0 ch\gamma l_2}{Z_{SR} sh\gamma l + Z_0 ch\gamma l} I_{load} \\ I_{SS} = \dfrac{Z_{RES}}{Z_{RES} + Z_{SS}} I_{SR} = (\dfrac{Z_{RES}}{Z_{RES} + Z_{SS}})(\dfrac{Z_0 ch\gamma l_2}{Z_{SR} sh\gamma l + Z_0 ch\gamma l}) I_{load} \end{cases} \tag{7}$$

In this case, the harmonic Q_{am}' is,

$$Q_{am}' = \frac{I_{ss}}{I_{load}} = (\frac{Z_{RES}}{Z_{RES} + Z_{SS}})(\frac{Z_0 ch\gamma l_2}{Z_{SR} sh\gamma l + Z_0 ch\gamma l}) \tag{8}$$

According to Formula (8), the image of Q_{am}' changing with locomotive location and frequency is drawn, as shown in Fig. 4b. It can be seen from the image that when resonance occurs, the peak value of Q_{am}' decreases obviously, and the resonant frequency shifts to the right, and the resonance occurs near the 27th and 87th harmonics. In addition, a slight harmonic current amplification occurs near the 5th harmonic.

In order to verify the theoretical analysis in Sect. 2.2, a simulation model is established in Sect. 3.

3 Simulation Model of FCTPSS

3.1 Simulation Model of Traction Network

The traction network is a parallel multi-conductor system. In terms of topology, one power supply arm of the traction network forms a chain network. The chain network is composed of longitudinal series impedance elements and transverse parallel admittance elements, as shown in reference [13]. The parallel multi conductor system is cut into different segments by the admittance elements in transverse parallel. And the node voltage equation was proposed in [13], which can obtain the voltage and current at each node and the corresponding impedance at a specific frequency.

No matter what power supply mode is adopted for the traction network, and the line is single line or double line, the essence of the traction network is a chain network composed of multi conductor transmission lines. When modeling, the traction network is generally divided evenly according to the natural cutting of the transverse parallel elements. In order to simplify the modeling process, the wires are merged according to the parallel connection. The contact line and catenary of the same traction network are equivalent to one conductor (T), and the rails of the same line are combined into one (R). When calculating the self impedance, mutual impedance, self admittance and mutual admittance coefficient matrix, the influence of other merged wires is also taken into account, so the accuracy of the model will not be reduced by merging wires for modeling. According to the distributed parameter method of traction network proposed in reference [13], the simulation model of T-R+NF power supply mode for double line traction network is established in Matlab/Simulink software with the unit of 1km.

3.2 Simulation Model of RES and HESS

RES and HESS converters for FCTPSS are similar to converters commonly used in power systems. The core part is the single-phase DC/AC converter, which can convert the DC bus voltage to 27.5 kV AC voltage for the locomotive load. The converter used in this simulation system is a voltage-type bidirectional DC/AC converter, and its topology is shown in Fig. 6a.

DC/AC converter adopts double closed loop control strategy, and its control block diagram is shown in Fig. 7a. The outer voltage loop is used to stabilize the DC bus voltage, and the inner current loop is used to control the converter output grid connected current. The voltage outer loop compares the measured DC bus voltage with the reference voltage, and the error obtained is output as a DC flow I_{cp}* through the PI regulator. The

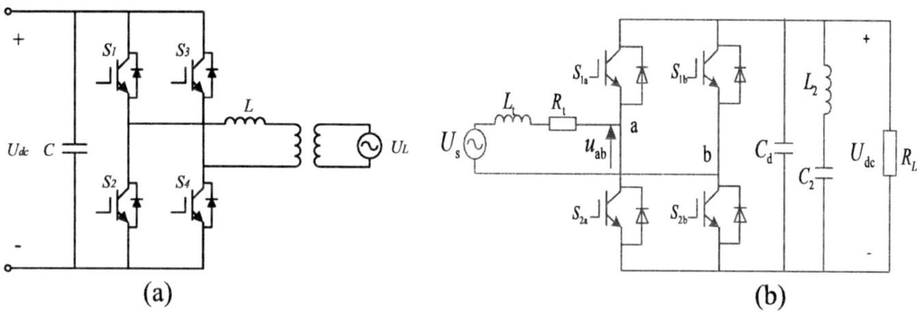

Fig. 6. **a** Topology of voltage-type bidirectional DC/AC converter; **b** simulation model of CRH3 EMU.

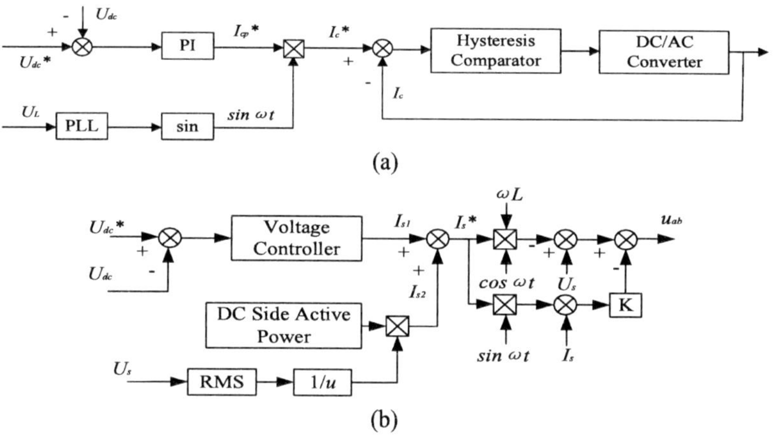

Fig. 7. **a** Double closed loop control strategy diagram of DC/AC converter; **b** Block diagram of TCCS for CRH3.

reference current $\boldsymbol{I_c}$* is obtained by multiplying $\boldsymbol{I_{cp}}$* by the AC network voltage $\boldsymbol{U_L}$ through the phase locked loop processing.

This paper only explores the influence of the introduction of RES and HESS on the harmonic current transmission characteristics of the locomotive. In order to avoid the influence of the harmonic generated by RES, the DC side of the inverter (RES and HESS) is replaced by a constant DC source.

3.3 Simulation Model of Locomotive Load

The locomotive is the main harmonic source of the traction power supply system. The simulation model of CRH3 EMU is established in this paper, as shown Fig. 6b. The electric drive system of CRH3 EMU mainly includes pantograph, on-board transformer, four quadrant converter, intermediate DC link, pulse width modulation (PWM) inverter, traction motor, etc. The intermediate DC link can effectively suppress the influence of the inverter side on the grid side system, that is, the high order harmonics on the locomotive grid side are mainly generated by the four quadrant converter modulation.

CRH3 EMU traction rectifier adopts transient current control strategy (TCCS), and the control block diagram is shown in Fig. 7b. And the principle of transient current control is shown in Formula (10). Where, U_{dc}* is the given value of DC side voltage, and U_{dc} is the actual value of DC side voltage. Is* is the command value of network side current; is(t) is the actual current value of the network side; Kp, K and Ki are proportional integral coefficients; Idc indicates the DC load current. U_s, Is and U_{dc} are input signals of the control system.According to the control principle, the locomotive model is established in Matlab/Simulink.

$$\begin{cases} I_{s1} = K_P(U_{dc}^* - U_{dc}) + 1/K_i \int (U_{dc}^* - U_{dc})dt \\ I_{s2} = I_{dc}U_{dc}/U_s \\ I_s^* = I_{s1} + I_{s2} \\ u_{ab}(t) = U_s(t) - \omega L I_s^* \cos(\omega t) - R_t I_s^* \sin(\omega t) - K[I_s^* \sin(\omega t) - i_s(t)] \end{cases} \tag{9}$$

4 Simulation Verification and Analysis

According to Part 2.2, when the length of traction network and other parameters are fixed, the harmonic transmission characteristics of traction power supply system are related to the access position of RES and HESS and the operating position of locomotive. Therefore, this part verifies the influence law of related factors of TPSS and FCTPSS harmonic transmission characteristics by simulation. The verification process and results are as follows.

(1) Simulation and verification of harmonic transmission characteristics of TPSS.

The change of locomotive location will cause the change of voltage and current distribution of traction network. A 25 km long traction network simulation model is established with 1 km traction network as the basic unit. By changing the location where the locomotive is connected to the traction network, the harmonic current amplification characteristics of the traction power supply system at different locations are obtained as shown in Fig. 8a.

It can be seen from Fig. 8a that when the traction network is 25 km long, $\boldsymbol{Q_{am}}$ will have a peak value near the 15th and 87th harmonics, that is, harmonic resonance occurs; When the locomotive runs at 5 km, the $\boldsymbol{Q_{am}}$ (near the 15th harmonic) is 25.78; When the locomotive runs at 25 km, the $\boldsymbol{Q_{am}}$ (near the 15th harmonic) is 28.42; The farther the locomotive is from the substation, the larger the $\boldsymbol{Q_{am}}$ becomes. As the locomotive is far away from the substation, the longer the transmission distance of current on the traction network is, the more likely the resonance will occur. The simulation results are consistent with the theoretical verification in 2.2.

(2) Simulation and verification of harmonic transmission characteristics of FCTPSS.

According to the theoretical analysis in Sect. 2.2, when RES and HESS are connected, the resonant frequency of the system will shift to the right, and the amplification of harmonic current will decrease. In this section, RES and HESS will be accessed at 5, 10, 15, 20 and 25 km respectively, and the locomotive will be placed at a distance of 10 km from the substation. Figure 8b shows the harmonic current amplification characteristics of the FCTPSS.

As can be seen from Fig. 8b, the harmonic resonance phenomenon has been significantly weakened, and the harmonic resonance frequency has shifted to the right, transferring to the vicinity of the 23rd harmonic frequency, avoiding the harmonic characteristic range of the locomotive. In FCTPSS, the maximum Q_{am} is only 7.88 times, which is about 3.6 times lower. In addition, it can be seen from the Fig. 8b that under the FCTPSS, a slight resonance will be generated near the 5th harmonic, and the maximum Q_{am} is 4.94 times, which is also consistent with the theoretical analysis in Sect. 2.2.

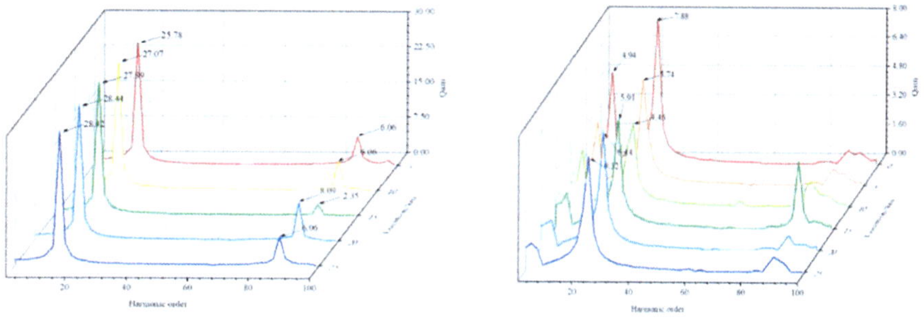

(a) (b)

Fig. 8. **a** Change of Q_{am} with locomotive location under TPSS; **b** Change of Q_{am} with RES location under TPSS.

5 Conclusions

In this paper, the harmonic transmission characteristics of TPSS and FCTPSS are analyzed theoretically, and the simulation models of TPSS and FCTPSS are established in Matlab/Simulink, and the accuracy of the models is verified. The change of harmonic current amplification with the locomotive access position under TPSS are simulated and analyzed. In addition, the variation of harmonic current amplification is simulated and analyzed when RES and HESS are connected at different locations in FCTPSS. The conclusions are as follows.

The frequency spectrum of locomotive harmonic current does not change with the locomotive power and running state, but under the traction condition, the harmonic content decreases with the increase of locomotive power; When the parameters and length of the traction network are fixed, the system resonance frequency is fixed, independent of the location of the locomotive. But the farther the locomotive is from the substation, the stronger the resonance phenomenon will be; In FCTPSS, the resonant frequency of the system will shift to the right, and the amplitude of harmonic current amplification will be reduced.

References

1. Roccotelli, M., Mangini, A.M., Fanti, M.P.: Smart district energy management with cooperative microgrids. IEEE Access **10**, 36311–36326 (2022)
2. Hayashiya, H.: Recent Trend of Regenerative Energy Utilization in Traction Power Supply System in Japan. In: Urban Rail Transit 3, 183–191 (2017)
3. Fei, S., Wu, M.: Photovoltaic DC microgrid with hybrid energy storage system connected to electrified railway traction power supply system. In: Proceedings of the 5th International Conference on Electrical Engineering and Information Technologies for Rail Transportation (EITRT) 2021. EITRT 2021. Lecture Notes in Electrical Engineering, vol 864. Springer, Singapore
4. Mingliang, W., Weiying, W., Wenli, D., Huabo, C., Chaohua, D., Weirong, C.: Back-to-back PV generation system for electrified railway and its control strategy. In: 2017 IEEE Transportation Electrification Conference and Expo, Asia-Pacific (ITEC Asia-Pacific), 2017, pp. 1–6
5. Deng, W., Dai, C., Chen, W., Gao, S.: Experimental investigation and adaptability analysis of hybrid traction power supply system integrated with photovoltaic sources in AC-fed railways. IEEE Trans. Transp. Electrif. **7**(3), 1750–1764 (2021)
6. Aguado, J.A., Sánchez Racero, A.J., de la Torre, S.: Optimal operation of electric railways with renewable energy and electric storage systems. IEEE Trans. Smart Grid **9**(2), 993–1001 (2018)
7. Ying, Y., Liu, Q., Wu, M., Zhai, Y.: The flexible smart traction power supply system and its hierarchical energy management strategy. IEEE Access **9**, 64127–64141 (2021)
8. Hu, H., Shao, Y., Tang, L., Ma, J., He, Z., Gao, S.: Overview of harmonic and resonance in railway electrification systems. IEEE Trans. Ind. Appl. **54**(5), 5227–5245 (2018)
9. Xie, S., Zhang, Y., Wang, H.: A novel co-phase power supply system for electrified railway based on V type connection traction transformer. Energies **14**, 1214 (2021)
10. Liu, Y., Chen, M., Cheng, Z., Chen, Y., Li, Q.: Robust energy management of high-speed railway co-phase traction substation with uncertain PV generation and traction load. IEEE Trans. Intell. Transp. Syst. **23**(6), 5079–5091 (2022)
11. Chang, G.W., Lin, H.-W., Chen, S.-K.: Modeling characteristics of harmonic currents generated by high-speed railway traction drive converters. IEEE Trans. Power Deliv. **19**(2), 766–773 (2004)
12. Rangarajan, S.S., Collins, E.R., Fox, J.C.: Harmonic resonance repercussions of PV and associated distributed generators on distribution systems. In: 2017 North American Power Symposium (NAPS), 2017, pp. 1–6
13. Liu, Q., Zhang, W., Cao, G., Liu, J., Ye, J., Wu, M., Yang, S.: Influence of the catenary distributed parameters on the resonance frequencies of electric railways based on quantitative calculation and field tests. Energies **15**, 3752 (2022)

An Energy Management Strategy for Traction Substation with ESS Based on Reinforcement Learning

Jiaming Luo(✉), Xiaoguang Wei, Shibin Gao, Xiaoran Guo, and Tianbao Zhong

Southwest Jiaotong University, Chengdu 611756, China
carrot9191@my.swjtu.edu.cn

Abstract. The traction power supply system (TPSS) consumes a large amount of electrical energy for locomotive traction every year. To effectively utilize the regenerative braking energy of locomotives and reduce the overall traction energy consumption, the reinforcement learning (RL) method is introduced into energy management strategy in this paper. A reinforcement-learning framework for energy management is constructed to enhance energy utilization efficiency in traction scenario and regenerative braking scenario respectively, and to achieve the effect of adaptive charging and discharging by setting multiple reward functions. Meanwhile, due to the model-free energy management strategy, which avoids modeling the traditional circuit model and has the ability to update online, it is more flexible compared to the rule-based class approach. The simulation test results show that compared with the traditional rule-based energy management method, the RL-based energy management strategy can effectively improve the utilization efficiency in different electric locomotive operating scenarios, reduce the power impact of the traction power supply system, and improve the system operation economy.

Keywords: Energy management · Reinforcement learning · Energy storage system

1 Introduction

With the continuous expansion of the electrified railroad scale, electricity is used as the only source of energy for electrified railroad locomotives, and the rapidly expanding size of railroads has also led to a rapid increase in electrical energy. In order to reduce energy consumption for traction, how to improve energy utilization efficiency has become urgent issue [1]. During the braking process of an electric locomotive, the motor works under generator conditions, so, the kinetic energy of the locomotive can be converted into electrical energy and returning that energy to the TPSS [2]. However, this portion of the energy cannot be stored or utilized unless there are other electric locomotives in the same feeding section. Therefore, utilizing regenerative braking energy and improving its efficiency have become one of the effective ways to reduce the energy consumption of the TPSS [3].

L. Jia et al. (Eds.): EITRT 2023, LNEE 1135, pp. 11–19, 2024.
https://doi.org/10.1007/978-981-99-9307-9_2

Fortunately, the development of energy storage technology provides new ideas for the utilization of regenerative braking energy, and the installation of energy storage equipment in traction substations to absorb regenerative braking energy has become a novel way to save energy [4]. The structure of a typical traction substation with energy storage system (ESS) is shown in Fig. 1. With the operation of the railway power conditioner (RPC), it is possible to achieve a bidirectional flow of energy between the left and right feeding sections. The ESS is directly connected to the DC bus of RPC, thus realizing the energy exchange between the TPSS and the ESS. However, the capacity of ESS is limited, and to maximize the energy consumption of the ESS, reasonable energy management strategies are needed to enhance operational efficiency [5].

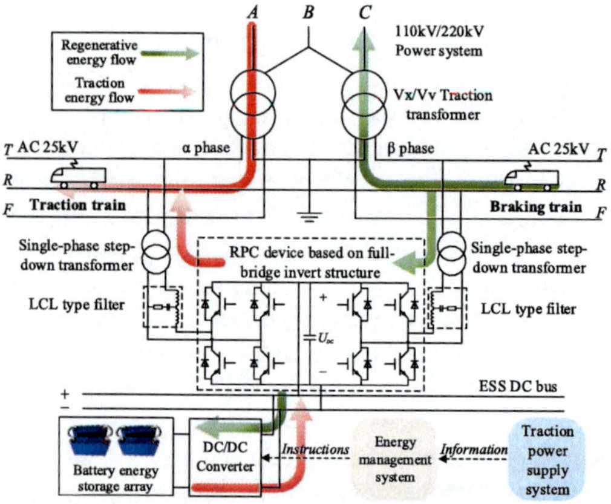

Fig. 1. The structure of energy storage TPSS.

From existing research, rule-based energy management strategies are mainstream, usually setting multiple thresholds to accommodate various work situations according to traction power or catenary voltage [6]. In the field of DC-TPSS, because of the fully parallel TPSS with through connection and the low power level of locomotives, the power fluctuation of trains will directly respond to the catenary voltage, and the adoption of voltage-based energy management strategy can achieve better results [7]. However, the voltage fluctuation of the electrified railroad is drastic and the voltage reference value is unstable, so the charging and discharging thresholds are usually set according to the locomotive power to realize the control of ESS [8]. Nevertheless, such methods require artificially set charging and discharging rules and need to be readjusted to maintain efficiency when the operating environment changes, so they are not flexible enough. Although there are current strategies based on the model predictive control (MPC) and the fuzzy logic control (FLC) that enable adaptive energy exchange [9, 10], they do not have the capability of system migration key parameters still rely on the manual setting.

To overcome the shortcomings of the above strategies, a reinforcement-learning-based (RL) energy management strategy is proposed. By building a RL model of the energy management system (EMS), adaptive energy management is achieved by realizing the autonomous exploration and adjustment of experience in the offline state, while the reward function table is also updated in real time. The algorithm also can be updated online and adjusts the system parameters in real-time during operation to adapt to different power environments.

2 RL-Based Energy Management Strategy Framework

The essence of the energy management strategy is a Markov decision process problem, in which reinforcement learning is used to select the optimal action to improve the efficiency of the system operation. Therefore, it is necessary to establish relevant mathematical models and build algorithmic frameworks for different scenarios to achieve the expected results.

2.1 System Basic Constraints and Objective Functions

The ESS's main purpose is to improve energy utilization efficiency. On the one hand, it is necessary to improve the absorption efficiency of regenerative braking energy to reduce energy loss as much as possible; on the other hand, it is required to actively return energy to reduce the average traction power of locomotives as well. Based on these two basic principles, the objective function is illustrated as follows:

$$\begin{aligned} \max\ e_{\%} &= a \cdot e_{tr\%} + (1-a) \cdot e_{ab\%} \\ &= a \cdot \left(\sum_{i=0}^{m} \int_0^{\Delta t} P^{re}_{ess,t} dt / (\sum_{i=0}^{m} \int_0^{\Delta t} (P^{tr}_{lc,t} + P^{re'}_{ess,t}) dt) \right) \\ &\quad + (1-a) \cdot \left(\sum_{i=0}^{m} \int_0^{\Delta t} P^{ab}_{ess,t} dt / \sum_{i=0}^{m} \int_0^{\Delta t} P^{re}_{lc,t} dt \right) \end{aligned} \quad (1)$$

where $e_{\%}$ is the overall energy use efficiency, a is the weighted parameter, $e_{\mathrm{tr}\%}$ represents the energy efficiency, $e_{\mathrm{ab}\%}$ denotes the absorption efficiency, $P^{ab}_{ess,t}$ and $P^{re}_{ess,t}$ indicate the absorption power and return power of the ESS respectively, $P^{tr}_{lc,t}$ and $P^{re}_{lc,t}$ denote the traction power and regenerative braking power of the locomotive, $P^{re'}_{ess,t}$ is the return power of the ESS under the idle state, m represents the total length of the training data series. The higher value of $e_{\mathrm{tr}\%}$ indicates the more active energy return of the ESS, while the higher value of $e_{\mathrm{ab}\%}$ represents the higher efficiency of regenerative braking energy utilization of the ESS.

In the cooperation of the ESS and TPSS, the basic constraints also need to be satisfied, which include power balance constraints (2)–(3), power variation constraint (4), maximum power constraint (5)–(6) and state of charge constraints (7).

$$s.t.\ P^{tr}_{lc,t} = P^{re}_{ess,t} + P_{ps,t} - P_{loss,t} \quad (2)$$

$$P^{re}_{lc,t} = P^{ab}_{ess,t} + P_{ps,t} + P_{loss,t} \quad (3)$$

$$P^{down}_{ess} \Delta t \le P_{ess,t} - P_{ess,t-\Delta t} \le P^{up}_{ess} \Delta t \quad (4)$$

$$0 \leq P_{ess,t}^{re} \leq P_{ess}^{re,\max} \tag{5}$$

$$P_{ess}^{ab,\max} \leq P_{ess,t}^{ab} \leq 0 \tag{6}$$

$$SOC^{\min} \leq SOC_t \leq SOC^{\max} \tag{7}$$

where $P_{ps,t}$ denotes the exchange power between TPSS and the utility power grid, $P_{loss,t}$ indicates the wasted power caused by transmission loss, Δt is the time step, P_{ess}^{down} and P_{ess}^{up} are the maximum short-time energy storage power and energy feedback power of the ESS respectively, $P_{ess}^{ab,\max}$ and $P_{ess}^{re,\max}$ are the maximum absorb power and return power of the ESS, SOC_t represents the state of charge (SOC)at the time t, $SOC^{\max}$ and $SOC^{\min}$ are the maximum and minimum constraints of the SOC.

2.2 A Reinforcement Learning Framework for Energy Management Processes

The essence of the RL-based EMS is a Markov decision process solving problem, that is, the state of the system at the next moment depends only on the current action and state, independent of the historical data. Therefore, in this paper, the optimal decision is obtained using reinforcement learning. A complete reinforcement learning framework needs to include the basic elements of state $\mathcal{S}$, action $\mathcal{A}\ominus$, reward $\mathcal{R}$, return G and strategy π.

State $\mathcal{S}$: In terms of the operation of the ESS, the SOC is used to characterize the remaining charge, while the power exchange affects the SOC at the next moment. Therefore, the state of RL is as follows

$$S_t = \{P_{lc,t-\Delta t}, P_{ess,t-\Delta t}, P_{ps,t-\Delta t}, SOC_t\} \tag{8}$$

where $P_{lc,t\text{-}\Delta t}$ denotes the locomotive traction power at t-Δt, $P_{ess,t\text{-}\Delta t}$ is the exchange power of ESS at t-Δt, $P_{ps,t\text{-}\Delta t}$ is the exchange power between TPSS and the utility power grid at t-Δt.

Action $\mathcal{A}$: In order to control the ESS for power exchange, the exchange power of the ESS is set as the action of the RL. Because the ESS has three basic power exchange states of energy storage, idle and energy-feedback, it's necessary to divide the storage power and feedback power into b gradients to reduce the dimension of the empirical space, the action of RL a_c can be expressed as follows.

$$a_c = \begin{cases} (b-c)P_{ess}^{ab,\max}/b \text{ (energy-storage state)} \\ 0 \text{ (idle state)} \\ (c-b-1)P_{ess}^{re,\max}/c \text{ (energy-feedback state)} \end{cases} \tag{9}$$

Reward $\mathcal{R}$: The reward function for RL directly determines the agent's action selection. Therefore, the rewards set in this paper mainly include work condition selection rewards r_{con}, SOC range rewards r_{soc}, idle work condition charging rewards r_{cr}, peak power reduction rewards r_{peak} and energy absorption rewards r_{en}, and the reward function $R(S_t,a_t)$ can be expressed as follows:

$$R(S_t, a_t) = r_{con}(S_t, a_t) + r_{soc}(S_t) + r_{cr}(S_t, a_t) + r_{peak}(S_t, a_t) + r_{en}(a_t) \tag{10}$$

Return G: According to the definition of the reward function, the reward G can be set as the cumulative sum of the discounts of the reward $R(S_t,a_t)$. Let the discount factor be γ, then the reward G can be expressed in (11).

$$G = \mathrm{E}\left(\sum_{t=0}^{\infty} \gamma^t R(S_t, a_t)\right) \tag{11}$$

Policy $\pi : \mathcal{S} \rightarrow \mathcal{A}$: The essence of a strategy for reinforcement learning is the probability distribution of actions. The initial distribution of actions in this paper is a random uniform distribution. The purpose of RL is to select the optimal sequence of actions, i.e., the maximum return G_0 under the strategy π, where the agent makes the decision, as shown in (12). Then the maximum expectation of the reward is the optimal value function, which can be expressed as (13). According to the Bellman optimal equation, Eq. (13) can be rewritten as (14), as follows.

$$G_0 = \max \mathrm{E}\left(\sum_{t=0}^{\infty} \gamma^t R(S_t, a_t)\right) \tag{12}$$

$$Q^*(S) = \max_{\pi} \mathrm{E}(G_0/S_0 = S, a_0 = a, \pi) \tag{13}$$

$$Q^*(S_t, a_t) = R(S_t, a_t) + \max_{a_t \in \mathcal{A}\ominus} \gamma \sum_{S_{t+1} \in \mathcal{S}} P_{S_t, S_{t+1}}(a_t) Q^*(S_{t+1}, a_{t+1}) \tag{14}$$

where $P_{S_t,S_{t+1}}$ is the transition probability from the state S_t at time t to the next state S_{t+1} taking action a_t. However, the power of the train is difficult to predict accurately, so an accurate value of the transfer probability cannot be obtained.

2.3 Off-Line Training and Online Update Process

According to (14), the transfer probability $P_{S_t,S_{t+1}}$ must be obtained for the solution of the Bellman equation. However, this value is difficult to obtain because the power at the next moment is difficult to predict. To solve this problem, the Q-learning algorithm is used to solve this equation to obtain the optimal action sequence. The Q-learning agent will observe the environment parameters, then select the optimal action, and interact with the environment through the action. In this process, the Q value function will be updated for the next decision, which is shown in (15). The maximum Q-value function can be obtained under the optimal action sequence, as in (16). The action can be expressed as (17).

$$Q_{k+1}(S_k, a_k) = Q_k(S_k, a_k) + \alpha[R(S_k, a_k) + \gamma \max_{a^* \in A} Q_k(S_{k+1}, a^*) - Q_k(S_k, a_k)] \tag{15}$$

$$Q_k\left(s_{t+1}, a^*\right) = \max_{a^* \in \mathcal{A}} Q_k\left(s_{t+1}, a^*\right) \tag{16}$$

$$a_t = a^* \left(Q_k(S_{t+1}, a^*) = \max_{a \in A} Q_k(S_{t+1}, a)\right) \tag{17}$$

where k denotes the number of iterations, α is the learning factor, γ represent and discount factor, and a^* is the optimal action.

In the offline training process, the RL algorithm first extracts training data fragments from the historical power data set to simulate the operation of a real locomotive. After the agent performs operations such as decision-making, interaction and update, the next state is calculated based on the training data and the current state, and the process is repeated. Different from offline training, the online update of RL can only rely on real-time locomotive power data, updating parameters and states in real-time after each interaction, and its algorithm can be expressed as in Table 1.

Table 1. The online update algorithm based on RL algorithm.

Algorithm: The online update algorithm based on RL algorithm			
1:	State initialization	*6:*	Calculate $R(S_t, a_t)$ using (10)
2:	Repeat (for each episode)	*7:*	Update the Q-value table
3:	Determine the current state S_t	*8:*	Update status parameters
4:	Choose the optimal action a_t by (17)	*9:*	Go to next stage S_{t+1}
5:	Control output/input power of ESS	*10:*	**Until** receive the stop command

3 Case Study

In order to verify the effectiveness of the algorithm, the 24 h of power data set of a traction substation is selected as an arithmetic example in this paper, as shown in Fig. 2. The daily traction energy is 37,154 kWh, the regenerative braking energy is 3,174 kWh, the maximum traction power is 16.167 MW, and the maximum regenerative braking energy is 6.62 MW. The capacity of the ESS used in this case study is 331.2 kWh, and the maximum exchanging power is 3.2 MW. The SOC is in the range of 0.25 to 0.85.

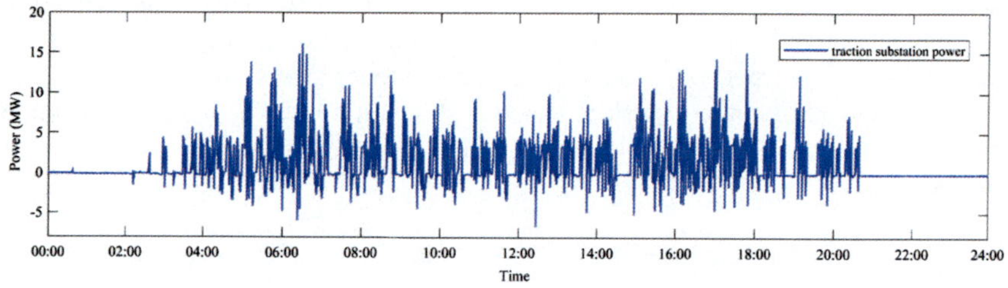

Fig. 2. One day power of a traction substation.

To study the convergence of the Q-value function matrix, one hundred simulations are conducted and averaged to offset the random sampling bias, and 3000 data windows are randomly drawn for each training with a data window of length 100. According to Fig. 3, it can be seen that with the increase in the number of iterations, the 2-parametric number of the Q-value function matrix gradually approximates the convergence value with good convergence.

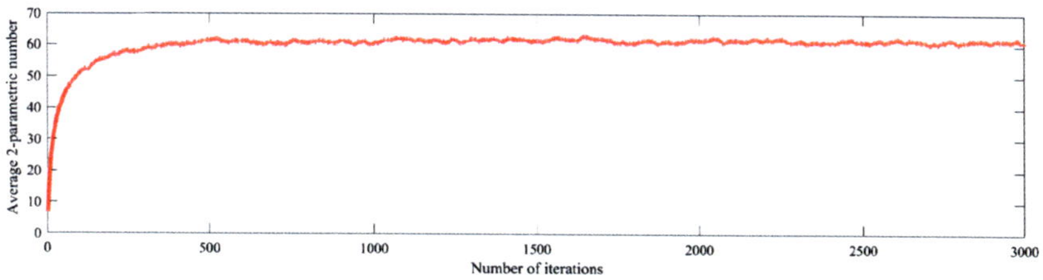

Fig. 3. Average 2-parametric number of the Q-value function matrix.

According to Fig. 4, the RL-based strategy can effectively achieve adaptive power exchange with a good power reduction effect. At the same time, the strategy can charge under idle conditions to maintain SOC. However, it should be noted that the SOC has been running at a low level because the train operation in this case is very intensive, which makes the idle working condition less, and the regenerative braking energy only accounts for about 10% of the total traction energy.

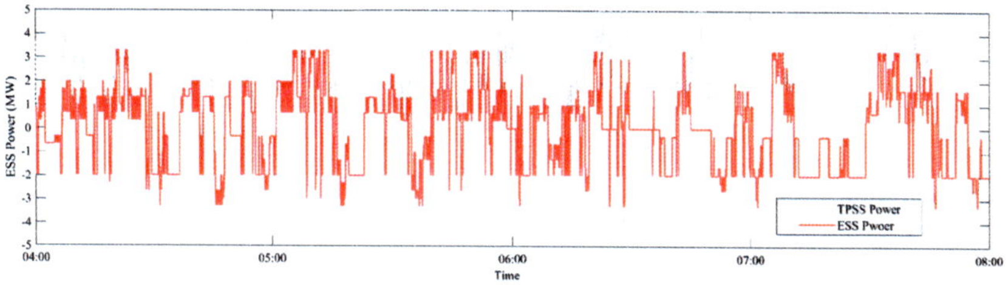

Fig. 4. The ESS exchange power.

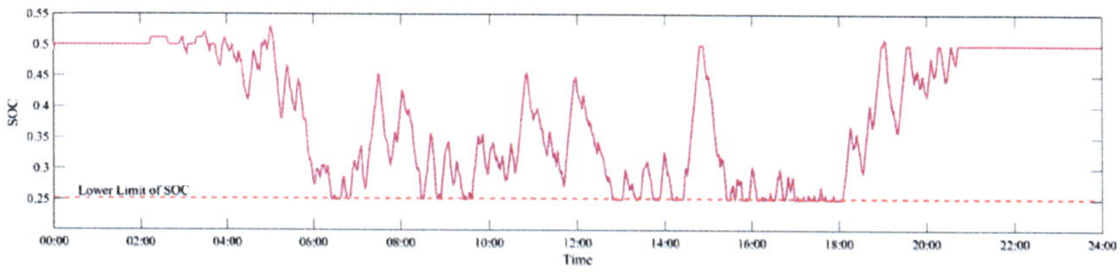

Fig. 5. The SOC of the ESS.

Further, to verify the performance of the energy management strategy based on RL algorithm, this paper compares the strategy with rule-based class methods such as fixed threshold control (FTC) strategy and fuzzy logic-based control (FLC) strategy, and the parameter metrics of each strategy are shown in Table 2.

Table 2. The parameter metrics of each strategy.

Parameter	RL	FLC	FTC
$e_{tr\%}$ (–)	26.23%	19.27%	15.43%
$e_{ab\%}$ (–)	99.45%	98.31%	98.31%
E_{ab} (kWh)	3157.71	3122.09	3120.88
E_{id} (kWh)	7320.62	5173.21	3098.56
E_{re} (kWh)	10576.50	8158.56	6210.17
P_{ave} (MW)	2.47	2.59	2.82

where $e_{tr\%}$ is the traction energy reduction ratio, $e_{ab\%}$ is the regenerative braking energy absorption efficiency, E_{ab} is the absorbed electricity, E_{id} is the energy storage electricity under idle conditions, E_{re} is the return electricity, and P_{ave} is the average traction power. From the results, the RL-based energy management strategy outperforms the rule-based energy management strategy in all metrics. Due to the adoption of a more accurate adaptive discharge strategy, it makes the ESS more active in discharging and effectively improves the economy of system operation.

4 Conclusion

This paper constructs an adaptive energy management strategy based on RL algorithm for TPSS with ESS. The paper first describes the objectives and constraints of the energy management model, based on which the framework of the RL-based algorithm is presented, and the algorithmic process of offline training and online optimization is described. On this basis, a case validation was carried out with the measured traction substation data on a certain day. The simulation results present that compared with the traditional fixed threshold-based strategy and fuzzy logic control-based strategy, the RL-based energy management strategy can effectively improve the utilization efficiency of regenerative braking energy and enhance the comprehensive energy efficiency of TPSS, which shows good economic value.

Acknowledgments. This work was supported by the National Key Research and Development Program of China under Grant 2021YFB2601500.

References

1. Luo, J., Wei, X., Gao, S., Huang, T., Li, D.: Summary and outlook of capacity configuration and energy management technology of high-speed railway energy storage system. Proc. CSEE **42**(19), 7028–7050 (2022). (in Chinese)
2. Luo, J., Gao, S., Wei, X., Zang, T., Zhang, J.: Research on "grid–source–storage–vehicle" dynamic threshold energy management based on fuzzy petri nets. Adv. Eng. Sci. **55**(1), 48–58 (2023). (in Chinese)

3. Luo, J., Gao, S., Wei, X., Tian, Z., Wang, X.: Parallel-reinforcement-learning-based online energy management strategy for energy storage traction substations in electrified railroad. IEEE Tran. Transp. Electrif. (Early Access)
4. Vazquez, S., Lukic, S., Galvan, E., Franquelo, L.: Carrasco.: Energy storage systems for transport and grid applications. IEEE Trans. Industr. Electron. **57**(12), 3881–3895 (2010)
5. Sengor, I., Kilickiran, H., Akdemir, H., Kekezoglu, B., Erdinc, O., Catalao, J.: Energy management of a smart railway station considering regenerative braking and stochastic behavior of ESS and PV generation. IEEE Trans. Sustain. Energy **9**(3), 1041–1050 (2018)
6. Deng, W., Dai, C., Han, C., Chen, W.: Back-to-back hybrid energy storage system of electric railway and its control method considering regenerative braking energy recovery and power quality improvement. Proc. CSEE **39**(10), 2914–2923 (2019). (in Chinese)
7. Ciccarelli, F., Iannuzzi, D., Kondo, K., Fratelli, L.: Line–voltage control based on wayside energy storage systems for tramway networks. IEEE Trans. Power Electron. **31**(1), 884–899 (2016)
8. Yang, H., et al.: Coordinated demand response of rail transit load and energy storage system considering driving comfort. CSEE J. Power Energy Syst. **6**(4), 749–759 (2020)
9. Chen, M., Cheng, Z., Liu, Y., Cheng, Y., Tian, Z.: Multitime-scale optimal dispatch of railway ftpss based on model predictive control. IEEE Trans. Transp. Electrif. **6**(2), 808–820 (2020)
10. Liu, Y., et al.: Adaptive threshold adjustment strategy based on fuzzy logic control for ground energy storage system in urban rail transit. IEEE Trans. Vehicul. Technol. **70**(10), 9945–9956 (2021)

A Global Optimized Energy Management Strategy for Reducing Fuel Consumption in Diesel-Electric Hybrid Power Supply System

Shuo Zhang[1](✉), Wanrui Ren[2], Cunguo Liu[2], and Xuhui Dai[2]

[1] Beijing China Railway Construction Electrification Design and Research Institute Co., Ltd., Beijing 100043, China
zhangshuo.dqh@crcc.cn

[2] China Railway 14Th Bureau Group the Electrification Engineering Co., Ltd., Beijing 100043, China

Abstract. Diesel-electric hybrid power supply system (HPSS) is widely used in railway vehicles due to the advantages of reducing fuel consumption and improving dynamic response. However, the complex control and output relationships among power sources are unfavorable for the effective management of power flow in the system. This problem can be solved by designing an effective energy management strategy (EMS). In this paper, the mathematical model of diesel generator/battery HPSS is firstly established. A detailed mathematical analysis was then conducted on common rule-based strategies and the dynamic programming strategy designed in this paper. Finally, the relevant performance of the HPSS at a full working conditions of railway vehicles under the two strategies is simulated in MATLAB/Simulink. It can be seen from simulation results that the HPSS energy management based on DP strategy optimization is more reasonable. The diesel generator can work in the best fuel consumption range as far as possible, so the power compensation of energy management strategy is more reasonable. Compared with the conventional RB strategy, DP strategy reduces the fuel consumption by 1.6145 kg, about 58.19%, which proves the effectiveness of the proposed EMS in fuel saving.

Keywords: Hybrid power supply system · Energy storage system · Energy management strategy · Dynamic programming

1 Introduction

As economies and societies continue to grow, transportation has emerged as a key technology that countries worldwide are competing to advance [1]. Unfortunately, this has led to a surge in the demand for coal fuel within the transportation industry, resulting in the emission of a significant amount of pollutants and greenhouse gases [2]. Although China has made progress in promoting clean energy usage, with clean energy consumption reaching 24.3% in 2020 according to the National Economic and Social Development Statistics Bulletin of the People's Republic of China, non-clean energy still accounts for a significant 75.7%, indicating the need for further energy conservation and emission reduction efforts [3].

L. Jia et al. (Eds.): EITRT 2023, LNEE 1135, pp. 20–27, 2024.
https://doi.org/10.1007/978-981-99-9307-9_3

Rail transit is a major consumer of energy for medium and long-distance transportation, representing a significant portion of total transportation energy consumption [4]. Given the continued implementation of China's "dual carbon" policy, it has become imperative to promote the adoption of green and low-carbon railway infrastructure. Despite this, most rail vehicles continue to rely on diesel generators (DGs) due to the demands of long-distance transportation [5]. To reduce pollution emissions resulting from fuel consumption and advance the transformation of railway transportation towards an environmentally friendly, low-carbon, and economically sustainable direction, it is essential to develop a new mode of green, low-carbon rail transit [6].

The emergence of energy storage technology has made the hybrid power supply system (HPSS) consisting of DGs and energy storage systems (ESS) the primary method for improving the fuel efficiency of rail vehicles [7]. By enabling real-time power compensation through the ESS, the DG can operate within its optimal fuel consumption range, thereby minimizing pollution emissions. As the most suitable energy storage option due to its high energy density and established application technology, batteries have become the preferred choice for ESS [8].

The diverse output characteristics of various power sources make the energy management of HPSSs more complex. To ensure the stable operation of the system and minimize fuel consumption, the development of a well-designed and effective energy management strategy (EMS) is crucial [9]. To meet the operational requirements for stable vehicle operation, the EMS should optimize the operation modes of each power source by combining the actual working status of the vehicle and real-time road information. Consequently, the dynamics and efficiency of the HPSS are directly influenced by the effectiveness of its EMS. The research field related to EMS is broad. Khan et al. [10] designed an energy management system based on a PI controller for a medium voltage DC ship to control the charging and discharging of the hybrid energy storage system. According to the fuel utilization rate of the hybrid power three-wheeler diesel generator, Maddumage et al. [11] artificially formulated a logical RB strategy to control the power output of DGs and batteries, thereby reducing the fuel consumption of the three-wheeler. He formulated an RB EMS with frequency rules for the DG and battery of diesel-electric locomotive. Compared with the single diesel generator drive, the HPSS reduces the fuel consumption by 25%. In conclusion, RB strategy is currently the most widely used strategy for HPSS EMS. Since the formulation of rules mainly depends on engineering experience, the interval design of rules is mostly not optimal [12]. To overcome this defect, this paper proposes an optimization strategy based on dynamic programming (DP) algorithm to optimize the power output of diesel generators and batteries, thereby reducing the fuel consumption of vehicles. According to the power allocation result after DP strategy, it can be used as a reference for the subsequent RB strategy formulation.

Compared with other papers on energy management strategies for vehicle diesel electric multi source power systems, the contributions and innovations of this paper lie in:

1. Aiming at the energy management problem of a diesel generator/battery hybrid system, a mathematical model of the system was established to more accurately describe the performance of the system.

2. This paper conducts a detailed mathematical analysis of common rule based strategies and the dynamic planning strategy designed in this article to compare the performance of the two strategies.
3. In this paper, MATLAB/Simulink is used to simulate the performance of hybrid systems under two strategies under given railway conditions. By comparing the simulation results of the two strategies, it is proved that the dynamic planning strategy is effective in saving fuel.

Therefore, this paper proposes an effective optimization method for diesel generator/battery hybrid systems, which achieves more effective energy management through optimization of dynamic planning strategies, and has achieved significant results in fuel saving, with high practical value and promotion significance. The present study is organized as follows. Section 2 builds the mathematical models of the HPSS. Section 3 describes the energy management strategy including RB strategy and DP strategy. Section 4 provides the results and discussion, and Sect. 5 draws conclusions.

2 HPSS for Railway Vehicles

The HPSS of rail vehicles adopts the hybrid power supply mode of diesel generator and battery [7], which is mainly composed of diesel generator system (DGS), energy storage system (ESS), and load, as shown in Fig. 1. Due to space limitations, the specific subsystem modeling process in this article will focus on energy management.

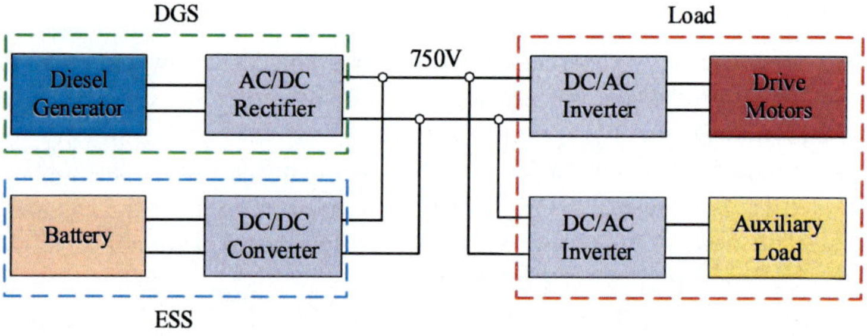

Fig. 1. Simple structure of HPSS for a railway vehicle.

3 Energy Management Strategy for HPSS

EMS is the top control mode of vehicle power flow. By developing appropriate EMS, the working model of the vehicle's power source can be managed, so as to realize the stable operation of the vehicle and reduce fuel consumption. Based on HPSS model, this paper studies two vehicle energy management strategies: logic rule-based (RB) energy management strategy and dynamic programming (DP) optimization-based energy management strategy.

3.1 Logic Rule-Based Strategy

The RB strategy is essentially to manage energy in real time according to established logic rules. In this paper, the different values of load demand power are used as the difference areas for rule-making. The specific logic rules are shown in Fig. 2.

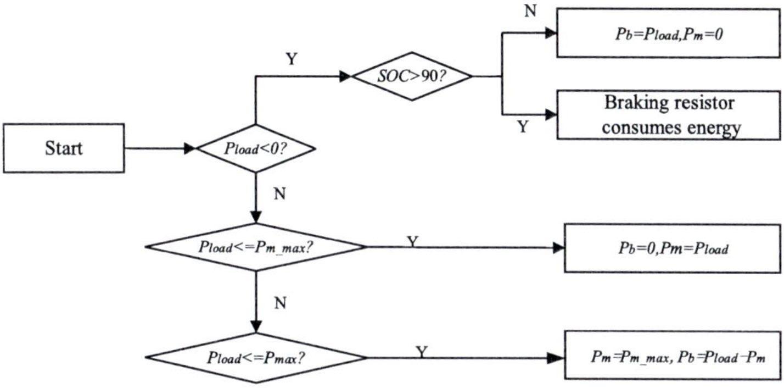

Fig. 2. RB energy management strategy logic diagram.

According to the actual working conditions of the vehicle, load demand and the SOC value of the power battery pack, the instantaneous power control rules are shown in Table 1, where P_{m_max} is the maximum output power of the diesel generator; P_{max} is the maximum load power; P_b is the output power of the battery pack; P_m is the output power of diesel generator; P_{load} is the load power.

Table 1. Logical rule table of RB strategy.

Condition	Control
$0 < P_{load} < P_{m\_max}, 0 < = SOC < = 100$	$P_b = 0, P_m = P_{load}$
$P_{m\_max} < P_{load} < P_{max}, 0 < = SOC < = 100$	$P_b = P_{load}\text{-}P_m, P_m = P_{m\_max}$
$P_{load} < 0, SOC > = 90$	$P_{bat} = 0, P_m = 0$
$P_{load} < 0, SOC < 90$	$P_{bat} = P_{load}, P_m = 0$

The RB strategy has a small amount of calculation and fast response speed of energy management, which can basically realize the stable cooperative work of diesel generator and battery. However, the formulation of RB strategy logic rules mainly depends on human experience, and it is difficult to ensure the optimal distribution of power flow and achieve the minimum fuel consumption RB.

3.2 Logic Rule-Based Strategy

In order to solve the technical bottleneck that RB policy rules are excessively dependent on experience, DP strategy with global optimization effect is adopted as the energy management strategy of diesel generator and battery. Compared with the traditional RB strategy, the DP strategy proposed in this paper can achieve the best fuel consumption and provide a benchmark for the formulation of RB strategy logic rules. The energy management optimization of DP strategy is shown in Fig. 3.

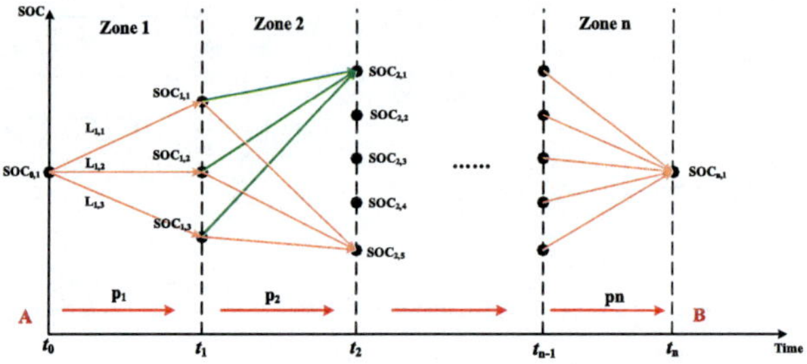

Fig. 3. Energy management optimization diagram of DP strategy.

The optimization calculation objective of the DP strategy is to obtain the minimum fuel consumption of the system. It can be seen from the figure that when the train is divided into n cells from position A to B of the line. The vertical axis represents the SOC of the battery, that is, to achieve the best battery power compensation by optimizing the SOC, thus reducing the total fuel consumption. The calculation process of DP strategy is as follows:

(1) Collect the fuel consumption at each time from the beginning to the first interval, and record it as L, and all the sum values are the total fuel consumption.
(2) Record the value of L at each time from zone 1 to zone 2, and only keep the data path from the initial point to the small total fuel consumption in zone 2.
(3) Repeat step 2 in the subsequent interval to ensure the minimum fuel consumption in the whole operating cycle. The optimal power allocation result obtained by DP strategy is this set of discrete sequences. In the optimization calculation, it mainly includes: variables of state, variables of decision, transfer equations, single-region objective functions and cost functions.

The goal of the single-region objective function is to minimize the fuel consumption of the vehicle. The fuel consumption calculation method proposed in this paper includes the actual fuel consumption of the diesel generator and the equivalent fuel consumption of the battery, as shown in (1).

$$\begin{cases} L(t) = F(t) = F_{DG}(x(t), y(t)) + F_b(x(t), y(t)) \\ F_b(t) = \dfrac{E_b}{H_l \times s \times \eta_{DG}} \end{cases} \tag{1}$$

where F_{DG} is the real fuel consumption of diesel generator. F_b is the equivalent fuel consumption of battery. H_l is the calorific value of fuel. s is the conversion factor of electric-fuel.

The optimization process of DP strategy is divided into n intervals, and the objective function is the sum of the optimization functions of n intervals, as shown in (2).

$$D = \min \sum_{t=1}^{n} L(t)\Delta t = \min \sum_{t=1}^{n} F(t)T_s \tag{2}$$

4 Results

In order to verify the optimization effect of the DP strategy on the HPSS, this paper adopts a typical operating condition (about 300 s) of the vehicle to carry out relevant simulation tests. Under this operating condition, the power output results of the vehicle based on RB strategy and DP strategy are shown in Figs. 4 and 5 respectively. The blue line represents the load power, the red line and the yellow line represent the output power of the DG and the battery respectively. It can be seen from Fig. 4 that the DG mainly provides energy, and the battery only works when the diesel generator's output power is insufficient to meet the load demand and braking, which is obviously not enough to save energy. According to the energy management results after DP strategy optimization in Fig. 5, the DG works in the power range with minimum fuel consumption as far as possible, and the remaining energy is output by the battery. Compared with RB strategy, the power allocation optimized by the DP strategy is more reasonable.

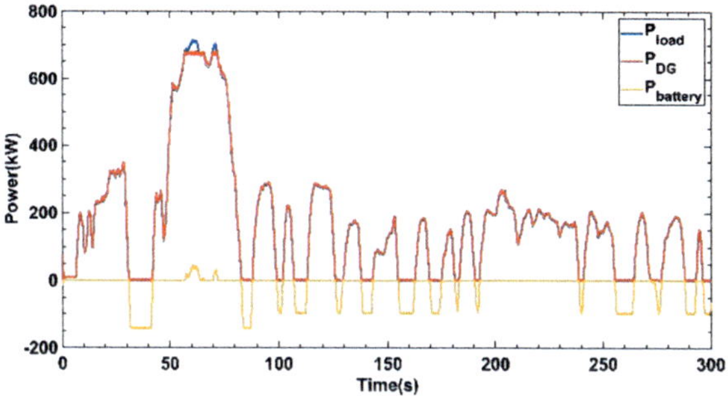

Fig. 4. Power-split managed by RB strategy.

Finally, the comparison results of vehicle fuel consumption based on the RB strategy and DP strategy are shown in Fig. 6. Under this typical operating condition, the vehicle consumes 2.7744 kg of fuel under RB strategy and 1.1599 kg under DP strategy. These results verify that the vehicle optimized by DP strategy can reduce the fuel consumption of 1.6145 kg, about 58.19%, thus verifying the superiority of DP strategy.

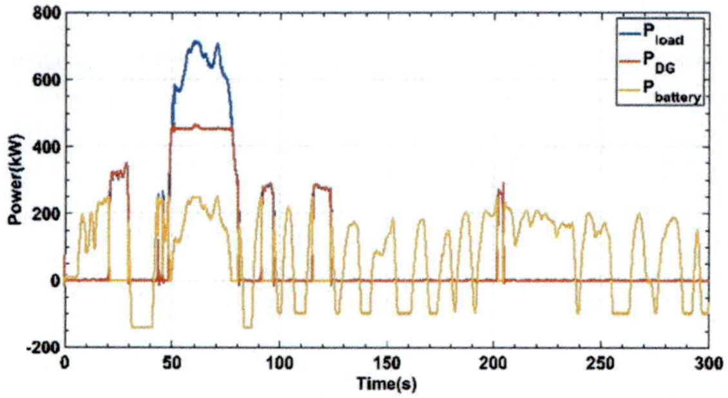

Fig. 5. Power-split managed by DP strategy.

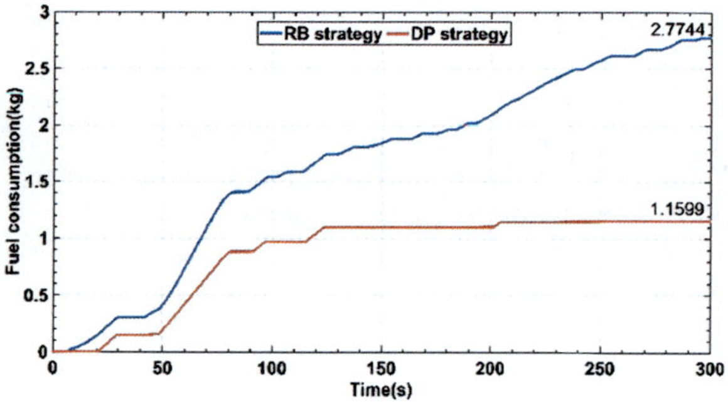

Fig. 6. Fuel consumption of the railway vehicle under two strategies.

References

1. Kapetanović, M., Vajihi, M., Goverde, R.M.P.: Analysis of hybrid and plug-in hybrid alternative propulsion systems for regional diesel-electric multiple unit trains. Energies **14**(18). https://doi.org/10.3390/en14185920
2. Boretti, A.: High-efficiency internal combustion engine for hybrid hydrogen-electric locomotives. Int. J. Hydrogen Energy **48**(4), 1596–1601 (2023)
3. He, H., Shou, Y., Wang, H.: Fuel economy optimization of diesel engine for plug-in hybrid electric vehicle based on equivalent operating points. Control Eng. Pract. **123**, 105162 (2022)
4. Al-Thani, H., Koç, M., Isaifan, R.J., Bicer, Y.: A review of the integrated renewable energy systems for sustainable urban mobility. Sustainability **14**(17). https://doi.org/10.3390/su141710517
5. Zhao, J., et al.: Modeling and integrated optimization of power split and exhaust thermal management on diesel hybrid electric vehicles. Energies **14**(22). https://doi.org/10.3390/en14227505
6. Kapetanović, M., Núñez, A., van Oort, N., Goverde, R.M.P.: Reducing fuel consumption and related emissions through optimal sizing of energy storage systems for diesel-electric trains. Appl. Energy **294**, 117018 (2021)
7. Zhang, Y., et al.: An optimization of new energy hybrid configuration parameters based on GA method. J. Mar. Sci. Eng. **10**(11). https://doi.org/10.3390/jmse10111747
8. Din, T., Tian, Z., Li, K., Hillmansen, S., Roberts, C.: Operation and energy evaluation of diesel and hybrid trains with smart switching controls. Control Eng. Pract. **116**, 104935 (2021)

9. Kusakaka, K., Phiri, S.F., Numbi, B.P.: Optimal energy management of a hybrid diesel generator and battery supplying a RTG crane with energy recovery capability. Energy Rep. **7**, 4769–4778 (2021)
10. Khan, M.M.S., Faruque, M.O.: Management of hybrid energy storage systems for MVDC power system of all electric ship. North American Power Symposium, Denver, 2016
11. Maddumage, W., Perera, M., Attalage, R., Kelly, P.: Power management strategy of a parallel hybrid three-wheeler for fuel and emission reduction. Energies **14**(7). https://doi.org/10.3390/en14071833
12. Lü, X., et al.: Energy management of hybrid electric vehicles: a review of energy optimization of fuel cell hybrid power system based on genetic algorithm. Energy Convers. Manage. **205**, 112474 (2020)

Check for updates

A Dynamic Tuning Method of Wireless Power Transfer System Based on Phase-Controlled Inductor Circuit

Shan He[1](✉), Xiaona Li[2], Yuhao Peng[2], Weige Zhang[2], Guohong Zeng[2], and Xiaogong Du[3]

[1] Shenzhen Power Supply Co. Ltd, Shenzhen 518001, China
heshan@sz.csg.cn

[2] School of Electrical Engineering, Beijing Jiaotong University, Beijing 100044, China
{20121469,22121492,wgzhang,ghzeng}@bjtu.edu.cn

[3] Xunfeng Electronics Co. Ltd, Jinan 250200, China
duxiaogong@hxrtdl.com

Abstract. Wireless power transfer (WPT) technology has increasing potential as the charging infrastructure in railway transportation and the interaction between large-scale or electric vehicles and power grid. To solve the problem of receiver side detuning caused by changes in coil spacing and equivalent DC load in wireless electric vehicle charging system, an LCC-S WPT system with receiver side dynamic tuning method is designed. This paper first analyzes the transmission characteristics of the LCC-S WPT system and the working principle of the phase-controlled inductor circuit. The phase-controlled inductor circuit is applied to the series compensation capacitor at the receiver side, and the equivalent capacitance value is dynamically adjusted to make the receiver side always resonate. The experimental results demonstrated the effectiveness of the proposed dynamic tuning method, which can restore resonance at the receiver side and improve the output power when the coil spacing and equivalent DC load change.

Keywords: Wireless Power Transfer · LCC-S · Phase-controlled Inductor Circuit · Dynamic Tuning

1 Introduction

In recent years, multiple energy source railway vehicles have been a research hotspot. Meanwhile, there is undoubtedly increasing promotion of Electric Vehicles (EVs) as well as large-scale charging and discharging interaction between EVs and the power grid. WPT technology has the advantages of good safety, flexible power supply and low maintenance costs, which has great potential in railway transportation as well as EV charging applications [1–3]. To improve the transmission performance of the system, a compensation network needs to be added to the system to make it in a resonant state. In practical applications, changes in coil spacing and DC side load can cause changes in the equivalent impedance of the line, leading to system detuning, reducing system output power, and affecting system stability [4]. Therefore, the WPT system tuning technology is very important.

L. Jia et al. (Eds.): EITRT 2023, LNEE 1135, pp. 28–35, 2024.
https://doi.org/10.1007/978-981-99-9307-9_4

At present, the main tuning methods for WPT system include frequency tracking control and dynamic compensation tuning. Frequency tracking control restores resonance of the system by adjusting the switch frequency in real-time to the resonance frequency point of the system [5]. Paper [6] addresses the issue of detuning in WPT system by measuring the effective value of the transmitting coil current in real-time and the average output current of the DC power supply. The frequency tracking control tuning method does not require additional tuning devices and has a relatively simple structure. However, it is only applicable to single transmitting and receiving devices. In applications where multiple sets of transmitting and receiving coils are used, due to the different resonant frequencies of different circuits, this method is difficult to achieve the resonance state of multiple sets of coils at the same time, and may occur frequency bifurcation phenomenon, leading to system tuning failure.

Dynamic compensation tuning replaces the compensation capacitors in the system with adjustable capacitor structures such as capacitor arrays and phase-controlled inductor circuits [7]. However, there are issues with the capacitance arrays, such as the inability to continuously adjust the equivalent capacitance value, the need for a large number of capacitors and switches, and the increase in equipment volume. Paper [8–10] address the issue of detuning in the WPT system by using a phase-controlled inductor circuit as the compensation capacitor. Based on the principle of minimum voltage and maximum current in the system, the equivalent capacitance value of the phase-controlled inductor circuit is dynamically adjusted to achieve dynamic tuning of the system. However, the control algorithm is relatively complex.

This paper proposes a simple dynamic tuning method based on phase-controlled inductor circuit. Firstly, the transmission characteristics of the LCC-S WPT system are analyzed. At a fixed switching frequency, the phase-controlled inductor circuit is applied to the series compensation capacitor of the system's receiver side. Based on the principle of in-phase control, the equivalent capacitance value of the phase-controlled inductor circuit is dynamically adjusted to make the receiver side circuit resonate at all times and improve the system's output power. Finally, experimental verification was conducted on the proposed method.

2 Dynamic Tuning Method at the Receiver Side

2.1 Analysis of the Principle of Phase-Controlled Inductor Circuit

The main circuit of the LCC-S WPT system is shown in Fig. 1. The switches T_1-T_4 form a full bridge inverter circuit. L_f, C_f, and C_1 are the series compensation inductor, parallel compensation capacitor, and series compensation capacitor at the transmitting end, respectively. L_1 and L_2 represent the self-inductance of the transmitting coil and the receiving coil, respectively. M represents the mutual inductance between two coils, C_2 is the series compensation capacitor at the receiver side. Diodes D_1-D_4 form an uncontrolled rectifier circuit, C_d is the DC output filtering capacitor, R is the DC side load resistance, they can be equivalent to rectifier load Z_o. $U_{in\text{-}dc}$ and $U_{out\text{-}dc}$ are the system's DC input voltage and DC output voltage, respectively. U_{in} and u_{out} are the output square wave voltage of the inverter circuit and the input square wave voltage of the rectifier bridge, respectively. i_1 and i_2 are the current of the transmitting coil and the current of the receiving coil, respectively. u_{C2} represents the voltage at both ends of capacitor C_2.

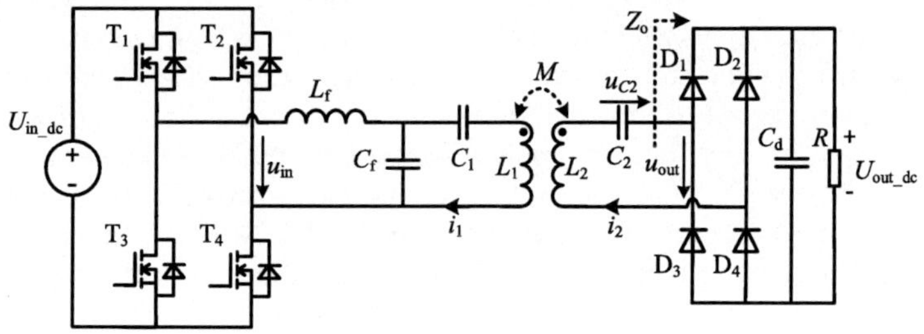

Fig. 1. Main circuit of LCC-S WPT system.

Assuming that L_f and C_f always resonate:

$$\frac{1}{\omega C_f} = \omega L_f \tag{1}$$

$\boldsymbol{I}_{1-1}$, the fundamental component of the transmitting coil current i_1, will be constant

$$\boldsymbol{I}_{1-1} = \frac{\boldsymbol{U}_{\text{in - 1}}}{\text{j}\omega L_f} \tag{2}$$

Among them, $\boldsymbol{U}_{\text{in-1}}$ is the fundamental component of the square wave voltage u_{in}. $\boldsymbol{I}_{2-1}$, the fundamental component of the receiving coil current i_2, is

$$\boldsymbol{I}_{2\text{-}1} = \frac{\text{j}\omega M \boldsymbol{I}_{1\text{-}1}}{\text{j}\omega L_2 + \frac{1}{\text{j}\omega C_2} + Z_{\text{o-1}}} \tag{3}$$

Among them, $Z_{\text{o-1}}$ is the fundamental equivalent impedance of the rectifier load Z_{o}. $\boldsymbol{U}_{\text{out-1}}$, the fundamental component of square wave voltage u_{out}, is

$$\boldsymbol{U}_{\text{out-1}} = \boldsymbol{I}_{2\text{-}1} Z_{\text{o-1}} = \frac{\frac{M}{L_f}\boldsymbol{U}_{\text{in-1}} Z_{\text{o-1}}}{\text{j}\omega L_2 + \frac{1}{\text{j}\omega C_2} + Z_{\text{o-1}}} \tag{4}$$

According to Fourier decomposition, the numerical relationship between the amplitude of $\boldsymbol{U}_{\text{out-1}}$ and $U_{\text{out-dc}}$ is

$$U_{\text{out - }\mathbf{1}} = \frac{4U_{\boldsymbol{out-dc}}}{\boldsymbol{\pi}} \tag{5}$$

Therefore, the output power $P_{\text{out-dc}}$ on the DC side is

$$P_{\text{out - dc}} = \frac{U_{\text{out-dc}}^2}{R} = \frac{\boldsymbol{\pi}^2 U_{\text{out-1}}^2}{16R} \tag{6}$$

When C_2 meets

$$\mathrm{Im}\left(\mathrm{j}\omega L_2 + \frac{1}{\mathrm{j}\omega C_2} + Z_{\text{o-1}}\right) = 0 \tag{7}$$

The receiving end will resonate. According to Eqs. (4) and (6), the output voltage and output power of the system are at their maximum at this time; According to Eq. (3), $\boldsymbol{I}_{2\text{-}1}$ lead $\boldsymbol{I}_{1\text{-}1}$ 90°, Combining Eq. (8), it can be seen that at this point, $\boldsymbol{I}_{1\text{-}1}$ and $\boldsymbol{U}_{C2\text{-}1}$ are in phase.

$$U_{c2-1} = \frac{I_{2-1}}{j\omega C_2} \tag{8}$$

During the actual operation of the system, changes in coil spacing and DC side load can cause Eq. (7) to fail, resulting in detuning of the receiver side and a decrease in the output voltage and power of the system. To ensure that Eq. (7) remains true, it is necessary to correspondingly change the value of the series compensation capacitor C_2 at the receiver side in the system.

2.2 Control Strategy

In order to continuously adjust the series compensation capacitance value of the receiver side, this paper adopts a phase-controlled inductor circuit to replace the original capacitor C_2. A typical phase-controlled inductor circuit is shown in Fig. 2. The static inductor L_T is connected in series with two MOSFET switches S_1 and S_2 in reverse series to form a phase-controlled inductor branch, and a static capacitor C_T is connected in parallel at both ends of the phase-controlled inductor branch. By controlling the triggering conduction angle of switch tubes S_1 and S_2, the equivalent capacitance C_{eq} of the phase-controlled inductor circuit can be continuously adjusted. The effective range of the triggering conduction angle α of the switch tube S_1 is [π/2, π]. Within a switching cycle, S_2 and S_1 complement each other and are only turned on for half a cycle.

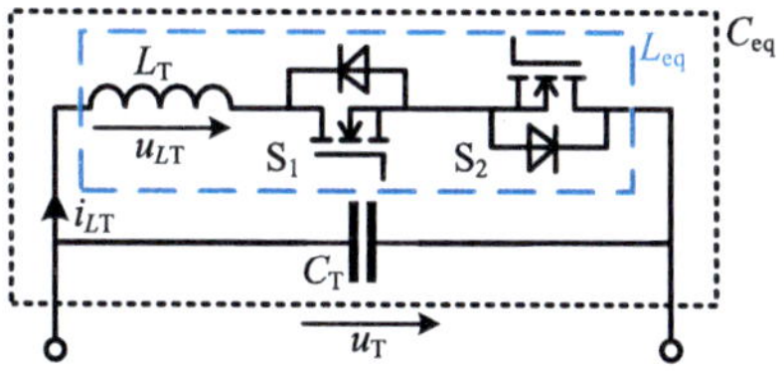

Fig. 2. Typical phase-controlled inductor circuit

Set the voltage u_T at both ends of the capacitor C_T as a sine wave

$$u_T(t) = U_T \sin(\omega t) \tag{9}$$

The switch sequence and typical working waveform of the phase-controlled inductor circuit are shown in Fig. 3. Among them, u_{LT} is the voltage at both ends of the inductor L_T, and i_{LT} is the current flowing through the inductor L_T.

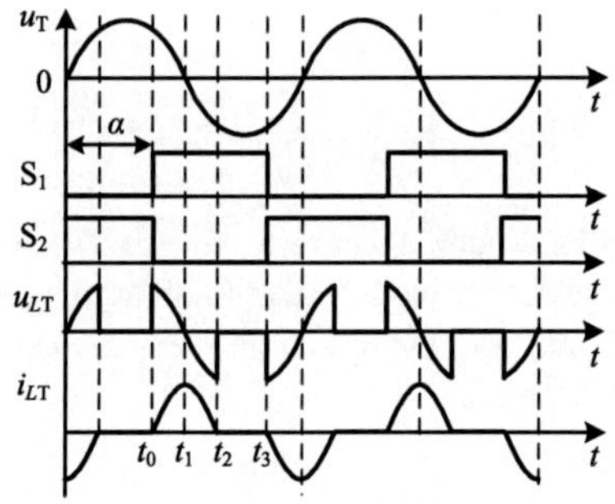

Fig. 3. Switch sequence and typical working waveform of phase-controlled inductor circuit.

The current i_{LT} is decomposed by Fourier series, and the amplitude of its fundamental component is

$$I_{LT1} = \frac{U_T}{\omega L_T} \frac{[2\pi + \sin(2\alpha) - 2\alpha]}{\pi} \tag{10}$$

Therefore, the equivalent inductance L_{eq} of the phase-controlled inductor branch can be obtained as

$$L_{eq} = \frac{U_T}{\omega I_{LT1}} = \frac{\pi L_T}{2\pi + \sin(2\alpha) - 2\alpha} \tag{11}$$

The equivalent capacitance C_{eq} of the overall phase-controlled inductor circuit is

$$C_{eq} = C_T - \frac{1}{\omega^2 L_{eq}} = C_T - \frac{2\pi + \sin(2\alpha) - 2\alpha}{\pi\omega^2 L_T} \tag{12}$$

According to the analysis in Sect. 1, the characteristic of resonance at the receiver side of the system when the switching frequency remains constant is that the fundamental wave of the transmitting coil current i_1 is in phase with the fundamental wave of the compensating capacitor voltage u_{C2} at the receiver side. Therefore, this paper is based on the principle of in-phase for dynamic tuning control.

The dynamic tuning control strategy of the system receiver is shown in Fig. 4. Firstly, the current i_1 and voltage u_{C2} are sampled, and then the fundamental components of i_1 and u_{C2} are obtained through bandpass filter circuits. Then, the corresponding zero crossing comparison square wave signals are obtained through zero crossing comparison circuits, The phase difference φ between these two square wave signals, i.e. the phase difference between $\boldsymbol{I}_{1\text{–}1}$ and $\boldsymbol{U}_{C2\text{–}1}$, is calculated internally in the control chip FPGA. Subtract the reference phase difference $\varphi_{ref} = 0$ and φ to obtain the error value. To obtain the error value. Input this error value into the PI controller to obtain the conduction trigger angle α, Finally, using u_{C2} as the synchronous signal, the driving signal of the switch tube is obtained by switch drive signal modulation, and C_{eq} is adjusted to achieve dynamic tuning of the receiver side circuit.

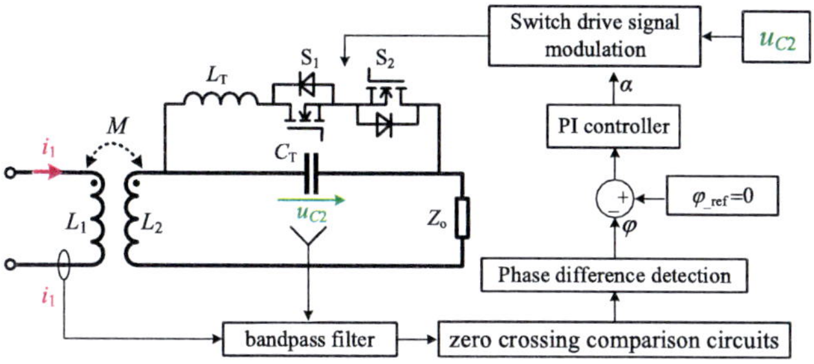

Fig. 4. Dynamic tuning control strategy at the receiver side.

3 Experiment

To verify the effectiveness of the dynamic tuning method mentioned above, an LCC-S WPT system experimental platform was constructed based on Figs. 1 and 4, as shown in Fig. 5. Configure the system circuit parameters to ensure that the system is in a resonant state when the tuning device is not turned on, the coil spacing d is 6cm, and the DC side load R is 8Ω. The specific circuit parameters are shown in Table 1.

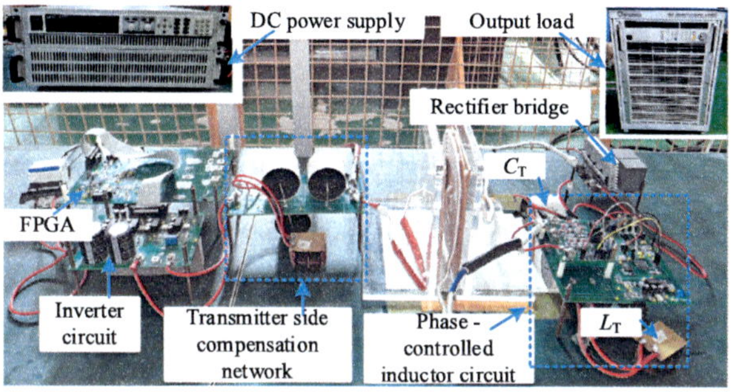

Fig. 5. Experimental platform.

When the tuning device is not turned on, $d = 6$ cm, and the $R = 8$ Ω, the experimental waveforms of i_1 and u_{C2} are shown in Fig. 6a. At this point, i_1 and u_{C2} are in phase, and the receiver side is in a resonant state. The output voltage is 48 V and the output power is 288 W. Adjust d to 3 cm and R to 16 Ω. When the tuning device is not turned on, the waveforms of i_1 and u_{C2} are shown in Fig. 6b, i1 leads u_{C2}, and the receiver side is out of tune. The output voltage of the is 103.93 V and the output power is 675 W. After turning on the tuning device, the waveforms of i_1 and u_{C2} are shown in Fig. 6c, i1 and u_{C2} are in phase, and the receiver side resonates. The output voltage is 108.3 V, and the output power is 733 W. The typical waveform of the phase-controlled inductor circuit is shown in Fig. 6d, and the triggering conduction angle α is 1.687. Figure 7 summarizes the experiment results which proved the satisfying tuning effects.

Table 1. Circuit parameters.

Parameters	Values	Parameters	Values
U_{in_dc}/V	300	f/kHz	85
d/cm	2–6	R/Ω	8–16
L_f/μH	36	C_f/nF	97
L_1/μH	56.3–50.2	C_1/nF	247
L_2/μH	12.18–10.9	M/μH	15.96–6.05
L_T/μH	36	C_T/nF	248

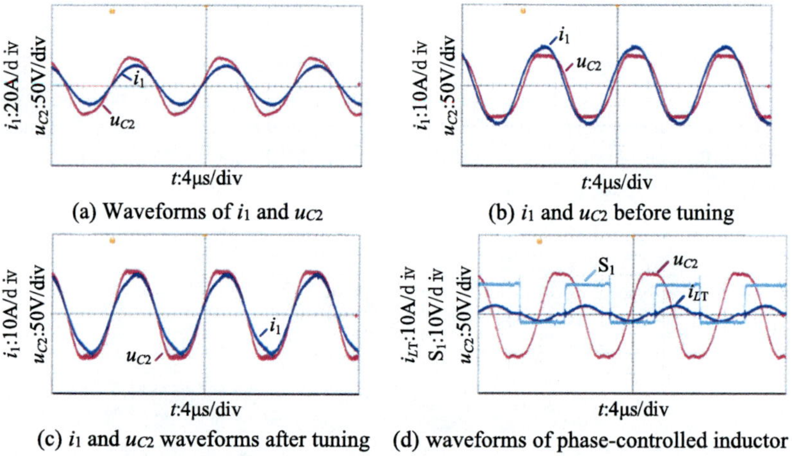

Fig. 6. Experimental waveforms at $d = 6$ cm and $R = 8\ \Omega$ for (**a**) and $d = 3$ cm and $R = 16\ \Omega$ for (**b–d**).

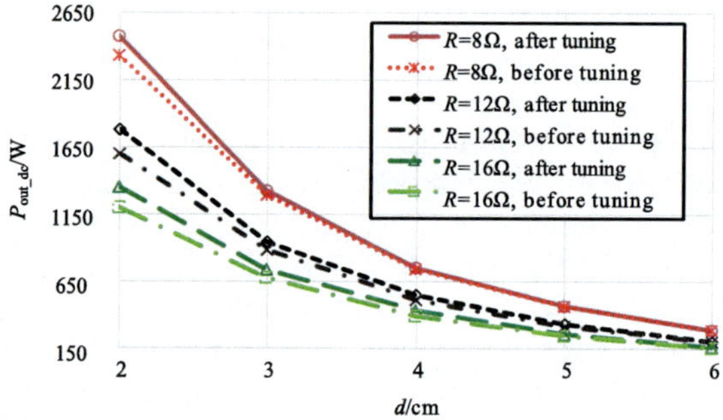

Fig. 7. Comparison of output power experimental values before and after receiver side tuning.

4 Conclusion

This paper analyzes the transmission characteristics of the LCC-S WPT system in railway or EV transportation applications. The equivalent capacitance expression of the phase-controlled inductor circuit is derived, and the phase-controlled inductor circuit to the series compensation capacitor at the receiver side of the system is adopted. By controlling the current of the transmitting coil and the voltage of the compensation capacitor at the receiver side to maintain the same phase, the dynamic tuning of the system at the receiver side is achieved. The experimental results demonstrate that the proposed receiver dynamic tuning method can improve the system output power effectively.

Acknowledgments. This work is supported by Science and Technology Project of China Southern Power Grid Co., Ltd (Grant Number. 090000KK52210132): Research and Demonstration of Key Technologies for Interaction between Large-scale Electric Vehicles and Power Grid (Phase I).

References

1. Cheng, L., Wang, Y., Peng, Y.: Research on risk assessment of high-speed railway operation based on network ANP. Smart Resilient Transp. **3**(1), 37–51 (2021)
2. Qi, J., et al.: A multiple-modes resonant switched capacitor DC/DC converter with variable voltage ratios. IEEE Trans. Power Electron. **38**(6), 7428–7443 (2023)
3. Wang, H., Cheng, K.W.E.: Analysis, design, and validation of a decoupled double-receiver wireless power transfer system with constant voltage outputs for industrial power supplies. IEEE Trans. Industr. Inf. **19**(1), 362–370 (2023)
4. Xu, W., Li, X., Peng, Y.: Review and new developments of adaptive tunning technologies for wireless power transfer. In: 2022 IEEE 9th International Conference on Power Electronics Systems and Applications, Hong Kong (2022)
5. Zhou, H., Yao, G., Ma, L.: A new method of frequency tracking control for ICPT system. Power Electron. **47**(9), 77–79 (2013). (in Chinese)
6. Mai, R., Xu, D., Yang, M.: Dynamic tuning method of frequency tracking based on the minimum current ratio for IPT system. Trans. China Electrotech. Soc. **33**(6), 1276–1284 (2018). (in Chinese)
7. Wu, Y., Zhang, J., Zhang, S.: An adaptive impedance matching method based on radial basis function neural network in multi-load wireless power transfer systems. Trans. China Electrotech. Soc. **36**(19), 3969–3977 (2021). (in Chinese)
8. Qiang, H., Huang, X., Tang, L.: Maximum power transmission of inductive coupling radio energy transmission system based on dynamic tuning. Sci.. China Press **42**(7), 830–837 (2012). (in Chinese)
9. Mai, R., Lu, L., Li, Y.: Dynamic resonant compensation approach based on minimum voltage and maximum current tracking for IPT system. Trans. China Electrotech. Soc. **30**(19), 32–38 (2015). (in Chinese)
10. Wang, H., Cheng, K.W.E., Yang, Y.: A new resonator design for wireless battery charging systems of electric bicycles. IEEE J. Emerg. Selected Topics Power Electron. **10**(5), 6009–6019 (2022)

Energy-Efficient Train Control Incorporating Realistic Traction Characteristics Using Convex Optimization

Yang Peng, Fuwang Chen, and Shaofeng Lu(✉)

Shien-Ming Wu School of Intelligent Engineering, South China University of Technology, Guangzhou 511442, China
lushaofeng@scut.edu.cn

Abstract. The study proposes an energy-efficient train control (EETC) model that utilizes the convex optimization (CO) method to incorporate more practical and complex traction system characteristics. These characteristics, referred to as RMeB-RP, include regenerative and mechanical braking at low speeds and reduced power at relatively high speeds. In many previous studies on the EETC problems, the traction system characteristics often incorporated the simplified pure regenerative braking and constant power characteristics during high speeds, known as RB-CP, to reduce the complexity of the model. We applied convexification and linearization to model such nonlinear characteristics during traction and braking operation over various speeds and applied convex optimization to achieve the optimal solution efficiently. Our simulation results showed that the train with RMeB-RP has a higher total energy consumption, cruising speeds, and speeds at the beginning of the braking stage than the train with RB-CP. Furthermore, the train with RB-CP in the EETC model often leads to underestimating energy consumption since the partial loss of traction and regenerative braking effort. As for the real-world scenario in this model incorporating RB-CP, energy consumption was underestimated by 17% compared to incorporating RMeB-RP, with a difference of 0.9136 kWh/km in energy consumption per kilometer. In addition, applying CO to quickly address the EETC problems incorporating the complex traction system characteristics has shown to be effective in the real case, in which the total computational time of interstation operation by using the CO model is only within the range of [0.52, 1.2] seconds.

Keywords: Energy-efficient Train Control · Reduced Power Characteristics · Regenerative and Mechanical Braking Characteristics · Convex Optimization · Urban Rail Transit

1 Introduction

In modern rail transportation, energy-efficient train control (EETC) is concerned with the optimal train speed profile or control strategy to achieve the minimum net energy consumption under different traction system constraints. Massive studies aim to minimize net energy consumption and obtain optimal train control through indirect or direct

L. Jia et al. (Eds.): EITRT 2023, LNEE 1135, pp. 36–45, 2024.
https://doi.org/10.1007/978-981-99-9307-9_5

methods. Pontryagin's maximum principle (PMP), which is a classical optimal control theory, was utilized to indirectly obtain the optimal train control strategies comprising maximum traction, cruising, coasting, and full braking by analyzing the necessary conditions for optimality [1–3]. As another critical method in optimal control, dynamic programming (DP) is also utilized to tackle the EETC problem [4, 5]. Besides that, more applicable mathematical programming methods, i.e., the Pseudospectral method [6, 7], mixed-integer linear programming [6, 8], and convex optimization [9], are applied to obtain the optimal train operation.

Despite the integration of these technologies in practical industry scenarios, some highly energy-consuming and optimal train profile-associated realistic traction characteristics such as hybrid braking characteristics [10, 11], varying traction efficiency [12], and reduced power characteristic (RP) at relatively high speed [13], etc., have not been extensively studied. Most of the reviewed papers on the EETC problem usually study simplified traction and braking characteristics [14], i.e., incorporating pure regenerative braking characteristics and constant power at relatively high speed, referred to as RB-CP. However, there is a reduced power region when the train operates at high speeds. When the train runs at low speed during the braking stage, there is a switching process from regenerative braking to mechanical braking. We refer to these two characteristics mentioned above as RMeB-RP. Ignoring RMeB-RP during train control may lead to inaccurate energy cost modeling and an infeasible train control strategy. It is thus necessary to incorporate more realistic traction system characteristics RMeB-RP to obtain a more practical and precise EETC for the train.

This paper investigated an EETC model based on the CO method incorporating RMeB-RP and the more commonly adopted simplified RB-CP. The real-life urban rail transit line is used for numerical experiments to analyze the impact of these traction characteristics, such as RMeB-RP and RB-CP, on optimal train control. The contributions of this paper are as follows:

1. A CO model to address the EETC problem that incorporates RMeB-RP. The achieved optimal train control strategy leads to more accurate energy consumption evaluation and more practical train optimal control.
2. Real-world urban rail transit line data are utilized for comparative case studies on the EETC model for the trains with RMeB-RP and RB-CP.

2 Problem Formulation

In this section, the typical traction system characteristics are presented. The mathematical model to tackle the EETC problem with realistic traction system characteristics RMeB-RP is given based on the CO method, in which the nonconvex traction/braking forces are solved by relaxation and convexification techniques.

The traction system characteristics incorporating more practical RMeB-RP characteristics and simplified RB-CP characteristics are introduced to analyze the impact on energy consumption and optimal speed profiles. Figure 1 demonstrates the constant power (CP) and reduced power characteristics (RP) when the speed is within $[v_2, v_3]$ and regenerative and mechanical braking characteristics (RMeB).

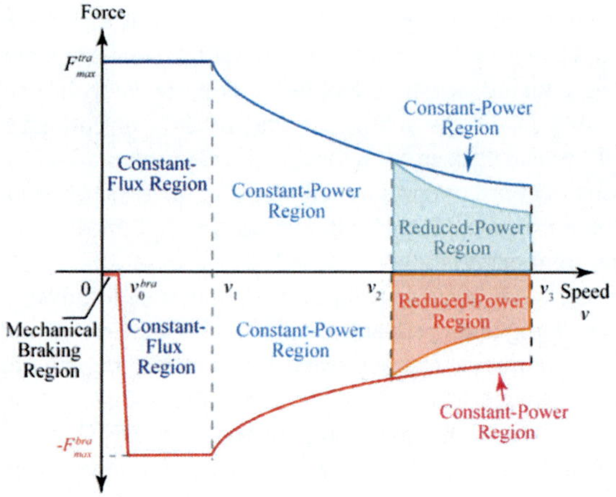

Fig. 1. It shows a typical induction motor traction and braking characteristics diagram. The traction characteristics diagram includes the Constant-Flux, Constant-Power, and Reduced-Power regions. The braking characteristics diagram consists of the constant-effort, hyperbolic-effort, and zero-effort regions.

2.1 Energy-Efficient Train Control Model

In this paper, the model is constructed discretely, in which i represents the index of the discrete distance segments along the journey. The total distance D between two adjacent stations is divided equally into N segments with a length of Δd_i satisfies

$$\sum_{i=1}^{N} \Delta d_i = D. \tag{1}$$

Due to the punctuality requirement, the sum of elapsed time in each distance segment should be equal to the total journey time T

$$\sum_{i=1}^{N} \frac{\Delta d_i}{v_i} = T. \tag{2}$$

The model of train operation can be expressed as follows

$$(f_i - R_i^{drag}) \cdot \Delta d_i = \frac{1}{2} \cdot M \cdot (1 + \rho) \cdot (v_{i+1}^2 - v_i^2) + Mg \cdot \Delta H_i \tag{3}$$

where M is the static train mass, ρ is the rotary allowance, f_i is the traction or braking force, g is the gravitational acceleration ΔH_i is the altitude difference in the ith segment, and R_i^{drag} is the train resistance, which can be formulated as

$$R_i^{drag} = A + B \cdot v_i + C \cdot v_i^2 \tag{4}$$

where A,B,C are the Davis coefficients. The energy e_i consumed during the traction stage and regenerative energy recovered during the braking stage can be expressed as

$$e_i \geq f_i \cdot \Delta d_i / \eta^{tra} \tag{5}$$

$$e_i \geq f_i \cdot \Delta d_i \cdot \eta^{bra} \tag{6}$$

where η^{tra} and η^{bra} are the motor traction and regenerative braking efficiency, respectively. Considering the motor characteristics, traction or braking force f_i is bounded by

$$-F^{b,\max} \leq f_i \leq F^{t,\max} \tag{7}$$

$$-P^{b,\max}/v_i \leq f_i \leq F^{t,\max}/v_i \tag{8}$$

where $F^{b,max}$ and $P^{b,max}$ is the maximum braking force and power, $F^{t,max}$ and $P^{t,max}$ is the maximum traction force and power. The objective function of the EETC problem is to minimize the sum of the electrical energy in each Δd_i, which is defined by

$$\min \cdot f_{obj}(v, e, f) = \sum_{i=1}^{N} e_i \tag{9}$$

Subject to: (1)–(8)

It can be seen that the EETC model (9) is nonconvex caused by Eqs. (2) and (8) with $\frac{1}{v_i}$, Eqs. (3) and (4) with v_i^2. Therefore, to address the nonconvex equations, we introduce auxiliary variables α_i and β_i into a convex optimization model.

2.2 Traction/Braking Characteristics

As shown in Fig. 2, during the traction stage, the expressions of the traction system characteristics that incorporate RP and CP are as follows

$$f_i^{tra} = \begin{cases} F_{\max}^{tra} & 0 \leq v_i \leq v_1^{tra} \\ F_{\max}^{tra} \cdot v_1^{tra}/v_i = f_2^{tra} & v_1^{tra} \leq v_i \leq v_2^{tra} \\ f_2^{tra} \cdot (v_2^{tra})^2/v_i^2, f_2^{tra} \cdot v_2^{tra}/v_i & v_2^{tra} \leq v_i \leq v_3^{tra} \end{cases} \tag{10}$$

where f_i^{tra} is the traction force in the ith segment, $f_2^{tra} = F_{max}^{tra} \cdot \frac{v_1^{tra}}{v_i}, f_3^{tra,RP} = f_2^{tra} \cdot \frac{v_2^{tra}}{v_3^{tra}}$ is the expression of traction force, in which the train with RP when the speed is within $[v_2^{bra}, v_3^{bra}]$. $f_3^{tra,CP} = f_2^{tra} \cdot (\frac{v_2^{tra}}{v_3^{tra}})^2$ is the expression of traction force, in which the train with CP when the speed is within $[v_2^{bra}, v_3^{bra}]$. During the braking stage, the expression of the train with RMeB-RP can be expressed as

$$f_i^{RMeB-RP} = \begin{cases} 0 & 0 \leq v_i \leq v_0^{bra} \\ -170.34 \cdot v_1^{bra} - 11 & v_0^{bra} \leq v_i \leq v_0^{bra} + \delta \\ F_{\max}^{RMeB} & v_0^{bra} + \delta \leq v_i \leq v_1^{bra} \\ F_{\max}^{RMeB} \cdot v_1^{bra}/v_i & v_1^{bra} \leq v_i \leq v_2^{bra} \\ f_2^{bra} \cdot (v_2^{bra})^2/v_i^2 & v_2^{bra} \leq v_i \leq v_3^{bra} \end{cases} \tag{11}$$

where, $\overline{C_1D_1} = 170.34 \cdot v_1^{bra} - 11$ represents the straight line segment from zero braking force to maximum RB force, δ is the speed difference between point C_1 and D_1. v_0^{bra} is the critical speed, that is, the speed point between pure MeB and RB. F_{max}^{RMeB} is a negative constant, which is the maximum braking force, the braking force at the speed of v_1^{bra} is $f_2^{bra} = F_{max}^{RMeB} \cdot \frac{v_1^{bra}}{v_2^{bra}}$, and the expression of traction force at the speed of v_3^{tra} is $f_3^{bra,RP} = F_{max}^{RMeB} \cdot \left(\frac{v_2^{tra}}{v_3^{tra}}\right)^2$. Moreover, the expression of the train with RB, and with CP when the speed is within $[v_2^{bra}, v_3^{bra}]$ can be expressed as

$$f_i^{RB-CP} = \begin{cases} F_{\max}^{RB} & 0 \le v_i \le v_1^{bra} \\ F_{\max}^{RB} \cdot v_1^{bra}/v_i = f_2^{bra} & v_1^{bra} \le v_i \le v_2^{bra} \\ f_2^{bra} \cdot v_2^{bra}/v_i & v_2^{bra} \le v_i \le v_3^{bra} \end{cases} \tag{12}$$

where F_{max}^{RB} is a negative constant, which is the maximum braking force of RB mode.

2.3 Relaxations and Convexification

To address the nonconvexity, two auxiliary variables α_i and β_i were introduced to solve the convexification of the nonconvex problems, in which nonconvex constraints are replaced with relaxed but convex constraints. In the relaxation of Eqs. (3), (9), and (12), $\frac{1}{v_i}$ is replaced by auxiliary variable α_i

$$\sum_{i=1}^{N} \Delta d_i \cdot \alpha_i = T \tag{13}$$

$$-P^{b,\max} \cdot \alpha_i \le f_i \le F^{t,\max} \cdot \alpha_i \tag{14}$$

$$f_i^{RB-CP} = \begin{cases} F_{\max}^{RB} & 0 \le v_i \le v_1^{bra} \\ F_{\max}^{RB} \cdot v_1^{bra} \cdot \alpha_i & v_1^{bra} \le v_i \le v_2^{bra} \\ f_2^{bra} \cdot v_2^{bra} \cdot \alpha_i & v_2^{bra} \le v_i \le v_3^{bra} \end{cases} \tag{15}$$

Meanwhile, the auxiliary variable α_i should satisfy the constraint (16) as below to maintain the convexity of the model.

$$\alpha_i \ge 1/v_i^2 \tag{16}$$

Besides that, in the relaxation of Eqs. (3) and (4), v_i^2 is replaced by auxiliary variable β_i which can be expressed by

$$\beta_i \ge v_i^2 \tag{17}$$

$$(f_i - R_i^{drag}) \cdot \Delta d_i = \frac{1}{2} \cdot M \cdot (1+\rho) \cdot (\beta_{i+1} - \beta_i) + Mg \cdot \Delta H_i \tag{18}$$

$$R_i^{drag} = A + B \cdot v_i + C \cdot \beta_i \tag{19}$$

In Eqs. (10) and (11), $\frac{1}{v_i}$ is replaced by auxiliary variable α_i, and v_i^2 is replaced by auxiliary variable β_i. For the above formula (10), the straight line $\overline{A_1B}$ instead of arc $\widehat{A_1B}$ for convexity process, which can be expressed as

$$f_i^{tra} = \begin{cases} F_{\max}^{tra} & 0 \le v_i \le v_1^{tra} \\ F_{\max}^{tra} \cdot v_1^{tra} \cdot \alpha_i & v_1^{tra} \le v_i \le v_2^{tra} \\ f_2^{tra} \cdot (v_2^{tra})^2 \cdot \alpha_i, or, -12.56 \cdot v_1^{tra} + 403 & v_2^{tra} \le v_i \le v_3^{tra} \end{cases} \tag{20}$$

The straight line $\overline{E_1F_1}$ instead of arc $\widehat{E_1F_1}$ for the convexity process for (11), as shown in Fig. 2, which can be expressed as

$$f_i^{RMeB-RP} = \begin{cases} 0 & 0 \le v_i \le v_0^{bra} \\ -170.34 \cdot v_1^{bra} - 11 & v_0^{bra} \le v_i \le v_0^{bra} + \delta \\ F_{\max}^{RMeB} & v_0^{bra} + \delta \le v_i \le v_1^{bra} \\ F_{\max}^{RMeB} \cdot v_1^{bra} \cdot \alpha_i & v_1^{bra} \le v_i \le v_2^{bra} \\ 16.96 \cdot v_1^{bra} - 544.26 & v_2^{bra} \le v_i \le v_3^{bra} \end{cases} \tag{21}$$

Introducing the above relaxations transform model (9) to a convex model.

$$\min .f_{obj}(v, e, f, \alpha, \beta) = \sum_{i=1}^{N} e_i \tag{22}$$

Subject to: (1), (5)–(7), (13)–(21).

3 Numerical Experiments

In this section, optimal train operation and speed profiles of the EETC model of the train with realistic traction system characteristics RMeB-RP and simplified characteristics RB-CP are presented based on real-world data. The case incorporating RMeB-RP is taken as the baseline case to evaluate the energy underestimate rate and different train operation points of other studied cases. The parameters of urban rail transit are shown in Table 1. The gradient and speed limits of real-world data from the Beijing Changping line and the realistic traction/braking characteristics are shown in Fig. 3 (Fig. 4).

Under the same journey time, compared to the case with RB-CP, the energy consumption of the train with RMeB-RP is higher, the time for the traction and braking stages is longer, and the variation of acceleration is more significant. In addition, the train with RMeB-RP has a higher cruising speed, a higher speed at the beginning of the braking phase, and a more traction and braking distance.

The left subfigure of Fig. 5 shows the distance and operation time of the traction systems characteristics with RMeB-RP and RB-CP under different operations of the train. Results demonstrate that the train with RB-CP has a longer coasting distance, a shorter traction distance, and a shorter braking distance. It also indicates that when the train with the characteristics of RMeB-RP, requires a longer time for traction and braking stages, consumes more traction energy, and recovers more braking energy. For the train with RMeB-RP, the traction energy and the recovered braking energy are 170.65 kWh

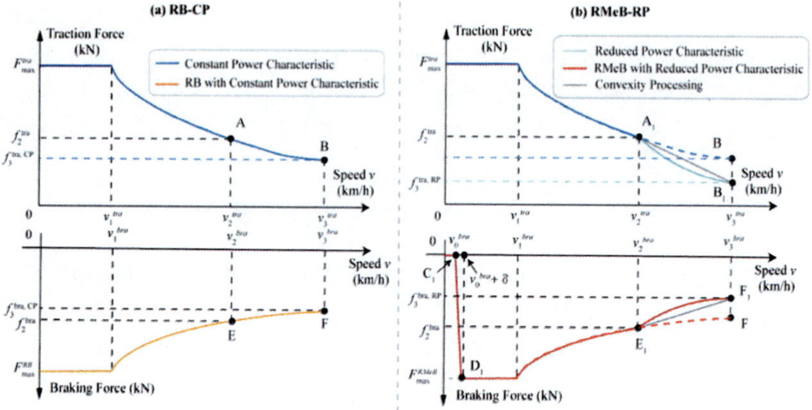

Fig. 2. Schematic diagram of the traction system characteristics incorporating **a** RB-CP (Left) and **b** RMeB-RP (Right). For RMeB-RP, the power reduced curve $\widehat{A_1B_1}$ and $\widehat{E_1F_1}$ are directly replaced by straight line $\overline{A_1B_1}$ and $\overline{E_1F_1}$ in this model, to simplify the complexity of the CO model.

Table 1. Basic parameters of urban rail transit.

Train Mass [t]	Maximum Traction/braking Power [kW]	Maximum Traction/braking Force [kN]	Motor Traction/braking Efficiency	Rotary Allowance [%]	Maximum Acceleration [m/s^2]
287	3144/4327	289/−390	0.9/0.6	8	$\mp$ 1.2

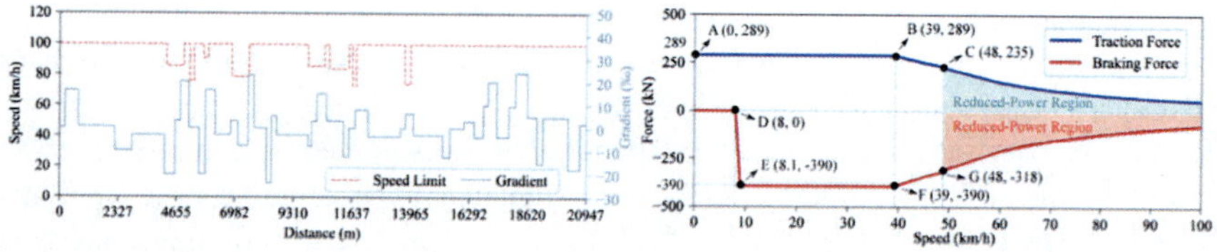

Fig. 3. The speed limit/gradient-distance profile of the Beijing Changping line (Left), and the realistic traction/braking force curve from the AC motor of the DKZ32 model train in Beijing Metro (Right) of the case studies. The Davis coefficients A, B, and C are 3.48 kN, 0.14 kN · s/m, and 0.01 kN · s^2/m^2, respectively.

and 62.03 kWh, respectively. For the train with RB-CP, they are 134.29 kWh and 46.17 kWh, respectively.

The right figures of Fig. 5 show the train operation conditions for the traction system characteristics with RMeB-RP and RB-CP, where different colors represent four various operating states: traction, cruising, coasting, and braking. It can be observed that the operating conditions for the train with RMeB-RP change frequently, and switch between multiple operation states. Regarding the fourth section of the speed trajectory, due to

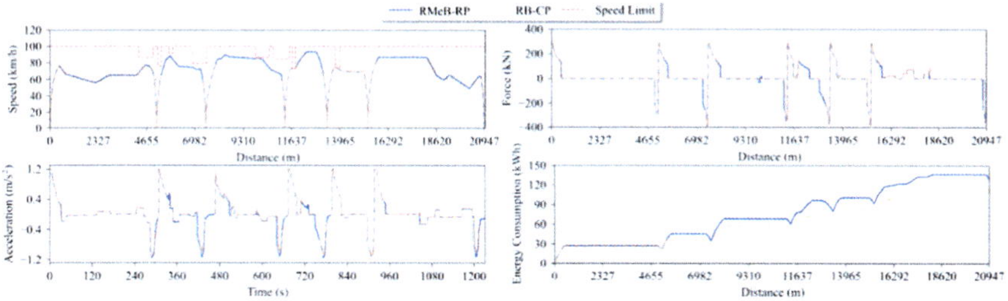

Fig. 4. Simulation results for real-world scenarios based on the Beijing Changping line using the two proposed EETC models, i.e., RMeB-RP and RB-CP. The speed-distance, force-distance, acceleration-operation-time, and energy-distance profiles are presented.

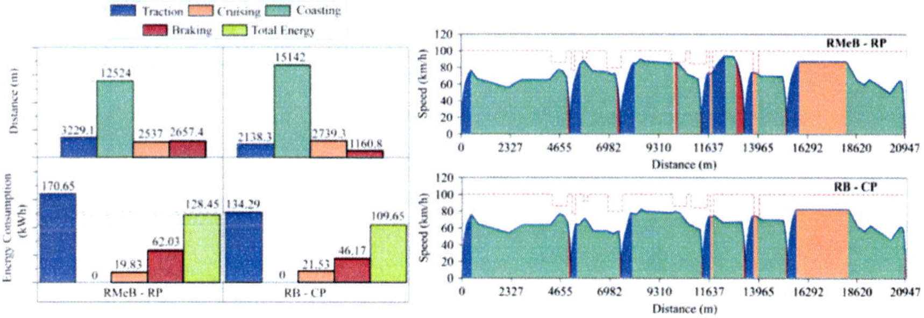

Fig. 5. The comparison of the operation distance and energy consumption with different traction characteristics of RMeB-RP and RB-CP (Left). Speed profiles under different operation conditions of the traction systems incorporating RMeB-RP and RB-CP (Right).

Table 2. Optimization result of EETC under real-world case scenarios considering the traction system characteristics incorporating RMeB-RP and RB-CP.

No	Length [m]	Energy Consumption [kWh]		Time [s]	Distance Segment	EUR [%]	Computation Time [s]
		RMeB-RP	RB-CP				
1	5144.7	19.374	18.739	308	514	3.389	0.76
2	2367.3	14.855	9.335	159	472	59.132	0.52
3	3800	25.647	22.524	206	380	13.865	1.20
4	2025	16.323	11.527	123	505	41.607	0.69
5	1964	11.399	11.359	119	491	0.352	0.89
6	5646.6	37.711	36.174	316	564	4.249	1.17
Total	20947.6	128.795	109.658	1231	2926	17.452	5.23

the RP at high speeds and the inability of regenerative braking below the critical speed

v_0^{bra}, the train with RMeB-RP consumes more traction energy and recovers more braking energy during higher speeds to meet operation constraints.

Table 2 shows the energy consumption of the train with RMeB-RP and RB-CP. The energy consumption of the train with RMeB-RP is higher, so the total operating energy consumption of the train with the simplified traction characteristic of RB-CP is often underestimated. In this case, the energy consumption per kilometer of the train with RB-CP is underestimated by 0.9136 kWh/km than that of the train with RMeB-RP. In this case, the energy consumption per kilometer for the train with RMeB-RP is 6.148 kWh/km, while that for the train with RB-CP is 5.235 kWh/km.

Table 3. Energy consumption for EETC of the trains with RMeB-RP and RB-CP in the fourth interstation with a length of 2025 m

Characteristics		Journey Time (s)			
		123	129	136	148
Energy Consumption [kWh]	RMeB-RP	16.323	14.856	12.124	9.419
	RB-CP	11.527	10.136	8.944	7.428
EUR [%]		41.607	46.567	35.554	26.804

In addition, due to the time limitation, the energy consumption difference is significant under different train operation times. As the train operation time increases, the time that the train operates at speeds within the range of $[v_2, v_3]$ decreases, and the energy consumption difference caused by RP also decreases. Therefore, the energy consumption difference between the train with RMeB-RP and RB-CP also decreases. As shown in Table 3, with the increase of train operation time, the EUR of the train with RB-CP compared to RMeB-RP decreases from 41.607% to 26.804%.

4 Conclusion

This paper proposes an optimization model based on CO to study the EETC problem with more realistic traction system characteristics. In summary, the traction system characteristics that incorporate RMeB-RP results in higher total energy consumption, higher cruising speed, higher speed at the beginning of the braking stage, and longer traction and braking distances, compared to the one incorporates the simplified RB-CP used in many existing literatures. Wwhen the traction forces are reduced, the train with RMeB-RP needs to take more time to increase to the cruising phase. Additionally, due to the reduction in braking force, the train with RMeB-RP also needs to spend more time and distance in the braking phase. For the real-world case scenarios considering the traction system characteristics incorporating RMeB-RP and RB-CP, the energy consumption of the train with RMeB-RP is 0.9136 kWh/km higher than that of the train with RB-CP on a per-kilometer basis. Furthermore, the total energy consumption when considering the traction characteristics with RB-CP is underestimated by 17% compared to the energy consumption when considering the realistic traction characteristics of RMeB-RP. In the future, we will further consider how to validate the above models on an experimental platform to obtain more realistic train operation strategies.

Funding Information. This work was supported in part by the National Science Foundation of Guangdong Province, China (Grant No.2023A1515012949), in part by the 2022 Fundamental and Applied Fundamental Research Project of Guangzhou Municipal Basic Research Program (Grant No.202201010716) and the Featured Creative Project of Higher Education Institution of Guangdong Province, China (No. 2021KTSCX001).

References

1. Khmelnitsky, E.: On an optimal control problem of train operation. IEEE Trans. Autom. Control **45**(7), 1257–1266 (2000)
2. Ichikawa, K.: Application of optimization theory for bounded state variable problems to the operation of train. Bull. JSME **11**(47), 857–865 (1968)
3. Scheepmaker, G.M., Goverde, R.M.P., Kroon, L.G.: Review of energy-efficient train control and timetabling. Eur. J. Oper. Res. **257**(2), 355–376 (2017)
4. Lu, S., Hillmansen, S., Ho, T.K., Roberts, C.: Single-train trajectory optimization. IEEE Trans. Intell. Transp. Syst. **14**(2), 743–750 (2013)
5. Su, S., Wang, X., Cao, Y., Yin, J.: An energy-efficient train operation approach by integrating the metro timetabling and eco-driving. IEEE Trans. Intell. Transp. Syst. 1–17 (2019)
6. Wang, Y., Schutter, B.D., van den Boom, T.J., Ning, B.: Optimal trajectory planning for trains - a pseudospectral method and a mixed integer linear programming approach. Transp. Res. Part C Emerg. Technol. **29**, 97–114 (2013)
7. Goverde, R.M., Scheepmaker, G.M., Wang, P.: Pseudospectral optimal train control. J. Oper. Res. **292**, 353–375 (2021)
8. Tan, Z., Lu, S., Bao, K., Zhang, S., Wu, C., Yang, J., Xue, F.: Adaptive partial train speed trajectory optimization. Energies **11**, 3302 (2018). http://www.mdpi.com/1996-1073/11/12/3302, https://doi.org/10.3390/en11123302
9. Jibrin, R., Hillmansen, S., Roberts, C., Zhao, N., Tian, Z.: Convex optimization of speed and energy management system for fuel cell hybrid trains (2021). arXiv:2109.13833
10. Scheepmaker, G.M., Goverde, R.M.P.: Energy-efficient train control using nonlinear bounded regenerative braking. Transp. Res. Part C Emerg. Technol. **121**, 102852 (2020). ISSN 0968-090X
11. Lu, S., Weston, P., Hillmansen, S., Gooi, H.B., Roberts, C.: Increasing the regenerative braking energy for railway vehicles. IEEE Trans. Intell. Transp. Syst. **15**(6), 2506–2515 (2014)
12. Novak, H., Lešić, V., Vašak, M.: Energy-efficient model predictive train traction control with incorporated traction system efficiency. IEEE Trans. Intell. Transp. Syst. **23**(6), 5044–5055 (2022)
13. Ghaviha, N., Bohlin, M., Holmberg, C., Dahlquist, E., Skoglund, R., Jonasson, D.: A driver advisory system with dynamic losses for passenger electric multiple units. Transp. Res. Part C Emerg. Technol. **85**, 111–130 (2017). ISSN 0968-090X
14. Peng, Y., Wu, C., Lu, S.: Optimal energy management strategy for fuel-cell hybrid trains with different types of energy storage devices. In: Proceedings of the 5th International Conference on Electrical Engineering and Information Technologies for Rail Transportation (EITRT) 2021. Singapore: Springer Singapore, pp. 417–425 (2022)

Design and Modeling of Subway Train Traction System Based on MBSE

Jiahao Zhao[1], Baomin Wang[1](✉), Lujie Zhou[1], Tingli Huang[1], and Qingyong Wang[2]

[1] College of Automotive and Transportation, Tianjin University of Technology and Education, Tianjin 300222, China
1142031254@qq.com, wangbaomin1982@126.com, 792321186@qq.com
[2] Tianjin Rail Transit Operation Group Co., Ltd., Tianjin 300392, China
tjmetro08@163.com

Abstract. In order to apply model-based system engineering (MBSE) to the study of metro train traction system, SysML was chosen as the modeling language, M-Design was adopted as the modeling tool, and OFAP (Operational, Functional, Architecture, Physical) method was used to carry out the modeling work. From the perspective of the stakeholders of the metro train traction system, the forward design idea is used to view the traction system as a black box, a gray box and a white box in turn, and the concrete implementation of business, functional, architectural and physical modeling for realizing the metro train traction system is analyzed. The results show that by applying the MBSE idea and adopting the OFAP method, the business analysis, functional decomposition, architectural design and unified description of physical components of the metro train traction system can be realized, and the whole chain tracing between requirements, functions and physics can be realized.

Keywords: MBSE · Subway train · Traction system · System modeling

1 Introduction

The development of systems engineering has gone through 3 phases: traditional, complex, and complex organism systems engineering. Model-based system engineering (MBSE) came into being to accommodate the significant growth in system scale and complexity in design problems [1]. In recent years, MBSE has carried out a lot of theoretical research and practice at home and abroad [2], and has been widely used in the design and development process of complex systems in the nuclear industry [3], aerospace [4], and automotive [5], and has achieved significant application results. The MBSE method was adopted by NASA and applied in the design of CubeSat [6] and Fire Warning Satellite [7] using Magic Draw modeling tool, which efficiently solved the user requirements and reflected the advantages of MBSE method in the design and development of complex spacecraft. Aditya Akundi and Viviana Lopez [8], a literature review is presented highlighting the current need for an industry-wide transition from document-centric systems engineering to Model-Based Systems Engineering (MBSE). Currently, researchers and

L. Jia et al. (Eds.): EITRT 2023, LNEE 1135, pp. 46–55, 2024.
https://doi.org/10.1007/978-981-99-9307-9_6

developers in the field of rail transportation are also exploring ways to combine MBSE with complex system design [9]. Based on the adoption of MBSE ideas, Yuan Wenqiang [10] et al. take the high-voltage system of a rolling stock as the research object, and propose a set of methods applicable to the intelligent and efficient generation of the architecture of the high-voltage system of a rolling stock for the problem of automatic generation of architectural solutions in the innovative design of complex products. Wang Baomin [11] et al. proposed the OFAP (Operational, Functional, Architecture, Physical) approach to establish the requirements, use cases, functions, architecture, and physical models of the entire logic of the rolling stock in order to apply MBSE to the study of the entire logic of the rolling stock, and the entire logic of the rolling stock that exists in the form of existing text The modeling transformation is carried out. From domestic and international research and application experience, model-based system engineering has obvious advantages for large-scale projects with variable requirements and complex system composition, and is an important direction for future industrial development. The subway train system is a more complex system, and few research results at home and abroad have taken the subway train as the research object. The specific application of using MBSE method to complete the design of subway train-related systems has not been reported yet, so this paper applies MBSE to the research of subway train traction system. The modeling language, modeling tools and modeling methods are the basis for the implementation of the metro train traction system research. In this paper, the SysML modeling language, the M-Design modeling tool developed by Zhejiang University and the OFAP modeling method are used.

The OFAP modeling approach consists of four abstraction levels: business, functional, architectural and physical. The modeling process starts from the stakeholders of the metro train traction system, and adopts the forward design idea to treat the traction system as black box, gray box and white box in turn. Based on this, the concrete implementation of the business, functional, architectural and physical modeling of the metro train traction system is analyzed, and the four levels of business, functional, architectural and physical layers are developed.

2 Business Layer Modeling

Business analysis is critical for manufacturers of metro train systems because of their complex structure, logical processes, and the many disciplines and departments involved in their development. The business analysis of a metro train traction system includes stakeholder analysis, requirements analysis and use case analysis.

2.1 Stakeholder Analysis

A stakeholder is an external entity that influences or interacts with a system, not necessarily a person, but usually connected to the system, has a need for the system or limits the system. Stakeholder analysis is used to establish the context of the system and to understand the connections and influences between the system and the external environment.

Through analysis, the subway train traction system stakeholders (see Fig. 1). The driver is responsible for controlling the joystick to control the traction and braking of the traction system. The contact network is responsible for the basic power of the traction system. ATO drives the train automatically and ATP prevents the train from running over speed. The traction system provides kinetic energy to the bogie to ensure the normal operation of the train.

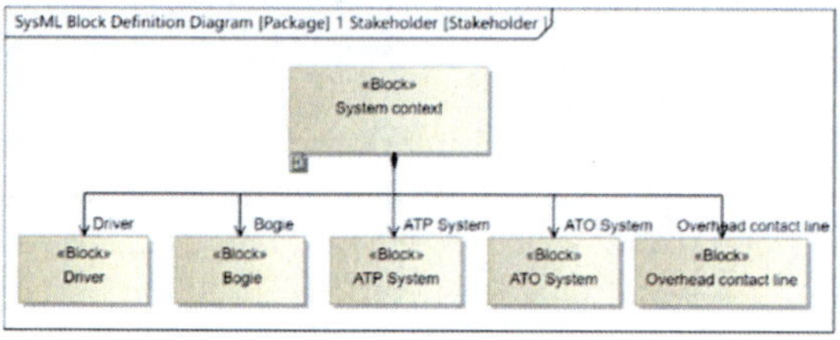

Fig. 1. Stakeholders

2.2 Demand Analysis

The top-level user requirements for the metro train traction system are "The metro traction system is an integral part of the train drive system, the main purpose of which is to reverse the DC voltage on the contact network into a three-phase voltage with variable amplitude and frequency to provide the right energy for traction motor operation. The subway train traction system is the core system of the subway vehicle, providing power for the train. The performance of the traction system is directly related to the performance of the vehicle and the comfort of the passengers".

After analysis, the user requirements are to satisfy the traction characteristics, make the traction system of metro trains have the best dynamic performance, ensure the safety and comfort of metro train operation, make full use of equipment capacity and reduce energy consumption. The top-level requirements are decomposed, and the user requirements of metro train traction system can be decomposed into the following five aspects, traction performance requirements, braking performance requirements, on-time and safe delivery requirements, fault rescue requirements and green travel requirements.

(1) Traction/braking performance requirements, the acceleration at which the train can start at the maximum ramp should not be less than 0.083m/s2, the performance indicators under the working conditions of flat and dry track, semi-abrasive state of the wheels and rated voltage are shown in Table 1 [12].
(2) On-time and safe delivery requirements, control the running time and maximum speed of trains to safely deliver passengers to their destinations.
(3) Fault rescue requirements. The train shall have fault runner rescue capability and meet the following requirements, under the overcrowding load (AW3) condition, when 1/4 or less power is lost, the train with a dynamic drag ratio greater than 1:1 shall be able to maintain all-day operation, and the train with a dynamic drag ratio of 1:1 shall be able to maintain operation to the end. Under the overcrowding load

(AW3) condition, when 1/3 power is lost, the train shall be able to start on the main line at the maximum Under the overcrowding load (AW3) condition, when 1/2 power is lost, the train shall be able to start and maintain operation until the end of the line; under the overcrowding load (AW3) condition, when 1/2 power is lost, the train shall be able to prevent overfilling protection function.

(4) Green travel demand, the traction system can work to achieve the effect of energy saving and emission reduction, in order to respond to the national green travel policy call.

Table 1. Traction/braking performance indicators

Maximum vehicle speed (km/h)	Average acceleration under constant load (m/s^2)		Average deceleration rate (m/s^2)	
			Common brakes	Emergency braking
80	(0–40) km/h, $\geq$ 1.0	(0–80) km/h, $\geq$ 0.6	$\geq$1.0	$\geq$1.2
100	(0–45) km/h, $\geq$ 1.0	(0–100) km/h, $\geq$ 0.6	$\geq$1.0	$\geq$1.2
120	(0–50) km/h, $\geq$ 1.0	(0–120) km/h, $\geq$ 0.5	$\geq$1.0	$\geq$1.2

Note When the towing ratio is 1:1, (0–40) km/h average acceleration $\geq$ 0.83m/s^2, (0–80) km/h average acceleration $\geq$ 0.5 m/s^2

Note Normal load conditions (AW0–AW3), the average deceleration of braking from the highest running speed of the vehicle (including response time) to zero

The structured design of user requirements (see Fig. 2). Each requirement block diagram includes two main attributes, "ID" and "Text", where "ID" indicates the unique identification of the requirement, which is usually assigned automatically by the system. "Text" indicates the text description of the requirement [13]. In addition, "Derived Requirements" is chosen to represent the "derived requirement relationship" between the five requirements decomposed from the user requirements and the user top-level requirements.

2.3 Use Case Analysis

Use cases are a refinement of user requirements. Through use case analysis, the application scenarios of the traction system for metro trains, the interaction between the traction system and stakeholders, and thus the functions of the traction system can also be introduced. When a subway train is in operation, the traction system inverts the DC voltage on the contact network into a three-phase voltage with variable amplitude and frequency to provide the right energy for the traction motor operation. Its main tasks in subway train operation include electrical energy conversion, traction and braking. When the power

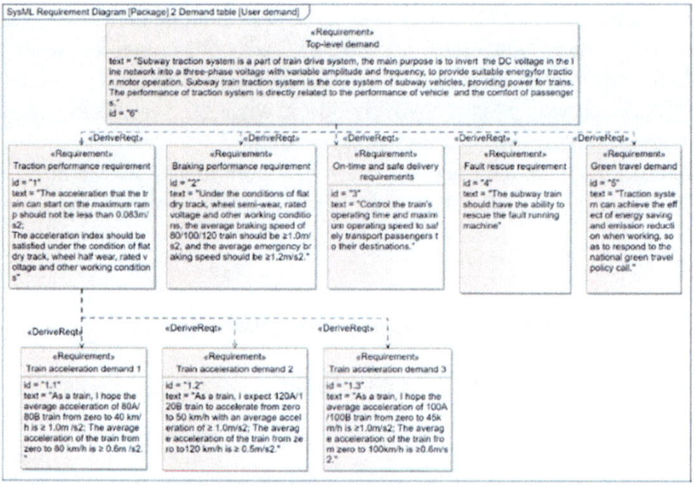

Fig. 2. User requirements

of the traction system is normal, traction braking can be carried out normally and the train operates normally; when the traction system loses part of its power, fault rescue is activated to ensure that the train travels to the next station or vehicle section.

Based on the above use case analysis, the use cases of metro train traction system are clarified in the business layer according to the use case analysis module, which are mainly divided into four use case scenarios: power conversion, traction control, brake control and fault operation (see Fig. 3).

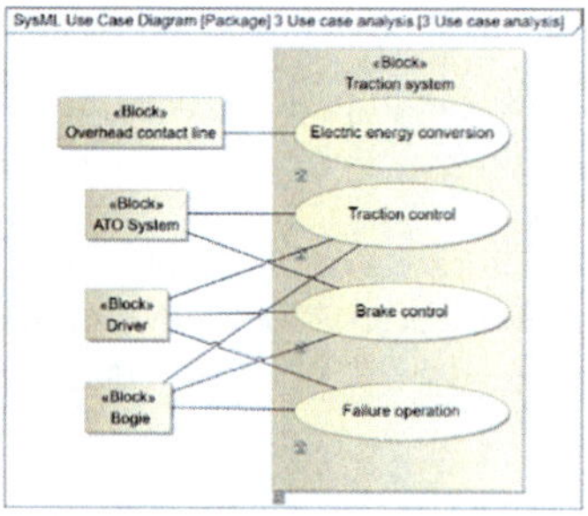

Fig. 3. Use case analysis

3 Functional Layer Modeling

An important task in the system concept design phase of a subway train traction system is to analyze what functions need to be designed to meet the business requirements of the target system, so a reasonable and feasible functional analysis is essential. The traction system is initially opened as a gray box to view the functions it has. The highest level is divided into four categories of functions, electrical energy conversion function, traction control function, brake control function and fault operation function, where the traction control function is divided into starting function, torque control function and speed control function (see Fig. 4).

Under the electric energy conversion function, the traction system converts the electric energy provided by the contact network into the electric energy used by the traction motor to supply the traction motor, and the traction motor works as an electric motor to convert the electric energy into mechanical energy to realize the electric energy conversion of the train; under the traction control function and brake control function, the traction system controls the torque and speed of the traction motor according to the driver's or ATO's instruction to realize traction and braking; under the fault operation function, the power battery acts as a power source to provide energy to the traction system to realize the fault operation of the train.

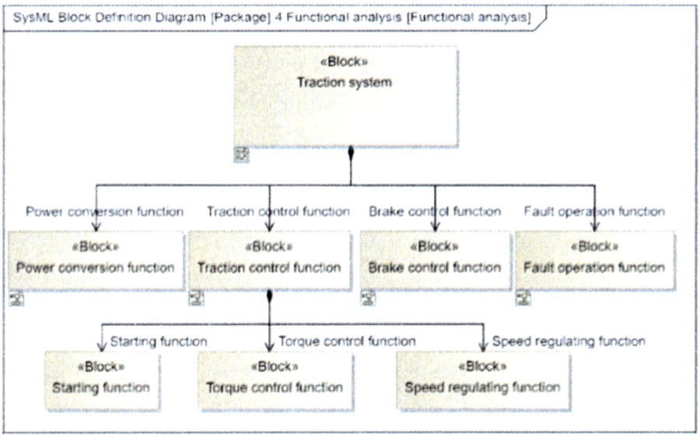

Fig. 4. Functional analysis

4 Architecture Layer Modeling

The metro train traction system is white-boxed on the basis of its business and functional layers. The system architecture model includes a complex structure obtained from different levels of abstraction and different perspectives of analysis, which contains various elements in addition to the interaction between different elements, so as to establish the metro train traction system architecture model, including High voltage unit, Inverter unit, Traction control unit, PWM encoder, Brake module, Power battery, AC traction motor, grounding device, DC/ DC converter and main controller (see Fig. 5).

At the same time, in order to refine the input-output relationship between the subsystems in the system during the operation of the device, the input-output relationship of the systems within the traction system is designed based on the black box use case. The Traction inverter converts the DC power into three-phase AC power with voltage and frequency regulation and sends it to the traction motor to make the train run. If the

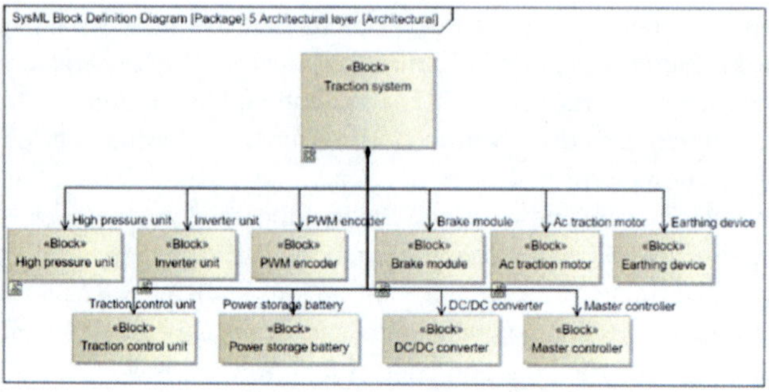

Fig. 5. System architecture

ATP system detects that the train is overspeeding, it will perform a traction blockade and disconnect the traction motor (see Fig. 6).

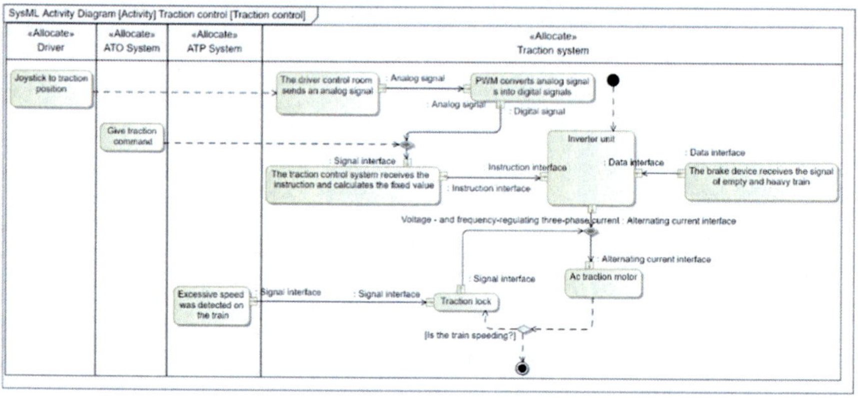

Fig. 6. Traction control activities

5 Physical Layer Modeling

The mapping of the logical architecture of the metro train traction system to the physical architecture of the subsystems in the traction system is a further refinement of the logical components. The main equipment of High voltage system includes Pantograph, Surge arrester, Main disconnecting switch, High-speed circuit breaker, etc. The main equipment of Traction inverter includes DC voltage sensor, Brake chopper, Filter capacitor, Brake current sensor, etc. In order to meet the demand of green travel, AC traction motor choose permanent magnet traction synchronous motor main equipment including stator, rotor, end cover, sensor, junction box, bearings, etc. The physical components (see Fig. 7).

Based on the physical architecture of the system, the functional activities of each component module of the system are further refined, and an internal module diagram is

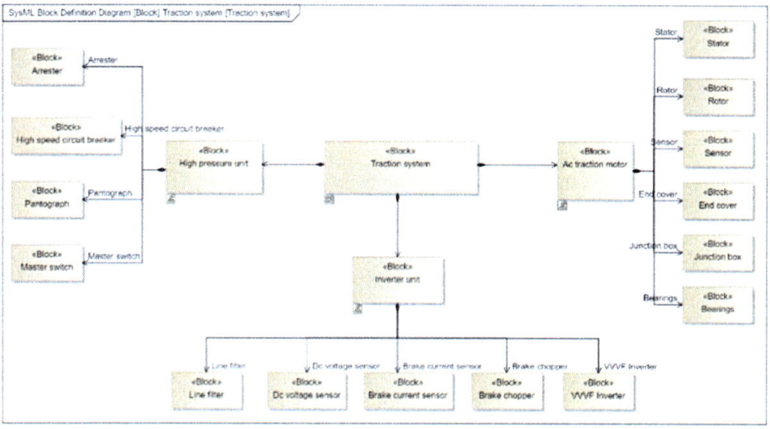

Fig. 7. Subsystem physical architecture

established to further clarify the interaction between the sub-modules of the system and refine the information interaction of the traction system (see Fig. 8).

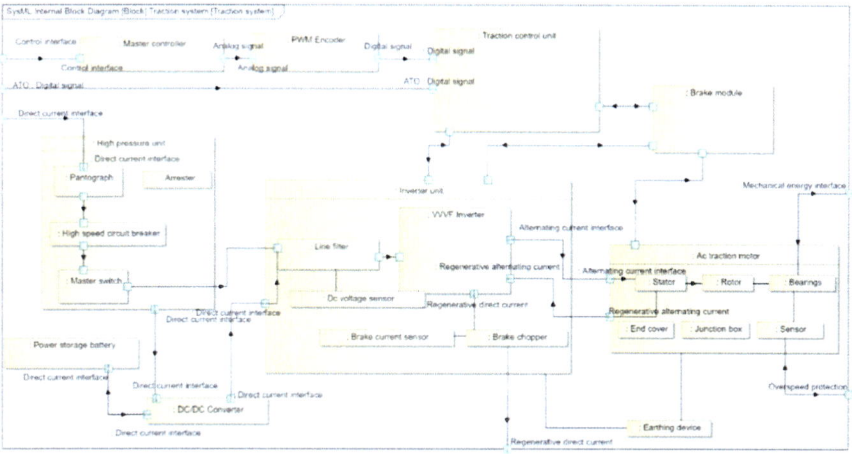

Fig. 8. Internal modules

6 Trace Matrix

After completing the design of the metro train traction system, it is necessary to verify that the user requirements are met. The traceability matrix from user requirements to use cases, functions and system logical architecture design (see Fig. 9). In Fig. 9a, shows the traceability from requirements to use cases. In Fig. 9b, shows the traceability from use cases to functions. In Fig. 9c, shows the traceability from functions to logical architecture. In Fig. 9d, shows the traceability from logical architecture to physical components. Realizes the mapping from user requirements to use cases, functions and system logical architecture design.

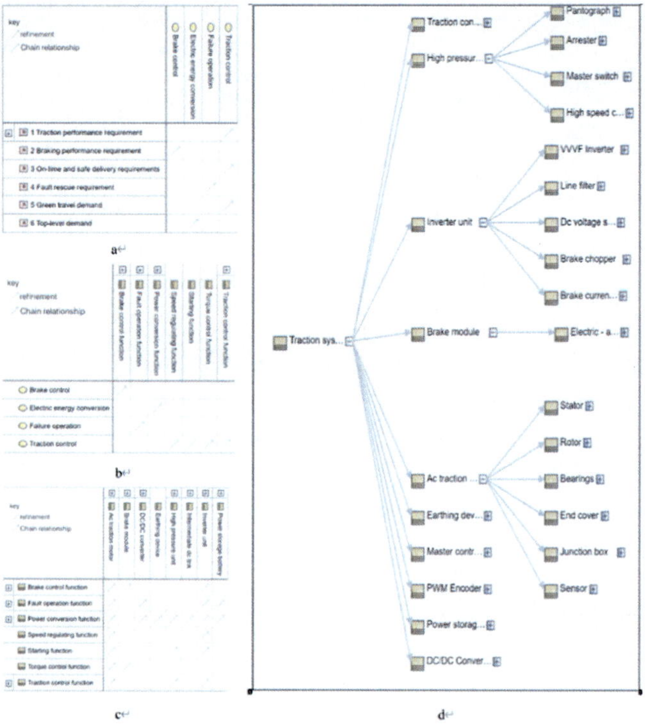

Fig. 9. Requirements traceability matrix

7 Concluding Remarks

In this paper, a model-based design method for metro train traction system is applied to the metro train traction system as the research object. First, with user requirements as input, the traction system is refined layer by layer to perform requirement use case analysis, functional decomposition and architecture design at the business, functional, architectural and physical layers, respectively. Second, after the traction system is designed, the mapping of user requirements to use cases, functions and logical architectures is realized according to the traceability matrix, which facilitates the rapid identification of affected layers and rapid iteration when requirements change, effectively ensuring the traceability and consistency of the metro train traction system design.

Acknowledgments. This study is supported by the Tianjin Municipal Education Commission Research Program Grant (2020KJ121) and the Tianjin University of Technology and Education Research Initiation Grant (KRKC012215).

References

1. Wang, B.M., Chen, B., Zhang, S.C.: Research method and application of MBSE-based whole vehicle logic for motive power trains. China Railway Sci **42**(06), 143–151 (2021). (in Chinese)
2. Daniel, R., Wibbe, N., Roberto, F.: Model-based systems engineering approach for the development of the science processing and operations center of the NASA OSIRIS-Rex asteroid sample return mission. ActaAstronaut **115**, 147–159 (2015)
3. Zhu, J.Z., Yang, Y., Wan, L., et al.: Preliminary application study of MBSE in nuclear power design. Nucl. Power Eng. **43**(01), 163–168 (2022). (in Chinese)
4. Cheng, L.: Model based system engineering and birtual iron bird solution. Aeronaut. Manuf. Technol. **3**, 100–102 (2013)
5. Li, D.C.: Research and Implementation of PREE Vision-Based Electrical Architecture Modelling for Automotive Electronics. Nanjing University of Posts and Telecommunications, Nanjing (2021). (in Chinese)
6. Spangelo, S.C., Cutler, J., Anderson, L., et al.: Model based systems engineering (MBSE) applied to radio aurora explorer (RAX) CubeSat mission operational scenarios. In: Proceedings of the IEEE Aerospace Conference (2013)
7. Larson, W.J.: Applied Space Systems Engineering. McGrew-Hill Learning Solution, New York (2009)
8. Aditya, A., Viviana, L.: A review on application of Model based systems engineering to manufacturing and production engineering systems. Proc. Comput. Sci. 185 (2021)
9. Liu, Y.X., Liu, X.X., Wang, B.M. et al.: Application of MBSE in the architecture design of bogie instability detection system for moving trains. Mech. Sci. Technol. 1–9. https://doi.org/10.13433/j.cnki.1003-8728.20230049. (in Chinese)
10. Yuan, W.Q., Wang, B.M., Chen, B., et al.: MBSE enables intelligent and efficient generation scheme in the field of high voltage system for rolling stock. J. Comput. Aided Des. Graph. **32**(06), 979–988 (2020). (in Chinese)
11. Wang, B.M., Chen, B., Zhang, S.C.: Research method and application of vehicle logic for EMU based on MBSE[J]. China Railway Sci. **42**(06), 143–151 (2021)
12. Liu, X.: Serialized Chinese standard metro train traction system design [J]. Locomotive electric drive, 2022, No. 285 (02): 55–66. https://doi.org/10.13890/j.issn.1000-128X.2022.02.009. (in Chinese)
13. Gao, J.Y., Wang, L.Y., Pan, Z.S., et al.: MBSE architecture modeling for mars maintenance and management device. Syst. Eng. Electron. **45**(05), 1441–1450 (2023). (in Chinese)

Research on Electromagnetic Characteristics of Linear Synchronous Motor of Superconducting High-Speed Maglev Train

Yuhao Zheng[1], Jingyu Huang[1,2(✉)], and Xiaonong Wang[2]

[1] Institute of Rail Transit, TongJi University, Shanghai 201804, China
huangjingyu@tongji.edu.cn
[2] National Maglev Transportation Engineering R&D Center, Tongji University, Shanghai 201804, China

Abstract. Taking the Japanese superconducting high-speed maglev train as the research object, a finite element model of the superconducting linear synchronous motor is established, the armature magnetic field characteristics of the superconducting linear synchronous motor under different air gap conditions are analyzed, and the variation law of the electromagnetic force in different directions of the superconducting linear synchronous motor with the amplitude of the armature current and the electric angle is studied. Simultaneously, by establishing a mathematical model, the characteristics of mutual inductance change between the superconducting magnet and track coil with the change in train position, the relationship between the vertical and lateral forces, the power angle of the linear synchronous motor, and the pulsation law of the lateral and vertical forces are analyzed. The results showed that the armature magnetic field density gradually converged to a sinusoidal wave with an increase in the air gap, and the vertical and lateral forces generated by the superconducting linear motor significantly influenced the operational stability of the magnetic levitation train.

Keywords: Finite element model · Linear synchronous motors · Electromagnetic properties · Armature magnetic field

1 Introduction

With economic development and progress, the demand for fast travel is gradually increasing, and high-speed magnetic levitation trains, as a new type of ultra-high-speed transportation, are characterized by energy conservation, low environmental impact, and strong hill-climbing ability. Superconducting magnetic levitation trains, a high-speed magnetic levitation transportation system, are represented by the Japanese MLX01 type EDS magnetic levitation train, which has the advantages of a simple structure and high reliability [1–4]. The traction of Japanese EDS maglev trains is driven by a long-stator linear synchronous motor, and the running section of the train is divided into several partitions. Only when the train enters the current partition is a three-phase alternating

L. Jia et al. (Eds.): EITRT 2023, LNEE 1135, pp. 56–67, 2024.
https://doi.org/10.1007/978-981-99-9307-9_7

current applied to the traction winding to generate a traveling wave magnetic field moving forward along the track. Because high-speed superconducting magnetic levitation trains use superconducting linear synchronous motors to obtain traction, it is important to study the electromagnetic force characteristics of high-speed superconducting magnetic levitation trains, which are important for designing and optimizing the structures of linear motors.

G. Martinelli proposed a three-dimensional analytical method to calculate the magnitude of levitation, drag and guiding forces of EDS magnetic levitation system [5]. Liu J. calculated the electromagnetic force along the vertical direction of superconducting magnets using semi-analytical model and finite element method respectively, and the results were in good agreement with the numerical calculations [6]. Ribani, He, Su and others calculated the levitation and guiding forces of the "8" shaped coil using dynamic circuit theory [7–10]. Lv G. et al. proposed a linear synchronous motor integrated with PLG and calculated the magnitude of the three-phase electromagnetic force generated by LSM at different speeds and gaps [11]. Li G. et al. proposed a current compensation-based thrust control method for the problem of thrust fluctuation of hollow-core motors [12]. Tang J. et al. studied the superconducting linear motor by numerical simulation and experimental measurement and analyzed the efficiency of the superconducting linear motor and other related problems [13]. Nonaka S et al. analyzed the end effect of the linear motor of an electric maglev train and studied the magnetic force line of a linear motor distribution [14]. From the above literature, it can be seen that scholars at home and abroad have conducted a lot of research on the electromagnetic characteristics of superconducting maglev systems, but the analysis of superconducting linear synchronous motors has not provided an in-depth analysis of their armature magnetic field characteristics and the effect of armature current on the electromagnetic force characteristics of linear synchronous motors of superconducting maglev trains. At the same time, the vertical and lateral forces of the linear synchronous motor of superconducting maglev trains have not been investigated further.

To compensate for the above research deficiencies, this paper takes the Japanese MLX01 superconducting high-speed maglev train as the research object, establish-es a finite element model of the superconducting linear synchronous motor, and analyzes its armature magnetic field characteristics. On the basis of this study, the variation law of electromagnetic force with electric angle and current amplitude in different directions of superconducting linear synchronous motor is quantified numerically, and the relationship between vertical and lateral forces and power angle of superconducting linear synchronous motor and the pulsation law of lateral force and vertical force are further analyzed and studied to provide data sup-port for the design of linear motor of magnetic levitation train and the study of traction characteristics of linear motor.

2 Modeling of Linear Motors for Superconducting Magnetic Levitation Trains and Study of Electromagnetic Force Characteristics

2.1 Finite Element Modeling of Linear Motors for Superconducting Magnetic Levitation Trains

Based on the relevant parameters of the Japanese MLX01 superconducting high-speed magnetic levitation train, a linear synchronous motor finite element model of the superconducting high-speed magnetic levitation train was established, as shown in Fig. 1. The linear motor finite element model mainly consists of three structural layers: the stator layer, the air gap layer, and the kinematic layer [15]. The electromagnetic relationship of the superconducting magnet is described by Eq. (1), where H is the magnetic field strength, μ is the magnetic conductivity. $E(J)$ is the electric field associated with the current, and the expression is calculated as in Eq. (2), where J is the current density vector in the primary coil, J_c is the critical current density, α and E_0 is the nonlinear process constant affecting the resistance jump to zero resistance. The electromagnetic force on the superconducting magnet in the finite element model is given by Eq. (3), where V is the air domain and B is the flux density vector.

$$\nabla \times \vec{E}(\vec{J}) = -\mu \frac{\partial \vec{H}}{\partial t} \tag{1}$$

$$E(J) = \begin{cases} 0 \, |J| < J_c \\ E_0 (\frac{|J| - J_c}{J_c})^{\alpha} \frac{J}{|J|} \, |J| \geq J_c \end{cases} \tag{2}$$

$$F = \iiint_V B \times J dv \tag{3}$$

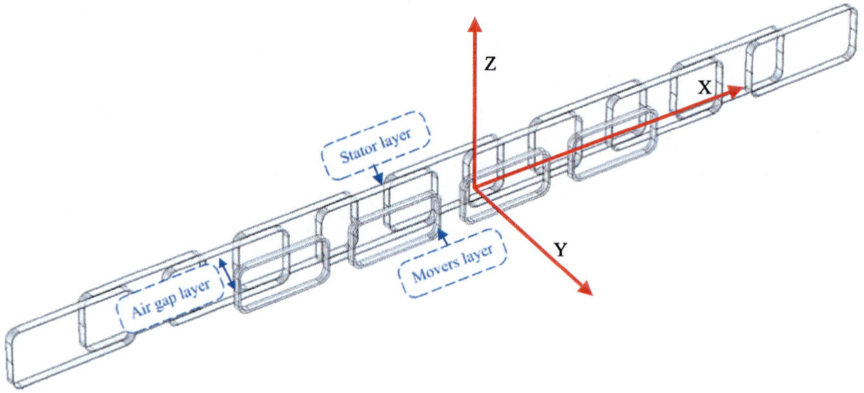

Fig. 1. Finite element model of a superconducting linear synchronous motor

2.2 Analysis of Armature Magnetic Field Characteristics of Superconducting Linear Synchronous Motor

The positions of the superconducting electric maglev train LSM armature winding and the superconducting magnet are shown in Fig. 2. To investigate the superconducting linear synchronous motor armature magnetic field characteristics under different space air gap conditions, based on the finite element model of a superconducting linear synchronous motor built in Sect. 2.1, a three-phase AC cur-rent with a frequency of 50 Hz and RMS size of 1000 A is passed into the traction coil, and the SC coil current is shown in Fig. 3. When the air gap increased, the maximum values of the armature magnetic field flux density Y-component are 0.05 T, 0.012 T and 0.009 T, and the armature magnetic field flux density was almost two times lower than that of the SC coil. The flux density generated by the armature magnetic field is almost two orders of magnitude lower than that generated by the SC coil, that is, the low flux density generated by the traction coil on the SC coil is conducive to reducing the eddy current and hysteresis loss inside the SC coil, which is in line with the design principle of the superconducting LSM. The harmonic spectral decomposition of the armature flux density at different air gaps using the numerical integration method is shown in Fig. 4. Figure 4 shows that the harmonic amplitude of the armature magnetic field density improved significantly with an increase in the air gap and gradually converged to a sinusoidal waveform.

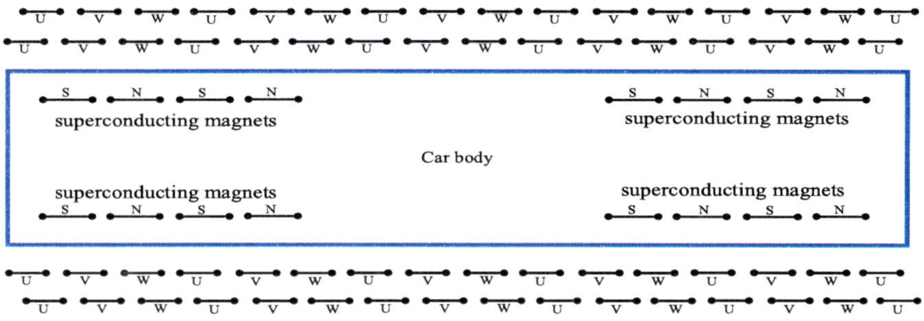

Fig. 2. Armature winding and superconducting magnet position

2.3 Analysis of Traction Characteristics of Linear Motors for Superconducting Maglev Trains

Owing to the fast running speed of superconducting high-speed magnetic levitation trains, to ensure safe and accurate running of trains, the operating conditions and speed of the vehicles, such as acceleration, deceleration, and stopping, are completely controlled directly by the ground control center. The ground control center realizes the control of train running speed by adjusting the frequency (phase) of the current delivered to the rail winding by the substation, and the control of train driving force by changing the size (amplitude) of the current, so based on the synchronous linear motor model of superconducting high-speed maglev train in Sect. 2.1, when the current size through

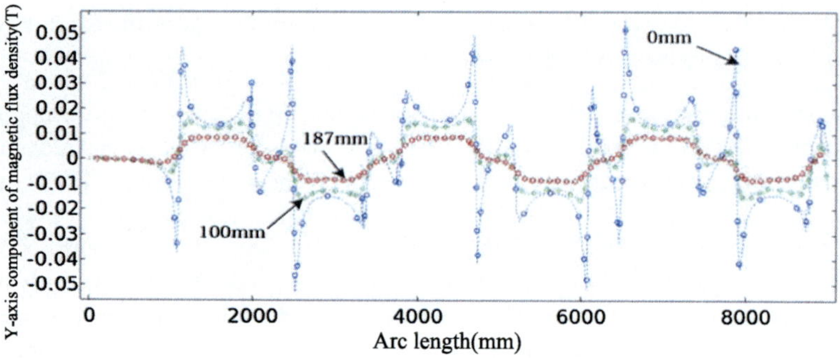

Fig. 3. Trend of the y-component of armature flux density along the running direction of the train at different centre distances

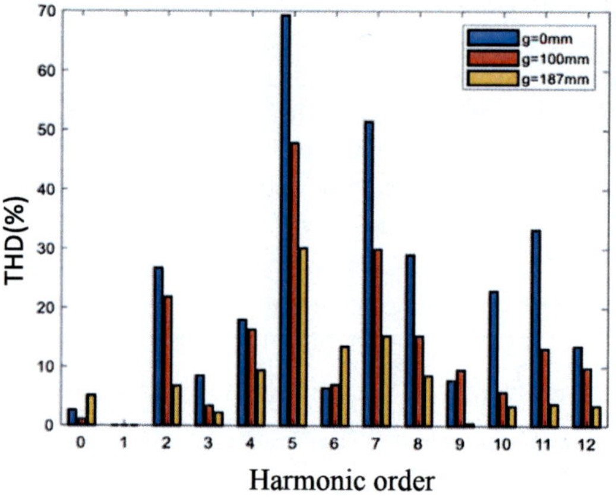

Fig. 4. Superconducting LSM armature magnetic field density harmonic spectrum

the rail coil is 1000 A, 1100 A, 1200 A, 1300 A, 1400 A, through the coupling effect of the magnetic field of the track side and superconducting coil, the change of tractive force along the running direction of superconducting maglev train, i.e., X-direction, is shown in Fig. 5a, and it can be seen that the tractive force along the X-direction increases linearly with the increase of the current amplitude, which is in line with the tractive characteristic requirement of superconducting linear synchronous motor, and the magnitude of tractive force in one cycle is not 0 in the root mean square value There-fore, the magnitude of traction force accumulates during the cycle, which is beneficial to the starting of the superconducting coil on the secondary side. The maximum traction force was 15.712KN when the current size was 1000 A. The lateral and vertical forces in the Y and Z directions are generated during the superconducting magnetic levitation train, and their magnitude changes are shown in Fig. 5b, c, respectively. The lateral force affects the guidance system of the train, and the vertical force affects the levitation system of the train during operation of the superconducting magnetic levitation train. Therefore, the vertical and lateral forces generated by the superconducting linear motor should be considered in the design of superconducting magnetic levitation trains to affect their operational stability.

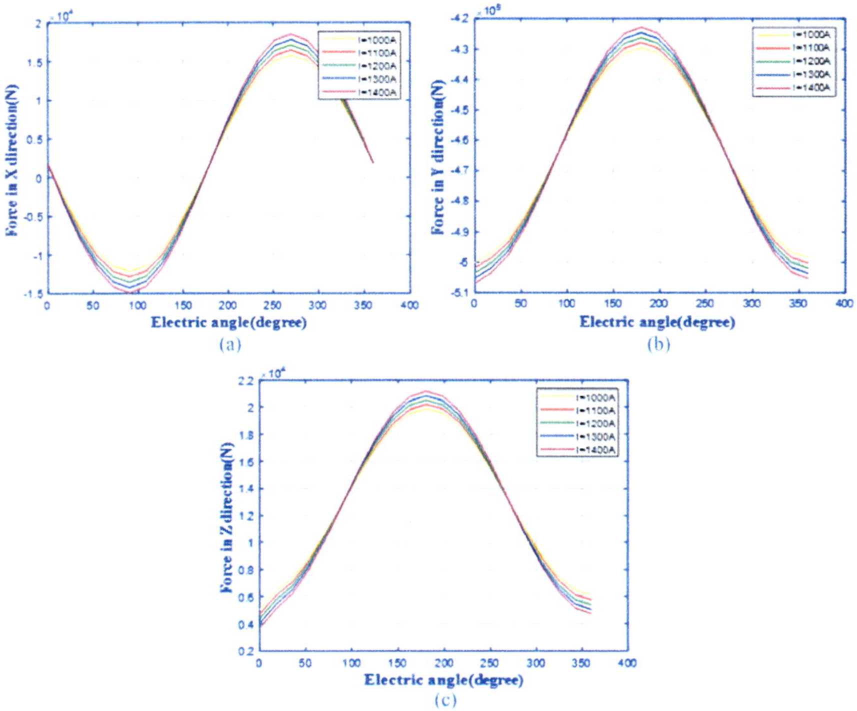

Fig. 5. Superconducting linear synchronous motor force magnitude at different armature currents

3 Analysis of Vertical and Lateral Forces in Linear Synchronous Motors of Superconducting Maglev Trains

In Sect. 2.3, the simulation and study of the vertical and lateral forces of superconducting linear synchronous motors are presented. Because the magnitudes of the lateral and vertical forces affect the stability of the superconducting high-speed magnetic levitation guidance system, and the magnitudes of the vertical and lateral forces are related to the change in power angle, the magnitudes of the lateral and vertical forces are considered while considering the magnitude of the traction force in the control process of the magnetic levitation vehicle operation to further improve the stability of the high-speed magnetic levitation system. Assuming that the primary-side traction coil is rectangular, the expressions for calculating the vertical and lateral forces are as shown in Eqs. (4) and (5) for the a-phase current in the three-phase circuit system, where I_0 is the current in the superconducting coil (A), i_a is the A-phase current in the track coil (A), and M_{st}

is the mutual inductance between the superconducting coil and the track coil (a-phase).

$$F_z = NI_0 i_a \frac{\partial M_{st}}{\partial z} \tag{4}$$

$$F_y = NI_0 i_a \frac{\partial M_{st}}{\partial y} \tag{5}$$

The relative positions and parameters of the superconducting magnet coil and the a-phase orbital coil are shown in Fig. 6a, b, respectively, and the expressions for the mutual inductance of the superconducting coil and the a-phase orbital coil can be obtained according to the Riemann formula as in Eq. (6).

$$M = \frac{\mu_0}{4\pi} \int_{c1} \int_{c2} \frac{dxdx' + dydy' + dzdz'}{\left[(x - x')^2 + (y - y')^2 + (z - z')^2\right]^{1/2}} \tag{6}$$

where (x, y, z) and (x', y', z') are the coordinates of two points, one is on the orbital coil and the other on the magnetic coil, c_1 and c_2 are the border of the two coils. Suppose $(x'_{0i}, y'_{0i}, z'_{0i}, i = 1, 2, 3, 4)$ that are the coordinates of A', B', C', D', and $(x_{0j}, y_{0j}, z_{0j}, j = 1, 2, 3, 4)$ are the coordinates of A, B, C, D, and so Eq. (6) can be transformed into Eq. (7).

$$M = \frac{\mu_0}{4\pi} \sum_1^4 \sum_1^4 (-1)^{i+j} \left[\psi(x_{0i}, y_{0i}, 0; x'_{0i}, y'_{0i}, z_0) + \psi(y_{0j}, x_{0j}, 0; y'_{0i} x'_{0i}, z_0)\right] \tag{7}$$

where: $\psi(x, y, z; x', y', z') = -(x - x')\ln[(x - x') + \sqrt{(x - x')^2 + (y - y')^2 + (z - z')^2}]$ $+\sqrt{(x - x')^2 + (y - y')^2 + (z - z')^2}$

Because the superconducting high-speed magnetic levitation train has infinite mutual inductance between the track coils of alternating polarity and magnet coils, the total mutual inductance expression is given by Eq. (8).

$$\begin{aligned} M_{total} =& (-1)^{n+1} \frac{\mu_0}{4\pi} \sum_{n=0}^{\infty} \sum_1^4 \sum_1^4 (-1)^{i+j} \left[\psi(x_{nj}, y_{nj}, 0; x_{0i}, y_{0i}, z_0)\right. \\ &\left. + \psi(y_{nj}, x_{nj}, 0; y'_{0i} x_{0i}, z_0)\right] \\ &+ (-1)^n \frac{\mu_0}{4\pi} \sum_{n=0}^{\infty} \sum_1^4 \sum_1^4 (-1)^{i+j} \left[\psi(x_{-nj}, y_{-nj}, 0; x'_{0i}, y'_{0i}, z_0)\right. \\ &\left. + \psi(y_{-nj}, x_{-nj}, 0; y'_{0i} x_{0i}, z_0)\right] \end{aligned} \tag{8}$$

As the superconducting high-speed railroad will consist of a number of ground propulsion windings in series with each other for each partition, the ground winding of each partition is only connected to power when the vehicle passes through the partition, and the length of each partition is generally 1200 m. Therefore, it is assumed that the number of track coils in each partition is 1335, the speed of the vehicle is 135 m/s, and the mutual inductance change value between the track coils and superconducting coils is shown in Figs. 7a, which shows that the mutual inductance is a periodic function

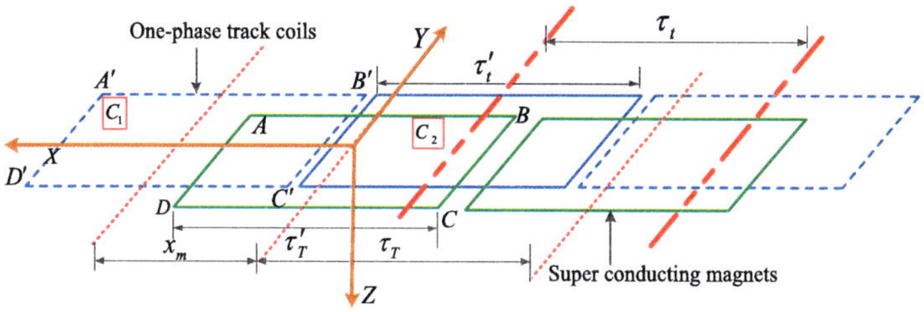

(a) Perpendicular position and coordinates of the superconducting magnet and the a-phase orbital coil

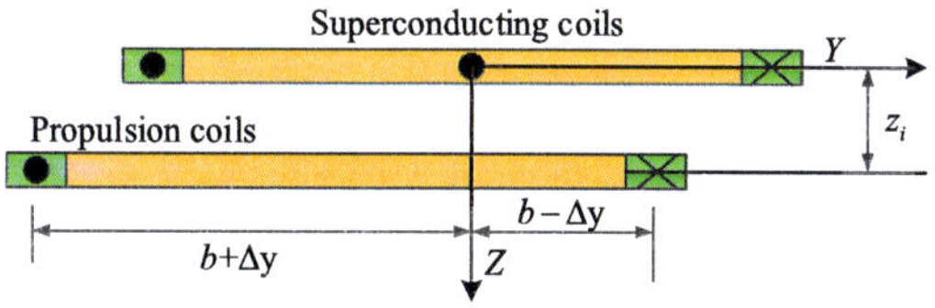

(b) Horizontal coordinates of the superconducting magnet and the a-phase orbital coil

Fig. 6. Superconducting coil and a-phase orbital coil coordinates

of time. Therefore, in the superconducting linear synchronous motor, the fundamental and third-harmonic components of the track current are further considered, as shown in Eqs. (9) and (10), respectively.

$$i_1 = \sqrt{2}I\cos(wt-\varphi) \tag{9}$$

$$i_3 = -\sqrt{2}I_3\cos(3wt-3\delta-\varphi_3) \tag{10}$$

where $\varphi_3 = \arctan[3w(L/R)]$. Because the orbital voltage is triangular connection, the expressions for the total vertical and lateral forces of the three phases can be obtained using Eqs. (11) and (12). The variation characteristics of the total mutual inductance in the Y and Z directions are shown in Fig. 7b, c, where N is the number of magnets, M_{st1}, M_{st3}, M_{st5} are the first harmonic, third harmonic and fifth harmonic of the mutual inductance of the superconducting magnet coil and the orbital coil, respectively. The Y and Z directions of Eq. (8) are used to find the partial derivatives, and the values of different directions and harmonics of different orders can be obtained by numerical integration.

$$\begin{aligned} F_{zr} =& 3NI_0\left\{I\frac{\partial M_{st1}}{\partial z}\sin(\delta-\varphi) - I_3\frac{\partial M_{st3}}{\partial z}[\sin\varphi_3 - \sin(6wt-6\delta-\varphi_3)]\right. \\ &\left. + I\frac{\partial M_{st5}}{\partial z}\sin(6wt-5\delta-\varphi)\right\} \end{aligned} \tag{11}$$

$$F_{zr} = 3NI_0\left\{I\frac{\partial M_{st1}}{\partial y}\sin(\delta-\varphi) - I_3\frac{\partial M_{st3}}{\partial y}[\sin\varphi_3 - \sin(6wt-6\delta-\varphi_3)]\right.$$

$$+I\frac{\partial M_{st5}}{\partial y}\sin(6wt-5\delta-\varphi)\Bigg\} \tag{12}$$

When the input voltage is 4200 V and the input frequency is 50 Hz, the relationship between the vertical and lateral forces of the superconducting linear synchronous motor and the power angle is shown in Fig. 8. It can be seen that the attractive vertical force of the superconducting linear synchronous motor changes to a repulsive vertical force and the restoring lateral force changes to a centrifugal lateral force when the power angle, thus, the restoring lateral force always accompanies the appearance of the attractive vertical force, and conversely, the centrifugal lateral force is always accompanied by repulsive a vertical force. When the pulsation curve of the lateral force in one cycle is shown in Fig. 9a, it can be observed that the pulsation frequency of the magnitude of the vertical and lateral force values of the superconducting linear synchronous motor is six times the fundamental frequency. When the vertical displacement is 0.03 m, the value of the mutual inductance change is calculated sequentially using the above method. The pulsation curve of the vertical force at time is shown in Fig. 9b, and the average vertical force in one cycle is 775.5629 N, which shows the repulsive vertical force.

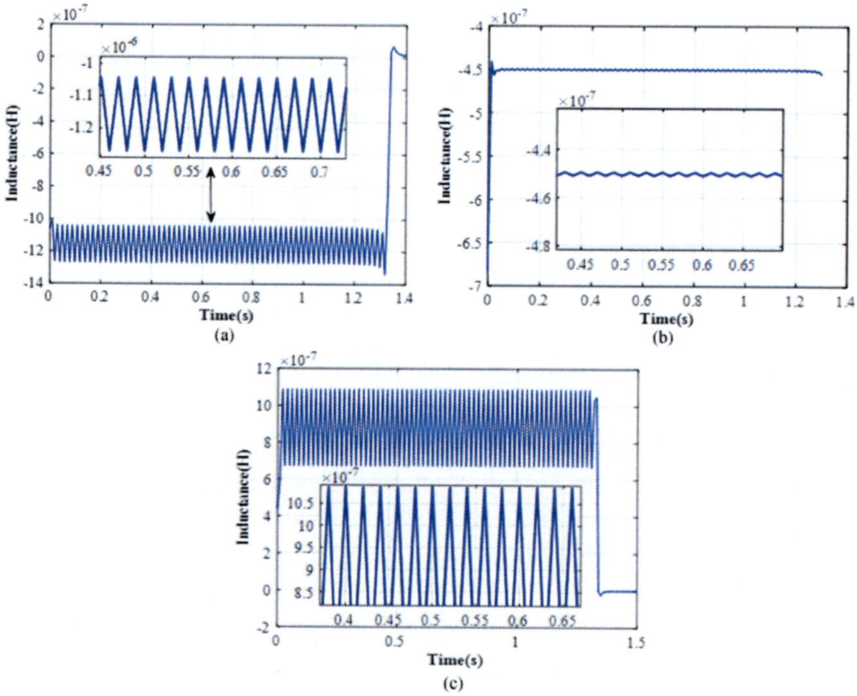

Fig. 7. Mutual inductance variation characteristics of superconducting magnets and orbital coils

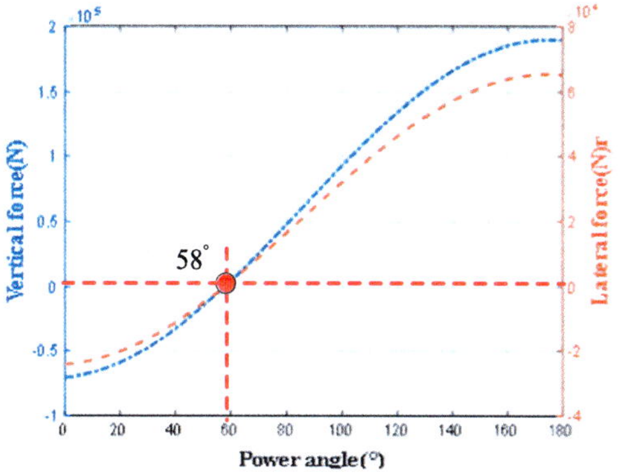

Fig. 8. The relationship between vertical and lateral forces and power angle

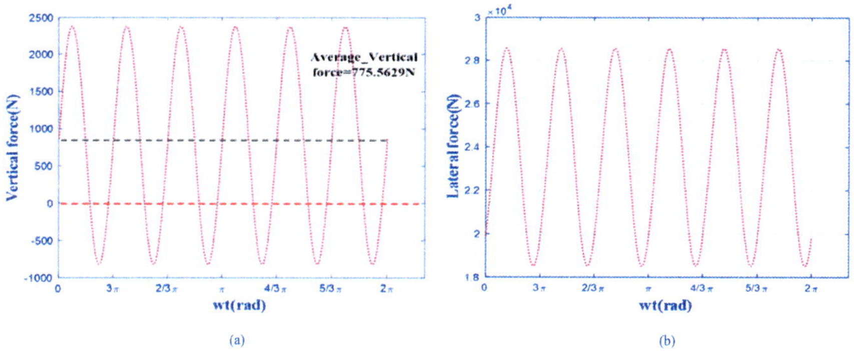

(a) Lateral force pulsation curve at a power angle of 70 degrees

(b) Vertical force pulsation curve when vertical displacement is 0.03m and power angle is 70 degrees

Fig. 9. Lateral and vertical force pulsation curve at a power angle of 70°

4 Conclusion

In this paper, the superconducting high-speed magnetic levitation train of Japan was used as the research object, and the armature magnetic field and electromagnetic force characteristics of superconducting linear motors were quantitatively analyzed and explored. The vertical and lateral forces generated by superconducting linear synchronous motors are discussed in detail, and the following conclusions are drawn:

(1) The armature magnetic field produces a flux density almost two orders of magnitude lower than that produced by the superconducting coil, which helps reduce the eddy current and hysteresis losses inside the SC coil. Simultaneously, the armature magnetic field density gradually converged to a sine wave as the air gap increased.

(2) The magnitude of the train traction force increased linearly with an increase in the armature current amplitude, and the root-mean-square value of the traction force was not zero in one cycle. However, the lateral force of the superconducting linear synchronous motor affects the guiding system of the train, whereas the vertical force affects the suspension system of the train. Therefore, the influence of the vertical and lateral forces generated by the superconducting linear motor on the operational stability of the maglev train should be considered.
(3) The recovered lateral force of a superconducting linear synchronous motor is always accompanied by attractive vertical force. Conversely, the centrifugal lateral force was always accompanied by the push-repulsion vertical force, and the pulsation frequency of the vertical and lateral forces values was six times the fundamental frequency.

References

1. Fujie, J.: Current status of superconducting Maglev system. In: Advances in Superconductivity III: Proceedings of the 3rd International Symposium on Superconductivity (ISS'90), November 6–9, 1990, Sendai. Springer Japan, 1991, pp. 1263–1268
2. Ono, M., Koga, S., Ohtsuki, H.: Japan's superconducting Maglev train. IEEE Instrum. Meas. Mag.Instrum. Meas. Mag. **5**(1), 9–15 (2002)
3. Deng, Z., Zhang, W., Wang, L., et al.: A high-speed running test platform for high-temperature superconducting maglev. IEEE Trans. Appl. Supercond.Supercond. **32**(4), 1–5 (2022)
4. Ge, Q., Kou, B., Zhang, Y., et al.: 3-D electromagnetic force characteristics and modeling of double-sided air-cored superconducting linear synchronous motor for EDS train. IEEE Trans. Transp. Electrif. **8**(3), 3411–3426 (2022)
5. Carbonari, N., Martinelli, G., Morini, A.: Calculation of levitation, drag and lateral forces in EDS-MAGLEV transport systems. Archiv für Elektrotechnik (Berlin) **71**(2), 138–148 (1988)
6. Liu, J., Wang, Q., Yan, L.: Analysis of force characteristics of a superconducting ball in a given magnetic field. Phys. C **469**(13), 756–759 (2009)
7. Ribani, P.L., Urbano, N.: Study on figure-eight-shaped coil electrodynamic suspension magnetic levitation systems without cross-connection. IEEE Trans. Magn.Magn. **36**(1), 358–365 (2000)
8. He, J.L., Rote, D.M., Coffey, H.T.: Applications of the dynamic circuit theory to maglev suspension systems. IEEE Trans. Magn.Magn. **29**(6), 4153–4164 (1993)
9. Su, Z., Ma, G., Luo, J.: Analytical Modeling of the closed-loop operation and quench behavior for superconducting electrodynamic suspension train. IEEE Trans. Trans. Electrif. **8**(1), 1026–1039 (2021)
10. Burkhardt, E.E., Schwartz, J., Nakamae, S.: Analysis of superconducting magnet (SCM)-ground coil interactions for EDS Maglev coil configurations. IEEE Trans. Appl. Supercond.Supercond. **3**(1), 430–433 (1993)
11. Lv, G., Zhang, Z., Liu, Y., et al.: Characteristics analysis of linear synchronous motor integrated with propulsion, levitation, and guidance in high-speed maglev system. IEEE Trans. Transp. Electrif. **7**(4), 3185–3193 (2021)
12. Li, G., Li, J.: Design and control of a superconducting DC linear motor for electromagnetic launch system. Wirel. Pers. Commun.. Pers. Commun. **102**, 611–628 (2018)
13. Tang, J., Li, J., Li, X., et al.: Numerical studies on the force characteristic of superconducting linear synchronous motor with HTS bulk magnet. Phys. C Superconduct. Appl. **546**, 44–49 (2018)

14. Nonaka, S., Hirosaki, T., Kawakami, E.: Analysis of characteristics of repulsive magnetic levitated train using a space harmonic technique. Trans. Inst. Electr. Eng. Japan B **100**(10), 601–608 (1980)
15. Wang, X., Huang, J., Fang, Z. (2022) Research on EDS propulsion characteristics of superconducting high speed maglev train. In: Proceedings of the 5th International Conference on Electrical Engineering and Information Technologies for Rail Transportation (EITRT) 2021: Novel Traction Drive Technologies of Rail Transportation, pp. 20–28. Springer Singapore, Singapore (2022)

NSGA-II Genetic Algorithm Based Optimization of Parameters of Multiphase Interleaving Buck/Boost DC-DC Converter

Haotian Yang, Qing Lv, and Keling Song(✉)

China North Vehicle Research Institute, Beijing 100072, China
autumnscene18@gmail.com, sk172@vip.sina.com

Abstract. Buck/Boost DC-DC converters can be used in grids with bidirectional power flows. In the actual designing process of a distribution box, power loss would cause converter to heat, which might cause damage to components in the converter if the heat generated greater than the thermal dispersion capacity of the converter. While ripple currents and the volume are also considered in distribution box designing. This paper considers the above-mentioned factors, establishes objective functions, and use NSGA-II genetic algorithm to optimize the parameter design of multiphase interleaving Buck/Boost DC-DC converters. Then several relatively better outputs have been selected, scored and weighted to estimate working performance under the parameter sets. Finally, the very best parameter set of the converter can be obtained.

Keywords: NSGA-II Algorithm · Multi-objective Optimization · Multiphase interleaving Buck/Boost DC-DC Converter

1 Introduction

Buck/boost converter is a single-switch un-isolated DC-DC converter with an output voltage that can either be lower or be higher than the input voltage. While the input voltage polarity is the same as the output voltage.

When the IGBT of the converter operates in hard switching mode, there will be switching losses and conduction losses, when the operating frequency of IGBT increases, the switching loss would increase. For the inductors, due to hysteresis effects, eddy current effects, and their own resistance, heat would generate during operation of the converter, while improper selection of inductance parameters can cause a sharp increase in inductance loss, what's more, when the temperature of the inductance increases, the inductance value will correspondingly decrease, even causing saturation of the inductance. These effects can cause an increase in local internal temperature, which may cause permanent damage to the components in the converter. In the actual designing, the heat dissipation capacity of the device is limited, so it is necessary to limit the heat generation of the converter to a certain range, to avoid a low efficiency or damage to the converter due to heating during operation.

L. Jia et al. (Eds.): EITRT 2023, LNEE 1135, pp. 68–79, 2024.
https://doi.org/10.1007/978-981-99-9307-9_8

Power loss have long been a significant parameter in converter designing [1, 2]. in terms of power loss analysis of a interleaving converter, Fan et al. [3] made analysis to the bidirectional Buck/Boost circuit operating in hard switching mode, and made physical testing and verification. Various algorithms including PSO algorithm [4, 5], NSGA based algorithms [6, 7], Grey Wolf optimization algorithm [8] have been taken into consideration into converter designing.

At the same time, in the actual design process, the overall occupied volume of the converter and the stability of the generated voltage are also very important indicators.

2 Multi-phase Interleaving Buck/Boost Circuit

The solution process for multi-objective optimization problems in multiphase interleaving Buck/Boost circuit is divided into three parts: analysis of the circuit, optimization function analysis, and multi-objective parameter optimization, the process is shown in Fig. 1.

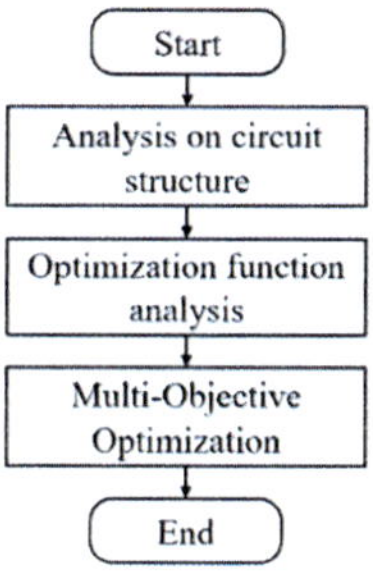

Fig. 1. Flowchart of optimization process of Buck/Boost circuit

2.1 Structure of Buck/Boost Circuit

Figure 2 shows the basic structure of a Buck/Boos circuit. In Fig. 2, U_1 is the voltage of high-voltage side, U_2 is the voltage of low-voltage side, VQ_1 and VQ_2 are IGBT switching tubes, D_1 and D_2 are parasitic diodes of the IGBTs, L is the inductor, C_2 is the output filtering capacitor.

During the operation process of the circuit, taking example of the circuit working in Buck mode with a hard switch and continuous current mode, the IGBTs conduct alternately. The conduction waveform of VQ_2 and the current waveform flowing through the inductor L are shown in Fig. 3:

Multiphase interleaving converter, refers to the parallel connection of both input and output of N single-phase converters, and the phase difference of each PWM drive signal is $\theta = 2\pi/N$. The use of multiphase interleaving can reduce ripple current and improve the stability of the output, improving the power that the bidirectional Buck/Boost converter can transfer. Compared to other methods, it can reduce the volume and weight of the converter and improve system reliability. The topology structure of a three-phase interleaving converter is shown in Fig. 4.

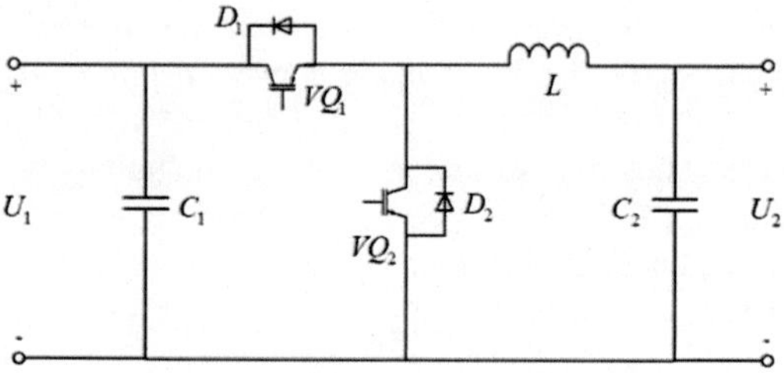

Fig. 2. Buck/Boost circuit structure

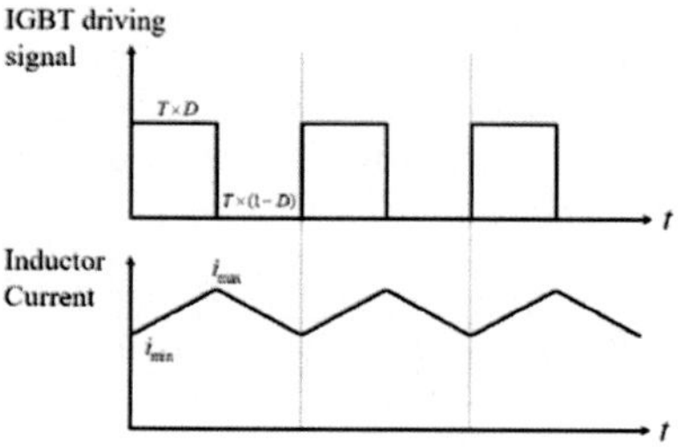

Fig. 3. IGBT driving signal waveform and inductive current waveform

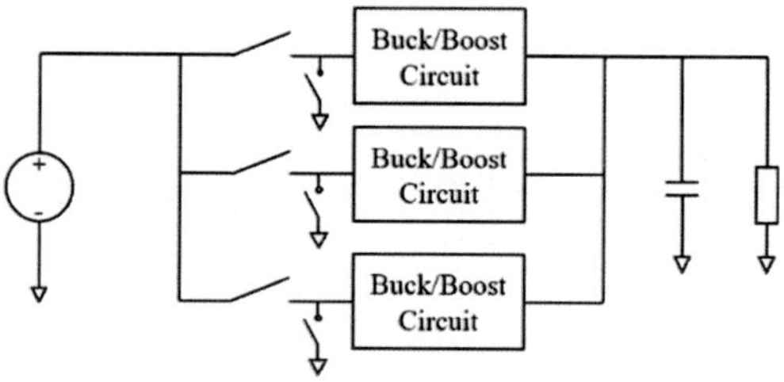

Fig. 4. Topological structure of a three-phase interleaving converter

2.2 Optimization Objectives and Constraints

The optimization objectives of the Buck/boost converter discussed in this article are divided into two aspects: one is the power loss of the converter during operation, the other is to evaluate the volume and ripple current of the converter.

(1) Power loss functions

The losses of the Buck/Boost circuit are mainly IGBT losses and inductance losses, and there is also a small amount of loss for the capacitors, which is small compared to the others and is ignored in calculation.

(a) Power loss of IGBTs

The losses of IGBT are mainly conduction losses and on-off losses. The commonly used IGBT in practice refers to integrated IGBT with auxiliary freewheeling diode. Diodes are used to protect IGBTs or achieve ZVS switches for IGBTs. When reverse recovery occurs, the diodes will generate losses.

In hard switching mode, when the converter is operating at rated power, there is no circulating current for the Buck/Boost converter, and the working current of the auxiliary transistor of the IGBT module is small. Considering the ideal situation, ignoring the auxiliary tube loss to the general Buck/Boost circuit, the total conduction loss during IGBT conduction and diode conduction is shown in Eq. (1).

$$P_1 = V_Q I_o \frac{T_{on}}{T} + V_D I_o \frac{T_{off}}{T} \tag{1}$$

During the open and shut-down periods of IGBT, the loss of IGBT is:

$$P_2 = \frac{1}{6} V_{DC} I_o \times \frac{T_{qon} + T_{qoff}}{T} \tag{2}$$

For a single IGBT, the total loss of IGBT is the sum of P_1 and P_2:

$$P_{\text{IGBT}} = P_1 + P_2 \tag{3}$$

In these formulas, V_Q is the conduction voltage drop of IGBT, V_D is the on-state voltage drop of the diode, I_0 is the average current flowing through IGBT, T is the switching period, $T = 1/f$, T_{on} is the turn-on time in a single period, T_{off} is the turn-off time in a single period, V_{DC} is the voltage value on the high-voltage side, $T_{q,on}$ and $T_{q,off}$ are the opening and closing time during switching.

(b) Power loss of inductors

The loss of an inductor mainly includes the magnetic core loss of the inductor, including losses caused by hysteresis and eddy current effects, as well as losses caused by resistance.

The sum of hysteresis loss and eddy current loss is calculated using the empirical formula of a certain type of magnetic material:

$$P_{\text{loss,core}} = \left(\frac{\Delta B}{2}\right)^{\beta} (af + bf^{\alpha}) \tag{4}$$

where $P_{loss,core}$ is the power loss per unit volume of the iron core, with a unit of mW/cm^3, ΔB is change in magnetic flux density in *kGauss*, f for switching frequency in *kHz*, α and β are empirical parameters which may not be an integer, a and b are constants.

ΔB can be calculated using the following equation, where the inductance unit is in μH, cross-sectional area of iron core in mm^2.

$$\Delta B = \frac{L \times \Delta I}{A_e \times N} \tag{5}$$

For the current variation ΔI in the equation, taking the converter running in Buck mode as an example, when operating in the mode that the inductance current is continuous, and the inductance current changes linearly during circuit operation, the following formula is approximately used for calculation:

$$\Delta I_{buck} = \frac{V_{o,buck}}{Lf}(1 - D) \tag{6}$$

For ΔI in Boost mode, there is:

$$\Delta I_{boost} = \frac{V_{o,boost}}{Lf}D(1 - D) \tag{7}$$

In the above equations: V_0 stands for high side voltage, f for switching frequency, D for duty cycle, A for cross-sectional area of iron core.

The formula for calculating the inductance value is:

$$L = A_L \times N^2 \tag{8}$$

where N is the number of windings of the inductor.

Comprehend formula (4)–(8), the loss of the iron core can be obtained.

The loss due to resistance of the inductor could be calculated using the following equation set:

$$\begin{cases} P_{loss,Cu} = I_{rms}^2 \times R_{Cu} \\ R_{Cu} = \rho_{Cu}\frac{length}{S} \end{cases} \tag{9}$$

The total length of the wire in the formula is taken as a fixed value.

(c) Total power loss function

The total loss function is the sum of IGBT loss and total inductance loss, and the total loss is described as follows:

$$P_{loss} = P_{IGBT} + P_{loss,Core} + P_{loss,Cu} \tag{10}$$

In the formula of total power loss, the cross-sectional area of iron core and wire, the high side voltage V_0, the duty cycle under steady-state operating condition D, and the four parameters in the empirical formula for iron core power loss calculation are constants. The variables are the switching frequency f, the number of turns of inductance N_{wind}, and the number of phases N_{phase}.

(2) Evaluation function

(a) Volume evaluation function

The internal local structure of an actual two-phase interleaving converter is shown in Fig. 5, where the circuit board is an IGBT module, and the black part in the middle is the filtering capacitor. It can be seen that the capacitor needs to occupy a large volume.

When the material and withstand voltage level of the capacitor are the same, the capacity of the capacitor is positively correlated with its volume. Interleaving converter, which means parallel connection of multiple circuits with the same structure. Therefore,

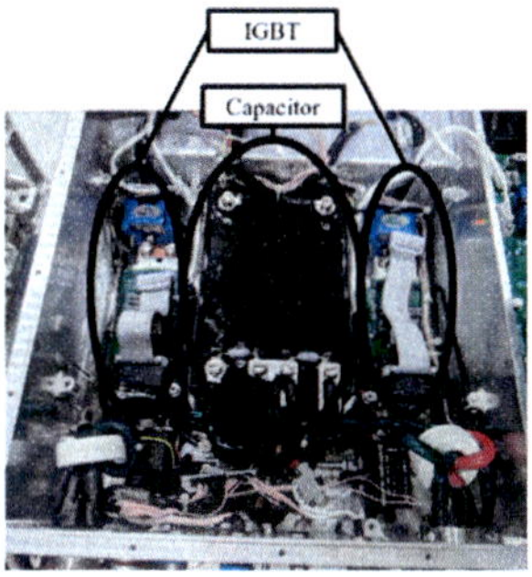

Fig. 5. Photograph of a two-phase interleaving converter

we will roughly evaluate the volume of the converter by multiplying the sum of input capacitance and output capacitance with the number of phases as part of the expression, then square root it to not make the function differences to obvious during optimization, which finally is:

$$f_{volume} = f\left(C_1, C_2, N_{phase}\right) = \sqrt{(C_1 + C_2)N_{phase}} \tag{11}$$

The unit of capacitance in the equation is in μF.

(b) Ripple current limitation and related evaluation functions

During the operation of the circuit, the periodic opening and closing of the switch tube will generate ripple current, affecting the quality of power supply, and burden higher current stress to the capacity, which will increase losses and shorten the life of the DC bus capacitor [9]. For multi-phase interleaving converters, different number of phases have different ripple calculation method. When the circuit is in single-phase, the circuit has no backup capability and the current flowing through a single IGBT is too large, when the number of phases is too big, the total volume would exceed an allowed limit which would not meeting design requirements. Therefore, the number of phases from 2 to 4 are only discussed.

For a converter of a certain phase, in Buck mode, the current change rate is obtained from Eq. (6). In parameter design, it is necessary to limit the magnitude of current change in the single phase to not less than a certain value, that is, the expression is:

$$N^2 f \geq \frac{V_o(1 - D)}{A_L \Delta I_{max}} \tag{12}$$

For interleaving converters, the phase difference of the conduction angle between each phase is $\theta = 2\pi/N$, the total output current is the sum of each single-phase current. Song et al. [10] made ripple current analysis on a four-phase interleaving converter.

Here we take a two-phase interleaving converter as an example, the relationship and waveform between the total output current and the output current of each phase are shown in Fig. 6.

Based on this superposition principle, the relationship between the total ripple current under different phase numbers can be obtained according to deduction. In the parameter

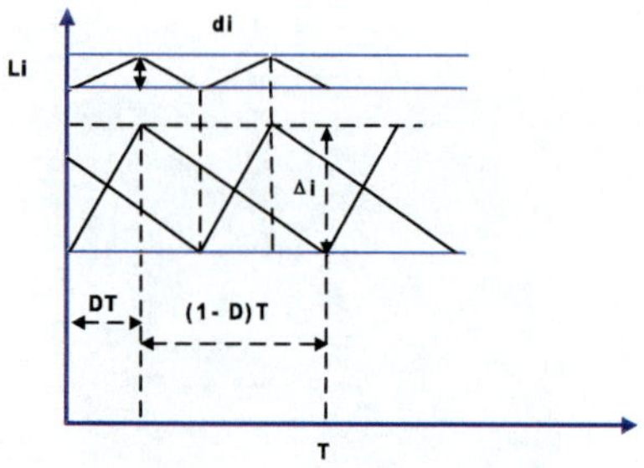

Fig. 6. The waveform of output current for every phase and total output current of a two-phase interleaving converter

setting, the duty cycles are between 1/3 and 1/2 in Buck and Boost modes, and this specific situation is analyzed during ripple analysis.

In the two-phase situation, it has been derived as follows:

$$\begin{cases} di = 2k_{down}(\frac{1}{2} - D)T \\ k_{down} = \frac{\Delta I}{(1-D)T} \end{cases} \tag{13}$$

As:

$$di = \left(1 - \frac{D}{1-D}\right)T \tag{14}$$

Similarly, in three-phase situation, when the duty cycle is between 1/3 and 2/3:

$$di = \left(2 + \frac{1-3D}{3(1-D)} - \frac{2}{3D}\right)T \tag{15}$$

In four-phase situation, when the duty cycle is between 1/4 and 1/2:

$$di = \left(1 + \frac{1-4D}{2(1-D)} + \frac{2D-1}{2D}\right)T \tag{16}$$

The current variation in the above formulas is obtained from Eqs. (6) and (7).

Finally, due to the input and output capacitors can smoothen the current, so we introduce a parameter that is negatively correlated with the sum of input capacitance and output capacitance. The final expression is:

$$f_{ripple} = f\left(N_{phase}, N, f, C_1, C_2\right) = di \times e^{1-\frac{\ln(C_1+C_2)}{10}} \tag{17}$$

The unit of capacitance in the equation is in μF.

(3) Evaluation function

Due to the significant correlation between the volume evaluation function and the ripple current and variables, so we weighting and adding each function, which is divided into Buck mode and Boost mode, to analyze the optimization. The weighted function for volume evaluation and ripple current has the following expression.

3 Multi-objective Optimization and Decision Process

3.1 Multi-objective Optimization of NSGA-II Algorithm

This paper uses the NSGA-II genetic algorithm [11] for optimization, and the specific optimization process is shown in Fig. 7:

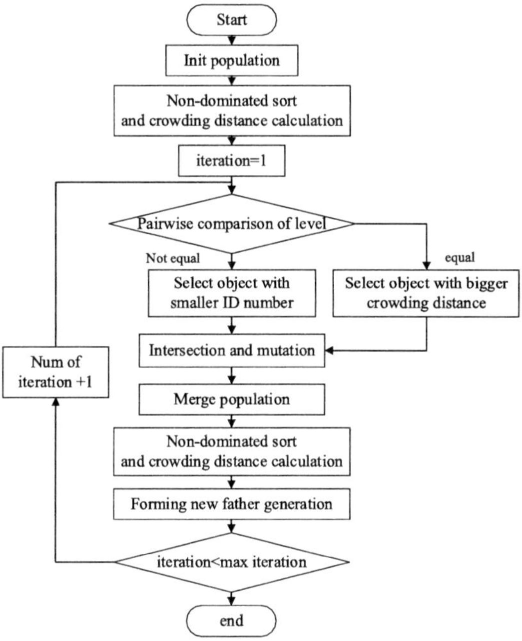

Fig. 7. The Calculation Process of NSGA-II Genetic Algorithm

The algorithm introduces non dominated sorting, distance sorting, and elite strategy methods on the basis of genetic algorithm. Fast non dominated sorting compares the objective functions in the generated population, stratifies the strengths and weaknesses of individuals in the population, and selects them for the next iteration. The computational complexity of the algorithm is lower than that of general genetic algorithms.

After non dominated sorting, in order to solve the internal comparison problem of data in the same dominant level of the population, the program sort the solutions of the same rank of the non-dominated sorted population in descending order of aggregation distance, maintaining population diversity.

The elite strategy is manifested in the algorithm by selecting the optimal individual as the new parent population after non dominated sorting and crowding calculation for each generation of population. After multiple generations of selection, crossover, and mutation, the resulting population will converge to an optimal front-end, improving the convergence of the algorithm.

In the actual program process, during the generation of the population. It is necessary to determine the limiting conditions for the magnitude of the current change, and when the current change exceeds a certain set value, the data point will be regenerated.

After completing the program, it is necessary to select data and rate each performance to the targets, after weighting, the performance of the target is comprehensively evaluated, to select the results that are more advantageous.

3.2 Parameter Setting

(1) Parameter setting of functions

In the calculation of IGBT and diode losses, the parameters of the converter in steady-state mode with low power output are set as follows:

In buck mode, high-voltage side voltage is set to 600V, low side voltage 270V, in constant power mode. Power for low voltage side $P_{buck,low} = 2$kW, therefore the steady state average current $I_{buck} = 7.41A$, duty cycle $D_{buck} = 0.45$.

In boost mode, there's a battery in the low-voltage side that can charge some components on the high-voltage side bus. Where, high-voltage side voltage can be fix to around 560V, low side voltage 330V, in constant current mode, with a current $I_{boost} = 20A$. duty cycle $D_{boost} = 1 - 330/560 = 0.41$.

The parameters of IGBT are set to: in both buck and boos mode, IGBT and diode voltage drop during conduction $V_q = V_d = 2V$, conduct time in a single cycle of IGBT is approximately taken as $T_{on} = T \cdot D$, $T_{off} = T \cdot (1 - D)$. Turn-on and turn-off time of IGBT is fixed to 0.5μs. With the total IGBT power loss be:

$$P_{IGBT,buck} = 3.7f + 74.07 \tag{18}$$

$$P_{IGBT,boost} = 19.0f + 67.88 \tag{19}$$

To calculate iron core power loss, according to the parameters of a certain magnetic material, $a = 5.101$, $b = 0.1561$, $\alpha = 1.822$, $\beta = 2.150$, volume of the iron core is $V_e = 50.69cm^3$, formula (4) is changed to the following form:

$$P_{\text{loss,core}} = 50.69 \times \left(\frac{\Delta B}{2}\right)^{2.150} (5.101f + 0.1561f^{1.822}) \tag{20}$$

In the calculation of magnetic flux variation ΔB, $A_L = 143nH/N^2$, $A_e = 310mm^2$, when the unit of frequency is in *kHz*, for ΔB there would be:

$$\Delta B_{buck} = 95.81 \times \frac{1}{Nf} \tag{21}$$

$$\Delta B_{boost} = 87.44 \times \frac{1}{Nf} \tag{22}$$

When calculating the wire loss of the inductor, the wire winding is set as $0.3\text{mm} \times 160$ with a length of 6.1 m and a material of copper, then the resistance would be:

$$R_{Cu} = 0.0173 \times \frac{6.1}{\pi \times (0.3/2)^2 \times 160} = 9.3m\Omega \tag{23}$$

For the weighting section for volume function and ripple current in both buck and boost mode, we take $x_1 = 0.45$, $x_2 = 0.7$.

(2) Input parameters

The input variable parameters include: working frequency, inductance size, input capacitance, output capacitance, and the number of phases for the interleaving converter. In the calculation of inductance, the number of windings directly related to the inductance, and the number of windings needs to be taken as an integer. It is roughly estimated that when the inductance is between 100 and 600 μH, the minimum value of windings is 27, the maximum value is 64, and the step size is taken as 1.

The range and step size of each parameter are shown in Table 1:

Table 1. Parameter range and step size

Parameter	Minimum	Maximum	Step size
Working frequency f/kHz	10	30	0.5
Number of windings N_{wind}	270	64	1
Input capacity $C_1/\mu F$	50	600	10
Output capacity $C_2/\mu F$	50	600	10
Interleaving phases N_{phase}	2	4	1

(3) Ripple current restriction

In terms of limiting ripple current, the maximum ripple current is limited to 120 A, when other conditions are the same, taking the larger value of the square of the turns and the frequency product in buck and boost mode, which can be equivalent to:

$$N^2 f \geq 19231 \tag{24}$$

3.3 Optimization Results

The parameters of the NSGA-II genetic algorithm are set as follows: population size of 160, crossover ratio of 0.8, mutation probability of 0.1, and maximum number of iterations of 40. After several generations, most of the data in the population can no longer be completely dominated by other data. Due to the fact that the converter is in a more current supply state in practice, which is in buck mode, the losses in Buck mode are sorted in ascending order and some results are selected as shown in Table 2:

Subsequently, the output results are weighted and scored, as in more cases the converter runs in buck mode, weight for power loss in buck mode is 0.07, for power loss in boos mode in 0.03, evaluation function in buck mode in 0.7, evaluation function in boost mode in 0.3, the weighted values obtained based on the 10 data in Table 2 are shown in Table 3. The smaller the value, the better the overall performance.

Table 2. Selected output results

Order	$[f, N, C_1, C_2, N_{phase}]$	$P_{loss,buck}$	$P_{loss,boost}$	f_{buck}	f_{boost}
1	[28.5, 64, 230, 280, 4]	964.65	739.87	13.21	12.19
2	[29.5, 63, 210, 270, 4]	971.05	759.29	12.81	11.83
3	[30, 62, 240, 340, 3]	988.30	771.18	12.20	11.26
4	[29.5, 62, 280, 90, 4]	998.18	763.49	11.25	10.38
5	[27.5, 63, 160, 480, 2]	1013.2	728.96	10.46	9.65
6	[26, 64, 300, 410, 3]	1021.1	702.53	13.49	12.46
7	[25.5, 64, 580, 530, 2]	1034.0	695.31	13.78	12.72
8	[29, 61, 180, 600, 4]	1037.4	760.36	9.07	8.35
9	[25, 64, 300, 470, 2]	1047.5	688.19	11.47	10.59
10	[27, 62, 230, 450, 2]	1054.3	726.10	10.78	9.95

Table 3. Weighted value of the selected output results

Order	Weighted Value
1	102.62
2	103.27
3	104.23
4	103.77
5	103.01
6	105.74
7	106.70
8	104.29
9	105.18
10	106.12

Based on the above data, we think the data that order number 1 represents, with: $f = 28kHz$, $N_{wind} = 64$, $C_1 = 230\mu F$, $C_2 = 280\mu F$, $N_{phase} = 4$, is the relatively optimal solution generated during the parameter optimization design process.

4 Conclusion

Based on the losses of multiphase Buck/Boost converters, while also considering inductance current ripple and design volume factors. This article proposes a parameter design multi-objective optimization method based on the NSGA-II algorithm.

Based on this method, this article achieved optimal values for the parameter design of the Buck/Boost circuit under specific parameter condition, and evaluated the output results of the algorithm that were not dominated by other parameters.

However, there are still shortcomings in the algorithm, such as insufficient processing and optimization function design for the number of windings of inductors in the function, as well as insufficient optimization for the evaluation function and weighted values in the function. In subsequent work, further improvement for these aspects should be made.

References

1. Liu, Q., et al.: An improved quadrangle control method for four-switch buck-boost converter with reduced loss and decoupling strategy. IEEE Trans. Power Electron. **36**(9), 10827–10841 (2021)
2. Wei, C.-L., et al.: Design of an average-current-mode noninverting buck–boost DC–DC converter with reduced switching and conduction losses. IEEE Trans. Power Electron. **27**(12), 4934–4943 (2012)
3. Fan, L., et al.: Loss analysis of bi-directional buck/boost converter. Vehicle and Power Technology (2022). (in Chinese)
4. Veerachary, M., Saxena, A.R.: Optimized power stage design of low source current ripple fourth-order boost DC–DC converter: a PSO approach. IEEE Trans. Ind. Electron. **62**(3), 1491–1502 (2014)
5. Vadi, S., et al.: Optimization of pi based buck-boost converter by particle swarm optimization algorithm. In: 2021 9th International Conference on Smart Grid (icSmartGrid). IEEE (2021)
6. Huang, X., Zhang, X., Li, X.: Multi-objectiv1 1e optimization for smaller, efficient and better performed design of buck-boost converters. In: 2020 IEEE 11th International Symposium on Power Electronics for Distributed Generation Systems (PEDG). IEEE (2020)
7. Chang, W., Wang, J., Chen, Q.: Multi-objective optimization of synchronous buck converter based on NSGA-U algorithm. In: 2016 IEEE International Conference on Mechatronics and Automation. IEEE (2016)
8. Sreedhar, R., et al.: Design and validation of a single-phase buck–boost inverter with Grey Wolf optimization algorithm under partial shaded conditions. Int. J. Inf. Technol. 1–11 (2022)
9. Wang, Z., Yu, X.-P.: Analysis and inhibition studies of the DC-bus ripple of inverter power supply. Power Electron. **3**, 93–95 (2013). (in Chinese)
10. Ke-ling, S.O.N.G., et al.: Study on the influence of duty cycle on ripple current in four-phase interleaving circuits. Vehicle Power Technol. **3**, 15–19 (2019)
11. Deb, K., et al.: A fast and elitist multiobjective genetic algorithm: NSGA-II. IEEE Trans. Evol. Comput. **6**(2), 182–197 (2002). (in Chinese)

Research on Optimal Power Flow Control Strategy for Urban Rail Flexible Traction Power Supply System

Junjie Zhang[1(✉)], Shuai Jiang[1], Zhigang Liu[1,2], and Jie Chen[1,2]

[1] School of Electrical Engineering, Beijing Jiaotong University, Beijing, China
15801351357@163.com, {21121532,22126177,zhgliu, jiechen}@bjtu.edu.cn

[2] Rail Transit Electrical Engineering Technology Research Center, Beijing, China

Abstract. In order to solve the problems of uncontrollable DC voltage and low utilization of regenerative braking energy in traditional traction power supply system, a bi-directional converter is used instead of diode rectifier to form a flexible traction power supply system. Firstly, the AC network and DC network of the system, as well as the working mode are introduced. Then a system hierarchical control strategy is proposed to study the modeling, dual closed-loop control of the bidirectional converter, and the optimization objectives, variables, and Gray Wolf optimization algorithm are introduced to achieve flexible control of the system. Finally, a simulation example is given to verify the proposed control strategy. The results show that the control strategy can better improve the regenerative braking energy distribution of the train and realize the energy-saving optimization of the system.

Keywords: Urban rail transit · Flexible traction power supply system · Optimal power flow · Energy saving

1 Introduction

As a major consumer of energy and electricity, urban rail transit has the advantages of energy-saving, pollution-free, large volume, safety and convenience [1]. By the end of 2022, 55 cities in our country have urban rail transit, with the total operating mileage exceeding 10,000 km [2]. Under this background, it is of great significance to study the energy-saving strategy of the traction power supply system of metro. The energy generated during train braking accounts for 10–20% of the total energy consumption of electricity, so an important way to achieve energy saving is to efficiently use the regenerative braking energy of the subway trains [3].

The traditional traction power supply system uses diode rectifier, the energy can only flow in one direction and the DC side voltage can not be controlled. With the advantages of low cost and high power factor, reversible converter has been widely concerned [4]. According to the operation condition and network structure of metro traction power

L. Jia et al. (Eds.): EITRT 2023, LNEE 1135, pp. 80–88, 2024.
https://doi.org/10.1007/978-981-99-9307-9_9

supply system, the optimal power flow [5] is used to optimize the energy of the system, so that the system can run in a stable state with the optimal preset target. With the rapid development of computer technology, intelligent optimization algorithm [6] is gradually applied to power system optimization. In [7] presents a power flow algorithm for DC traction power supply system including AC network based on diode rectifier. In [8] based on adaptive particle swarm optimization algorithm of fuzzy control theory, the active power loss and voltage quality in DC power network are optimized. Most of these papers do not consider the inverter feedback device, and do not carry out optimization research from the system point of view.

In this paper, the optimal power flow of flexible traction power supply system is studied. The second part of this paper introduces the concept and working mode of the flexible traction power supply system. The third part discusses the flexible control technology from the steady state model, hierarchical control strategy. In the fourth part, the energy-saving effect of the proposed method is verified by a numerical example. The fifth part is a summary.

2 Flexible Traction Power Supply System

2.1 The Concept of Flexibility

This paper uses bi-directional converter instead of diode rectifier unit to build a flexible traction power supply system, its structure as shown in Fig. 1. As shown in Fig. 1, the system consists of a main substation, a medium-voltage ring network, a traction substation, a station load, a DC network and a train. Taking the bi-directional converter as the boundary, the whole system is divided into AC network and DC network.

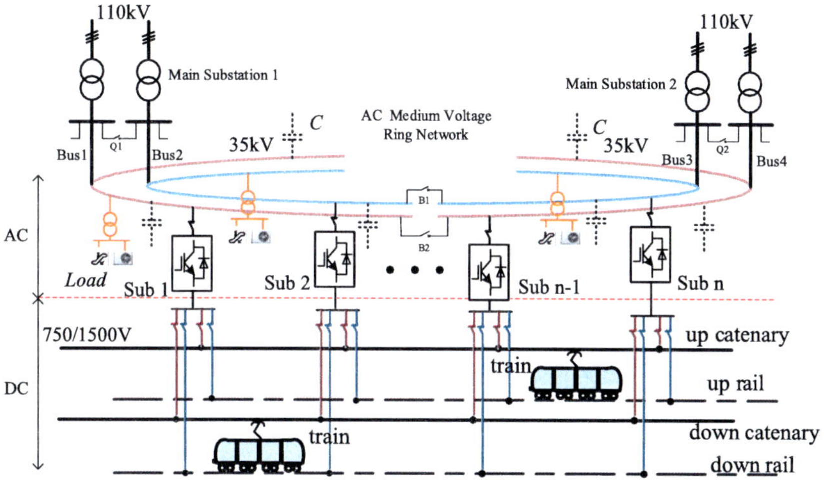

Fig. 1. The topology of flexible traction power systems.

(1) AC network

The main transformer outlet Bus1-Bus4 constitutes the middle voltage ring network of the system. When the system is in normal operation, the contact switches B1, B2, Q1, Q2 are disconnected, and the AC bus Bus1-Bus4 operates independently.

(2) DC network

The DC network is composed of catenary, train and return rail. Because the train has traction, uniform speed, inertia, braking a variety of working conditions, so its position and power changes with time.

2.2 Mode of Operation

The working mode and function of the flexible traction power supply system are introduced below.

(1) Train traction

In Fig. 2a, the main substation reaches the medium voltage ring network through the AC bus to supply power to each substation. Through the bidirectional converter in the traction substation to provide power for train traction.

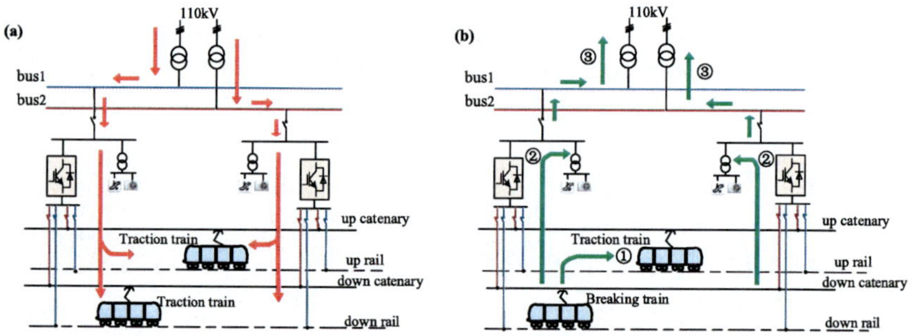

Fig. 2. Flexible traction power supply system operating mode **a** train traction **b** train brake.

(2) Train brake

Figure 2b as shown, this braking energy will first be used on other trains in a nearby traction state. Then feds back to the medium voltage ring network through the traction substation for other loads on the AC bus; the last feds back to the city grid through the main substation.

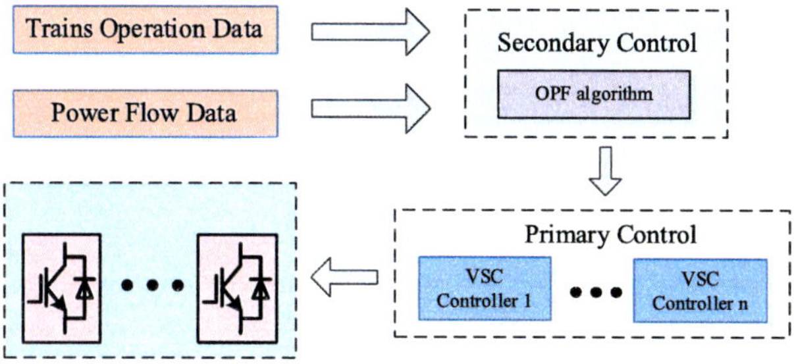

Fig. 3. Hierarchical control structure.

3 Flexible Control Strategy

In this paper, a hierarchical control structure based on optimal tides is proposed, as shown in Fig. 3. The first control layer is located in traction substation. The second level control is located in the control center of the traction power supply system.

3.1 Steady-State Model

Figure 4 shows the equivalent circuit of the bidirectional converter. The bi-directional converter injected power is derived in [9].

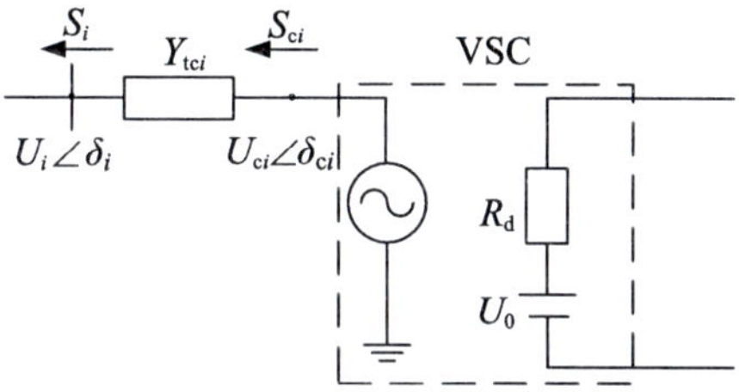

Fig. 4. Equivalent circuit of bidirectional converter.

3.2 First Layer Control

In the first layer of control, the bidirectional converter uses a voltage and current dual closed-loop feedforward decoupling control, which consists of two parts: DC voltage droop control and current control. The role of the voltage outer loop is to control the DC voltage through the voltage command to optimally distribute the regenerative braking power to the train. The role of the current inner loop is to control the current according to the current command output from the voltage outer loop. The bidirectional converter control system is shown in Fig. 5.

When the bidirectional converter uses constant voltage control, this control method is not conducive to the uniform distribution of regenerative braking power among substations, when the train is braking.

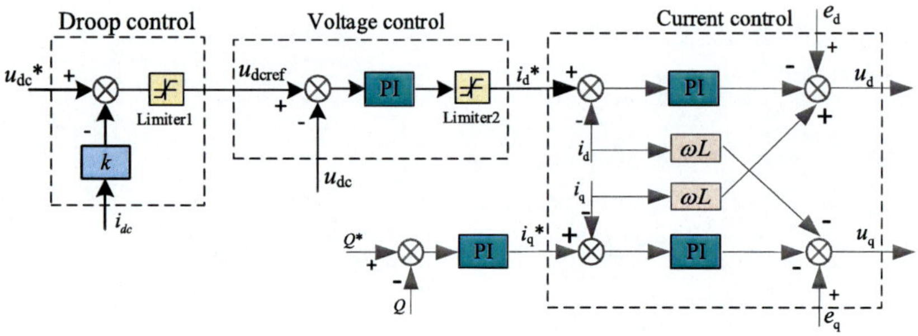

Fig. 5. First layer control structure diagram.

When the bidirectional converter uses DC voltage droop control, multiple substations can simultaneously participate in regulating the DC voltage and improve the distribution of regenerative braking energy.

3.3 Second Layer Control

In order to realize the energy-saving operation of the system, the optimal power flow algorithm is added to the second level control [1]. The train's real-time position and power are calculated as the input of system power flow simulation, taking into account the running route data, train parameters, speed limit and schedule.

A. Optimization objectives

The total incoming power of all main substations is minimum. The objective function can be written as follows:

$$\mathrm{Min} f(x) = \mathrm{Min} \sum_{i=1}^{N} P_i \tag{3}$$

$$\boldsymbol{x} = \boldsymbol{R}_{\mathrm{droop}} \tag{4}$$

Among them, N is the number of main substation, P_i is the instantaneous active power of the incoming line of the first main substation, x is the optimal variable, and R_{droop} is the droop coefficient.

B. Equality constraints

(1) AC power flow equation

$$\Delta P_{ci} = P_{\mathrm{loss}i} + U_i U_{ci}[G_{\mathrm{tc}i}\cos(\delta_i - \delta_{ci}) - B_{\mathrm{tc}i}\sin(\delta_i - \delta_{ci})] - U_{ci}^2 G_{\mathrm{tc}i} = 0 \tag{5}$$

$$\Delta Q_i = Q_{\mathrm{ref}i} - U_i U_{ci}[G_{\mathrm{tc}i}\sin(\delta_i - \delta_{ci}) - B_{\mathrm{tc}i}\cos(\delta_i - \delta_{ci})] - U_i^2 B_{\mathrm{tc}i} = 0 \tag{6}$$

$$\Delta F_{\mathrm{P}i} = P_{\mathrm{G}i} - P_{\mathrm{L}i} - U_i^2 G_{\mathrm{tc}i} + U_i U_{ci}[G_{\mathrm{tc}i}\cos(\delta_i - \delta_{ci})$$

$$+ B_{tci} \sin(\delta_i - \delta_{ci})] - P_i(U, \delta) = 0 \tag{7}$$

$$\begin{aligned} \Delta F_{Qi} = & Q_{Gi} - Q_{Li} + U_i^2 B_{tci} + U_i U_{ci} [G_{tci} \sin(\delta_i - \delta_{ci}) \\ & - B_{tci} \cos(\delta_i - \delta_{ci})] - Q_i(U, \delta) = 0 \end{aligned} \tag{8}$$

Among them, P_{Gi} and Q_{Gi} are the active and reactive generation power of AC node *i*, respectively; P_{Li} and Q_{Li} are the active and reactive load power of AC node *i*, respectively; Q_{refi} is the initial value calculated by OPF algorithm; ΔF_P is the active power deviation vector of VSC; ΔF_Q is the reactive power deviation vector of VSC; P_{lossi} is the equivalent loss of PWM voltage source converter.

(2) DC power flow equation

$$\Delta P_{dci} = P_{ti} - U_{dci} \sum_{j=1, j \neq i}^{n} G_{dcij}(U_{dcj} - U_{dci}) = 0 \tag{9}$$

Among them, *n* is the number of DC nodes; U_{dci} is the DC node *i* voltage; P_{dci} is the power of the traction substation or train power at node *i*; G_{dcij} is the element of the i-th row and j-th column of the DC node derivative matrix.

C. Inequality constraints

(1) AC voltage limit

$$U_{\min} \leq U_i \leq U_{\max} \tag{11}$$

Among them, U_{min} and U_{max} are the minimum and maximum values of AC voltage amplitude of traction substation nodes, respectively.

(2) DC voltage limit

$$U_{dc\min} \leq U_{dci} \leq U_{dc\max} \tag{12}$$

Among them, U_{dcmin} is the value of current limiting protection and U_{dcmax} is the value of overvoltage protection.

(3) Main substation capacity limit

$$P_i^2 + Q_i^2 \leq S_N^2 \tag{12}$$

D. Gray Wolf Optimization Algorithm

The GWO algorithm achieves optimization by simulating the predatory behavior of gray wolf packs in nature, based on the mechanism of wolf pack collaboration. The social hierarchy of gray wolves plays an important role in the group hunting process, i.e., under the leadership of α, β, and δ, ω follows these three wolves in the process of predation [10]. Using the Gray Wolf optimization algorithm for moisture prediction of sintered return powders from smelters in [11].

4 Simulation Verification

Taking a subway line as an example, a simulation model of flexible traction power supply system is established based on MATLAB simulation software, and the control strategy proposed in this paper is verified. There are 21 stations on the whole line, including 11 traction substations. The simulation parameters are shown in Table 1. Traction network using DC 1500V power supply, trains using B-6 vehicle formation. In order to give full play to the regenerative braking capacity of the train, the maximum train voltage is limited to 1800V.

Table 1. Simulation system parameters

Parameter	Value
35 kV Cable resistance (Ω/km)	0.0778
35 kV Cable reactance (Ω/km)	0.1720
35 kV Cable susceptance (10^{-6}s/km)	108.016
Catenary resistance (Ω/km)	0.014
Rail resistance (Ω/km)	0.025
Nominal dc voltage (V)	1650
Diode rectifiers equivalent resistance (Ω)	0.05
Capacity of reversible converters (kVA)	6000
Auxiliary load (kVA)	3000
Headway interval (s)	300

Set the departure interval 300 s, the two-way converter of the flexible system adopts sag control, compared with the traditional traction power supply system using diode rectifier power supply, the two-way converter only inverts and adopts constant voltage control.

Table 2 shows the comparison of power consumption of the system within 300 s. E1 represents the power taken by the system from the city grid, E2 is the power returned by the train braking to the main substation, the power consumption of the flexible system using sag control is reduced by 3.67%, and the power returned is reduced by 85.17%. Because the regenerative braking energy of train will be absorbed by several traction substations in droop control, the transmission path of braking energy will increase, so the transmission loss of flexible droop is larger than that of conventional constant voltage control.

As shown in Fig. 6, the traction energy consumption of the train is reduced and the regenerative braking energy distribution is improved by using flexible sag compared with traditional constant pressure control.

Using gray wolf optimization algorithm, the system power consumption as the optimization objective, droop coefficient as the optimization variable, after optimization,

Table 2. Simulation results within 300s.

Control method	U_0 (kV)	k	E_1 (kW·h)	E_2 (kW·h)	Ploss (kW·h)
Constant voltage	1.65	/	1116.77	−2.36	−30.25
Flexible droop	1.70	by GWO	1075.71	−0.35	−33.91

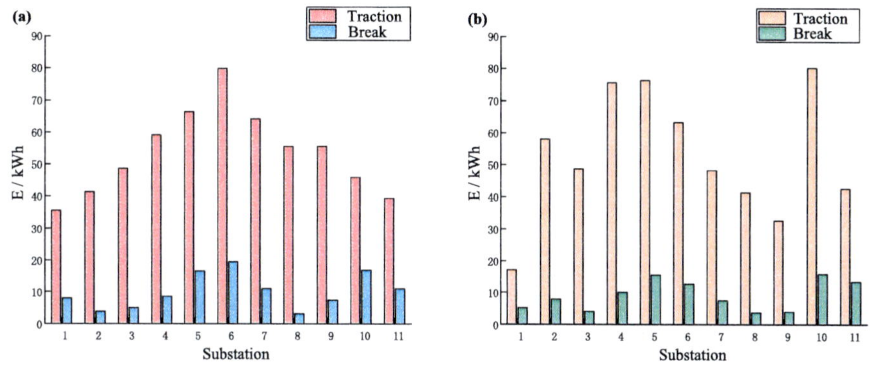

Fig. 6. Traction substations energy consumption comparison **a** traditional constant voltage control, **b** flexible droop control.

set the initial population number to 50 and iterations to 50, the system power down by 1.28%.

The simulation results show that the control strategy based on the optimal power flow algorithm can realize the energy-saving optimization of the system.

5 Conclusion

In this paper, based on the structure of urban rail transit flexible traction power supply system, the optimal power flow control strategy is studied, a hierarchical control strategy based on grey wolf optimization algorithm is proposed to optimize the SAG coefficient of bi-directional converter in traction substation. The simulation results show that compared with the traditional traction power supply system, the control strategy proposed in this paper can better improve the distribution of regenerative braking energy and reduce the regenerative braking energy of the train returning to the main station, significantly reduce the cost of electricity system.

References

1. Dai, H., Li, Z., Song, J.: Energy consumption status of urban rail transit in Beijing and suggestions on energy saving measures. Railway Tech. Innovat. 77–80 (2016). (in Chinese)
2. Wang, F., Feng, A.: Statistics and development analysis of urban rail transit in China in 2022. Tunnel Construct. **43**(3), 521–528 (2023). (in Chinese)

3. Zhang, G., Tian, Z., Tricoli, P., Hillmansen, S., Wang, Y., Liu, Z.: Inverter operating characteristics optimization for DC traction power supply systems. IEEE Trans. Vehicul. Technol. **68**(4), 3400–3410 (2019)
4. Khodaparastan, M., Dutta, O., Saleh, M., Mohamed, A.A.: Modeling and simulation of DC electric rail transit systems with wayside energy storage. IEEE Trans. Vehicul. Technol. **68**(3), 2218–2228 (2019)
5. Li, P., Sheng, W., Duan, Q.: Optimal power flow calculation method for AC/DC hybrid distribution network based on power router. In: IEEE 6th Asia Conference on Power and Electrical Engineering, pp. 1694–1699 (2021)
6. Yu, L., Li, L., Wang, Z., Zhang, Y.: Optimization algorithm of power marketing AI response system in the era of intelligent technology. In: IEEE 6th Advanced Information Technology, Electronic and Automation Control Conference, 2008–2013 (2022)
7. Pires, C.L., Nabeta, S.I., Cardoso, J.R.: DC traction load flow including AC distribution network. IET Electr. Power Appl. **3**(4), 289–297 (2009)
8. Liu, X., Wen, J., Pan, Y., Wu, P., Li, J.: OPF control of DC-grid using improved PSO algorithm. Power Syst. Technol. **41**(3), 715–720 (2017). (in Chinese)
9. Beerten, J., Cole, S., Belmans, R.: Generalized steady-state VSC MTDC model for sequential AC/DC power flow algorithms. IEEE Trans. Power Syst. **27**(2), 821–829 (2012)
10. Zhang, X., Wang, X.: Comprehensive review of grey wolf optimization algorithm. Comput. Sci. **46**(3), 30–38 (2019). (in Chinese)
11. Liu, Z., Wang, Z., He, J, Shen, L., Shao, X., Jiang, X.: Moisture prediction of lead zinc sintering recycled-fine-ore based on multi-hidden-layer radial basis function neural network optimized by gray wolf optimizer. J. Central South Univ. (Science and Technology), **54**(2), 754–764 (2023). (in Chinese)

Research on Harmonic Noise Characteristics of Train Four-Quadrant Converter Using SiC Devices

Yongxiang Wang[1], Kan Liu[1](✉), Fanyang Meng[2], and Jianqiang Liu[2]

[1] State Key Laboratory of Traction and Control System of EMU and Locomotive, China Academy of Rail-Way Sciences Corporation, Beijing, China
{wangyongxiang,liukan}@zemt.cn
[2] School of Electrical Engineering, Beijing Jiaotong University, Beijing, China
{21126167,liujq}@bjtu.edu.cn

Abstract. High frequency devices represented by SiC have the advantages of low switching loss, high switching frequency, and high junction temperature. Consequently, SiC have emerged as the dominant direction for the development of the new generation of traction converter technology. This study specifically examines the harmonic noise characteristics of the four-quadrant converter (4QC), which utilizes SiC devices in trains. Firstly, the influential factors affecting the harmonic characteristics on the grid side of the SiC 4QC are investigated through theoretical derivation. Secondly, the common-mode (CM) conducted interference coupling path of the 4QC is analyzed, and an equivalent circuit for the CM electromagnetic interference (EMI) is constructed. By studying the spectrum of the CM voltage source, the factors influencing the CM EMI in the SiC 4QC are analyzed. Then, an optimal selection method for selecting the switching frequency of the SiC 4QC is proposed, taking into account both the harmonic characteristics on the grid side and the CM EMI characteristics. Finally, a low-power experimental platform is established, and the research content is experimentally verified.

Keywords: Silicon carbide · Four-quadrant converter · Harmonic · Common-mode electromagnetic interference · Switching frequency

1 Introduction

The four-quadrant converter (4QC) is one of the important components of traction converter, which is responsible for the conversion of AC current to DC current. With the reduction of switching losses in SiC power modules, the 4QC employing SiC devices can operate at higher switching frequencies, facilitating the development of high-frequency traction drive systems. On the one hand, the high-frequency operation effectively optimizes the harmonic performance of the traction drive system. On the other hand, it introduces more severe electromagnetic interference (EMI) issues to the traction drive system [1, 2].

L. Jia et al. (Eds.): EITRT 2023, LNEE 1135, pp. 89–98, 2024.
https://doi.org/10.1007/978-981-99-9307-9_10

Extensive research has been conducted by domestic and international scholars regarding EMI issues. References [3] and [4] investigate the common-mode (CM) EMI noise in electric vehicle power systems using SiC MOSFET power devices. Reference [5] analyzes the influence of the switching performance of SiC MOSFET and the severity of switching ringing on EMI in motor drive systems. Reference [6] focuses on 4QC in high-speed trains, analyzing conducted EMI in traction drive systems. Reference [7] investigates the EMI mechanism in high-speed trains and proposes an overall EMI model. Reference [8] presents a frequency-domain calculation method for SiC motor drive systems that considers both equivalent voltage and current sources. Reference [9] conducts frequency-domain analysis on the output waveform of synchronous Buck converters based on SiC MOSFET modules. Reference [10] quantitatively analyzes the impact of wide bandgap semiconductor devices on CM EMI noise in PWM inverter motor drive systems.

This study investigates the harmonic noise characteristics of 4QC for trains using SiC devices. By analyzing the grid-side harmonic components and studying the CM EMI characteristics of the system, the impact of SiC device application on grid-side harmonics and CM EMI is analyzed. On this basis, an optimal method for selecting the switching frequency of the SiC 4QC is proposed, considering both the grid-side harmonic characteristics and CM EMI characteristics, and simulation verification is conducted to validate the proposed approach. Finally, experiments are conducted on a low-power SiC experimental platform to verify the results of the analysis.

2 Analysis of Harmonic Noise Characteristics of 4QC

2.1 Harmonic Characteristics on the Grid Side

Under unipolar SPWM modulation, the AC voltage of the 4QC is the difference between the voltages of the two bridge arms. The harmonic components of the voltage are the main cause of current harmonics. According to the KVL equation on the AC side of the 4QC, the expression for the grid-side current i_s can be obtained as shown in formula (1). Here, U_{dc} represents the DC bus voltage, M represents the modulation index, ω_m is the modulation wave angular frequency, θ_0 is the initial phase of the modulation wave, ω_c is the carrier wave angular frequency, and J_n is the Bessel function.

$$
\begin{aligned}
i_s = {} & \frac{\sqrt{(U_{dc}M)^2 - 2U_s^2}}{\omega_m L}\sin(\omega_m t) + \\
& \frac{U_{dc}}{L}\sum_{n=1}^{\infty}(-1)^{\frac{n}{2}}\frac{4}{n\pi}\left\{\sum_{k=1,3,5}^{\infty} J_k\left(\frac{Mn\pi}{2}\right)\left[\begin{array}{l}\dfrac{1}{k\omega_m + n\omega_c}\cos(k\omega_m t + n\omega_c t + k\theta_0) + \\ \dfrac{1}{k\omega_m - n\omega_c}\cos(k\omega_m t - n\omega_c t + k\theta_0)\end{array}\right]\right\}
\end{aligned}
\tag{1}
$$

The harmonic components of the grid side current correspond to the harmonic components of the AC side voltage and only the harmonic components whose angular frequency is $n\omega_c \pm n\omega_m$ are included as shown in formula (2) for their amplitude.

$$A = \frac{U_{dc}}{L(n\omega_c \pm k\omega_m)} \frac{4}{n\pi} J_k \left(\frac{Mn\pi}{2} \right) \tag{2}$$

2.2 CM EMI Characteristics

The CM conducted path on the 4QC can be depicted as shown in Fig. 1. It includes the high-frequency equivalent models of the traction transformer, long transmission cables, and power switching devices. The power switching devices in the upper arms of the 4QC are equivalently represented by current sources with the same current as the branch current, while the power switching devices in the lower arms are equivalently represented by voltage sources with the same voltage as the branch voltage. After these substitutions, the external voltage and current characteristics remain unchanged [11].

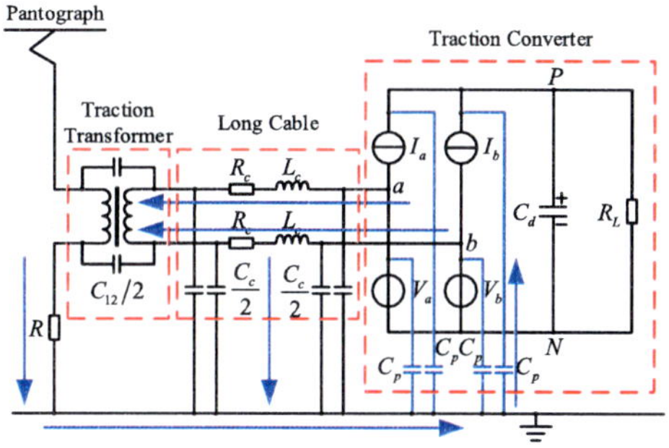

Fig. 1. CM conduction path of 4QC.

In the analysis of CM EMI, the influence of the equivalent current source on CM EMI can be neglected, and only the CM interference current generated by the equivalent voltage source is considered [8]. The voltage interference source in the system can be replaced with V_{CM}, and due to the symmetry of the output circuit structure, the circuit can be treated in parallel, establishing an equivalent circuit for the CM conducted interference in the 4QC system, as shown in Fig. 2. The CM voltage source expression is $V_{CM} = (V_a + V_b)/2$.

Taking into account the transient processes during the turn-on and turn-off of the switching devices, the midpoint voltage of each individual bridge arm in the 4QC can be equivalently represented as a series of trapezoidal pulse waveforms with certain rise and fall times. The spectrum expression of the CM voltage source is obtained as shown in formula (3). This expression is derived under the unipolar SPWM modulation. In the formula, A represents the DC bus voltage, T represents the switching period, τ_r and τ_f characterize the switching speed of the devices.

$$S_n = \frac{AT}{2n^2\pi^2} \left| \frac{1 - e^{-j2n\pi\tau_r f_s}}{\tau_r} + (-1)^k \frac{\left(e^{-j2n\pi\tau_f f_s} - 1\right)e^{-jn\pi f_s(\tau_r - \tau_f)}}{\tau_f} \right|$$

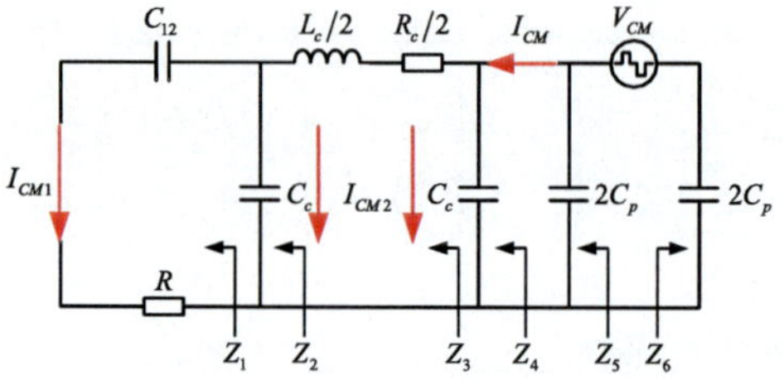

Fig. 2. CM conducted interference equivalent circuit of 4QC.

$$\times \left| \sum_{m=0}^{m_s-1} e^{j\frac{n}{2}\pi M_0 \sin(2\pi m/m_s)} / m_s \right| \quad (3)$$

3 Analysis of Harmonic Noise Characteristics of 4QC

3.1 Operating Frequency Range of 4QC

The SiC MOSFET power module consists of a SiC MOSFET and an anti-parallel SiC SBD. The power loss of the 4QC was calculated using PLECS software to determine the potential operating frequency range. Figure 3a shows the comparison of losses between SiC 4QC and traditional Si 4QC at a switching frequency of 500 Hz, and Fig. 3b shows the SiC 4QC losses at different switching frequencies.

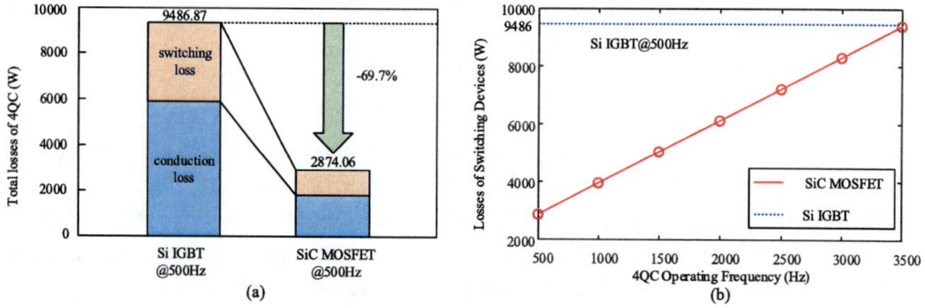

Fig. 3. Calculation of losses in SiC 4QC.

When SiC devices operate at the same switching frequency as Si devices, the losses of the switching devices in the 4QC are reduced by approximately 69.7%. Moreover, when the switching frequency of the SiC devices is 3500 Hz, the losses are roughly equivalent to those at a switching frequency of 500 Hz for Si devices. This implies that, without making any changes to the cooling system's heat dissipation capacity, the possible operating frequency range for the 4QC using SiC devices is from 500 Hz to 3500 Hz.

3.2 Impact of Switching Frequency on Grid Side Harmonics

The adoption of SiC devices in the 4QC for trains leads to an increase in the switching frequency, resulting in optimized grid-side harmonics. Figures 4 and 5 present the simulated waveform and FFT analysis of the grid-side current at a switching frequency of 500 Hz and 1000 Hz, respectively.

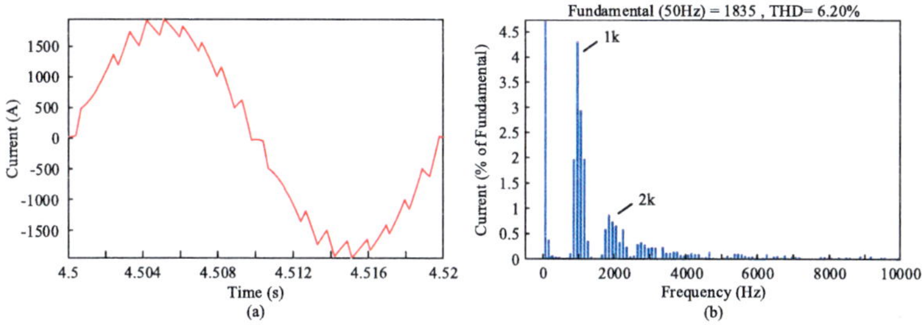

Fig. 4. Network side current waveform and FFT analysis at switching frequency of 500Hz.

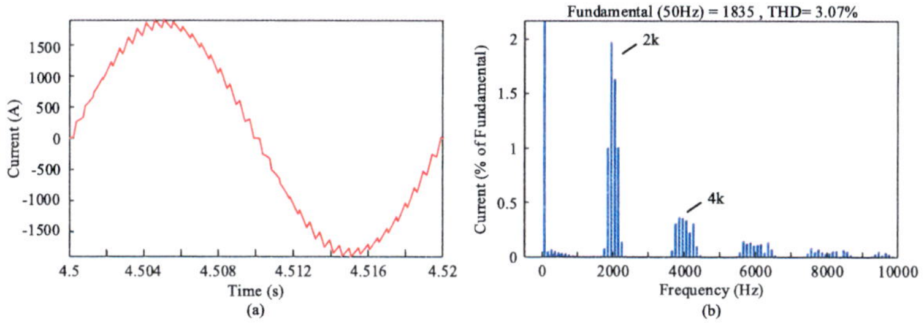

Fig. 5. Network side current waveform and FFT analysis at switching frequency of 1000 Hz.

Figure 6 illustrates the THD of the grid-side current at switching frequencies ranging from 500 Hz to 3500 Hz, obtained through simulations. It can be observed that the magnitude of grid-side current THD decreases as the switching frequency increases. After increasing the switching frequency from 500 Hz to 3500 Hz, the current THD decreases by approximately 85.6%. The use of SiC in the 4QC results in significant improvement in the waveform quality of the grid-side current.

3.3 CM EMI Characteristics

The increase in switching frequency of the 4QC using SiC devices amplifies the CM voltage source in the system, thereby exacerbating the CM EMI noise. Therefore, the magnitude of the CM EMI in the system can be characterized by the CM voltage source spectrum. Figure 7 presents the CM voltage source spectrum at three different switching frequencies. It is evident from the graph that as the switching frequency increases, the voltage spectrum exhibits an overall increase across the entire frequency range, with the increment being nearly constant within each frequency band.

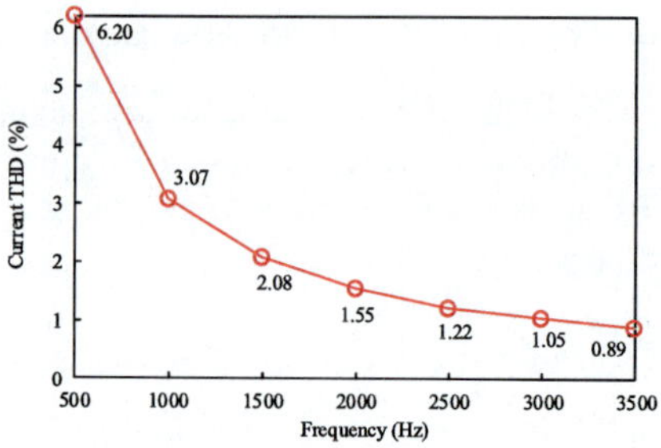

Fig. 6. Grid side current THD at different switching frequencies.

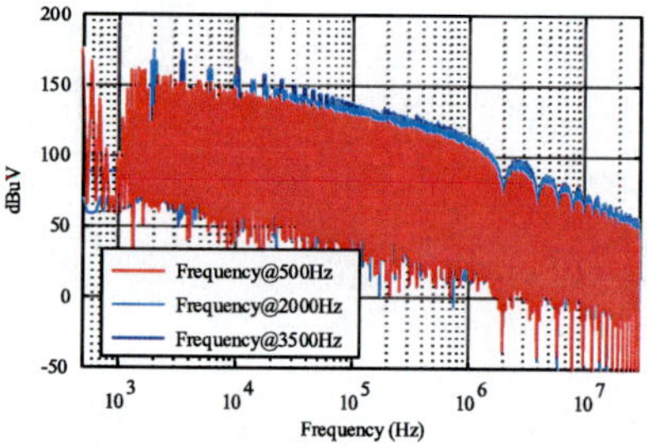

Fig. 7. System CM voltage source spectrum.

To quantify the magnitude of CM EMI at a specific switching frequency, the maximum value of the CM voltage source spectrum within the frequency range of 1–3 MHz is selected as the metric for assessing the CM EMI level. Figure 8 illustrates the simulated CM EMI for switching frequencies ranging from 500 Hz to 3500 Hz. When the switching frequency is increased from 500 Hz to 3500 Hz, the CM EMI in the system is increased by about 8 dB. The results demonstrate that as the switching frequency increases, the CM EMI noise in the system becomes more severe.

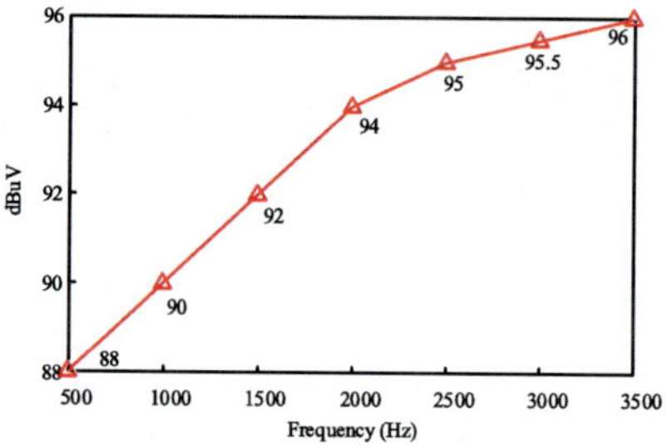

Fig. 8. CM EMI at different switching frequencies.

3.4 Optimal Switching Frequency Selection Method

To optimize both the current THD and CM EMI, a linear weighting model is employed. Assuming the variation of current THD with switching frequency follows function *thd*(*f*), and the variation of CM EMI with switching frequency follows function *emi*(*f*). By scaling *thd*(*f*) and *emi*(*f*) using formula (4) and formula (5), *THD*(*f*) and *EMI*(*f*) are obtained, ensuring their function values lie between 0 and 1.

$$THD(f) = (thd(f) - thd(f)_{\min})/(thd(f)_{\max} - thd(f)_{\min}) \tag{4}$$

$$EMI(f) = (emi(f) - emi(f)_{\min})/(emi(f)_{\max} - emi(f)_{\min}) \tag{5}$$

Formula (6) represents the objective function for optimization, where λ_1 and λ_2 are the weights assigned to functions *THD*(*f*) and *EMI*(*f*), respectively. The point where function *THD*(*f*) and *EMI*(*f*) intersect are considered as the optimal solutions.

$$\min z = \lambda_1 THD(f) + \lambda_2 EMI(f) \tag{6}$$

Figure 9 illustrates the functions of current THD and CM EMI, denoted as *THD*(*f*) and *EMI*(*f*), respectively. The switch frequency at the intersection of the curves is approximately 1200 Hz. When the traction drive system operates at this switch frequency, the magnitude of the current THD on the grid side is around 2.67%, and the magnitude of the CM EMI is approximately 90.8 dBuV. This intersection represents the optimal switch frequency that considers both the harmonic characteristics of the grid-side current in the traction drive system and the CM EMI characteristics.

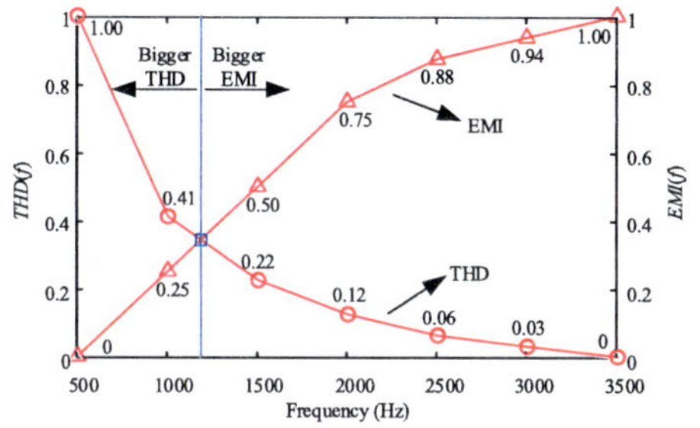

Fig. 9. Function curve of current THD and CM EMI.

4 Experimental Research

The SiC low-power experimental platform was built to simulate the 4QC using SiC devices. The 4QC operates under unipolar SPWM modulation. The operating frequency range for the power devices is assumed to be from 1 kHz to 4 kHz. The experimental results of harmonic characteristics are depicted in Fig. 10.

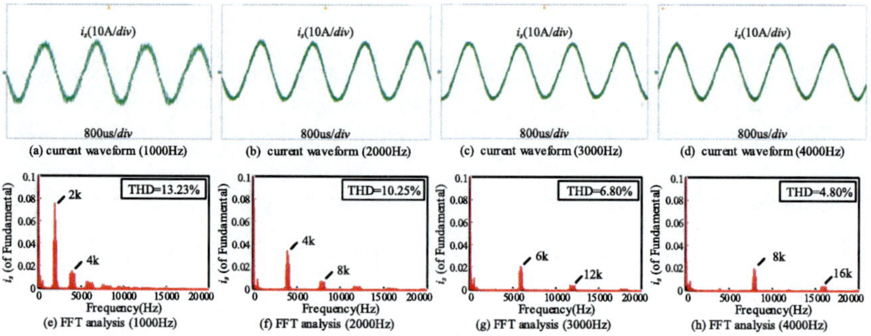

Fig. 10. Harmonic characteristics of grid side under different switching frequencies.

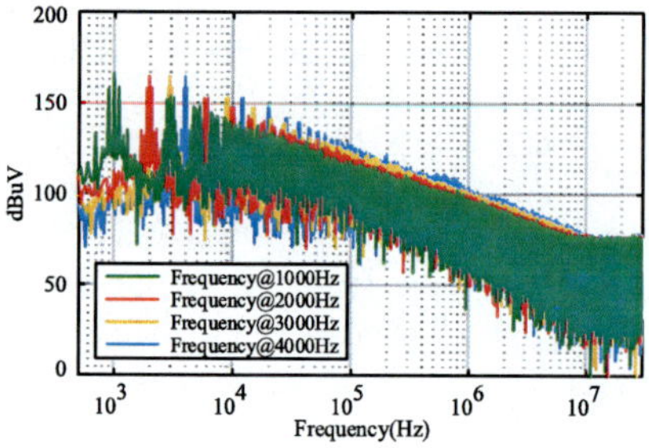

Fig. 11. CM voltage source spectrum at different frequencies.

The experiment measured the midpoint voltages, V_a and V_b, of the two bridge arms of the 4QC at switch frequencies ranging from 1 kHz to 4 kHz. The CM voltage source spectrum was then calculated, as shown in Fig. 11.

Figure 12 illustrates the plots of functions $THD(f)$ and $EMI(f)$. The two curves intersect between switching frequencies of 2000 Hz and 2500 Hz, with the intersection occurring at approximately 2300 Hz. When the system operates at this switching frequency, the grid-side current THD is approximately 9.22%, and the CM EMI is approximately 96.5 dBuV. This switching frequency is considered the optimal choice that balances the harmonic characteristics of the grid-side current and the CM EMI characteristics of the system.

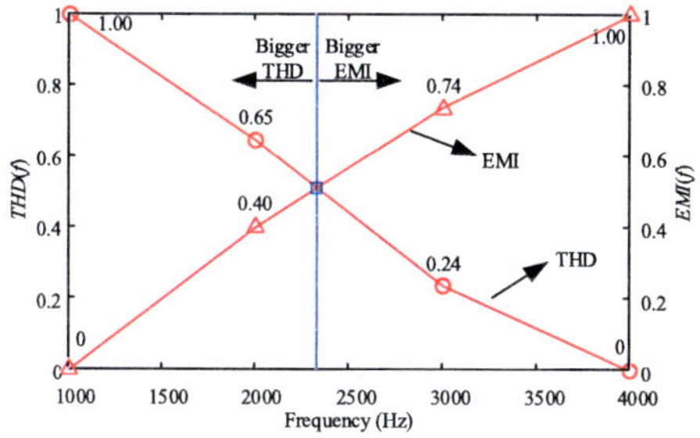

Fig. 12. Function curve of current THD and CM EMI.

5 Conclusion

This paper investigates the harmonic noise characteristics of a train four-quadrant converter using SiC devices. The study focuses on the grid-side current harmonic characteristics of the 4QC under unipolar SPWM. The switching frequency of the power devices affects the harmonic content of the grid-side current, and the adoption of SiC devices leads to an increase in the switching frequency, effectively reducing the harmonic content of the grid-side current and improving its waveform quality. Furthermore, the CM EMI conduction path in the 4QC system is examined. An equivalent circuit for CM EMI in the 4QC system is proposed, and the analysis of the CM voltage source is conducted. The higher switching frequency of SiC devices results in an overall increase in CM EMI across the frequency range. Additionally, the possible operating frequency range of the SiC 4QC is analyzed, and an optimal switching frequency selection method is proposed to achieve a balance between grid-side harmonic characteristics and CM EMI characteristics. Finally, a SiC low-power experimental platform is built and the research content is experimentally verified. Under the selected switching frequency, the grid-side current THD and common mode EMI are at a low level at the same time, which can meet the normal operation requirements of the system.

Acknowledgment. This work is supported by the State Key Laboratory of Traction and Control System of EMU and Locomotive Open Project 2021YJ279.

References

1. Zhi, Y., Yang, D., Zhu, B., Yuan, K., Gao, Z., Li, P.: High-frequency negative effect analysis and countermeasures of traction inverter based on all SiC MOSFET for rail transit. Eltric Drive Locomot. **5**, 49–55 (2020). (in Chinese)
2. Zhang, X., Luo, Z.: Research on EMC application of traction inverter based on silicon carbide MOSFET power module. Rolling Stock **60**(2), 83–88 (2022). (in Chinese)
3. Jia, X., Hu, C., Dong, B., He, F., Wang, H. and Xu, D.: System-level conducted EMI model for SiC powertrain of electric vehicles. In: 2020 IEEE Applied Power Electronics Conference and Exposition, pp. 885–892 (2020)

4. Jia, X., Hu, C., Dong, B., He, F., Wang, H., Xu, D.: Influence of system layout on CM EMI noise of SiC electric vehicle powertrains. CPSS Trans. Power Electron. Appl. **6**(4), 298–309 (2021)
5. Wu, Y., Yin, S., Liu, Z., Li, H., See, K.Y.: Experimental investigation on electromagnetic interference (EMI) in motor drive using silicon carbide (SiC) MOSFET. In: 2020 International Symposium on Electromagnetic Compatibility - EMC EUROPE, pp. 1–6 (2020)
6. Yang, Z., Li, H., Yang, Z., Xu, Y., Lin, P.: Conducted electromagnetic interference analysis method for traction drive system in high speed trains. In: 2017 IEEE Transportation Electrification Conference and Expo, pp. 1–5 (2017)
7. Ji, J., Liu, Y., Chen, W., Li, J., Yang, X.: Modeling and simulation on overall-train EMI of china standardized high-speed train. In: 2018 IEEE International Symposium on Electromagnetic Compatibility and 2018 IEEE Asia-Pacific Symposium on Electromagnetic Compatibility, pp. 953–955 (2018)
8. Duan, Z., Zhang, D., Fan, T.: Modeling and prediction of electromagnetic interference in SiC motor drive systems. Trans. China Electrotech. Soc. **35**(22), 4726–4738 (2020). (in Chinese)
9. Cheng, L., Ou, H., Bi, C., Feng, S., Wu, J.: Trans. China Electrotech. Soc. **36**(S2), 627–634+643 (2021). (in Chinese)
10. Han, D., Li, S., Wu, Y., Choi, W., Sarlioglu, B.: Comparative analysis on conducted CM EMI emission of motor drives: WBG versus Si devices. IEEE Trans. Industr. Electron. **64**(10), 8353–8363 (2017)
11. Han, D., Sarlioglu, B.: Comprehensive study of the performance of SiC MOSFET-based automotive DC-DC converter under the influence of parasitic inductance. IEEE Trans. Ind. Appl. **52**(6), 5100–5111 (2016)

Research on the Common Mode EMI Suppression Method for SiC Traction Converters

Kan Liu[1], Donghui Liu[1](✉), Yongxiang Wang[1], Fanyang Meng[2], and Jianqiang Liu[2]

[1] State Key Laboratory of Traction and Control System of EMU and Locomotive, China Academy of Rail-Way Sciences Corporation, Beijing, China
{Liukan,Liudonghui,wangyongxiang}@zemt.cn

[2] School of Electrical Engineering, Beijing Jiaotong University, Beijing, China
{21126167,liujq}@bjtu.edu.cn

Abstract. The SiC traction converter possesses the advantages of low power loss and high operating frequency. However, the faster switching speed and higher switching frequency of SiC devices also lead to more severe electromagnetic interference (EMI) issues in traction drive systems. This paper proposes a common-mode (CM) EMI suppression method specifically for SiC traction converters. Firstly, the CM conducted interference paths in the traction drive system are studied, and an equivalent circuit modeling method for CM EMI applicable to SiC traction converters is presented. The influence of applying SiC devices on the magnitude of CM EMI in the traction drive system is analyzed. Furthermore, a CM EMI suppression measure is proposed by introducing nanocrystalline chokes at the input and output ports of the SiC traction converter, along with a design method for the parameters of the nanocrystalline choke. Finally, a low-power experimental platform for the SiC traction converter is established, and experimental verification is conducted. The results demonstrate that the proposed suppression method effectively suppresses CM currents in the system.

Keywords: Silicon carbide · Traction converter · Common-mode electromagnetic interference · Nanocrystalline choke · Suppression method

1 Introduction

The higher switching frequency and faster switching speed of SiC devices give rise to more severe common-mode (CM) electromagnetic interference (EMI) issues in traction drive systems. High-frequency current harmonics propagate through the complex conduction paths within the traction drive system, posing significant electromagnetic compatibility challenges [1, 2].

In terms of EMI equivalent circuit modeling, References [3] and [4] have investigated the CM EMI noise in electric vehicle power systems using SiC MOSFET power devices. Reference [5] analyzes the influence of SiC MOSFET switching performance on EMI in motor drive systems and also examines the effect of motor impedance on

L. Jia et al. (Eds.): EITRT 2023, LNEE 1135, pp. 99–107, 2024.
https://doi.org/10.1007/978-981-99-9307-9_11

CM EMI noise. Reference [6] proposes a frequency-domain calculation method for SiC motor drive systems that simultaneously considers the equivalent voltage source and equivalent current source replacing the power switches in SiC inverter circuits. Reference [7] analyzes the output waveform of a synchronous Buck converter based on SiC MOSFET modules in the frequency domain.

In the study of CM EMI suppression methods, reference [8] designs and analyzes the dimensions and performance of CM chokes required at 200 kHz and 20 kHz switching frequencies. Reference [9] proposes a CM circuit modeling method suitable for SiC motor drives and, based on this equivalent circuit model, analyzes and calculates the required parameters for CM chokes. Reference [10] designs a CMIS choke for suppressing CM noise in two-level voltage source inverters. Reference [11] focuses on the saturation issue of CM chokes and analyzes the saturation mechanism caused by CM voltage.

This paper investigates the CM EMI suppression method for SiC traction converters. In Sect. 2, an analysis of the CM conducted paths in the traction converter is conducted. Based on this analysis, a CM EMI equivalent circuit modeling method suitable for SiC traction cnnverters is proposed, and an analysis of the CM voltage and current spectrum in the system is presented. In Sect. 3, a CM suppression method based on nanocrystalline chokes is proposed, along with a design process for the parameters of the CM chokes. Finally, in Sect. 4, experimental verification is performed on a low-power experimental platform based on the SiC traction converter, and the experimental results are presented.

2 CM EMI Equivalent Circuit Modeling of Traction Converter

2.1 CM EMI Conduction Path

The conduction path of CM EMI is highly complex, with the CM conducted current in the traction converter system primarily forming a conduction loop through distributed parameter coupling. Figure 1 illustrates the CM conducted paths in the traction drive system, highlighting four main paths for CM current circulation.

2.2 Common Mode EMI Equivalent Circuit

Based on the high-frequency equivalent models of key components and the CM EMI conduction paths in traction drive systems, a CM EMI model for the traction drive system is established, as shown in Fig. 2.

The CM conduction equivalent circuit of traction drive system is shown in Fig. 3. The positive and negative DC terminals, P and N, of the traction converter busbar can be considered shorted, and the CM current in the positive and negative busbars is the same. The voltage interference source in the system can be replaced by V_{CM1} and V_{CM2}, and due to the symmetry of the output circuit structure, the circuit can be subjected to parallel processing. The expression for the equivalent CM voltage source of the four-quadrant converter (4QC) is $V_{CM1} = (V_{\mathrm{a}} + V_{\mathrm{b}})/2$, and the expression for the equivalent CM voltage source of the traction inverter is $V_{CM2} = (V_{\mathrm{a}} + V_{\mathrm{b}} + V_{\mathrm{c}})/3$.

Since the CM equivalent circuit consists only of linear components and independent sources, according to the superposition theorem, the equivalent two-port network is

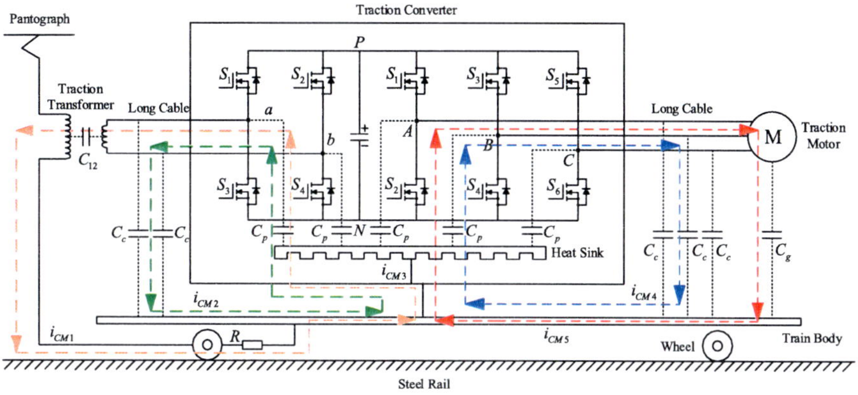

Fig. 1. CM conduction path of traction drive system.

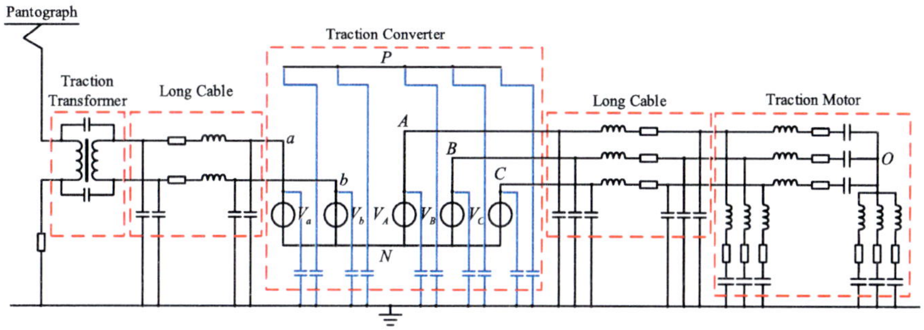

Fig. 2. CM EMI model of traction drive system.

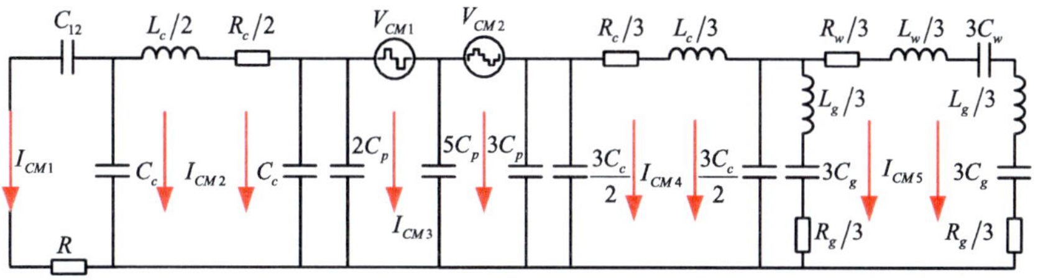

Fig. 3. CM conduction equivalent circuit of traction drive system.

obtained as shown in Fig. 4. $H_1(\omega)$ and $H_2(\omega)$ are functions related to the angular frequency and are uniquely determined by the CM conduction path. Therefore, the CM current I_{cm} on the impedance Z is equal to the sum of the current I_{cm_1} and the current I_{cm_2}. In the case where the frequency response of the conduction path is uniquely determined, the magnitude of the CM current is determined by CM voltage sources V_{CM1} and V_{CM2}.

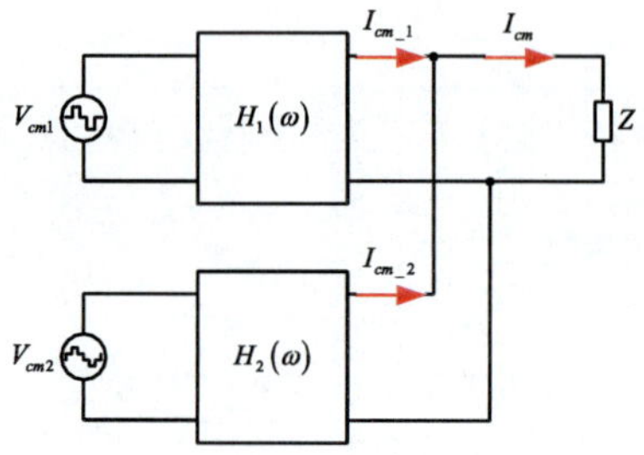

Fig. 4. Equivalent two port network.

The expressions for the amplitude of each harmonic of the CM current are given by formula (1), where C_{1k} and C_{2k} represent the spectrum of CM voltage sources V_{CM1} and V_{CM2}, respectively. ${H_1}^{(k)}$ and ${H_2}^{(k)}$ denote the amplitude-frequency response of the kth harmonic.

$$I_{cm}^{(k)} = C_{1k}H_1^{(k)} + C_{2k}H_2^{(k)} \tag{1}$$

2.3 CM Voltage Spectrum and CM Current Spectrum

The CM voltage spectrum in the traction drive system are presented below. Figure 5a shows the CM voltage source spectrum of the 4QC, and Fig. 5b shows the CM voltage source spectrum of the traction inverter. The spectrum of the CM currents flowing into the grid side and the traction motor, as depicted in Fig. 6a and b, respectively.

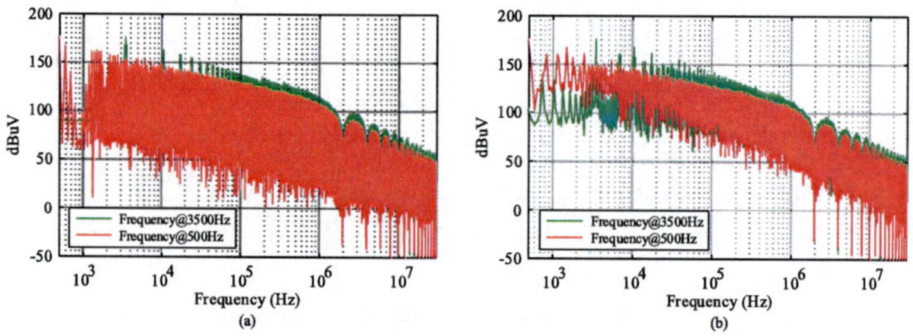

Fig. 5. System CM voltage spectrum.

It is observed that the higher the switching frequency, the larger the CM currents in the system, with an increase of over 10 dB across the entire frequency range. This poses a more severe EMI challenge for the traction drive system.

3 CM Suppression Method Based on Nanocrystalline Chokes

3.1 Nanocrystalline Choke

To mitigate the more severe CM EMI noise resulting from the use of SiC devices, interventions can be made on the system's conducted path by increasing the impedance along the path to reduce the system's CM current. Inserting filters can be an effective

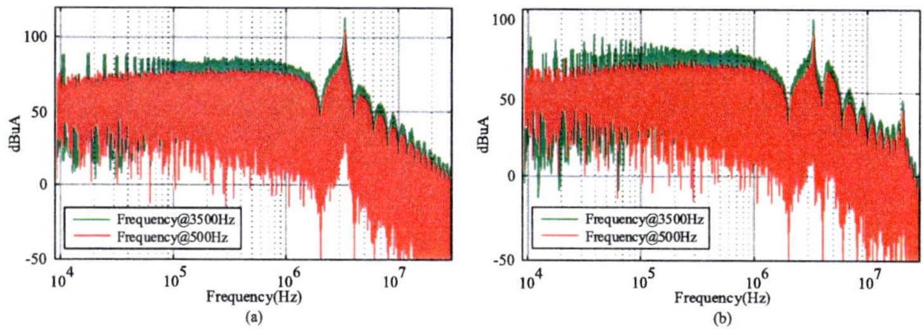

Fig. 6. System CM current spectrum.

approach to attenuate conducted interference to a certain extent. Adding CM chokes at the input and output terminals of the traction converter is a method to reduce the CM current in the system, as illustrated in Fig. 7.

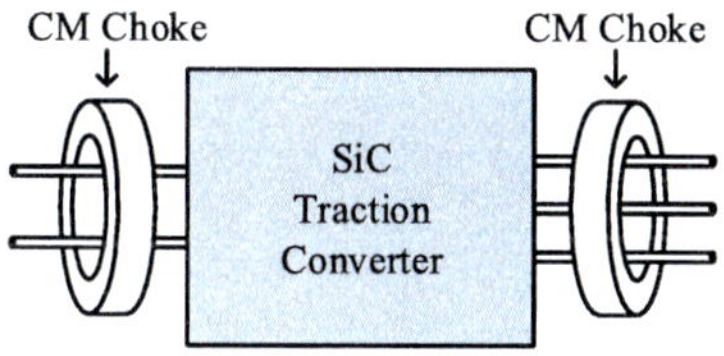

Fig. 7. Schematic diagram of adding CM chokes to the traction converter.

The recommended core material for the CM choke is iron-based nanocrystalline material. Nanocrystalline cores exhibit higher saturation magnetic induction compared to traditional ferrite cores. With a Curie temperature as high as 570 °C, nanocrystalline cores demonstrate stable performance even in high-temperature environments. Therefore, nanocrystalline cores are well-suited for suppressing CM interference in high-power traction converters.

3.2 Design Method for CM Chokes Parameters

The design process for CM chokes includes the following steps: determination of the original EMI noise, determination of insertion loss, determination of the CM choke inductance value, verification of suppression effectiveness, and selection of the CM choke design.

The insertion loss requirement is equal to the original CM EMI noise subtracted by the EMC standard limit, with an additional 6dB suppression margin, as shown in formula (2). Here, IL_{need} is the insertion loss requirement, CM_{ori} is the original CM EMI noise, and CM_{lim} is the EMC standard limit.

$$IL_{need} = CM_{ori} - CM_{lim} + 6dB \tag{2}$$

Insertion loss is the most direct indicator used to measure the suppression performance of passive EMI filters. Insertion loss is defined as the logarithmic ratio of power loss in the load before and after the EMI filter is connected to the system. After inserting the CM choke, the insertion loss of the filter is given by formula (3).

$$IL = 20\lg\left|1 + \frac{j\omega L_{cm}}{Z_s + Z_L}\right| \tag{3}$$

This type of L-shaped filter provides an insertion loss of 20 dB/Dec. The insertion loss of the filter can be simplified as a slope of 20 dB/Dec insertion loss curve, with the intersection point of the curve and the 0 dB line representing the cutoff frequency f_c of the filter. The value of f_c can be expressed as:

$$f_c = \frac{|Z_L + Z_s|}{2\pi L_{cm}} \tag{4}$$

In order to achieve EMI suppression, it is necessary to ensure that the insertion loss of the CM choke is greater than the insertion loss requirement (Fig. 8).

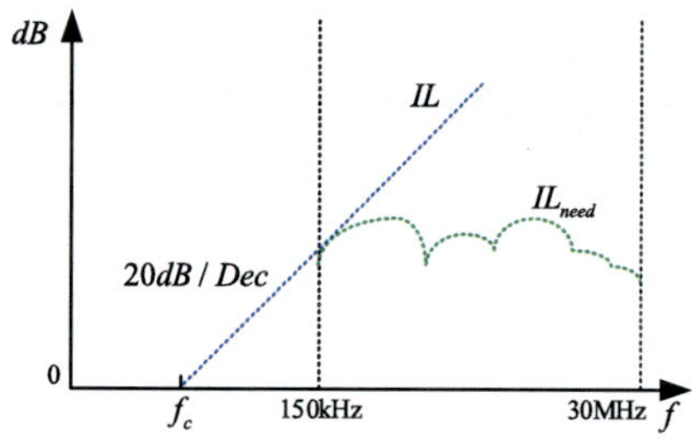

Fig. 8. Insertion loss demand curve and the choke insertion loss curve.

Once the value of the CM choke inductance is determined, the structure of the CM choke needs to be designed. The value of the CM choke inductance can be determined using formula (5), where A_e is the effective cross-sectional area of the core, l_e is the effective magnetic path length of the core, μ_0 is the vacuum permeability, μ_r is the relative permeability, and N is the number of turns in the winding. The CM choke can be considered as an inductance with only one turn.

$$L_{cm} = \frac{\mu_0 \mu_r A_e N^2}{l_e} \tag{5}$$

To prevent saturation of the CM choke, an analysis of the CM volt-second stress on the CM choke needs to be conducted. The CM volt-second stress on the CM choke can be determined using the peak amplitude of the CM current component I_{cmpk}, as described in formula (6).

$$\lambda_{cm} = \int V_L dt \approx 2L_{cm} I_{cmpk} \tag{6}$$

The magnetic flux density on the choke must satisfy the requirement in formula (7). Higher volt-second stress results in a larger cross-sectional area of the CM choke.

$$B_{\max} \geq \frac{\lambda_{cm}}{2NA_c} \tag{7}$$

Taking the design of the CM choke on the output side of the traction inverter as an example, Fig. 9a presents the CM EMI noise and the corresponding standard limits at a switching frequency of 3500 Hz. Figure 9b illustrates the insertion loss requirement, which is determined using formula (2). By determining a 20 dB/Dec slope insertion loss curve that tangentially intersects the insertion loss requirement, the value of the magnetic choke inductance can be determined. Without considering the volume and weight of the choke, a larger magnetic choke inductance value provides better suppression performance.

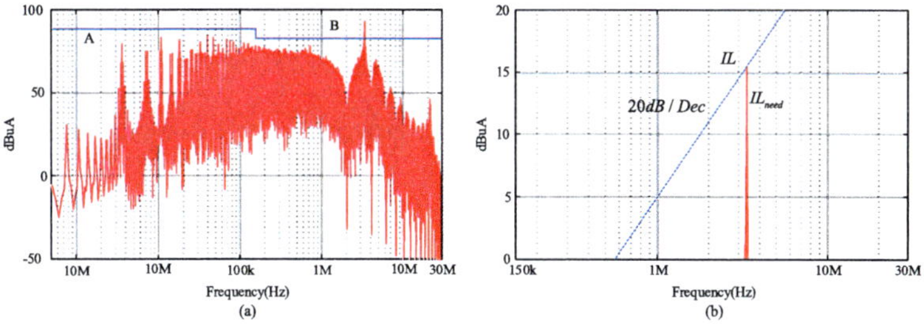

Fig. 9. CM EMI noise and standard limits.

The CM EMI noise after inserting a CM choke with 0.1 mH inductance value is shown in Fig. 10a. It is evident that the insertion of the CM inductance increases the impedance characteristics of the load circuit, particularly in the high-frequency range. Figure 10b compares the CM EMI noise before and after the insertion of the CM choke, demonstrating a significant reduction in EMI intensity within the 2–30 MHz frequency range.

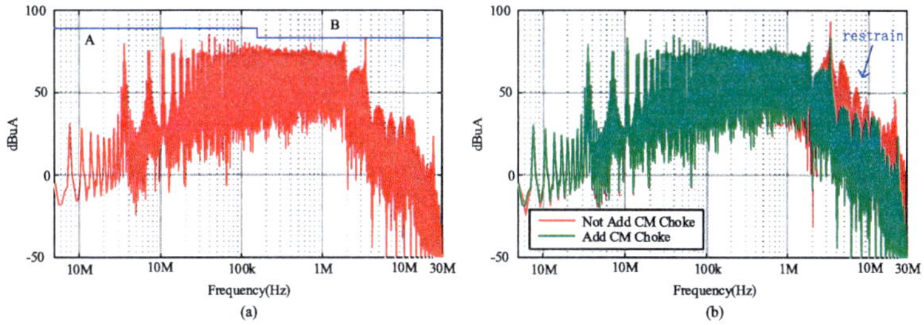

Fig. 10. Comparison of CM EMI before and after inserting a CM choke.

4 Experimental Research

To analyze the impact of high-frequency SiC devices on the system CM EMI, the CM current spectrum flowing into the grid-side and the asynchronous motor were measured and are presented in Fig. 11a and b, respectively. The higher switch frequency results in increased CM interference sources, and under unchanged conduction paths, the CM current spectrum also exhibit an increase.

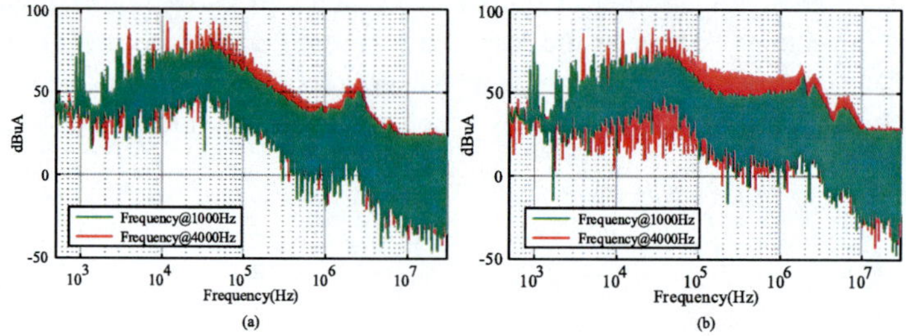

Fig. 11. System CM current spectrum.

To validate the effectiveness of using nanocrystalline CM chokes to suppress the CM currents in the system, CM chokes with an inserted inductance value of 0.12 mH were employed on the rectifier side and inverter side of the SiC experimental platform. Figure 12 illustrates the suppression effect before and after the installation of CM chokes. These results highlight the effectiveness of using CM chokes to suppress the CM currents in the system and effectively suppress the electromagnetic interference within the frequency range of 600 kHz to 10 MHz.

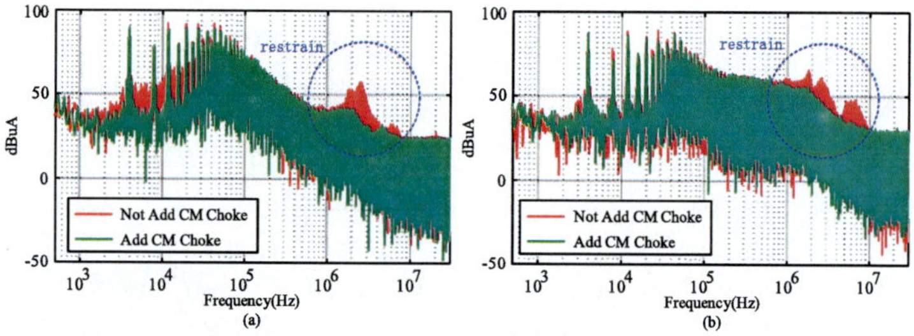

Fig. 12. Comparison of suppression effects before and after installing CM chokes.

5 Conclusion

This paper focuses on the study of CM EMI suppression methods for SiC traction converters. The equivalent circuit of the traction converter's CM EMI is first modeled, analyzing the paths of CM currents in the system. The application of SiC devices exacerbates the EMI issues across the entire frequency range of the traction converter. The CM voltage spectrum exhibits an approximate 10 dB increase, when frequency is increased from 1 to 4 kHz. Subsequently, a CM EMI suppression method is proposed, which involves inserting nanocrystalline CM chokes at the input and output ports of the traction converter to reduce the CM currents in the system. A design method for the CM choke parameters is provided, determining the required choke inductance based on insertion loss, where a larger inductance value leads to better suppression performance. Finally, experimental validation is conducted on a SiC low-power experimental platform. The CM current of the grid-side and motor-side is effectively reduced within the frequency range of 600 kHz to 10 MHz after the installation of the chokes.

Acknowledgment. This work is supported by the State Key Laboratory of Traction and Control System of EMU and Locomotive Open Project 2021YJ279.

References

1. Zhi, Y., Yang, D., Zhu, B., Yuan, K., Gao, Z., Li, P.: High-frequency negative effect analysis and countermeasures of traction inverter based on all SiC MOSFET for rail transit. Electric. Drive Locomot. **5**, 49–55 (2020). (in Chinese)
2. Zhang, X., Luo, Z.: Research on EMC application of traction inverter based on silicon carbide MOSFET power module. Rolling Stock **60**(2), 83–88 (2022). (in Chinese)
3. Jia, X., Hu, C., Dong, B., He, F., Wang, H. and Xu, D.: System-level conducted EMI model for SiC powertrain of electric vehicles. In: 2020 IEEE Applied Power Electronics Conference and Exposition, pp. 885–892 (2020)
4. Jia, X., Hu, C., Dong, B., He, F., Wang, H., Xu, D.: Influence of system layout on CM EMI noise of SiC electric vehicle powertrains. CPSS Trans. Power Electron. Appl. **6**(4), 298–309 (2021)
5. Wu, Y., Yin, S., Liu, Z., Li, H., See, K.Y.: Experimental investigation on electromagnetic interference (EMI) in motor drive using silicon carbide (SiC) MOSFET. In: 2020 International Symposium on Electromagnetic Compatibility—EMC EUROPE, pp. 1–6 (2020)
6. Duan, Z., Zhang, D., Fan, T.: Modeling and prediction of electromagnetic interference in SiC motor drive systems. Trans. China Electrotechnical Soc. **35**(22), 4726–4738 (2020). (in Chinese)
7. Cheng, L., Ou, H., Bi, C., Feng, S., Wu, J.: Trans. China Electrotechnical Soc. **36**(S2), 627–634+643 (2021) (in Chinese)
8. Han, D., Morris, C.T., Lee, W., Sarlioglu, B.: Comparison between output CM chokes for SiC drive operating at 20- and 200-kHz switching frequencies. IEEE Trans. Ind. Appl. **53**(3), 2178–2188 (2017)
9. Han, D., Morris, C.T., Lee, W., Sarlioglu, B.: Determination of CM choke parameters for SiC MOSFET motor drive based on simple measurements and frequency domain modeling. In: 2016 IEEE Applied Power Electronics Conference and Exposition, pp. 2861–2867 (2016)
10. Kumar, M., Jayaraman, K.: Common mode impedance shaping choke to attenuate the conducted EMI in three phase drive. In: 2020 IEEE International Conference on Power Electronics, Drives and Energy Systems, pp. 1–5 (2020)
11. Fang, L., Boroyevich, D., Mattevelli, P., Gazel, N.: A comprehensive design for high power density common mode EMI inductor, pp. 1861–1867 (2011)

Inductor Current Ripple Modeling and Mitigation Method of Cascaded Half-Bridge DC/DC Converter Under Unbalanced Conditions

Tianmu Chen, Guohong Zeng(✉), Fen Tang, and Luoqi Wu

National Energy Active Distribution Network Technology R&D Center, Beijing Jiaotong University, Beijing 100044, China
{19117002,Ghzeng,Fent,22121528}@bjtu.edu.cn

Abstract. For cascaded half-bridge DC/DC converter based on phase-shifted carrier modulation, if input voltages and references of each module are different, the current ripple on inductor will increase, which could cause more power loss and larger size of the filter. In this paper, the problems of traditional fixed-angles phase-shifting technology are revealed through analyzing the model of current ripple in different conditions. In order to express the ripple characteristics under unbalanced conditions, a frequency domain model is established, which explain the relation among the harmonics in ripple, input and output voltages and phase-shifted angles. Based on the model, a current ripple mitigation strategy by varying the angles is proposed to remove the selective harmonic in ripple. Simulation results on a three-cell modular multilevel DC/DC converter is presented to verify the feasibility and effectivity of proposed models and modulation method.

Keywords: Cascaded Half-bridge · DC/DC converter · Current ripple

1 Introduction

In the application of electric locomotive regenerative braking, energy storage devices are used to recover the energy when breaking, and releasing the energy when traction, which could reduce the impact of traction load on the system [1]. Battery as a kind of flexible energy storage method is widely adopted in locomotive power supply system. Since each battery's voltage and capacity is on a low level, to structure high voltage energy storage system, the conventional method is placing the batteries in long series strings, but it is difficult to control every battery's state [2].

Cascaded Half-bridge DC/DC converter (CHDC), which has the feature of flexible control method, high degree of modularization, low voltage stress of switching devices and small volume of output filter components, is widely used in regenerative braking [3, 4], battery energy storage system [5] and renewable energy generation [6–8]. In CHDC topology, a large amount of half-bridge cells are in series connected to transfer the energy.

L. Jia et al. (Eds.): EITRT 2023, LNEE 1135, pp. 108–116, 2024.
https://doi.org/10.1007/978-981-99-9307-9_12

In this topology, high voltage energy storage system and battery status monitoring can be conveniently implemented.

Inductor current ripple is key factor affecting the operation of the converter. Excessive current ripple not only increases the loss of inductance, but also accelerates the aging of filter components. Although current ripple can be suppressed by increasing the switching frequency or inductance value, these methods could bring large switching loss and volume of the filter [9, 10]. To solve this problem, phase-shifted carrier (PSC) modulation which sets the angles as $2\pi/N$ is usually adopted [3]. With this strategy in N-cell CHDC, 1st to N-1th harmonics are removed in current ripple, when all modules' input voltage and reference are the same. With this method, the current ripple is reduced $1/N^2$ of the ripple based on the modulation method without phase shifting. However, in practical applications, due to the influence of varied system parameters or control strategies, there may be significant differences on the input and output voltages of various modules. Traditional PSC method could not effectively suppress the ripple.

To address the issue, this paper proposes a detailed analysis of current ripple characteristic in unbalanced conditions. The ripple mathematic model in frequency-domain is established, which is used to figure out the relation among ripple and modules' parameters. Then, a variable angles phase-shifted modulation based on the model is presented under unbalanced condition. Finally, the simulation results verify the proposed modeling and ripple suppression method.

2 Current Ripple Analysis of CHDC Converter Using PSC-PWM

2.1 Description of CHDC

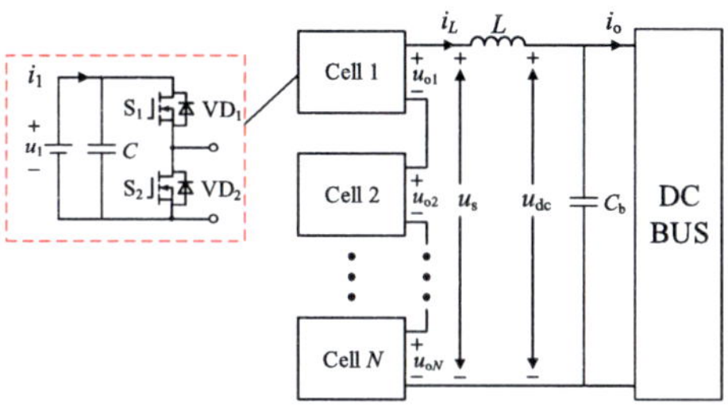

Fig. 1. Topology of CHDC system

The diagram of CHDC is shown in Fig. 1, which is composed with N DC/DC cells in series. In the jth half-bridge DC/DC cell, u_j is the battery voltage, i_j represents the battery current, u_{oj} is the output voltage. u_s and u_{dc} are the multilevel voltage and the average output voltage, i_L and i_o are the inductor current and average output current. L and C_b compose the filter. S_1 and S_2 are the switching device.

CHDC usually uses phase-shifted method to convert each module's reference signal into pulses, as shown in Fig. 2a, where T_s is switching period and φ is phase-shifted angle.

If ignoring the effect of dead-time, the output voltage of *j*th module can be expressed as

$$u_{oj}(t) = s_j(t)u_j \tag{1}$$

where s_j(t) is the switching function of the *j*th module. s_j(t) = 0 represents S_1 is off while S_2 is on, and s_j(t) = 1 represents S_1 is on while S_2 is off. The output pulses of modules are added up to form the multilevel voltage u_s(t), which can be expressed as

$$u_s(t) = \sum_{j=1}^{N} u_{oj}(t) \tag{2}$$

u_s(t) is consisted of DC and AC component, where the DC part is average value of the expected output of u_s(t), which is equal to u_o, and the expression is

$$\overline{u}_s = u_o = \sum_{j=1}^{N} d_j u_j \tag{3}$$

In (3) d_j represents *j*th module' duty.

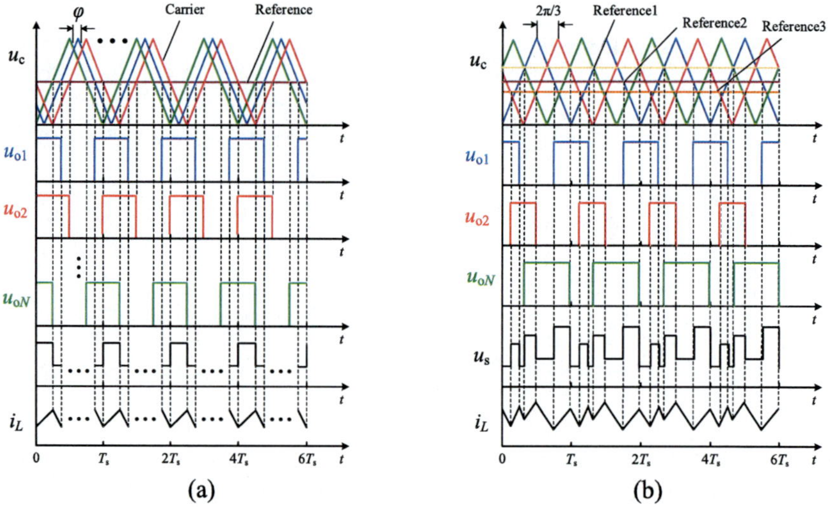

Fig. 2. Diagram of PSC. **a** Balanced condition. **b** Unbalanced conditions

The AC part in (2) is the difference between u_s(t) and u_o, which brings the ripple on inductor, and it can be written as

$$u_{ac}(t) = u_s(t) - u_o \tag{4}$$

the current ripple can be expressed as

$$\Delta i_L(t) = \frac{1}{L}\int u_{ac}(t)\mathrm{d}t \tag{5}$$

Equation (5) is used to calculate the ripple value at every sampling time.

2.2 Ripple Analysis Under Balanced Condition

In balanced condition, where $u_1 = u_2 = \ldots = u_N = u_0$ while $d_1 = d_2 = \ldots = d_N = d_0$, the waveform of u_s(t) is a series two-level pulses with same phase, pulse width and amplitude based on fixed angle PSC in Fig. 2a. The expression of u_s(t) is

$$u_s(t) = \begin{cases} (k+1)u_0 & l\varphi \le t < l\varphi + M \\ ku_0 & l\varphi + M \le t(l+1)\varphi \end{cases} \tag{6}$$

where $k = 0,1,2, \ldots, N-1$; $l = 0,1,2, \ldots, N$; $\varphi = 2\pi/N$; $M = d_0T_s - l\varphi$.

The pulses on the inductor bring current ripple, where high voltage level increases the current while low voltage level decreases it. Assume Δi_{L0} is the ripple amplitude generated by any cell puts into operation, the current ripple can be expressed as

$$\Delta i_L = \left[(k+1)\frac{\Delta i_{L0}}{dT_s} - (l-k-1)\frac{\Delta i_{L0}}{(1-d)T_s}\right]M \tag{7}$$

From (7), the normalized expression of ripple r is

$$r = \frac{\Delta i_L}{\Delta i_{L0}} = \frac{(ld-k)(1+k-ld)}{ld(1-d)} \tag{8}$$

The diagram of r varying is shown in Fig. 3, which can be seen that with the number of operating cells increases, the ripple performance is improved. Thus, the inductance value could be reduced when putting more cells into work.

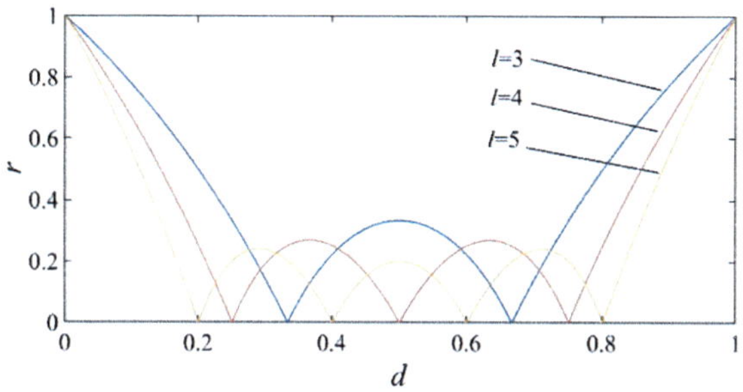

Fig. 3. Relation between normalized current ripple and cell duty

2.3 Ripple Model in Frequency-Domain Under Balanced Condition

When CHDC is under unbalanced condition, the voltage level of u_s(t) and its shifting time is unpredictable, as shown in Fig. 2b. It is difficult to express the current ripple as in balanced condition.

To suppress the ripple in this complicated condition, frequency domain analysis is adopted, which regards ripple as sum of multiple harmonics.

Applying Fourier series, the pulse of jth module can be written as

$$u_{oj}(t) = \frac{a_0}{2} + \sum_{n=1}^{\infty} (a_n \cos(n\omega t) + b_n \sin(n\omega t)) \tag{9}$$

where $n(n \in \mathbf{N})$ indicates the nth harmonic in ripple, and $\omega = 2\pi/T_s$.

Based on even symmetric modulation, coefficients a_{0j}, a_{nj} and b_{nj} are obtained as

$$\begin{cases} a_0 = \dfrac{2}{\pi}\displaystyle\int_0^{\pi d_j} u_j(\omega t)\mathrm{d}(\omega t) = 2u_j d_j \\ a_n = \dfrac{2}{\pi}\displaystyle\int_0^{\pi d_j} u_j(\omega t)\cos(n\omega t)\mathrm{d}(\omega t) \\ \quad = \dfrac{2u_j}{n\pi}\sin(n\pi d_j) \\ b_n = 0 \end{cases} \tag{10}$$

With the consideration of (10) and phase-shifted angel φ_j, the jth module's output voltage is

$$u_{oj}(t) = u_j d_j + \sum_{n=1}^{\infty} \frac{2u_j}{n\pi} \sin(n\pi d_j)\cos(n\omega t + n\varphi_j) \tag{11}$$

With the sum of N modules' pulses, the multilevel voltage is

$$u_s(t) = \sum_{j=1}^{N} u_j d_j + \sum_{j=1}^{N}\sum_{n=1}^{\infty} \frac{2u_j}{n\pi} \sin(n\pi d_j)\cos(n\omega t + n\varphi_j) \tag{12}$$

Combine (5) and (12) with the assumption that initial value is 0, the current ripple can be expressed as

$$\Delta i_L(t) = \sum_{j=1}^{N}\sum_{n=1}^{\infty}\left[\frac{2u_j}{L\omega n^2\pi}\sin(n\pi d_j)\sin(n\omega t + n\varphi_j)\right] \tag{13}$$

From (13), the expression of amplitude A_n for nth harmonic is

$$A_n = \sqrt{\left(\sum_{j=1}^{N}\frac{2u_j}{L\omega n^2\pi}\sin(n\pi d_j)\sin(n\varphi_j)\right)^2 + \left(\sum_{j=1}^{N}\frac{2u_j}{L\omega n^2\pi}\sin(n\pi d_j)\cos(n\varphi_j)\right)^2} \tag{14}$$

3 Current Ripple Suppressing Strategy Based on Variable Angle PSC

In order to suppress the ripple, the efficiency method is to eliminate the ripple harmonic. It can be seen from (14), the coefficient of nth harmonic is $1/n^2$ of the fundamental harmonic, which is in switching frequency. As a result, it is necessary to remove A_1.

To eliminate the harmonic, A_n in (13) should be suppressed to 0, and the sufficient condition is

$$\begin{cases} \sum_{j=1}^{N} h_j \sin(n\varphi_j)=0 \\ \sum_{j=1}^{N} h_j \cos(n\varphi_j)=0 \end{cases} \tag{15}$$

where h_j is

$$h_j = \frac{2u_j}{L\omega n^2 \pi} \sin(n\pi d_j) \tag{16}$$

To simplify the analysis in this article, a modulation strategy for eliminating fundamental harmonic was calculated and validated in CHDC with a cascaded module number of $N = 3$. Assume cell 1 is the phase reference, which means $\varphi_1 = 0$, angles φ_2 and φ_3 need to be solved. From (16), to remove fundamental harmonic, the following requirement needs to met

$$\begin{cases} h_{1,1}+h_{2,1}\cos(\varphi_2)+h_{3,1}\cos(\varphi_3)=0 \\ h_{2,1}\sin(\varphi_2)+h_{3,1}\sin(\varphi_3)=0 \end{cases} \tag{17}$$

The solutions of (17) are

$$\begin{cases} \varphi_2 = \arccos\left(\frac{h_{3,1}^2 - h_{1,1}^2 - h_{2,1}^2}{2h_{1,1}h_{2,1}}\right) \\ \varphi_3 = \arccos\left(\frac{h_{2,1}^2 - h_{1,1}^2 - h_{3,1}^2}{2h_{1,1}h_{3,1}}\right) \end{cases} \tag{18}$$

Because of the arccos function in the equations, the method should be limited by the relation among h_j as

$$\begin{cases} \frac{h_{3,1}}{h_{1,1}} \le \frac{h_{2,1}}{h_{1,1}}+1 \\ \frac{h_{2,1}}{h_{1,1}} \le \frac{h_{3,1}}{h_{1,1}}+1 \\ \frac{h_{2,1}}{h_{1,1}} + \frac{h_{3,1}}{h_{1,1}} \ge 1 \end{cases} \tag{19}$$

Based on the above analysis, the optimized PSC modulation strategy is proposed in this article. The method calculates the harmonic coefficients $h_{j,n}$ of each cell in modulation cycle, and then calculating the angles using Eq. (18). Compared with traditional fixed angle PSC, this strategy can effectively suppress current ripple under unbalanced conditions.

4 Simulation Results

To verify the effectiveness of the proposed current ripple model and mitigation strategy, simulation tests for a three-cell CHDC based on MATLAB/Simulink are presented in this section.

For system parameters, the output voltage is 110 V; switching frequency is 100 kHz; inductance value is 33 uH; The resistance value of load is 25 Ω. In Case I, where the converter is running under balanced condition, the input voltages are the same as 60 V while the duties are the same as 0.611. In In Case II, where the converter is running under unbalanced condition, the input voltages are 55, 60 and 65 V while the duties are 0.509, 0.611 and 0.713.

For Case I which is in balanced condition, the ripple peak-to-peak value is 0.88 A in Fig. 4. From the harmonic spectrum, the 1st and the 2nd order harmonics are almost 0, which is same to theoretical calculation based on the proposed model in Table 1. The 3rd order harmonic is 8% of DC, which amplitude is 0.343 while the theoretical value is 0.308.

For Case II, the unbalanced condition increases the ripple peak-to-peak value to 1.89A in Fig. 5. From the harmonic spectrum, the 1st to the 3rd order harmonics are 5.5, 11.5 and 5% of DC, which amplitudes are 0.24, 0.495 and 0.22, while the theoretical values are 0.196, 0.472 and 0.179. As can be seen from the comparison, the theoretical and simulation values are basically consistent. When using the proposed ripple suppression method, ripple value is reduced to 1.57A in Fig. 6. The 1st harmonic is removed.

Table 1. Harmonic amplitude comparison

Case	In theory [1st, 2nd, 3rd]	Simulation result
I	[0, 0, 0.308]	[0, 0, 0.343]
II	[0.196, 0.472, 0.149]	[0.24, 0.495, 0.22]

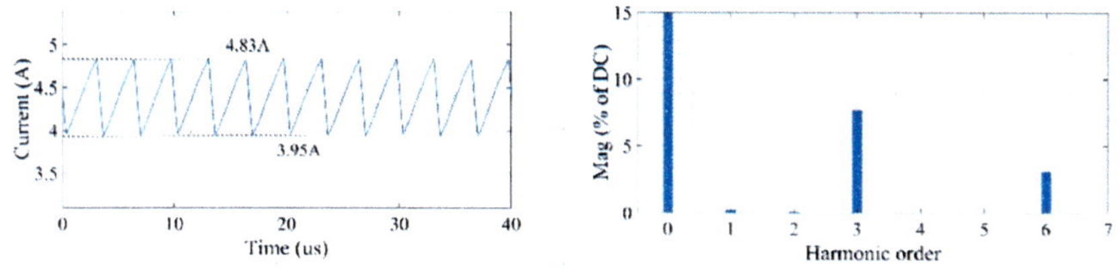

Fig. 4. Ripple performance for Case I with traditional PSC

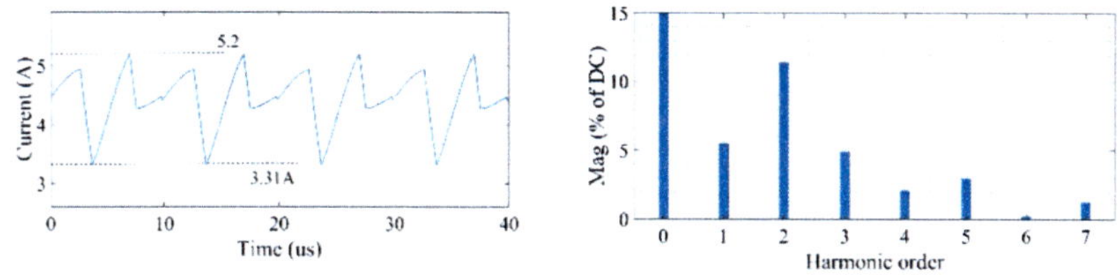

Fig. 5. Ripple performance for Case II with traditional PSC

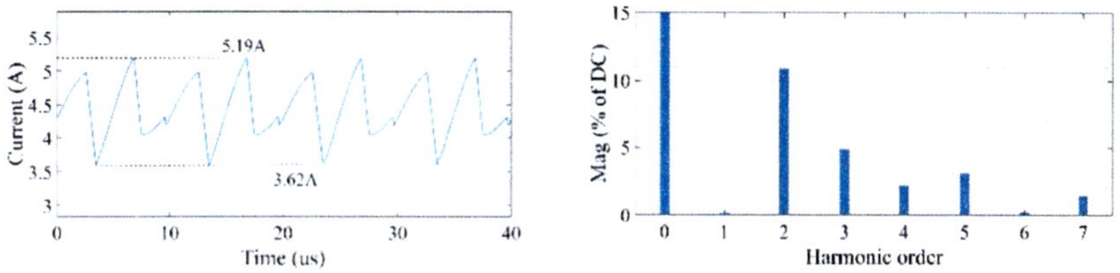

Fig. 6. Ripple performance for Case II with proposed PSC

5 Conclusion

In this paper, a current ripple model and mitigation method is proposed for CHDC under unbalanced conditions. Firstly, the problems in current ripple analysis were explained through time-domain modeling. Then, a frequency-domain model is established to analyze the relation between harmonics in ripple and other cells' parameters. Finally, a variable angle phase-shifted modulation is proposed to suppressing the ripple. The simulation of harmonic value is tested to evaluate the effectivity of proposed modeling and modulation method. The results show that the frequency-domain model could calculate harmonics in different conditions and the modulation method can effectively remove the harmonic.

References

1. Bao, G.N., Lu, C., Yuan, Z.C., Han, Y,D.: Load shift real-time optimization strategy of battery energy storage system based on dynamic programing. Autom. Electric Power Syst. **36**(12), 11–16 (2012) (in Chinese)
2. Soong, T., Lehn, P.W.: Evaluation of emerging modular multilevel converters for BESS applications. IEEE Trans. Power Deliv. **29**(5), 2086–2094 (2014)
3. Montesinos-Miracle, D., Massot-Campos, M., Bergas-Jane, J., Galceran-Arellano, S., Rufer, A.: Design and control of a modular multilevel DC/DC converter for regenerative applications. IEEE Trans. Power Electron. **28**(8), 3970–3979 (2013)
4. Bi, K., Sun, L., An, Q., Duan, J.: Active SOC balancing control strategy for modular multilevel super capacitor energy storage system. IEEE Trans. Power Electron. **34**(5), 4981–4992 (2019)
5. Mukherjee, N., Strickland, D.: Analysis and comparative study of different converter modes in modular second-life hybrid battery energy storage systems. IEEE J. Emerg. Sel. Top. Power Electron. **4**(2), 547–563 (2016)
6. Lu, Y., Sun, K., Wu, H., Dong, X., Xing, Y.: A three-port converter based distributed DC grid connected PV system with autonomous output voltage sharing control. IEEE Trans. Power Electron. **34**(1), 325–339 (2019)
7. Zhuang, Y., Liu, F., Huang, Y., Zhang, X., Zha, X.: A voltage-balancer-based cascaded DC–DC converter with a novel power feedforward control for the medium-voltage DC grid interface of photovoltaic systems. IEEE Access **7**, 178094–178107 (2019)
8. Majeed, Y.E., Ahmad, I., Habibi, D.: A multiple-input cascaded DC–DC converter for very small wind turbines. IEEE Trans. Ind. Electron. **66**(6), 4414–4423 (2019)
9. Nupur, N., Nath, S.: Unifying inductor current ripples and inductor design in coupled SIDO converters by forming sectors of duty ratios. IEEE Trans. Ind. Appl. **58**(3), 3830–3839 (2022)
10. Kang, T., Suh, Y.: Optimized coupling factor design of multiple-phase coupled inductor for minimum inductor current ripple operation in EV charger systems. In: 2017 IEEE 3rd International Future Energy Electronics Conference and ECCE Asia (IFEEC 2017—ECCE Asia), pp. 1178–1183, Kaohsiung, Taiwan (2017)

Weak Magnetic Control of Automotive Permanent Magnet Synchronous Motor Based on Sliding Mode Control

Huchuan Yang[1,2](✉), Jing Chen[1,2], Meng Zhang[1], Zhuo Wang[1], and Wei Wu[1]

[1] Wuhan University of Technology, Wuhan 430079, China
y13687194094@163.com

[2] Foshan Xianhu Laboratory of the Advanced Energy Science and Technology Guangdong Laboratory, Foshan 528200, China

Abstract. In order to solve the problems of large torque speed pulsation and waveform distortion in the traditional PI control method for weak magnetic control of PMSM of electric vehicles, a new weak magnetic control method for PMSM was proposed based on sliding mode control in this work. Firstly, the mathematical model of the motor is established, then the controller is designed according to the weak magnetic control theory and the principle of sliding mode control, and finally the simulation analysis is established by MATLAB/Simulink. The results of simulation evaluation turned out that compared with the traditional PI weak magnet control, the response speed, anti-interference ability and robustness of control system were improved when the proposed control method was used.

Keywords: Permanent magnet synchronous motor · Sliding mode control · Weak magnetic control

1 Introduction

In order to improve the energy utilization, driving range of electric vehicles (EVS) and reduce the braking energy loss, the weak magnetic speed control technology is widely used in the permanent magnet synchronous motor (PMSM). Apart from it, it could reduce the energy consumption and emission of EVS, extend the battery life, and also improve the driving experience and ride comfort. Besides, the PMSM is a system with multiple variables, strong coupling, non-linearity and complex characteristics, the control strategy of PMSM is current research hotspot.

Currently, many scholars have studied the weak magnetic control of PMSM at home and abroad. Domestic scholar Niu ZZ proposed a weak magnetic control method of PMSM based on sliding mode speed control, which effectively improved the adjustable range of speed [1] and the stability of speed is enhanced [1]. By analyzing the weak magnetism principle of PMSM, Luo Xiang used the Maximum Torque Per Ampere

L. Jia et al. (Eds.): EITRT 2023, LNEE 1135, pp. 117–124, 2024.
https://doi.org/10.1007/978-981-99-9307-9_13

(MTPA) control strategy to improve the response speed and stability of the system [3]. Jia et al. proposed a space vector pulse width modulation control strategy based on current feedforward control, which compensated the d-q axis reference current components in traditional vector control separately, effectively suppressing the motor torque pulsation and enabling faster and stable control of the motor [4]. The above control methods still rely on the traditional PI control, while sliding mode control (SMC) has stronger robustness, faster response, better model independence, lower steady-state error and better real-time performance compared to the traditional PI control.

In summary, a weak magnetic control method with sliding mode control was proposed in this work which was used both the speed and current loops. Firstly, the mathematical model of PM synchronous motor was constructed. And then, the block diagram of PM synchronous motor control system structure was given. Again, through the weak magnetic control and vector control, the sliding mode controller (SMC) was designed. Finally, the superiority of the proposed method was verified by MATLAB/Simulink simulation results.

2 Mathematical Model of PMSM

The PMSM is a system with strong coupling, non-linearity and complex characteristics, and so as to match the design of its control system, its mathematical model needs to be established in the d-q rotation coordinate system.

The stator voltage expression of the mathematical model in the d-q rotating coordinate system is shown in (1):

$$\begin{cases} u_d = R_s i_d + L_d \frac{\mathrm{d}i_d}{\mathrm{d}t} - p_\mathrm{n}\omega_\mathrm{m} L_\mathrm{q} i_\mathrm{q} \\ u_q = R_s i_q + L_q \frac{\mathrm{d}i_q}{\mathrm{d}t} + p_\mathrm{n}\omega_\mathrm{m} L_d i_d + p_\mathrm{n}\omega_\mathrm{m}\psi_f \end{cases} \tag{1}$$

where u_d and u_q are the stator winding d-axis and q-axis voltages, i_d and i_q are the stator winding d-axis and q-axis voltages, L_d and L_q are the stator winding d-axis and q-axis inductances, R_s is the stator winding resistance, P_n is the number of pole pairs, ω_m is the mechanical angular velocity, and ψ_f is the permanent magnet chain.

Equation of motion:

$$T_e = T_L + B\omega_m + J\frac{d\omega_m}{dt} \tag{2}$$

where T_e is the electromagnetic torque, T_L is the load torque, B is the damping factor, and J is the rotor inertia.

Motor torque equation:

$$T_\mathrm{e} = 1.5P_\mathrm{n}\left[\psi_\mathrm{f} + \left(L_\mathrm{d} - L_\mathrm{q}\right)i_\mathrm{d}\right]i_\mathrm{q} \tag{3}$$

3 PMSM Control System

The block diagram of SMC-based vector control of a PMSM is shown in Fig. 1. The reference torque output from the speed loop SMC is used as the input to the MTPA module, and the d-axis reference current and q-axis reference current output from the weak magnet module are used as the inputs to the d-axis current loop SMC and q-axis current loop SMC, respectively.

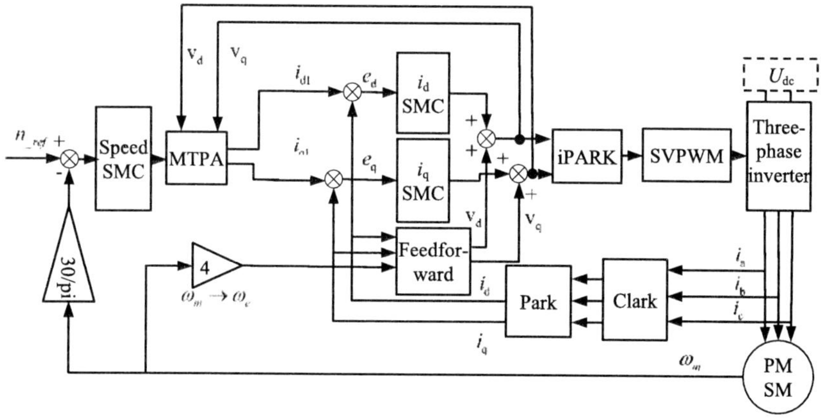

Fig. 1. Block diagram of SMC based weak magnetic control of PMSM.

4 Design of SMC

SMC is a kind of nonlinear controller, whose core idea is to slide the system state from the initial state to the target state by introducing a sliding mode surface to achieve the control purpose of the system. The design of the SMC for this system is divided into two parts, one of which is the design of the SMC for the velocity loop and the other is the design of the SMC for the current loop.

4.1 Speed Ring Design

When designing the SMC for speed, considering that the input parameter of the MPTA module is torque, the state variable of the sliding mode control method is defined according to the motor equation of motion.

$$\begin{cases} x1 = \omega_{ref} - \omega_m \\ x2 = \dot{x}1 = -\frac{d\omega_m}{dt} \end{cases} \quad (4)$$

where ω_{ref}, is a reference speed of the motor, and in general ω_{m} is a constant, which is the actual speed of the motor.

The sliding mode surface function can be defined as

$$s = cx_1 + x_2 \tag{5}$$

In the above equation, c is the parameter need to be designed in this work, and $c > 0$, the derivative of the above equation can be obtained

$$\dot{s} = c\dot{x}1 + \dot{x}2 = cx2 + \dot{x}2 = cx2 - \dot{T}_e/J \tag{6}$$

For the sake of ensuring that the control system has good dynamic quality, the exponential convergence law method is used here, and the resulting controller expression is

$$\dot{s} = -\varepsilon \mathrm{sgn}(s) - qs \tag{7}$$

where $\varepsilon > 0, q > 0$, is the parameter to be designed. The expression of the SMC can be obtained as

$$T_e = J\int_0^t [cx2 + \varepsilon \mathrm{sgn}(s) + qs]dt \tag{8}$$

The expression of the controller contains an integral term, so it can weaken some jitter phenomenon, and also reduce the steady-state error of the system and improve the dynamic quality of the system.

4.2 Current Loop Design

Define the current error as a state variable:

$$e_d = i_{d1} - i_d \tag{9}$$

Define the sliding mode surface function as:

$$\begin{cases} s_1 = c_1 e_d + \dot{e}_d \\ \dot{e}_d = \dfrac{di_{\mathrm{d}1}}{dt} - \dfrac{u_{\mathrm{d}}}{L_{\mathrm{d}}} + \dfrac{R_s}{L_{\mathrm{d}}} i_{\mathrm{d}} - \dfrac{P_n \omega_m L_{\mathrm{q}}}{L_{\mathrm{d}}} i_{\mathrm{q}} \end{cases} \tag{10}$$

where $c_1 > 0$, are parameters to be designed, the derivative of the above equation can be obtained.

$$\dot{s}_1 = c_1 \dot{e}_d + \ddot{e}_d \tag{11}$$

Using the exponential convergence law method, the resulting controller expression is

$$\dot{s}_1 = -\varepsilon_1 \mathrm{sgn}(s_1) - q_1 s_1 \tag{12}$$

Combining Eqs. (11) and (12), the expressions u_{d}, u_{q} can be obtained as follows:

$$u_d = L_d \int_0^t \left(c_1 \left(\frac{di_{dd}}{dt} + \frac{R_s}{L_d} i_d - \frac{P_n \omega_m L_q}{L_d} i_q \right) + \varepsilon_1 |e_d| \mathrm{sgn}(s_1) + q_1 s_1 \right) dt \quad (13)$$

Similarly, we can obtain:

$$u_q = L_q \int_0^t \left(c_2 \left(\frac{di_{qd}}{dt} + \frac{R_s}{L_q} i_q + \frac{P_n \omega_m L_d}{L_q} i_d + \frac{P_n \omega_m}{L_q} \varphi_f \right) + \varepsilon_2 |e_q| \mathrm{sgn}(s_2) + q_2 s_2 \right) dt \quad (14)$$

When $\varepsilon_1 > 0,\ q_1 > 0,\ \varepsilon_2 > 0,\ q_2 > 0$, Eqs. (13) and (14) satisfy the Lyapunov stability condition, the system will eventually slide to the slipform surface.

$$\begin{cases} s_1 \dot{s}_1 \le 0 \\ s_2 \dot{s}_2 \le 0 \end{cases} \quad (15)$$

5 Simulation Results and Analysis

So as to verify the superiority of the control system, firstly, the simulation model of control system is built in the Matlab/Simulink environment, and then the parameters of motor are shown in Table 1 below.

Table 1. Motor parameters table.

Parameters	Numerical value	Parameters	Numerical value
U_{dc} (V)	311	L_{d} (mH)	0.00002
Ψ_{f} (Wb)	0.062	L_{q} (mH)	0.00047
P_{n}	4	J (kg·m^2)	0.003
$Rs(\Omega)$	0.025	B (N·m·s)	0

The initial value of the load torque of the motor is 5 N·m, which changes to 25 N·m at 0.2 s.

The Simulink simulation model based on sliding mode weak magnetic control system is built as shown in Fig. 2.

In order to verify the performance of the control system designed in this work, a simulation scheme of the PMSM control strategy is designed as shown in Table 2.

The control system in this paper contains a weak magnetic control module, so the speed regulation range can reach more than 7000 rpm, set different given speed and conduct simulation test, the speed response simulation results are shown in Fig. 3.

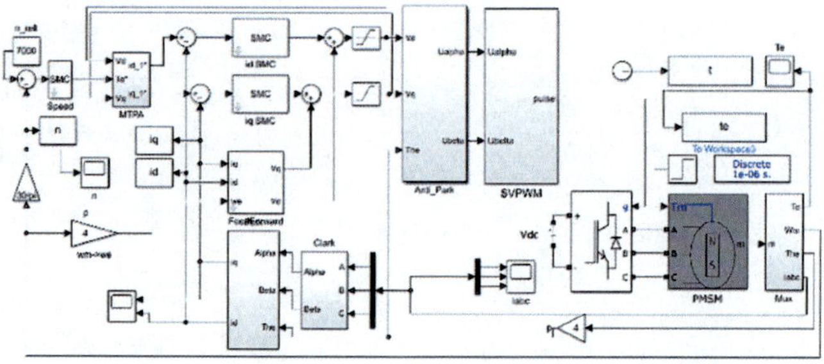

Fig. 2. Simulink simulation model based on sliding mode weak magnetic control system

Table 2. Simulation scheme of PMSM control strategy

Simulation number	Category	Test content
CM1	Slipform speed control	Only the speed loop is controlled by sliding mode
CM2	The sliding mode control proposed in this paper	Sliding mode control for both speed and current loops

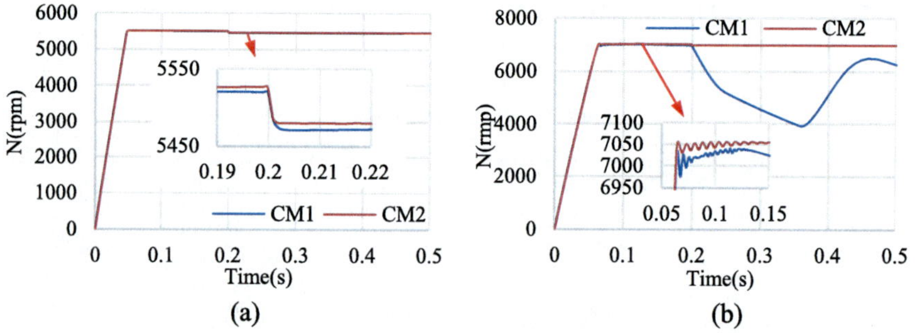

Fig. 3. CM1/CM2 system speed profile at different given speeds. **a** Given speed 5500 rmp. **b** Given speed 7000 rmp

As shown in Fig. 3a, the control effects of CM1 and CM2 are almost the same during the process of starting the motor and reaching the given speed. When the system is stable again after 0.2 s torque change, the regulation time of CM1 is 0.004 s and the stable speed is 5471 rpm. The regulation time of CM2 is 0.002 s and the stable speed is 5480 rpm, and the comparison shows that CM2 has 0.002 s faster regulation time compared with CM1, with faster response and less steady-state error.

From Fig. 3b, it is obvious that CM1 is still able to run stably and reach the target speed when no torque change occurs when the speed rises to 7000 rpm, but the pulsation of the speed curve is larger than CM2 at this time. When the torque changes at the moment of 0.2 s, CM1 is no longer able to stabilize the control system speed, and a large speed fluctuation occurs, with the highest deviation reaching 1486 rpm. At this point CM2 is still able to stabilize the control system and respond quickly to changes in torque that occur in the system.

When the given speed is 7000 rmp, the d-axis and q-axis current waveforms are shown in Fig. 4a and b, respectively.

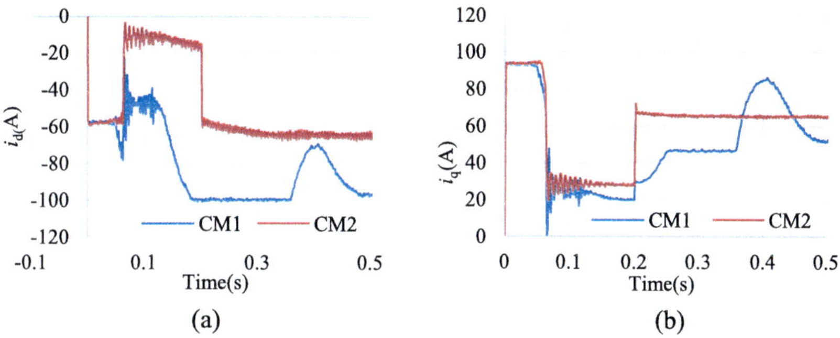

Fig. 4. CM1/CM2 d-axis and q-axis current waveform. **a** d-axis current waveform. **b** q-axis current waveform

Analysis of Fig. 4a and b shows that the current waveform of CM2 is more regular and flat compared to CM1 before and after the torque change, and in CM1, the current is distorted at 0.35 s with a larger overshoot. The results shows that the control effect of CM2 on the current is significantly better than CM1 in both stability and response speed.

When the given speed is 7000 rpm, the torque response simulation results are shown in Fig. 5.

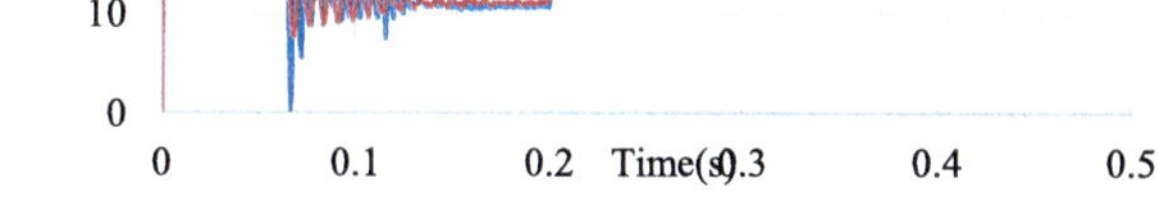

Fig. 5. CM1/CM2 Torque Curve

Comparing the torque curves of CM1 and CM2, it can be seen that the former has a larger torque curve pulsation and a larger overshoot after 0.2, indicating that CM2 can effectively reduce the torque pulsation of the system.

6 Concluding Remarks

In this paper, we proposed a sliding mode control method for the weak magnetic control of PMSM based on the traditional PI control method in the weak magnetic control of PMSM of EVS, which not only adopts the sliding mode control in the speed loop, but also adopts the sliding mode control in the current loop, and designs the SMC for the speed loop and current loop respectively.

In contrast with the traditional PI control, the sliding mode control strategy proposed could effectively deal with the problems of large motor torque and speed pulsation and waveform distortion in this work. In addition, it could improve the dynamic response speed and anti-interference capability, and could provide a feasible solution for the high performance control of the motor.

References

1. Niu, Z.Z., Li, S.P., Jia, Q.Y.: Weak magnetic control of permanent magnet synchronous motor based on sliding mode control. Ind. Technol. Innov. **8**(4), 119–123, 148 (2021) (in Chinese)
2. Jiang, W., Yang, J., Gao, J., Lou, T.: Sliding mode observer-based weak magnetic control of automotive permanent magnet synchronous motor. Electrotechnology **15**, 59–63 (2022) (in Chinese)
3. Xiang, L.: Research on vector control method of permanent magnet synchronous motor based on MTPA. Automot. Ind. Res. **8**, 25–29 (2018). (in Chinese)
4. Jia, N., Cheng, Y., Tian, Y.: Vector control of permanent magnet synchronous motor for electric vehicles based on current feedforward. Foreign Electr. Meas. Technol. **40**(12), 61–66 (2021) (in Chinese)
5. Zhao, Y., Liu, B.: Weak magnetic vector control of built-in permanent magnet synchronous motor based on sliding mode variable structure. Inf. Control **46**(4), 428–436, 442 (2017) (in Chinese)
6. Rongjun, D., Kan, L.: Perspectives on key technologies of new energy vehicle motor drive system. China Eng. Sci. **21**(03), 56–60 (2019). (in Chinese)
7. Fu, X., Chen, R., Dong, T., et al.: A review on MTPA control of permanent magnet synchronous motor considering parameter uncertainty. Chin. J. Electr. Eng. **42**(02), 796–808 (2022) (in Chinese)
8. Li, X.H., Huang, S.R., Zhang, Q.: Analysis of the inherent frequency of the stator structure of permanent magnet synchronous motors for electric vehicles. Chin. J. Electr. Eng. **37**(08), 2383–2391 (2017). (in Chinese)
9. Yang, Y., He, Q., Fu, C., et al.: Efficiency improvement of permanent magnet synchronous motor for electric vehicles. Energy (2020) (prepublish)
10. Mohd Zaihidee, F., Mekhilef, S., Mubin, M.: Robust speed control of PMSM using sliding mode control (SMC)—a review. Energies (Basel) **12**(9), 1669 (2019)
11. Suman, K., Mathew, A.T.: Speed control of permanent magnet synchronous motor drive system using PI, PID, SMC and SMC plus PID Controller, IEEE (2018)
12. Yuhao, W., Li, S., Zhongxian, C.: An improved sliding mode control method to increase the speed stability of permanent magnet synchronous motors. Energies **15**(6313), 6313 (2022)
13. Alnami, H., Pang, C., Wang, Q.: A novel sliding mode control method of interior-mounted PMSM, IEEE (2021)
14. Liu, X., Yu, H., Yu, J., et al.: Combined speed and current terminal sliding mode control with nonlinear disturbance observer for PMSM drive. IEEE Access **6**, 29594–29601 (2018)

Fuzzy Sliding Mode Control of DC/DC Converter for Hydrogen Vehicles

Zhuo Wang[1,2](✉), Jing Chen[1,2], Wei Wu[1], Huchuan Yang[1], and Meng Zhang[1]

[1] Wuhan University of Technology, Wuhan 430079, China
19687979212@qq.com

[2] Foshan Xianhu Laboratory of the Advanced Energy Science and Technology Guangdong Laboratory, Foshan 528200, China

Abstract. Aiming at the problems that DC/DC converters require low voltage ripple and low current ripple in high-power applications, a fuzzy sliding mode control method for DC/DC converters was proposed to apply in hydrogen fuel vehicles in this work. And a four-phase interleaved boost converter topology was built in this control system by using Matlab/Simulink. The results of simulation evaluation turned out that utilizing the present proposed control algorithm could effectively accelerate the system response speed and reduce the output current ripple and voltage ripple compared with the traditional double closed-loop PID algorithm. Moreover, before and after changing the load, the system showed good dynamics and strong robustness, which could better meet the requirements of automotive DC/DC converter.

Keywords: High-power · Interleaved · DC/DC converter · Fuzzy sliding mode control · Ripple

1 Introduction

Hydrogen fuel cell system has the characteristics of soft output properties, large range of output voltage variation, large output current ripple, etc. Through DC/DC converter, it can be used as the power source of the car to achieve voltage boost and ripple reduction, thereby increasing the durability of the hydrogen fuel cells [1–6]. Therefore, the design of high-performance DC/DC converter is extremely important. The interleaved boost converter can effectively reduce the output voltage and the current ripple of the boost converter, and is often used in high-power situations with boost demand [7–11].

Therefore, researchers began to apply nonlinear control such as fuzzy control and sliding mode control to converter control, so as to effectively reduce the voltage ripple and current ripple of the system [12–15]. Among them, sliding mode control has the advantages of simple design, strong robustness, fast dynamic response, etc. It is often used in nonlinear systems such as boost converter. However, sliding mode control has chattering phenomenon, so it is usually not used alone but combined with fuzzy control to improve the control effect of the system.

L. Jia et al. (Eds.): EITRT 2023, LNEE 1135, pp. 125–133, 2024.
https://doi.org/10.1007/978-981-99-9307-9_14

Based on the above merits of fuzzy sliding mode control, we proposed to apply it to the DC/DC converter of hydrogen vehicles to achieve low voltage ripple, low current ripple and boost voltage in high-power situations. Additionally, considering that interleaved boost converter can effectively reduce the output voltage and current ripple of the Boost converter, we employed a four-phase interleaved Boost converter. Resultantly, the simulation evaluation indicated that utilizing the proposed algorithm could effectively accelerate the system response speed and reduce the output current ripple and voltage ripple when comparing with the traditional double closed-loop PID algorithm. Moreover, before and after changing the load, the system has good dynamics and strong robustness, which verified the superiority of the proposed control algorithm.

2 Model Building

In the proposed algorithm, the DC/DC converter is a four-phase interleaved boost converter. And its topology is shown in Fig. 1, where Vin is the input voltage source, Vo is the output voltage of the DC/DC converter, C is the internal capacitance of the DC/DC converter, R is the internal load resistance of the DC/DC converter, and PCPL is the constant power load.

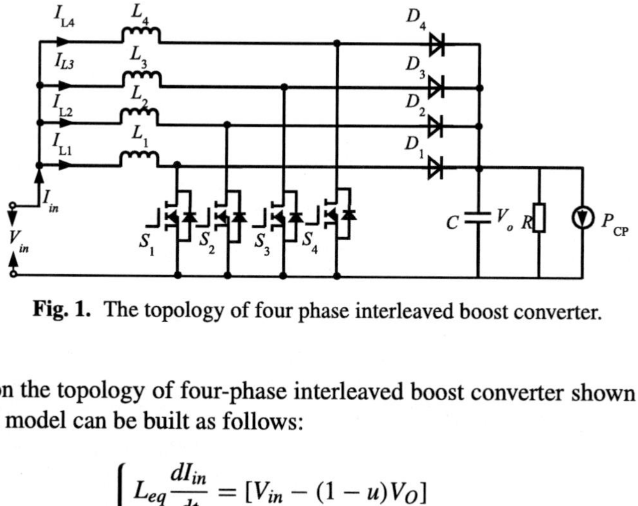

Fig. 1. The topology of four phase interleaved boost converter.

Based on the topology of four-phase interleaved boost converter shown in Fig. 1, a generalized model can be built as follows:

$$\begin{cases} L_{eq}\dfrac{dI_{in}}{dt} = [V_{in} - (1-u)V_O] \\ C\dfrac{dV_o}{dt} = I_{in}(1-u) - \left(\dfrac{V_O}{R} + \dfrac{P_{CPL}}{V_O}\right) \end{cases} \tag{1}$$

where L_{eq} is the equivalent input inductance, I_{in} is the total input current, dI_{in}/dt is the rate of change of the total input current with time t, dV_o/dt is the rate of change of the output voltage with time t, and u is the switch control signal obtained. To simplify the model, let $P_{CPL} = 0$.

3 Algorithm Optimization

The control schematic of the control system is shown in Fig. 2. The outer loop is a voltage loop, which uses fuzzy sliding mode control. The inner loop is a current loop, which uses PID control. The fuzzy sliding mode controller in this work includes a sliding mode controller and a fuzzy controller.

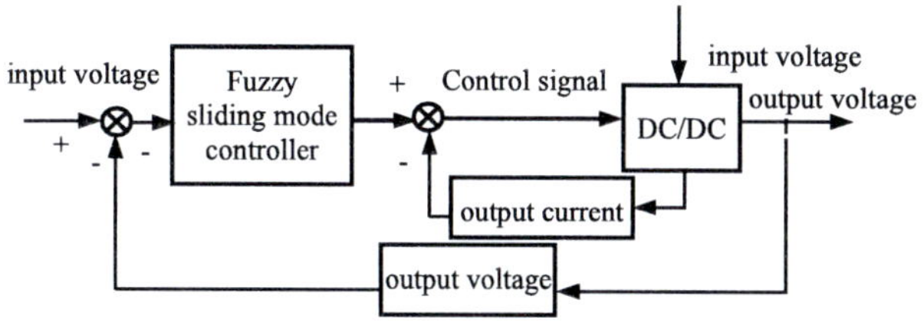

Fig. 2. The principle diagram of control system.

In order to design the sliding surface, the energy conservation formula and interference term are obtained according to Eq. (2):

$$\begin{cases} \dot{x}_1 = x_2 + d_1 \\ \dot{x}_2 = k + d_2 \\ x_1 = L_{eq}I_{in}^2/2 + CV_o^2/2 \\ x_2 = V_{in}I_{in} - V_o^2/R_o \\ d_1 = -V_o^2/R + V_o^2/R_o \\ d_2 = 2(V_o^2/R - V_o^2/R_o)/(CR_o) \end{cases} \tag{2}$$

where x_1 is the total energy of the DC/DC converter, x_2 is the rate of change of the total energy, $\dot{x}_1$ is the first derivative of the total energy, $\dot{x}_2$ is the first derivative of the rate of change of the total energy, k is the virtual control law, d_1 is the matched interference term, and d_2 is the unmatched interference term.

According to the matched and unmatched interference terms, the nonlinear interference observer is designed as follows:

$$\begin{cases} \hat{d}_1 = k_{d1}x_1 + \beta_1, \dot{\beta}_1 = -k_{d1}(x_2 + \hat{d}_1) \\ \hat{d}_2 = k_{d2}x_2 + \beta_2, \dot{\beta}_2 = -k_{d2}(k + \hat{d}_2) \end{cases} \tag{3}$$

where $\hat{d}_1$ is the output value of the first interference observer matching the interference term d_1, is the output value of the second interference observer not matching the interference term d_2, k_{d1} is the gain of the first interference observer, k_{d2} is the gain of the second interference observer, β_1 is the variable state of the first interference observer, and β_2 is the variable state of the second interference observer.

According to Eq. (5), the system dynamic error is obtained:

$$\begin{cases} \dot{e}_{d1} = -k_{d1}e_{d1} + \dot{d}_1 \\ \dot{e}_{d2} = -k_{d2}e_{d2} + \dot{d}_2 \end{cases} \tag{4}$$

where $e_{d1} = d_1 - \hat{d}_1$, $e_{d2} = d_2 - \hat{d}_2$, represent the estimation error.

According to the results of observation and estimation, the sliding surface is designed as:

$$s = a(x_1 - \hat{x}_{1ref}) + (x_2 - \hat{x}_{2ref}) - \dot{\hat{x}}_{1ref} \tag{5}$$

In order to make the sliding surface S converge to zero, the control law corresponding to the designed sliding mode controller is determined as:

$$\begin{aligned} k = &-a(e_{x2} - \dot{\hat{x}}_{1ref}) + \ddot{\hat{x}}_{1ref} - \dot{\hat{d}}_1 - \hat{d}_2 \\ &- k_{s1}\mathrm{sgn}(s) - k_{s2}s \end{aligned} \tag{6}$$

where K_{s1} and K_{s2} are both designed positive sliding mode control signal gains.

It can be obtained that:

$$\dot{s} = ae_{d1} + e_{d2} + k_{s1}sgn(s) - k_{s2}s \tag{7}$$

Lyapunov operator is defined:

$$V(s) = \frac{1}{2}s^2 \tag{8}$$

There is:

$$\begin{aligned} \dot{V}(s) &= s \cdot \dot{s} = -k_{s1}|s| + k_{s2}s^2 + (ae_{d1} + e_{d2})s \\ &\leq -ks1|s| - k_{s2}s^2 + (ae_{d1\max} + e_{d2\max})|s| \\ &-\sqrt{2}[k_{s1} + k_{s2}|s| - (ae_{d1}\max + e_{d2}\max)]V^{\frac{1}{2}} \end{aligned} \tag{9}$$

When $(k_{s1} + k_{s2}|s|) > (ae_{d1\max} + ed_{2\max})$, the system is asymptotically stable and the state error of the sliding surface is zero in a finite time, then:

$$\begin{cases} e_{x2} = -ae_{x1} + \dot{\hat{x}}_{1ref} \\ \dot{e}_{x1} = ex2 + ed1 - \dot{\hat{x}}_{1ref} \end{cases} \tag{10}$$

The fuzzy controller takes the sliding surface s and the derivative of the sliding surface as the input of the fuzzy controller, and the output of the fuzzy controller adjusts the observed value of the fuzzy controller. The fuzzy word set symbol comparison table is shown in Table 1. The membership function is Gaussian function, the fuzzy rule is the if then rule of expert experience, the fuzzy reasoning is Mamdani type, and the defuzzification is the centroid method.

According to formula (11), under the condition of stability, the fuzzy controller in this work can be designed. The fuzzy rules are shown in Table 2.

Table 1. Fuzzy word set symbol table

Fuzzy word set	Symbol	Fuzzy word set	Symbol
Positive big	PB	Negative small	NS
Positive small	PS	Negative big	NB
Zero	ZO	/	/

Table 2. Fuzzy rule table

s					
$\dot{s}$	NB	NS	ZO	PS	PB
PB	ZO	ZO	PS	PB	PB
PS	ZO	ZO	ZO	PS	PB
ZO	NS	ZO	ZO	ZO	PS
NS	NB	NS	ZO	ZO	ZO
NB	NB	NB	NS	ZO	ZO

4 Simulation and Optimization

The DC/DC converter in this work is a 120 kW of high-power converter. According to Figs. 1 and 2, Simulink was used to complete the construction of the simulation model of the our-phase interleaved boost converter system as shown in Fig. 3. The parameters used in the simulation circuit are as follows: $L = 270\ \mu$H, $C = 500$ uF $V_{ref} = 550$ V, $R = 2.5\ \Omega$, $V_{in} = 220$ V. Load switching shall be carried out at 0.05 s.

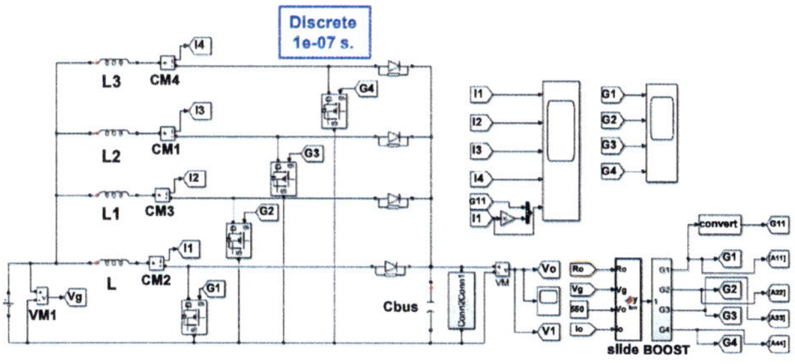

Fig. 3. The system simulation model of four-phase interleaved Boost converter

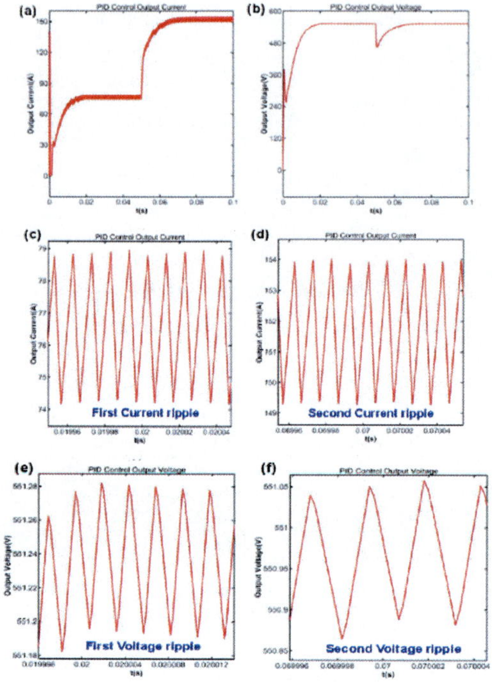

Fig. 4. The simulation results without using fuzzy sliding mode algorithm optimization.

The simulation results without using fuzzy sliding mode algorithm optimization are shown in Fig. 4, and the corresponding results from fuzzy sliding mode algorithm optimization are shown in Fig. 5.Through analyzing these two groups of data, the comparison of output current performance indicators and output voltage performance indicators are obtained as shown in Tables 3 and 4, respectively.

Through the analysis and comparison of the above simulation results, it could be concluded that the proposed algorithm could effectively accelerate the system response speed and reduce the output current ripple and voltage ripple compared with the traditional double closed-loop PID algorithm. Moreover, before and after changing the load, the system showed good dynamics and strong robustness, which could better meet the requirements of automotive DC/DC converter as shown in Table 5.

According to the data in Tables 3 and 4, it is found that compared with the non-optimized double closed-loop PID control algorithm, after using fuzzy sliding mode control optimization, the system response speed is faster, the adjustment time is reduced, the current ripple and voltage are reduced, and the corresponding stability is better.

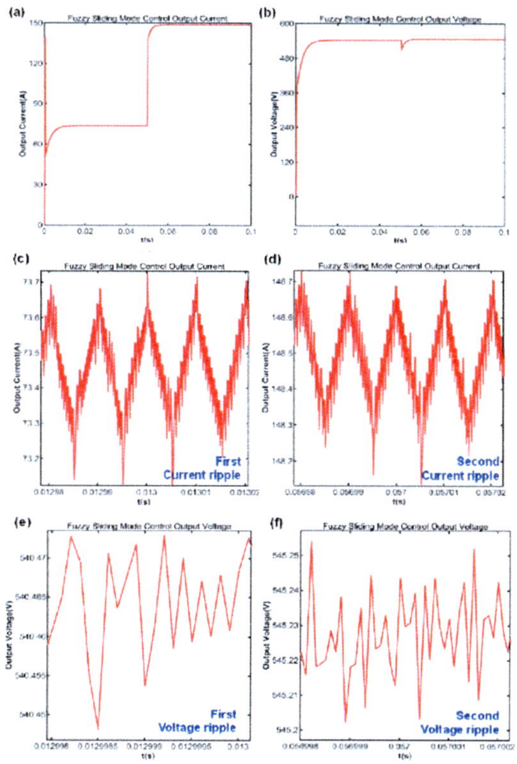

Fig. 5. The simulation results using fuzzy sliding mode optimization.

Table 3. The comparison of output current performance indexes

Performance indicators	Without using fuzzy sliding mode algorithm optimization	Using fuzzy sliding mode algorithm optimization
Adjustment time 1	0.020 s	0.013 s
Current ripple 1	5 A	0.5 A
The rate of current ripple 1	6%	0.68%
Adjustment time 2	0.070 s	0.057 s
Current ripple 2	5 A	0.6 A
The rate of current ripple 2	3%	0.4%

Table 4. The comparison of output voltage performance indexes

Performance indicators	Without using fuzzy sliding mode algorithm optimization	Using fuzzy sliding mode algorithm optimization
Adjustment time 1	0.020 s	0.013 s
Voltage ripple 1	0.1 V	0.03 V
The rate of voltage ripple	0.018%	0.006%
Adjustment time 2	0.070 s	0.057 s
Voltage ripple 2	0.2 V	0.06 V
The rate of voltage ripple 2	0.036%	0.012%

Table 5. The Design requirements of DC/DC converter

Name of the performance	Indexes
Rated power	120 kW
Input voltage range	70 V ~ 200 V
Output voltage rating	550 V
Input current ripple	<8%
Output voltage Ripple	<1%
Output current ripple	<1%

5 Conclusion

In this paper, a fuzzy sliding mode control method for DC/DC converter used in hydrogen vehicles was proposed. By using Simulink, a simulation model of the topology control system of four-phase interleaved Boost converter was build, and the control effect of the proposed control method was verified. Through the analysis of simulation results, it turned out that: (1) the system response speed was accelerated and the system could be stabilized after load shedding; (2) the adjustment time was reduced, and the output voltage ripple and current ripple were reduced; (3) the output current and voltage of Boost converter could be stabilized at the set value; and (4) the output current ripple and output voltage ripple could be kept in a good range. These results well prove the efficiency of the proposed fuzzy sliding mode control method for DC/DC converters used in hydrogen vehicles. And this work might provide guidance for the experimental investigation of converters for hydrogen vehicles.

References

1. Bairabathina, S., et al.: Review on non-isolated multi-input step-up converters for grid-independent hybrid electric vehicle. Int. J. Hydrogen Energy **45**, 21687 (2020)
2. Farhani, S., et al.: Design and practical study of three phase interleaved boost converter for fuel cell electric vehicle. J. Power. Sources **479**, 228815 (2020)
3. Gao, D., et al.: An interleaved step-up/step-down converter for fuel cell vehicle applications. Int. J. Hydrogen Energy **41**, 22422 (2016)
4. Jyotheeswara Reddy, K., Natarajan, S.: Energy sources and multi-input Dc-Dc converters used in hybrid electric vehicle applications – a review. Int. J. Hydrogen Energy **43**, 17387 (2018)
5. Sagar Bhaskar, M., et al.: Survey of Dc-Dc non-isolated topologies for unidirectional power flow in fuel cell vehicles. IEEE Access **8**, 178130 (2020)
6. Shang, J., et al.: Hybrid hydrogen pem fuel cell and batteries without Dc–Dc converter. Int. J. Low-Carbon Technol. **11**, 205 (2016)
7. Akhlaghi, B., et al.: Efficient Zvt cell for interleaved Dc–Dc converters. IET Power Electr. **13**, 1925 (2020)
8. Alajmi, B.N., et al.: Multiphase interleaved converter based on cascaded non-inverting buck-boost converter. IEEE Access **10**, 42497 (2022)
9. Babazadeh, Y., et al.: A new continuous input current nonisolated bidirectional interleaved buck-boost Dc-Dc converter. Int. Trans. Electr. Energy Syst. **2022**, 1 (2022)
10. Farooq, A., et al.: A three phase interleaved boost converter with L & C voltage extension mechanism. Tehnicki Vjesnik-Technical Gazette **25**, 52 (2018)
11. Granados-Luna, T.-R., et al.: Two-phase, dual interleaved buck-boost Dc–Dc converter for automotive applications. IEEE Trans. Ind. Appl. **56**, 390 (2020)
12. Garza-Arias, E., et al.: The fourth-order single-switch improved super-boost converter with reduced input current ripple. Electronics **10**, 2379 (2021)
13. Gorji, S.A., et al.: Topologies and control schemes of bidirectional Dc–Dc power converters: an overview. IEEE Access **7**, 117997 (2019)
14. Guo, L., et al.: Comparative evaluation of sliding mode fuzzy controller and Pid controller for a boost converter. Electric Power Syst. Res. **81**, 99 (2011)
15. Lin, G., Zhang, Z.: Low input ripple high step-up extendable hybrid Dc-Dc converter. IEEE Access **7**, 158744 (2019)

Research on Observer-Based Vector Control Strategy for Battery Vehicle Drive System

Congding Yang[1,2], Jing Chen[1,2](✉), Chengbo Xu[1], and Junjie Zhao[1,2]

[1] Wuhan University of Technology, Wuhan 430079, China
310393164@qq.com
[2] Foshan Xianhu Laboratory of the Advanced Energy Science and Technology Guangdong Laboratory, Foshan 528200, China

Abstract. The main research content of this paper is as follows: Firstly, the observer-based MTPA vector control strategy is adopted, and the observer is used to improve the anti-interference ability of the system. On the Matlab simulation platform, the simulation model of the drive control system of hydrogen fuel cell vehicle is constructed, and the test scheme of the drive control system of hydrogen fuel cell vehicle is designed. The simulation results show that the design of the hydrogen fuel cell vehicle drive control system in this paper can meet the requirements of hydrogen fuel cell vehicle drive response time, speed control precision and other technical indicators at start-up, acceleration and deceleration, and constant speed operating conditions. And in the working condition, the motor can quickly follow the given value, and has a strong anti-interference ability; Relative to the stable value of stator current, the speed and torque of the motor can maintain high stability, the speed of the motor, torque and stator current fast follow and strong anti-interference ability, to ensure the stable operation of the system.

Keywords: Hydrogen fuel cell vehicle · Drive control system · Vector control · MTPA · Observer

1 Development Status of Hydrogen Fuel Cell Vehicles at Home and Abroad

At present, hydrogen fuel cell vehicle has become one of the development directions of the global automobile industry. Many countries and regions around the world are vigorously promoting the development of hydrogen fuel cell vehicles [1]. In terms of technology, the key technologies of hydrogen fuel cell vehicles have been maturing, and the production cost has been decreasing. Currently, the cost of hydrogen fuel cell vehicles on the market is decreasing, and more hydrogen fuel cell vehicles are expected to come out in the future. Although the sales volume of hydrogen fuel cell vehicles is relatively small compared with that of ordinary fuel vehicles, more and more automobile manufacturers have invested in the research and development and production of hydrogen fuel cell vehicles, and the market competition is becoming more and more fierce. In order to solve the problem of hydrogen filling infrastructure construction, some countries and regions have started to actively promote related work, such as Japan, South Korea, Europe and other places have built a certain scale of hydrogen filling stations [2].

L. Jia et al. (Eds.): EITRT 2023, LNEE 1135, pp. 134–140, 2024.
https://doi.org/10.1007/978-981-99-9307-9_15

The fuel cell technology and related automobile technology are developing rapidly under the support of the government, and all the core components have realized the independent production. At present, there is a certain gap between the domestic level and the world's top level, but this gap is gradually narrowing [3–5].

2 Mathematical Model of Permanent Magnet Synchronous Motor

2.1 Modeling of Three-Phase Rest Coordinate System

The voltage and flux equation are as follows:

$$\begin{bmatrix} u_A \\ u_B \\ u_C \end{bmatrix} = \begin{bmatrix} R_S & 0 & 0 \\ 0 & R_S & 0 \\ 0 & 0 & R_S \end{bmatrix} \begin{bmatrix} i_A \\ i_B \\ i_C \end{bmatrix} + \frac{d}{dt} \begin{bmatrix} \varphi_A \\ \varphi_B \\ \varphi_C \end{bmatrix} \tag{1}$$

$$\begin{bmatrix} \varphi_A \\ \varphi_B \\ \varphi_C \end{bmatrix} = \begin{bmatrix} L_{AA} & L_{AB} & L_{AC} \\ L_{BA} & L_{BB} & L_{BC} \\ L_{CA} & L_{CB} & L_{CC} \end{bmatrix} \begin{bmatrix} i_A \\ i_B \\ i_C \end{bmatrix} + \varphi_f \begin{bmatrix} \cos\theta_e \\ \cos(\theta_e - 120°) \\ \cos(\theta_e + 120°) \end{bmatrix} \tag{2}$$

2.2 Mathematical Model of Two Phase Synchronous Rotating Coordinate System

According to the transformation principle, Clark transformation was used to convert the ABC three-phase stationary coordinate system into α-β two-phase stationary coordinate system, and Park transformation was used to convert the two-phase stationary coordinate system into d - q the two-phase synchronous rotation coordinate system, thus obtaining the mathematical model under the two-phase synchronous rotation coordinate system.

The flux equation is:

$$\begin{bmatrix} \varphi_d \\ \varphi_q \end{bmatrix} = \begin{bmatrix} L_d & 0 \\ 0 & L_q \end{bmatrix} \begin{bmatrix} i_d \\ i_q \end{bmatrix} + \begin{bmatrix} \varphi_f \\ 0 \end{bmatrix} \tag{3}$$

The voltage equation is:

$$\begin{bmatrix} u_d \\ u_q \end{bmatrix} = R_s \begin{bmatrix} i_d \\ i_q \end{bmatrix} + \begin{bmatrix} L_d \frac{di_d}{dt} \\ L_q \frac{di_q}{dt} \end{bmatrix} + \omega_e \begin{bmatrix} -L_q i_q \\ L_d i_d \end{bmatrix} + \begin{bmatrix} 0 \\ \omega_e \varphi_f \end{bmatrix} \tag{4}$$

The torque equation is:

$$T_e = \frac{3}{2} n_p \left[\varphi_f i_q + \left(L_d - L_q \right) i_d i_q \right] \tag{5}$$

3 Observer Based Vector Control Strategy Principle

According to the drive control strategy scheme, based on the space vector control based on MTPA [6], the speed controller adopts PI controller +MTPA [7]. An observer based MTPA vector control system is obtained, and its structure is shown in Fig. 1.

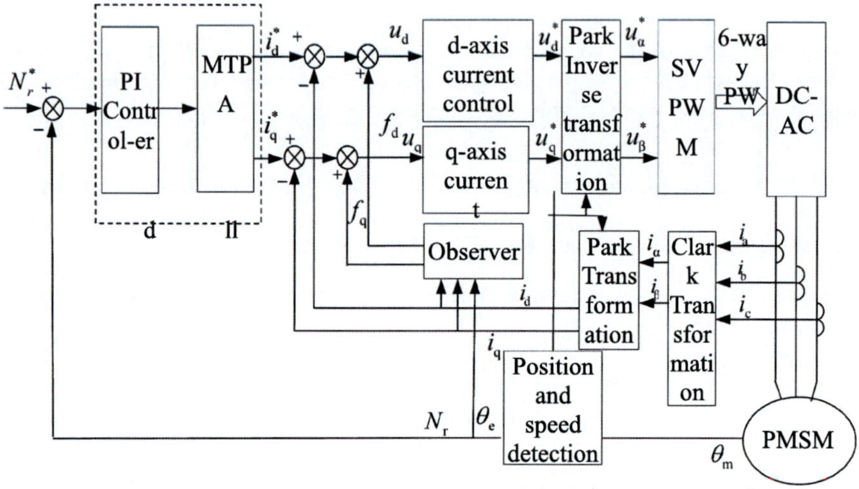

Fig. 1. Structure diagram of observer based MTPA vector control system

3.1 Maximum Torque Current Ratio Strategy (MTPA)

The maximum torque current ratio control optimizes the configuration of d-q axis current id and i_q to minimize stator current and electromagnetic torque T_e when the PMSM output torque is known.

The effective values Id of current id and the combined values I_q of i_q are:

$$I_s = \sqrt{I_d^2 + I_q^2} \tag{6}$$

Order $L_s = L_q - L_d$, by introducing Eqs. (2, 3, 4, 5), we obtain:

$$T_e = 1.5P_n i_q(\varphi_f - L_s i_d) \tag{7}$$

Using Lagrange number multiplication to construct auxiliary functions:

$$F = \sqrt{i_d^2 + i_q^2} + \lambda\left\{T_e - \frac{3}{2}P_n\left[\varphi_f \cdot i_q + \left(L_d - L_q\right) \cdot i_d \cdot i_q\right]\right\} \tag{8}$$

3.2 Observer Principle

The structural diagram of the observer is shown in Fig. 2, taking the d-axis current as an example. $G_0(s)$ is the transfer function of the actual controlled object, $G_n^{-1}(s)$ is the model of the controlled object, $Q(s)$ is a low-pass filter, d is the equivalent interference, $\hat{d}$ is the observed interference, ξ is the measurement noise, i_d^* and i_d are the system control input u and output y, respectively.

When the observer does not include a filter $Q(s)$, it indicates that $Q_{uy}(s) = Q_n(s)$, $Q_{dy}(s)= 0$, the controlled object will fully exhibit the characteristics of the nominal model $Q(s)$, and external interference will be completely suppressed. The observer has no effect on measurement noise [8].

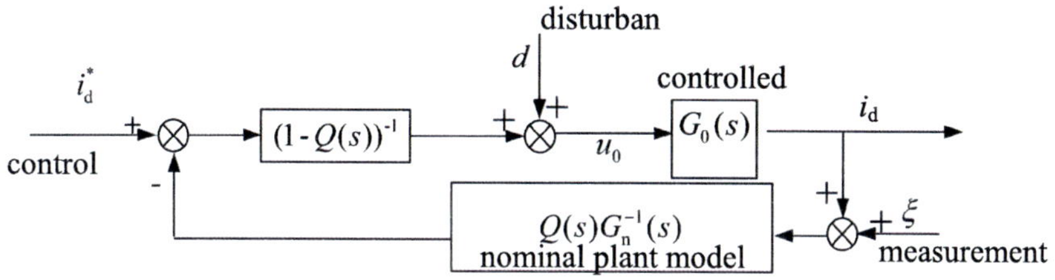

Fig. 2. Equivalent Observer Structure Box

4 Simulation Test of Hydrogen Fuel Cell Vehicle Drive Control System

4.1 Simulation Model Construction

Based on the structure of the hydrogen fuel cell vehicle drive control system [9], a simulation model of the hydrogen fuel cell vehicle drive control system is constructed (Figs. 3, 4, and 5). The model mainly consists of MTPA simulation model, observer simulation model, etc.

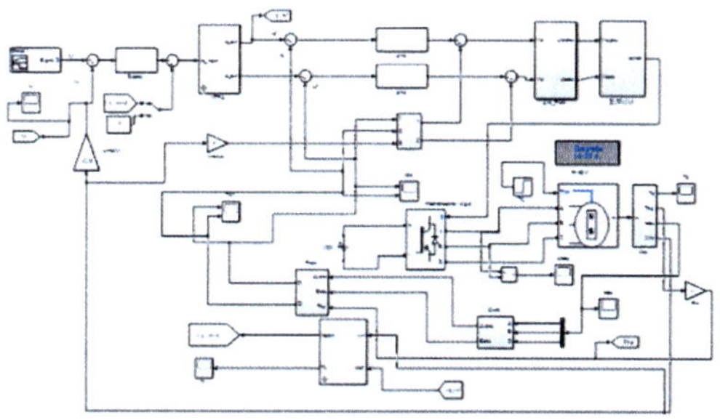

Fig. 3. Simulation Model of the Drive Control System for Hydrogen Fuel Cell Vehicles

4.2 Design of Simulation Test Plan

In order to verify the feasibility of the driving performance and control strategy of the hydrogen fuel cell vehicle designed in this article, a simulation scheme was designed. The simulation model of the driving control system of hydrogen fuel cell vehicles can be used to design solutions for hydrogen fuel cell vehicles under different driving conditions.The driving conditions of hydrogen fuel cell vehicles include normal operating conditions such as start, acceleration, constant speed, deceleration, etc. The simulation experimental scheme of the drive control system is shown in Table 1.

The simulation experimental scheme of the drive control system mainly verifies that hydrogen fuel cell vehicles can better maintain stability, accuracy, and speed in certain situations.

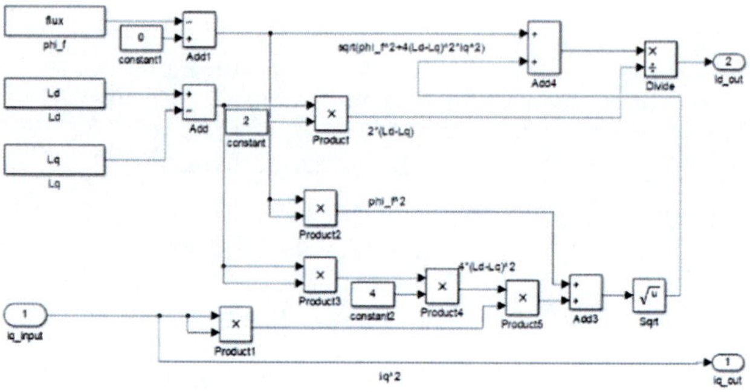

Fig. 4. MTPA Controller Simulation Model

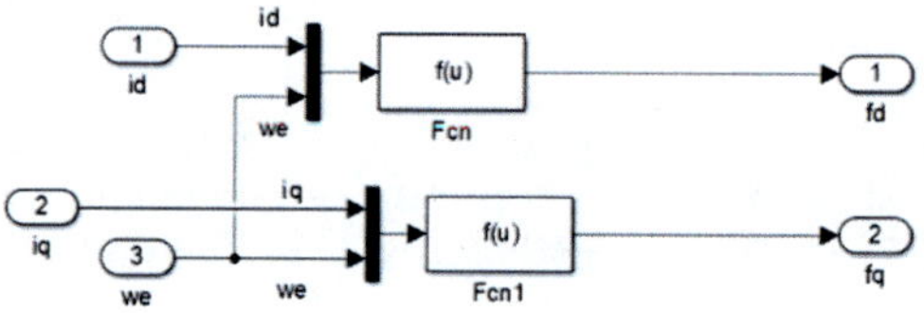

Fig. 5. Observer Simulation Model

Table. 1. Simulation Experiment Plan for Drive Control System.

Working status	Load torque T_L(N · m)	Theoretic speed N_r^*(r/min)	Content validation
Start acceleration and deceleration of constant speed running	0 ~ 0.7s: 8	0s: 800, 0.1s: 1100 0.2s: 1200 0.4s: 1100, 0.5s: 800	Speed accuracy and system response time under various working conditions

4.3 Analysis of Simulation Results

In the simulation, the time is set to 0.7 s, and the given speed is set to 800, 1100, 1200, 1100, and 800 r/min at 0 s, 0.1 s, 0.2 s, 0.4 s, and 0.5 s, respectively. The load torque is constant at 8 N·m from 0 to 0.7 s, thus obtaining the simulation results of speed, electromagnetic torque, and dq axis current.

The simulation curves of speed, torque, and current are shown in Figs. 4 to 6, respectively (Figs. 7 and 8).

The following conclusions can be drawn from Figs. 4, 5, and 6:

(1) Control the motor to use a lower current during startup and achieve the rated speed faster, thereby reducing energy consumption and mechanical pressure during startup;
(2) Smooth torque output is achieved during acceleration and deceleration, thereby reducing wear and vibration of mechanical components.

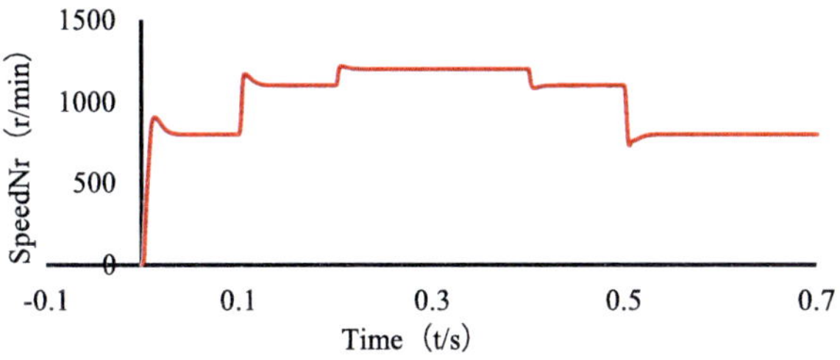

Fig. 6. Speed Simulation Result Curve

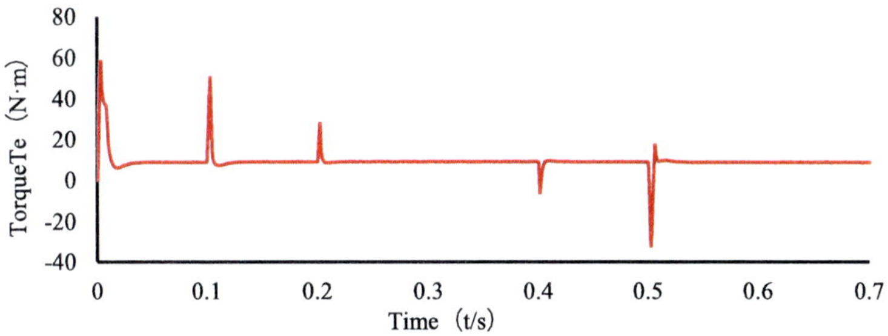

Fig. 7. Torque Simulation Result Curve

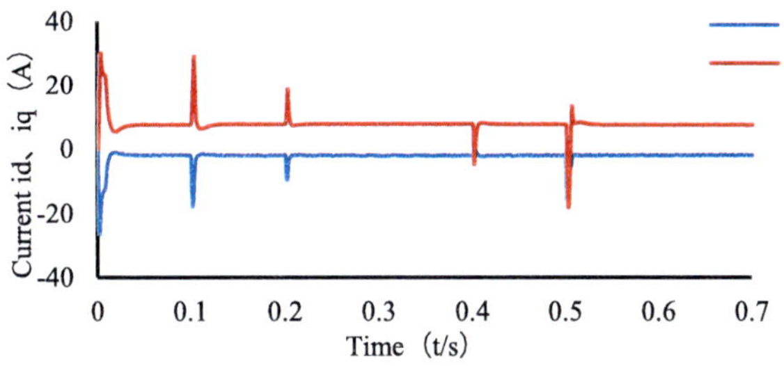

Fig. 8. Current Simulation Result Curve

In addition, better control of the motor's torque output during rapid acceleration and deceleration allows for more precise control of the motion process;

(3) Achieving more stable torque output during constant speed operation can better maintain the stable operation of the motor;
(4) The speed, torque, and stator current of the motor can quickly follow the given values and have strong anti-interference ability. The stable value of the stator current is much smaller [10]. The entire system has fast response speed and good following ability, which can meet the requirements of hydrogen fuel cell vehicles for driving under working conditions and meet the control requirements, namely speed control accuracy $\leq$8% and response time $\leq$49 ms.

References

1. Yanzhong, H., Bo, D.: Development status and prospect analysis of hydrogen fuel cell vehicles in China. Sci. Technol. Wind **4**, 5–6 (2021). (in Chinese)
2. Hongliu, X.: Application status and development trend analysis of hydrogen fuel cell technology. Sci. Technol. Innov. **2**, 160–161 (2019). (in Chinese)
3. Hongxue, Z., Xiao, L., Zhifei, P., et al.: Brief analysis on the development status of hydrogen fuel cell vehicles. Transp. Energy Conserv. Environ. Prot. **16**(4), 11–15 (2020). (in Chinese)
4. Jiaheng, W., Zhen, H.: Development status and trend of hydrogen fuel cell technology in China. Automot. Pract. Technol. **22**, 20–23 (2019). (in Chinese)
5. Ping, Z., Wang, T., Huang, Y., et al.: Internal model control of PMSM position servo system: theory and experimental results. IEEE Trans. Industr. Inf. **16**(4), 2202–2211 (2020)
6. Xingdi, H.: Research on MTPA Control Strategy Engineering of Permanent Magnet Synchronous Motor. Huazhong University of Science and Technology, Wuhan (2019). (in Chinese)
7. Liu, K., Hou, C., Hua, W.: A novel inertia identification method and its application in PI controllers of PMSM drives. IEEE Access **7**, 13445–13454 (2019)
8. Liu, X., Yu, H., Yu, J., et al.: Combined speed and current terminal sliding mode control with nonlinear disturbance observer for PMSM drive. IEEE Access **6**, 29594–29601 (2018)
9. Yachuan, H.: Research on Electric Vehicle Drive System Based on Permanent Magnet Synchronous Motor. Beijing University of Technology, Beijing (2008). (in Chinese)
10. Chow, H.W., Cheung, N.C.: Disturbance and response time improvement of submicrometer precision linear motion system by using modified disturbance compensator and internal model reference control. IEEE Trans. Industr. Electron. **60**(1), 139–150 (2013)

Simulation Study of Bidirectional DC/DC Converter for New Energy Vehicles Based on RBF Neural Variable Structure

Junjie Zhao[1,2], Jing Chen[1,2(✉)], Congding Yang[1], Jiawei Gao[1], and Mu Yang[1]

[1] Wuhan University of Technology, Wuhan 430079, China
15271740727@163.com

[2] Foshan Xianhu Laboratory of the Advanced Energy Science and Technology Guangdong Laboratory, Foshan 528200, China

Abstract. To make the bidirectional DC/DC converter have better performance and stability, this paper presents a RBF neural network-based variable structure control approach. Mathematically describe the working principle of bi-directional DC/DC converter and construct a variable structure controller using RBF neural network according to the design requirements. The variable structure control is optimized by the RBF neural network algorithm and the chattering phenomenon in the variable structure control is reduced. The simulation model is built and simulated in MATLAB/Simulink, and results of the simulation show that the dynamic response performance and robustness of the control system are improved after using the RBF neural variable structure control optimized voltage controller.

Keywords: Bidirectional DC/DC · Variable structure control · RBF neural network

1 Introduction

Bidirectional DC/DC converters allow energy to be transferred to each other and efficiently utilized between two batteries (or energy storage devices), which has wide application in new energy vehicles [1–3]. However, the bidirectional DC/DC converters also need to have reliability, stability, and dynamic performance. By reducing the switching frequency as well as current and voltage ripple, reducing switching losses and filter size, the three-phase interleaved parallel bidirectional DC/DC converter has the benefits [4, 5]. Therefore, the bidirectional DC/DC converter control strategy is a popular research topic.

Bilalovic et al. proposed to apply variable structure control to DC/DC converters to achieve fast tracking and stabilization of the output voltage [6]. However, jittering phenomenon exists in the variable structure control, which affects the stability and lifetime of the system [7, 8]. To address the limitations of variable structure control, domestic and foreign scholars proposed many intelligent algorithms, such as fuzzy logic, genetic algorithms, and neural networks, to achieve optimization of bidirectional DC/DC converter control [9, 10]. Domestic scholars added RBF neural network to the variable

L. Jia et al. (Eds.): EITRT 2023, LNEE 1135, pp. 141–148, 2024.
https://doi.org/10.1007/978-981-99-9307-9_16

structure control, and used the approximation ability of RBF neural network to optimize the variable structure control, simplify the design difficulty of the sliding mode surface, reduce the system jitter, and enhance the robustness and dynamic response performance of the system, which is an effective control strategy [11, 12].

Therefore, this paper uses the RBF neural variable structure control strategy to design a voltage controller for a three-phase interleaved parallel bidirectional DC/DC converter, which is used for new energy vehicles. The RBF neural network algorithm optimizes the variable structure control and reduces the chattering phenomenon in the variable structure control. The paper also conducts a simulation test by MATLAB/Simulink to verify the effectiveness of the proposed control strategy.

2 Mathematical Model of Bidirectional DC/DC Converter

The topology of a three-phase interleaved parallel bidirectional DC/DC converter is shown in Fig. 1, which is composed of three identical bidirectional half-bridge converters connected in parallel, with the two adjacent phases staggered by 120° when on.

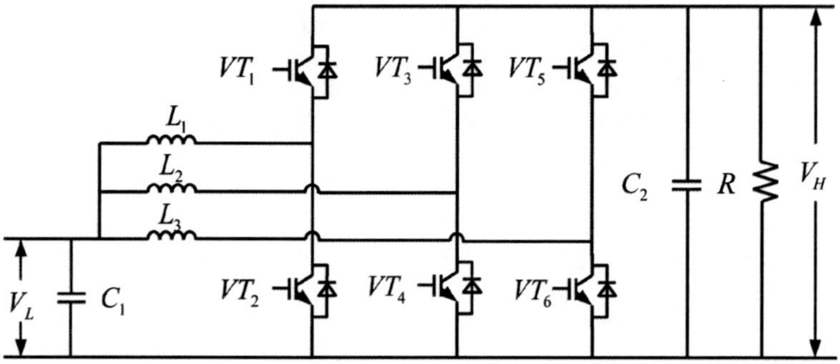

Fig. 1. Three-phase interleaved parallel bidirectional DC/DC converter topology

Where $VT_1 \sim VT_6$ is the switching element of the converter, $L_1 \sim L_3$ is the energy storage inductor, C_1 is the primary side filter capacitor, C_2 is the secondary side filter capacitor, V_L is the primary side voltage, and V_H is the secondary side voltage.

Since the three-phase interleaved parallel bidirectional DC/DC converter is derived from three identical bidirectional half-bridge converters connected in parallel, the circuit analysis can be performed for a single circuit. To simplify the theoretical analysis, the capacitive inductors in the circuit structure are ideal devices, and the power switching devices are free from parasitic parameters. Where, u_i is the switching function of the ith switching tube, u_i defined as:

$$u_i = \begin{cases} 1, & \text{on} \\ 0, & \text{off} \end{cases} \tag{1}$$

In boost mode, the power switching tubes VT_2, VT_4 and VT_6 operate, and the voltage equation of the nth phase circuit can be derived from Kirchhoff's voltage law as follows:

$$L_n \frac{di_{Ln}}{dt} = V_L - V_H u_i (n = 1, 2, 3) \tag{2}$$

where i_{Ln} denotes the nth phase inductance current.

The current equation for the load end node on the secondary side can be derived from Kirchhoff's current law:

$$C_2 \frac{dV_H}{dt} = i_{Ln} u_i - \frac{V_H}{R} \tag{3}$$

Similarly, the voltage equation and current equation in buck mode can be obtained.

Using the state-space averaging modeling method, the state variables are the inductor currents of each phase i_{L1}(t), i_{L2}(t), i_{L3}(t) and the secondary side voltage V_H, and the primary side voltage V_L is the input quantity, small signal perturbations are introduced, and the second-order minutiae term is ignored, and the state expression of the three-phase interleaved parallel bidirectional DC/DC converter in boost mode is derived as shown in Eq. (4).

$$\begin{bmatrix} \dot{i}_{L1} \\ \dot{i}_{L2} \\ \dot{i}_{L3} \\ \dot{V}_H \end{bmatrix} = \begin{bmatrix} 0 & 0 & 0 & \frac{1}{L_1} \\ 0 & 0 & 0 & \frac{1}{L_2} \\ 0 & 0 & 0 & \frac{1}{L_3} \\ \frac{1-d_2}{C_2} & \frac{1-d_4}{C_2} & \frac{1-d_6}{C_2} & -\frac{1}{C_2 R} \end{bmatrix} \begin{bmatrix} i_{L1} \\ i_{L2} \\ i_{L3} \\ V_H \end{bmatrix} + \begin{bmatrix} \frac{d_2-1}{L_1} \\ \frac{d_4-1}{L_2} \\ \frac{d_6-1}{L_3} \\ 0 \end{bmatrix} V_L \tag{4}$$

where the duty cycle of the power switching device VT_2 is d_2, VT_4 is d_4, and $VT6$ is d_6.

Similarly, in the buck mode, the state expression of the three-phase interleaved parallel bidirectional DC/DC converter is shown in Eq. (5).

$$\begin{bmatrix} \dot{i}_{L1} \\ \dot{i}_{L2} \\ \dot{i}_{L3} \\ \dot{V}_H \end{bmatrix} = \begin{bmatrix} 0 & 0 & 0 & -\frac{1}{L_1} \\ 0 & 0 & 0 & -\frac{1}{L_2} \\ 0 & 0 & 0 & -\frac{1}{L_3} \\ \frac{1}{C_2} & \frac{1}{C_2} & \frac{1}{C_2} & -\frac{1}{C_2 R} \end{bmatrix} \begin{bmatrix} i_{L1} \\ i_{L2} \\ i_{L3} \\ V_H \end{bmatrix} + \begin{bmatrix} \frac{d_1}{L_1} \\ \frac{d_2}{L_2} \\ \frac{d_3}{L_3} \\ 0 \end{bmatrix} V_L \tag{5}$$

where the duty cycle of the power switching device VT_1 is d_1, VT_3 is d_3, and VT_5 is d_5.

3 Controller Design for Bidirectional DC/DC Converter

3.1 Design Scheme of Three-Phase Interleaved Parallel Bidirectional DC/DC Converter

This paper uses a RBF neural network-based variable structure control strategy for the three-phase interleaved parallel bidirectional DC/DC converter. The voltage loop adopts RBF neural variable structure control to optimize the variable structure control and weaken the jitter phenomenon in the variable structure control by RBF neural network algorithm, and the current loop adopts PI control strategy to achieve fast tracking and stabilization of the current and improve the system accuracy. The structure of RBF neural variable structure control based on double closed-loop is shown in Fig. 2.

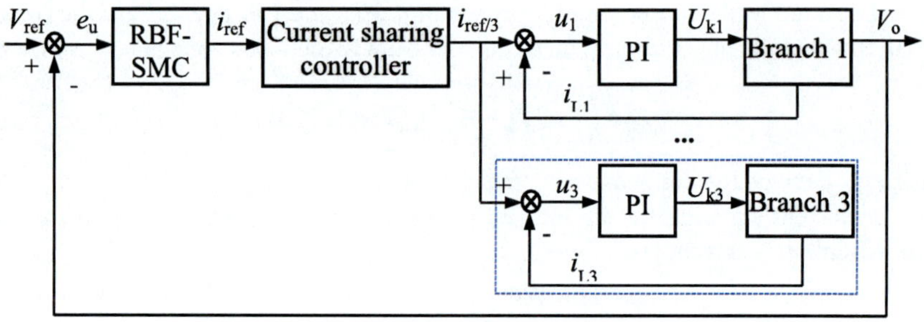

Fig. 2. Structure of RBF neural variable structure control based on double closed loop

3.2 Variable Structure Controller Design Based on RBF Neural Network

The output voltage of the three-phase interleaved parallel bidirectional DC/DC converter applied in this paper is V_H, while the desired output voltage is V_r. Due to the output command and the actual output of the system, the system error is defined as:

$$e = V_r - V_H \tag{6}$$

Establish the variable structure function and switch the sliding mode surface design as:

$$s(t) = ce(t) + \dot{e}(t) \tag{7}$$

Using the switching slip surface function s and as the input to the RBF neural network, defined as:

$$X = \begin{bmatrix} s \ \dot{s} \end{bmatrix}^T \tag{8}$$

In designing the control system, its stability should satisfy the Lyapunov theory stability requirements, i.e. satisfy $s(t)\dot{s}(t) \leq 0$, and set the weight adjustment index of the RBF neural network as:

$$E = s(t)\dot{s}(t) \tag{9}$$

then the weight of each neuron is:

$$dw_j = -\eta \frac{\alpha E}{\alpha w_i(t)} = -\eta \frac{\alpha s(t)\dot{s}(t)}{\alpha w_j(t)} = -\eta \frac{\alpha s(t)\dot{s}(t)}{\alpha u(t)} \frac{\alpha u(t)}{\alpha w_j(t)} \tag{10}$$

The RBF neural network weight learning algorithm is:

$$dw_j = \gamma s(t) e^{-\frac{\|s - c_j\|^2}{b_j}} = \gamma s(t) h_j(s) \tag{11}$$

The variable structure controller is the output of the RBF neural network, and the parameters of the variable structure controller are adjusted by the RBF neural network:

$$u = \sum_{j=1}^{m} w_j e^{-\frac{\|s - c_j\|^2}{b_j}} \tag{12}$$

where, there are m such neurons in the hidden layer, and c_j represents the center vector of the jth node of the network, and b_j represents the base width parameter of node j.

4 Simulation Results and Analysis

The corresponding simulation model is constructed in Matlab/Simulink to validate the effectiveness of the control approach proposed in this paper. The parameters used in the simulation circuit are as follows: inductor $L_1 = L_2 = L_3 = L = 5\,mH$, capacitor $C_1 = 30\,\mu F$, $C_2 = 100\,\mu F$, reference voltage $V_r = 380\ V$, and input voltage $V_{in} = 200\ V$.

The RBF-SMC controlled bidirectional DC/DC converter simulation model built by packaging the trained and tested RBF neural network into sub-modules in Simulink is shown in Fig. 3.

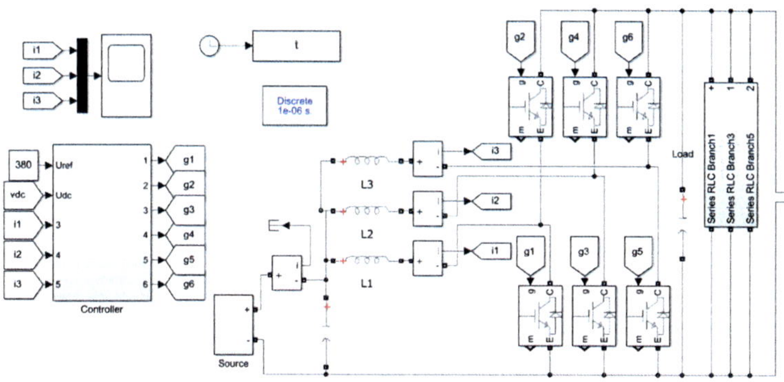

Fig. 3. Bi-directional DC/DC converter simulation model

This paper uses a dual-loop control scheme based on voltage mode, in which the voltage controller adopts the RBF-SMC control technique, and the current controller adopts PI control, (set parameters) the simulation model of the control module is shown in Fig. 4.

The simulation time is set to 0.3 s, and the constant voltage provided by the CVS is changed at 0.1 s to jump from 370 V to 390 V, and the ideal switching state is changed at 0.2 s to let the ideal switch from the on state to the off state to simulate different load conditions. The simulation results of the three-phase interleaved parallel bidirectional DC/DC converter using the conventional PID double closed-loop control strategy are shown in Fig. 5, and the simulation results of the RBF-SMC optimization algorithm are shown in Fig. 6.

According to Figs. 5 and 6, when the load conditions change, the optimized three-phase interleaved parallel bi-directional DC/DC converter output voltage overshoot is smaller and more responsive when compared to the unoptimized dual closed-loop PID control algorithm, and can reach the steady state faster.

According to the data analysis in Table 1, compared with the unoptimized dual closed-loop PID control algorithm, after using RBF-SMC control optimization, the system response speed is faster, the adjustment time is reduced by about 57.14%, greatly

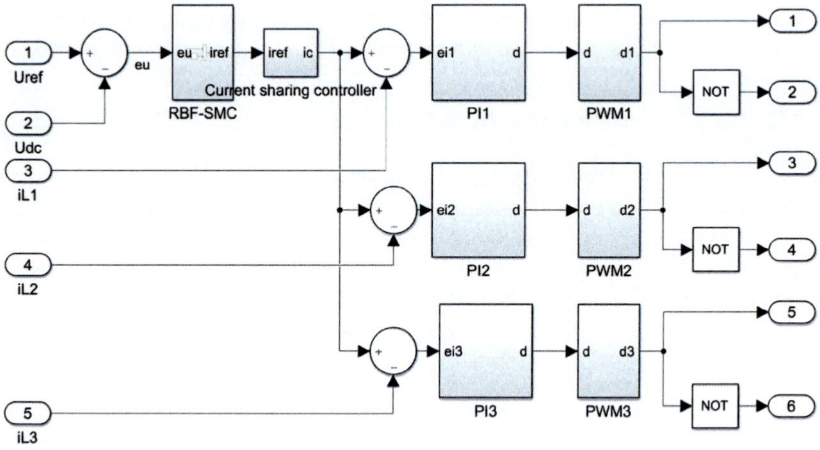

Fig. 4. Simulation model of control module

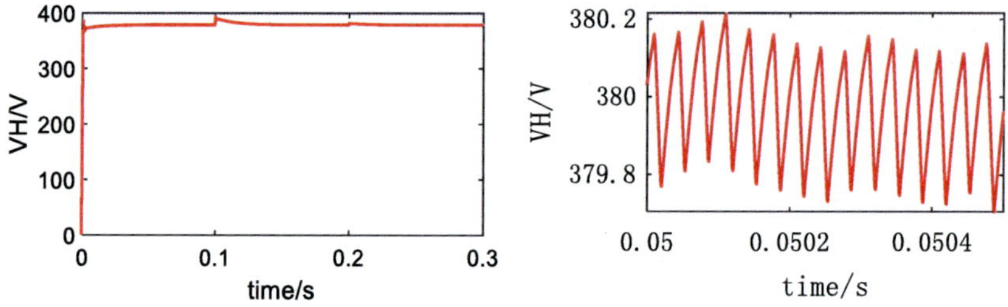

Fig. 5. Simulation results of traditional PID double closed-loop control strategy

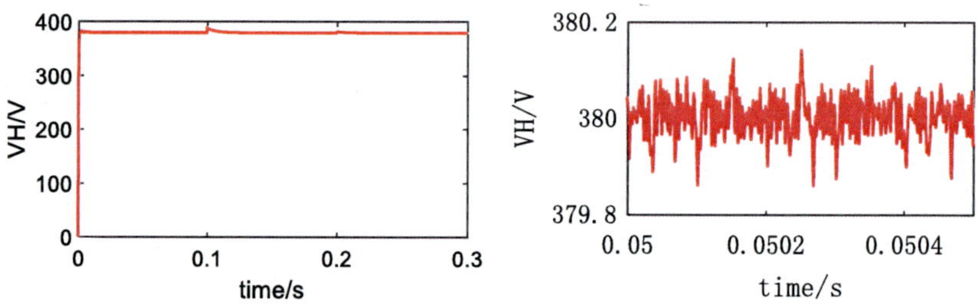

Fig. 6. Simulation results of RBF-SMC optimization algorithm

shortening the time to reach a stable state, and the overshoot is also slightly reduced, the voltage ripple is reduced by 0.09%, the performance is optimized, and the stability is better.

Table 1. Output voltage performance comparison

Performance index	Unoptimized double closed-loop PID control	RBF-SMC control optimization
Overshoot amount	4.42%	3.01%
Adjustment time	35 ms	15 ms
Peak time	3 ms	1 ms
Voltage ripple	0.6 V	0.3 V
Voltage ripple rate	0.16%	0.07%

5 Conclusion

To enhance the dynamic performance and robustness of the bidirectional DC/DC converter for new energy vehicles, this paper proposes a variable structure control strategy based on RBF neural network. The voltage outer loop is controlled by RBF neural variable structure control, and the current inner loop is controlled by PI control, using a three-phase interleaved parallel bidirectional DC/DC converter. The accuracy and stability of voltage and current control are successfully achieved. It is verified by a simulation model in MATLAB/Simulink. The results show that the system response speed is faster, the adjustment time is greatly shortened, reduced by about 57.14%, the overshoot is also slightly reduced, the output voltage ripple is reduced by about 33.3%, and even if the load conditions change, it can reach a stable state more quickly, which shows the superiority of the RBF-SMC controller, meeting the requirements of new energy vehicles for high performance and high reliability of bidirectional DC/DC converters.

References

1. Zhang, W., Xiang, H.: Fuel cell vehicle basic technology and development overview. Smart Power **48**(04), 36–41 (2020). (in Chinese)
2. Xiaogang, W., Wang, J., Zhang, Y., et al.: Review of DC-DC converter topologies based on impedance network with wide input voltage range and high gain for fuel cell vehicles. Autom. Innov. **4**(4), 351–372 (2021)
3. Xiong, J., Yang, T., Zhang, J.: A power system design based on electric-electric hybrid full-power hydrogen fuel cell vehicle. Electromech. Eng. Technol. **51**(06), 72–77 (2022). (in Chinese)
4. Celik, M., Naci, G., Uzmus, H., et al.: Experimental verification of interleaved hybrid DC/DC boost converter. J. Power Electron. **22**(10), 1–11 (2022)
5. Liu, Z., Du, J., Yu, B.: Research on triple switch dual boost high gain DC/DC converter. Power Technol. **45**(08), 1082–1086 (2021). (in Chinese)
6. Bilalovic, S.K.: Enhanced design of an indirect adaptive fuzzy sliding mode power system stabilizer for multi-machine power systems. Int. J. Electr. Power Energy Syst. **54**(54), 425–431 (2014)
7. Zou, R., He, R., Gao, W.: Application of sliding mode control in three-phase interleaved parallel bidirectional DC-DC converters. J. Qiqihar Univ. (Nature Science Edition), **36**(01), 1–5 (2020). (in Chinese)

8. Alsmadi, Y., Utkin, V., Haj-ahmed, M., et al.: Sliding mode control of power converters: DC/DC converters. Int. J. Control **91**(11), 2472–2493 (2018)
9. Wu, Y., Qian, Q., Sun, W.: Genetic algorithm-based efficiency optimization of two-stage DC/DC systems. Commun. Power Technol. **36**(06), 58–62 (2019). (in Chinese)
10. Zhang, M., Chen, Q., Xie, J.: Research on controlling switching power supply based on improved neural network algorithm. Optoelectron. ·Laser **33**(04), 349–355 (2021). (in Chinese)
11. Qin, B., Zhang, J., Wang, X., Zhang, T.: RBF network-based sliding mode control system for automotive supercapacitors. Electr. Drive **48**(8), 65–69 (2018). (in Chinese)
12. Li, L.: Research on the control strategy of PEMFC converter based on RBF neural network. Microcomput. Appl. **36**(2), 135–137 (2020). (in Chinese)

Rotor Position Estimation Method for PMSM Based on Improved Sliding Mode Observer

Jing Wang[1], Jiale Chen[2](✉), Guohong Zeng[2], and Jingdou Liu[2]

[1] Shenzhen Power Supply Company, Shenzhen 518028, China
489671806@qq.com
[2] National Active Distribution Network Technology Research Center, Beijing Jiaotong University, Beijing 100044, China
{20121417,Ghzeng,jdliu}@bjtu.edu.cn

Abstract. To address the problems of high-frequency chattering and phase delay in traditional sliding mode observer, an improved sliding mode observer combining sinusoidal switching function with back electromotive force observer is proposed. Firstly, the sinusoidal switching function is employed as a substitute for the conventional sign function, thereby effectively suppressing the chattering problem. Secondly, an adaptive observer is constructed to process the equivalent back electromotive force, which reduces the harmonic content and avoids the phase delay issue that arises from the use of a low-pass filter. Finally, the experimental analysis is carried out with 1kW surface-mounted permanent magnet synchronous motor. Experimental findings indicate the improved sliding mode observer effectively addresses concerns related to high-frequency chattering, phase delay, and harmonic content in the equivalent back electromotive force. This confirms the practicality and credibility of the proposed improved sliding mode observer.

Keywords: Permanent magnet synchronous motor · Sensorless control · Sliding mode observer

1 Introduction

The permanent magnet synchronous motor (PMSM) finds extensive application in diverse domains such as rail transit, aerospace, electric vehicles and other fields. This is primarily attributed to its compact size, impressive efficiency, remarkable power density, and extensive range of speed regulation [1]. Vector control is the primary PMSM control technique, to attain optimal vector control performance, it is crucial to precisely sense the motor's rotor position and speed. This is typically accomplished through the utilization of mechanical sensors such as encoders and resolvers. However, for one thing, the installation of mechanical sensors increases the volume and cost of the motor, for another, the sensor accuracy is easily influenced by external environmental factors, which reduces the system reliability. Hence, to address the aforementioned issues caused by the use of mechanical sensor, sensorless control strategies have emerged as a research hotspot of PMSM [2].

L. Jia et al. (Eds.): EITRT 2023, LNEE 1135, pp. 149–156, 2024.
https://doi.org/10.1007/978-981-99-9307-9_17

There are two types of PMSM sensorless control technology, depending on the speed range. The first type is applied to the low-speed stage where back electromotive force (EMF) is not significant, mainly including: IF control [3], VF control [4], high-frequency signal injection method [5]. The second type is based on the back EMF model, and is applicable to the medium and high-speed stage, including: extended Kalman filter method (EKF) [6], model reference adaptive system (MRAS) [7], extended state observers, (ESO) [8], sliding mode observers (SMO) [9–12]. Among them, EKF requires complex matrix operations, which requires high chip computing level; although MRAS and ESO are relatively straightforward, the observation precision of these two methods rely heavily on motor parameters and have poor robustness. Compared with other methods, SMO exhibits high robustness, remaining unaffected by alterations in motor parameters and external disturbances. Its straightforward structure and ease of implementation make it a popular choice for use in PMSM sensorless control systems.

Regrettably, as a result of the intrinsic traits of sliding mode control (SMC), the estimated back EMF is subject to high-frequency chattering and harmonics, leading to a substantial margin of error in rotor position observation value. To mitigate the chattering problem, the utilization of the hyperbolic tangent function is implemented as switch function [9]. The saturation function is utilized in [10] as a substitute for the conventional sign function, resulting in a reduction of high-frequency chattering. In conventional SMO, a low-pass filter (LPF) is utilized to filter the equivalent back EMF, however, this use of LPF introduces a phase delay that can negatively impact accuracy. In order to get rid of using LPF, while reducing the harmonic content in equivalent back EMF. Jiang et al. [11] proposed a proportional resonant filter, which can effectively eliminate the harmonic elements present in the equivalent back EMF and suppress rotor position pulsation, but its resonance frequency is determined by the estimated speed, which will cause a large rotor position observation error when the speed changes abruptly. In [12], an adaptive filter is constructed using the least mean square algorithm to eliminate the specified subharmonics in equivalent back EMF, which availably weakens the pulsation error in the rotor position observation.

This paper presents an improved SMO that incorporates a sinusoidal switching function and a back EMF observer. The conventional sign function is replaced with a continuous and smooth sinusoidal switching function to mitigate the high-frequency chattering. Additionally, a back-EMF observer is utilized instead of a LPF, resulting in reducing harmonic content of the equivalent back EMF and eliminating phase delay. Finally, the efficacy of the proposed improved SMO is confirmed through experimentation on the 1 kW SPMSM sensorless control platform.

2 Mathematical Model of SPMSM

Assuming that the iron saturation, eddy currents and hysteresis losses are disregarded, the mathematical model of SPMSM in αβ coordinate system is

$$\begin{bmatrix} u_\alpha \\ u_\beta \end{bmatrix} = \begin{bmatrix} R_s + pL_s & 0 \\ 0 & R_s + pL_s \end{bmatrix} \begin{bmatrix} i_\alpha \\ i_\beta \end{bmatrix} + \begin{bmatrix} E_\alpha \\ E_\beta \end{bmatrix} \tag{1}$$

where u_α and u_β are αβ-axis stator voltages, i_α and i_β are αβ-axis stator currents, R_s is stator resistance, L_s is stator inductance, $p = \mathrm{d}/\mathrm{d}t$ is the differential operator, E_α and E_β are αβ-axis back EMFs.

The back EMF of αβ-axis is as follows

$$\begin{bmatrix} E_\alpha \\ E_\beta \end{bmatrix} = \omega_e \psi_f \begin{bmatrix} -\sin\theta_e \\ \cos\theta_e \end{bmatrix} \tag{2}$$

where ω_e is the electrical angular velocity, ψ_f is the flux linkage of permanent magnet, θ_e is the electrical rotor angle.

3 Design of an Improved SMO

3.1 Design of Sliding Mode Current Observer

Traditional SMO utilizes the sign function as its switching function. However, the inherent nonlinearity associated with the sign function leads to significant chattering, thereby compromising the precision of rotor position estimation. To address this issue, a sinusoidal switching function is proposed as a substitute for the sign function. The expression is as stated below

$$F(s) = \begin{cases} \mathrm{sgn}(s) & |s| > a \\ \sin\left(\frac{\pi s}{2a}\right) & |s| \le a \end{cases} \tag{3}$$

where sgn(·) denotes the sign function, a is the boundary layer. Figure 1 shows the sinusoidal switching function curve.

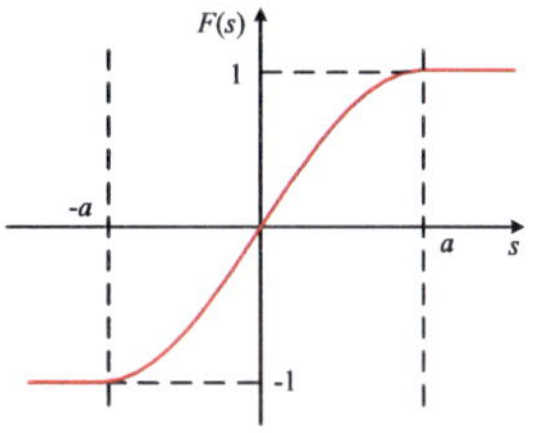

Fig. 1. The sinusoidal switching function curve.

According to (3), the error expression of current of sliding mode current observer with sinusoidal switching function can be constructed as

$$\frac{\mathrm{d}}{\mathrm{d}t}\begin{bmatrix} \hat{i}_\alpha - i_\alpha \\ \hat{i}_\beta - i_\beta \end{bmatrix} = -\frac{R_s}{L_s}\begin{bmatrix} \hat{i}_\alpha - i_\alpha \\ \hat{i}_\beta - i_\beta \end{bmatrix} + \frac{1}{L_s}\begin{bmatrix} E_\alpha - kF\left(\hat{i}_\alpha - i_\alpha\right) \\ E_\beta - kF\left(\hat{i}_\beta - i_\beta\right) \end{bmatrix} \tag{4}$$

where $\hat{i}_\alpha$ and $\hat{i}_\beta$ are αβ-axis observed currents, k is the sliding mode gain.

The sliding surface $\boldsymbol{S}$ is defined as follows

$$\boldsymbol{S} = \begin{bmatrix} \tilde{i}_\alpha \\ \tilde{i}_\beta \end{bmatrix} = \begin{bmatrix} \hat{i}_\alpha - i_\alpha \\ \hat{i}_\beta - i_\beta \end{bmatrix} \tag{5}$$

where $\tilde{i}_\alpha = \hat{i}_\alpha - i_\alpha$ and $\tilde{i}_\beta = \hat{i}_\beta - i_\beta$ are the αβ-axis stator current observation errors.

When the system enters the sliding surface ($\tilde{i}_\alpha = \dot{\tilde{i}}_\alpha = 0, \tilde{i}_\beta = \dot{\tilde{i}}_\beta = 0$), the expression for equivalent back EMF can be given as

$$\begin{bmatrix} E_\alpha \\ E_\beta \end{bmatrix} = \begin{bmatrix} kF\left(\hat{i}_\alpha - i_\alpha\right) \\ kF\left(\hat{i}_\beta - i_\beta\right) \end{bmatrix} \tag{6}$$

To guarantee the stability of the sliding mode current observer, it is necessary to satisfy the given relational expression.

$$k > \max\left(|E_\alpha|, \left|E_\beta\right|\right) \tag{7}$$

3.2 Design of Back EMF Observer

Conventional SMO typically employs LPF to eliminate high-frequency harmonic components from the equivalent back EMF. Nevertheless, although LPF effectively removes high-order harmonics, it unavoidably introduces phase delay issues, which ultimately diminish the precision of rotor position estimation. Thus, to overcome this issue and enhance the accuracy of rotor position estimation, a back EMF observer is devised to replace LPF.

Due to the high sampling frequency, it can be considered that $\dot{\omega}_e = 0$ in a control period, differentiating (2) yields

$$\begin{cases} \frac{dE_\alpha}{dt} = -\omega_e E_\beta \\ \frac{dE_\beta}{dt} = \omega_e E_\alpha \end{cases} \tag{8}$$

Based on (8), the construction of the back EMF observer can be accomplished by utilizing the given expression.

$$\begin{cases} \frac{d\hat{E}_\alpha}{dt} = -\hat{\omega}_e \hat{E}_\beta - l\left(\hat{E}_\alpha - E_\alpha\right) \\ \frac{d\hat{E}_\beta}{dt} = \hat{\omega}_e \hat{E}_\alpha - l\left(\hat{E}_\beta - E_\beta\right) \\ \frac{d\hat{\omega}_e}{dt} = \left(\hat{E}_\alpha - E_\alpha\right)\hat{E}_\beta - \left(\hat{E}_\beta - E_\beta\right)\hat{E}_\alpha \end{cases} \tag{9}$$

where $\hat{E}_\alpha$ and $\hat{E}_\beta$ are α-β axis observed back EMFs, $\hat{\omega}_e$ is observed electrical angular velocity, l is the observer gain.

Subtracting (8) from (9) results in

$$\begin{cases} \frac{d\tilde{E}_\alpha}{dt} = -\tilde{\omega}_e \hat{E}_\beta - \omega_e \tilde{E}_\beta - l\tilde{E}_\alpha \\ \frac{d\tilde{E}_\beta}{dt} = \tilde{\omega}_e \hat{E}_\alpha + \omega_e \tilde{E}_\alpha - l\tilde{E}_\beta \\ \frac{d\tilde{\omega}_e}{dt} = \tilde{E}_\alpha \hat{E}_\beta - \tilde{E}_\beta \hat{E}_\alpha \end{cases} \tag{10}$$

where $\tilde{E}_\alpha = \hat{E}_\alpha - E_\alpha$ and $\tilde{E}_\beta = \hat{E}_\beta - E_\beta$ are α-β axis back EMF observation errors, $\tilde{\omega}_e = \hat{\omega}_e - \omega_e$ is the speed observation error.

By utilizing Lyapunov stability theory, it is evident that the constructed back EMF observer can ensure stability when the gain $l > 0$.

Figure 2 illustrates the improved SMO that incorporates a sinusoidal switching function and back EMF observer. Additionally, the use of PLL aids in accurately calculating the rotor position and speed, mitigating calculation errors that may arise due to motor parameter variation.

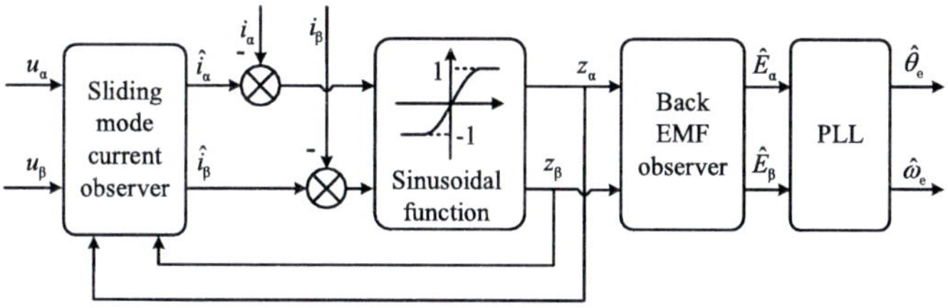

Fig. 2. Block diagram of improved SMO.

4 Experimental Validation and Results Comparison

Figure 3a shows the block diagram for sensorless control of PMSM using $i_d = 0$ vector control strategy, where rotor position and speed are estimated by the improved SMO. The efficiency of the proposed observer has been verified with a 1 kW SPMSM sensorless experimental platform based on dSPACE control system. The experimental setup is shown in Fig. 3b. The inverter's PWM switching frequency is 10 kHz, an incremental encode is implemented to obtain the actual rotor position and speed information of the motor, which is only used for the experimental comparison.

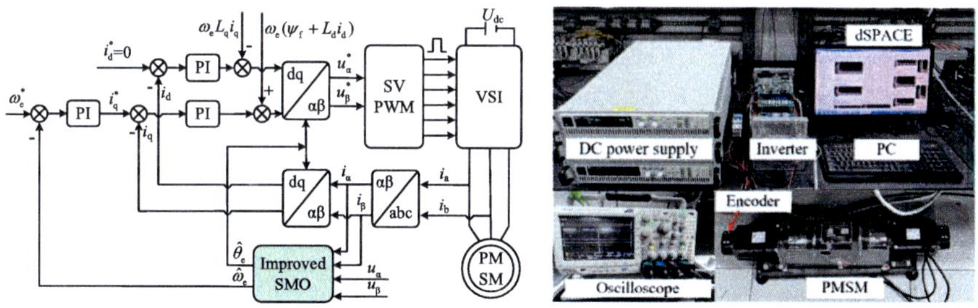

(a) Control block diagram (b) Experimental setup

Fig. 3. Block diagram of PMSM sensorless control system and experimental setup. **a** Control block diagram. **b** Experimental setup

The PMSM parameters are listed as follows: stator inductance $Ls = 3.22$mH, stator resistance $Rs = 1.58\ \Omega$, rotor inertia $J = 3.7 \times 10^{-4}$ kg·m^2, flux $\psi f = 0.1181$ Wb, pole pairs $pn = 4$, rated speed $N = 2500$ r·min^{-1}, rated current $IN = 4$ A, rated torque $TN = 4$ N·m, rated power $PN = 1$ kW.

The comparison results of rotor position observation between traditional SMO and improved SMO at 750 r·min^{-1} are presented in Fig. 4, where θ_e is the actual position, $\hat{\theta}_e$ is the observed position, $\Delta\theta$ is the observation error. In Fig. 4a, it is evident that the rotor position observed by traditional SMO exhibits noticeable chattering, resulting in an observation error of 30°. Additionally, there exists a considerable phase delay between rotor estimated position and actual position, with a delay angle of 0.16 π. Conversely, in Fig. 4b, the improved SMO yields a smoother waveform for rotor position observation, reducing the observation error to 5.4° from 30°. Furthermore, the phase delay is significantly improved, with the delay angle reduced from 0.16 π to 0.05 π. Experimental findings demonstrate that the proposed SMO offers higher accuracy in rotor position observation.

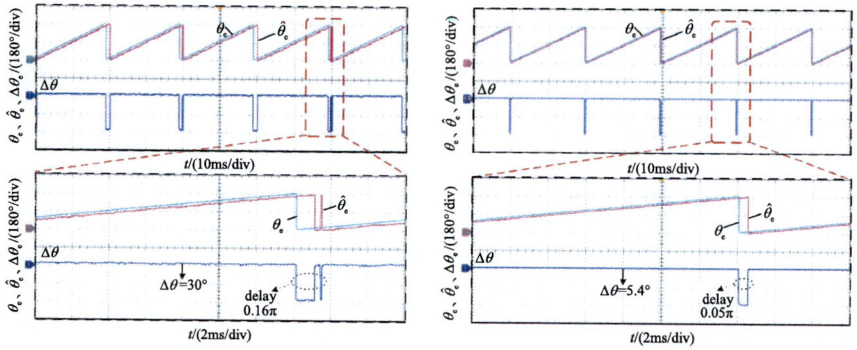

(a) Traditional SMO (b) Improved SMO

Fig. 4. Experimental waveforms of observed rotor position and observation error. **a** Traditional SMO. **b** Improved SMO

Figure 5 displays the waveforms of the observed back EMF and its FFT analysis results under both control techniques at 750 r·min^{-1}. Evidently, the implementation of improves SMO results in the observation waveforms of the back EMF being more sinusoidal, thereby reducing the total harmonic distortion (THD) from 8.69 to 2.21% compared to the conventional control. Experimental findings demonstrate the effectiveness of the improved SMO in reducing harmonic content in the observed back EMF, and thus improve the observation accuracy of system.

Figure 6 shows the observed speed and observation error waveforms under the control of traditional SMO and improved SMO when the speed varies continuously at a rate of 100 r·min^{-1}/s. Where n is actual speed, $\hat{n}$ is observed speed, Δn is speed observation error. Obviously, the observed speed under the control of traditional SMO has a large chattering, and the observation error fluctuation amplitude can reach 20 r·min^{-1}. In contrast, the speed observation waveform under the improved SMO control is smooth, and the waveform chattering is significantly suppressed, the margin of error of observation is decreased from 20 to 12 r·min^{-1}, resulting in a reduction of 40% in the range of

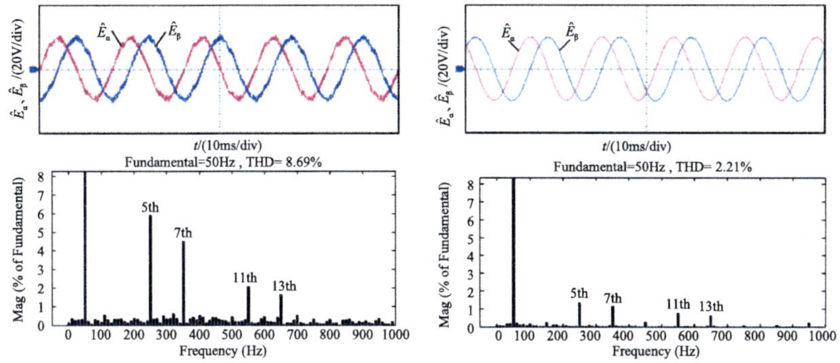

(a) Traditional SMO (b) Improved SMO

Fig. 5. Experimental waveforms of estimated back EMF and its FFT analysis. **a** Traditional SMO. **b** Improved SMO

fluctuation. Experimental findings indicate that improved SMO can markedly reduce the fluctuation amplitude of speed and have higher speed observation accuracy.

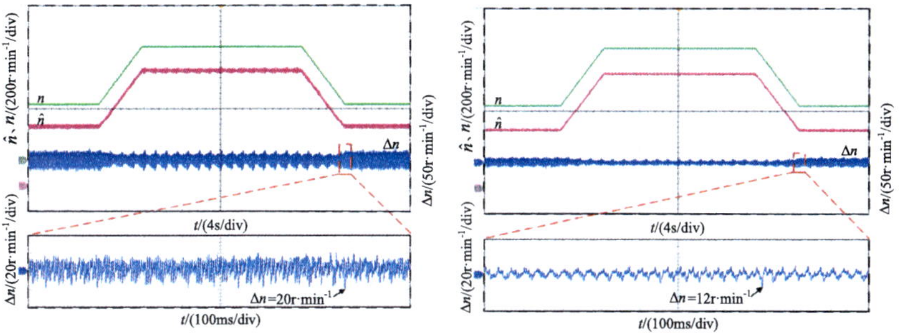

(a) Traditional SMO (b) Improved SMO

Fig. 6. Experimental waveforms of estimated speed and estimation error. **a** Traditional SMO. **b** Improved SMO

5 Conclusion

In this paper, an improved SMO using sinusoidal switching function and back EMF observer is proposed. Instead of utilizing the sign function, the continuous smooth sinusoidal switching function is employed, which effectively improves the chattering problem. Subsequently, a back EMF observer is devised as a substitute for LPF, thereby diminishing the presence of harmonics and circumventing the phase delay that arises from the implementation of LPF. The validity of the proposed method was verified with a 1 kW sensorless SPMSM tested platform. Experimental results show that the proposed SMO can effectively improve the problems of high-frequency chattering and phase delay, reduce the harmonic content in back EMF, and have higher observation accuracy.

Acknowledgments. This work is supported by Science and Technology Project of China Southern Power Grid Co., Ltd (Grant Number. 090000KK52210134): Research and Application of Key Technologies of Interaction between Urban Buildings and Power Grid for Large-scale Renewable Energy Consumption.

References

1. Zhang, Z., Wang, C., Zhou, M., et al.: Parameters compensation of permanent magnet synchronous motor in flux-weakening region for rail transit. IEEE Trans. Power Electron. **35**(11), 12509–12521 (2020)
2. Zhang, J.H., Liu, G.H., Chen, Q.: MTPA control of sensorless IPMSM drive system based on virtual and actual high-frequency signal injection. IEEE Trans. Transp. Electrif. **7**(3), 1516–1526 (2021)
3. Tang, Q.P., Chen, D.X., He, X.N.: Integration of improved flux linkage observer and I-f starting method for wide-speed-range sensorless SPMSM drives. IEEE Trans. Power Electron. **35**(8), 8374–8384 (2019)
4. Tang, Z.Y., Li, X., Dusmez, S., et al.: A new V/f-based sensorless MTPA control for IPMSM drives. IEEE Trans. Power Electron. **31**(6), 4400–4415 (2016)
5. Wu, X.M., Zhu, Z.Q., Freire, N.M.: High frequency signal injection sensorless control of finite-control-set model predictive control with deadbeat solution. IEEE Trans. Ind. Appl. **58**(3), 3685–3695 (2022)
6. Quang, N.K., Hieu, N.T., Ha, Q.P.: FPGA-based sensorless PMSM speed control using reduced-order extended Kalman filters. IEEE Trans. Ind. Electr. **61**(12), 6574 (2014)
7. Sun, X.D., Zhang, Y., Tian, X., et al.: Speed sensorless control for IPMSMs using a modified MRAS with gray wolf optimization algorithm. IEEE Trans. Transp. Electrif. **8**(1), 1326–1337 (2022)
8. Xu, Y.J., Zhu, Y.H., Zhou, F., et al.: SPMSM speed sensorless control based on fuzzy sliding mode and new expanded state observer. Electric Mach. Control **26**(07), 133–146 (2022). (in Chinese)
9. Xu, W.J., Qu, S.C., Zhao, L., et al.: An Improved adaptive sliding mode observer for middle- and high-speed rotor tracking. IEEE Trans. Power Electron. **36**(1), 1043–1053 (2021)
10. Lu, W.Q., Zheng, D.Y., Lu, Y.J., et al.: New sensorless vector control system with high load capacity based on improved SMO and improved FOO. IEEE Access **9**(1), 40716–40727 (2021)
11. Jiang, Y., Li, B.W., Wu, Y., et al.: An improved rotor position observer for permanent magnet synchronous motors based on proportional resonant filtering. Trans. China Electrotechnical Soc. **35**(17), 3619–3630 (2020). (in Chinese)
12. Zhang, G.Q., Wang, G.L., Xu, D.G., et al.: ADALINE-network-based PLL for position sensorless interior permanent magnet synchronous motor drives. IEEE Trans. Power Electron. **31**(2), 1450–1460 (2016)

Application Research of Photovoltaic Power Generation Technology in Railway System

Yajing Han[1](✉), Jie Jin[1], Feng Lin[1], Dawei Li[1], and Jianfan Wu[2]

[1] National Railway Test Center, China Academy of Railway Sciences Co., Ltd, Beijing 100015, China
hanyajing1997@hotmail.com

[2] Beijing Jiaotong University, Beijing 100081, China

Abstract. Photovoltaic power generation is one of the most promising renewable energy utilization methods in the world, but there are few related researches in the field of railway photovoltaic power generation. In this paper, the construction conditions of photovoltaic power generation, main equipment selection, energy storage equipment, energy control platform, combined with the national railway test center, to carry out relevant research.

Keywords: Photovoltaics · Railway · Equipment types

1 Introduction

Solar energy is an inexhaustible and inexhaustible renewable energy for human beings. Photovoltaic power generation has become one of the most mature and widely used technologies for solar energy utilization. It is clean, renewable, safe, noiseless, and flexible in application. It is a typical green power with significant environmental protection and economic benefits. Solar photovoltaic (PV) technology is indispensable for realizing a global low-carbon energy system and, eventually, carbon neutrality [1]. Photovoltaic (PV) power is expected to play an important role in reducing global warming and improving energy security [2]. "Opinions of the National Development and Reform Commission and the National Energy Administration on Improving the System, Mechanism, and Policy Measures for Energy Green and Low-Carbon Transformation" (Fagai Energy [2022] No. 206) encourages the application of new energy: support new energy power construction, Consolidate and develop as much as possible [3, 4].

The National Railway Testing Center is located in Beijing Chaoyang District, covering an area of 2723 acres. Since its opening in 1958, it has completed over 900 types of tests and international cooperation projects in various fields such as high-speed, conventional, heavy-haul, and urban rail transportation. It is China's only comprehensive railway testing center. Based on the testing base, the distributed photovoltaic power generation system test project is carried out. Distributed photovoltaic power generation has the characteristics of "local generation and local use", which is the best form of solar energy application [5]. The pilot demonstration section of the Anting Photovoltaic

L. Jia et al. (Eds.): EITRT 2023, LNEE 1135, pp. 157–169, 2024.
https://doi.org/10.1007/978-981-99-9307-9_18

Power Generation Project adopts domestic high-efficiency solar energy panels and connects them in series to the photovoltaic inverter. The photovoltaic inverter is connected to a combiner box, which then enters the grid distribution box set up in the park distribution substation. The storage system unit is connected after the grid distribution box is combined, and the park distribution substation is completed for grid-connected power generation.

2 Feasibility of Constructing Distributed Photovoltaic Power Generation Facilities on the Side Slopes of Railway Tracks and Railway Tunnels

Because of the large amount of solar radiation power that is clean and pollution free, solar energy resources occupy an important status in the modern energy system [6]. The characteristic of the railway is its linear structure, variable track direction, scattered available land blocks along the line, small area per unit, resulting in insufficient single-unit conversion of electricity. Some areas near the photovoltaic power station have low power demand, and limited utilization, and if the converted electricity is delivered to the station area for utilization, the transmission distance may be long and may result in poor economic efficiency. Therefore, it is necessary to conduct a comprehensive on-site survey to determine the available photovoltaic land blocks for use.

The first step is to confirm the solar energy resource conditions. Beijing is a typical temperate subtropical monsoon climate zone, according to the distribution of solar energy resources in China. Beijing belongs to Class II solar energy resources regions. Solar radiation in Beijing is abundant [7]. The average annual solar radiation in Beijing is between 112 and 136 kilocalories per square centimeter. The annual average sunshine time in Beijing is between 2,000 and 2,800 h, and from January to May, the monthly total radiation increases, with the highest increase in March and April. From May to June, the total radiation reaches its highest value, and from June onward, it begins to decrease. In December, it reaches its lowest value, and therefore, it has the conditions to carry out photovoltaic power generation (Table 1).

Table 1. Beijing solar monthly radiation

Month station name	1	2	3	4	5	6	7	8	9	10	11	12	Full year
Qixiangtai	6.8	8.1	12.2	13.7	16.6	16.1	13.9	12.9	11.6	9.5	6.6	5.7	131.9
Gubeikou	7.1	8.4	12.4	13.5	16.2	15.9	13.9	13.2	11.7	9.6	6.8	6.2	136.3
Yanqin	7.2	8.6	12.5	13.4	16.4	15.9	13.8	12.8	11.7	9.6	6.8	6.2	136.3
Changpin	6.8	8	12	13.2	15.9	15.6	13.1	12.4	11.5	9.4	6.4	6	130.7
Fanshan	6.6	7.9	11.7	13.1	15.9	15.4	13.1	12.4	11.4	9.1	6.3	5.7	129.2
Chaoyang	6.5	7.7	11.7	12.8	15.7	15.4	12.8	12.1	11.4	9	6.2	5.5	127.8
Xiayunlin	5.6	6.8	9.8	11.5	16.2	13.5	11.4	10.7	9.6	7.6	5.5	4.9	112.3

The project mainly utilizes land owned by the strip line and rail intersection to carry out distributed photovoltaic system construction. As shown in Fig. 1, in Zone 1, approximately 300 square meters of unused ground are utilized, while in Zone 2, approximately 1,300 square meters of rail slope are utilized to build a distributed photovoltaic system, and a storage system is constructed as well.

Fig. 1. Photovoltaic layout of the Loop Track Test Center

3 Selection of Major Equipment for Photovoltaic Power Generation

The main photovoltaic equipment includes: solar photovoltaic support foundations, solar photovoltaic supports, solar photovoltaic batteries, inverters, etc.

3.1 Selection of Solar Photovoltaic Support Foundation

The main types of supporting foundations for expansion include independent foundations, horizontal beam-type foundations (horizontal beam-type longitudinal and horizontal beams), grid foundations, and pile foundations. Expansion foundations have good adaptability, low cost, and are commonly used in ground solar photovoltaic power plants and concrete-reinforced waterproof rooftop photovoltaic power plants. Pile foundations mainly include precast pile foundations and cast-in-place pile foundations. There are various forms of pile foundations, such as H-type steel pile foundations, spiral steel pile foundations, PHC pile foundations, cast-in-place concrete pile foundations, etc. Pile foundations are commonly used in ground solar photovoltaic power projects and waterproof rooftop photovoltaic power projects with poor geological conditions [8]. Based on the comparison table in (Table 2), it is recommended to use spiral steel column foundations.

3.2 Comparison of Photovoltaic Support Forms

Currently, photovoltaic support can be divided into fixed types, tilt adjustable types, horizontal single-axis, skewed single-axis, azimuth single-axis, and dual-axis types, and so on, with a total of six types.

Table 2. Photovoltaic support foundation

Bracket form	Advantages	Disadvantages
Reinforced concrete strip foundation	1. Shallow burial depth, low soil bearing capacity requirement;	1. Requires extensive site leveling, causing significant environmental impact;
	2. Construction does not require specialized equipment, and the construction process is simple	2. Long construction period
Helical steel piles	1. Fast construction, no need for site leveling, adjustable bracket height;	1. Requires specialized construction machinery;
	2. Minimal damage to existing vegetation;	2. Demands higher requirements for soil layers, including a certain degree of compaction;
	3. Helical steel piles can be reused during recovery	3. Poor corrosion resistance of helical steel piles
Prefabricated reinforced concrete pile column	1. Good anti-corrosion performance of the structure;	1. Difficult to control the elevation of the pile top;
	2. Fast construction speed;	2. Challenging to reserve embedded components at the pile top
	3. The soil around the piles is compacted, providing strong resistance against pull-out forces	
Cast-in-place reinforced concrete pile column	1. Material-saving, low cost, and fast construction	1. Requires high-quality soil layers, suitable for moderately compacted silty soil or plastic and stiff plastic loam

Fixed-type support structures (Fig. 2) are simple in structure and cannot adjust the tilt angle and azimuth angle after installation. After deployment, there is no need for a large amount of maintenance work. They are suitable for deployment near railways. Fixed-type support structures commonly adopt structural forms such as trusses, single-span structures, and single-beam structures. Through comparative analysis, it is recommended to adopt truss-type support structures as photovoltaic support forms (Table 3).

Fig. 2. Photovoltaic bracket (Truss-type fixed bracket, Single-column fixed bracket, Single-row column fixed bracket).

Table 3. Photovoltaic bracket selection

Bracket form	Advantages	Disadvantages
Truss-type fixed bracket	1. Clear force distribution, simple fabrication	1. Large foundation engineering quantity
Single-column fixed bracket	1. Only one row of foundations, causing minimal disturbance to the ground during construction;	1. The structure is more complex, with low installation efficiency;
	2. Better adaptability to terrain compared to truss-type fixed brackets	2. Compared to truss-type brackets, it has lower resistance to wind and snow loads, and the structure is prone to swaying
Single-row column fixed bracket	1. Strongest adaptability to terrain;	1. Fewer photovoltaic modules can be installed on a single set of brackets;
	2. Occupies less land area;	2. There is more overhang
	3. Fast construction speed	

3.3 Selection of Solar Photovoltaic Cells

Solar photovoltaic cells are mainly classified based on the substrate material as follows: silicon solar photovoltaic cells, compound semiconductor solar photovoltaic cells, organic semiconductor solar photovoltaic cells, and thin-film solar photovoltaic cells. Performance technology comparisons are made between current mainstream solar photovoltaic cells (namely crystalline silicon solar cells, amorphous silicon solar cells, and thin-film solar cells), as shown in the table below.

Table 4. Solar cell characteristics

Battery types	Crystalline silicon		Thin film category		
	Monocrystalline silicon	Polycrystalline silicon	Amorphous silicon	Cadmium telluride	Copper indium selenide
Commercial efficiency	14% ~ 18%	14% ~ 16%	6% ~ 9%	7% ~ 9%	5% ~ 8%
Laboratory efficiency	24.70%	20.30%	13%	16.40%	19.50%
Lifespan	25 years	25 years	25 years	25 years	25 years
Thickness of cell layers	Thick layer	Thick layer	Thin layer	Thin layer	Thin layer
Scale production	Already formed	Already formed	Already formed	Already formed	Proven to be feasible
Environmental issues	Neutral	Neutral	Neutral	Yes (uses cadmium)	Neutral, except for the use of cadmium
Energy payback time	2 ~ 3 years	2 ~ 3 years	1 ~ 2 years	1 ~ 2 years	1 ~ 2 years
Primary raw materials	Medium	Medium	Abundant	Both cadmium and telluride are rare metals	Indium is an expensive rare metal
Production cost	High	High	Lower	Relatively lower	Relatively lower
Key advantages	High efficiency, mature technology	High efficiency, mature technology	Good weak light performance, lower cost	Good weak light performance, relatively lower cost	Good weak light performance, relatively lower cost

As can be seen from Table 4, the biggest difference between the monocrystalline silicon battery and polycrystalline silicon battery in the silicon-based photovoltaic battery is that the photoelectric conversion efficiency of the monocrystalline silicon battery is slightly higher than that of the polycrystalline silicon battery. For a power of the same battery module, the area of the monocrystalline silicon battery module is smaller than that of the polycrystalline silicon battery module [9].

3.4 Inverter Selection

According to the application form of the photovoltaic power station [10], the development of the photovoltaic inverter has been relatively rapid, and there are three main inverter solutions.

Centralized inverter solution refers to the grid-connected solar photovoltaic power generation system that converts the direct current output of the solar panel array into alternating current with the same frequency, phase, and amplitude as the low voltage power grid, and achieves balanced three-phase alternating current. The string inverter is based on the modular concept, which inputs each photovoltaic string in the photovoltaic panel array into a small power inverter, and multiple photovoltaic strings and inverters are modularly combined. All inverters are parallel connected in the AC output end, and then sent to a double-winding transformer on site for step-up. The decentralized inverter solution is a grid-connected sub-station solution that adopts a centralized inverter, distributed tracking approach. It adds DC/DC boost conversion hardware modules and MPPT control software modules to the traditional PV aggregation box inside, and achieves distributed tracking functions for every 2 ~ 4 string PV components corresponding to 1 line MPPT (Table 5).

Table 5. Inverter selection

Inverter types	Advantages	Disadvantages
Centralized	1. Fewer inverters, easier to manage 2. Fewer components in the inverter, higher reliability 3. Lower harmonic content and DC components, resulting in higher power quality 4. High integration and power density of the inverter, leading to lower costs 5. Inverters equipped with various protection functions, ensuring high safety of the power station 6. Active power factor regulation and low voltage ride-through capability, providing good grid regulation	1. The failure rate of DC combiner boxes is relatively high 2. Centralized inverters have a narrow MPPT voltage range 3. With the string array going through two combiners before reaching the inverter, the MPPT cannot monitor the operation of each string individually, making it impossible to achieve the optimal operating point for every string

(continued)

Table 5. (*continued*)

Inverter types	Advantages	Disadvantages
String	1. Each photovoltaic string corresponds to one inverter, which is not affected by module differences between the strings or shading, reducing the mismatch between the optimal operating point of the PV modules and the inverter, thereby maximizing the power generation 2. String inverters have a wide MPPT (Maximum Power Point Tracking) voltage range	1. There are more electronic components, resulting in slightly lower reliability 2. The design does not include an isolation transformer, leading to slightly lower electrical safety and a larger DC component, which has a greater impact on the grid 3. With a larger number of inverters, the overall failure rate increases, making system monitoring more challenging 4. However, when multiple inverters are connected in parallel, it becomes more difficult to implement functions such as zero voltage ride-through, reactive power regulation, and active power regulation
Distributed	The distributed inverter solution combines the advantages of centralized inverter solution and string inverter solution	In the distributed inverter solution, the boosting unit and control unit of the combiner box, as well as the structure and functions of the combiner box itself, are more complex compared to the DC combiner box in the centralized inverter solution and the AC combiner box in the string inverter solution. The heat dissipation and adaptability to harsh environments of the combiner box deserve attention. Currently, there are few manufacturers in China that support the distributed inverter solution, and there are relatively few application cases. The design advantages of the distributed inverter solution need further validation in practical applications

Through analyzing the advantages and disadvantages of the three types of inverters, it is recommended that the string inverter be selected as the main selection.

4 Storage System

Photovoltaic power generation output power varies greatly with changes in irradiance and temperature, and it is highly nonlinear, making it difficult for the power generated by the power system to be effectively controlled to ensure the safety and reliability of power supply. The storage system in the photovoltaic power generation system can solve the power imbalance problem in the power generation system, and the storage system is crucial to the stable operation of the photovoltaic power station.

4.1 Configuration Method of Energy Storage Systems

The electricity generated by the photovoltaic battery module is managed by the photovoltaic controller and controlled automatically by the photovoltaic controller. The photovoltaic inverter converts the direct current into alternating current to supply power for alternating loads. The system has automatic DC balance with the battery at the front end of the photovoltaic inverter, and the main characteristics of this mode are high system efficiency, the output capacity of the power station can be controlled internally by the photovoltaic power plant, and seamless connection can be achieved, with good quality of output electricity, very small output fluctuations, etc., which can greatly improve the smoothness, stability, and adjustability of the output of photovoltaic power generation. However, this type of converter has some disadvantages: it is designed for specific use and cannot be used for upgrading existing photovoltaic power plants that have been installed in many places. Another disadvantage is that the energy storage unit of the photovoltaic battery module only serves as a charging source for the electrical energy of this unit, and other nearby photovoltaic power stations or units do not have enough electricity to charge, which lacks large-scale grid power regulation functions within the power station (Fig. 3).

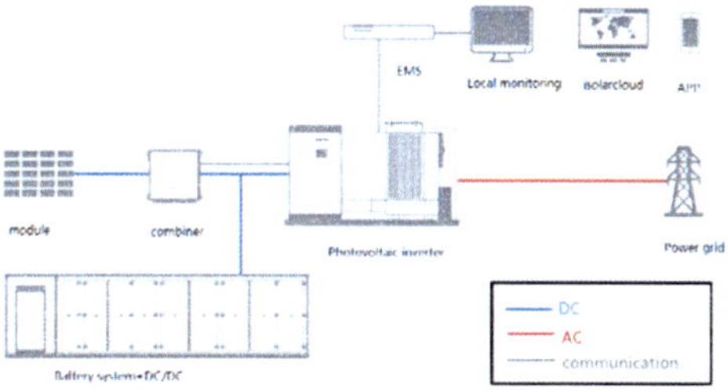

Fig. 3. Structure diagram of the DC side of the power supply

The configuration scheme on the AC side of the power supply is also known as a common AC busbar scheme. It uses a separate charging and discharging controller and inverter to charge or invert the battery, which actually means attaching a storage device to an existing photovoltaic power system, which can be upgraded and installed in any photovoltaic power station, wind power station, or other power station currently available. This mode overcomes the problem of the DC side of the storage system not being able to fully centrally control the scheduling of non-fuel energy, and it can be charged or discharged intelligently by the control system or centralized scheduling by the power grid. It can even quickly and effectively collect cheap surplus energy from the entire low-voltage side of the storage system and dispatch it to the power grid for charging, even if the surplus energy is distributed from neighboring photovoltaic power stations or stations that lack sufficient electricity (Fig. 4).

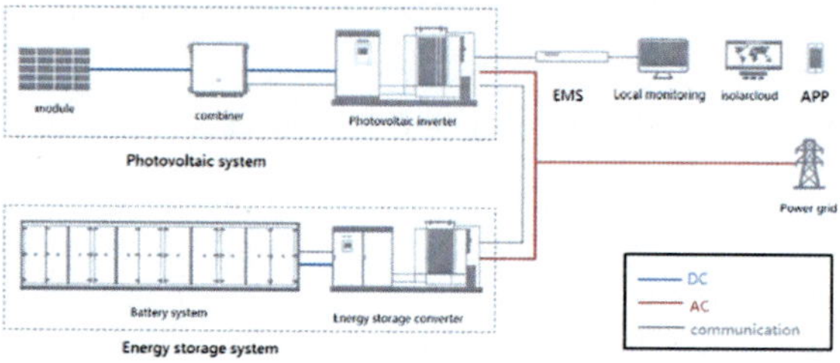

Fig. 4. Structure diagram of the AC power supply

4.2 Selection of Batteries for Energy Storage Systems

The main types of batteries used in power applications include lead-acid batteries, fuel cells, nickel-metal hydride batteries, and lithium-ion batteries (LiBs). LiBs can be divided into three categories based on their positive and negative electrode materials: lithium iron phosphate batteries, lithium-cobalt-nickel oxide batteries, and lithium titanate batteries.

The densities of lithium iron phosphate batteries and compacted density are relatively low, resulting in lower energy density compared to lithium-cobalt-nickel oxide batteries. However, due to its low cost and excellent safety characteristics, LiFePO4 batteries are widely used in power applications due to their high cost-performance ratio. Based on the comparison of different energy storage modes, it is recommended to use LiFePO4 batteries as energy storage elements due to their advantages such as high energy density, long cycle life, and low self-discharge rate (Table 6).

4.3 Energy Storage Inverter

Energy storage inverter (as shown in Fig. 5) adopts a primary conversion main circuit structure, and the AC side should be configured with AC switches, contactors, EMI filters, and so on. The DC output unit includes EMI filters, precharge circuits, and DC switches. An energy storage inverter with bidirectional requirements needs to achieve bidirectional flow of electricity. This project adopts a three-level power conversion topology with high power control response speed, high accuracy, high reliability, high stability, and excellent performance in meeting the functional and performance requirements of energy storage inverters.

The energy storage inverter should have the following functions: active and reactive power control, protection and alarms, communication management, controlling power supply, fault recording, etc.

Table 6. Characteristics of different energy storage components

Energy storage methods	Energy storage methods				
	Supercapacitor	Ternary lithium battery	Lithium iron phosphate battery	Lithium titanate battery	Flywheel
Advantages	High power density Low internal resistance	High energy density	High safety level	Long cycle life, high power density, high safety	Good energy-saving effect
Disadvantages	Low energy density	Lower safety level	Lower energy density compared to ternary lithium batteries	Higher cost	Large in size Immature technology

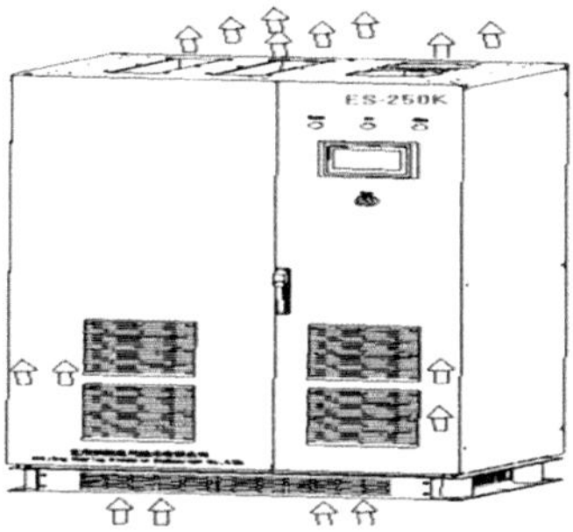

Fig. 5. Energy storage converter diagram

5 Photovoltaic Energy Storage Comprehensive Energy Management Platform

The photovoltaic energy storage comprehensive energy management platform aims to achieve interoperability between energy-related information flows and integrate various available resources. It reasonably monitors and allocates the supply, consumption, and services of energy resources in the ring of iron and around it, ensuring safe and reliable energy supply while increasing the utilization of clean energy and reducing unnecessary energy consumption, thereby reducing energy costs and improving energy quality.

5.1 Operation Monitoring Function

The platform has monitoring functions such as SCADA monitoring, meter monitoring, equipment monitoring, power quality monitoring, and video monitoring (optional), which can present real-time data in diverse page formats to show the system and equipment in a comprehensive way.

5.2 Warning Management Function

The platform has but is not limited to remote monitoring and warning, remote sensing displacement, equipment entry and exit, remote operation, fault location, and classification and hierarchical alarms. The warning methods include alarm bell flashing, alarm sound, alarm push notification, text message, and app push notification.

5.3 Other Functions

The platform has functions such as related operation and maintenance, statistical analysis, mobile connectivity, dynamic monitoring, etc., which can achieve safe, reliable, stable, and high-quality power supply and demand while providing a safe, comfortable, agile, and efficient working environment for operation and maintenance managers (Fig. 6).

Fig. 6. Schematic diagram of comprehensive energy control platform

6 Conclusion

The rail iron solar photovoltaic construction built a railway solar photovoltaic testing environment by using railway embankment slope and railway sandwiched land for building solar photovoltaic power generation. By comparing and analyzing major photovoltaic equipment, this provides a test environment for future research on railway solar photovoltaic applications.

Acknowledgments. This work is supported by the National Railway Track Test Center, China Academy of Railway Sciences Corporation Limited (No. 2022YJ003).

References

1. Li, Z.: Prospects of photovoltaic technology. Engineering **21**, 28–31 (2023). ISSN 2095-8099
2. Zanxin, W., Wenrui, F.: Economic and environmental impacts of photovoltaic power with the declining subsidy rate in China. Environ. Impact Assess. Rev. **87** (2021)
3. Opinions of the National Development and Reform Commission and the National Energy Administration on Enhancing the Energy-saving and Environmental Protection Policies and Measures for Energy Supply and Demand Coordination Platforms, Notice No. 206. https://www.ndrc.gov.cn/xxgk/zcfb/tz/202202/t20220210_1314511_ext.html. (2022). Accessed 10 Aug 2023 (in Chinese)
4. Jingdong, W., Zenghai, Z., Yanxing, G., et al.: 2021 China's PV power development status and prospects. Water Power **48**(10), 4–8 (2022). (in Chinese)
5. Qiang, L., Li-Juan, L., Ming, Y., et al.: Choice of the distributed photovoltaic power generation operating mode for a manufacturing enterprise: Surrounding users vs a power grid. J. Clean. Prod. **293** (2021) (in Chinese)
6. Zhenyu, Z., Yujia, Y.: Research on spatio-temporal coupling suitability of solar resources and distributed PV power projects in Beijing. Sol. Energy **4**, 16–25 (2021)
7. Jianghua: Current situation and trends of domestic and overseas PV industry development. Solar Energy **12**, 15–17 (2016) (in Chinese)
8. Tingting, Z., Shaixiong, J.: Study on selection and design optimization of PV support foundation. J. Build. Struct. **52**(S1), 2353–2357 (2022). (in Chinese)
9. Yan, Z.: Optimization Design of Distributed Photovoltaic Power Generation Systems. Tianjin Business University, 2017 (in Chinese)
10. Jianguo, Z., Jianting, D.: Discussion on the selection of inverters for centralized photovoltaic power plants. Energy Saving **39**(03), 26–28 (2020). (in Chinese)

Improvement on Command Current Extraction of Railway Static Power Conditioner

Yunpeng Zhang[1(✉)], Ying Wang[1,2], Xiaoqiang Chen[1,2], and Chenyu Song[1]

[1] Lanzhou Jiaotong University, Lanzhou 730070, China
2461295015@qq.com

[2] Key Laboratory of Opto-Technology and Intelligent Control Ministry of Education, Lanzhou 730070, China

Abstract. The Railway Static Power Conditioner (RPC) can effectively overcome the shortcomings of traditional treatment devices for negative sequence and harmonic compensation, and provide a new idea for the treatment of high-speed railway power quality. The low-pass filter (LPF) plays a key role in the extraction of the command current signal. Traditional command current signal extraction faces challenges in terms of speed and accuracy in balancing techniques. To address this issue, this paper proposes an improved technique to extract the command current signal using the phase shifting capability of the second order generalized integrator (SOGI). Firstly, this method effectively improves the speed and accuracy of the instruction current signal extraction. Secondly, the elimination of low-order harmonic signals allows for higher cutoff frequency of the filter, ensuring both filtering speed and waveform smoothness. Finally, simulation experiments have verified the feasibility and effectiveness of the improved current signal extraction technique.

Keywords: Railway power static regulator · Second-order generalized integrator · Low-pass filter

1 Introduction

With the rapid development of electrified railways, while promoting economic and social development and facilitating people's travel, it has also brought serious power quality problems to the public power grid. Electric locomotives have the characteristics of high power, large load fluctuation, and single-phase power supply, which cause the power quality problems caused by the traction power supply system to the public power grid mainly include negative sequence, reactive power, and harmonic problems, which reduce the power quality of the power grid [1]. Therefore, it is necessary to control the power quality of electrified railways. The Railway Static Power Regulator (RPC) can almost solve the power quality problems such as negative sequence, reactive power, and harmonic distortion in the traction power supply system. RPC has excellent comprehensive compensation performance, port extension and other advantages, so it has received widespread attention. Research on the topology, instruction current, compensation principle and capacity optimization of RPC provides a corresponding basis for its

L. Jia et al. (Eds.): EITRT 2023, LNEE 1135, pp. 170–177, 2024.
https://doi.org/10.1007/978-981-99-9307-9_19

application. At the same time, the application of RPC has also promoted the development and application of static power regulation technology.

Through the study of the control strategy of RPC, it can be found that the Low-pass filter (LPF) technology runs through the entire command current signal extraction process, and its quality plays an important role in the extraction of the command signal, directly determining the performance indicators of RPC. A series-connected LPF structure was given in [2], which can effectively improve the extraction speed and quality of active and reactive command signals. However, it is affected by the cutoff frequency of the filter, which has a certain impact on the extraction speed. By adding the proportional link, phase shift link and summation link to the signal, the extraction speed of the command current is improved and the filtering effect is improved. However, the phase shift link requires additional time and storage space, and increases as the sampling accuracy is improved in [3]. An improved ip-iq harmonic detection algorithm based on second-order generalized integrator is proposed in [4], but it is only applicable to three-phase systems. It mainly studies the phase-locked loop (PLL) technology in command current extraction in [5]. Other literature mainly studies the RPC control method and topology, and the command current extraction technology adopted is an extraction method based on instantaneous reactive power theory.

Therefore, in this paper, the speed and accuracy of instruction extraction are analyzed, and a new improved method based on second-order generalized integrator (SOGI) is proposed for the extraction of instruction current, which reduces the limitation of LPF on extraction technology, and can quickly and accurately extract the amplitude of instruction current. Finally, the proposed method is verified by semi-physical simulation platform.

2 Compensation Structure and Control Strategy of V/v Traction Power Supply System

2.1 Topology of RPC

The RPC compensation structure is shown in Fig. 1. The 220 kV three-phase high voltage is converted into two 27.5 kV single-phase electricity by a three-phase V/V traction transformer, which are connected to two feeder segments. RPC is composed of two AC/DC converters sharing DC capacitors, two converters are connected in parallel with large-capacity DC capacitors to ensure their stable performance, the two converters are connected to the single-phase step-down transformer through two output inductors, and connected to the traction power supply arm through the step-down transformer, and the power flow of the two power supply arms is controlled by controlling the IGBT bridge of the converter, and a certain amount of active power is transferred from one feeder segment to another. Both converters can also provide harmonic suppression and reactive power compensation [6]. Therefore, the system can realize the comprehensive compensation of negative sequence current, harmonic current and reactive power, and improve the power quality of the primary side of the traction network.

2.2 Control Strategy of RPC

The RPC control block diagram is shown in Fig. 2, which mainly consists of two parts: the first part is the RPC signal detection and instruction current extraction part, which

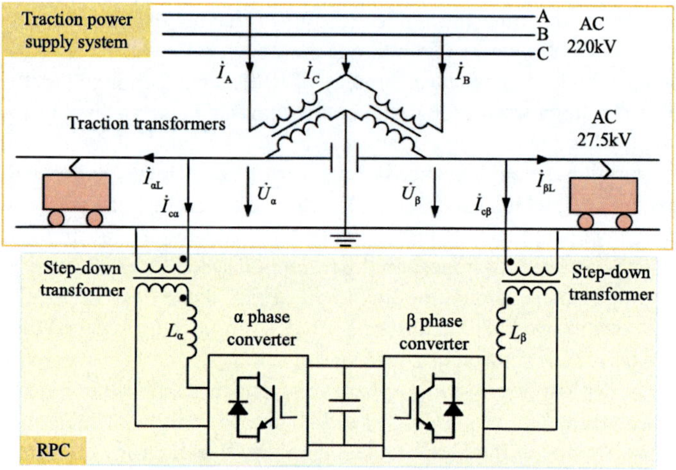

Fig. 1. Construction of RPC compensation

mainly completes the detection and extraction of reactive power, active power, and harmonics; the second part is the RPC control algorithm, which consists of a double-loop control composed of voltage outer loop and current inner loop. The voltage outer loop control includes quasi proportional integral (PI) control, composite PI control, and composite PI control based on mean filtering. The current inner loop control includes composite PI control, quasi-proportional resonance (PR), hysteresis current control, repetitive prediction of zero-crossing control, fuzzy tracking control, and fuzzy recursive PI control, etc., to complete RPC grid connection and governance of the power quality of the traction power grid of electrified railways [7, 8]. As shown in Fig. 2, the extraction of RPC active power, reactive power, and harmonics, RPC grid connection control, etc. are closely related to the grid phase-locked loop, harmonic extraction, and LPF, which directly determine the RPC control speed and performance.

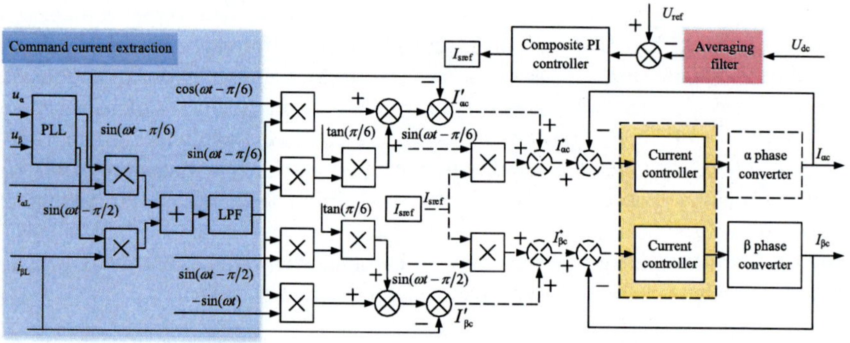

Fig. 2. Control strategy block diagram of RPC

3 Improved Instruction Signal Detection Method

In Fig. 3, the second-order generalized integrator (SOGI) module is an orthogonal signal generator, which filters the input signal and retains the signal in the specified frequency range [9, 10]. At the same time, it generates another signal that is orthogonal to the input signal. The discrete domain state space expression of the SOGI module is as follows:

$$\begin{cases} ID(s) = \frac{I'}{I} = \frac{k\omega s}{s^2 + k\omega s + \omega^2} \\ IQ(s) = \frac{qI'}{I} = \frac{k\omega^2}{s^2 + k\omega s + \omega^2} \end{cases} \tag{1}$$

where k is the gain of the transfer function; ω is the resonant frequency.

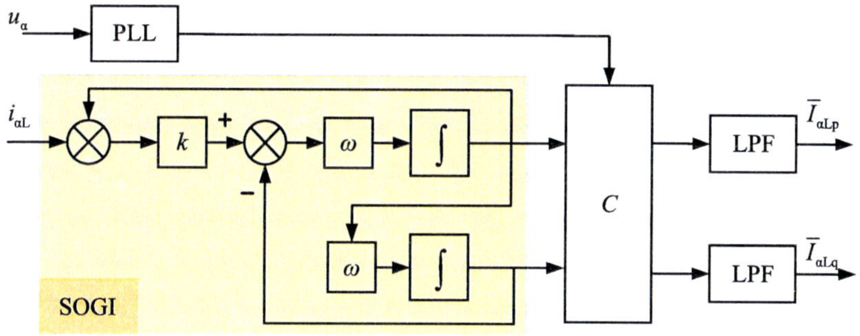

Fig. 3. Block diagram of the improved instruction signal detection method

As the gain of the transfer function decreases, the filter bandwidth also changes in the same direction. At this time, the filtering effect becomes better, but the response speed becomes slower in contrast. The phase shift at the resonant point is always 90 degrees, which can achieve the purpose of phase shifting.

In order to verify the implementation method, the A-phase voltage is taken as the reference vector $\dot{U}_A = U\sin(\omega t)$, The fundamental current of Phase α and Phase β power supply arms is expressed as:

$$\begin{cases} i_{\alpha L}(t) = \sqrt{2}I_{\alpha LF}\sin\left(\omega t - \frac{\pi}{6} + \varphi_\alpha\right) + \sum_{k=2}^{\infty} i_k(t) \\ i_{\beta L}(t) = \sqrt{2}I_{\beta LF}\sin\left(\omega t - \frac{\pi}{2} + \varphi_\beta\right) + \sum_{k=2}^{\infty} i_k(t) \end{cases} \tag{2}$$

where ω represents the fundamental frequency; $i_k(t)$ represents the harmonic component.

The transformation matrix in Fig. 3 is as follows:

$$C = \begin{bmatrix} \sin\left(\omega t - \frac{\pi}{6}\right) & \cos\left(\omega t - \frac{\pi}{6}\right) \\ \cos\left(\omega t - \frac{\pi}{6}\right) & -\sin\left(\omega t - \frac{\pi}{6}\right) \end{bmatrix} \tag{3}$$

After the current signal passes through the SOGI module, it outputs orthogonal signals that are 90 degrees out of α phase with each other, and can filter out some of the high-order harmonics. Assuming that the output of phase a becomes:

$$\begin{cases} i_{\alpha \mathrm{Ld}}(t) = \sqrt{2} I_{\alpha \mathrm{LF}} \sin\left(\omega t - \frac{\pi}{6} + \varphi_\alpha\right) + \sum_{k=2}^{\infty} i'_k(t) \\ i_{\alpha \mathrm{Lq}}(t) = \sqrt{2} I_{\alpha \mathrm{LF}} \cos\left(\omega t - \frac{\pi}{6} + \varphi_\alpha\right) + \sum_{k=2}^{\infty} i'_k(t) \end{cases} \tag{4}$$

where $i'_k(t)$ is the harmonic component after passing through the SOGI module.

Formula (5) results from the multiplication by the transformation matrix as follows:

$$\begin{cases} I_{\alpha \mathrm{Lp}} = i_{\alpha \mathrm{Ld}}(t) \sin\left(\omega t - \frac{\pi}{6}\right) + i_{\alpha \mathrm{Lq}}(t) \cos\left(\omega t - \frac{\pi}{6}\right) \\ I_{\alpha \mathrm{Lq}} = i_{\alpha \mathrm{Ld}}(t) \cos\left(\omega t - \frac{\pi}{6}\right) - i_{\alpha \mathrm{Lq}}(t) \sin\left(\omega t - \frac{\pi}{6}\right) \end{cases} \tag{5}$$

The simplified result of Formula (6) is as follows:

$$\begin{cases} I_{\alpha \mathrm{Lp}} = \sqrt{2} I_{\alpha \mathrm{Lf}} \cos \varphi_1 + \sum\limits_{k=2}^{\infty} i'_k(t) \sin\left(\omega t - \frac{\pi}{6}\right) + \sum\limits_{k=2}^{\infty} i'_k(t) \cos\left(\omega t - \frac{\pi}{6}\right) \\ I_{\alpha \mathrm{Lp}} = \sqrt{2} I_{\alpha \mathrm{Lf}} \sin \varphi_1 + \sum\limits_{k=2}^{\infty} i'_k(t) \sin\left(\omega t - \frac{\pi}{6}\right) + \sum\limits_{k=2}^{\infty} i'_k(t) \cos\left(\omega t - \frac{\pi}{6}\right) \end{cases} \tag{6}$$

The direct current component and high-order harmonic components of the reference current can be extracted through calculations. The Formula (7) can be filtered by a LPF to remove the harmonic signal, resulting in the following formula:

$$\begin{cases} \overline{I}_{\alpha \mathrm{Lp}} = \sqrt{2} I_{\alpha \mathrm{Lf}} \cos \varphi_\alpha \\ \overline{I}_{\beta \mathrm{Lq}} = \sqrt{2} I_{\alpha \mathrm{Lf}} \sin \varphi_\alpha \end{cases} \tag{7}$$

Similarly, the reference current signal for β phase can be obtained.

The specific calculation method is described in [11]. Further, the α and β phase converters can be controlled in such a way that the two converters dynamically track the reference values, thus achieving a comprehensive compensation function. Furthermore, in addition to performing multi-index compensation, it is still effective to perform specific single-index complete compensation by separating the reference components of harmonics, negative sequence, and reactive current from the total current reference quantity under specific operating conditions. This detection method can provide various optional compensation objects for RPC.

4 Instruction Current Extraction Simulation

To verify the improved instruction current extraction technology, a high-speed railway power supply system and an RPC simulation model were constructed using the MT6020 real-time simulation system, and the RPC model was adjusted using a composite PI control method of voltage outer loop and current inner loop [12]. The MT6020 real-time simulation system is shown in Fig. 4.

Assuming that at the initial time, the power of the A phase supply arm is 4800 kW and the power of the B phase supply arm is 2400 kW. At 1.5 s, the power load of the

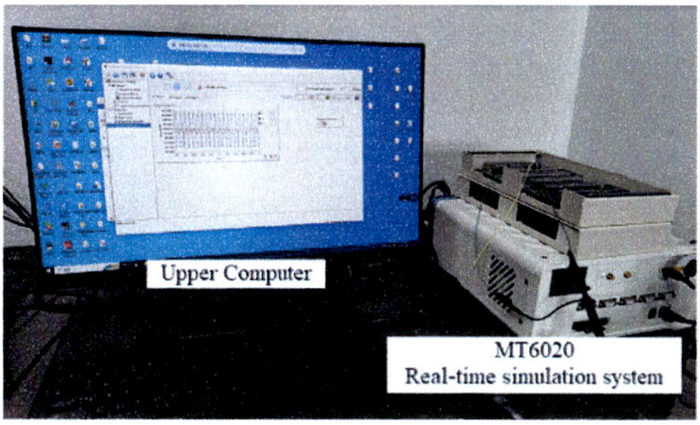

Fig. 4. Real-time simulation system

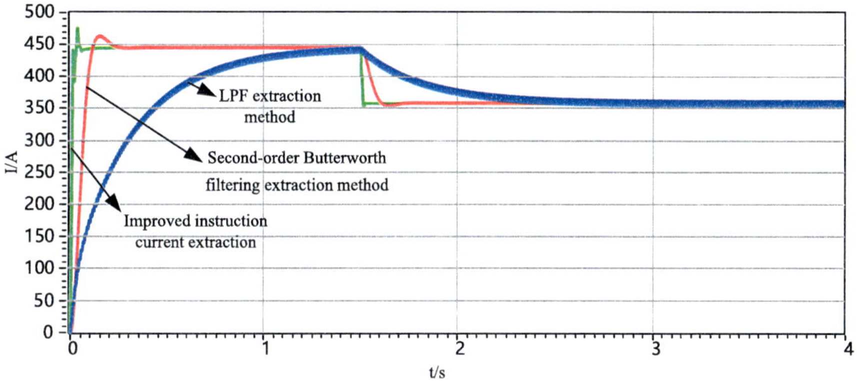

Fig. 5. Output instruction current waveform

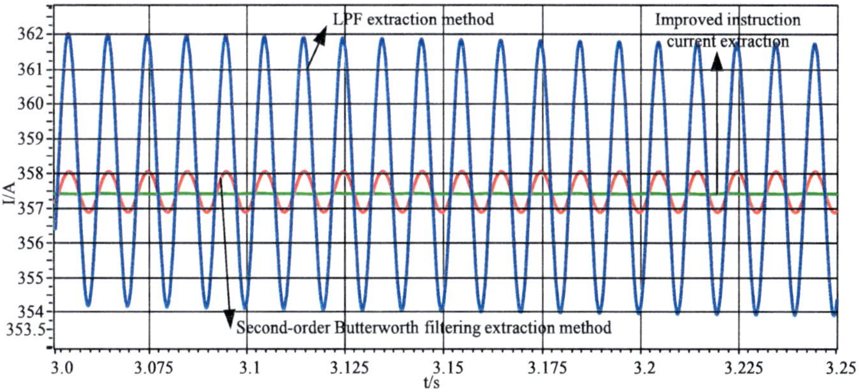

Fig. 6. The local amplification waveform of output instruction current

A phase supply arm changed to 2400 kW, and the resulting waveform of the instruction current extraction is shown in Fig. 5. In Fig. 5, the improved instruction current extraction method shows a significantly faster response speed compared to the methods based on low-pass filters and second-order Butterworth filters, and it can quickly respond to load changes. The local amplification waveform of the output instruction current is shown in Fig. 6. The improved current extraction technology shows a significantly better waveform quality compared to the other two methods.

5 Conclusion

LPF have a significant impact on the instruction current signal extraction technique. To address this issue, an improved technique for extracting instruction current signal using the phase shifting capability of SOGI was proposed. This technique eliminates low-order harmonic signals, which allows for both improved response speed and waveform smoothness.

Acknowledgments. This research is supported by National Natural Science Foundation of China Regional under Grant (Grant No. 52067013), Natural Science Key Foundation of Gansu Provincial Department of Science and Technology under Grant (Grant No. 21JR7RA280), Natural Science Key Foundation of Gansu Provincial Department of Science and Technology under Grant (Grant No. 22JR5RA318), and Natural Science Foundation of Gansu Provincial Department of Science and Technology under Grant (Grant No. 22JR11RA162).

References

1. Hu, H., Ge, Y., Huang, Y., et al.: "Source-network-train-storage" integrated power supply system for electric railways. Proc. CSEE **42**(12), 4374–4391(2022) (in Chinese)
2. Jiang, Y., Wang, W., Zhao, L.: The research and improvement of the signal extraction for railway static power conditioner. Power Electron. **53**(03), 59–63 (2019) (in Chinese)
3. Li, K.: Study of Railway Static Power Conditioner. Harbin Institute of Technology (2013) (in Chinese)
4. Gu, C., Zhang, X., Tong, T., et al.: An improved ip-iq harmonic detection algorithm based on second-order generalized integrator. Electr. Meas. Instrum. **57**(10), 88–93 (2020) (in Chinese)
5. Jiang, Y., Wang, H., Zhao, L., et al.: Research on command current extraction and phase-locked technology of railway power conditioner. Power Electron. **56**(05), 17–20 (2022). (in Chinese)
6. Luo, A., Wu, C., Shen, J., et al.: Railway static power conditioners for high-speed train traction power supply systems using three-phase V/V transformers. IEEE Trans. Power Electron. **26**(10), 2844–2856 (2011)
7. Deng, W., Dai, C., Chen, W., et al.: Research progress of railway power conditioner. Proc. CSEE **40**(14), 4640–4655+4742 (2020) (in Chinese)
8. Guo, P., et al.: Analysis and control of modular multilevel converter with split energy storage for railway traction power conditioner. IEEE Trans. Power Electron. **2**(35), 1239–1255 (2020)
9. Zhang, J., Chen, Q., Li, G., et al.: Research on improved second order generalized integrator phase-locked loop in non-ideal power grid. Power Electron. **55**(07), 7–11 (2021). (in Chinese)

10. Smadi, I.A., Bany Fawaz, B.H.: DC offset rejection in a frequency-fixed second-order generalized integrator-based phase-locked loop for single-phase grid-connected applications. Prot. Control Mod. Power Syst. **1**(7), 1–13 (2022)
11. An, B., Li, Y., Liu, F., et al.: An asymmetrical connection balance transformer-based hybrid railway power conditioning system with cost-function optimization. IEEE Trans. Transp. Electrification **4**(2), 577–590 (2018)
12. Ma, Q., Luo, H., Yang, S.: Study of the strategy of composite control based on V/v transformer RPC study of the strategy of composite control based on V/v transformer RPC. Power Electronis Technol. **49**(09), 88–92 (2015). (in Chinese)

Research on High Frequency Current Measurement Method for the SiC MOSFET

Shuai Liang(✉), Yuming Liu, Chunhui Miao, and Ke Sun

CSSC Systems Engineering Research Institute, Beijing 100094, China
liangshuaigz@126.com

Abstract. Silicon carbide MOSFET can operate stably at high voltage level and high switching speed with lower switching power loss and has been widely used in electric vehicles and other fields. The measurement of current signal is very important in device protection and condition monitoring. With the increase of the current change rate in the transient process, traditional current measurement methods face more severe challenges. Based on the principle of double pulse test, the drain current of the SiC MOSFET is divided into different parts to calculate the equivalent frequency of transient current. Based on the double pulse test platform, the characteristics of coaxial resistance, Pearson coil, clamp current probe and Rogowski coil in high frequency current signal measurement are compared and analyzed from the Angle of bandwidth range and mechanical size. Finally, according to the actual application condition of the SiC MOSFET, the selection of current measurement method is proposed.

Keywords: The SiC MOSFET · Double pulse test · Current measurement method

1 Introduction

Power semiconductor devices have been widely used in power, energy, machinery, transportation and other traditional industries, and highlight their irreplaceable importance in aerospace [1], rail transit [2], artificial intelligence and other high-tech industries [3]. With the increasingly complex application conditions, power semiconductor devices have higher and higher requirements for voltage level, switching speed, power loss and other performance. Si power semiconductor devices are gradually approaching the theoretical limit of their materials. It is difficult to meet the higher requirements for power device performance. As a new generation of wide bandgap semiconductor materials, silicon carbide (SiC) materials have more excellent performance in width of forbidden band, breakdown field strength [4, 5], among which, silicon carbide MOSFET can operate stably under high voltage level and high switching speed conditions with lower switching power loss, has been gradually applied to electric vehicles, new energy power generation and other industries. It is a powerful substitute for silicon-based devices in the field of medium and high voltage [6]. Due to the fast switching speed of the SiC MOSFET and the large rate of current change in the transient process, the equivalent frequency of the transient current is also relatively large, which puts forward a more severe

L. Jia et al. (Eds.): EITRT 2023, LNEE 1135, pp. 178–187, 2024.
https://doi.org/10.1007/978-981-99-9307-9_20

test for the current measurement method of the SiC MOSFET. Obtaining accurate test currents is a key issue in the study of the switching characteristics of the SiC MOSFET devices [7, 8].

Firstly, the principle of double pulse test is described in detail, and then the drain current of SiC MOSFET is divided into different parts and the equivalent frequency is calculated. Secondly, the double pulse test platform is built, and the characteristics of four different current measurement methods in high frequency current signal measurement are compared and analyzed through experiments. Finally, according to the actual application condition of the SiC MOSFET, the selection of current measurement method is proposed.

2 Double Pulse Test

At present, the common test method for switching characteristics is the double pulse test, based on the typical test circuit in the IEC60747-8 field effect transistor standard, as shown in Fig. 1. The test circuit adopts diode clamp inductive load circuit, which is a common application circuit of MOSFET, and can reflect the switching characteristics of the device under actual application conditions. As shown in Fig. 2, the driving control timing of the test circuit is double pulse signal. After the first pulse is triggered, the gate-source voltage V_{GS} rises to the drive forward bias voltage V_{GH}, and the drain-source voltage V_{DS} drops to the on-state voltage. During the first pulse, the energy stored in the DC bus capacitor is charged to the load inductor to establish the required test current for the device under test, through the switched device. After the first pulse, the inductive current continues to flow through the diode and begins to slowly decay. Then, the second pulse is triggered, the device under test is turned on again, and the continuing diode enters the reverse recovery stage. The inductance current flows to the device under test again, and the DC power supply charges the load inductance again. Until the end of the second pulse, the device under test is turned off again. To compare the on-off and off-off characteristics of the device reasonably, ensure that the final steady-state current is consistent with the initial steady-state current before the device is turned off, after the device is turned on. Therefore, the off transient characteristics of the device are usually captured at the end of the first pulse, and the on transient characteristics of the device at the beginning of the second pulse.

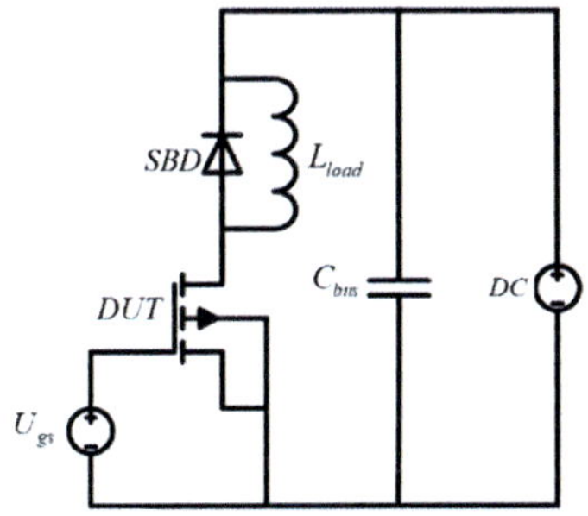

Fig. 1. Standard text circuit.

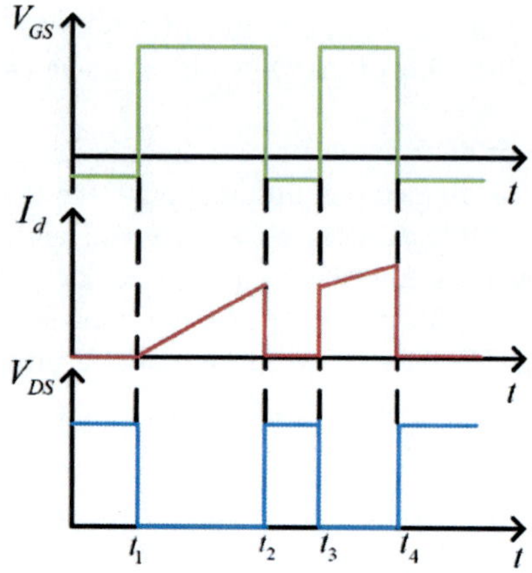

Fig. 2. Schematic of typical waveforms.

3 The Frequency Calculation of Drain Current

The measurement of drain current is an important basis for judging the switching transient characteristics of the SiC MOSFET. As shown in the figure, the i_d of drain current is divided into five parts in this paper, namely, the on transient part, the off transient part, the on oscillation part, the off oscillation part and the steady state part. In the figure, ① and ③ are the open transient and open oscillation parts, ② and ④ are the off transient and off oscillation parts, and the steady state parts are between ① and ② and ③ and ④. Meanwhile, a discrete the SiC MOSFET with voltage and current class of 1200 V/36 A is selected as the measured device. The key parameters can be obtained from the data manual provided on the company's official website, and specific parameters are shown in Table 1. Finally, equivalent frequencies of each part of drain current id are calculated according to device related parameters provided in the data manual (Fig. 3).

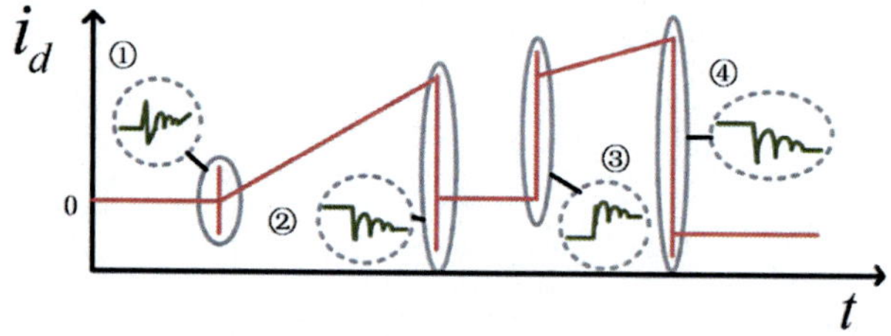

Fig. 3. Drain current of the SiC MOSFET switching process.

In the opening stage of the SiC MOSFET, when the drain current i_d reaches the peak value, the drain current attenuates from the peak value to the steady-state value in the form of second-order damped oscillation, due to the existence of LC oscillation circuit formed by stray inductance, stray capacitance of the continuing loop and stray resistance of the loop. At this time, the SiC MOSFET is fully switched on, and the

Table 1. Measured the SiC MOSFET discrete device parameters.

Parameters	Data	Parameters	Data
Datasheet	C2M0080120D	C_{DS}/pF	100
U_{DSmax}/V	1200	t_r/ns	20
I_D(@25 °C)/A	36	t_f/ns	19

Table 2. Equivalent frequency of each part of drain current.

Parameters	Data	Formula
Turn on the transient frequency	12.5 MHz	$f_{on} = \frac{1}{T_{on}} = \frac{0.35}{t_r}$
Switch off transient frequency	13.2 MHz	$f_{off} = \frac{1}{T_{off}} = \frac{0.35}{t_f}$
Open oscillation frequency	65.1 MHz	$f_{osc(on)} = \frac{1}{T_{osc(on)}} = \frac{1}{2\pi\sqrt{L_{loop}C_F}}$
Turn off the oscillation frequency	91.9 MHz	$f_{osc(off)} = \frac{1}{T_{osc(off)}} = \frac{1}{2\pi\sqrt{L_{loop}C_{D-S}}}$
Steady state frequency	12.5 MHz	$f = \frac{1}{t_{on}}$

voltage at both ends of the drain-source electrode can be considered as zero, and the drain-source capacitor C_{DS} is bypassed. Since the impedance of the common source inductor is far less than that of the drive branch, its switched oscillation frequency can be equivalent to:

$$f_{osc(on)} = \frac{1}{T_{osc(on)}} = \frac{1}{2\pi\sqrt{L_{loop}C_F}} \tag{1}$$

At the end of the SiC MOSFET off phase, the load inductance current continues to flow through the SiC Schottky relay diode. At this time, the relay diode is fully switched on, and the diode junction capacitor C_d and the load inductance stray capacitor C_{load} are bypassed. When the SiC MOSFET is completely turned off, the LC oscillation circuit is formed by the stray inductance L_{loop} and stray capacitor C_{DS} of the continuing loop, and the drain source voltage will decay from peak to steady state under the influence of the stray resistance R_{loop} of the loop. The oscillation frequency of the turn-off waveform can be equivalent to:

$$f_{osc(off)} = \frac{1}{T_{osc(off)}} = \frac{1}{2\pi\sqrt{L_{loop}C_{D-S}}} \tag{2}$$

To sum up, equivalent frequencies of each part of drain current can be expressed in Table 2. As can be seen from the Table 2, the turn-off oscillation part of the drain current of the dynamic characteristics of the SiC MOSFET has the highest frequency, and the turn-off oscillation part is an important aspect of current measurement evaluation.

4 The SiC MOSFET Dynamic Characteristic Test Platform

In this paper, the double pulse experimental platform for the SiC MOSFET is designed and built. The second generation the SiC MOSFET of Cree is 1600 V/36 A, and the SiC Schottky diode C4D20120D of Cree is selected as the continuing diode for the device to be tested. At the same time, digital signal processor (DSP) is used to generate the double pulse signal needed for driving control. In addition, IXDI609 of IXYS company is used as the driver chip, which can isolate and amplify the driver signal generated by DSP. The positive and negative bias required by the driver is provided by low-voltage DC power supply. Yokogawa DLM2054 4-channel 500 M oscilloscope was used to measure and record the double-pulse test waveform.

The main power circuit and driving circuit of the platform designed in this paper are based on PCB circuit board. Because PCB is very convenient and highly controllable in the design of current path, in order to minimize the influence of parasitic parameters on the current value of dynamic characteristics of the SiC MOSFET, the DC capacitance bus is designed as a layered structure in this paper. In order to reduce the influence of parasitic inductance on the SiC MOSFET dynamic current measurement, the current flow path is shortened as much as possible on the premise of matching the measurement size of 100 MHz high-frequency current probe and Rogowski coil (Fig. 4).

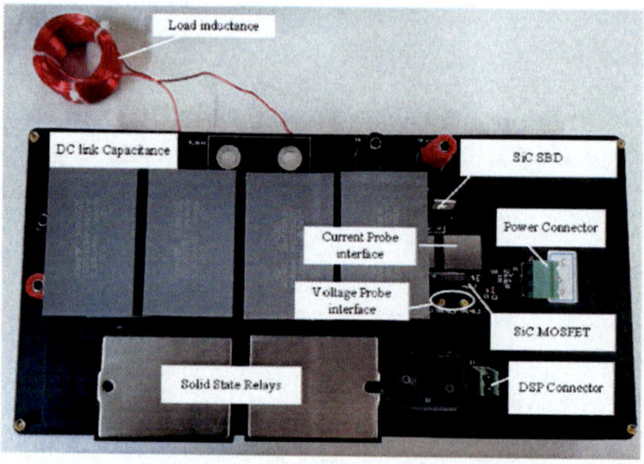

Fig. 4. The SiC MOSFET dynamic characteristic test platform.

5 The SiC MOSFET Current Measurement Method

For the SiC devices, sufficient bandwidth and appropriate mechanical size are the key factors affecting the accuracy of pulse current measurement. On the one hand, because the SiC MOSFET switching speed is fast, the current change rate in the transient process is large, the equivalent frequency of the transient current is also relatively large, so the bandwidth of the current measurement equipment is correspondingly high; On the other hand, due to the parasitic inductance of the test platform, the mechanical size also has an effect on the measurement results. There are four methods widely used in the field of current measurement: coaxial resistance, Rogowski coil, Pearson coil, and clamp current probe.

5.1 Coaxial Resistance

Coaxial resistance has high bandwidth, with a maximum bandwidth range of several hundred MHz. But the coaxial resistance is actually a low-sensing precision resistor, the measurement must be connected in series to the current loop, so it does not have the function of electrical isolation, easy to interfere with the switching transient characteristics of the SiC MOSFET. Therefore, this method is generally not used in the actual measurement process of the SiC MOSFET current (Fig. 5).

Fig. 5. Coaxial resistance.

5.2 Clamp Current Probe

Hall principle of clamp current probe, the highest measurement bandwidth of 100 MHz, is a relatively ideal measurement mode, can achieve high reduction degree of the SiC MOSFET drain current. However, before use, it is necessary to introduce a certain length of extra loop on the double-pulse experimental platform for measurement, which will produce large parasitic inductance in the power loop. This method is not conducive to the measurement of switching transient characteristics of the SiC MOSFET (Figs. 6 and 7).

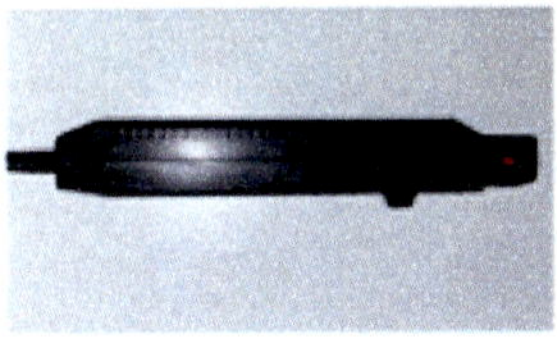

Fig. 6. Clamp Current Probe.

5.3 Pearson Coil

Pearson coils also have higher bandwidth, but they have a larger mechanical size than the clamp current probe, which is not conducive to installation and use. Some literatures

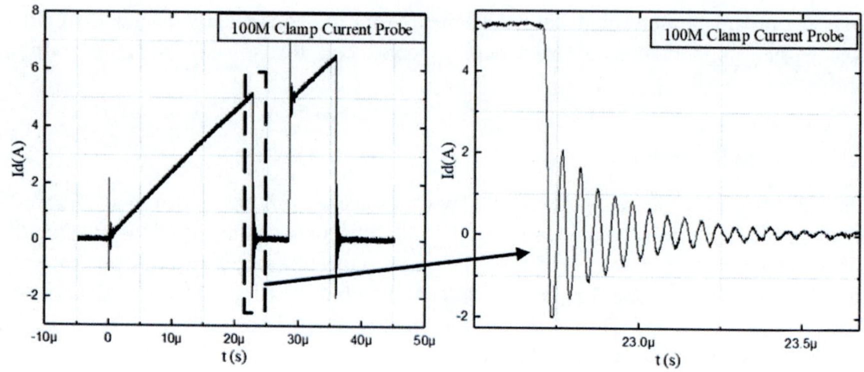

Fig. 7. Turn-off drain current measurement results (clamp current probe).

suggest that a primary current transformer composed of ferrite magnetic ring with a ratio of 1:5 can be added to solve the problem of mechanical size. However, this method will increase the parasitic inductance of the measurement loop of the coil, thus reducing the measurement bandwidth (Figs. 8 and 9).

Fig. 8. Pearson Coil.

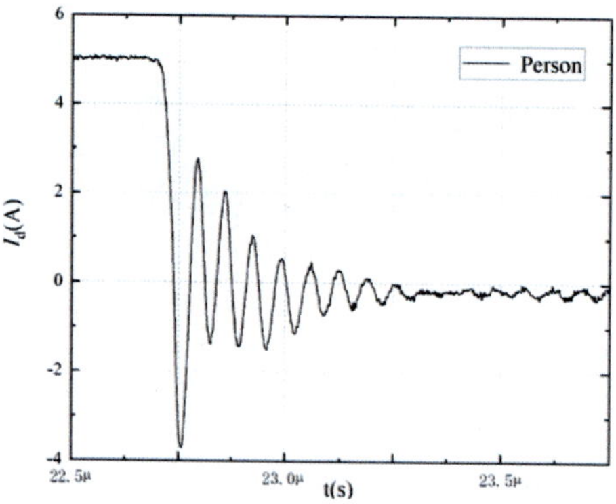

Fig. 9. Turn-off drain current measurement results (Pearson coil).

5.4 Rogowski Coil

As a main current measurement method, Rogowski coil has great advantages in low intrusion and electrical isolation. The measuring end is an extremely thin coil that can be opened or closed for easy installation and measurement. But the current commercial Rogowski coil maximum bandwidth can only reach 30 MHz (Figs. 10 and 11).

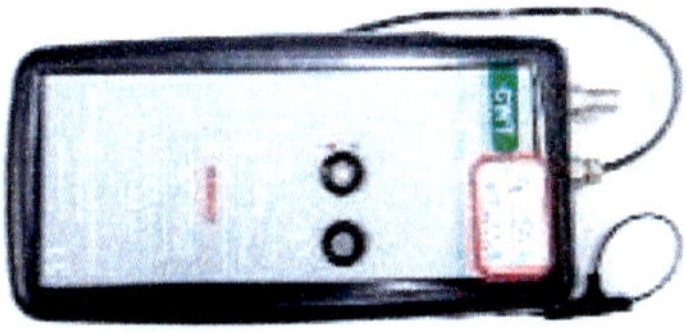

Fig. 10. Rogowski coil.

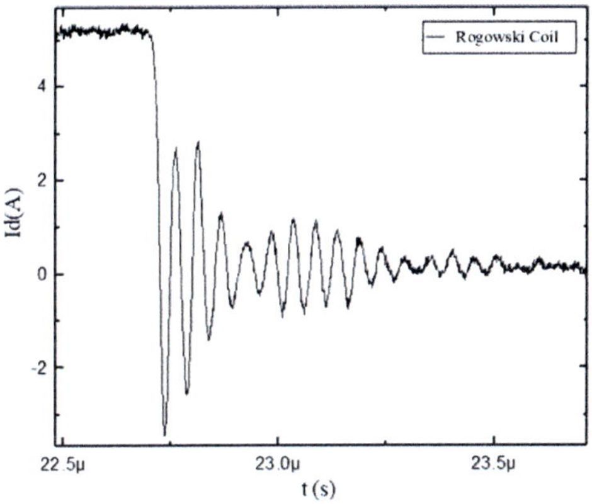

Fig. 11. Turn-off drain current measurement results (Rogowski coil).

5.5 Summary

Pearson coil, clamp current probe and Rogowski coil are commonly used in high frequency current signal measurement. Pearson coil is mechanically large and suitable for large installation space in large power electronic equipment. The clamp current probe is stable in measurement and does not require as much space as Pearson coil, but it is often used for laboratory measurement.

As the main measurement method, Rogowski coil has good universality, and the problem of insufficient bandwidth can be solved by using many methods. Through the active integrator at the output end of the Rogowski coil, the distortion of the measured pulse current waveform can be eliminated to correct the output voltage value, and the accuracy of current measurement at high frequency can be optimized [9]. Literature [10, 11] use the ratio of the peak value of the output and input pulse of the Rogowski coil to obtain the constant coefficient used to calibrate the output voltage to correct the peak value of the measured signal. The figure below shows the measurement results of the Rogowski coil after correction and compensation. It can be seen that the instantaneous current overshot measured by the Rogowski coil after compensation is smaller, but there are also problems of overcompensation and noise compensation, which need further study (Fig. 12).

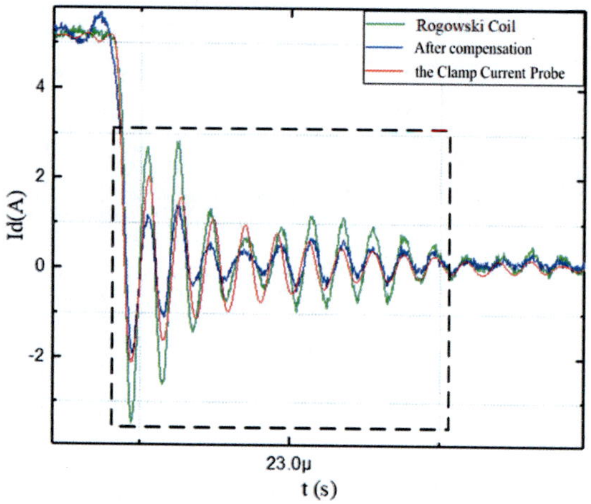

Fig. 12. Turn-off drain current measurement results (using compensated Rogowski coils).

6 Conclusion

In this paper, the drain current of the SiC MOSFET is analyzed in stages, and the double pulse test platform is built. The characteristics of coaxial resistance, Pearson coil, clamp current probe and Rogowski coil in high frequency current signal measurement are described from the Angle of bandwidth range and mechanical size, and the application of different current measurement methods are analyzed. In order to solve the problem of insufficient bandwidth of Rogowski coil, it is proposed that measurement correction can be used to compensate, but some additional problems of noise compensation need further study.

References

1. Linder, S., De, L., Xi, X.: Power Semiconductors—Devices and Applications. China Machine Press, Beijing (2009). (in Chinese)
2. Zhaoming, Q.: Current status and development of power electronic devices and their applications. Inverter World J. **34**(009), 34–37 (2014). (in Chinese)
3. Shidong, C.: High Voltage and High Current SiC MOSFET Series Parallel Module. Zhejiang University, Hangzhou (2014). (in Chinese)
4. Kuang, S.: Application prospect of SiC power electronic devices in power system. Proc. CSEE, J. **32**(30), 1–7 (2012). (in Chinese)
5. Matsunami H.: Current SiC technology for power electronic devices beyond Si. Microelectron. Eng. J. **83**(1), 2–4 (2006)
6. Xuemei, W.: Research and application of wide band gap SiC power devices in electric vehicles. Proc. CSEE J. **34**(3), 371–379 (2014). (in Chinese)
7. Chengzhan, L.: Research progress of SiC devices for rail transit. Electron. Packag. J. **22**(06), 17–30 (2022). (in Chinese)
8. Song, B.: Development progress of high voltage and high power SiC power electronic devices. Sci. Technol. Guide J. **39**(14), 56–62 (2021). (in Chinese)
9. Chen, Q.: Design and characteristics of two Rogowski coils based on printed circuit board. IEEE Trans. Instrum. Meas. J. **55**(3), 939–943 (2006)
10. Shi Y.: A review of traditional helical to recent miniaturized printed circuit board Rogowski coils for power-electronic applications. IEEE Trans. Power Electron. J. **35**(11), 12207–12222 (2020)
11. Lei, M.: Structure and modelling of four-layer screen returned PCB Rogowski coil with very few turns for high-bandwidth SiC current measurement. IET Power Electron. J. **13**(4), 765–775 (2020)

Bandwidth Compensation Method for Rogowski Coil for the SIC MOSFET

Shuai Liang(✉), Huiqing Du, Yuming Liu, and Xiran Sun

CSSC Systems Engineering Research Institute, Beijing 100094, China
falseliangshuaigz@126.com

Abstract. Compared with Si based devices of the same grade, the SiC MOSFET has a higher switching frequency, and its speed can reach ns level in the switching transient process. Therefore, current measuring devices with higher bandwidth should be selected for its transient current measurement. Currently, the bandwidth of commercial Rogowski coil is only 30 MHz, which cannot meet the switching frequency of the SiC MOSFET, resulting in inaccurate measurement results. In order to solve the problem of accurate measurement of transient high equivalent bandwidth current in the SiC MOSFET switching, a high frequency bandwidth compensation method for low bandwidth Rogowski coils is proposed in this paper. Based on the analysis of deconvolution theory, the error correction curve of Rogowski coil is obtained on the experimental platform of measurement reference system, and then the high frequency compensation is realized for the bandwidth of Rogowski coil. Finally, the experimental verification is completed on the double-pulse test platform. Experimental results show that the method proposed in this paper can effectively improve the measurement accuracy and the available frequency range of Rogowski coil.

Keywords: The SiC MOSFET · Rogowski Coil · Bandwidth Error Correction Curve

1 Introduction

In power electronic system, the current information of power semiconductor devices is very important in the research of switch characteristics, short circuit protection and condition monitoring [1, 2]. The faster switching speed of the SiC MOSFET results in higher requirement on the bandwidth range of current sensor [3, 4]. As a mainstream current measuring device, Rogowski coil has the advantages of flexibility, low cost, and good transient response ability and safety [5]. However, at present, the highest bandwidth of commercial Rogowski coil is about 30 MHz, and for the SiC devices, the current rise time can reach 20 ns or even less than 10 ns [6]. Therefore, the current commercial Rogowski coil cannot meet the bandwidth requirements of the SiC devices for current testing devices, so that accurate current measurement values cannot be obtained [7, 8].

In view of the insufficient bandwidth of Rogowski coil in the SiC MOSFET transient current testing, some studies have been carried out. Literature [9] proposes and verifies

L. Jia et al. (Eds.): EITRT 2023, LNEE 1135, pp. 188–197, 2024.
https://doi.org/10.1007/978-981-99-9307-9_21

the reasons for the influence of high frequency on the measurement current of Rogowski coil by establishing the transmission line model of Rogowski coil. It shows that with the increase of frequency, coil transmission time is delayed, thus leading to inaccurate measurement results of Rogowski coil high frequency. Literature [10] established the lumped parameter model of Rogowski coil and concluded that the bandwidth of Rogowski coil could be increased by reducing the inductance and capacitance of the coil. Moreover, it is verified through experiments that reducing the lumped inductance and capacitance of Rogowski coil could optimize the high-frequency measurement results of Rogowski coil. Literature [11] studies the effect of transmission delay of Rogowski coil caused by parasitic parameters of the probe on accurate measurement of the SiC MOSFET current. Literature [6] proposes a current probe with smaller parasitic parameters and higher bandwidth through novel structural optimization design.

In this paper, the bandwidth error compensation method for low bandwidth Rogowski coils is proposed, considering the effect of noise on the compensation results, to solve the problem of insufficient bandwidth for measuring the the SiC MOSFET switching current. Then, the error correction curve of Rogowski coil is obtained by setting up the experimental platform of measurement reference system. Finally, a double pulse test platform is built and the correctness of the proposed compensation method is verified.

2 Deconvolution Theory Analysis

2.1 Frequency Domain Representation of Deconvolution

The original signal value can not be obtained after the signal is transmitted by the measuring system or sensor. In mathematics, a convolution integral can be used to describe this change.

$$y(t) = \int_{-\infty}^{\infty} h(t - \tau)x(\tau)d\tau + \xi(t) \tag{1}$$

In the formula, $x(\tau)$is the original signal, $y(t)$ is the signal obtained by the measuring system, $h(t)$ is the unit impulse response of the measuring system, $\xi(t)$ represents noise. We really want to get the original signal, that is, without losing the true physical quantity $x(\tau)$. Deconvolution is a common technique to estimate the original signal $x(\tau)$ based on the output $y(t)$ and the statistical properties of the noise obtained from the measurement system. Without considering noise interference, $x(\tau)$ can be obtained by knowing the unit impulse response $h(t)$ and output $y(t)$ of the system.

The frequency domain expression of deconvolution can be obtained by Fourier transform of formula (1).

$$X(f) = \frac{Y(f)}{H(f)} \tag{2}$$

In the deconvolution process, $H(f)$ is not usually measured directly. It is usually necessary to determine the specific input according to reasonable assumptions, and on this premise to measure the output signal as a reference, to obtain the system transfer function *H(f)*. Firstly, the transfer function is calculated according to the reference signal:

$$H_1(f) = \frac{Y_1(f)}{X_1(f)} \tag{3}$$

where, $X_1(f)$ and $Y_1(f)$ are the representation of reference input signal and output signal respectively in the frequency domain. The original input signal can then be calculated from the actual output signal:

$$X_2(f) = Y_2(f) / H_1(f) \tag{4}$$

where, $X_2(f)$ is the original input signal, and $Y_2(f)$ is the actual output signal.

2.2 Pathological Solutions to Deconvolution

Deconvolution computation is generally ill-conditioned. In practical problems, the bandwidth of input signal *X(f)* is limited, and its amplitude in one or more frequency regions is small or even zero. Therefore, the deconvolution errors in these regions will be amplified during the *H(f)* calculation, resulting in a series of spikes similar to δ function. These spikes have large amplitudes and may become the main components of the time domain signal after Fourier inversion, thus masking the real input signal and causing the failure of deconvolution in the frequency domain. In order to obtain the best approximate solution, *H(f)* needs to be modified to reduce the spikes in the frequency component, so that the inverse Fourier transform result is close to the real signal. The concrete implementation is generally to add a filter *F(f)* to *H(f)* to obtain a new transfer function $H_2(f)$.

$$H_2(f) = H_1(f) \times F(f) \tag{5}$$

$$H_2(f) = Y_1(f) / [X_1(f) / F(f)] \tag{6}$$

Due to the insufficient bandwidth when Rogowski coil measures the dynamic current of the SiC MOSFET, the measurement value of the SiC MOSFET dynamic current has bandwidth error. According to the principle of deconvolution, by calculating the transfer function of the current measurement system, the current measurement value bandwidth error can be corrected. At the same time, due to the general morbidity of deconvolution calculation, the dynamic current of the SiC MOSFET is mainly composed of effective current signals and noise, and noise is the cause of the failure of deconvolution in frequency domain. Therefore, in the deconvolution calculation of the dynamic current of the SiC MOSFET, it is necessary to de-noise the actual input dynamic current of the SiC MOSFET. Only effective current signals are extracted for deconvolution calculation.

3 Bandwidth Error Correction Curve

Before compensating the SiC MOSFET dynamic current bandwidth error by using the back-winding principle, it is necessary to obtain the transfer function *H(f)* of the measurement system through the reference system, also known as the bandwidth error correction curve.

$$H(f) = I_m(f) / I_0(f) \tag{7}$$

where $I_m(f)$ is the frequency domain value of the reference system output current $i_m(t)$.

$$I_0(f) = U_0(f)/Z_A(f) \tag{8}$$

where $I_0(f)$ is the frequency domain value of the reference system input current $i_0(t)$, $U_0(f)$ is the frequency domain value of the reference system input current $u_0(t)$, $Z_A(f)$ is the frequency domain impedance characteristic of the reference system. The reference system uses a voltage probe with a bandwidth greater than 100MHz, and it is approximately considered that the output voltage value of the system is equal to the input voltage value, as follows:

$$I_0(f) = U_m(f)/Z_A(f) \tag{9}$$

Therefore, the transfer function of the measurement system can be expressed as:

$$H_1(f) = \frac{I_m(f)}{U_m(f)/Z_A(f)} = \frac{U_m(f)/I_m(f)}{Z_A(f)} \tag{10}$$

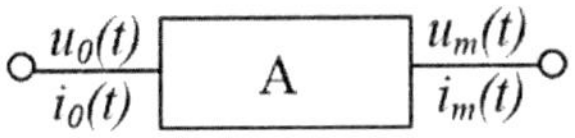

Fig. 1. Measure reference system input and output.

4 Bandwidth Compensation Method for Rogowski Coil

The measured current $i(t)$ varying with time t can be obtained by using the Rogowski coil. Then the measured current can be compensated in the frequency domain by combining with the bandwidth error correction curve that characterizes the measurement accuracy of the current measurement reference system at different frequencies, so as to obtain the real current value. In the process of testing the SiC MOSFET switch characteristics, the measured current signal is aperiodic signal $i(t)$. By de-noising this aperiodic signal after Fourier decomposition, the test current $I(f)$ in the frequency domain can be obtained. Combined with the bandwidth error correction curve, the bandwidth error correction of the SiC MOSFET dynamic characteristic current is carried out. Finally, the time-domain current waveform after bandwidth error correction is obtained by Fourier inverse transformation.

5 Experimental Verification

5.1 Acquisition of Bandwidth Error Correction Curve

As shown in the figure below, the experimental platform of measurement reference system based on high-frequency signal generator is built in this paper. Using the PCB circuit board, the signal generator is connected to the 10 Ω load resistor, and the frequency of the

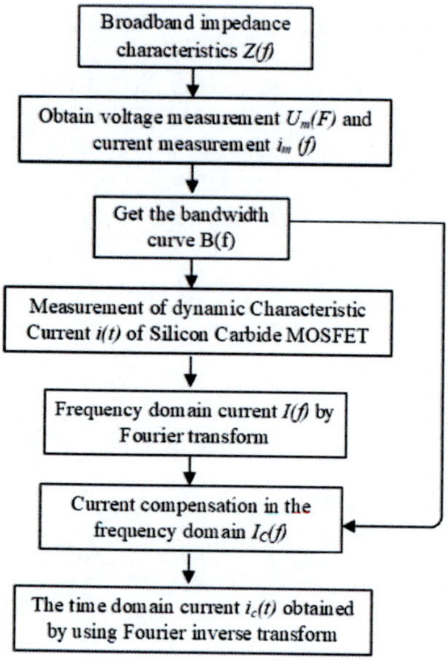

Fig. 2. Block diagram of bandwidth compensation for Rogowski coil.

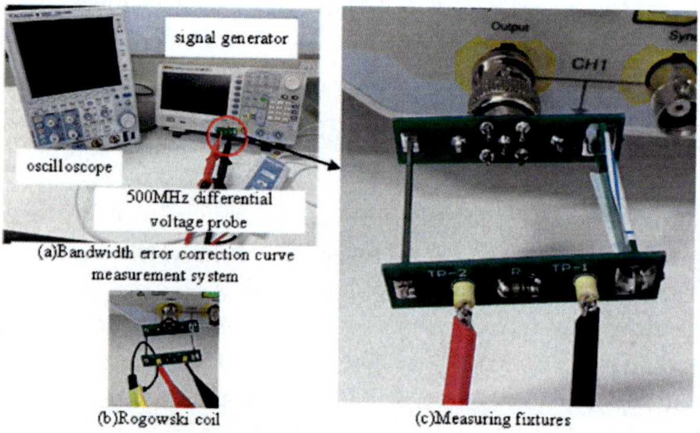

Fig. 3. The experimental platform of measurement reference system.

measured voltage and current is adjusted by controlling the output voltage frequency of the signal generator. In the figure, the signal generator can send out a sinusoidal voltage signal of up to 100 MHz, and the 10 Ω resistor is the load of the output voltage of the signal generator. The voltage $u_m(t)$ at both ends of the load resistor is measured by the 500 MHz differential probe, and the load current $i_m(t)$ flowing through the 10 Ω resistor is measured by the Rogowski coil.

Because of the sensitivity of Rogowski coil itself, easy to be affected by external environment such as electromagnetic interference, vibration caused by the change of position and other factors, resulting in distortion of the output inductance parameters, sensitivity and accuracy are greatly reduced. Therefore, in order to ensure the stability and accuracy of the measurement of the Rogowski coil, the use of PCB circuit board fixed circuit, excluding the influence of the external environment on the sensitivity and accuracy of the Rogowski coil, in addition to the PCB circuit board can ensure that the parasitic parameters of the loop will not change, so as to ensure the stability of the electrical parameters measured at both ends of the test loop.

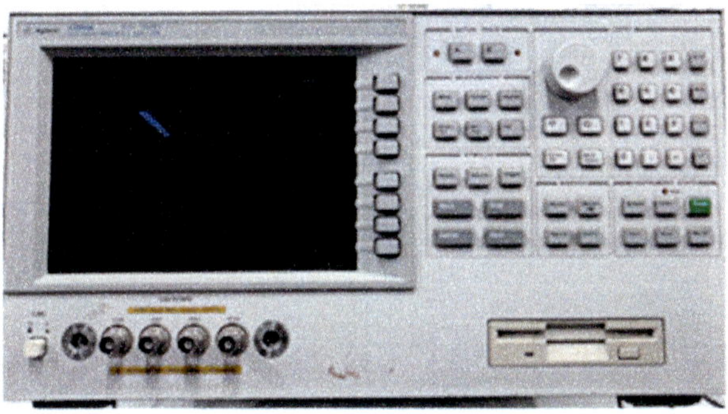

Fig. 4. Impedance analyzer.

At the frequency of tens or even hundreds of MHz, the high frequency characteristics of the load resistance can not be ignored, and it can not be regarded as a pure resistance. Figure 4 shows the impedance analyzer of Model Agilent 4249a. The broadband impedance characteristic *Z(f)* of a 10 Ω resistance PCB board can be obtained through the impedance analyzer test.

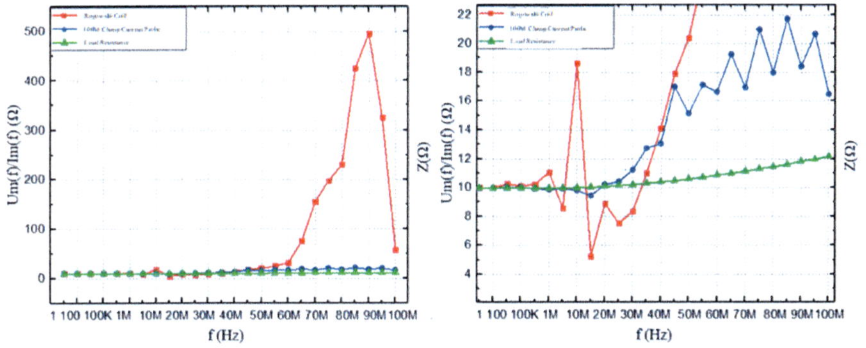

Fig. 5. Wide-band impedance characteristics of Rogowski coil and load resistor.

According to formula (10), the bandwidth error correction curve of Rogowski coil can be obtained, as shown in Fig. 6. As can be seen from the figure, the measurement value of Rogowski coil has a large deviation when it is above 10M, and the measurement error of current when it is above 40M exceeds plus or minus 3dB.

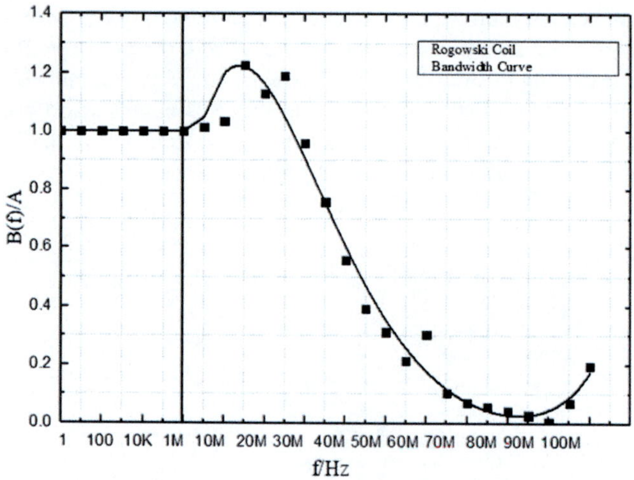

Fig. 6. Rogowski coil bandwidth error correction curve.

5.2 Double Pulse Test Results and Analysis

The SiC MOSFET dynamic characteristics test platform is built and a 1200 V silicon carbide MOSFET is selected for double pulse test. The test platform, as shown in the figure below, is mainly composed of the SiC SBD, load inductance, DC bus capacitor and drive module. The load inductance is 0.2 mH hollow inductance, and the DC bus capacitor is 900 V/50 uF thin film capacitor.

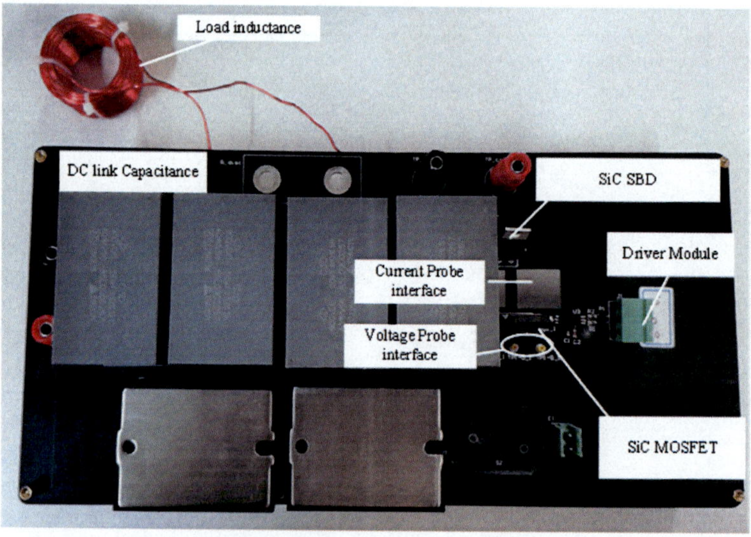

Fig. 7. Double pulse test platform.

The current data measured by Rogowski coil is decomposed by fast Fourier transform to obtain the current waveform in the frequency domain. After filtering, the current waveform is compensated in the frequency domain and Fourier inverse transform. Since the test current includes two parts: effective information and noise, when compensating the test current by using the error correction curve, the measured value of the current needs to be processed in sections to determine the frequency band where the effective information is located and compensate it. The current waveform after compensation is shown in Fig. 8.

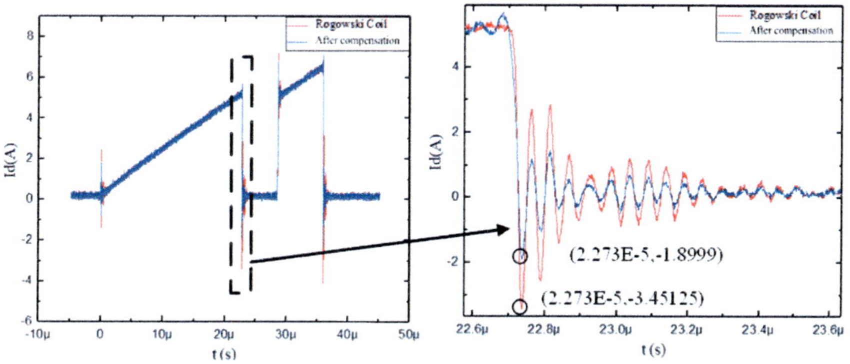

Fig. 8. Measurement results of Rogowski coil before and after compensation.

As can be seen from the figure, after compensation, the measurement current switch oscillation of Rogowski coil decreases significantly. The bandwidth compensation method can reduce the measurement error of Rogowski coil and make the switching oscillation closer to the real current value. In order to realize comparative verification, the bandwidth compensation method is applied to the 100MHz clamp current probe, and the waveform is almost unchanged, which proves that the compensation method can realize the realization of increasing the bandwidth range without introducing other additional errors, and proves the effectiveness of the method.

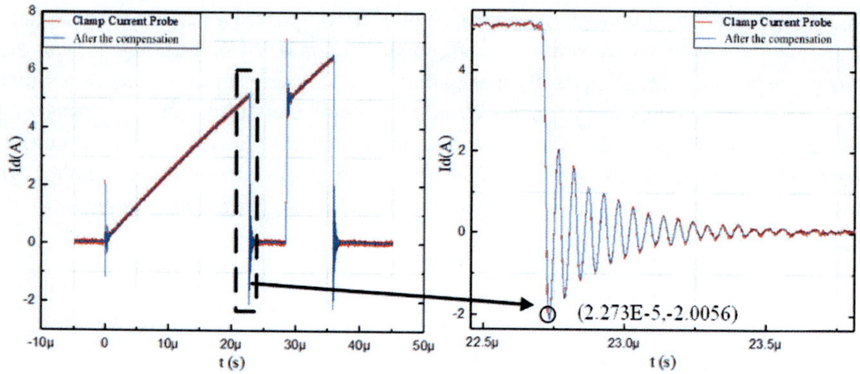

Fig. 9. Measurement results of clamp current probe before and after compensation.

6 Conclusion

In this paper, a bandwidth error compensation method for low bandwidth Rogowski coil is proposed, considering the effect of noise on the compensation results, to solve the problem of insufficient bandwidth for measuring the SiC MOSFET switching current. The bandwidth error correction curve required by this method can be obtained on the experimental platform of the measurement reference system, which is composed of a signal generator and a self-made fixture. Experimental results show that the measurement current switch oscillation of Rogowski coil is reduced by about 50% after compensation, and this compensation method can effectively expand the available frequency range of Rogowski coil.

References

1. Xuemei, W.: Reliability and stability of SiC power mosfets and next-generation. Proceedings of the CSEE, J. 34(3):371–379 (2014). (in Chinese)
2. Nawaz, M.: Replacing Si to SiC: Opportunities and challenges. European Solid-State Device Research Conference, pp. 472–475 (2016)
3. Hull, B.: Reliability and stability of SiC power mosfets and next-generation SiC MOSFETs. Wide Bandgap Power Devices and Applications, pp. 139–142 (2014)
4. She, X.: Review of silicon carbide power devices and their applications. IEEE Transactions on Industrial Electronics, Journal **64**(10), 8193–8205 (2017)
5. Jalebi, E.: High-performance low-cost Rogowski transducers and accompanying circuitry. IEEE Transactions on Instrumentation and Measurement, Journal **59**(2), 753–759 (2007)
6. Ray, W.: High performance Rogowski current transducers. Proceedings of IEEE International Conference, Journal **5**, 3083–3090 (2000)
7. Chengzhan, L.: Research progress of SiC devices for rail transit. Electronics and packaging, Journal **22**(06), 17–30 (2022). (in Chinese)
8. Song, B.: Development progress of high voltage and high power SiC power electronic devices. Science and technology guide, Journal **39**(14), 56–62 (2021). (in Chinese)
9. Shi, Y.: A review of traditional helical to recent miniaturized printed circuit board Rogowski coils for power-electronic applications. IEEE Transactions on Power Electronics, Journal **35**(11), 12207–12222 (2020)

10. Lei, M.: Structure and modelling of four-layer screen returned PCB Rogowski coil with very few turns for high-bandwidth SiC current measurement. IET Power Electronics, Journal **13**(4), 765–775 (2020)
11. Nibir, S.: Magneto resistor with planar magnetic concentrator as wideband contactless current sensor for power electronics applications. IEEE Transactions on Industrial Electronics, Journal **65**(3), 2766–2774 (2018)

An Interleaved Control Strategy for Input-Series Output-Parallel System with Three-Level Boost Converter+Full-Bridge LLC Converter

Ruiqi Ma[1,2], Xinbo Liu[1,2], Shuiyuan He[1,2], Jiepin Zhang[3], Yingtao Ma[3], Zheming Jin[1,2], and Lijun Diao[1,2](✉)

[1] Beijing Jiaotong University, Beijing, China
ljdiao@bjtu.edu.cn

[2] Key Lab. of Vehicular Multi-Energy Drive Systems (VMEDS), Ministry of Education, Beijing, China

[3] Locomotive and Car Research Institute, China Academy of Railway Sciences Corporation Ltd, Beijing, China

Abstract. LLC resonant converters are known for their advantages such as soft switching technology, high power density, and high efficiency. In order to achieve wide gain range under wide input voltage, reduce the voltage stress on the switching devices, and ensure that the output voltage is stable, this paper proposes an input-series and output-parallel (ISOP) system composed of a three-level boost (TLB) converter and a full-bridge LLC (FB-LLC) resonant converter in cascade. The three-level boost converter provides high voltage conversion capability for output voltage regulation. The LLC resonant converter achieves ZVS (Zero Voltage Switching) and ZCS (Zero Current Switching) within the full load range, then reducing switching losses and improving system efficiency. In order to improve the dynamic response of the system, this paper adopts an interleaved control strategy. By employing dual-loop pulse phase-shift control in the three-level boost converter and switch interleaving control in the LLC resonant converter, the output voltage ripple can be reduced, and the stability and response speed of the system can be enhanced. Finally, the feasibility of the ISOP system under interleaved control is verified through Simulink simulations.

Keywords: ISOP System · Three-Level Boost Converter · LLC Resonant Converter · Interleaved Control

1 Introduction

With the progress of power electronics topology and the emergence of new requirements, wide-range input/output power converters are gradually increasing [1]. Input series output parallel dc-dc conversion systems are widely used in high input voltage and high output current applications. Series-parallel power conversion system has become a hot topic of power electronics system integration [2]. LLC converters are widely used in data centers, aerospace and distributed power systems [3]. However, there are still some

L. Jia et al. (Eds.): EITRT 2023, LNEE 1135, pp. 198–206, 2024.
https://doi.org/10.1007/978-981-99-9307-9_22

problems to be solved in practical application, such as light load regulation voltage problem and dynamic response problem [4]. In order to reduce the high voltage stress of power semiconductors and overcome limitations of LLC, three-level boost DC/DC converters have been proposed [5]. The voltage stress of the switches is half of the output voltage, and the size of the filter inductance is reduced, as well as the power capacity of the system is increased [6]. For LLC, PSM, PFM, PWM and other control strategies are usually adopted. For TLB, a double closed-loop based on output capacitors is usually adopted [7].

In this paper, an ISOP topology based on three-level boost converter and full-bridge LLC resonant converter is proposed. This paper adopts a Boost dual-loop switch interleaving control strategy to stabilize the input voltage of the LLC resonant converter. Additionally, pulse phase-shift control is employed in the upper and lower full-bridge LLC resonant converters to enhance the system's efficiency and reliability.

2 Operation Principle

2.1 Proposed ISOP Topology

Figure 1 illustrates the isolated DC/DC topology consisting of a three-level boost converter and a full-bridge LLC resonant converter, forming the ISOP system with input series and output parallel configuration. The three-level boost circuit can regulate output voltage, which can achieve a wide range of input voltage [8]. Both modules are composed of a three-level boost converter and a full-bridge LLC resonant converter. To improve system efficiency, the switching frequency and duty-ratio of the FB-LLC resonant converter are fixed. In this configuration, the FB-LLC resonant converter primarily serves the purpose of isolation and transformation, while the output voltage is primarily regulated by the preceding three-level boost converter. In this paper, the two three-level boost converters adopt interleaved control to reduce the input inductance of the Boost converter and enhance system stability and response speed.

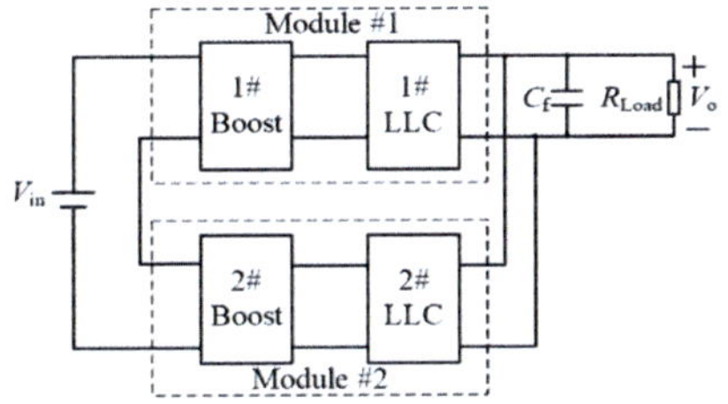

Fig. 1. The proposed ISOP topology consisting of a three-level boost converter and a full-bridge LLC resonant converter.

2.2 Three-Level Boost Working Principle

Figure 2 shows the topology of a TLB converter. In this topology, V_{in} represents the input voltage, L_b is the boost inductor, Q_1 and Q_2 are the switching devices, D_1 and D_2

are diodes, C_{b1} and C_{b2} are the output filter capacitors, R is the load, V_1 and V_2 are the voltages across the filter capacitors, and V_o is the output voltage. The purpose of this topology is to control the output voltage (V_o) to reach the desired set value and maintain the balanced voltage potential at the midpoint of capacitors C_{b1} and C_{b2}.

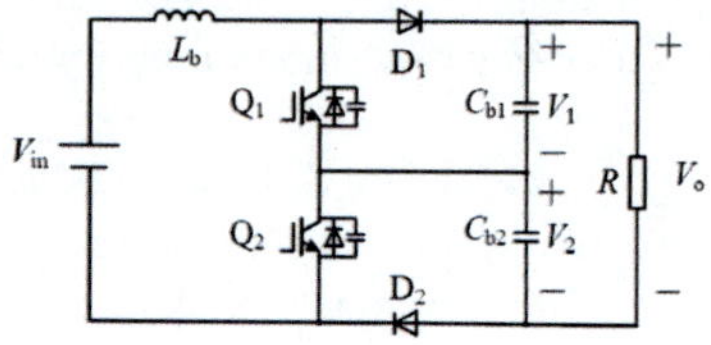

Fig. 2. Schematic of three-level boost converter.

When the voltage at the midpoint of the capacitors is balanced, $V_1 = V_2 = 0.5V_o$. Based on the relationship between V_{in} and V_o, the duty-ratio D of the three-level boost converter can be determined in two different modes [9].

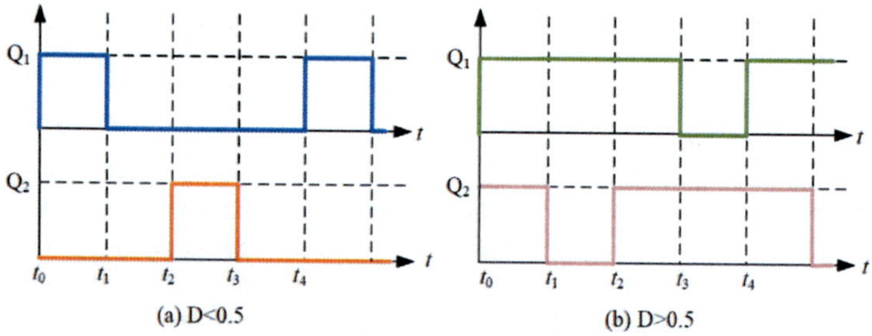

Fig. 3. Key switch operating waveforms of TLB.

(1) $V_{in} > 0.5V_o$

In this case, the duty-ratio is D < 0.5, and the switch working waveform is as shown in the Fig. 3a. The circuit operating status diagram is illustrated in Fig. 4.

- Mode 1 (t_0, t_1) corresponding to Fig. 3a
  During this mode, Q_1 is turned on while Q_2 is turned off. The voltage across the boost inductor is V_{in}-0.5V_o. At this time, the current through the boost inductor linearly increases, and its increment is given by:

$$\Delta I_{L_b} = \frac{V_{in} - 0.5V_o}{L_b}(t_1 - t_0) = \frac{V_{in} - 0.5V_o}{L_b}DT_{sb} \tag{1}$$

- Mode 2 (t_1, t_2) corresponding to Fig. (b)
  During this mode, both Q_1 and Q_2 are turned off. The voltage across the boost inductor is V_{in}-V_o. At this time, the current through the boost inductor linearly

decreases, and its decrement is given by:

$$\Delta I_{L_b} = \frac{V_{in} - V_o}{L_b}(t_2 - t_1) = \frac{V_{in} - V_o}{L_b}\left(\frac{T_{sb}}{2} - DT_{sb}\right) = \frac{V_{in} - V_o}{L_b}T_{sb}\left(\frac{1}{2} - D\right) \tag{2}$$

- Mode 3 (t_2, t_3) corresponding to Fig. (c).

  During this mode, Q_1 is turned off while Q_2 is turned on. The voltage across the boost inductor is V_{in}–$0.5V_o$. At this time, the current through the boost inductor linearly increases, similar to Mode 1.
- Mode 4 (t_3, t_4) corresponding to Fig. (d)

During this mode, both Q_1 and Q_2 are turned off, similar to Mode 2.

When the circuit reaches steady-state, the sum of Eqs. (1) and (2) is equal to 0, yielding:

$$\frac{V_o}{V_{in}} = \frac{1}{1 - D} \tag{3}$$

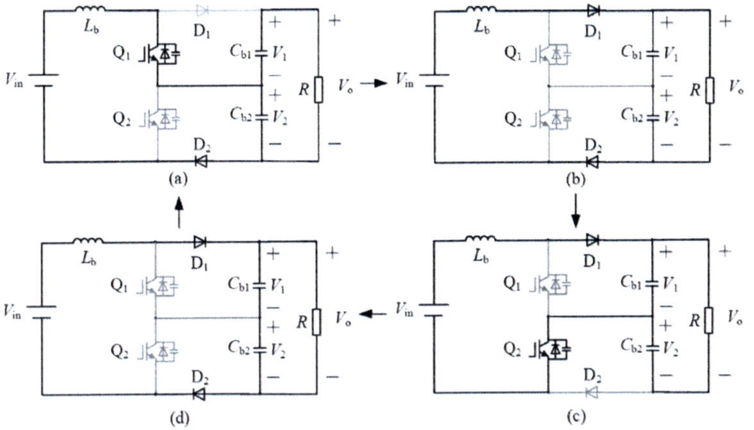

Fig. 4. TLB operations in different switching steps ($V_{in} > 0.5V_o$).

(2) $V_{in} < 0.5V_o$

In this case, the duty-ratio is D > 0.5 and the switch working waveform is as shown in the Fig. 3b. However, compared with Fig. 4, the working condition in which Q_1 and Q_2 are turned off at the same time will not appear. The specific derivation process of the working condition in which Q_1 and Q_2 are added at the same time on is similar to the above, and the final input-output relationship still satisfies Eq. (3).

2.3 Full-Bridge LLC Working Principle

The primary side of the FB-LLC converter consists of four power switches, Q_1 to Q_4, each including a body diode and a junction capacitor. It also includes the resonant inductor L_r, resonant capacitor C_r, and excitation inductor L_m, as shown in the Fig. 5. The secondary side consists of a rectifier bridge, output filter capacitor C_f, and load resistor *R*. The operation of the primary switches is as follows: Q_1 and Q_4 are controlled to switch on and off simultaneously, while Q_2 and Q_3 switch on and off together. Q_1 and Q_3 are driven with a 50% duty-ratio signal, and a certain dead time is set to avoid simultaneous conduction of the bridge arms and prevent damage to the switches. FHA (first harmonic approximation analysis) is commonly utilized for circuit analysis [10].

The topology has two different switching frequencies. When L_r and C_r are in resonance, the resonant frequency is f_r. When L_r, C_r, and L_m are in resonance, the resonant frequency is f_m. Based on the relationship between the switching frequency f_s and the resonant frequencies, there are three operating modes.

- Mode 1 ($f_s < f_m$)

In this mode, the converter operates in the capacitive region. The current flowing through the excitation inductor reverses before the switch turns off, preventing ZVS from being achieved.

- Mode 2 ($f_s > f_m$)

In this mode, the converter operates in the inductive region. The primary side switches can achieve ZVS during turn-on.

- Mode 3 ($f < f_s < f_m$)

In this mode, the secondary side diodes can achieve ZCS during their turn-off.

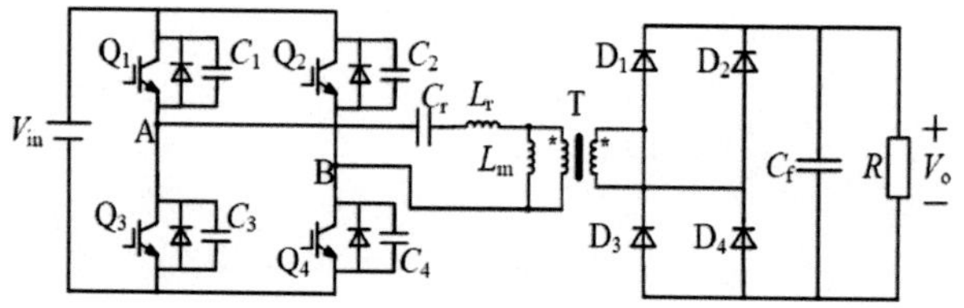

Fig. 5. Basic circuit diagram of the full-bridge LLC converter.

3 Operation Principle of the Proposed Control Strategy

3.1 Principle of Three-Level Boost Control

According to Sect. 2.2, when there is a deviation in the duty-ratio of the two switches, V_1 and V_2 will have a mismatch, causing an increase in the voltage and current stress on certain circuit components. To achieve balance in the midpoint voltage, the voltage difference across the output capacitors is measured, and the duty-ratios of the two switches are adjusted accordingly. This is achieved through a dual-loop control system consisting of an outer voltage loop and an inner capacitor voltage balancing loop.

The control strategy employed in this paper involves the series connection of two three-level boost converters at the input side, as depicted in Fig. 6. The strategy utilizes 180° phase-shifted control for the switches in the same bridge arm and 90° phase-shifted control for the switches in the upper and lower bridge arms. The carrier waveforms are illustrated in Fig. 7a.

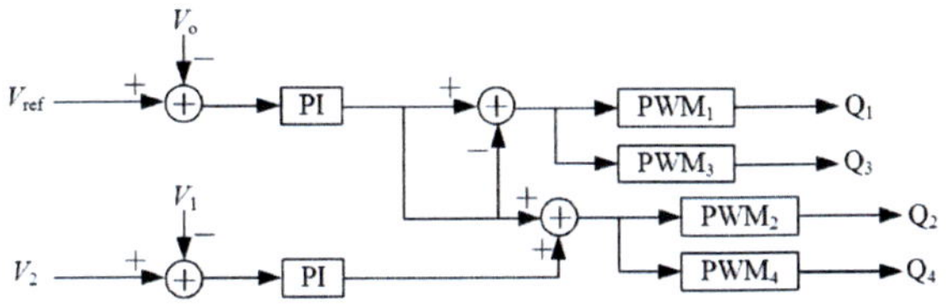

Fig. 6. The proposed three-level boost interleaving control strategy.

3.2 Principle of Full-Bridge Control

LLC resonant converter is operated as a dc transformer (DCX) for electrical isolation and voltage matching [11]. The parallel connection of two LLC converters at the output side. The direct duty-ratio control method is employed. During operation, both LLC converters receive identical drive signals, and the upper and lower switches in the same bridge arm conduct in a complementary manner. The control signals of the two LLC converters are phase-shifted by 90°, as shown in the Fig. 7b.

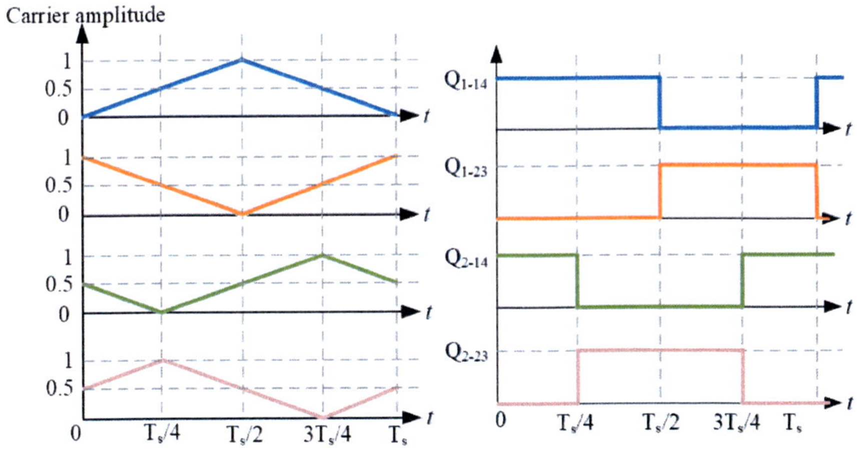

(a) The carrier waveforms of TLB. (b) The control signals of LLC.

Fig. 7. Switch control waveform.

4 Simulation Results of the Interleaved Control Strategy

Given input voltage 1800V, output voltage 700V, single TLB output voltage 1100V, power 300 kW.The simulation results are as follows:

- The control signals of the switches in the three-level boost converter are shown in Fig. 8a. The switches in the same bridge arm conduct in a complementary manner, and the two TLB converters operate with a 90° phase shift.
- The simulation results for the voltage balancing in the TLB converter in the Fig. 9a shows the voltage balancing results of the two output capacitors in the upper and lower Boost converters, and Fig. 9b shows the voltage balancing result of the output voltages of the two boost converters. Because of the Interleaved control, the voltage equalization waveforms fluctuate and overlap in a certain period.
- The control signals of the switches in the full-bridge LLC converter are shown in Fig. 8b. The switches in the same bridge arm conduct in a complementary manner, and the two LLC converters operate with a 90° phase shift.
- The input and output voltage results are shown in the Fig. 10. The input voltage is 1800V DC, and the output voltage is approximately 700V.

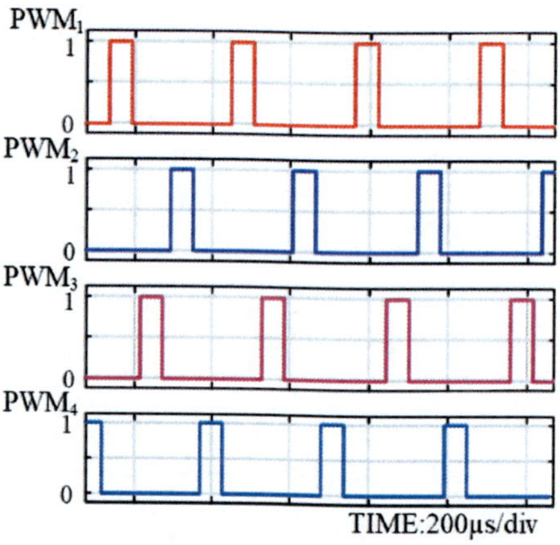

(a) The control signals of TLB.

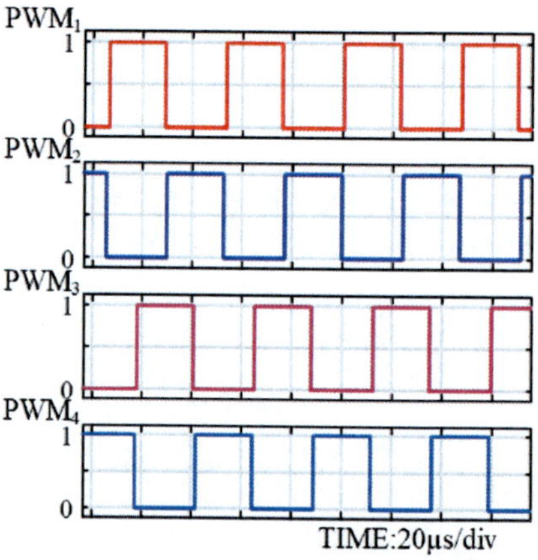

(b) The control signals of LLC.

Fig. 8. Simulation waveform of converter switches.

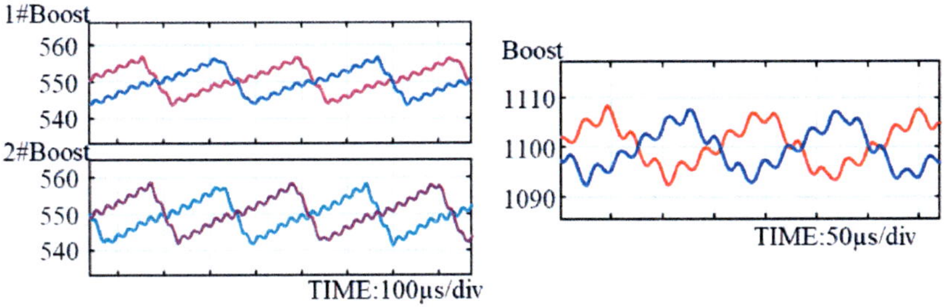

(a) The upper and lower boost converters. (b) Two boost converters.

Fig. 9. The simulation results for the voltage balancing in the three-level boost converter.

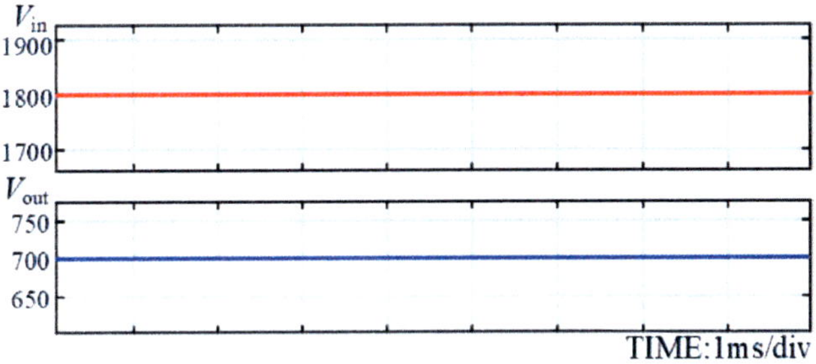

Fig. 10. The input and output voltage results.

5 Conclusion

This paper combines the principles of TLB and FB-LLC to form an ISOP system. TLB utilizes a dual-loop control based on voltage outer loop and capacitor voltage balancing. Two three-level boost converters are connected in series at the input, with the upper and lower switches of the same bridge controlled in a 180° staggered manner, and the two bridges controlled with a 90° phase difference. The two full-bridge LLC converters are connected in parallel at the output. The drive signals for the switches in a single LLC converter are consistent, and the upper and lower switches of the same bridge are complementary and staggered by 180°. The control signals for the two LLC converters are staggered by 90°. Simulation results confirm that the ISOP system achieves stable and fast response of the TLB converter output voltage, while the full-bridge LLC converter also achieves stable output voltage.

Acknowledgements. This work was supported in part by the Science and Technology Research and Development Plan of China National Railway Group Company Ltd., under Grant L2021J001.

References

1. Ma, D., Chen, W., Ruan, X.: A review of voltage/current sharing techniques for series—parallel-connected modular power conversion systems. IEEE Trans. Power Electron. **35**(11), 12383–12400 (2020)
2. Chen, W., Zhu, X., Wang, G., Jiang, W., Yao, L.: A novel input voltage sharing control strategy for input-series output-parallel system with high reliability. In: IEEE Energy Conversion Congress and Exposition (ECCE), pp. 1207–1212 (2014)
3. Wang, Y., Chen, C., Chen, B., Wang, Z., Ji, R., Zhang, M.: Transformer integration and winding design for ISOP-LLC converter. In: IEEE Energy Conversion Congress and Exposition (ECCE), pp. 1–5 (2022)
4. Huang, Y.C., Hsieh, Y.C., Lin, Y.C., Chiu, H.R., Lin, J.Y.: Study and implementation on start-up control of full-bridge LLC resonant converter. In: IEEE Transportation Electrification Conference and Expo, Asia-Pacific (ITEC Asia-Pacific), pp. 1–5 (2018)
5. Zhang Y., et al.: Wide input-voltage range boost three-level DC–DC converter with Quasi-Z source for fuel cell vehicles. IEEE Trans. Power Electron. **32**(9), 6728–6738 (2017)
6. Zhang, C., Xu, M., Wang, W., Meng, X., Kan, Z.: A high-gain floating-interleaved three-level boost converter. In: International Conference on Intelligent Green Building and Smart Grid (IGBSG), pp. 126–131 (2019)
7. Oulad-Abbou, D., et al.: Combined control of MPPT, output voltage regulation and capacitors voltage balance for three-level DC/DC boost converter in PV–EV charging stations. In: International Symposium on Power Electronics, Electrical Drives, Automation and Motion (SPEEDAM), pp. 372–376 (2018)
8. Xu, M., Wang, Z., Wang, N., Jiang, T., Chen, W., Ge, Y.: A new integrated topology design of auxiliary power supply for metro. In: IEEE 1st International Power Electronics and Application Symposium (PEAS), pp. 1–5 (2021)
9. Lee, M., Kim, J.W., Lai, J.S.: Derivation of DCM/CCM boundary and ideal duty-ratio feed-forward for three-level boost rectifier. In: IEEE Applied Power Electronics Conference and Exposition (APEC), pp. 1173–1179 (2019)
10. Israr, M., Samuel, P.: Study and design of DC-DC LLC full bridge converter for electric vehicle charging application. In: International Conference on Power, Control & Embedded Systems (ICPCES), pp. 1–6 (2023)
11. Liu, F., Ruan, X., Huang, X., Qiu, Y.: Second harmonic current reduction for cascaded inverter with pre-regulator + LLC converter as front-end DC–DC converter. In: IEEE Energy Conversion Congress and Exposition (ECCE), pp. 686–690 (2019)

Design and Simulation Analysis of Hybrid Pulse Width Modulation for Urban Rail Traction System

Jianyu Yang[1], Xiaoxi Lu[1], and Yilin Wang[2](✉)

[1] Changchun Rail Transit, ChangChun Jilin, China
[2] Shanghai Rail Transit Equipment Development Co Ltd., Shanghai, China
ylwang0403@hotmail.com

Abstract. In the urban railway industry, in order to achieve the stability and reliability of the permanent magnet motor control in urban railway vehicles, the on-off control of the switching device of the traction inverter circuit is used to obtain a series of pulses with the same threshold value at the three-phase output terminals, and these pulses can be used to replace the sine wave and the desired waveform. Then input an AC current of a certain frequency into the permanent magnet motor. In order to solve the phenomenon of unstable control of permanent magnet motors at different speeds, this paper proposes a specific harmonic elimination pulse width modulation (SHEPWM) hybrid pulse width modulation method, and uses Simulink software for modeling and simulation simulation waveforms, which proves that the mixed pulse width modulation method proposed in this paper Stability and reliability of the modulation output.

Keywords: SHEPWM · Permanent magnet motor · AC current

1 Introduction

At present, in order to ensure redundant and reliable operation of the urban railway vehicles, most of the traction systems use permanent magnet motors as the train power source, and the hybrid pulse width modulation method is widely used in the traction inverter software of urban railways [1]. In order to prevent the input current instability caused by the permanent magnet motor at different speeds, a hybrid pulse width modulation is used to ensure the stability and reliability of the three-phase output current of the inverter.

In the urban railway traction system, the power of the transmission system maintains a large xvalue, and on the contrary, most of the switching devices in the inverter work at low frequencies (usually only a few hundred Hz), so the hybrid pulse width modulation strategy are often used in this system. In urban railway traction system, hybrid pulse width modulation adopts different synchronous modulation strategies in the mid-frequency band. It includes regular sampling segmental synchronous modulation and special synchronous PWM mode, and special synchronous PWM mode of optimized

L. Jia et al. (Eds.): EITRT 2023, LNEE 1135, pp. 207–216, 2024.
https://doi.org/10.1007/978-981-99-9307-9_23

PWM mode (such as specific harmonic elimination, hereinafter referred to as SHE); in fact, the mid-frequency band also A pulse width modulation mode that combines regular sampling segmental synchronous modulation with special synchronous modulation is often used [2].

This article will first analyze the principle selection of the power module, and briefly describe the modulation method for SHE synchronous modulation. And the simulation analysis of the modulation methods and output current stability is carried out.

2 Organization of the Text

Permanent magnet motor control is divided into three parts: torque given, current calculation, voltage calculation. Torque reference is responsible for determining the given torque of the current motor through parameters such as level command, network voltage, power module temperature, and motor speed, and providing instructions for the core algorithm of motor control; the current calculation obtains the given torque through the look-up table method The corresponding given torque current and given excitation current. The voltage calculation is responsible for converting the given torque and excitation current into corresponding given voltage values through certain calculations. Furthermore, the voltage feedforward technology can solve both the dq-axis current coupling problem and the system response rate problem, which can make the system have good dynamic performance and robustness. Finally, the corrected given modulation degree can be obtained through voltage calculation, which provides control instructions for the subsequent pulse width modulation algorithm. The pulse width modulation algorithm can provide the stable and three-phase current value required by the permanent magnet motor. The block diagram of the permanent magnet motor vector control is shown in Fig. 1.

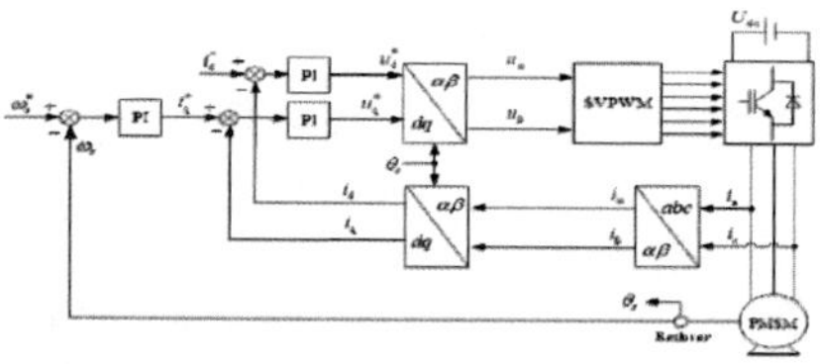

Fig. 1. Block diagram of permanent magnet synchronous motor vector control

2.1 Analysis of SHEPWM Control Principle

In the permanent magnet synchronous motor control system of rail urban railway, the voltage or current output by three-phase UVW generally contains a large number of harmonics, which will interfere with the operation of the motor control algorithm to a certain extent.

Specific harmonic elimination pulse width modulation technology SHEPWM is a more optimized pulse width modulation technology with SPWM. Its working principle is to achieve specific harmonics by optimizing the switching time of the inverter output phase voltage pulse. Elimination can effectively reduce the harmonics in the output voltage and current.

According to the mathematical model of the three-phase output phase voltage UN, VN, WN of the inverter, and then perform Fourier calculation on the output phase voltage UN, VN, WN of the urban railway traction inverter, the output phase of the inverter is obtained. The pulse switching angle corresponding to the specified sub-harmonic in the voltage being zero. Then, the inverter is controlled by the switching angle trigger pulse obtained above, and the elimination of the specified sub-harmonic can be realized in the output phase voltage UN, VN, WN of the traction inverter.

The advantage of specific subharmonic elimination PWM (SHEPWM) is that it can eliminate unwanted harmonic orders in the output voltage, and at the same time keep the output fundamental wave amplitude following the command value. In the urban rail traction system, when the carrier ratio is not greater than 10, SHEPWM is the best pulse width modulation method.

SHEPWM realizes the elimination of some harmonic components on the basis of meeting the requirements of the fundamental wave, thereby improving the performance of the modulation method as a whole [6]. The waveform of the SHEPWM modulation mode has the characteristics of waveform symmetry, which are half-wave symmetry and cycle symmetry. Then the even harmonics in the harmonics in these two symmetrical ways will be completely eliminated. The waveform of the SHEPWM modulation method has two different forms, taking the U phase as an example, as shown in the figure below.

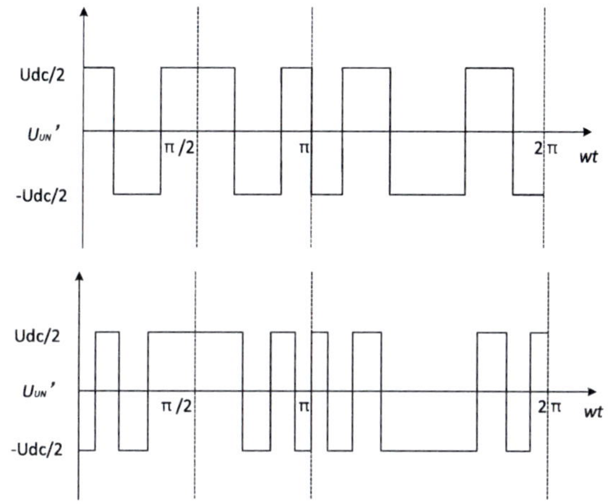

Fig. 2. SHEPWM output U-phase voltage waveform

Some characteristics of the SHEPWM modulation waveform can be obtained from Fig. 2.

The output phase voltage waveform has the characteristics of 1/4 cycle axis symmetry and 1/2 cycle center symmetry. According to the symmetry characteristics of the periodic axis and the periodic center, it can be seen that the output waveform of the SHEPWM modulation mode must have an opening and closing condition at 0 degrees, 180 degrees and 360 degrees respectively, so as to ensure the symmetrical characteristics of the output waveform; so when The switching angle of the modulated output waveform in 1/4 cycle is obtained, and the remaining switching angles can be calculated through the above-mentioned axisymmetric and cycle symmetric.

According to the number of switching angles, the waveforms output by SHEPWM modulation have two forms, as shown in a and b in Fig. 5 above. In Fig. 4, the output waveform of type a begins to complete a fundamental wave cycle with a high level, while in Fig. 5, the type b waveform begins to complete a fundamental wave cycle with a low level. When the number of switching angles is an even number, the output waveform is a type waveform, and when the number of switching angles is an odd number, the output waveform is a type b waveform.

The Fourier series expansion of SHEPWM output phase voltage can be written as the following form:

$$U_u(t) = \frac{a_0}{2} + \sum_{n=1}^{\infty} [a_n \sin(nt) + b_n \cos(nt)](n = 1, 2, 3 \ldots)$$

$$a_0 = \frac{1}{\pi} \int_0^{2\pi} U_u(t) d(t)$$

$$a_n = \frac{1}{\pi} \int_0^{2\pi} U_u(t) \sin(nt) d(t)(n = 1, 2, 3 \ldots)$$

$$b_n = \frac{1}{\pi} \int_0^{2\pi} U_u(t) \cos(nt) d(t)(n = 1, 2, 3 \ldots)$$

In the above formula, a_0 is the DC component, a_n is the sine component, and b_n is the cosine component.

Among them, $U_u(t) = U'_{un} = \pm\frac{U_1}{2}$ is a centrally symmetrical odd function with 1/4 period axisymmetric properties.

Based on the central symmetry of the odd function, the voltage waveform of SHEPWM only contains sine harmonic components (the DC component and the cosine component do not have central symmetry. Therefore, it is impossible to become part of the output voltage, so a, b are all 0, This conclusion can also be obtained from the perspective of mathematical calculation).

In the following, only Class A waveform is taken as an example to expand.

As a_0 and b_n are equal to 0, so $U_n(t) = \sum_{n=1}^{\infty} a_n \sin(nt)$.

Only need to analyze and solve the sine component a_n, according to the integral characteristics of the product of two odd functions in the symmetrical interval, there are:

$$a_n = \frac{1}{\pi}\int_0^{2\pi} U_u(t)\sin(nt)d(t)(n = 1, 2, 3\ldots)$$
$$a_n = \frac{2}{\pi}\int_0^{\pi} U_u(t)\sin(nt)d(t)(n = 1, 2, 3\ldots)$$
$$a_n = \frac{2}{\pi}\int_0^{\pi} \pm\frac{U_1}{2}\sin(nt)d(t)(n = 1, 2, 3\ldots)$$
$$a_n = \pm\frac{U_1}{\pi}\int_0^{\pi} \sin(nt)d(t)(n = 1, 2, 3\ldots)$$
$$a_n = \pm\frac{U_1}{n\pi}\int_0^{\pi} d - \cos(nt)(n = 1, 2, 3\ldots)$$
$$a_n = \pm\frac{U_1}{n\pi}[1 - \cos(nt)](n = 1, 2, 3\ldots)$$

It can be seen that when n is an even number, $\cos(n\pi) = 1$, the above formula is equal to 0; when n is odd, $\cos(n\pi) = -1$, the above formula is equal to $\pm\frac{2U_{dc}}{n\pi}$; therefore, there is no even-order sine harmonic component in the output voltage of SHEPWM modulation, so it can be deduced as: $a_n = \pm\frac{2U_1}{n\pi}(n = 2k - 1, k = 1, 2, 3\ldots)$.

$$U_u(t) = \pm\frac{2U_1}{n\pi}\left[\sin t + \frac{1}{3}\sin 3t + \frac{1}{5}\sin 5t + \frac{1}{7}\sin 7t + \ldots + \frac{1}{n}\sin(nt)\right]$$

Thus,

$$U_u(t) = \sum_{n=1}^{\infty} \pm\frac{2U_1}{n\pi}\sin(nt)(n = 2k - 1, k = 1, 2, 3\ldots)$$

e $U_n(t)$ to represent the nth harmonic of the SHEPWM output phase voltage, considering the 1/4 cycle axisymmetric characteristics, then:

$$U_n(t) = \sum_{n=1}^{\infty} a_n\sin(nt) = \frac{1}{\pi}\int_0^{2\pi} \pm\frac{U_1}{2}\sin(nt)d(t)$$

When ω is any real

$$U_n(t) = \frac{4}{\pi}\int_0^{2\pi} \pm\frac{U_1}{2}\sin(nt)d(t)(n = 2k - 1, k = 1, 2, 3\ldots)$$

number, the above formula should be written as:

$U_n(t) = \frac{4}{\pi}\int_0^{2\pi} \pm\frac{U_1}{2}\sin(n\omega t)d(\omega t)(n = 2k-1, k = 1, 2, 3\ldots)$ Let the switching angles be $\alpha_1, \alpha_2\alpha_3\ldots\alpha_k$ in sequence, and solve it by integral by parts:

$$U_n(t) = \pm\frac{2U_1}{\pi}\begin{bmatrix} \int_0^{a_1}(+1)\sin(n\omega t)d(\omega t) \\ +\int_{a_1}^{a_2}(-1)\sin(n\omega t)d(\omega t) \\ +\int_{a_2}^{a_3}(+1)\sin(n\omega t)d(\omega t) \\ +\ldots \\ \int_{a_{k-1}}^{a_k}(-1)^{k-1}\sin(n\omega t)d(\omega t) \\ +\int_{a_k}^{\frac{\pi}{2}}(+1)\sin(n\omega t)d(\omega t) \end{bmatrix}$$

$$= \pm\frac{2U_1}{\pi}\left[1 + 2\sum_{i=1}^{k}(-1)^i\cos(na_i)\right]$$

In Fig. 2. Above, a and b correspond to + and - respectively. If the number of switching angles within 1/4 cycle is set to k, the reference fundamental wave voltage U_{1s}* can be completely equal to the fundamental wave equation, and the remaining (k − 1) The second equation can be used to eliminate 2k ± 1 harmonics, and the corresponding equations are as follows:

$$U_{1'} = \pm\frac{2U_1}{\pi}\left[1 + 2\sum_{i=1}^{k}(-1)^i\cos(a_i)\right] = U_{1s}^*$$

$$U_n = \frac{2U_1}{\pi}\left[1 + 2\sum_{i=1}^{k}(-1)^i\cos(na_i)\right] = 0(k-1, n = 2k \pm 1)$$

If the number k of switching angles is an odd number and is 3, 2, and 1 respectively, then it is the type b waveform in Fig. 4. Its output waveform is a low level as the initial level, and on the contrary, when k is an even number, it is a type a waveform in Fig. 4, and its output waveform is a high level as the initial level. Normally, in the line voltage of a 3-phase symmetrical circuit, the 3rd harmonic contained in the phase voltage cancels each other out. So$U_3 = 0$, and there is no need to consider the 3rd harmonic when doing harmonic cancellation.

According to the above formula, the corresponding objective equation can be obtained as follows.

k = 3, SHE7PWM:

$$U_1 = -\frac{2U_1}{\pi}[1 - 2\cos a_1 + 2\cos a_2 - 2\cos a_3] = U_{1s}^*$$

$$U_5 = -\frac{2U_1}{\pi}[1 - 2\cos(5a_1) + 2\cos(5a_2) - 2\cos(5a_3)] = 0$$

k = 2, SHE5PWM:

$$U_7 = -\frac{2U_1}{\pi}[1 - 2\cos(7a_1) + 2\cos(7a_2) - 2\cos(7a_3)] = 0$$

$$U_1 = \frac{2U_1}{\pi}[1 - 2\cos a_1 + 2\cos a_2] = U_{1s}^*$$

$$U_5 = -\frac{2U_1}{\pi}[1 - 2\cos(5a_1) + 2\cos(5a_2)] = 0$$

k = 1, SHE3PWM:

$$U_1 = -\frac{2U_1}{\pi}[1 - 2\cos a_1] = U_{1s}^*$$

According to the above formula, when k = 1, there is only one equation, so it can only meet the basic requirement that the amplitude of the output waveform is equal to the command value, so it does not have the ability to eliminate harmonics.

According to the above three sets of objective equations, substitute $\frac{U_{1s}{}^*}{\frac{2U_{dc}}{\pi}} = m$ into,

By calculating them separately, we can get the change law of each switching angle with the modulation degree m under SHE7PWM frequency division, SHE5PWM frequency division and SHE3PWM frequency division. The change range of m is 0–1. As follows:

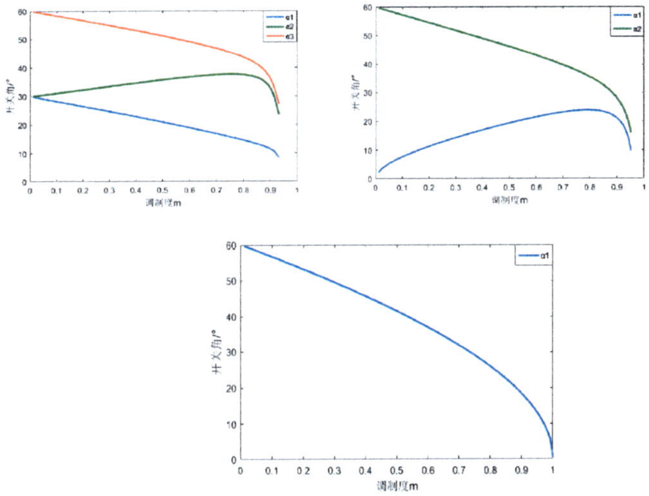

Fig. 3. Change of SHEPWM switching angle with modulation degree (SHE7, 5, 3)

From the curve of the angle value of the switching angle versus the degree of modulation in Fig. 5, it can be clearly seen that the angle value of the switching angle under the frequency division of SHE7PWM, SHE5PWM and SHE3PWM corresponds to a certain transcendental equation within the modulation degree of 0–1. The range of the relationship, among which the available modulation range of the 7-frequency SHEPWM pulse width modulation is 0–0.93, the 5-frequency available modulation range is 0–0.95, and the 3-frequency available modulation range is 0–1.

The calculation of the switching angle of SHEPWM modulation mode involves the solution of transcendental equations. In the practical application of urban railway traction systems, the offline calculation table look-up method or the real-time calculation of polynomial fitting (such as the least square method) are often used. The biggest advantage of offline calculation is that it is easy to implement, but the defect is also very obvious, that is, it needs to occupy a large amount of storage space on the board while calculating other algorithms. If the requirement for switching angle accuracy is relatively high, the large storage space will be further consumed. Occupancy; the real-time calculation method of polynomial fitting performs curve fitting on the obtained switching angle to realize real-time online calculation, but the error of the fitting curve is inevitable. The simpler the fitting method, the greater the error, but the smaller the error The fitting method is more complicated to implement. In order to meet the application requirements, usually more than 3 times of fitting are required. For the application of multiple SHEPWM modes, segmental fitting is also required, which increases the complexity of the program implementation. Therefore, in the actual application of the rail industry, the off-line look-up table method is mostly used to realize the calculation and processing of the switching angle.

When realizing the SHEPWM modulation mode, the most typical way is to take the output end of the traction inverter as a whole, and divide a time period into several states at the same time, and the output of the inverter remains stable in these several states. Change, and output three-phase power through the acquisition state table and send out PWM pulses. However, this method also has corresponding disadvantages. First, if there are many switching angles, at the same time, the number of states in a fundamental cycle will increase rapidly, which will cause the time occupied by each state to decrease rapidly; secondly., the change of the switching angle follows the change of the modulation degree, which will cause the state table to change under different modulation degrees, especially when the modulation degree tends to 1, which will bring a great deal of difficulty to the algorithm implementation. Difficulty.

2.2 Hybrid PWM Switching and Basic Principles of Implementation

The pulse width modulation algorithm is implemented in two parts, DSP and FPGA, among which the period count value calculation, comparison value calculation and modulation mode selection are realized in DSP, and the PWM module function is realized in FPGA.

The calculation function of the cycle count value is to provide the cycle count value for the carrier required for PWM pulse generation. Different motor frequencies, different modulation modes, and carrier frequency (period) are also different. Therefore, when calculating, it is necessary to properly correct the motor frequency according to the modulation mode, and use different calculation formulas to calculate the PWM period count value, and even when the carrier mode changes, it is necessary to correct the actual period count value that should be used. In order to reduce the current or voltage impact generated by the pulse mutation during the switching process.

The calculation function of the comparison value is based on the difference of the modulation mode, and the calculation method of the comparison value also changes accordingly. SHE synchronous modulation (including 7, 5 and 3 frequency division), use look-up table method to calculate switching angle in 1/4 cycle (not less than a) corresponding to the comparison value.

Modulation mode selection is to initiate a mode request according to the motor frequency or voltage modulation degree to determine the current modulation mode.

3 Experimental Simulation Analysis

The simulation results of the permanent magnet motor under the SHEPWM pulse condition are shown in the figure below.

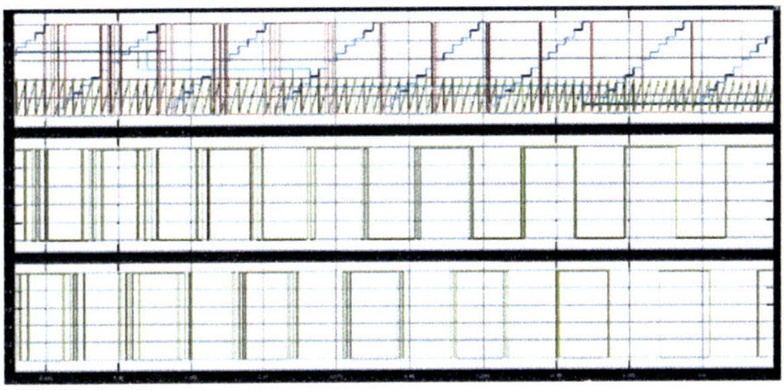

Fig. 4. Switching between SHEPWM pulses 7- > 5- > 3

The simulation results of the three-phase current output of the inverter under the condition of mixed pulse modulation are shown in the figure below.

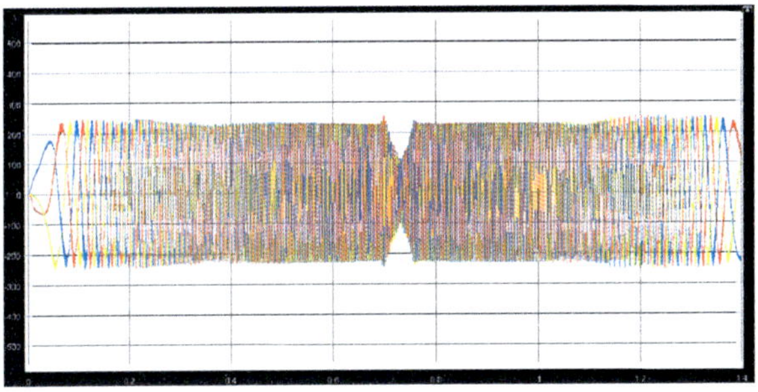

Fig. 5. Three-phase output current value of the inverter

It is not difficult to see from the simulation results that the SHEPWM waveform achieves the elimination of specific harmonics by optimizing the switching time of the inverter output phase voltage pulse. And as shown in Fig. 5, the three-phase current input of the permanent magnet motor tends to be stable at different speeds through the mixed modulation method of SHEPWM.

4 Conclusion

This paper analyzes the principle of power module selection and the principle and characteristics of permanent magnet motor hybrid pulse width modulation control. This modulation method of the SHEPWM. The simulation results demonstrate that the hybrid pulse width modulation of the traction inverter on rail vehicles can control the permanent magnet motor well.

References

1. Wang Jun, Influence of Inverter Carrier Frequency on Motor Operation. Qilu Petrochemical, 2011
2. Yuan Jialiang. Research on Optimal Pulse Width Modulation Method for Permanent Magnet Synchronous Traction Motor. Beijing Jiaotong University, 2014
3. Zhao Tao, Zhang Yonghao. Research on PMSM vector control servo system based on SHPWM. Control Technology, 2011
4. Gong Dongan. Research on Voltage Control Method of Auxiliary Inverter of Electric Locomotive. Beijing Jiaotong University, 2012
5. Li Ruifu. Research on vector control of asynchronous motor under synchronous modulation. Beijing Jiaotong University, 2011
6. Zhou Minglei. Research on Control Strategy of Electric Locomotive Traction Motor in Full Speed Range. Beijing Jiaotong University, 2013
7. Diao, L., Tang, J., Loh, P.C., Yin, S., Wang, L., Liu, Z.: An Efficient DSP–FPGA-Based Implementation of Hybrid PWM for Electric Rail Traction Induction Motor Control. IEEE Trans. Power Electron. **33**(4), 3276–3288 (April 2018)
8. Current-Source Inverter Integrated Motor Drives Using Dual-Gate Four-Quadrant Wide-Bandgap Power Switches[J]. Torres Renato Amorim,Dai Hang,Lee Woongkul,Sarlioglu Bulent,Jahns Thomas. IEEE TRANSACTIONS ON INDUSTRY APPLICATIONS
9. A 7-level inverter with less number of switches for grid-tied PV applications[J]. J. N. Bhanutej,Rani Chinnappa Naidu. International Journal of Advanced Technology and Engineering Exploration (IJATEE)
10. Seven-Level Symmetrical Series/Parallel Multilevel Inverter with PWM Technique Using Digital Logic[J]. Motaparthi Nagaraju,Kumar Malligunta Kiran. International journal of electrical and computer engineering systems
11. XMEGA-based implementation of four-switch, three-phase voltage source inverter-fed induction motor drive[J]. Boloor Kashani E.,Halvaei Niasar A.. International Journal of Power Electronics and Drive Systems
12. Variable switching frequency hybrid PWM technique for switching loss reduction in a three-phase two-level voltage source inverter[J]. Anas Ibrahim,Muhamad Zahim Sujod. Measurement
13. Suboptimal switching strategies for microprocessor-controlled PWM inverter drives[J]. Bowes S.R.,Midoun A.. IEE Proceedings B Electric Power Applications
14. Suboptimal switching strategies for microprocessor-controlled PWM inverter drives[J]. Halász S.,Bowes S.R.,Midoun A.. IEE Proceedings B Electric Power Applications

A Review of Hydrogen Fuel Cell Applications in Rail Transit

Suyao Liu, Yifei Zhang, Chunmei Xu(✉), and Lijun Diao

School of Electrical Engineering, Beijing Jiaotong University, Beijing 100044, China
chmxu@bjtu.edu.cn

Abstract. The application of hydrogen fuel cell trains in rail transportation is gradually gaining attention due to the continuous development of contactless grid-powered trams and high-power fuel cell technology. Hydrogen fuel cell trains, as a new environmentally friendly transportation method, utilize the reaction between hydrogen and oxygen to generate electrical energy for train operation. The findings indicate that hydrogen fuel cell trains offer advantages such as zero emissions, efficient energy utilization, and quiet operation. These benefits contribute to the effective reduction of environmental pollution and energy consumption, making hydrogen fuel cell trains a crucial direction for the future development of urban rail transportation. Nevertheless, hydrogen fuel cell trains encounter various challenges related to hydrogen storage and transportation, construction costs, and policy considerations. To achieve the goal of sustainable development, further efforts should be directed towards enhancing research and promoting applications.

Keywords: Hydrogen Fuel Cell · Rail Transportation · Contactless grid-powered trams

1 Introduction

Trams in China have experienced rapid development in recent years. The power supply methods of trams include contact network power supply and contactless network power supply. The contact network power supply method is extensively utilized due to its mature technology and extended range. However, this method is associated with high construction costs, an unfavorable urban outlook, and specific safety hazards. Contactless network power supply technology offers an effective solution to the aforementioned problems, leading to the emergence of contactless network power supply trams as the prevailing trend in modern tram development.

Hydrogen fuel cells directly convert hydrogen and oxygen into electricity and water through electrochemical reactions, exhibiting high energy conversion efficiency, minimal environmental pollution, and low noise levels. They offer broad prospects for development. Hydrogen fuel cells directly convert hydrogen and oxygen into electricity and water through electrochemical reactions, exhibiting high energy conversion efficiency, minimal environmental pollution, and low noise levels. They offer broad prospects for development. In contrast to existing catenary-free trams, hydrogen fuel cell trains possess an independent power system and do not depend on ground traction for power

L. Jia et al. (Eds.): EITRT 2023, LNEE 1135, pp. 217–224, 2024.
https://doi.org/10.1007/978-981-99-9307-9_24

supply. Moreover, their performance is comparable to that of conventional power trains. Consequently, the application of hydrogen fuel cell trains has garnered increasing attention in recent years and has been piloted as a new form of environmentally friendly transportation in certain cities.

2 Hydrogen Fuel Cell Train Working Principle

Since hydrogen fuel cells have a slow dynamic response, cannot cope with high-frequency large load changes, and cannot perform braking energy recovery, hydrogen fuel cells are generally combined with supercapacitors, batteries, and other energy storage power sources to form hybrid power systems for rail vehicles. The operating principle of a hydrogen fuel cell train is shown in Fig. 1. Hydrogen fuel cell train working principle diagram. When a hydrogen fuel cell train works, hydrogen gas from a high-pressure hydrogen storage tank is delivered to the fuel cell system and combined with oxygen in the air to produce chemical energy through an electrochemical reaction, which is directly converted into electrical energy to drive the train. During vehicle start-up and emergency conditions, there is a high demand for instantaneous power, which requires the use of power batteries or supercapacitors to provide additional power output; When the vehicle is braked, the energy generated can be recovered through the power cell or supercapacitor. During the operation of a hydrogen fuel cell train, the only emission produced is water, which greatly reduces environmental pollution.

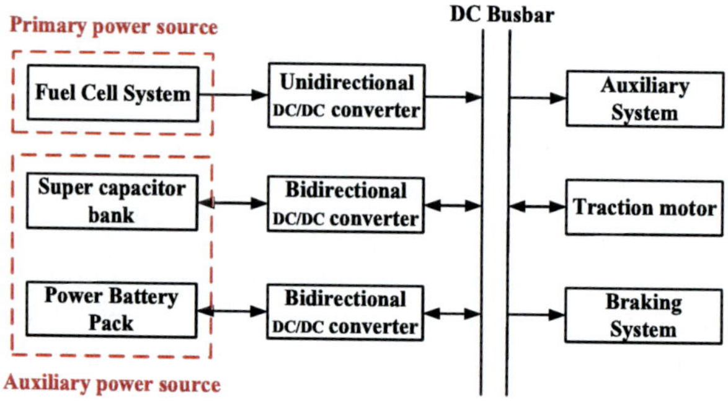

Fig. 1. Hydrogen fuel cell train working principle diagram

3 Hydrogen Train Development Status at Home and Abroad

3.1 Foreign Hydrogen Train Development Status

3.1.1 United States

In 2002, Vehicle Projects developed the first fuel cell-powered mining locomotive prototype, the world's first hydrogen fuel cell-powered locomotive. The vehicle is powered entirely by fuel cells without any auxiliary power source. The vehicle uses 3kg of metal hydride hydrogen storage, which allows the vehicle to run continuously for 8h at 17kW of power [1].

In 2007, the company began collaborating with BNSF [2] in the United States to develop fuel-cell hybrid shunting locomotives. The locomotive uses fuel cells as the main power source and is equipped with lead-acid batteries as auxiliary power sources to provide power output. The locomotive uses 14 35Mpa high-pressure hydrogen storage cylinders to store hydrogen gas, which can store a total of 68kg and provide continuous operation for 11 h. The hydrogenation time of the locomotive is 10 min to 45 min.

In 2019, the first commercial hydrogen powered passenger train (Flirt H2) project in the United States was launched. The vehicle consists of two carriages with 108 passenger seats and ample standing room, with a central power module in the middle of the two carriages to house the fuel cell and hydrogen storage tank. The vehicle can reach a maximum speed of 130 km/h. The vehicle is expected to be delivered in 2023 and enter service in 2024 [3].

3.1.2 France

In 2016, the French company Alstom developed the world's first hydrogen-powered passenger train, the Coradia iLint [4]. The vehicle has a fuel cell and hydrogen storage tank mounted on the roof and a battery mounted under the floor. The train is powered by a hybrid of fuel cells and batteries. The vehicle is equipped with 24 35Mpa high-pressure hydrogen storage cylinders to store hydrogen, and each vehicle can store 94 kg of hydrogen. The maximum speed of the vehicle can reach 140 km/h, the hydrogen refueling time is 15min, and the range is 1000 km. The maximum passenger capacity of the vehicle can reach 300 people. Compared to internal combustion locomotives, Coradia iLint's hydrogen technology can reduce CO_2 emissions by 700t/year.

3.1.3 Germany

Siemens is developing the hydrogen-powered train Mireo Plus H. The vehicle is scheduled for initial test runs in 2023 and will be in operation in 2024. The train can reach a maximum speed of 160 km/h, has a tractive power of 1,700 kW, provides acceleration of up to 1.1 m/s2 and can be refueled in. The two-pack train has 120 seats with a range of 600 km, and the three-pack train provides 160 seats with a range of 800–1000 km. Compared to internal combustion powered trains, Mireo Plus H hydrogen fuel trains will reduce carbon dioxide emissions by approximately 330 tons per year. In addition, Siemens has built a mobile hydrogen refueling station specifically for the train, making it possible to refuel hydrogen-powered trains as fast as diesel trains [5].

3.1.4 Spain

In 2013, the rail vehicle manufacturer CAF delivered 21 "Urbos 3" fuel cell trams to Zaragoza for operation. The vehicle uses the PEMFC fuel cell system as the main energy source and the power cell and supercapacitor as auxiliary energy sources. The overall length of the vehicle is 32.3 m, the actual passenger capacity is 275, and the maximum speed of the vehicle is 70 km/h [6].

3.1.5 United Kingdom

The UK is the first country in the world to legislate for net zero emissions by 2050. In June 2019, the UK exhibited its first hydrogen-powered train, the Hydro Flex [7]. The vehicle is equipped with four high-pressure hydrogen tanks that can store 44 lb of liquid hydrogen. A single hydrogen refill provides a range of 800 km and a maximum operating speed of 120 km/h. The train is in dual power mode, with the vehicle running on either hydrogen fuel cell power or switching to overhead contact network power.

3.1.6 Japan

East Japan Railway Company developed a diesel hybrid NE train in 2003 and converted it to a fuel cell hybrid in 2006. The maximum speed of the vehicle is 100 km/h. The vehicle is equipped with two 65kW fuel cell systems and one 340 kW power cell pack, with a hydrogen storage tank capacity of approximately 410L and a pressure of 35 MPa. Since the train is not equipped with a high-pressure hydrogen storage tank, and the performance of hydrogen fuel cell and lithium battery is also low, the longest distance the vehicle can travel with full hydrogen is only about 50–70 km, which is difficult to reach the standard of commercialization [8].

On the basis of the development of the NE Train train, JR East Japan Railway Company announced on June 4, 2019, that it would invest 4 billion yen to develop an FV-E991 series train equipped with high-pressure hydrogen tanks [9]. The train consists of a moving car and a trailer, with the moving car equipped with two sets of 25 KWh lithium-ion batteries and the trailer car equipped with a high-pressure hydrogen storage tank (mounted on the roof) and two sets of 180kW fuel cells. When the maximum pressure of hydrogen stored in the vehicle is 70MPa, the maximum distance the train can travel is 140 km; when the maximum pressure is 35MPa, the maximum distance the train can travel is 80 km. The train has a total of 108 seats and a maximum travel speed of 100 km/h.

3.2 Domestic Hydrogen Train Development Status

In recent years, domestic research related to fuel cell rail vehicles has flourished. Southwest Jiaotong University started to research fuel cell rail vehicles in 2008 and designed the first proton exchange membrane fuel cell locomotive "Blue Sky" in China in 2013. The rated power is 150kW and the hydrogen storage tank can store about 23 kg of compressed hydrogen. 13.5m in length, 45t in weight, 21 km/h continuous speed and 65 km/h maximum allowable speed [10].

In April 2016, the world's first commercially available hybrid tram rolled off the production line at CNR Tangshan. The car adopts the world's most advanced 100% low-floor technology, the carriage floor height from the rail surface is only 0.35 m. The vehicle adopts "two moving and one towing" three vehicle grouping, the vehicle adopts fuel cell as the main power source traction power supply. When the vehicle brakes and stops, the super capacitor and the battery absorb the braking energy, and the braking energy recovery rate can reach more than 30%. The vehicle can carry 12 kg of hydrogen in one filling, and the hydrogen refueling time is 15 min, which can be used for the train to travel continuously for more than 40 km, with a maximum speed of 70 km/h and

Fig. 2. Diagram of the appearance of foreign hydrogen fuel cell trains

a maximum passenger capacity of 336 people. In addition, the vehicle uses the entire line without contact network operation, the overall construction cost of the line than the contact network tram savings of more than 30,000 yuan/km. On October 26, 2017, the vehicle started passenger operation on Tangshan Tangxu Railway.

In December 2017, the first hydrogen tram developed and designed by China Motor Qingdao Sifang was launched in Foshan, and will be officially put into service in Gaoming, Foshan in November 2019. The vehicle adopts 100% low-floor technology and applies 2 world-leading technologies of synchronous permanent magnet motor direct drive and articulated bogie. The first phase of the vehicle demonstration line is 6.57 km long, with 10 stations, 1 interchange station, 1 vehicle base and 1 hydrogen refueling station each. The vehicle uses a modular design and can be freely grouped in 2–5 sections, and is currently designed to carry up to 285 passengers in a 3-car formation. In terms of vehicle configuration, the vehicle is equipped with a 200kW proton exchange membrane fuel cell system and large-capacity lithium titanate batteries. Each train carries six 35MPa hydrogen storage tanks with a hydrogen storage capacity of 20 kg, which can fill up a tram in 15 min on average, and can continue to travel 100km after each refill, with a maximum operating speed of 70km/h [11].

Blue Sky Train **Foshan Tramway** **Tangshan Hybrid Train**

Fig. 3. Domestic hydrogen fuel cell train appearance schematic

4 Development Trend of Hydrogen Fuel Cell Trains

4.1 Advantages of Hydrogen Fuel Cell Trains

(1) Low pollution, good environmental friendliness; The only emission of hydrogen fuel cell trains during operation is water, which has almost zero environmental pollution and helps reduce dependence on limited resources such as oil.
(2) Low noise; The weak noise generated by auxiliary equipment such as refrigeration systems during the operation of hydrogen fuel cell trains improves passenger comfort and effectively reduces noise interference to surrounding residents;
(3) No power grid constraints, good environmental adaptability; The vehicle adopts catenary free power supply, getting rid of the constraint of catenary pantograph, and increasing the flexibility and reliability of the vehicle. At the same time, it has improved the city's appearance and reduced infrastructure costs. Suitable for operation in areas where it is difficult to lay contact networks such as high plateaus.
(4) Fast hydrogenation speed, capable of quickly replenishing fuel; At present, the refueling speed of hydrogen fuel cell trains is close to that of diesel trains, reducing the parking time during vehicle operation.
(5) Good performance and convenient transition; At present, hydrogen powered trains can meet the requirements of the railway system. They can be modified from the original diesel trains to become hydrogen fuel cell trains, which is very convenient to replace traditional trains for transition.

4.2 Bottlenecks in the Development of Hydrogen Fuel Cell Trains

(1) There are safety issues with hydrogen storage and transportation; At present, high-pressure gaseous hydrogen storage technology is relatively mature, widely used, and cost-effective, making it the main method of hydrogen storage application. However, it also has safety hazards such as leakage and explosion, and its safety needs to be further improved.
(2) The cost of using hydrogen is high; The current technological level in China is not advanced enough, and materials such as proton exchange membranes and catalysts mainly rely on imports, resulting in high costs for hydrogen fuel cell systems; In addition, currently hydrogen is mostly stored and transported in compressed gas or low-temperature liquid state, and the transportation cost of hydrogen is also high.

(3) There are problems with the construction of hydrogen refueling stations; The construction of hydrogen refueling stations is the foundation of the development of the hydrogen energy industry. Currently, most of the supporting equipment for hydrogen refueling stations relies on imports, resulting in high investment and construction costs for hydrogen refueling stations; In addition, the technical standards and policies and regulations related to hydrogen refueling stations in China are still not sound, increasing the difficulty of promoting hydrogen refueling stations.
(4) The adaptability of the vehicle power system needs to be improved; The onboard hydrogen storage equipment for hydrogen fuel cell trains, due to the presence of a hydrogen storage system, increases the weight of the vehicle body and occupies a large amount of space. Its spatial design will affect the adaptability of the entire system.
(5) Imperfect policies related to the hydrogen energy industry chain; The relevant supporting facilities and approval standards for the application of hydrogen fuel cells in the field of rail transit in China are still incomplete, and there are many problems in train operation, such as limited operational management experience and lack of effective regulatory authorities. These problems seriously limit the development and promotion of hydrogen fuel cell trains.

4.3 Market Prospects for Hydrogen Fuel Cell Trains

(1) The demand for energy structure transformation; The transformation of energy structure is the key factor to achieve China's carbon peak and carbon neutrality. The entire rail transit transportation system needs rapid decarbonization transformation, and hydrogen energy, as a green new energy source, has high utilization efficiency and a wide range of sources. The application of hydrogen fuel cells in the field of rail transit has a significant effect on energy conservation and carbon reduction, which helps to transform the energy structure.
(2) The market potential is huge; at present, the world is vigorously developing new energy trains, among which hydrogen fuel cell trains do not rely on catenary power supply, have little environmental pollution, and have broad development prospects in the field of rail transportation. In the future, with the continuous breakthrough of key core technologies and the continuous reduction of the cost of hydrogen energy industry chain, the global market scale of hydrogen fuel cell trains will also be gradually expanded.
(3) Strong support from the state; In China, the importance of hydrogen energy has attracted attention from the state and relevant departments, and the government is actively promoting the standards and specifications of hydrogen energy rail transit. CRRC Qingdao Sifang Co Ltd, CRRC Tangshan Co Ltd and other companies are actively promoting the application of hydrogen fuel cells in the field of rail transit. With the strong support of the country, hydrogen fuel cell trains will flourish in the field of rail transit.

5 Conclusion

As an emerging green transportation mode, hydrogen fuel cell train has obvious advantages such as environmental protection, high efficiency and energy saving, and has a broad application prospect. However, there are still some challenges in the process of promoting the application, mainly including technical difficulties in hydrogen preparation and storage, high manufacturing and operation costs, and the difficulty of transitioning the existing infrastructure. Therefore, policy support and technological breakthroughs are needed to accelerate the promotion and application of hydrogen fuel cell trains. However, it is believed that in the near future, with the continuous development of related technology and application practice, hydrogen fuel cell trains will become the mainstream of the future green transportation field.

Acknowledgments. This research has been supported by the Key R&D Projects of China Railway Corporation (No. L2022Z001).

References

1. Miller, A.R., Barnes, D.L.: Advanced underground vehicle power and control fuel cell mine locomotive. In: Proceedings of the 2002 US DOE Hydrogen Program Review, pp. 5–9 (2002)
2. Miller, A., Hess, K., Baenes, D., et al.: System design of a large fuel cell hybrid locomotive. J. Power. Sources **173**(2), 935–942 (2007)
3. Zhang Jing. Hydrogen powered trains with market prospects.Foreign Railway Locomotives and Multiple Trains, (6): 1–6,17(2021). (in Chinese)
4. M. Y. Mustafa, A. Al-Mahadin, et al. Fuel cell technology application for Dubai rail systems.Advances in Science and Engineering Technology International Conferences (ASET), IEEE:2018, pp.1–7(2018)
5. Palmer, C.: Hydrogen-Powered Trains Start to Roll. Engineering **11**, 9–11 (2022)
6. P. García, J.P, et al. Viability study of a FC-battery-SC tramway controlled by equivalent consumption minimization strategy. International journal of hydrogen energy, 37(11): 9368–9382(2012)
7. Calvert, C., Allan, J., Amor, P., et al.: Concept development and testing of the UK's first hydrogen-hybrid train (Hydro FLEX). Railway Engineering Science **29**, 248–257 (2021)
8. Kawasaki, J., Takeda, S., Furuta, R.: Development of the fuel cell hybrid railcar. Japanese Railway Engineering **48**(1), 6–8 (2008)
9. Iida, R., Lijie, L.: Development of fuel cell hybrid test multiple units. Foreign Railway Locomotives and Multiple Trains **479**(05), 1–4 (2021). (in Chinese)
10. Peng, F., Chen, W., Liu, Z., et al.: System integration of China's first proton exchange membrane fuel cell locomotive. Int. J. Hydrogen Energy **39**(25), 13886–13893 (2014)
11. Tian C, Yan K, Jia H. Hydrogen fuel cell systems and their applications in the field of rail transit. Railway Locomotives and Multiple Units, No.586(12):1–6+10+65(2022). (in Chinese)

Power Integration and Energy Management Strategy of CJ5 Hybrid EMU

Xuefei Li[1](✉), Wei Han[1], Suyao Liu[2], Lijun Diao[2], and Chengtao Wang[1]

[1] CRRC Changchun Railway Vehicles Co. Ltd., Changchun 130000, China
13844090307@163.com

[2] School of Electrical Engineering, Beijing Jiaotong University, Beijing 100044, China

Abstract. The CJ5 hybrid EMU is produced by CRRC Changchun Railway Vehicles Co., Ltd, and is a new type of EMU with multiple power sources independently developed by China. The EMU can not only meet the needs of cross-line operation between electrified and non-electrified railways, but also has a series of functions such as all-electric braking recovery, emergency traction for power failure, and non-reduction in excessive phases, which can simplify the catenary wiring of the EMU and realize Station-section shunting is environmentally friendly and has broad application prospects. This paper firstly introduces the CJ5 hybrid EMU platform and its derivative models. According to the selection and combination of the three power sources of catenary, power battery and internal combustion power pack (including diesel engine and permanent magnet generator), it has different functional characteristics and application advantages. An energy management control approach for hybrid EMU using droop control is proposed to realize instantaneous power distribution among various power sources. This strategy depends on the integrated model of hybrid EMUs. The service life of the power batteries can be increased, the fuel consumption of the train can be decreased, and the effectiveness of the hybrid system can be increased by continual iterative optimization of energy management control systems.

Keywords: Hybrid EMU · Power Battery · Energy Management

1 Introduction

Utilizing a hybrid power system can increase a vehicle's fuel efficiency, lower exhaust emissions, and increase the effectiveness of regenerative braking. It benefits from minimal energy use, low pollution, and low noise levels [1]. The hybrid mode is currently mostly used for locomotives and shunting in rail transit [2]. In 2011, Russia proposed to develop the first hybrid locomotive TEM9H Sinara Hybrid, using a hybrid power supply mode of engine and lithium battery, which can reduce fuel consumption by 40% and exhaust emissions by 55% [3]. Japan, France, and the United Kingdom are also developing their own hybrid locomotives through independent research and cooperation [4–6]. In 2009, CSR Ziyang Company started research on 2500 kW diesel and battery hybrid CKD6E locomotives. Compared with the same power combustion locomotive, the hybrid locomotive can reduce the use of fuel by more than 50%, the annual emission reduction of SO2, CO, NOX and other harmful substances more than 500 tons [7].

L. Jia et al. (Eds.): EITRT 2023, LNEE 1135, pp. 225–233, 2024.
https://doi.org/10.1007/978-981-99-9307-9_25

The research and development of hybrid EMUs is mostly seen in Japan. On March 25, 2010, Toshiba Corporation delivered a prototype HD300 shunting locomotive to Japan Freight Railway Company. The locomotive uses diesel-lithium-ion battery hybrid power, which can be charged by the engine and store regenerative energy during braking. Compared with the original DE10 diesel locomotive, the hybrid technology is expected to reduce noise by 10 dB and reduce harmful gas emissions by 40% [8]. To increase internal combustion train energy efficiency, Japan has successively developed E200 and HB-E300 diesel-electric series hybrid trains and developed a pure battery prototype EV-E301 in 2007 to reduce environmental hazards [9, 10].

Hybrid power systems have proliferated in the Chinese automobile industry in recent years, and numerous academic institutions and academics are investigating energy management for hybrid power systems for electric vehicles. In terms of rail transit vehicles, the research and related applications of energy management strategies for hybrid electric multiple units in China are still in the initial stage.

2 Introduction of CJ5 Hybrid EMU

As the name suggests, a hybrid EMU is an EMU with two types of power input, and there are two configuration methods: one is a hybrid EMU with "25kV catenary power supply + diesel generator power supply", hereinafter referred to as plan A; the other is "25kV catenary power supply + lithium battery power supply" hybrid EMU, hereinafter referred to as Plan B. The above two methods are uniformly designed, the main circuit and the car body structure have good compatibility and can be produced according to the needs of users. For scheme A and scheme B, the catenary power supply is the main power, and the diesel generator or power battery as auxiliary power source. Figure 1 introduces the energy transmission and conversion process of the hybrid EMU traction transmission system.

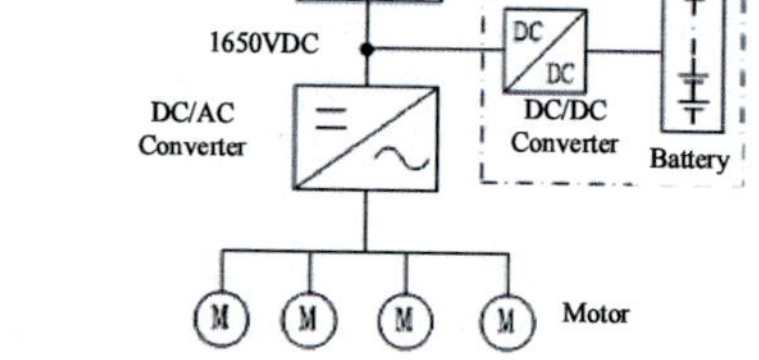

Fig. 1. Schematic diagram of energy transfer and conversion in the traction drive system of a hybrid EMU

For scheme A, the pantograph draws 25 kV 50 Hz single-phase alternating current from the catenary and reduces it to 900 V 50 Hz single-phase alternating current through the main transformer installed at the bottom of the vehicle (when there is no catenary, the diesel generator is used to directly generate 900 V 50 Hz three-phase Alternating current). Following this, it is transformed into 1650V DC direct current energy by the grid-side converter. The motor converter transforms the direct current energy into three-phase alternating current with variable amplitude and frequency before sending it to the traction motor to create mechanical energy for the traction train.

For plan B, the pantograph draws 25 kV 50 Hz single-phase AC power from the catenary, after being reduced to 900 V 50 Hz single-phase AC through the main transformer installed under the vehicle, it is converted into 1650V DC through the grid side converter. On the one hand, the motor converter receives the DC power, converts it into variable amplitude and frequency three-phase AC power, and then sends it to the traction motor to generate the mechanical energy required to traction the train. On the other hand, this direct current charges the power battery through bidirectional DC/DC. When there is no catenary, the electric energy stored in the lithium battery can be released to supply the motor converter so that the train can continue to run for a certain distance.

It is worth noting that: for plan A with diesel generators as auxiliary power, the diesel generators generate 900 V 50 Hz three-phase AC power, so the input of the grid-side converter is required to be switched from two single-phase 900V power supplies to one Three-phase 900V power supply (compatible with these two power supply systems); A bidirectional DC/DC converter must be connected in parallel with the grid-side converter's DC output in order to function as a lithium battery charging and discharging converter for Plan B, which employs lithium batteries as supplemental power. When the catenary has power, the DC/DC converter charges the lithium battery. When the catenary has no power, the DC/DC converter releases the energy stored in the lithium battery for use by the train.

3 Hybrid EMU Operating Indicators

The hybrid EMU uses a 2M1T (2 motors, 1 traction) configuration. Each EMU has four motors, a traction converter, and a hybrid system. The hybrid power system includes a power battery power supply system and a diesel generator set Power supply system with dual auxiliary converters on the trailer. The energy demand and power demand during train operation are shown in the table below.

4 Rule-Based Control Strategy for Energy Management of Hybrid Electric Multiple Units

The energy management system is the control and management hub of the train running state, how to optimize the energy management method is the key to achieve the goal of low fuel consumption and low emission of hybrid vehicles [12] This article selects the energy management method of droop control to real-time control the energy flow of hybrid electric multiple units.

Table 1. Emerfy demand in the different operating cycles of HEMU

System Indicators	Parameters			
	Traction process	Inert process	Braking process	Inbound Stops
Energy Sources	Diesel Generator Set/power battery pack	Diesel Generator Set/power battery pack	Regenerative braking energy	Power Battery Pack/diesel generator set
Peak power	838.72 kW	212.16 kW	–509.13 kW	85 kW
Running time	120.37 s	343.99 s	66.67 s	120 s
Operating mileage	2.554 km	11.335 km	1.111 km	0
Total energy consumption in the process	24.12 kWh	20.59 kWh	–7.78 kWh	2.83 kWh

4.1 Charge and Discharge Control of Power Batteries Based on Droop Characteristics

The battery charging and discharging control depending on droop characteristics is to calculate the droop voltage U_{drop} of the diesel generator at the current speed N_{dg} when outputting a certain power P_{dg} as the battery's desired voltage. At this time, the load of the diesel generator set can be equivalent to the power battery. By adjusting the charging and discharging of the battery, it is possible to adjust the generator's output power, and by working together, the two can achieve an even distribution of energy among the various power sources.

Figure 2 shows the system diagram for power battery droop control, which is primarily split into two parts: hysteresis control and droop control.

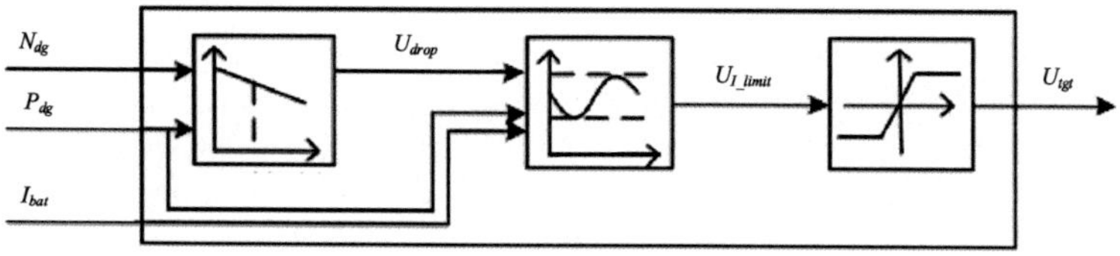

Fig. 2. Schematic of voltage reduction control system for power batteries

For example, if the speed of the diesel generator set is constant at 1800 rpm and its objective output power changes from 240 kW to 150 kW, the objective value of the DC side droop voltage is:

$$U_{tgt} = 1832 \times \frac{1800}{2000} - 0.4 \times P_{dg} \tag{1}$$

In order to simulate the above process, the target voltage control function of the power battery is written in Simulink with the m language, and the resulting DC voltage, power battery current, and diesel generator output power waveforms are shown in Fig. 3.

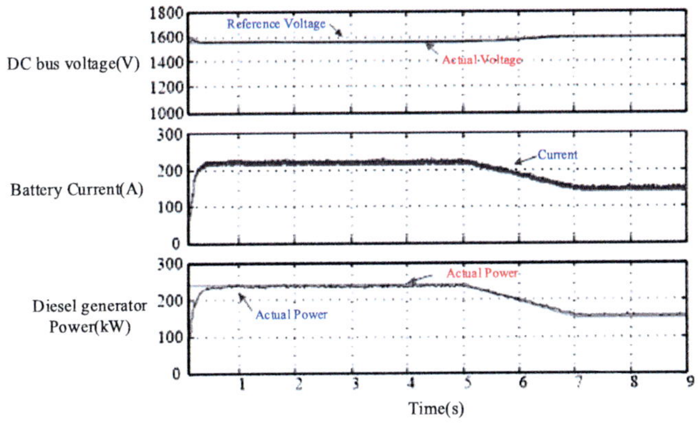

Fig. 3. Power battery droop control simulation of diesel generator output power change

As shown in Fig. 3, at a speed of 1800rpm, the target output power changes from 240 kW to 150 kW, and the target battery voltage increases from 1550 V to 1590 V, the diesel generator's output power reduces proportionately. According to the aforementioned simulation results, it is possible to achieve a one-to-one correspondence between the desired voltage and the power pack output power when controlling battery charge and discharge according to droop voltage.

Under the same speed conditions, the load power changes from 210 kW to 360 kW, requiring the power output of the diesel generator set to always be 240 kW. Figure 4 displays the corresponding simulation waveform.

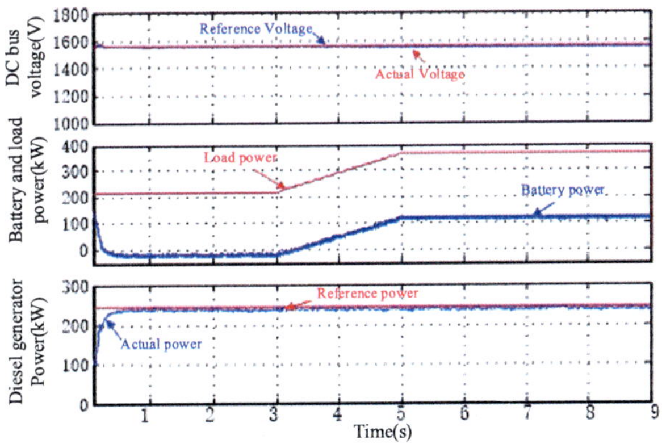

Fig. 4. Simulation of power battery droop control with load power variation

As observed in Fig. 4, the target voltage of the battery pack does not vary while the speed and target power of the diesel generator are both constant. The power battery pack is charged and the diesel generator set's output rises when the load is less than 240 kW. When the load increases to 360 kW, the power battery's charging current reduces and turns to discharging, supplying electricity alongside the diesel generator set. The DC voltage remained constant throughout the procedure at 1550 V, and the diesel generator set's output power was 240 kW, demonstrating that charging and discharging management according to droop characteristics can achieve distribution of power in the hybrid system.

During the operation of the power battery, except to controlling the size of the DC voltage, the size of the battery current should also be limited based on the battery status. In addition, under idling conditions, the diesel generator set also uses constant voltage control to charge the battery. On the basis of the drooping voltage, a certain hysteresis bandwidth is set according to the current limit, and the target voltage is adjusted to control the current.

4.2 Control Strategy Simulation Verification

Create a Simulink simulation model for a hybrid power system using a diesel generator and power batteries, the program to realize logic threshold control, diesel generator speed control, droop target voltage control, load reduction management, simulate hybrid power supply conditions, and verify hybrid power energy management With the performance of the control strategy, Fig. 5 displays the simulation model.

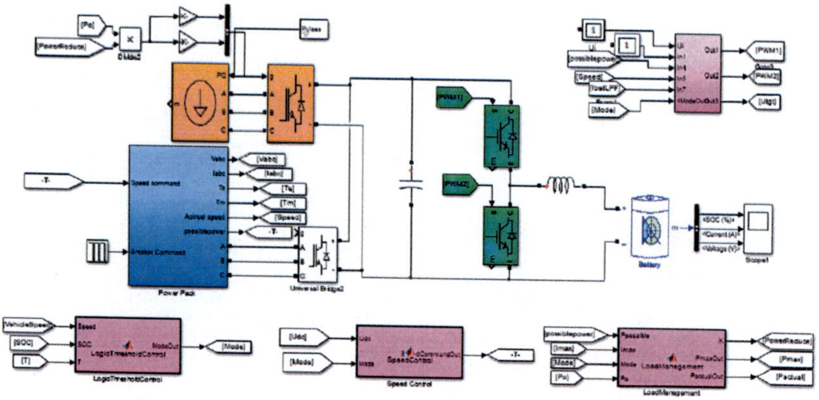

Fig. 5. Hybrid system simulation model

Simulate the start-traction-coasting-braking process, the maximum traction power is 640kW, the maximum braking power is 500 kW, and the maximum charging and discharging current of the power battery is 400 A. The diesel generator set's maximum power varies depending on speed. Among them, 2-3s refers to the power limitation of the simulated diesel generator set, with a maximum output power of 60% available. 7.3-9 s refers to the power limitation of the simulated power battery, with a maximum allowable current reduced from 400 A to 200 A. Figure 6 displays the simulation results.

The operating mode curve indicates that the hybrid energy management system can accurately identify the current operating mode of the system based on the actual operating parameters of the train, such as station start-up, hybrid traction, inert charge, or electric brake mode. Choose the appropriate speed control method according to the different operating modes of the diesel generator set, and the speed is 1800 rpm, 1700 rpm, 1400 rpm in the mixed traction, idle charging and braking modes respectively. During traction, the battery's objective voltage is converted to 1550 V based on the diesel generator's maximum output power. During coasting, the target voltage value is adjusted through the current hysteresis loop to achieve 85 A constant current charging, and the braking target voltage is 1750 V.

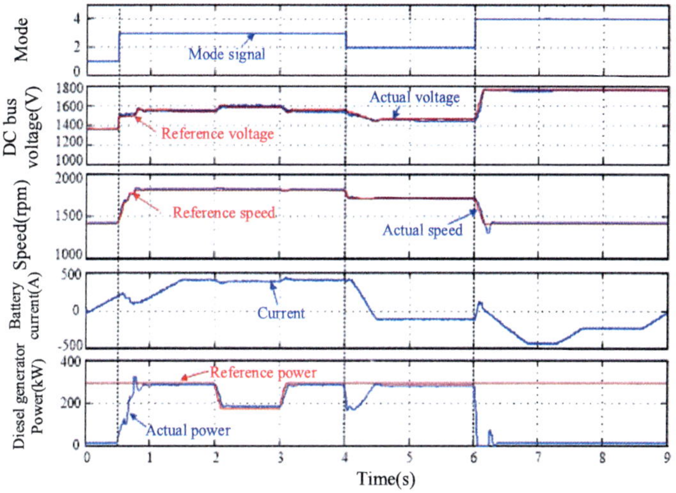

Fig. 6. Simulation result of hybrid power control strategy

Figure 7 displays the system power and power reduction factor simulation waveforms. In 2 ~ 3s and 7.3 ~ 9s, the power reduction coefficient is less than 1, and the output capacity of the power supply is reduced. The load power always operates based on the smaller value of the maximum output power and the command power. Energy management of the load is realized, avoiding overloaded operation of diesel generators or power batteries.

During the entire process of starting traction coasting braking, the speed of the diesel generator changes smoothly and the output power approaches the maximum available power. The system operates in the area of low fuel consumption and high efficiency. The DC side voltage and battery current are stable, able to quickly respond to changes in load and diesel generator power and adjust them, achieving reasonable energy flow.

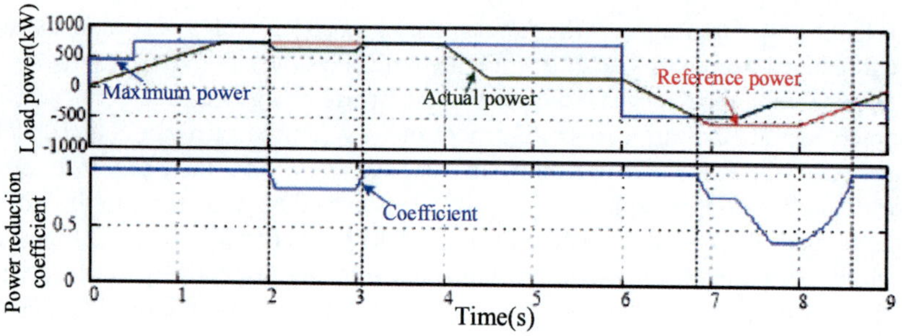

Fig. 7. Load reduction control simulation waveform

5 Conclusion

This paper introduces the CJ5 hybrid EMU platform and analyzes the energy flow relationship under different working conditions. In this paper, the operating characteristics of the diesel generator set and the power battery pack in the system are investigated for the hybrid train model, and the system energy management rules are formulated for the corresponding operating conditions, taking into account the actual operating conditions of the train. In order to achieve instantaneous power distribution between different power sources, the diesel generator set uses speed control, and the power battery pack uses charge and discharge control based on droop characteristics. Through the optimization of energy management strategies, battery life is extended, fuel consumption is decreased, and system efficiency is increased.

References

1. Fan, Y., Long, Y., Jiang, D., et al.: Li Wang. Overview of development status and key technologies of new energy hybrid locomotives . Electric locomotive and Urban Rail Vehicles, 46 (01): 1–11(2023) (in Chinese)
2. Yu, W.: DC side stability research and energy control management of hybrid Multiple unit. Beijing Jiaotong University (2017) (in Chinese)
3. Zubikhin, A., Kobylyanskiy, V., et al.: TEM9H SINARAHYBRID hybrid shafting technology at railway transport. Railway Equipment Magazine, no. 2, 06 (2012)
4. Li, L., Zhu, B., Ren, C.: Development and Application of Hybrid Electric Locomotives at Home and Abroad. Locomotive Electric Transmission, No.276 (05): 73–76+82(2020) (in Chinese)
5. Hou, R., Yang, Y., Emadi, A.: Hybrid electric locomotive powertrains.2014 IEEE Conference and Expo Transportation Electrification Asia-Pacific (ITEC Asia-Pacific). IEEE, pp. 1–6 (2014)
6. Zhang, C., Zeng, G., et al.: A bi-level optimization of speed and energy management for diesel-electric hybrid train. IEEE Trans. Vehicular Technol. pp. 1–13 (2023)
7. Xu, Y., Sheng, L.: China starts development of world's largest power hybrid locomotive. Science and Technology Daily (001)(2014) (in Chinese)
8. Tamada Masa, Liu, Y.: Japanese HD300 type hybrid shunting locomotive. Foreign railway vehicles, 52(01):31–34 (2015) (in Chinese)

9. Hayashiya, H., Kondo, K.: Recent trends in power electronics applications as solutions in electric railways. IEEJ Trans. Electr. Electron. Eng. **15**(5), 632–645 (2020)
10. Fedele, E., Iannuzzi, D., Del Pizzo, A.: Onboard energy storage in rail transport: Review of real applications and techno-economic assessments. IET electrical systems in transportation **11**(4), 279–309 (2021)
11. Wu, X.: Analysis of hybrid Multiple unit traction and auxiliary converter system. Urban Rail Transit Research **21**(11), 119–123 (2018). (in Chinese)
12. Wan, X., Xun, J., et al.: Energy management strategy for hybrid electric vehicles based on MPC. Journal of Beijing Jiaotong University **46**(05), 149–158 (2022). (in Chinese)

Design and Application of Complex Vector Current Controller for IM Based on Ripple-Free Minimum Beat System

Xuelei Yue(✉), Lei Wang, and Di Tian

Technology Center, CRRC Xian YongeJieTong Electric CO. Ltd., Xian 710016, Shaanxi Province, China
yuexl@hyee.com.cn

Abstract. In the AC drive system of urban rail transit, in order to reduce the loss of switching devices and improve the output power of the converter, the switching frequency of the traction converter is usually low, resulting in a considerable digital control delay, which intensifies the coupling between the motor excitation current and the torque current. To solve the above problems, the paper comprehensively analyzes the mathematical models of the traction induction motor (IM) in the S domain and Z domain, and the coupling factors caused by the digital control delay. Combining with the design method of ripple-free minimum beat system, an optimized complex vector current controller for the traction IM is proposed. In the discrete time domain, the current controller not only ensures an excellent dynamic response performance; Moreover, it can realize effective decoupling between the excitation current and torque current. Through Matlab/Simulink simulation and testing with CRH6F-A intercity Electric Multiple Unit (EMU) IM traction converter, the effectiveness and feasibility of the optimized complex vector current controller are proved.

Keywords: Urban rail transit · Traction IM · Ripple-free minimum beat system · Complex vector current controller · Digital control delay

1 Introduction

In the rail traction electric drive system, to reduce the loss of switching devices and improve the output power of the converter, the switching frequency of the traction converter is usually less than 1[KHz], for example, the switching frequency of CRH6F-A intercity EMU IM traction converter is only 520[Hz] [1–3]. When the switching frequency is low, there is cross-coupling between the d-axis and q-axis currents in the synchronous reference frame (SRF), especially when the speed is high, which will reduce the performance of the current controller [4–7].

In Ref. [8], to avoid the cross-coupling effect between the d-axis and q-axis currents in the SRF, the proportional resonance regulator is used to control the current in the stationary frame (SF). Although this method can avoid the coupling problem between the d-axis and q-axis currents, the dynamic response performance is poor. The feedforward

L. Jia et al. (Eds.): EITRT 2023, LNEE 1135, pp. 234–246, 2024.
https://doi.org/10.1007/978-981-99-9307-9_26

decoupling control and feedback decoupling control can reduce the coupling effect of the motor to a certain extent by adding the decoupling voltage term to the output voltage of the traditional PI controller. Still, they ignore the digital control delay and the coupling effect of the rotor side of the IM on the stator side. When the switching frequency is low, or the motor speed is high, the decoupling effect of the d and q axis currents is poor [9, 10]. Reference [11, 12] directly design the CVCC based on the discrete-time zero-order hold model of the symmetrical system like an induction motor and surface-mounted permanent magnet synchronous motor to improve the performance at a low ratio of sampling-to-fundamental frequency. In Ref. [13], a concurrent proportional integral (PI) controller design technique is addressed for indirect vector controlled induction motor (IVCIM). This method is implemented using a static output feedback scheme in which iterative linear matrix inequality (ILMI) based H∞ control technique is applied, it has strong robustness, but the digital control delay is not taken into account, and in the SRF, the current of each frequency cannot be fully decouple control. With the rapid development of modern control theory, intelligent control technologies such as sliding mode variable structure control [14], model reference adaptive control [15], and extended Kalman filter [16] are also applicable to the control system with solid coupling nonlinearity such as IM. However, the amount of calculation in the digital realization process is enormous, and it cannot be applied to the actual control system at present.

Aiming at the coupling problem between the d and q axis currents of the IM, this paper uses the concept of complex vector to establish the discrete mathematical model of the IM, including the digital control delay, and analyzes the substantial causes of the coupling between the d and q axis currents. At the same time, based on the design method of ripple-free minimum beat system, a complex vector current controller of IM, including digital control delay compensation, is proposed, and the difference equation for digital control is obtained. Finally, the performance of the designed complex vector current controller is verified by Matlab/Simulink simulation and combined test. The results show that both torque current and excitation current can track the command value stably, have good dynamic response performance, and achieve effective decoupling of torque current and excitation current.

This paper is organized as follows. The discrete mathematical model of IM is shown in Sect. 2. The discrete model of IM considering digital control delay is shown in Sect. 3. The controller design of IM discrete system is shown in Sect. 4. Simulation and experiments results are shown in Sect. 5. A conclusion is drawn in Sect. 6.

2 Discrete Mathematical Model of IM

The stator voltage equation of IM in two-phase SF can be expressed as

$$\begin{cases} u_{s\alpha} = (R_s + \dfrac{L_m^2}{L_r\tau_r})i_{s\alpha} + \delta L_s\dfrac{d}{dt}i_{s\alpha} - \dfrac{L_m}{L_r}\dfrac{1}{\tau_r}\psi_{r\alpha} - \dfrac{L_m}{L_r}\omega_r\psi_{r\beta} \\ u_{s\beta} = (R_s + \dfrac{L_m^2}{L_r\tau_r})i_{s\beta} + \delta L_s\dfrac{d}{dt}i_{s\beta} - \dfrac{L_m}{L_r}\dfrac{1}{\tau_r}\psi_{r\beta} - \dfrac{L_m}{L_r}\omega_r\psi_{r\alpha} \end{cases} \tag{1}$$

where, $u_{s\alpha}$, $u_{s\beta}$ are stator voltage in SF, $i_{s\alpha}$, $i_{s\beta}$ are stator current in SF, $\psi_{r\alpha}$, $\psi_{r\beta}$ are rotor flux linkage in SF, R_s, R_r are stator and rotor resistance respectively, ω_r is rotor angular frequency, L_m are excitation inductance, L_s, L_r are stator and rotor inductance respectively, $\delta = \frac{1-L_m^2}{L_s L_r}$, $\tau_r = \frac{R_r}{L_r}$.

The flux linkage equation in a two-phase stationary coordinate system is

$$\begin{cases} \psi_{s\alpha} = (1 - \frac{L_m^2}{L_r L_s}) L_s i_{s\alpha} + \frac{L_m}{L_r} \psi_{r\alpha} \\ \psi_{s\beta} = (1 - \frac{L_m^2}{L_r L_s}) L_s i_{s\beta} + \frac{L_m}{L_r} \psi_{r\beta} \end{cases} \tag{2}$$

The motor state equation is expressed in the form of complex vector $x = x_\alpha + jx_\beta$. Namely, $\psi_s^s = \psi_{s\alpha} + j\psi_{s\beta}$ is stator flux vector, $u_s^s = u_{s\alpha} + ju_{s\beta}$ is stator voltage vector, $i_s^s = i_{s\alpha} + ji_{s\beta}$ is stator current vector, and $\psi_r^s = \psi_{r\alpha} + j\psi_{r\beta}$ is rotor flux vector. Where, the subscript 's' represents the stator variable, 'r' represents the rotor variable; The superscript 's' stands for the SF and 'r' for the SRF. Then Eqs. (1) and (2) can be expressed as

$$u_s^s = \left(R_s + \frac{L_m^2}{L_r \tau_r} \right) i_s^s + \delta L_s \frac{d}{dt} i_s^s - \frac{L_m}{L_r} \frac{1}{\tau_r} \psi_r^s - j \frac{L_m}{L_r} \omega_r \psi_r^s \tag{3}$$

$$\frac{d}{dt} \psi_r^s = \frac{L_m}{\tau_r} i_s^s - \frac{\psi_r^s}{\tau_r} + j\omega_r \psi_r^s \tag{4}$$

By introducing Eq. (4) into Eq. (3), the voltage state equation with stator current i_s^s and rotor flux ψ_r^s as state variables is obtained.

$$u_s^s = R_s i_s^s + \delta L_s \frac{d}{dt} i_s^s - \frac{L_m}{L_r} \frac{d}{dt} \psi_r^s \tag{5}$$

In the rotor field-oriented vector control system, if the excitation current control is stable, the rotor flux is approximately constant, so the last item of Eq. (5) can be taken as the interference item, namely.

$$E^s = \frac{L_m}{L_r} \frac{d}{dt} \psi_r^s \tag{6}$$

Therefore, according to Eq. (5), the transfer function of IM with voltage as input and current as output in the SF can be achieved.

$$G_P^s(s) = \frac{I_s^s(s)}{U_s^s(s)} = \frac{1}{\delta L_s s + R_s} \tag{7}$$

The complex vector structure diagram of IM under the corresponding SF is shown in Fig. 1.

For a digitally controlled inverter system, the voltage command signal remains unchanged in the middle of every two controller action times (control cycle), equivalent to introducing a zero-order holder in the system loop.

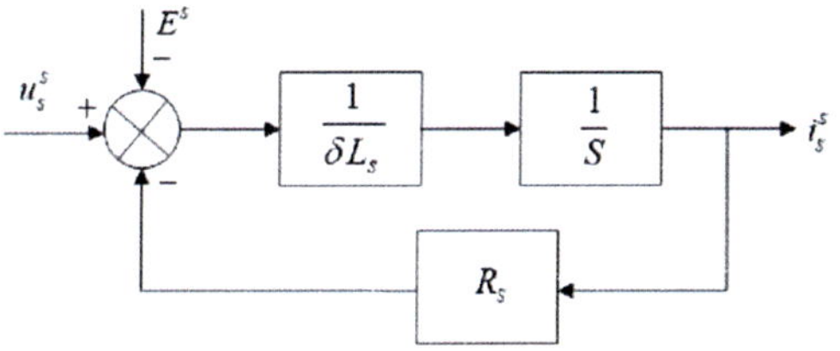

Fig.1. Complex Vector Structure Diagram of IM in Static Standard System.

Without considering the influence of harmonics and time delay, the digitally controlled inverter system can be equivalent to a zero-order holder, and its transfer function is

$$H(s) = \frac{1 - e^{-sT_s}}{s} \tag{8}$$

where T_s is the sample-control period.

The inverter open-loop discrete system with zero-order holder is shown in Fig. 2.

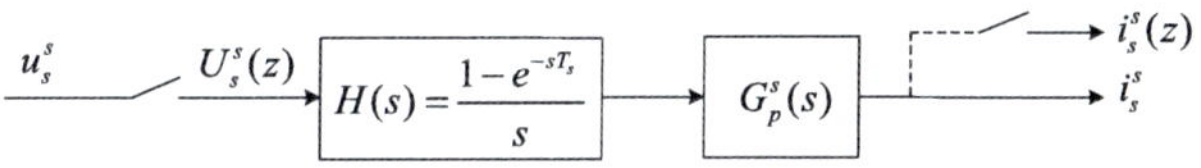

Fig. 2. Open loop discrete system of inverter with zero-order holder.

Then the pulse transfer function of the three-phase inverter fed IM system in SF is

$$G_P^S(z) = \frac{1 - e^{-\frac{R_S T_S}{\delta L_S}}}{(z - e^{-\frac{R_S T_S}{\delta L_S}})R_s} \tag{9}$$

Equation (9) is expanded into a difference Equation to obtain.

$$i_s^S(k+1) = e^{-\frac{R_S T_S}{\delta L_S}} i_s^S(k) + \frac{1 - e^{-\frac{R_S T_S}{\delta L_S}}}{R_s} u_s^S(k) \tag{10}$$

The complex vector Park coordinate transformation formula is

$$f_s^r(k) = f_s^s(k)e^{-j\theta(k)} \tag{11}$$

Assuming the Park coordinate transformation of the current i_s^s at the moment k + 1, the difference Equation in the SRF can be obtained.

$$i_s^S(k+1)e^{-j\theta(k+1)} = i_s^S(k)e^{-\frac{R_S T_S}{\delta L_S}} e^{-j\theta(k+1)} + \frac{1 - e^{-\frac{R_S T_S}{\delta L_S}}}{R_s} u_s^S(k)e^{-j\theta(k+1)} \tag{12}$$

For systems with large inertia, such as subway train and bullet train, the speed changes slowly, so it can be considered that the mechanical angular speed of the two sampling control periods before and after has not changed, then

$$e^{-j\theta(k+1)} = e^{-j[\theta(k)+\omega_s T_s]} \tag{13}$$

where ω_s is stator angular frequency.

Take Eq. (13) into Eq. (12)

$$i_s^s(k+1)e^{-j\theta(k+1)} = i_s^s(k)e^{-\frac{R_sT_s}{\delta L_s}}e^{-j[\theta(k)+\omega_sT_s]} + \frac{1-e^{-\frac{R_sT_s}{\delta L_s}}}{R_s}u_s^s(k)e^{-j[\theta(k)+\omega_sT_s]} \tag{14}$$

Rearrange Eq. (14) with Eq. (11) to obtain the current difference Equation in SRF.

$$i_s^r(k+1) = i_s^r(k)e^{-\frac{R_sT_s}{\delta L_s}}e^{-j\omega_sT_s} + \frac{1-e^{-\frac{R_sT_s}{\delta L_s}}}{R_s}u_s^r(k)e^{-j\omega_sT_s} \tag{15}$$

The discrete model of IM in SRF is obtained by converting it into a pulse transfer function.

$$G_P^r(z) = \frac{1-e^{-\frac{R_sT_s}{\delta L_s}}}{(ze^{-j\omega_sT_s} - e^{-\frac{R_sT_s}{\delta L_s}})R_s} \tag{16}$$

3 Discrete Model of IM Considering Digital Control Delay

When the control chip such as DSP is used as the main controller, the carrier PWM is usually realized by the way of continuous increase and decrease count of timer. Update PWM duty ratio output when timer overflows. The timing sequence of current sampling, operation and PWM update is shown in Fig. 3.

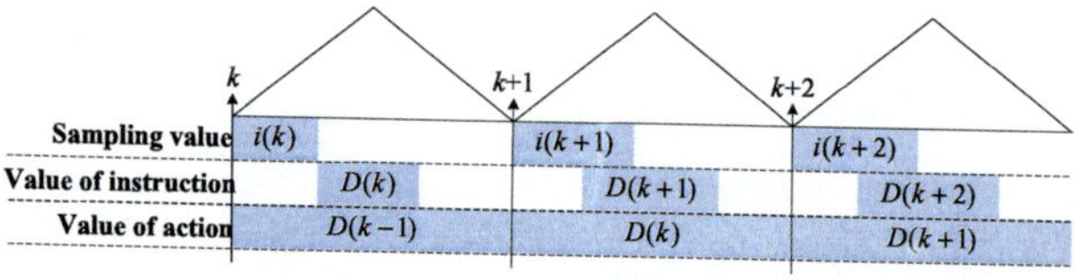

Fig. 3. Current sampling, operation and PWM update timing.

In Fig. 3, starting from the kth cycle, the duty cycle D(K-1) of PWM in the previous control period is updated first, and the current sampling is carried out at the same time. The sampling value i(k) is used to calculate the current controller and obtain the voltage instruction, and then the duty cycle D(k) of PWM is calculated. However, the update of duty cycle D(k) cannot be carried out until the end of this cycle, that is, the beginning of the k + 1 cycle, and the output of this instruction can be completed only after the end of the k + 1 cycle. To sum up, the whole digital control system has a one-beat lag, and the transfer function in the SF is z^{-1}.

According to Eq. (11), the transfer function of one beat lag link in SF can be converted into the transfer function in SRF.

$$G_d(z) = e^{-j\theta}z^{-1} \tag{17}$$

Since the speed of the front and rear two control cycles of the large inertia system is almost unchanged, Therefore, $\theta = \omega_s T_s$, the above Equation can be transformed into

$$G_d(z) = e^{-j\omega_s T_s} z^{-1} \tag{18}$$

Therefore, the pulse transfer function of traction IM with digital control delay in SRF is

$$G^r(z) = G_P^r(z) G_d(z) = \frac{1 - e^{-\frac{R_s T_s}{\delta L_s}}}{z e^{j\omega_s T_s}(z e^{j\omega_s T_s} - e^{-\frac{R_s T_s}{\delta L_s}}) R_s} \tag{19}$$

Convert Eq. (18) into a difference Equation to obtain Eq. (20).

$$u_s^r(k) = u_s^r(k-1)[\cos(\omega_s T_s) - j\sin(\omega_s T_s)] \tag{20}$$

It can be seen from Eq. (20) that there is a cross-coupling between the d and q axis voltages that finally act on the motor due to the existence of time delay. In particular, with the increase of synchronization frequency, or when the switching frequency is low, the coupling will become more serious, so the coupling part needs to be compensated.

4 Controller Design of IM Discrete System

4.1 Design Principle of Ripple-Free Minimum Beat System

If the generalized controlled object $G^r(z)$ of the system has no delay (if there is a delay link, it can be seen from the following formula (21) that it needs to be offset by $\Phi(z)$)and there are no zero poles on the unit circle in the z plane and outside the unit circle, it is required to select a closed-loop transfer function $\Phi(z)$ so that the system can make the steady-state error of the output sequence at each sampling time zero under the action of standard input after the minimum sampling period, to achieve complete tracking, Thus, the pulse transfer function $D(z)$ of the desired digital controller is determined(17).

$$D(z) = \frac{\Phi(z)}{G^r(z)\Phi_e(z)} \tag{21}$$

where $\Phi_e(z)$ is error transfer function.

4.2 Design of Ripple-Free Minimum Beat System

The IM control system with the digital controller can be regarded as a unit feedback linear time-invariant system, as shown in Fig. 4.

The generalized pulse transfer function with zero-order holder considering the digital control delay of IM is shown in Eq. (19), and the rewriting is shown in Eq. (22).

$$G^r(z) = \frac{\frac{1 - e^{-\frac{R_s T_s}{\delta L_s}}}{R_s e^{j2\omega_s T_s}} z^{-2}}{1 - e^{-\frac{R_s T_s}{\delta L_s}} e^{-j\omega_s T_s} z^{-1}} = \frac{A z^{-2}}{1 - B z^{-1}} \tag{22}$$

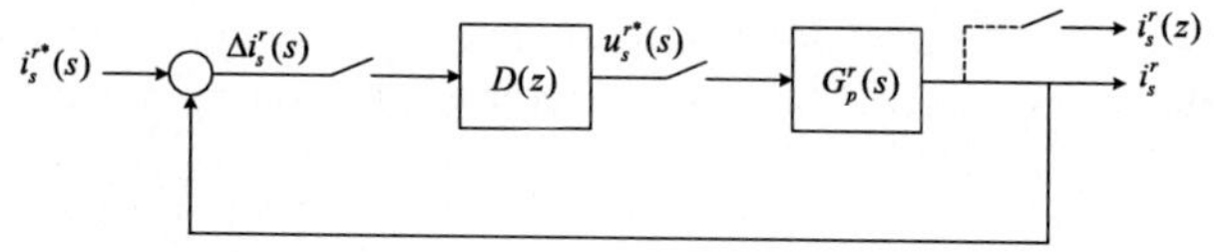

Fig. 4. Discrete system with the digital controller of IM.

where, $A = \frac{1-e^{-\frac{R_s T_s}{\delta L_s}}}{R_s e^{j2\omega_s T_s}}$, $B = e^{-\frac{R_s T_s}{\delta L_s}} e^{-j\omega_s T_s}$.

The switching frequency is 520[Hz], and the synchronous frequency changes from 0[Hz] to 150[Hz]. The pole and zero distribution diagram of the open-loop transfer function $G^r(z)$, including the digital control delay, can be obtained, as shown in Fig. 5. It can be seen from the figure that the poles and zero is distributed within the unit circle, and there is no poles and zero on or outside the unit circle, satisfying the design principle of the minimum beat system.

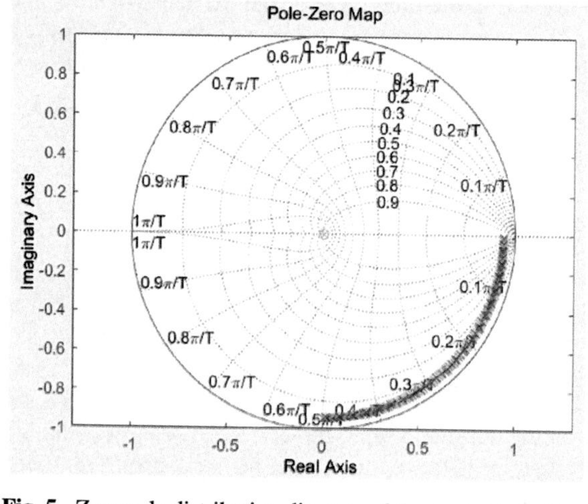

Fig. 5. Zero pole distribution diagram of the controlled object.

According to the design premise of the minimum beat system, the delay link z^{-2} in $G^r(z)$ should be included in the zero point of $\Phi(z)$. In addition, according to the additional conditions of the ripple-free minimum beat system, since $G^r(z)$ has no zero point, $\Phi(z)$ is not required to cancel. So let $M(z) = a + bz^{-1}$, where coefficients a and b are to be determined, select

$$\Phi(z) = z^{-2}M(z) = az^{-2} + bz^{-3} \tag{23}$$

Based on the minimum beat condition, the closed-loop transfer function and error transfer function under unit step input are

$$\begin{cases} \Phi(z) = z^{-1} \\ \Phi_e(z) = 1 - z^{-1} \end{cases} \tag{24}$$

Since $\Phi(z)$ is required to increase by one order when there is no ripple, and the realizable condition of $D(z)$ determined by Eq. (21) is that the number of zeros of $D(z)$ should not be greater than its pole points, so $\Phi_e(z)$ should also increase by one order. Under the condition that the minimum beat is required, select

$$\Phi_e(z) = (1 - z^{-1})(1 + cz^{-1}) \tag{25}$$

where c is to be determined, it can be obtained.

$$\Phi(z) = 1 - \Phi_e(z) = (1 - c)z^{-1} + cz^{-2} \tag{26}$$

Make Eq. (23) and Eq. (26) equal, and the coefficient can be solved.

$$\begin{cases} a = 1 \\ b = 0 \\ c = 1 \end{cases} \tag{27}$$

So

$$\begin{cases} \Phi(z) = z^{-2} \\ \Phi_e(z) = (1 - z^{-1})(1 + z^{-1}) \end{cases} \tag{28}$$

It can be seen from the above that $\Phi(z)$ already contains the transfer function delay z^{-1} of $G^r(z)$. Therefore, $\Phi(z)$ meets the delay link z^{-1} in cancellation $G^r(z)$, so $D(z)$ can be calculated according to Eq. (21), which can ensure that the given system becomes the minimum beat system under the action of $r(t) = 1(t)$. The pulse transfer function of the controller is

$$D(z) = \frac{R_s e^{-\frac{R_s T_s}{\delta L_s}} (e^{\frac{R_s T_s}{\delta L_s}} e^{j\omega_s T_s} - z^{-1}) e^{j\omega_s T_s}}{(1 - e^{-\frac{R_s T_s}{\delta L_s}})(1 - z^{-2})} \tag{29}$$

5 Matlab/Simulink Simulation and Test Results

To verify the performance of the IM complex vector current controller based on ripple-free minimum beat system, the algorithm is simulated and tested. The IM parameters used in the simulation are consistent with those used in the following tests. The specific parameters are shown in the following table:

5.1 Matlab/Simulink Simulation Experiment Results

In the Matlab/Simulink simulation environment, the control system simulation model based on asynchronous motor was built and the simulation test was carried out. The simulation System consists of two parts, the main circuit simulation module (Inverter & Motor) and the Control Algorithm simulation module (Control System), as shown in Fig. 6. The simulation system models are discretized and run in discrete time mode.

Table 1. Parameters of'the traction asynchronous motor.

Parameters and unit	Value
Rated power,KW	275
Rated voltage,V	1250
Rated current,A	150
Rated speed,Rpm	2100
Stator resistance,Ω	0.072
Stator leakage inductance,H	0.029
Rotor leakage inductance,H	0.03
Stator-rotor mutual inductance,H	0.029
Number of pole pairs	2

The main circuit simulation module is mainly composed of asynchronous motor model and three-phase two-level inverter model to realize mechanical and energy conversion simulation. Control algorithm simulation module, mainly composed of the custom M function model and basic calculation model, through the main circuit asynchronous motor three-phase current and speed signal, inverter bus voltage and output current for real-time sampling, after the custom function model calculation output PWM pulse, control inverter to complete the asynchronous motor traction, braking torque characteristics. The sampling rate of the simulation system is consistent with the simulation step size, that is, each sampling point is calculated once, and a PWM pulse is output, which belongs to the discrete step size simulation system.

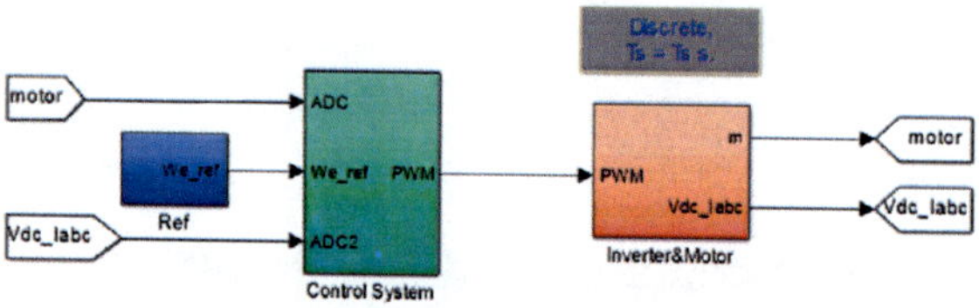

Fig. 6. Control algorithm simulation system.

Under the condition that the given excitation is unchanged, when the motor speed runs to 50[Hz], the torque current suddenly increases. When the motor speed runs to 60[Hz], the torque current decreases to the original value again. The typical PI current controller and the complex vector current controller with digital control delay compensation are used to compare the performance of the two current controllers. The torque current instruction fluctuates with the voltage fluctuation of DC side because the control item of DC side voltage oscillation is added to the torque current instruction.

Figure 7 shows the simulation waveform in the process of torque current rise (9.6–9.7[s]) and fall (10.3–10.5[s]) with the ordinary PI current controller. When the torque current surges or falls, both the torque current feedback value IqFdb and the excitation current feedback value IdFdb can track the instruction value (IqRef, IdRef) after control. However, in the process of the torque current rising and falling, the torque current feedback value IqFdb has a large overshooting, and the effect of following the instruction value is poor. When the excitation current instruction value IdRef is basically unchanged, the excitation current feedback value IdFdb fluctuates greatly. Therefore, it can be concluded that the performance of common PI current controller is poor in eliminating current coupling and dynamic response.

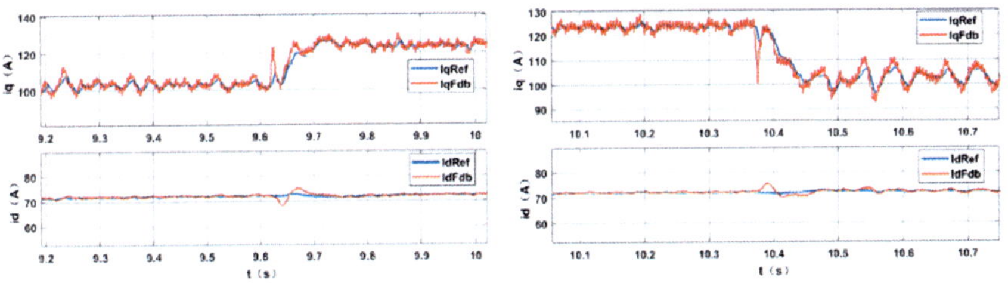

(a) Sudden increase of the motor torque current (b) Sudden decrease of the motor torque current

Fig. 7. Performance Test Waveform of Common PI Current Controller.

It can be seen from Fig. 8 that with the complex vector current controller, including time delay compensation, when the torque current suddenly increases or decreases, both the torque current feedback value IqFdb and the excitation current feedback value IdFdb can track the command value (IqRef, IdRef) through control. The torque current feedback value IqFdb can also track the command value well during the rise (9.6–9.7[s]) and fall (10.3–10.5[s]) of the torque current, with slight overshoot, The fluctuation of excitation current feedback value IdFdb is also tiny. From this, it can be seen that the complex vector current controller with time delay compensation effectively improves the dynamic response of the control system and eliminates the cross-coupling of d-axis and q-axis currents.

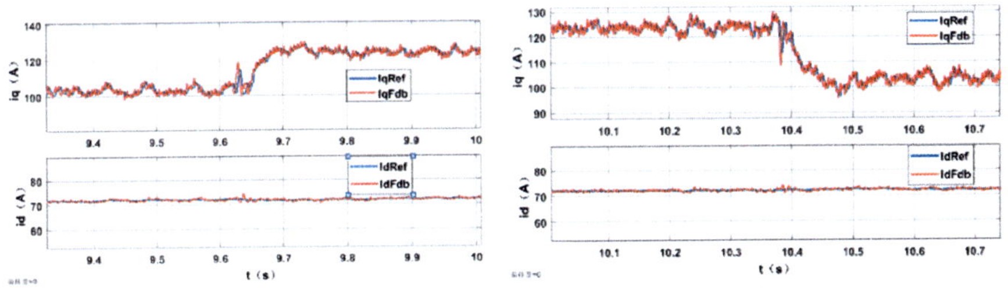

(a) Sudden increase of the motor torque current (b) Sudden decrease of the motor torque current

Fig. 8. Performance test waveform of current complex vector controller, including digital control delay compensation.

Compared with the simulation results in Figs. 7 and 8, it can be seen that the current complex vector controller, including digital control delay compensation, has better dynamic performance in the process of torque current change, and the current fluctuation is small, indicating that the current controller has better performance.

5.2 Combined Test Results

Using the CRH6F-A intercity EMU IM traction converter based on DSP TMS320F28335, the complex vector current controller algorithm, including digital control delay, compensation is tested and verified. The motor parameters are shown in Table 1, and the test equipment and measuring instruments are shown in Fig. 9.

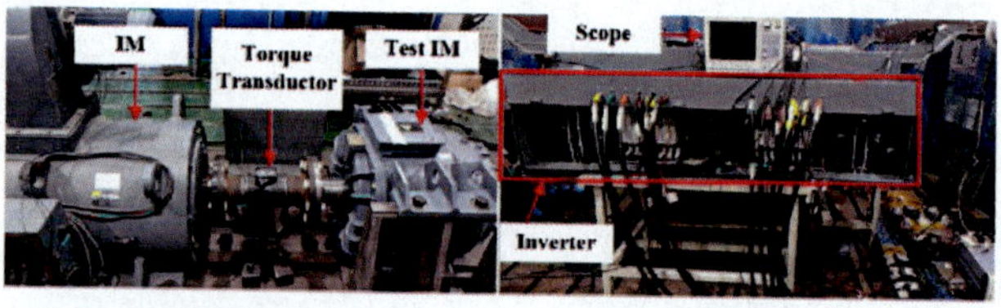

Fig. 9. Experimental prototype of IM.

Under the condition that the given excitation current remains unchanged, when the motor speed runs to 50[Hz], the given torque current suddenly increases. When the motor speed runs to 60 [Hz], the torque current decreases to the original value again. At the same time, the instruction value (IdRef) and feedback value (IdFdb) of excitation current and the instruction value (IqRef) and feedback value (IqFdb) of torque current are output to the DL850 recorder through optical DA for display, and the dynamic response and current decoupling effect when the given torque current changes are observed. The waveform is shown in Fig. 9.

Figure 10a shows the excitation current and torque current waveforms when the torque current suddenly increases from about 50[A] to about 76[A]. Figure 10b shows the excitation current and torque current waveforms when the torque current component suddenly increases from about 76[A] to about 50[A]. It can be seen that the excitation current has almost no change whether the torque current suddenly increases or decreases. In addition, the torque current feedback value IqFdb and the excitation current feedback value IdFdb can track the command value (IqRef, IdRef) through control, which not only has good dynamic response performance, but also realizes the effective decoupling of excitation current and torque current. Figure 10c, d show the waveform of UV line voltage and U-phase current when the torque current surges and dips. It can be seen that in the process of torque current surges and dips, the UV line voltage and U-phase current change stably without fluctuation, and the sinusoidal degree of current waveform is good, indicating that the current harmonics are small in the process of torque current surges or dips.

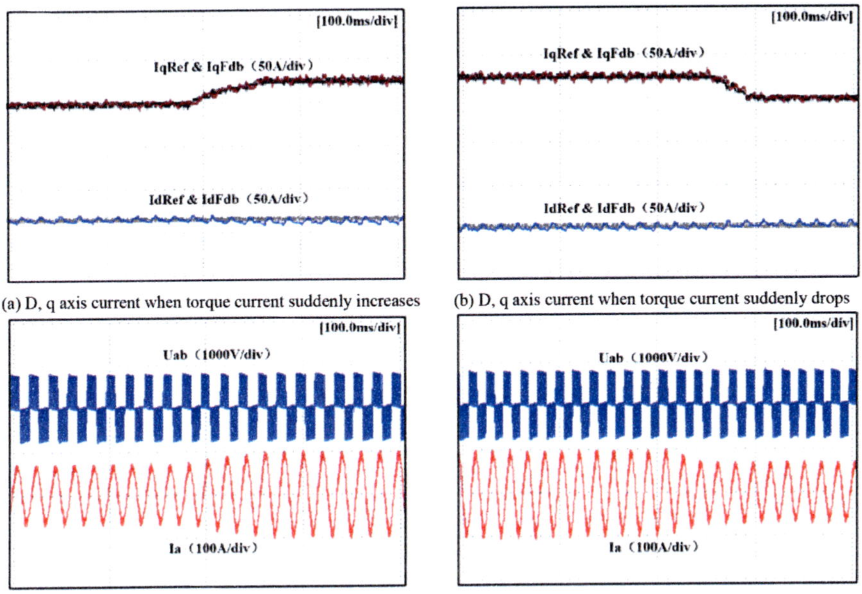

(a) D, q axis current when torque current suddenly increases (b) D, q axis current when torque current suddenly drops

(c) voltage and current when torque suddenly increases (d) voltage and current when torque drops suddenly

Fig. 10. Performance test waveforms of the current complex vector controller with digital control delay compensation.

6 Conclusions

In view of the shortcomings of conventional PI current controller at low switching frequency, the discrete mathematical model of IM, containing digital control delay, was analyzed based on the concept of complex vector, and the complex vector current controller, containing digital control delay compensation, was achieved. Through simulation test and test verification by using the CRH6F-A converter of intercity EMU, the following conclusions can be drawn: Simulation and experimental results show that the proposed complex vector current controller with digital control delay can effectively improve the dynamic coupling problem between the stator excitation current and torque current caused by low switching frequency. Disadvantages: Due to the tight test time, only the complex vector current controller algorithm including digital control delay compensation was verified on the ground test bench, and the conventional PI current controller algorithm was not verified and compared.

References

1. Oikonmou, N.: Control of medium-voltage drives at very low switching frequency, Wupperta. Logos Verlag, pp. 3–8 (2008)
2. Holtz, J., Xin, Q.: Optimal control of medium-voltage drives-an overview. IEEE Trans. Industr. Electron. **60**(12), 5472–5481 (2013)

3. Pontt, J., Rodriguez, J., Newman, P., Liendo, A., Holtz, J.: Network-friendly low-switching-frequency multipulse high-power three-level PWM rectifier. IEEE Trans. Industr. Electron. **56**(4), 1254–1262 (2009)
4. Holtz, J.: On the spatial propagation of transient magnetic fields in AC machines. IEEE Transactions on Induction Applications **32**(4), 927–937 (1996)
5. Holtz, J., Quan, J., Pontt, J., Rodriguez, J., Miranda, H.: Design of fast and robust current regulators for high-power drives based on complex state variables. IEEE Trans. Ind. Appl. **40**(5), 1388–1397 (2004)
6. Franklin G.F., Powell J.D., Naeini A.E.: Feedback control of dynamic systems. 6th Edition, New York. Prentice Hall, pp. 43–55 (2009)
7. Briz F, Degner M.W.: Analysis and design of current regulators using complex vectors, IEEE Transactions on Industry Applications, Vol. 36,No. 3, pp. 817–825 (2000)
8. Ghoshal, A., John, V.: Active damping of LCL filter at low switching to resonance frequency ratio. Power Electronics Iet **8**(4), 574–582 (2015)
9. Kwon, Y., Kim, S., Sul, S.: Six-Step operation of PMSM with instantaneous current control. IEEE Trans. Ind. Appl. **50**(4), 2614–2625 (2014)
10. Chattopadhyay, A.K., Meher, N.: Microprocessor implementation of a state feedback control strategy for a current source inverter-fed induction motor drive. IEEE Trans. Power Electron. **4**(2), 279–288 (2002)
11. Vukosavic, S.N., Peric, L.S., Levi, E.: Digital current controller with error-free feedback acquisition and active resistance. IEEE Trans. Ind. Electron. **65**(3), 1980–1990 (2018)
12. Liang, Y., et al.: Digital current controller design for SPMSM with low switching-to-fundamental frequency ratios. IEEE Trans. Ind. Appl. **58**(4), 4685–4697 (2022)
13. Jain, J.K., Ghosh, S., Maity, S.: Concurrent PI controller design for indirect vector controlled induction motor. Asian Journal of Control **22**(1), 1–13 (2020)
14. Utkin, V.I.: Sliding mode control design principles and applications to electric drives. IEEE Trans. Industr. Electron. **40**(1), 23–36 (2002)
15. Underwood, S.J., Husain, I.: Online parameter estimation and adaptive control of permanent-magnet synchronous machines. IEEE Trans. Industr. Electron. **57**(7), 2435–2443 (2010)
16. Jafarzadeh, S.: State Estimation of Induction Motor Drives Using the Unscented Kalman Filter. IEEE Trans. Industr. Electron.Industr. Electron. **59**(11), 4207–4216 (2012)
17. Powell, G.F., Franklin, J.D., Abbas, E.: Feedback Control of Dynamic Systems. 7th Edition, Pearson, pp. 601–613 (2015)

Design and Analysis of Energy Storage Converters for Regenerative Braking Energy Absorption in Urban Rail Transit

Yong Jin[1(✉)], Weijie Pu[1], Xianjin Huang[1], Zhongping Yang[1], and Seiki Igarshi[2]

[1] School of Electrical Engineering, Beijing Jiaotong University, Beijing 100044, China
{21121431,21121470,xjhuang,zhpyang}@bjtu.edu.cn
[2] Fuji Electric Co., Ltd, Tokyo, Japan
igarashi-seiki@fujielectric.com

Abstract. Nowadays, the development of urban rail transit is getting faster and faster, but its consequent electric consumption problem is getting more attention, and how to better solve the problem of regenerative braking energy absorption and utilization, which accounts for a larger portion of electric consumption, is gradually becoming one of the hot spots of research nowadays. This paper adopts energy storage technology and designs a wayside energy storage system (WESS). Firstly, the commonly used topologies of energy storage converters are compared, and finally a three-phase staggered parallel fly-across capacitor DC/DC converter is selected as the topology of energy storage converter; then its working characteristics are described from the three-level technology and staggered parallel technology respectively; then the WESS is added into the existing traction power supply system and MATLAB simulation is established. Finally, the simulation is carried out to verify the characteristics of the energy storage converter and the regenerative braking energy absorption effect. The results show that the designed WESS can be better adapted to the actual operation of urban railways and achieve the maximum utilization of energy, which provides good technical support for its energy-saving operation.

Keywords: Urban rail transit · Energy storage converter · Regenerative braking energy

1 Introduction

By the end of 2022, 55 cities nationwide had opened rail transit, with a total operational mileage of 10,291.95 km, increasing 1,085.17 km compared with that in 2021 [1]. The increasing scale of the urban rail network is also gradually making urban rail transit gradually become one of the preferred ways for people to take a short-distance travel. The increase of operating mileage also brings about the increase of power loss. In 2022, the total electricity consumption of national urban rail reached 227.92 kWh, with an increase of 6.39% year-on-year, among which traction energy consumption accounted for nearly 50% [2]. With the expansion of the road network afterwards, the energy consumption will further increase.

L. Jia et al. (Eds.): EITRT 2023, LNEE 1135, pp. 247–254, 2024.
https://doi.org/10.1007/978-981-99-9307-9_27

Due to the small station spacing of urban rail, coupled with the significant shortening of train intervals on the network during peak periods, it leads to frequent acceleration and deceleration of trains during operation. Literature [3] statistics the power consumption flow of urban rail transit, in which regenerative braking energy in traction energy consumption accounted for about 30–60%, if this part of the energy can be absorbed and used, it can effectively reduce the energy consumption of during operation.

Regenerative braking energy absorption and utilization of urban rail trains can be achieved in three main ways, such as train operation diagram optimization [4, 5], inverter feedback [6] and energy storage [7]. The specific optimization effect of train diagram optimization is not obvious, and also brings higher requirements for train operation organization. The inverter feed-back method has the problems such as poor quality of the power fed back to the AC grid and difficult quantification of energy saving effect statistics.

At present, energy storage technology is increasingly used in the field of urban rail, and its basic composition block diagram is shown in Fig. 1, including energy storage medium, energy storage converter and connection and other parts. Energy storage medium absorbs and releases energy according to the real-time changes of traction network pressure.

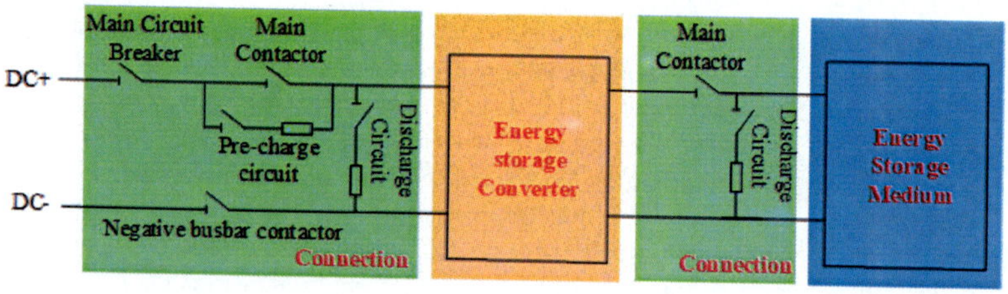

Fig. 1. Block diagram of basic components of WESS for urban rail transit.

According to the installation location of the energy storage system, it can be divided into urban rail WESS and urban rail OESS (on-board energy storage system) [8]. On-board energy storage systems are generally installed on the top or bottom of the train with low energy loss. However, due to its direct connection to the train, its weight and size are more limited, which also directly limits its configurable capacity; WESS are generally installed in traction substations. Because they are independent of each other and do not affect the trains, their volume and weight requirements are more relaxed, and installation and maintenance are relatively easy [9, 10]. Therefore, WESS is more widely used.

In this paper, supercapacitor is adopted as the energy storage medium and the three-phase interleaved parallel fly-across capacitor DC/DC converter is used as the energy storage converter topology. The operating principle and characteristics of this topology are analyzed in two dimensions: three-level technology and interleaved parallel technology. The simulation model of a complete WESS for urban rail on the basis of traction power supply system is built in MATLAB/Simulink, and the simulation is verified in terms of the operating characteristics of the energy storage converter and the effect of the WESS on regenerative braking energy absorption.

2 Design and Analysis of Bidirectional DC/DC Energy Storage Converters

Energy storage converter is the intermediate link of energy storage medium to the traction power supply system of urban rail, and undertakes the role of voltage level conversion and energy storage medium configuration and capacity management.

Energy storage converter can be divided into isolated and non-isolated converters according to whether the transformer is included in the converter topology. WESS used in urban rail transit have no isolation or insulation requirements and are of high power, so non-isolated bi-directional DC/DC converters are more suitable for the application in terms of device efficiency and economic efficiency.

A basic non-isolated bidirectional DC/DC converter includes: bidirectional Buck-Boost converter, bidirectional half-bridge converter, bidirectional Cuk converter and bidirectional Sepic converter. However, the outputs of the converters above are all two-level, which may bring about problems like static, dynamic, pressure and flow equalization, so the three-level converters are proposed, among which Buck-Boost with three levels and three levels with flying-across capacitors.

According to the application of energy storage converters, namely higher voltage in DC traction network as well as larger output current, finally the bidirectional three-levels with flying-across capacitors. Then combined with interleaved parallel technology, three phase bidirectional DC/DC with flying-cross capacitor using interleaved parallel technology is formed, which is shown in Fig. 2.

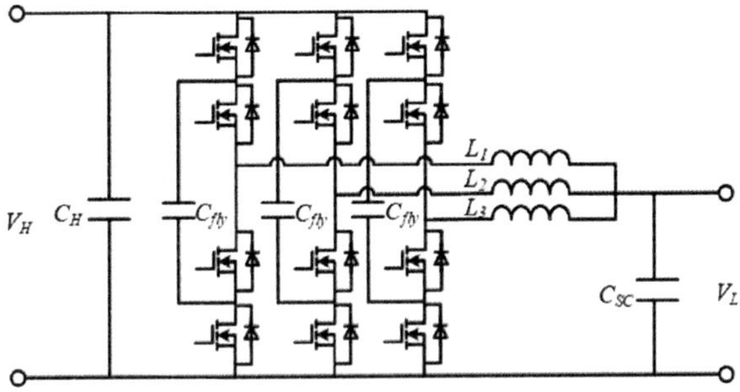

Fig. 2. Three phase bidirectional DC/DC converter with flying-cross capacitor using interleaved parallel technology.

The application of three-level technology reduces the inductor current ripple and the voltage stress on the switching tubes, but the ripple current injected into the divider capacitor is not reduced, making it necessary for the capacitor to absorb a larger ripple current. The advantage of the interleaved parallel technology is that it reduces the input and output current ripple after aggregation and extends the current output capability, but does not reduce the current ripple of individual inductors.

Combining the advantages of the three-level technology with the interleaved parallel technology make it easy to realize high power density DC-DC converters.

The following part will introduce the working principles in terms of the three-level technology and the interleaved parallel technology.

2.1 Working Principles and Characteristic Analysis of Single-Phase Three-Level DC/DC Converter with Flying-Across Capacitor

The topology and electrical parameters of the three phases are the same as Fig. 3, except that the switching tubes of each phase are interleaved, so one of the phases is used as an example for analysis.

The structure of the single-phase DC/DC converter with flying-across capacitors is shown as Fig. 3. V_H refers to DC voltage from the input; C_H refers to the support capacitors; L refers to output filters; T_1 ~ T_4 are the four power switches. T_1 and T_2, T_3 and T_4 are respectively in interleaved conduction, whose driving signals have a 180°phase shift. T_1 and T_4, T_2 and T_3 are in supplementary conduction. C_{fly} refers to the flying-cross capacitor, with the working voltage $U_{Cfly} = U_H/2$.

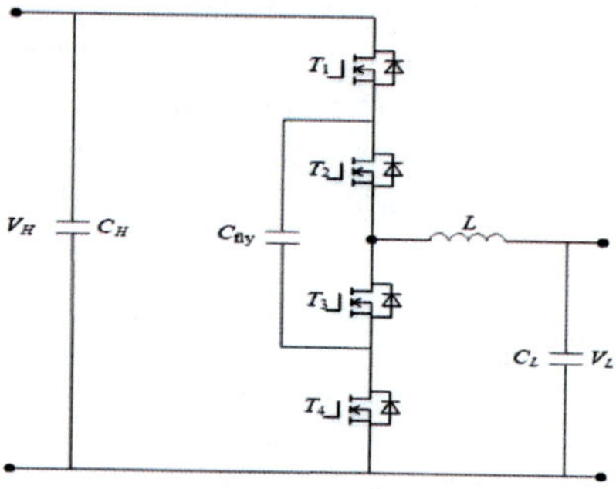

Fig. 3. Single phase three-level DC/DC converter with flying-across capacitor.

The working stages and working waveforms of the topology are shown respectively as Fig. 4. When the converter works in the stage of Buck and Boost, the only difference is the opposite direction of inductive current. Take the stage of Buck as the example to illustrate the working principles of the topology.

When the topology works in the stage of Buck with $D < 0.5$:

(1) t_0–t_1: T_1 and T_2 conduct, with an increase of inductor current, $U_{AB} = U_H$, C_{fly} does not work.
(2) t_1–t_2: T_1 and T_3 conduct, with a decrease of inductor current, $U_{AB} = U_H/2$, C_{fly} charges.
(3) t_3–t_4: T_2 and T_4 conduct, with a decrease of inductor current, $U_{AB} = U_H/2$, C_{fly} discharges.

When the topology works in the stage of Buck with $D < 0.5$:

(1) t_0–t_1: T_1 and T_3 conduct, with an increase of inductor current, $U_{AB} = U_H/2$, C_{fly} charges.

(2) t_1–t_2: T_3 and T_4 conduct, with a decrease of inductor current, $U_{AB} = 0$, C_{fly} does not work.

(3) t_1–t_2: T_1 and T_3 conduct, with an increase of inductor current, $U_{AB} = U_H/2$, C_{fly} discharges.

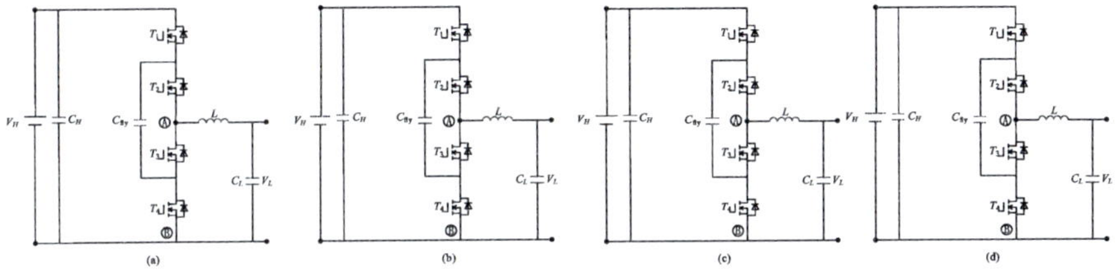

Fig. 4. Working stages of single-phase three-level converter with flying-across capacitor.

Make a deeper analysis of the main working waveforms of single-phase three-level DC/DC converter with flying-across capacitor, and it can be found that the external characteristic of single-phase three-level DC/DC converter with flying-across capacitor is the same as the conventional two-level bidirectional DC/DC converter, but the output equivalent switching frequency is twice that of the two-level DC/DC converter, so the three-level converter can effectively reduce the volume of inductor with constant output current ripples. Furthermore, three-level converter has less AC component of output voltage compared with two-level converter.

2.2 The Implementation of Interleaved Parallel Operation of Three-Phase Three-Level DC/DC Converter with Flying-Across Capacitor

Connect the three topologies shown in Fig. 3 in parallel with each other at the inputs and outputs, a three-phase topology is obtained as shown in Fig. 2. The three-phase topology remains the same in terms of structure, electrical parameters and control strategy, and the only difference is the phase of the three-phase carriers, which is achieved by applying interleaved parallel technology. Interleaved parallel technology means that the control signals remain the same, but the carrier phases are different. Unlike the synchronous operation mode, phase lag time T_s/n for each phase bridge arm in turn (where T_s refers to the switching period and n refers to the number of interleaved parallel bridge arms).

When the three-phase converter is operated in interleaved parallel, the phase angle of each phase bridge arm is lagged by 2π/3 in turn, i.e. Triangular carrier is shifted by 2π/3 in turn, and the total output current ripple is a interleaved addition of 120° phase difference in the output current ripple of the respective three-phase bridge arm, as is shown as Fig. 5. The amplitude of the total output current ripple is reduced and the ripple frequency of the total output current is three times the ripple frequency of the bridge arm current, reducing the design requirements for the output filter.

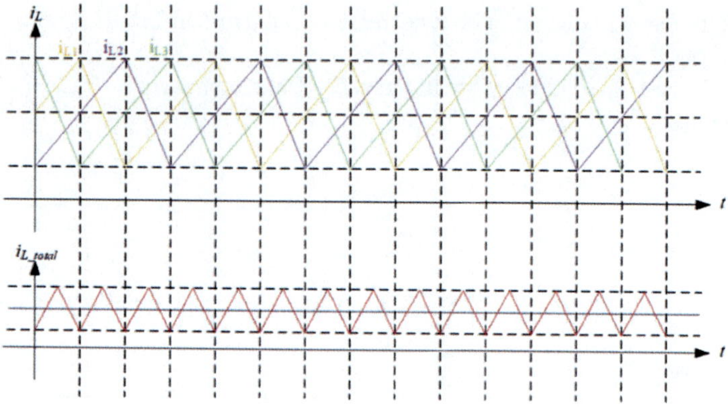

Fig. 5. The inductor current waveforms of three-phase three-level DC/DC converter with flying-across capacitor

3 Simulation Analysis of Urban Rail Wayside Energy Storage System

In order to verify the control effect of three-phase three-level DC/DC converter, a simulation model is built to verify the operation performance from the converter itself and the influence on the whole traction power supply system of urban rail transit.

3.1 Simulation of the Characteristics of a Three-Phase Interleaved Parallel Fly-Across Capacitor DC/DC Converter

According to the topology of the energy storage converter in Fig. 2, the main circuit topology of the WESS for urban rail transit is constructed.

In order to verify the performance of the three-phase three-level converter by simulation, the sizes of the main parameters of the converter in the simulation model are set according to the actual design values.

According to the above simulation model, the basic functions of the converter are simulated and verified. First, the current of the low voltage side is changed from 600 A to 400 A in Buck mode. Figure 6a, b show the voltage and current change curves in Buck mode and Boost mode respectively.

It can be seen from the simulation that the voltage of the flying capacitor is always stable at half of the voltage of the high voltage side when the converter cabinet is running, which reduces the voltage stress of the switching device. However, the current of the three-phase bridge arm is basically the same as one-third of the given value of the current, and the current equalization degree is high. Moreover, the three-phase interleaved parallel structure also reduces the size of the total charge and discharge current ripple.

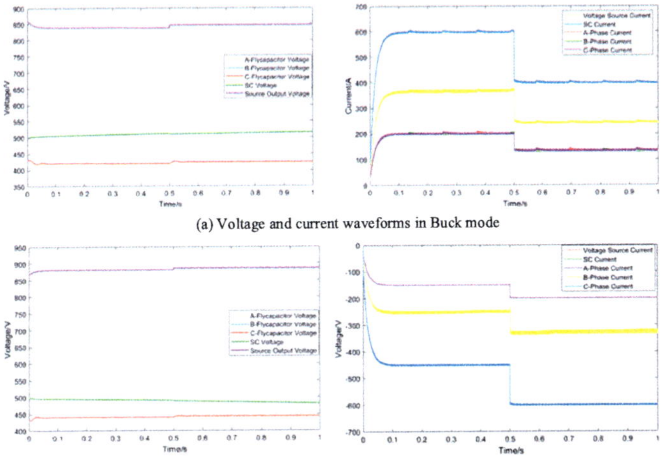

(a) Voltage and current waveforms in Buck mode

(b) Voltage and current waveforms in Boost mode

Fig. 6. Waveforms in Buck mode and Boost mode.

3.2 Simulation Analysis of WESS for Urban Rail Transit

In order to analyze the evaluation of the regenerative braking energy absorption effect after adding the WESS as accurately as possible, using the operating parameters and actual operating conditions of the actual line of Beijing Metro Batong Line.

From the simulation results shown in Fig. 7, it can be seen that the designed urban rail ground energy storage system can absorb and release energy according to the changes of train operating conditions and traction network pressure, reduce the fluctuation of network pressure, and further enhance the effect of regenerative braking energy absorption.

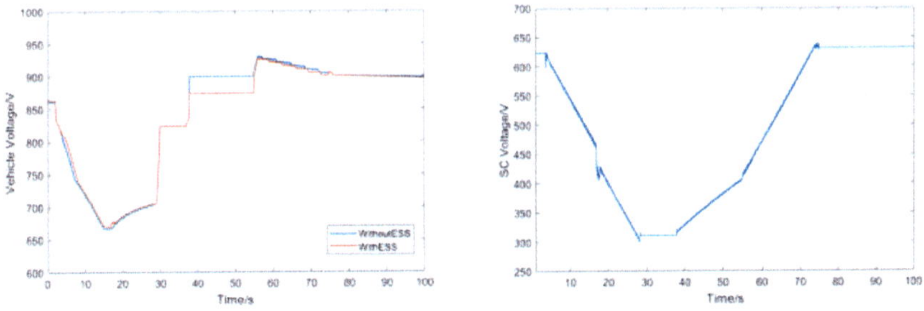

(a) Train voltage variation curves without WESS and with WESS.

(b) Voltage curve of wayside energy storage system.

Fig. 7. Simulation results of the application of WESS compared with that without WESS.

4 Conclusion

In this paper, the WESS is proposed to more effectively absorb and utilize the regenerative braking energy of urban rail transit. Three-phase interleaved parallel fly-across capacitor DC/DC converter topology is selected as the main circuit of energy storage converter, and its working principle is analyzed from the three-level technology and interleaved parallel technology respectively. In order to verify the effect of the WESS on regenerative braking energy absorption, a complete WESS model is built in MATLAB/Simulink. The result shows that the designed WESS for urban rail transit can effectively absorb and release energy according to the traction network pressure, reduce the abnormal fluctuation of traction network, and provide a method to guide the stable and energy-saving operation of urban rail.

Acknowledgement. This work was supported by the BJTU and Fuji Electric joint projects " Research on all SiC DC/DC converter for wayside hybrid energy storage system in urban rail transit".

References

1. Hou, X., Feng, C., Zuo, C., et al.: Statical Analysis of Urban Rail Transit in Chinese Mainland in 2022. Urban Rapid Rail Transit **36**(01), 9–13 (2023). (in Chinese)
2. China Urban Rail Transit Association: Annual Statistics and Analysis Report of Urban Rail Transit in 2022. China Metros **86**(04), 13–15 (2023). (in Chinese)
3. Ding, S.: Implementing six major actions to build green city rail— Interpretation of “the Action Plan for the Development of Green Urban Railways in Chinese Cities.” China Metros **10**, 36–40 (2022). (in Chinese)
4. Zheng, Y., Li, Y., Li, Y., et al.: Energy-saving optimization of subway train operating charts under regenerative braking conditions. Journal of South China University of Technology (Natural Science Edition) **49**(07), 1–7 (2021). (in Chinese)
5. Zhang, G., Tian, Z., Tricoli, P., et al.: Inverter Operating Characteristics Optimization for DC Traction Power Supply Systems. IEEE Trans. Veh. Technol. **68**(4), 3400–3410 (2019)
6. Yang, X., Li, X., Ning, B., et al.: A Survey on Energy-Efficient Train Operation for Urban Rail Transit[J]. IEEE Trans. Intell. Transp. Syst. **17**(1), 2–18 (2016)
7. Liu Y., Yang Z., Wu X., et al: Adaptive Threshold Adjustment Strategy Based on Fuzzy Logic Control for Ground Energy Storage System in Urban Rail Transit. IEEE Transactions on Vehicular Technology, (70–10) (2021)
8. Yang Z., Yang Z., Xia H, et al: Brake Voltage Following Control of Supercapacitor-Based Energy Storage Systems in Metro Considering Train Operation State. IEEE Transactions on Industrial Electronics, pp (99):1–1(2018)
9. Yang Z, Yang Z, Xia H, et al: Supercapacitor State Based Control and Optimization for Multiple Energy Storage Devices Considering Current Balance in Urban Rail Transit. Energies, 10(4):520-(2017)
10. Zhu, F., Yang, Z., Xia, H., et al.: Hierarchical Control and Full-range Dynamic Performance Optimization of Supercapacitor Energy Storage System in Urban Railway. IEEE Trans. Industr. Electron.Industr. Electron. **65**(8), 6646–6656 (2018)

Network-Side Power Optimization Control Considering Photovoltaic and Energy Storage Connection to RPC-Type Traction Power Supply System

Xiuqing Mu, Xiaoqiang Chen, Ying Wang(✉), Qiaoli Yang, and Yunpeng Zhang

Lanzhou Jiaotong University, Lanzhou 730070, China
wangying01@mail.lzjtu.cn

Abstract. Multi-source power supply for electrification of the railroad is essential for having a sustainable energy supply to drive critical EMU loads. Firstly, the working theory of railway power conditioner (RPC) is analyzed, and the "optical-storage-locomotive-network" model is built. Secondly, the specific working mode is proposed. Thirdly, the energy flow under various modes of the RPC converter are analyzed, and a cooperative control scheme for compensating the output power of the traction network by the photovoltaic storage system through the railway power conditioner is presented. Finally, the cooperative control scheme is verified and analyzed by Starsim HIL real-time test platform.

Keywords: Railway power conditioner (RPC) · Traction network output power · Photovoltaic and energy storage · Optimization control · Test verification

1 Introduction

With the high density and high speed development of electrified railways, it is urgent to carry out green and efficient transformation of its energy structure [1, 2]. Electrified railway relies on power electronic converter technology, and constructs a new "source-network-load-storage" consolidated power supply system [3]. Currently, the access methods are broadly categorized into two types: access to the battery device from the locomotive and from the traction network, both, can accomplish the recovery of braking energy to traction network voltage fluctuations [4, 5]. The paper [6] indicates that relying on RPC can help address the power quality issues in traction power supply systems. The literature [6–9] used traction power supply systems with different RPC topologies and conducted research on the issues of unbalanced power compensation and harmonic control. In this paper, firstly, an analysis of the topology of photovoltaic and energy storage traction power supply systems was conducted, and corresponding control strategies were designed for different modes. At the same time, an analysis of the different operating modes of the system was carried out, and a collaborative control scheme for compensating the traction network output power was proposed. Finally, cooperative control scheme effectiveness has verified by Starsim HIL real-time test platform.

L. Jia et al. (Eds.): EITRT 2023, LNEE 1135, pp. 255–263, 2024.
https://doi.org/10.1007/978-981-99-9307-9_28

2 Modeling of "Optical-Storage-Locomotive-Network" System

Figure 1 shows the topology of the " optical -Storage-Location-Network" system, which consists of four components: (1) The 220kV three-phase voltage is converted to 27.5kV two-phase voltage through the V/v traction transformer; (2) high-speed train and power arm system; (3) RPC composed of two AC/DC converters sharing DC capacitors, two converters are linked together via large-capacity DC capacitors to ensure their stable performance; (4) an innovative energy energy power supply system consisting in photovoltaic and energy storage. In Fig. 1, the function of the bidirectional DC/DC converter is to control the charging and discharging of the battery and to enable bidirectional energy flow between the battery and the power arm.

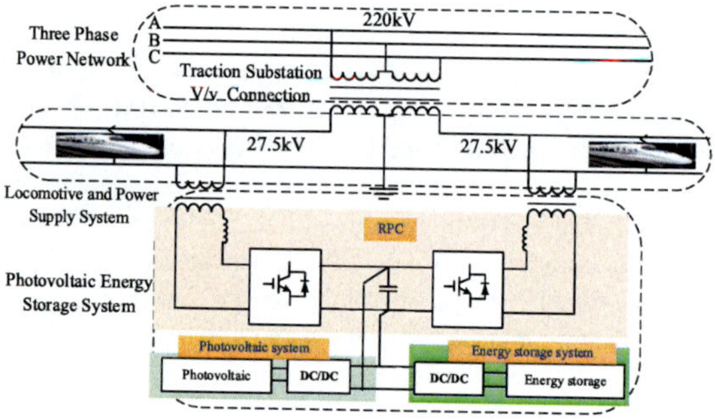

Fig. 1. Topology of the "optical-storage-locomotive-network" system

2.1 Modeling of Optical-Storage Subsystem

In Fig. 1, the optical-storage system is divided into a photovoltaic array and an energy storage system. In Fig. 2, the battery modules form the energy storage system. The output of a single solar cell is very small. The common photovoltaic array is to connect multiple photovoltaic modules in parallel. The photovoltaic cell is regarded as a nonlinear current source. By connecting several photovoltaic modules in parallel, the corresponding photovoltaic array can be obtained. In the case of a single PV module, for example, filter capacitors C_1 and C_2 are added to the photovoltaic array and DC side, respectively, in order to eliminate the effect of noise on the system. R_1 is used to equivalent the resistance of photovoltaic array, L_1 is the inductance, u_1 and i_1 are the output voltage and current of photovoltaic array respectively, i_2 is the low voltage side current, and i_3 is the high voltage side current.

The optical-storage system mentioned in Fig. 2 contains two DC/DC converters. Among them, the photovoltaic module boosts the DC voltage of the photovoltaic output through a DC/DC converter. For photovoltaic array side, the boost chopper circuit is used, which has small size and high conversion efficiency. For the conversion circuit on the battery side, the Buck-Boost converter is adopted. There are two cases: when the S_2 is locked, the freewheeling diode is turned on, the converter is boost mode, and the

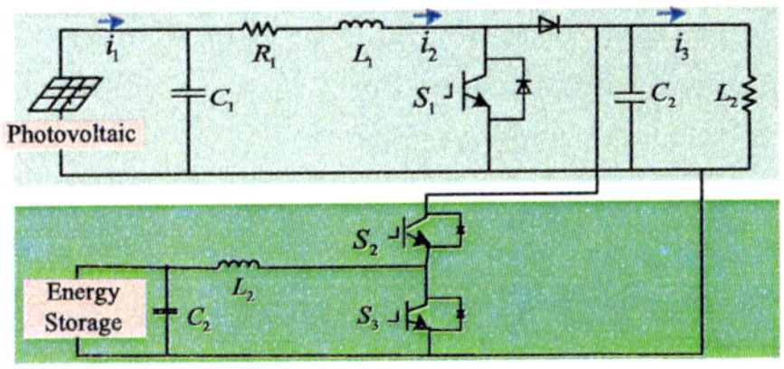

Fig. 2. Overall schematic diagram of optical-storage subsystem

outward discharge of storage batteries; when the S_3 is locked, the freewheeling diode is turned on, the converter is in Buck mode, and charging energy storage batteries.

2.2 Modeling of RPC Subsystem

The main structure of RPC consists of two back-to-back voltage source inverters. In Fig. 3, the two inverters use boost transformers to achieve their functions. The output voltage of the inverter is boosted and then inputted into the traction network.

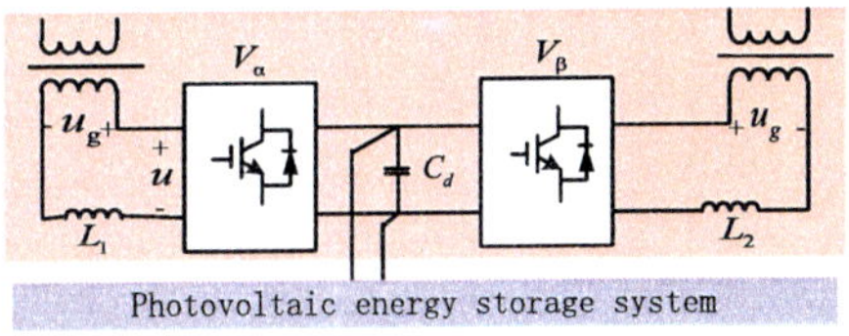

Fig. 3. RPC topology schematic diagram.

In Fig. 3, V_α and V_β are the inverters at both ends of RPC, u is inverter output voltage, and u_g is the voltage at the step-up transformer at both ends of the optical storage side. L_1 and L_2 is the inductance of inverter, C_d is DC side capacitance. The connection of RPC is achieved through two step-down transformers and two supply arms.

(1) Negative sequence current.

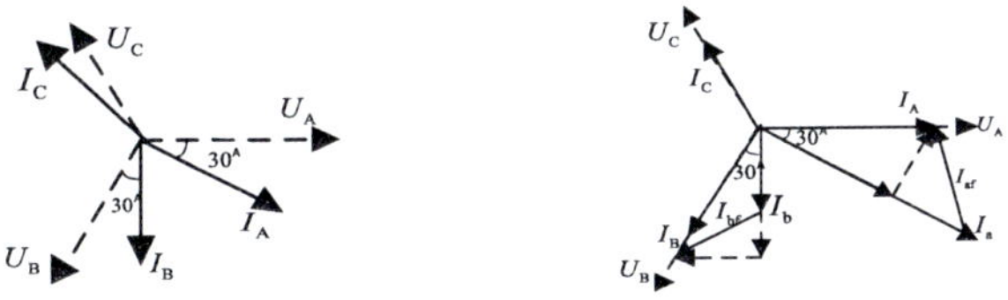

(a) Before compensation (b) After compensation

Fig. 4. Voltage and current phasor diagram of the primary side

In Fig. 1, the left and right sides represent the supply arms for Phase B and Phase A respectively. The high voltage side is a three-phase symmetrical voltage, and K represents the turns ratio of the three-phase transformer. From this, it can be inferred that the currents in each phase are not equal, and the voltages in each phase lag the respective phase currents by 30°. There is also negative sequence current in the three-phase currents. Figure 4(a) shows the phasor diagram of the voltage and current on the primary side before compensation.

For active power, RPC can complete the transfer of half of the difference between A and B two-phase load current from the heavy load side to the light load side. ΔI is represents the reactive current required for compensation.

$$\Delta I = (I_{\mathrm{af}} - I_{\mathrm{bf}})/2 \tag{1}$$

where I_{af} and I_{bf} are the fundamental currents in the supply arms of phase A and phase B, respectively. The two-phase current of A and B is expressed as follows.

$$\begin{cases} I_{\mathrm{A}'} = I_{\mathrm{A}} - \Delta I e^{-j30} \Big/ K \\ I_{\mathrm{B}'} = I_{\mathrm{B}} + \Delta I e^{-j90} \Big/ K \end{cases} \tag{2}$$

Sorting out formula (2), we can get

$$\begin{cases} I_{\mathrm{A}'} = (I_{\mathrm{af}} + I_{\mathrm{bf}}) e^{-j30} \Big/ 2K \\ I_{\mathrm{B}'} = (I_{\mathrm{af}} + I_{\mathrm{bf}}) e^{-j90} \Big/ 2K \end{cases} \tag{3}$$

From the formula (3), it can be concluded that the two-phase currents of A and B are equal, and the voltage leading current in phase A is 30°, the voltage lagging current of B-phase is 30°, and the voltage and current phases of C-phase overlap.

For reactive power, to obtain three-phase symmetrical current, the corresponding reactive current should be compensated, so that the phase of current and voltage of A and B phases is consistent. At this time, A-phase is capacitive, B-phase is inductive, A-phase emits reactive power, and B-phase absorbs reactive power. Figure 4(b) shows the phase diagram of the voltage and current on the primary side after compensating for both active and reactive power. After power compensation, complete symmetry of three-phase currents is achieved, as well as the alignment of voltage and current phases, resulting in a power factor of 1. The compensated current is obtained.

$$\begin{cases} I_{\mathrm{af}} = \left[(I_{\mathrm{bl}} - I_{\mathrm{al}}) \sin(\theta - 30^{\circ}) \right] \Big/ 2 + \left[(I_{\mathrm{al}} + I_{\mathrm{bl}}) \sin(\theta + 60^{\circ}) \right] \Big/ 2\sqrt{3} \\ I_{\mathrm{bf}} = \left[(I_{\mathrm{bl}} - I_{\mathrm{al}}) \sin(\theta - 90^{\circ}) \right] \Big/ 2 + \left[(I_{\mathrm{al}} + I_{\mathrm{bl}}) \sin(\theta - 180^{\circ}) \right] \Big/ 2\sqrt{3} \end{cases} \tag{4}$$

3 System Optimization Control Scheme

3.1 Cooperative Control and Four Working Conditions

When connecting the photovoltaic energy storage systemto the traction power supply system using RPC, it is necessary to consider the interactions and rela-tionships between photovoltaic array, energy storage modules, and the two power arms. Figure 5 illustrates

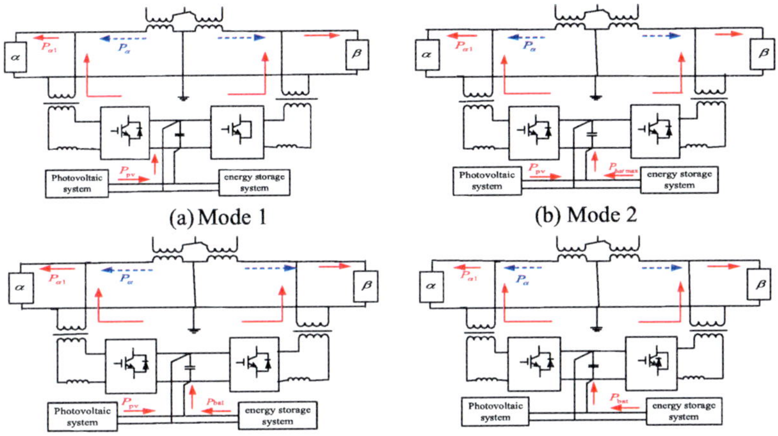

Fig. 5. Schematic diagram of energy flow from mode 1 to mode 4

the four operational states of the system. Among them, P_{dmax} is the maximum discharge power of the battery, P_{PV} is the output power of the photovoltaic, P_{bat} is the output power of the battery, P_{QY} is the traction power required for the load at both ends of the power supply arm, P_{QYW} is the output power of the traction network, and SOC_{min} is the minimum residual energy of the battery.

Mode 1: The electric locomotive works under the traction condition, the residual energy of the battery is equal to SOC_{min}, the specific mode is shown in Eq. 5.

$$\begin{cases} SOC = SOC_{\mathrm{min}} \\ P_{\mathrm{QY}} = P_{\mathrm{PV}} + P_{\mathrm{QYW}} \\ P_{\mathrm{bat}} = 0 \end{cases} \tag{5}$$

Mode 2: When the electric locomotive works in the traction condition, the residual energy of the battery is greater than SOC_{min}, the specific mode is shown in Eq. 6.

$$\begin{cases} SOC > SOC_{\mathrm{min}} \\ P_{\mathrm{QY}} = P_{\mathrm{PV}} + P_{\mathrm{QYW}} + P_{\mathrm{bat}} \\ P_{\mathrm{bat}} = P_{\mathrm{dmax}} \end{cases} \tag{6}$$

Mode 3: When the electric locomotive works in the traction condition, the residual energy of the battery is greater than SOC_{min}, the specific mode is shown in Eq. 7.

$$\begin{cases} SOC > SOC_{\mathrm{min}} \\ P_{\mathrm{QY}} = P_{\mathrm{PV}} + P_{\mathrm{QYW}} + P_{\mathrm{bat}} \\ P_{\mathrm{bat}} \neq P_{\mathrm{dmax}} \end{cases} \tag{7}$$

Mode 4: When the electric locomotive works in the traction condition, the residual energy of the battery is greater than $SOC_{\min}$. As shown in Eq. 8.

$$\begin{cases} SOC > SOC_{\min} \\ P_{QY} = P_{QYW} + P_{bat} \\ P_{PV} = 0 \end{cases} \tag{8}$$

3.2 Control of RPC Converter

In the above four modes, the braking modes are not considered in this paper, and the energy storage system only works in no power output. In the RPC model designed in this paper, VSCα is discussed here as an example. The command current form of the overlay of basic activity component, the fundamental reactive component and the characteristic harmonic component is shown in formula 9.

$$\begin{cases} i_{c\alpha}{'}(t) = k_T \times [I_{\alpha cp} \sin \omega t + I_{\alpha cq} \cos \omega t + i_{\alpha h}(t)] \\ i_{c\beta}{'}(t) = k_T \times [I_{\beta cp} \sin \omega t + I_{\beta cq} \cos \omega t + i_{\beta h}(t)] \end{cases} \tag{9}$$

In formula (9), $I_{\alpha cp}$ and $I_{\beta cp}$ are the effective values of active current components, $I_{\alpha cp}$ and $I_{\beta cp}$ are the effective values of reactive current components, and the characteristic harmonic components of the load current are represented by $i_{\alpha h}(t)$ and $i_{\beta h}(t)$. Figures 6 and 7 depict the specific control strategies. The voltage outer loop stabilizes the voltage on the DC side and corrects energy losses in the middle DC side. The current inner loop determines $i_{v\alpha}$ and is regulated through a PR controller.

Due to the existence of the filter inductance, it is necessary to normalize the obtained value with the reference value (U_{dc} is the base value), and then use it as the modulation wave to yield the pulse signal of converter. $i_{v\alpha}$ is the actual output current value of the converter, and $I'_{v\alpha ref}$ is the reference command current.

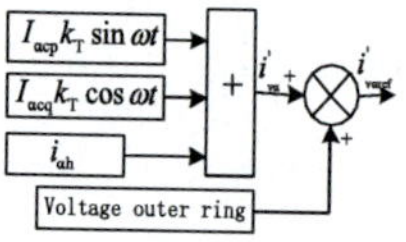

Fig. 6. Converter reference current

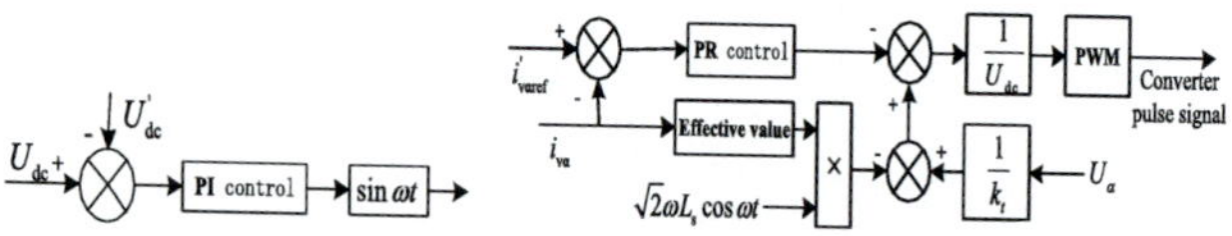

(a) Voltage outer loop (b) Current inner loop

Fig. 7. DC/DC control mode of energy storage system

3.3 DC/DC Control of Optical-Storage Energy System

DC/DC conversion of the optical-storage system is bidirectional. The DC/DC control of energy storage system is discussed here as an example. Here, the voltage outer loop and the current inner loop are used, and a novel sign variable is introduced here: charging and discharging sign current I_{savref}. When the value is higher than 0, the energy storage system is in the state of charge. The DC / DC control strategies are shown in Fig. 8 and Fig. 9.

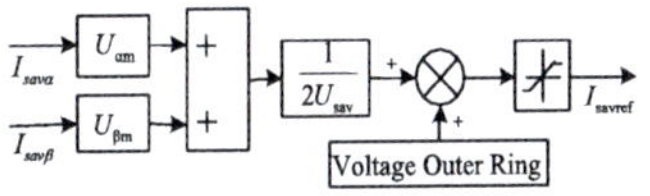

Fig. 8. DC/DC reference current

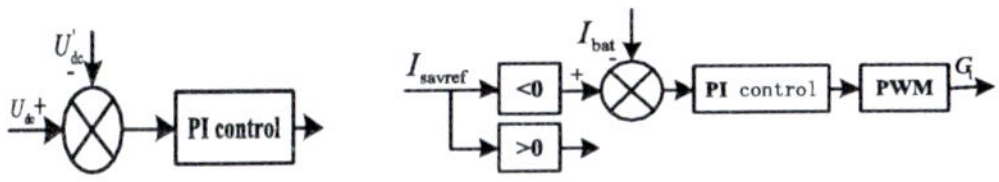

(a) Voltage outer loop (b) Current inner loop

Fig. 9. DC/DC control of energy storage system

4 Test and Analysis

In order to verify the validity of the proposed compensated output power control scheme, the test is conducted based on four modes. The specific test parameters can be found in the literature [5] published by our research team. The traction power required for the power supply arm is set to 3.23 MW, and the maximum output power of the energy storage system is 1 MW. The output power results of the optical storage system through the RPC compensated traction network are shown in Fig. 10. P_Z is the output power of the traction network before compensation, and P'_Z is the output power of the traction network after compensation.

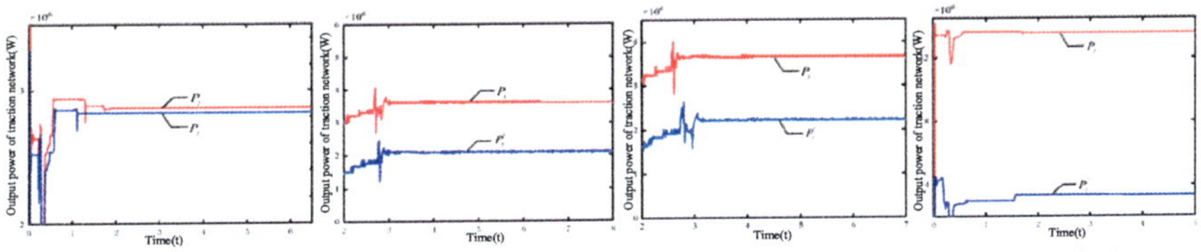

(a) Mode 1 (b) Mode 2 (c) Mode 3 (d) Mode 4

Fig. 10. Network-side output power before and after compensation under four modes

Figure 11 and Table 1 respectively display the output power and power flow curves of the optical storage under four different operating conditions.

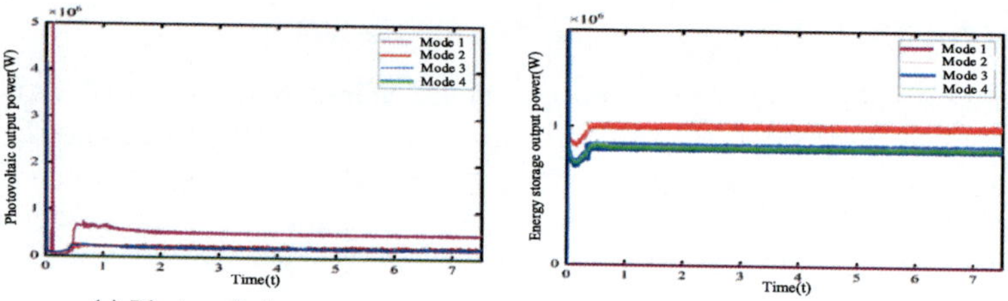

(a) Photovoltaic power (b) Energy storage power

Fig. 11. Output power diagram of optical storage under four modes after compensation

Table 1. Power flow distribution results (MW)

Mode	Output power of traction network	Tractive power	Photovoltaic output power	Energy storage output power
Mode 1	2.08	3.22	0.53	0
Mode 2	2.09	3.22	0.21	1.00
Mode 3	2.22	3.22	0.20	0.85
Mode 4	2.41	3.22	0	0.86

Taking the compensated load of the CRH5 EMU as an example, from Fig. 11, it can be observed that the control strategy proposed in this paper can compensate the output power of the optical storage system to the traction network under different operating conditions through RPC. This scheme is feasible and effective.

Figure 12 shows the further testing of the control strategy in a real environment using the Starsim HIL real-time testing platform from Shanghai Yuankuan Energy. This platform primarily consists of a host computer, real-time simulation system, oscilloscope, and other components, enabling semi-physical simulation.

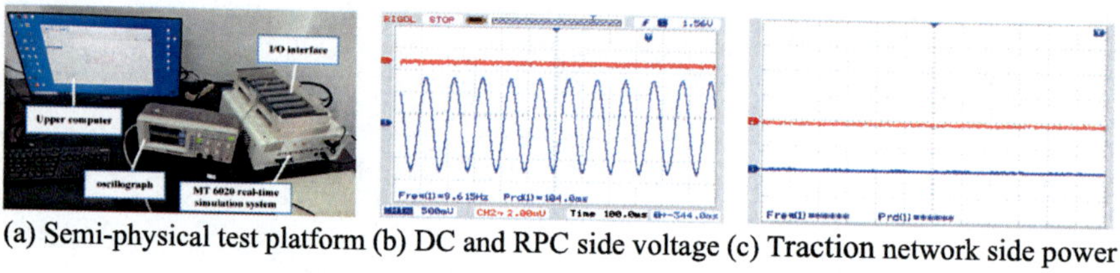

(a) Semi-physical test platform (b) DC and RPC side voltage (c) Traction network side power

Fig. 12. Starsim HIL test results (Red: before compensation; Blue: after compensation)

Taking the CRH5 EMU load as an example, in Fig. 12, the voltage wave forms on the DC side and RPC side are in line with expectations. The connection of RPC effectively compensates for the output power of the traction network. The output power of traction network after compensation is clearly lower than that before compensation.

5 Conclusion

In this paper, the issues of connecting an optical storage system and traction power supply system through RPC were investigated, and a cooperative control scheme was proposed to compensate for the output power of the traction network via RPC. The cooperative control scheme is tested and analyzed by means of power electronic real-time simulator. The study show that control program can make RPC compensate the output power of traction network under four working conditions, and the test results of power flow distribution are as expected, which demonstrates the effectiveness of the scheme.

Acknowledgments. This research was supported by China National Natural Science Foundation(Grant No. 52067013), Natural Science Key Foundation of Gansu Provincial Department of Science and Technology (Grant No. 21JR7RA280 and Grant No. 22JR5RA318), and Natural Science Foundation of Gansu Provincial Department of Science and Technology (Grant No. 22JR11RA162).

References

1. Huang, W.J, Zhang N., Cheng Y.H.: (2020). Multienergy Networks Analytics: Standardized Modeling, Optimization and Low Carbon Analysis. Proceedings of the IEEE,108(9):1411–1436. (in Chinese)
2. Xu. Y., Yan X.W., Li R.J.: Joint Optimization of Active and Reactive Powers in Active Distribution Network with "Generation-Grid-Load-Energy Storage" Interaction in power market environment. Power System Technology,43(10):3778–3789. (in Chinese) (2019)
3. Hu, H.T., et al.: A Novel, "Source-Network-Train-Storage" Integrated Power Supply System for Electric Railways. Proceedings of the CSEE **42**(12), 4374–4390 (2022). (in Chinese)
4. Chen, J., Wang, Y., Chen, X.: An on Board Energy Storage Solution Considering Voltage Fluctuations and Harmonic Control of Traction Networks under Mountain Area Conditions. J. Railw. Sci. Eng **18**, 1582–1594 (2021). (in Chinese)
5. Wang, Y., Xin, Y.Y., Xie, Z.Y., Mu, X.Q., Chen, X.Q.: Research on Low-Frequency Stability under Emergency Power Supply Scheme of Photovoltaic and Battery Access Railway Traction Power Supply System. Energies **16**, 1–32 (2023). (in Chinese)
6. Deng W.L.,Dai C.H.,Han C.B.(2019). Back-to-Back Hybrid Energy Storage System of Electric Railway and Its Control Method Considering Regenerative Braking Energy Recovery and Power Quality Improvement. Proceedings of the CSEE,39(10):2914–2923. (in Chinese)
7. Wei W.J., Hu H.T. ,Wang K.(2019).Energy Storage Scheme and .ontrol Strategies of High-Speed Railway Based on Railway Power Conditioner. Transactions of China Electrotechnical Society, 34(6):1290–1299. (in Chinese)
8. Guo, Y., Chen, X., Wang, Y.: Negative Sequence Optimization Compensation Strategy for Energy Storage Type Railway Power Conditioner Considering Grid Voltage Imbalance. High Volt. Eng **49**, 772–780 (2023). (in Chinese)
9. Barros, L.A.M., Martins, A.P., Pinto, J.G.: Balancing the Active Power of a Railway Traction Power Substation with RPC. Energies **16**, 3059–3074 (2023). (in Chinese)

Development of Double Suspension Frame Electromagnetic Permanent Magnet Hybrid Maglev Controller

Xiaolei Shi(✉), He Xi, Yuanzhi Li, Yanchao Wang, and Deyu Jiao

CRRC Dalian Research & Development Co., Ltd, Dalian 116200, China
sxl203@126.com

Abstract. In this paper, the electromagnetic permanent magnet hybrid levitation model was introduced, and the H-type four-quadrant hard switching chopping technology scheme was applied, and on the basis of analyzing the circuit principle and structure of the normally guided suspension controller chassis, the dual-suspension electromagnetic permanent magnet hybrid suspension controller was successfully developed to complete the engineering mass production application. The electrical performance test results show that the suspension controller fully meets the technical requirements, and has the advantages of reasonable structure layout and strong anti-electromagnetic interference ability.

Keywords: Hybrid maglev · H-bridge four-quadrant · IGBT gate driver · Performance test

1 Introduction

1.1 A Subsection Sample

A maglev train is a train that uses a contactless electromagnetic levitation, guide and drive system [1, 2].The maglev train is mainly composed of three parts: suspension system, propulsion system and guidance system. At present, there are three kinds of maglev trains in the world: Japanese MLU superconducting repulsive maglev trains; German TR type constant guided suction maglev train; Japanese HSST type constant conduction suction maglev train [3].

Permanent magnetic levitation technology has the characteristics of no contact, low noise and low energy consumption, and has broad prospects in the fields of light load magnetic bearings, low-frequency magnetic levitation vibration isolation, magnetic levitation gravity compensation and micro-moving machinery. Permanent magnet levitation technology mainly includes permanent magnet passive levitation and permanent magnet active levitation, among which the permanent magnet active levitation system can also be divided into air gap adjustment type, magneto resistive modulation type and magnetic source adjustment type according to the magnetic adjustment mode [4].

A large number of tests have shown that the EMS maglev system represented by the German TR series maglev train is technically mature and engineering feasible. The

L. Jia et al. (Eds.): EITRT 2023, LNEE 1135, pp. 264–275, 2024.
https://doi.org/10.1007/978-981-99-9307-9_29

levitation pole of the traditional EMS magnetic levitation system is simple and reliable in structure and is quite mature in technology, but the current flowing through the levitation pole is large when the levitation pole is stable levitation, so the power consumption of the levitation pole is large, and the pole heating is serious. The use of electromagnetic permanent magnet hybrid levitation technology can significantly reduce the power consumption of the levitation pole, optimize the structure of the EMS type magnetic levitation system, and reduce the system cost [6–8].

In this paper, the model of electromagnetic permanent magnet hybrid levitation system is analyzed, and then the engineering research of levitation chopper controller is carried out with H-type topology as the main circuit structure, and a levitation controller suitable for electromagnetic permanent magnet hybrid levitation frame is successfully developed, and the electrical performance test is completed.

2 Model of Maglev System

The basic structure of the electromagnetic permanent magnet hybrid levitation system is shown in Fig. 1. The permanent magnets are located in the middle of the bottom of the yoke, and they form a hybrid levitation magnet with the coil, and by changing the current in the coil, the magnetic force between the magnet and the track can be adjusted to control the gap to remain constant.

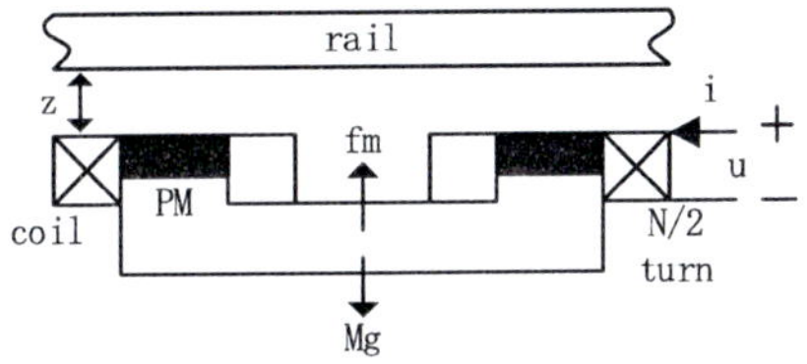

Fig. 1. Maglev system structure.

Where z is the length of the air gap of the suspended machinery; f_m is the Magnet attraction; M_g is the Gravity.

Air gap synthesis magnetic density B:

$$B = \mu_0 \cdot \frac{Ni + 2H_c h_{pm}}{2z + \frac{\mu_0}{\mu_{Fe}} l_{Fe} + \frac{2\mu_0 H_c}{B_r} h_{pm}} \tag{1}$$

The formula for calculating the suspension force of the hybrid suspension system is:

$$f = \frac{B^2 A}{\mu_0} \tag{2}$$

In the formula: N is the number of coil turns; i is the current; A is the active area of the magnetic circuit of the magnetic pole; l_{Fe} is the equivalent total length of the magnetic circuit in the pole core and the ferromagnetic track; u_{Fe} is the permeability of the iron core; u_0 is vacuum permeability; h_{pm} is the thickness of the permanent magnet; B_r is the residual magnetic density; H_c is coercivity.

3 Main Circuit Design

3.1 Model of Main Circuit

The main circuit topology is shown in Fig. 2, with R and L being equivalent circuits for the load electromagnets. When the switch tube T1 and T4 are turned on at the same time, if the electromagnet current is greater than zero, the positive direction will increase; if the electromagnet current is less than zero, the diodes on T1 and T4 freewheel, and the current will decrease in reverse. When the switch tube T2 and T3 are turned on at the same time, if the electromagnet current is less than zero, its current value will increase in reverse; if the electromagnet current is greater than zero, the diodes on T2 and T3 freewheel, and the current will decrease in the forward direction. This allows for fast performance when the suspension system rises and falls.

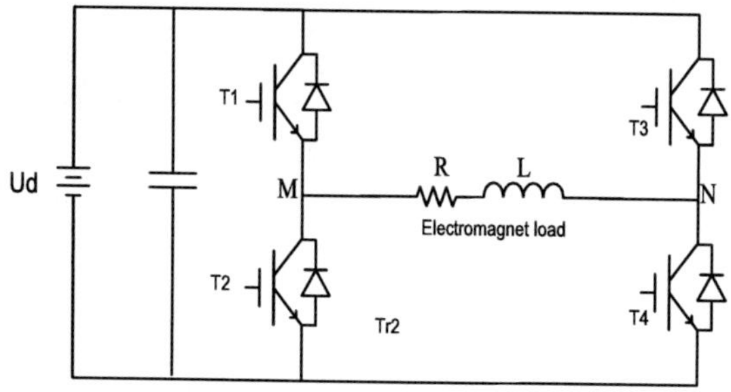

Fig. 2. Topology of the hovering controller.

3.2 Model of Main Circuit

The principle of the main circuit of the hovering controller is shown in Fig. 3. Under the rated state, the hybrid suspension controller input 480V DC, and the output loop is connected to the inductive load. Switching power supply module D1 outputs +5 V, +15 V, −15 V, 24 V1, 24 V2 control loop voltage. The main contactor KM1 is closed, the support capacitor is charged by the charging resistance and the voltage equalization resistance, and after 6 s, the KM2 contactor is closed, the capacitor supplies power to the IGBT, at this time the PWM pulse is sent to the IGBT driver board, and the IGBT will output rated DC after normal operation.

The basic principle of the suspension controller is to change the effective voltage on the electromagnet by controlling the on-off of the power switch, which is the core component of the suspension controller. Generally, IGBT is selected as the power switching element of the maglev controller, and the IGBT must have sufficient current capacity, and the appropriate gate resistance R should be selected to extend the IGBT off-time to reduce the $\frac{du}{dt}$ value and prevent the Optimus effect.

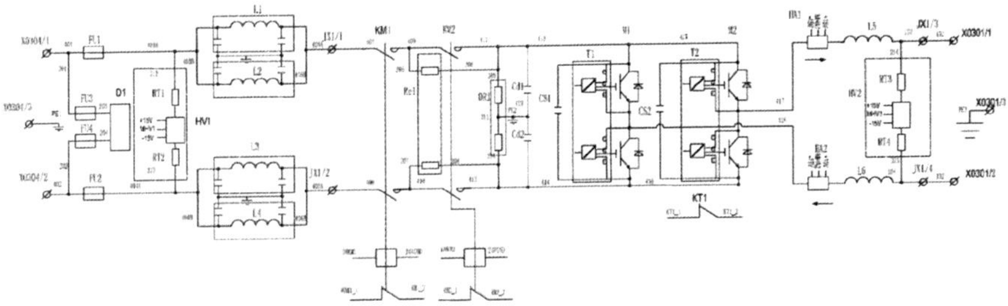

Fig. 3. Schematic diagram of the main circuit of the maglev controller.

3.3 Technical Parameters of the Main Circuit

The selection of the main circuit parameters of the suspension controller is based on the main technical conditions of the main suspension controller, and fully considers the working environment requirements of the maglev train to ensure the working performance and reliability of the product. The main technical parameters are shown in Table 1.

Table 1. Technical parameters of the maglev controller

Parameter	Metric value
Rated voltage U/V	480
Range of input voltage U/V	270–600
Rated output current I/A	40
Rated power P/VA	40000
Maximum starting current Imax/A	100
Maximum operating switching frequency f/Hz	15000

4 Design of Switching Power Supply

This power board takes a dedicated PWM control chip as the core, and the main circuit adopts fly-back and half-bridge topology, which provides multiple regulated DC power supplies for the maglev controller after high-frequency conversion. The power supply is internally protected against output overcurrent, overvoltage, and under-voltage.

4.1 Port Definition

The driver has a total of three external sockets, which have the following functions:

(1) JW1: DC power input socket;
(2) JW2: Output tap 1;
(3) JW3: Output tap 2;

4.2 The Principle of Power Conversion

After the rated 480V DC current is input to the power board, it is transformed by two fixed frequency adjustment inverter circuits. One inverter circuit outputs 24V, 5V, and the other inverter circuit outputs ± 15V and 24V with an output voltage accuracy of 5%.

In the two inverter circuits, the output voltage feedback can only take one of the taps as the main control and the other as the secondary control winding. At the same time, because the output filter loop adopts LC, if the multi-tap output is completely unloaded, the voltage may be higher than the design value. Therefore, the minimum load requirements for individual tap load output are as follows:

(1) 24V2 (Air Gap Sensor Power Supply): the minimum load current is 0.1A;
(2) +15V: Minimum load current 0.2A;
(3) −15V: Minimum load current 0.03A;
(4) 24V1: Minimum load current 0.05A.

5 Design of IGBT Driver Board

This driver board takes the IGBT special drive module 2SD315AI produced by Swiss Concept as the core, plus pulse shaping, input voltage detection, fault signal conversion and other circuits to form the drive circuit of four IGBTs of the maglev controller. The block diagram of the driver board is shown in Fig. 4.

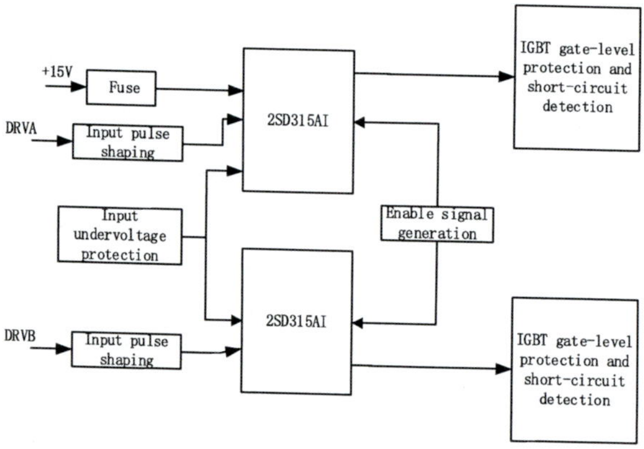

Fig. 4. Block diagram of IGBT driver board.

The driver board can be divided into the following parts of the functional circuit:

(1) IGBT special driver module 2SD315AI: This module is the core of the driver board;

(2) Input voltage detection circuit: When the input voltage exceeds 18V, the regulator tube WD1 breaks down, causing the fuse tube FU to blow;

(3) Input under voltage protection: When the input voltage is lower than 12V, the drive module blocks the gate drive signal to the IGBT, and the fault signal ERR changes from an open collector to a low level;

(4) Input pulse shaping circuit: DRIVA, DRIVB two driving signals pass through this circuit, filter out the interference signal during its transmission, and then shape the input to the INA pin of 2SD315AI;

(5) IGBT gate protection and short circuit detection: IGBT gate control voltage is not allowed to exceed $\pm$ 20V, in order to avoid the voltage applied to the IGBT gate due to pulse spikes or drive module failure exceeds the design value, the peripheral short circuit protection value is 500A (TVJ = 125 °C).

6 Design of Control Circuits

6.1 Functional Topology

As the monitoring and control equipment of the suspension controller, the control chassis collects the feedback information of the equipment status of the monitoring suspension controller in real time, and conducts/turns off the various control unit circuits of the suspension controller to ensure the safe operation of the equipment, and the functional topology is shown in Fig. 5.

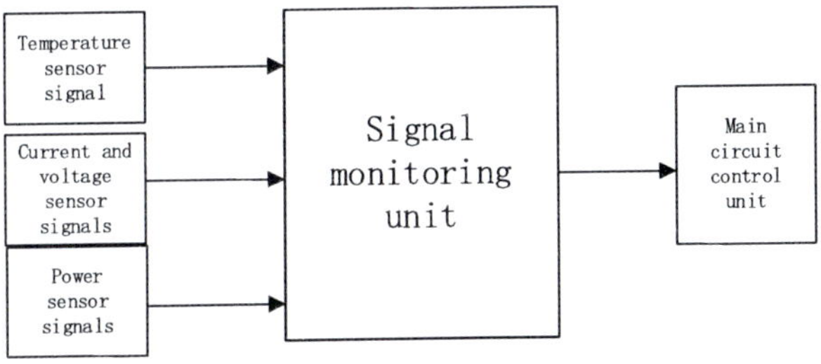

Fig. 5. Control circuit topology.

6.2 Control Program Algorithms

The control circuit uses a field programmable gate array (FPGA) to complete multi-channel signal acquisition filtering, status judgment and logic control, and the control program flow diagram is shown in Fig. 6.

7 Design of Hovering Controller

7.1 Functional Topology

The box structure of the maglev controller is divided into 7 modules: control chassis, support capacitor, contactor assembly, power board assembly, IGBT component, fuse component and filter component, which is convenient for installation, disassembly, maintenance and maintenance.

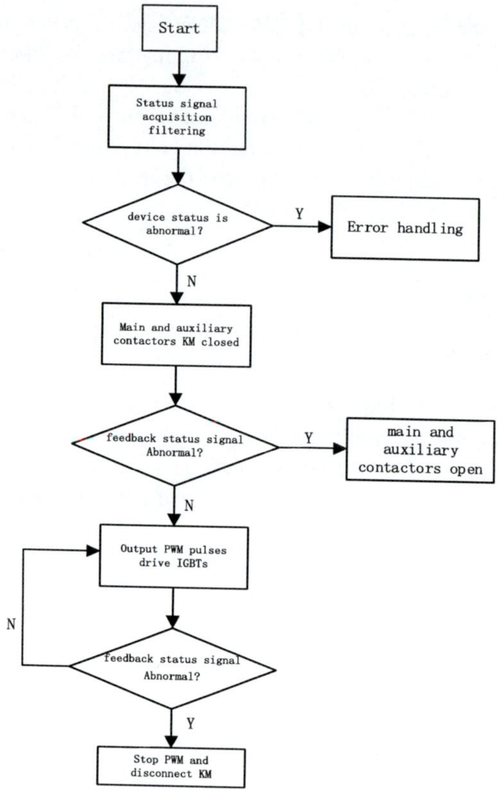

Fig. 6. Flowchart of Control program.

When considering the mutual interference factors of the devices in the circuit during the structural layout, the relative position of the device is adjusted to the best, and the air gap and creep age distance are fully guaranteed, so that the main circuit is routed smoothly. The main power line and the control line are separated and routed to increase electromagnetic compatibility and anti-interference ability. The three-dimensional rendering of the cabinet is shown in Fig. 7.

Fig. 7. 3D diagram of the maglev controller.

7.2 Mechanical Parameters

The mechanical and electrical parameters of the maglev controller cabinet are shown in Table 2.

Table 2. Mechanical parameters of the maglev controller box

Parameter	Metric value
Box size	1020mm(L) × 270mm(W) × 255mm(H)
Cooling method	Free cooling
Degree of protection	IP65
Weight	<40kg

8 Design of Hovering Controller

In order to verify the performance parameters of the suspension controller, the performance test is carried out according to IEC61287-1-2014 "Railway facilities - Power converters installed on railway locomotives Part I: characteristics and test methods" and GB/T 25119-2021 "Railway applications—Electronic equipment used on rolling stock" standards.

8.1 Sensor Test

When the hybrid suspension controller is rated for operation, the polarity and size of the output waveform of the current sensor and the voltage sensor are tested. The test results of current and voltage sensors are shown in Figs. 8 and 9, respectively.

8.2 Rated Load Test

In the case of input 270VDC, 480VDC and 600VDC in the main circuit, the output current of the hybrid maglev controller and the output voltage waveforms of both ends of IGBT CE are shown in Figs. 10, 11 and 12, respectively.

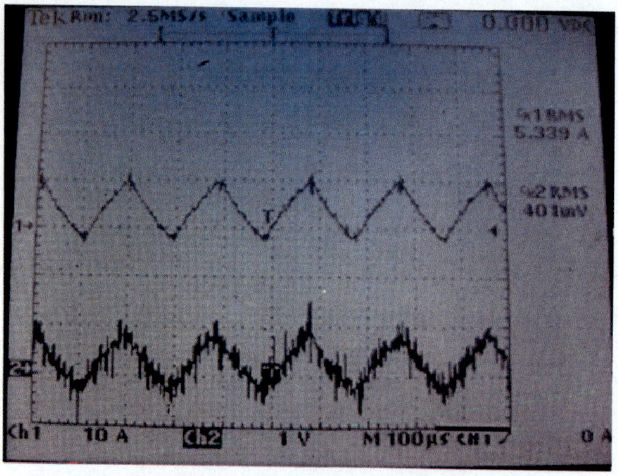

Fig. 8. Waveform of current sensor.

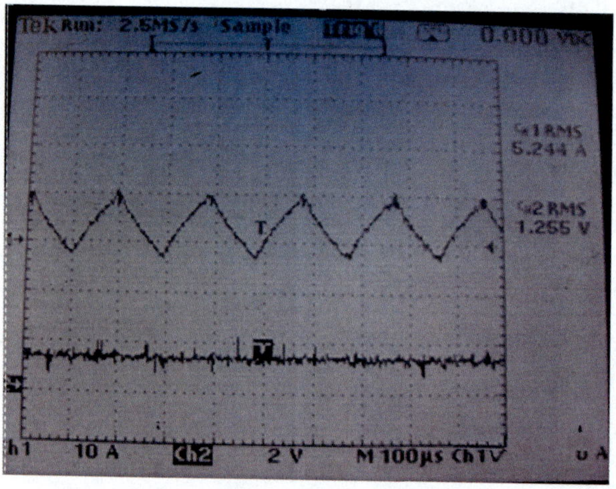

Fig. 9. Waveform of a voltage sensor.

In the figure above, the input voltage is within the range of 270VDC ~ 600VDC, the output PWM waveform has no distortion, the peak voltage when the IGBT is turned off is less than 30% of the bus voltage, the current waveform is continuous, there is a normal charging and discharging process, and the current waveform is a triangle wave, which meets the design requirements.

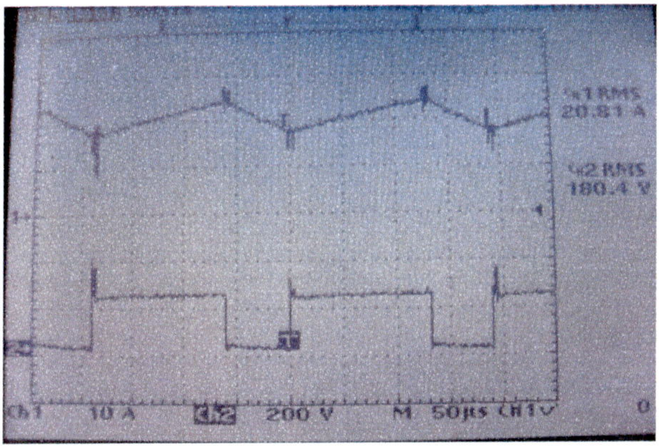

Fig. 10. Waveform of 270VDC IGBT voltage and current.

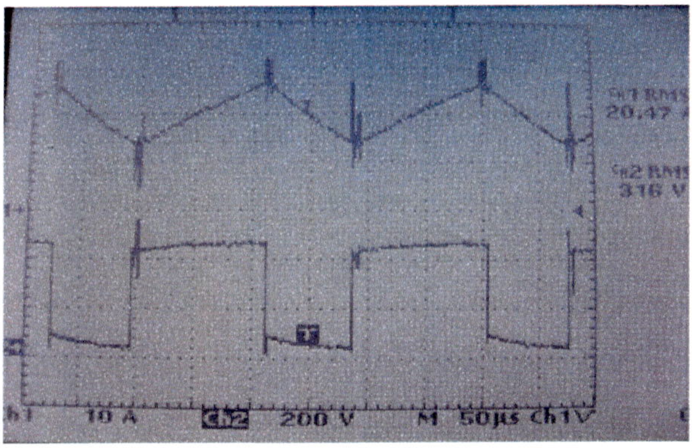

Fig. 11. Waveform of 480VDC IGBT voltage and current.

8.3 Inrush Current Test

When the main circuit input 480V rated voltage, the converter outputs 100A, 0.2s inrush current, the test waveform is shown in Fig. 13, and the suspension controller device is not damaged after the test.

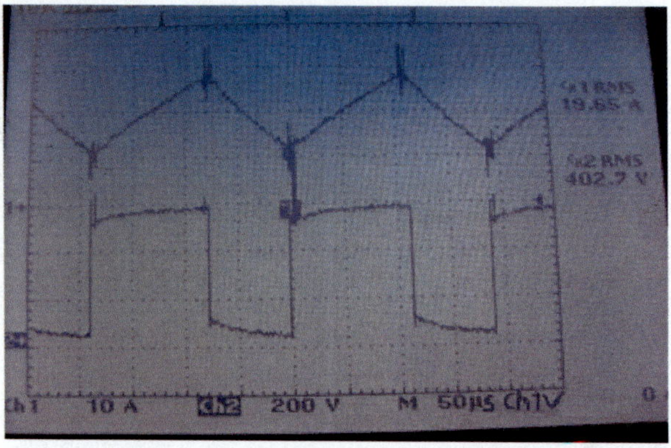

Fig. 12. Waveform of 600VDC IGBT voltage and current.

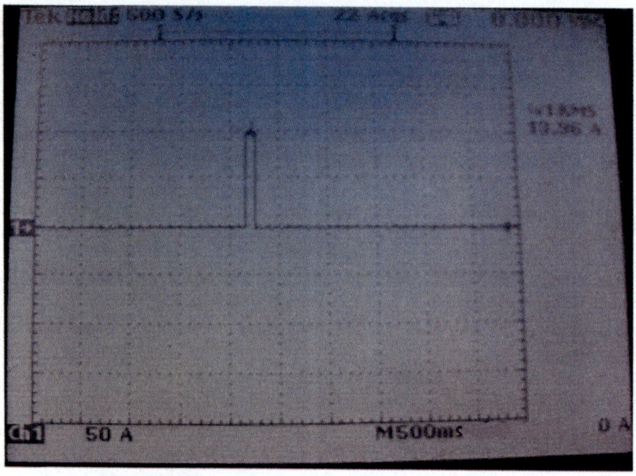

Fig. 13. Waveform for inrush current test.

9 Conclusion

In this paper, the four-quadrant H-type chopper circuit is studied and discussed, and an H-type four-quadrant suspension controller suitable for permanent magnet electromagnetic hybrid suspension control is developed. The test results show that the maglev controller has stable working performance, strong anti-interference ability, and fully meets the requirements of performance indicators. The successful development of this hybrid levitation controller has been applied to the Shanghai high-speed maglev demonstration line in small quantities, which provides certain guiding significance for the localization and engineering application of high-speed maglev trains.

References

1. Rui, T.A.N.G., Junquan, W.U.: Application prospect of medium and low speed maglev trains in urban rail transit in China. Urban fast rail transit **19**(2), 93–95 (2006)
2. Hongcheng, H.E., Yijing, X.U.: On-board electric traction system of medium and low speed maglev train. Electric Drive of Locomotive **20**(6), 35–42 (2010)
3. Jisan, L.I.A.N.: Principle and technical characteristics of maglev train. Electric Locomotive Technology **24**(3), 23–26 (2001)
4. Chuan, Z.H.A.O., et al.: Realization Mechanism and Development of Permanent Magnetic Levitation: A Review. JOURNAL OF MECHANICAL ENGINEERING **59**(1), 1–13 (2023)
5. Cheng, H., Xiao, Z.: LI Yungang: Characteristic analysis of electromagnetic permanent magnet hybrid suspension system. Electric Drive of Locomotive, Journal **20**(2), 18–21 (2010)
6. Weihua, M.A., Shihui, L.U.O.: ZHANG Min: Research review on medium and low speed maglev vehicle. Journal of Traffic and Transportation Engineering **21**(1), 199–216 (2021)
7. Li, H.T., Liu, D.: HONG: Modeling and identification of the hysteresis nonlinear levitation force in HTS maglev systems. Supercond. Sci. Technol. **33**(5), 054001 (2020)
8. Li, Y.J., Dai, Q.: ZHANG Y: Design and analysis of an electromagnetic turnout for the superconducting Maglev system. Physical C:Superconductivity and Its Applications **528**, 84–89 (2016)
9. Kim, K.B., Im, S.H.: UM D Y: Comparison of magnetic levitation systems using ring-shaped permanent magnets. IEEE Trans. Magn. **55**(7), 1–4 (2019)
10. Oka, K., Yamamoto, K.: HARADA A: Magnetic suspension mechanism using rotary permanent magnets. Int. J. Appl. Electromagnet Mech **64**(1–4), 977–983 (2020)
11. Taqavi, O.: TAGHAVI N: Development of a mixed solution of Maxwell's equations and magnetic equivalent circuit for double-sided axial-flux permanent magnet machines. IEEE Trans. Magn. **57**(4), 1–11 (2021)
12. Rodriguez, E., Sotelo, G.G., Oliveira, J.D., et al.: Designing, simulations and experiments of a passive permanent magnet bearing. Int. J. Appl. Electromagnet Mech **51**(2), 131–149 (2016)
13. Cho, H.W., Yu, J.S., Jang, S.M., et al.: Equivalent magnetic circuit based levitation force computation of controlled permanent magnet levitation system. IEEE Trans. Magn.Magn. **48**(11), 4038–4041 (2012)

Current Control Calculation and Simulation Analysis of Urban Railway Traction System

Yilin Wang(✉)

Shanghai Rail Transit Equipment Development Co Ltd., Shanghai, China
ylwang0403@hotmail.com

Abstract. In the urban railway industry, the stability and reliability of the permanent magnet motor control in urban railway vehicles is very important to the traction system and the overall vehicle operation. Control, so that the output of the inverter gets a series of pulses with the same threshold, and these pulses can be used to replace the sine wave and the desired waveform. Then input an AC current of a certain frequency into the permanent magnet motor. Among them, the current given closed-loop control is particularly important in the case of different rotational speeds in the permanent magnet motor algorithm control. This paper proposes maximum torque current ratio control (MTPA), field weakening control and maximum torque voltage ratio control (MTPV), and uses Simulink software to model and simulate simulation waveforms to verify the stability and reliability of the hybrid modulation output proposed in this paper.

Keywords: MAPA · Field weakening control · MTPV

1 Introduction

At present, for the system redundancy and reliable operation of urban railway vehicles, most of the traction systems use permanent magnet motors as train power sources. Different control strategies are adopted at different speeds of the motor to ensure the stability and reliability of the three-phase output current of the inverter [1].

In the urban railway traction system, during the operation of the motor, the back electromotive force of the motor will increase with the increase of the motor speed, so the three-phase stator voltage required by the motor should also increase accordingly, and the inverter output Due to the limitation of the bus voltage, the voltage will not increase indefinitely. At the same time, due to the limited thermal load capacity of the inverter and motor windings, the current of the motor cannot be too high or the drive system equipment will be damaged. Therefore, the motor drive The current of the system must also be limited. In order to achieve the maximum function of the motor, different control strategies need to be used when the speed of the motor is different.

The permanent magnet motor is divided into three areas when it is running at full speed: constant torque area, field weakening zone and field weakening zone II. This

L. Jia et al. (Eds.): EITRT 2023, LNEE 1135, pp. 276–286, 2024.
https://doi.org/10.1007/978-981-99-9307-9_30

article will briefly describe different control strategies for motors operating in three different regions, including MTPA, field weakening control and MTPV. And the simulation analysis of stator voltage and current, torque and speed, and AC and D axis currents is carried out.

2 Organization of the Text

Permanent magnet motor control is mainly divided into four parts: torque given, current calculation, voltage calculation and pulse width modulation. Torque reference is responsible for determining the given torque of the current motor through parameters such as level command, network voltage, power module temperature, and motor speed, and providing instructions for the motor control core algorithm; current calculation includes three parts: MTPA, weak For magnetic control and MTPV, a given current is obtained through the above methods. The voltage calculation is responsible for converting the given torque and excitation current into corresponding given voltage values through certain calculations. Furthermore, the voltage feedforward technology can solve both the dq-axis current coupling problem and the system response rate problem, which can make the system have good dynamic performance and robustness. Finally, the corrected given modulation degree can be obtained through voltage calculation, which provides control instructions for the subsequent pulse width modulation algorithm. The pulse width modulation algorithm can provide a stable and three-phase current value required by the permanent magnet motor. The overall control block diagram of the permanent magnet motor is shown in Fig. 1.

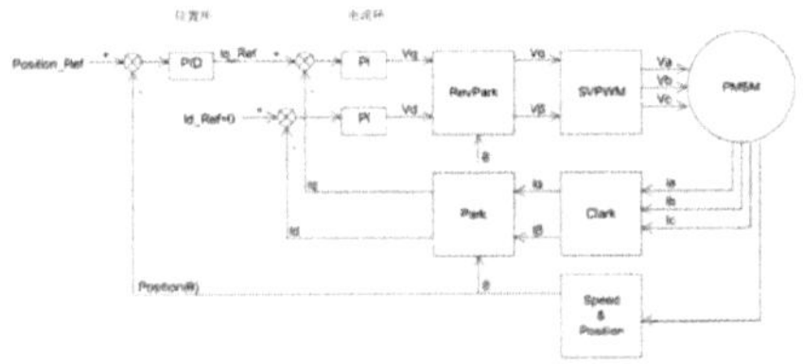

Fig. 1. Block diagram of permanent magnet synchronous motor vector control

2.1 Analysis of MTPA Control Principle

When the motor is running in the low-speed area, there is still a margin in the output voltage capability of the inverter, and the input voltage of the motor can be satisfied. At this time, the motor can reach the highest current operating point in the operating stage. The operating point of the motor is located inside the current and voltage limit circle. That is,$U_s < U_{max}$, $I_s < I_{max}$, the voltage and current of the motor meet the maximum sustainable output torque when the motor is running in this area, and the output power of the motor is getting higher and higher at this time.

During the operation of the motor, it can be obtained from the mathematical model of the motor that the back electromotive force of the motor will increase with the increase of the motor, so the heat treatment stator voltage required by the motor should also increase accordingly, and the output voltage of the sunshade device will increase accordingly. Restricted by the bus voltage, it will not increase indefinitely, namely: $u_d^2 + u_q^2 \leq U_{\max}^2$.

The MTPA control goal is to minimize the current of the motor under the same output current. The core problem of the maximum current ratio control is how to use the minimum stator current to exert the maximum action, which is converted into a mathematical problem is, Maximum: $\frac{\mathrm{Te}}{\mathrm{i_s}}$.

Since the target action time is fixed in the control process, the above problem can be transformed into the problem that the stator current is small under the control action condition. In the derivation process, in order to achieve parameter separation, the derivation formula is simplified, and the action formula is treated as per unit.

$$\mathrm{Te} = \mathrm{N_{pp}}\left[\psi_f \mathrm{i_{sq}} + \left(\mathrm{L_d} - \mathrm{L_q}\right)\mathrm{i_{sd}}\mathrm{i_{sq}}\right]$$

According to the torque formula above, in the per unitization process, the current per unitization coefficient is:

$$\mathrm{I}' = \frac{\psi_f}{L_q - L_d} > 0$$

The torque per unit coefficient is: $Te' = N\mathrm{pp}\psi_f i' > 0$.

Since i', Te' are all greater than 0, the change trend of the formula after unitization remains unchanged. After deriving the per-unit values of $\mathrm{Te}, \mathrm{i_{sq}}, \mathrm{i_{sd}}$, which can get:

$$\mathrm{te}^* = \frac{\mathrm{Te}}{\mathrm{Te}'} = \frac{\mathrm{Te}\left(\mathrm{L_q} - \mathrm{L_d}\right)}{\mathrm{N_{pp}}\psi_f{}^2}$$

$$\mathrm{i_{sq}}^* = \frac{\mathrm{i_{sq}}}{\mathrm{i}'} = \frac{\mathrm{i_{sq}}\left(\mathrm{L_q} - \mathrm{L_d}\right)}{\psi_f}$$

$$\mathrm{i_{sd}}^* = \frac{\mathrm{i_{sd}}}{\mathrm{i}'} = \frac{\mathrm{i_{sd}}\left(\mathrm{L_q} - \mathrm{L_d}\right)}{\psi_f}$$

The torque formula after per unit processing is: $\mathrm{te}^* = \mathrm{i_{sq}}^*(1 - \mathrm{i_{sd}}^*)$

The maximum torque current ratio control can be transformed into a mathematical problem of finding the minimum sum of squares of $\mathrm{i_{sq}}^*$ and $\mathrm{i_{sd}}^*$ when te^* it is a constant value. Using the optimization method to solve it, treat $\mathrm{i_{sq}}^*$ as x and $\mathrm{i_{sd}}^*$ as y, then you can get the mathematical relationship: $\min \mathrm{x}^2 + \mathrm{y}^2$ s.t. $\mathrm{te}^* = \mathrm{x}(1 - \mathrm{y})$

This problem can be solved using the Lagrange multiplier method, build function

$$\mathrm{fx,y} = \mathrm{x}^2 + \mathrm{y}^2 - \lambda_1[\mathrm{te}^* - \mathrm{x}(1 - \mathrm{y})]$$

Find the partial derivative of x with respect to fx, y, $2\mathrm{x} - \lambda_1(1 - \mathrm{y}) = 0$.

Find the partial derivative of y with respect to fx, y, $2\mathrm{y} - \lambda_1(-\mathrm{x}) = 0$

Restrictions untie: $\mathrm{te}^* = \mathrm{x}(1 - \mathrm{y})$

Transform the above formula into:

$$\mathrm{x} = \frac{\mathrm{te}^*}{(1 - \mathrm{y})}$$

$$y = 1 - \frac{te^*}{x}$$

Bring the above formula into:

$$\lambda_1 = \frac{te^*}{(y-1)^2}$$

Then bring the above formula into the solution

$$te^* = \sqrt{y(y-1)^3}$$

Similarly, the above formula can be brought into:

$$\lambda_1 = \frac{2te^* - 2x}{x^2}$$

Then bring the above formula into the solution

$$te^* = \frac{x}{2}(1 + \sqrt{1 + 4x^2})$$

Then the per unitized solution is:

$$te^* = \frac{i_{sq}{}^*}{2}(1 + \sqrt{1 + 4isq^{*2}})$$

$$te^* = \sqrt{i_{sd}{}^*(isd^* - 1)^3}$$

The formula can be obtained by removing per unitization:

$$te = \frac{N_{pp}\psi_f i_{sq}}{2}(1 + \sqrt{1 + 4 * \frac{i_{sq}{}^2(L_q - L_d)}{\psi_f{}^2}})$$

$$te = N_{pp}\sqrt{\frac{i_{sd}(i_{sd}(L_q - L_d) - \psi_f)^3}{(L_q - L_d)}}$$

At the same time the relationship of i_{sd} and i_{sq} is, $i_{sd} = \frac{-\psi_f + \sqrt{\psi_f^2 + 4(L_d - L_q)^2 i_{sq}^2}}{2(L_d - L_q)}$.

Since the relationship between te and i_{sd}、i_{sq} is a transcendental equation, it is difficult to solve it. Therefore, the usual practice is to calculate the corresponding values of i_{sd}, i_{sq}, and at each torque point off-line, and draw them into a table.

2.2 Analysis of the Principle of Negative Direct Axis Compensation Field Weakening Control

When the speed of the motor increases to the critical point of the field weakening zone I, due to the limitation of the inverter output and the motor itself, the voltage and current of the motor are limited by U_{max} and I_{max} and cannot continue to rise. In this case, only rely on MTPA Control strategy, it is no longer possible to increase the speed by increasing the terminal voltage of the motor. At this time, it is necessary to weaken the magnetic flux of the motor through negative direct axis compensation and other magnetic field weakening control methods, so that the motor can further speed up.

In the sinusoidal steady state, the magnitude of the motor stator voltage vector Us is directly related to the stator electrical angular velocity ωs. This means that the running speed of the motor is limited by the current and voltage limits of the inverter. Drawing these two limits into a curve is the current limit circle and voltage limit circle of motor control.

(1) Current limit circle.

When the motor is controlled with the maximum torque-to-current ratio, the motor current i_s becomes larger gradually. When i_s reaches the maximum current allowed by the motor, the current is no longer allowed to increase. Therefore, the value range of stator current and i_{sd}, i_{sq} is within a circle, see the figure below (green area in the figure) for details.

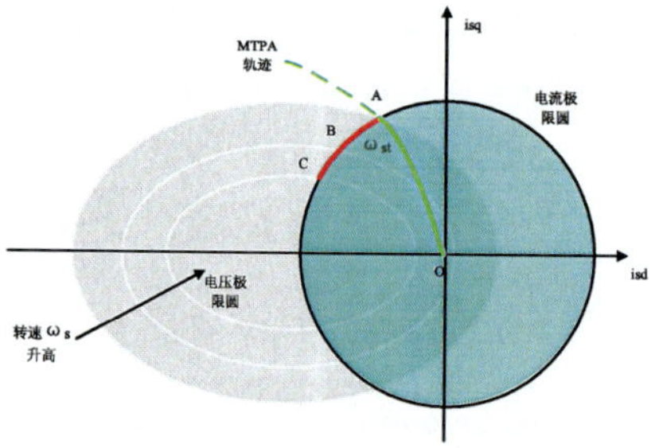

Fig. 3. Current limit circle

If the maximum torque-to-current ratio control is followed, there is a linear relationship between i_{sd} and i_{sq}, that is, curve OA.

(2) Voltage limit circle.

When i_s reaches the maximum current allowed by the motor, the current is no longer allowed to increase. In this range, the greater the speed, the greater the required voltage. The output power of the motor is $P = U_s I_s$. Because the motor cannot increase the output power infinitely. So when the power reaches P_{max}, that is, when Us reaches the maximum, the motor can no longer increase the power. In order to increase the motor speed, the output torque of the motor needs to be reduced at this time. The limit state curve for motor control can be drawn according to this procedure.

Decomposing the formula under the dq axis system can get the voltage decoupling equation

$$U_{sd} = R_s i_{sd} + p(L_d i_{sd}) - \omega_s L_q i_{sq}$$

$$U_{sq} = R_s i_{sq} + p(L_d i_{sd} + \psi_f) + \omega_s L_d i_{sd}$$

In the formula, the p differential operator represents the change of the corresponding parameter per unit time. Therefore, when the dq axis current of the motor is constant (the output torque is constant and the speed is variable), the system is in a steady state, and the formula can be obtained:

$$U_{sd} = R_s i_{sd} - \omega_s L_q i_{sq}$$

$$U_{sq} = R_s i_{sq} + \omega_s L_d i_{sd}$$

and have

$$|U_s| = \sqrt{{u_{sd}}^2 + {u_{sq}}^2}$$

Derive the voltage limit circle:

$${U_s}^2 = (R_s i_{sd} - \omega_s L_q i_{sq})^2 + (R_s i_{sq} + \omega_s (L_d i_{sd} + \psi_f))^2$$

and have:

$${U_s}^2 \leq {U_{sMax}}^2$$

where U_{sMax} is the maximum value of Us.

When running from the constant torque zone to the constant power zone, the control falls on point A. At this time, the stator current is the rated value, and $i_{sd} i_{sq}$ is the calculated value calculated from the maximum torque-current ratio.$|U_s|$ reaches limitU_{sMax}. The angular velocity of the rotor at this time is the turning angular velocity, denoted asω_{st}. When the speed is above the turning speed, the voltage drop caused by the resistance can be negligibly small (take this project as an example, the motor resistance is 0.012, and the maximum amplitude of $i_{sd} i_{sq}$ is below 300A and 500A respectively. The corresponding voltage drop is about 3 ~ 5V.Compared with the overall voltage drop of 400 ~ 500V, it is very small, so it can be ignored), and ω_{st} is:

$$\omega_{st} = \frac{|U_{smax}|}{\sqrt{(L_d * i_{sd} + \psi_f)^2 + (L_q * i_{sq})^2}}$$

After the turning speed, the motor runs at constant power. Maintain U_{sMax}, i_s at the maximum current limit circle. Dealing with Voltage Formulas

$$\sqrt{(L_d * i_{sd} + \psi_f)^2 + (L_q * i_{sq})^2} = \frac{|U_{smax}|}{\omega_s}$$

$$\omega_{st} < \omega_s$$

It can be found that if you want to get a speed greater than the turning speed, you need to change the i_{sd}, i_{sq} proportional relationship. The motor voltage limit circle formula is:

$$\frac{\left(i_{sd}+\frac{\psi_f}{L_d}\right)^2}{\left(\frac{U_{smax}}{\omega_s L_d}\right)^2}+\frac{\left(i_{sq}\right)^2}{\left(\frac{U_{smax}}{\omega_s L_q}\right)^2}=1$$

$$\omega_{st} < \omega_s$$

It can be obtained from the ellipse formula that as ωs increases, the radius $\frac{U_{smax}}{\omega_s L_d}$, $\frac{U_{smax}}{\omega_s L_q}$ of the voltage ellipse becomes smaller, and the voltage ellipse shrinks inward. Referring to Fig. 6, it can be seen that as the angular velocity gradually increases, the radius of the voltage limit circle gradually decreases, and the voltage limit circle shrinks inward (note again that the voltage remains unchanged, only the speed increases).

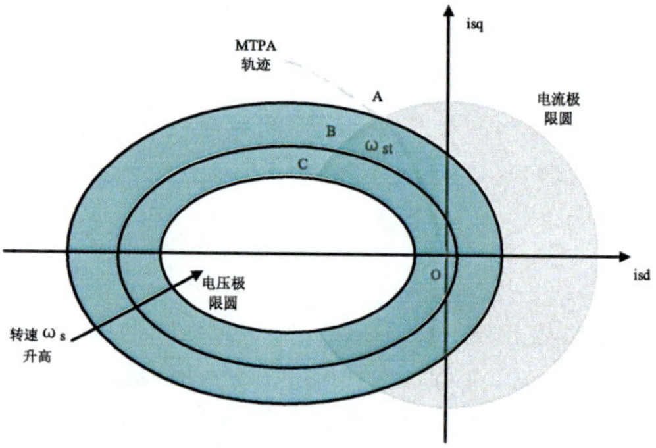

Fig. 4. Voltage limit circle

When $|U_s|$ reaches the limit $|U_{smax}|$, the modulation M of the inverter should be 1. Therefore, a method of field weakening control is that when the given modulation degree M does not exceed 1, the given torque is a constant value, and field weakening control is not performed.

When the given modulation degree M exceeds 1, the motor is controlled with constant power. The given torque is gradually reduced, and the field weakening control is performed. Adjusting $|i_{sd}|$ reduces $|U_s|$ so that $|U_s|$ does not exceed the maximum. Using PI adjustment to maintain $M = 1$ can meet the requirements.

2.3 Analysis of MTPV Control Principle

When the motor speed continues to rise after passing through the field weakening zone I, it will enter the deep field weakening zone. It can be seen from Fig. 6 that the motor enters the MTPV control area in the BC section, so point B is called the transition point from the field weakening zone I to the deep field weakening zone. The current information at point B can be obtained by combining the MTPV curve and the current limit circle, specifically:

$$\begin{cases} i_{dB} = -\dfrac{\psi_f}{L_d} + \dfrac{-\psi_f L_q + L_q\sqrt{\psi_f^2 + 4(L_d - L_q)^2 i_{qB}^2}}{2L_d(L_d - L_q)} \\ i_{qB} = \sqrt{i_{\max}^2 - i_{dB}^2} \end{cases}$$

From the above formula, the current information of the transition point B in the deep magnetic field weakening area can be calculated as:

$$\begin{cases} i_{dB} = -\psi_f L_d(2L_d - L_q) + M \\ i_{qB} = \sqrt{i_{\max}^2 - i_{dB}^2} \end{cases}$$

Then the current (i_{dB}, i_{qB}) at point B is brought into the voltage limit circle constraint equation, and the speed of the motor at point B can be obtained, that is, the critical speed of the motor transitioning from the field-weakening zone I to the field-weakening zone II ω_B, specifically:

$$\omega_B = \frac{U_{\max}}{\sqrt{(L_d i_{dB} + \psi_f)^2 + (L_q i_{qB})^2}}$$

The control frame of the permanent magnet synchronous motor based on MTPV and negative direct axis compensation full speed range compound field weakening control strategy is shown in the figure below. Below the turning speed, MTPA and negative direct axis compensation are used for control to complete the expansion of field weakening zone I; above the turning speed, MTPV curve is used for control.

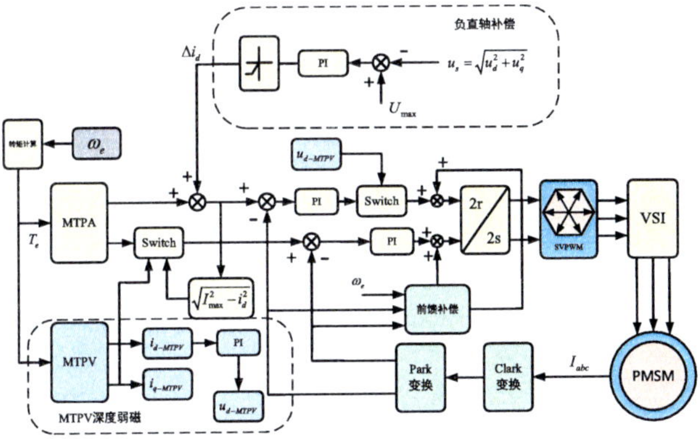

Fig. 5. Depth field weakening control strategy for MTPV and negative direct axis compensation

3 Experimental Simulation Analysis

The simulation results of the electronic voltage and current of the permanent magnet motor control algorithm under three different current control algorithm working conditions are shown in the figure below.

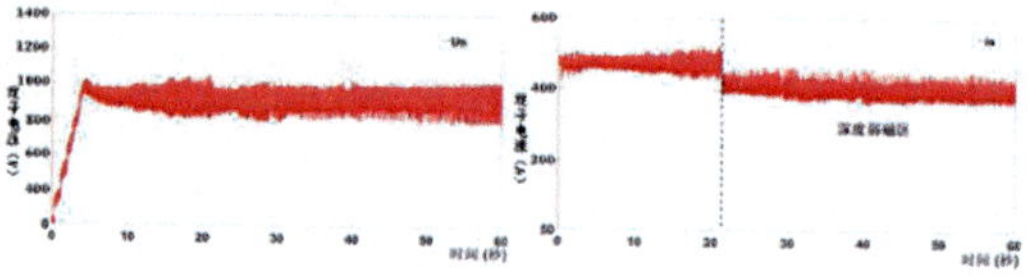

Fig. 6. Stator voltage and current

The simulation results of the torque and speed of the permanent magnet motor under the control algorithm under three different current control algorithm conditions are shown in the figure below.

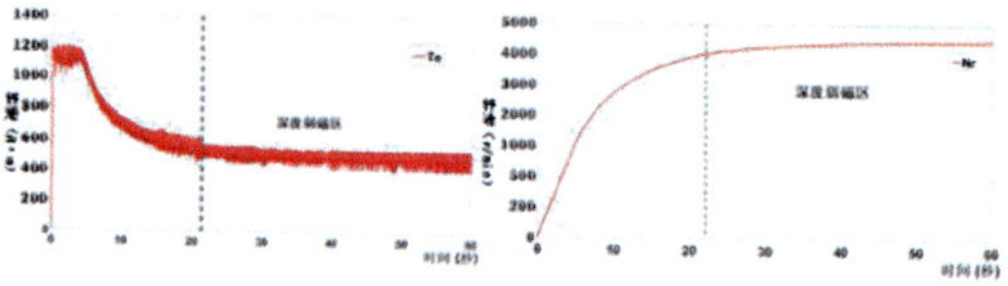

Fig. 7. Torque and speed

The simulation results of the AC and D axis currents of the permanent magnet motor under the control algorithm of three different current control algorithm conditions are shown in the figure below.

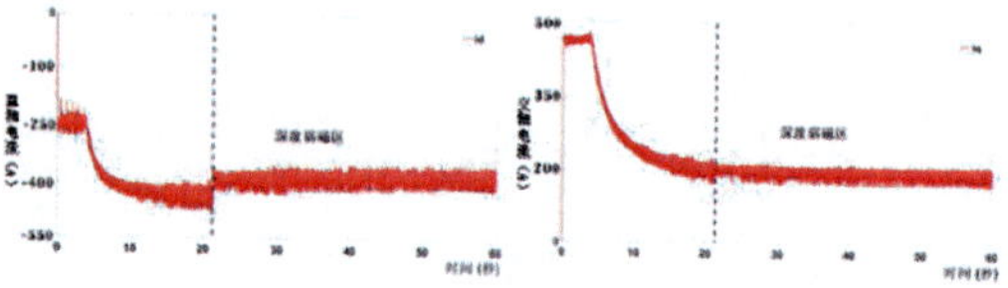

Fig. 8. AC-D axis current

It is not difficult to see from the simulation results that in the field weakening zone I/MC zone, the output capability of the inverter reaches the limit at this time, and the motor voltage cannot be further increased, that is,$u_s = U_{max}$, so the negative direct axis compensation control starts to work, and the MTPA control passes Perform reverse compensation on the direct axis current, and limit the motor to run on the voltage limit circle. At this time, the direct axis current compensation amount begins to increase in reverse, and at the same time, the stator voltage of the motor remains unchanged aroundU_{max}; in the direct axis current compensation Under the action of the quantity, the direct-axis current also increases in the opposite direction, while the quadrature-axis current will decrease positively due to the limitation of the current limit, as shown in Fig. 10; under the combined action of the direct-axis current, the electromagnetic The torque decreases gradually, and the acceleration of the rotational speed of the motor decreases gradually, as shown in Fig. 7.

After the motor enters the deep field weakening area, due to the limitation of the voltage limit circle, the stator voltage remains unchanged, that is, $u_s = U_{max}$, but the stator current will gradually decrease, as shown in Fig. 8; At point B, the actual AC-D axis current of the motor will intersect with the AC-D axis current curve of the motor under MTPV control at this time, and at this time the switching of the deep field weakening zone is completed, as shown in Fig. 6; Under the action, the torque of the motor continues to drop, the acceleration of the speed also continues to drop, and the speed rises more and more slowly, as shown in Fig. 6.

4 Conclusion

This paper analyzes the principle of power module selection and the principles and characteristics of MTPA, field weakening control and MTPV in the permanent magnet motor current control algorithm. This modulation method ensures the stability of the current and voltage output of the motor at different speeds and torques. The simulation results prove that the MTPA, field weakening control and MTPV control methods of the traction inverter on rail vehicles can control the permanent magnet motor very well.

References

1. Sepulchre, L., Fadel, M., Pietrzak-David, M., et al.: MTPV flux-weakening strategy for PMSM high speed drive. IEEE Trans. Ind. Appl. **54**(6), 6081–6089 (2018)
2. Dianov, A., Tinazzi, F., Calligaro, S., et al.: Review and classification of MTPA control algorithms for synchronous motors. IEEE Trans. Power Electron. **37**(4), 3990–4007 (2021)
3. Chen, Z., Yan, Y., Shi, T., et al.: An accurate virtual signal injection control for IPMSM with improved torque output and widen speed region[J]. IEEE Trans. Power Electron. **36**(2), 1941–1953 (2020)
4. Diao, L., Tang, J., Loh, P.C., Yin, S., Wang, L., Liu, Z.: An Efficient DSP–FPGA-Based Implementation of Hybrid PWM for Electric Rail Traction Induction Motor Control. IEEE Trans. Power Electron. **33**(4), 3276–3288 (2018)
5. Yang, R., Sun, T., Feng, W., et al.: Accurate online MTPA control of IPMSM considering derivative terms. Chinese Journal of Electrical Engineering **7**(3), 100–110 (2021)
6. Putri, A.K., Rick, S., Franck, D., et al.: Application of sinusoidal field pole in a permanent-magnet synchronous machine to improve the NVH behavior considering the MTPA and MTPV operation area. IEEE Trans. Ind. Appl. **52**(3), 2280–2288 (2016)
7. Mohamed, Y.A.R.I., Lee, T.K.: Adaptive self-tuning MTPA vector controller for IPMSM drive system. IEEE Trans. Energy Convers. **21**(3), 636–644 (2006)
8. Dianov A, Young-Kwan K, Sang-Joon L, et al. Robust self-tuning MTPA algorithm for IPMSM drives. In: 2008 34th Annual Conference of IEEE Industrial Electronics. IEEE, 2008: 1355–1360

9. Li, K., Wang, Y.: Maximum torque per ampere (MTPA) control for IPMSM drives using signal injection and an MTPA control law[J]. IEEE Trans. Industr. Inf. **15**(10), 5588–5598 (2019)
10. Bida V M, Samokhvalov D V, Al-Mahturi F S. PMSM vector control techniques—A survey[C]//2018 IEEE Conference of Russian Young Researchers in Electrical and Electronic Engineering (EIConRus). IEEE, 2018: 577–581
11. Sue, S.M., Pan, C.T.: Voltage-constraint-tracking-based field-weakening control of IPM synchronous motor drives. IEEE Trans. Industr. Electron. **55**(1), 340–347 (2008)
12. Lenke R U, De Doncker R W, Kwak M S, et al. Field weakening control of interior permanent magnet machine using improved current interpolation technique[C]//2006 37th IEEE Power Electronics Specialists Conference. IEEE, 2006: 1–5
13. Lei Z, Xuhui W, Feng Z, et al. Deep field-weakening control of PMSMs for both motion and generation operation[C]//2011 International Conference on Electrical Machines and Systems. IEEE, 2011: 1–5
14. Deng, T., Su, Z., Li, J., et al.: Advanced angle field weakening control strategy of permanent magnet synchronous motor[J]. IEEE Trans. Veh. Technol. **68**(4), 3424–3435 (2019)

Research on Wireless Power Transfer Systems Based on Semiactive Rectifier

Kangli Luo[1,2], Weiyao Mei[1,2], Jiangjun Yuan[1,2], Tongzhao Li[1,2], and Lijun Diao[1,2](✉)

[1] School of Electrical Engineering, Beijing Jiaotong University, Beijing 100044, China
{22121488,19117042,21121526,22121470,ljdiao}@bjtu.edu.cn

[2] Key Laboratory of Vehicular Multi-Energy Drive Systems (VMEDS), Ministry of Education, Beijing 100044, China

Abstract. Wireless power transfer (WPT) systems have become more and more popular for its advantages of operation safety, flexibility and convenience.In order to achieve constant current charging in wireless power transfer systems, control method based on the secondary side semiactive rectifier (SAR) has been proposed, which has more advantages compared to DC-DC converter and fully active bridge rectifier. In this paper, the dual-side LCC wireless power transfer system based on semiactive rectifier is introduced firstly. Second, the effect of the coil internal resistance on the charging current is investigated. Thirdly, the working principles of phase shift control (PSC), synchronous phase control (SPC), and duty cycle control (DCC) are analyzed in detail, and then the constant current output of wireless power transfer system are realized with these three control methods by simulation. Finally, the three control strategies are compared and analyzed in terms of control effects and implementation difficulty under the same simulation parameter conditions, and the advantages and disadvantages of these three control strategies are summarized.

Keywords: Wireless power transfer · Semiactive rectifier · Control method

1 Introduction

Compared with the traditional plug-in power electronic system, wireless power transfer (WPT) systems have become more and more popular for its advantages of operation safety, flexibility and convenience. WPT can be applied to portable devices powered by batteries or supercapacitors, biomedical implants, underwater devices, spacecraft, and electric vehicles. In practice, batteries or supercapacitor loads typically require constant current (CC) charging to achieve fast charging [1]. However, the change of the equivalent load resistance of battery or supercapacitor during charging can significantly affect the charging current, and in addition, the change of mutual inductance caused by coil misalignment can also affect the charging current.

The CC charging of wireless power transfer systems can be achieved in various ways. In [2], a PI controlled BUCK was applied to adjust the current to achieve CC charging under load changes. However, the BUCK makes the size of the system larger

L. Jia et al. (Eds.): EITRT 2023, LNEE 1135, pp. 287–296, 2024.
https://doi.org/10.1007/978-981-99-9307-9_31

and increases power losses. To simplify the secondary side, [3] proposed a BUCK on the primary side to achieve CC charging, but it needs extra wireless communication circuits to transmit the voltage and current signal of the secondary side to the primary side, which will lead to unstable control and malfunctions. In order to simplify the circuit topology and avoid bilateral communication, [4] proposed using a fully active bridge rectifier instead of uncontrolled rectifier to achieve rectification and constant charging current. The fully active bridge rectifier reduces system size and increases conversion efficiency compared with the buck converter. However, fully active bridge rectifier needs to control four switches, which is much more complex. Therefore, [5] proposed an optimized semiactive rectifier (SAR). Unlike fully active bridge rectifier, SAR reduces half of the MOSFETs, greatly simplifies the control method.

There are three main control strategies for SAR. The phase shift control for SAR is used for CC charging and impedance matching [5–7]. To eliminate the reactive impedance introduced by PSC, professor Chunbo Zhu's team proposed the synchronous phase control for SAR [8, 9]. To reduce the complexity of controling SAR, the duty cycle control strategy was introduced in [10].

In summary, using the SAR to achieve CC charging not only avoids applying additional DC-DC converter and wireless communication circuits, but also simplifies the circuit topology and control method compared to the full-bridge rectifier. Therefore, SAR is a simpler and more effective methods for achieving CC charging [11]. However, there is limited research on the characteristics of SAR control strategies, so this paper will compare three typical control strategies of SAR and explore their advantages and disadvantages.

2 System Structure and Theoretical Analysis

2.1 System Structure

The system structure is shown in Fig. 1. The input DC power is first converted into high-frequency AC power through a full bridge inverter, and then transmitted wirelessly from the primary side to the secondary side through a dual-side LCC compensation network by magnetic field coupling resonance. On the secondary side, a SAR is applied to be a rectifier and adjust output current.

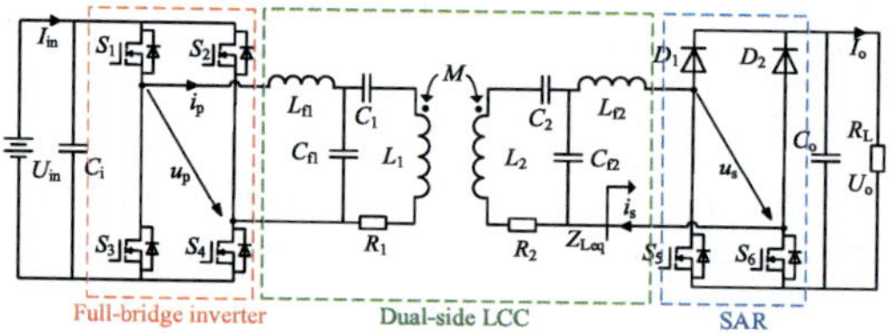

Fig. 1. Topology of dual-side LCC WPT system based the SAR.

In Fig. 1, the fully bridge inverter consists of four MOSFETs S_1-S_4, and the dual-side LCC compensation circuit consists of a transmitting coil L_1, a receiving coil L_2, a series compensation capacitor C_1/C_2, a parallel compensation capacitor C_{f1}/C_{f2}, and a series compensation inductor L_{f1}/L_{f2}. M is the mutual inductance of the coil, the SAR consists of two MOSFETs S_5, S_6, and two diodes D_1 and D_2, C_0 is a capacitor filter, and R_L is the equivalent variable resistance of the battery load.

2.2 Analysis of Dual-Side LCC Compensation Topology

The mutual inductance model of the dual-side LCC topology is shown in Fig. 2, where $i_1 - i_4$ is the grid current.

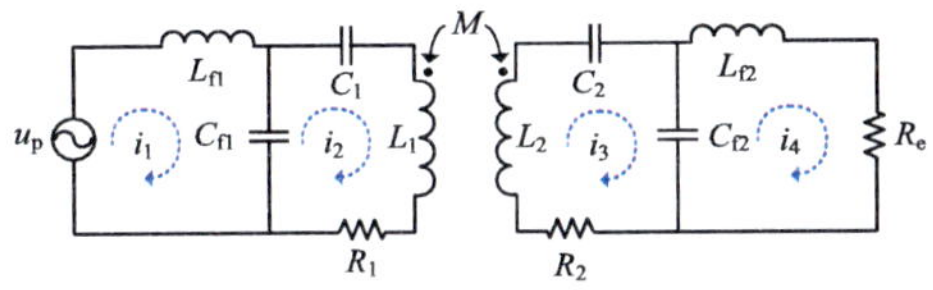

Fig. 2. Mutual inductance model of the dual-side LCC

Based on Kirchhoff's voltage law (KVL) and fundamental harmonic analysis (FHA), the following equation is derived as:

$$\begin{cases} \dot{U}_p = j\omega L_{f1} + \dfrac{1}{j\omega C_{f1}}(\dot{I}_1 - \dot{I}_2) \\ 0 = (j\omega L_1 + \dfrac{1}{j\omega C_1} + R_1)\dot{I}_2 + \dfrac{1}{j\omega C_{f1}}(\dot{I}_2 - \dot{I}_1) - j\omega M \dot{I}_3 \\ 0 = (j\omega L_2 + \dfrac{1}{j\omega C_2} + R_2)\dot{I}_3 + \dfrac{1}{j\omega C_{f2}}(\dot{I}_3 - \dot{I}_4) - j\omega M \dot{I}_2 \\ 0 = \dfrac{1}{j\omega C_{f2}}(\dot{I}_4 - \dot{I}_3) + (j\omega L_{f2} + R_e)\dot{I}_4 \end{cases} \tag{1}$$

According to [1], the dual-side LCC compensation network parameters meet the following principles:

$$\begin{cases} \omega_0^2 L_{f1} C_{f1} = 1 \\ \omega_0^2 L_{f2} C_{f2} = 1 \\ \omega_0^2 (L_1 - L_{f1}) C_1 = 1 \\ \omega_0^2 (L_2 - L_{f2}) C_2 = 1 \end{cases} \tag{2}$$

Therefore, the value of the equivalent system output current is:

$$I_s = \left|\dot{I}_4\right| = \frac{MU_p}{\omega L_{f1} L_{f2} + \frac{L_{f1}}{\omega L_{f2}} R_2 R_e} \tag{3}$$

where U_p is the output voltage of the full bridge inverter.

From Eq. (3), when considering the internal resistance of the coil, I_s is related to the load. It indicates that when the load resistance changes, the output current cannot be kept constant, so closed-loop control is needed to achieve precise current control.

3 Control Method of SAR

There are three main control methods for semiactive rectifier, phase shift control, synchronous phase control, and duty cycle control. The following will analyze the working principles of these three control methods.

3.1 Phase Shift Control

The PSC is the most traditional control method among the three control methods, and the working waveforms and switch state transition of its control principle are shown in Figs. 3 and 4. Typically, set the driving voltage pulse duty cycle of S_5 and S_6 to 50%. β is the pulse width of the rectifier input voltage. The operation of PSC in a cycle can be divided into the following four modes:

Mode 1 [t_0~t_1]: Switch S_5 is turned off and the current is greater than zero, so the current will pass through D_1 and D_6.

Mode 2 [t_1~t_2]: Switch S_6 is turned on and the current is smaller than zero, so the current will pass through S_6 and D_5.

Mode 3 [t_2 ~ t_3]: Switch S_6 is turned off and the current is smaller than zero, so the current will pass through D_2 and D_5.

Mode 4 [t_3 ~ t_4]: Switch S_6 is turned on and the current is greater than zero, so the current will pass through S_5 and D_6.

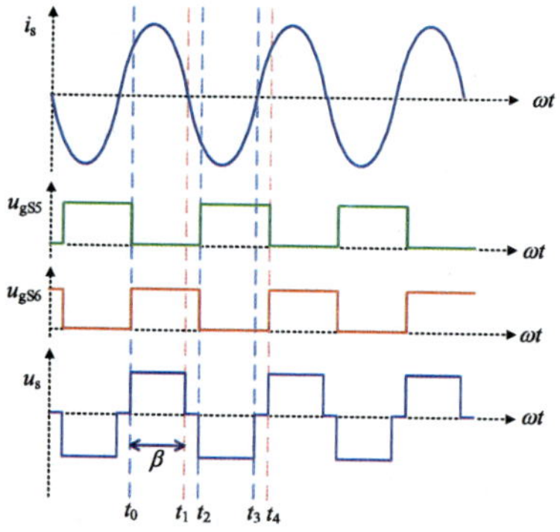

Fig. 3. The waveforms of the PSC strategy.

According to the Fourier Series and principles of electric circuits, the output current I_o can be obtained as follow:

$$I_o = \frac{8\sin\frac{\beta}{2}MU_{in}}{\pi^2\omega L_{f1}L_{f2}} \tag{4}$$

From the above equation, it can be seen that the PSC can control the output current I_o by adjusting the phase shift angle β.

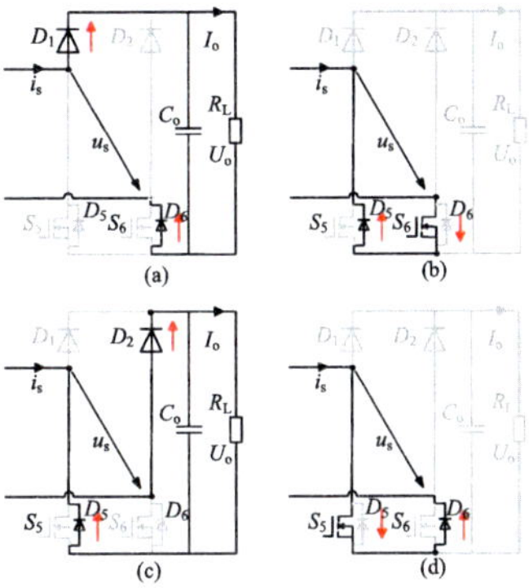

Fig. 4. Switch state transition and current path of PSC's control principle. **a** Mode 1. **b** Mode 2. **c** Mode 3. **d** Mode 4.

3.2 Synchronous Phase Control

The SPC is improved and optimized based on the PSC, and the working waveforms and switch state transition of its control principle are shown in Figs. 5 and 6. α is the pulse width of the gate drive voltage. Its operation in one cycle can be divided into the following four modes:

Mode 1: The current is greater than zero, S_5 is turned on and S_6 is turned off, so the current will pass through S_5 and D_6.

Mode 2: The current is greater than zero, S_5 is turned off and S_6 is turned off, so the current will pass through D_1 and D_6.

Mode 3: The current is smaller than zero, S_5 is turned off and S_6 is turned on, so the current will pass through D_5 and S_6.

Mode 4: The current is smaller than zero, S_5 is turned off and S_6 is turned off, so the current will pass through D_2 and D_5.

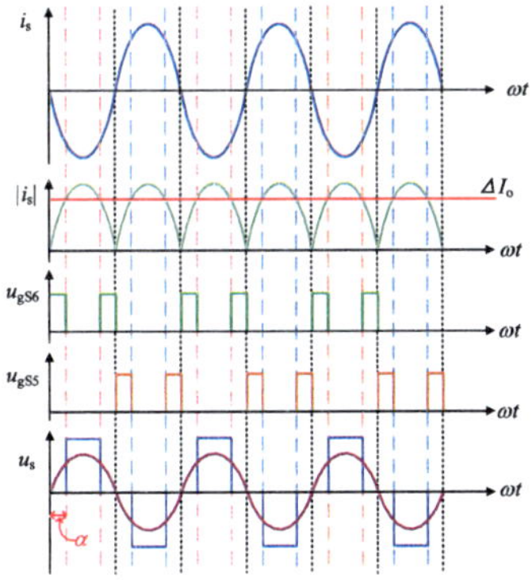

Fig. 5. Typical waveforms of the SPC strategy

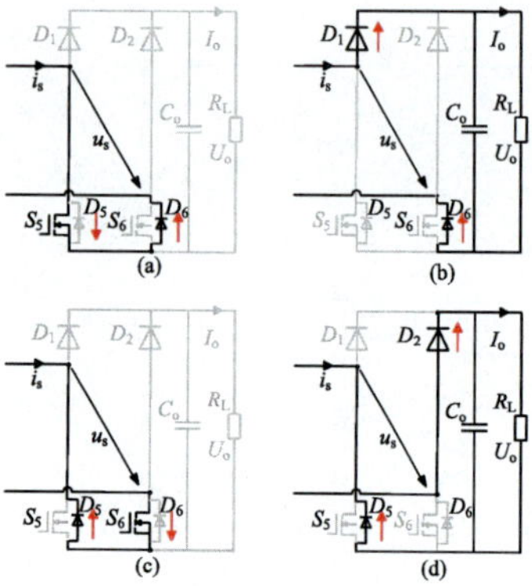

Fig. 6. Switch state transition and current path of SPC's control principle. **a** Mode 1. **b** Mode 2. **c** Mode 3. **d** Mode 4.

The output current I_o can be obtained as follow:

$$I_o = \frac{8\cos\frac{\alpha}{2}MU_{in}}{\pi^2\omega L_{f1}L_{f2}} \tag{5}$$

From the above equation, the SPC can control the output current I_o by adjusting the phase shift angle α.

3.3 Duty Cycle Control

The working waveforms and switch state transition of duty cycle control principle are shown in Figs. 7 and 8. Its operation in one cycle can be divided into the following two modes:

Mode 1 [t_0~t_1]: The switches S_5 and S_6 are turned off, so it is equivalent to uncontrolled rectifier during this time, and the output voltage is a square wave of alternating positive and negative.

Mode 2 [t_1~t_2]: The switches S_5 and S_6 are turned on, and the current flows through switch S_6 and D_5 or through switch S_5 and D_6, so the output voltage is zero.

The output current I_o can be obtained as follow:

$$I_o = \frac{8DMU_{in}}{\pi^2\omega L_{f1}L_{f2}} \tag{6}$$

From the above equation, the DCC can control the output current I_o by adjusting the duty cycle D.

4 Simulation Analysis

Based on the topology diagram of the wireless charging system circuit shown in Fig. 1, a simulation model was built in Matlab/Simulink, as shown in the Fig. 9.

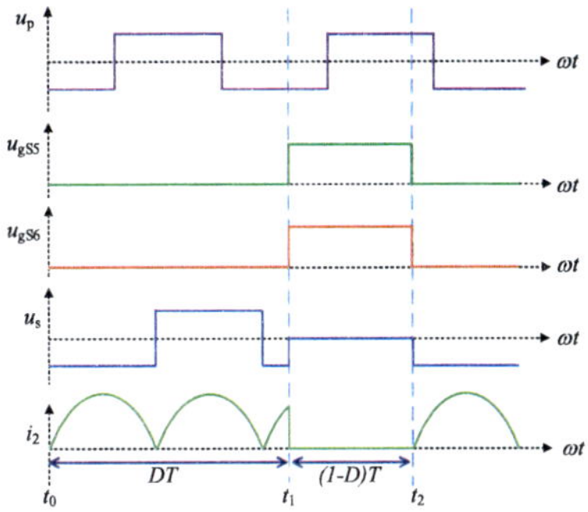

Fig. 7. Typical waveforms of the DCC strategy

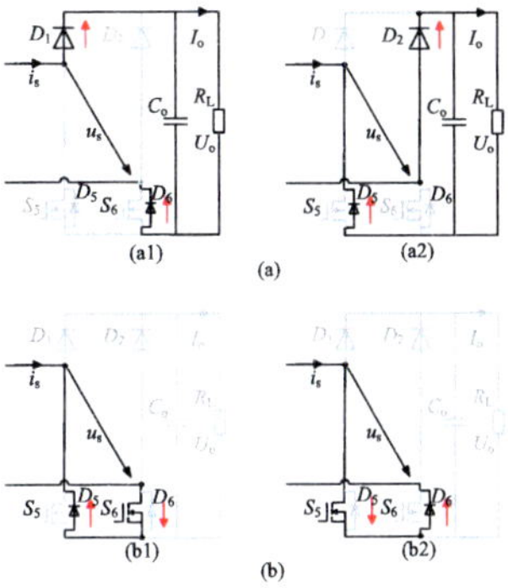

Fig. 8. Switch state transition and current path of DCC's control principle. **a** Mode 1. **b** Mode 2.

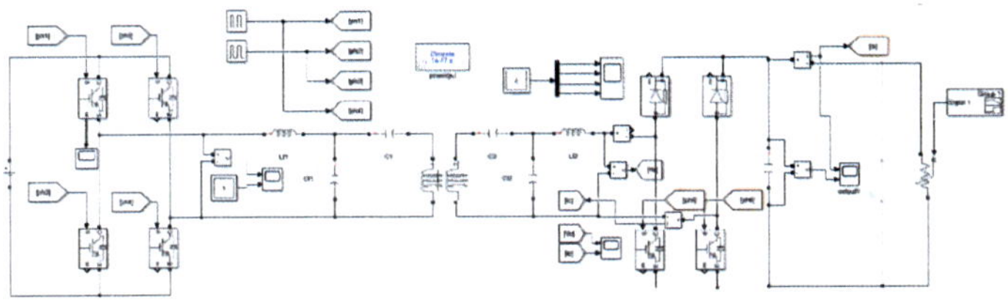

Fig. 9. Main circuit simulation model

Figure 10 shows the waveforms of system output voltage and current under open-loop control, PSC, SPC and DCC. The output current fluctuates with the changes of load resistance under open-loop control, so it cannot achieve CC charging without closed loop control. On the contrary, the output current remains at the expected current value when the load resistance changes under PSC, SPC and DCC, realizing accurate CC charging.

Figure 11 shows the waveforms of voltage u_s and current i_s, through comparative analysis, we can see that using traditional PSC to control the SAR cannot ensure that u_s and i_s are in the same phase. On the contrary, using SPC can ensure that the phase difference of u_s and i_s is zero, making the input impedance Z_{Leq} is pure resistive, which can improve the efficiency of the system. However, the control difficulty of SPC is the highest among the three control methods.

The difference between DCC and PSC or SPC is that the control process does not need to achieve synchronization technology, so the control of DCC is the simplest. However, according to the waveforms of the inverter output voltage u_p and current i_p under the DCC shown in Fig. 12, due to the irregular current, the ZVS on the primary side cannot be guaranteed, which may reduce the efficiency of the system.

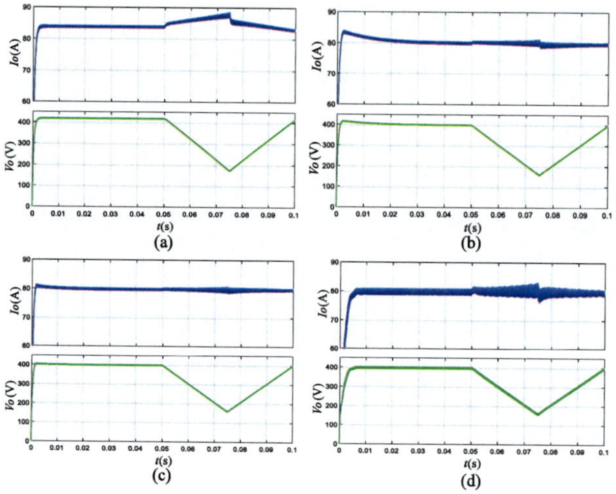

Fig. 10. Output voltage and current waveforms. **a** Open-loop control. **b** PSC. **c** SPC. **d** DCC.

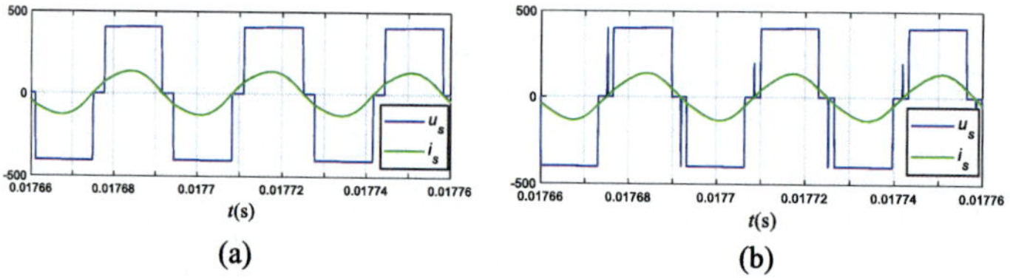

Fig. 11. Waveform of output voltage u_s and i_s.

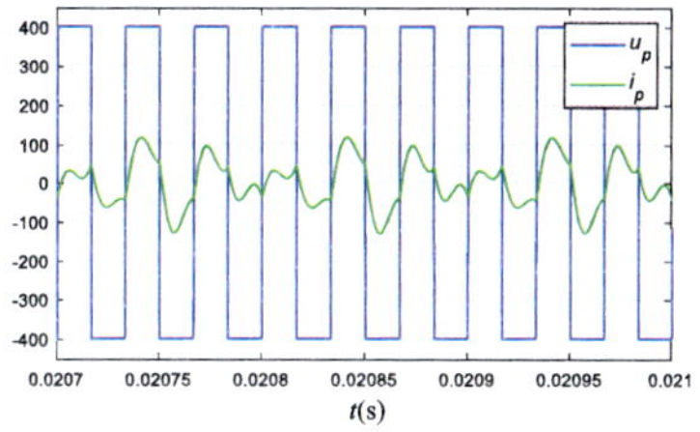

Fig. 12. Inverter output voltage u_p and current i_p waveforms under duty cycle control.

5 Conclusion

This paper research on the wireless power transfer systems based on semiactive rectifier, and analyse the working principles of PSC, SPC and DCC. Simulations verify that the SAR can achieve CC charging with all three control strategies. The control effect of SPC is the best, but it is the most complex to implement, and PSC is simpler but it will introduce reactive impedance. Besides, DCC is the simplest to implement, but the ZVS on the primary side cannot be guaranteed with DCC.

Acknowledgments. This research is supported by Magnetic Coupling Wireless Charging Technology for Unmanned Underwater Vehicles (E23JBZX00020).

References

1. Fu, N., Deng, J.J., Wang, Z.P.: A hybrid mode control strategy for LCC–LCC compensated WPT system with wide ZVS operation. IEEE Trans. Power Electron. **37**(2), 2449–2460 (2022)
2. Li, Z., Zhu, C., Jiang, J., Song, K., Wei, G.: A 3-kW wireless power transfer system for sightseeing car supercapacitor charge. IEEE Trans. Power Electron. **32**(5), 3301–3316 (2017)
3. Li, H., Li, J., Wang, K., Chen, W., Yang, X.: A maximum efficiency point tracking control scheme for wireless power transfer systems using magnetic resonant coupling. IEEE Trans. Power Electron. **30**(7), 3998–4008 (2015)
4. Berger, A., Agostinelli, M., Vesti, S., Oliver, J.A., Cobos, J.A., Huemer, M.: A wireless charging system applying phase-shift and amplitude control to maximize efficiency and extractable power. IEEE Trans. Power Electron. **30**(11), 6338–6348 (2015)
5. Cloak, K., Asa, E., Bojarski, M., Czarkowski, D., Onar, O.C.: A novel phase-shift control of semibridgeless active rectifier for wireless power transfer. IEEE Trans. Power Electron. **30**(11), 6288–6297 (2015)
6. Huang, Z., Lam, C.S., Mak, P.I.: A single-stage inductive-power-transfer converter for constant-power and maximum-effificiency battery charging. IEEE Trans. Power Electron. **35**(9), 8973–8984 (2020)
7. Huang, Z., Wong, S.C., Tse, C.K.: An inductive-power-transfer converter with high efficiency throughout battery-charging process. IEEE Trans. Power Electron. **34**(10), 10245–10255 (2019)
8. Li, Z., Song, K., Jiang, J., Zhu, C.: Constant current charging and maximum efficiency tracking control scheme for supercapacitor wireless charging. IEEE Trans. Power Electron. **33**(10), 9088–9100 (2018)

9. Song, K., Wei, R.Z., Yang, G.: Constant current charging and maximum system efficiency tracking for wireless charging systems employing dual-side control. IEEE Trans. Ind. Appl. **56**(1), 622–634 (2020)
10. Li, H., Xu, J., Gao, F., Zhang, Y., Yang, X., Tang, H.: Duty cycle control strategy for dual-side LCC resonant converter in wireless power transfer systems. IEEE Trans. Transp. Electrif. **8**(2), 1944–1955 (2022)
11. Zou, S.L., Omer, C.O., Veda, G.: Secondary active rectifier control scheme for a wireless power transfer system with double-sided LCC compensation topology. In: IECON 44th Annual Conference of the IEEE Industrial Electronics Society, pp. 2145–2150 (2018)

Resolver-To-Digital Conversion and Sensor Fault Identification of PMSM

Hao Zhu[1,2], Chunmei Xu[1,2](✉), Qiya Wu[1,2], Zichen Gao[1,2], and Jia Zhang[1,2]

[1] School of Electrical Engineering, Beijing Jiaotong University, Beijing 100044, China
{22121565,chmxu,20117040,19117007,20121524}@bjtu.edu.com

[2] Key Laboratory of Vehicular Multi-Energy Drive Systems (VMEDS), Ministry of Education, Beijing 100044, China

Abstract. The permanent magnet synchronous motor (PMSM) is the core equipment for next-generation trains in the rail transportation field, and the resolver is a commonly used position sensor for PMSM. The position of the motor is conversed from the output orthogonal waveform of the resolver. This paper presents a method for resolver-to-digital conversion (RDC) and a sensor fault identification method of PMSM. In the first part, the structure of the resolver is introduced, as the resolver modulates the angle value with the high-frequency excitation signal and outputs it to the system. So, the system first needs to demodulate the signal and then make resolver-to-digital conversion. This conversion uses a phase-locked loop to get the rotor angle. In the second part, a sensor fault identification method is presented. When the sensor gets a fault, there will be a pulsation in the stator's current, and the angle error is calculated from the motor current pulsation. According to the simulation results, the method has small identification errors and fast response speed.

Keywords: Resolver-to-digital conversion · Phase-locked loop · Sensor fault identification

1 Introduction

With the continuous maturity of power electronics and motor control technology, PMSM is widely used in industrial robots, aerospace, rail transit, electric vehicles, and other fields. PMSM is recognized as the future development direction of rail transit traction systems in rail transit due to its high-power factor, fast response speed, and low torque ripple [1].

Figure 1 shows the PMSM control system, to improve the operational stability of PMSM, the system uses suitable sensors to measure the position and speed of the motor. If the measurement deviates, the current will ripple after the coordinate transformation, and the torque will also oscillate because the electromagnetic torque is related to the current. Therefore, research on resolver-to-digital conversion for PMSM is needed.

Resolver is a kind of electromagnetic sensor, it is widely used in angle measurement [2], the structure of the resolver is Fig. 2. It is composed of a rotor side and stator side,

L. Jia et al. (Eds.): EITRT 2023, LNEE 1135, pp. 297–304, 2024.
https://doi.org/10.1007/978-981-99-9307-9_32

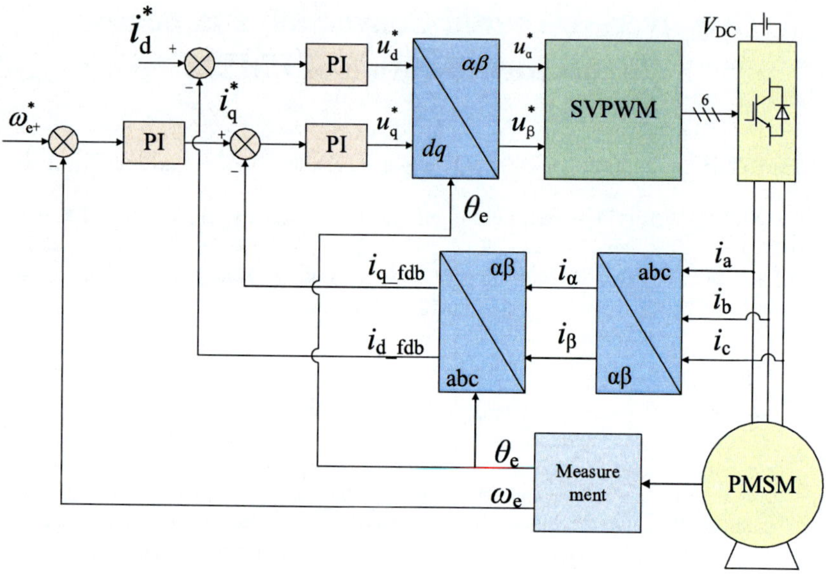

Fig. 1. PMSM control system.

primary excitation winding and a pair of secondary output windings are located on the stator side, primary output windings and secondary excitation windings are located on the rotor side [3]. The resolver is mechanically connected to the motor's rotor, when the motor rotates, the coupling between the windings also changes and is consistent with the change of motor angle [4].

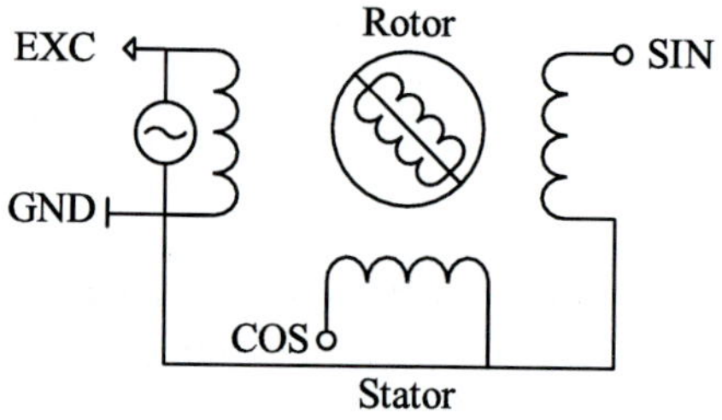

Fig. 2. Resolver's structure.

The angle can be obtained from the output signal by the decoding methods. The decoding methods include amplitude under-sampling and phase-sensitive demodulation. The amplitude under-sampling method samples once at each carrier's peak, and obtain the rotor angle sine cosine signal [5], but this method will deduce the signal accuracy, and in each carrier cycle, this method can only sample twice, which does not meet the Nyquist sampling theorem. Phase-sensitive demodulation method uses the principle that the excitation signal's frequency is higher than the rotor angle's frequency [6].

After demodulation, it is necessary to get the rotor angle from the demodulated results. The current mainstream RDC method is a phase-locked loop [7, 8].

Due to various reasons, the resolver has amplitude and phase fault [9], which impacts the system operation, but does not make the system fail immediately [10, 11], this kind of fault is the key part of fault identification.

For the resolver, this paper proposes a resolver-to-digital conversion and sensor fault identification method. Firstly, it obtains the angular sine cosine value from the output signal and then uses the phase-locked loop to obtain the rotor angle value. Secondly, this paper proposes an identification method, which uses the stator current to calculate the angle error. The simulation results verify the effectiveness of this method.

2 Resolver Decoding Method

The output signal of the resolver is as follows:

$$\begin{aligned} U_{\sin} &= A_{\text{out}} \sin(\omega t) \sin\theta \\ U_{\cos} &= A_{\text{out}} \cos(\omega t) \cos\theta \end{aligned} \tag{1}$$

where A_{out} is the output signal amplitude, ω is the frequency of the excitation signal and θ is the rotor angle.

Firstly, multiplied the resolver's output signal by the high-frequency excitation signal:

$$\begin{aligned} U_{\sin} * U_{\text{EXC}} &= A_{\text{out}}^2 \sin^2(\omega t) \sin\theta \\ U_{\cos} * U_{\text{EXC}} &= A_{\text{out}}^2 \cos^2(\omega t) \cos\theta \end{aligned} \tag{2}$$

In the demodulation, the output signal amplitude can be ignored. Then, separating the high-frequency part of the signal:

$$\begin{aligned} U_{\sin} * U_{\text{EXC}} &= \frac{1}{2}\left[\sin\theta + \frac{\sin(2\omega t + \theta)}{2} - \frac{\sin(2\omega t - \theta)}{2}\right] \\ U_{\cos} * U_{\text{EXC}} &= \frac{1}{2}\left[\cos\theta + \frac{\cos(2\omega t + \theta)}{2} - \frac{\cos(2\omega t - \theta)}{2}\right] \end{aligned} \tag{3}$$

Since the excitation signal frequency is much higher than the rotor angle frequency, a low-pass filter can filter out the high-frequency signal, and obtain the sine and cosine signal of the rotor angle:

$$\begin{aligned} U_{\sin} &= \frac{1}{2} \sin\theta \\ U_{\cos} &= \frac{1}{2} \cos\theta \end{aligned} \tag{4}$$

The phase-locked loop uses a PI controller to track the angle, when the tracking is successful, the rotor angle's difference is zero, and the tracking result is the rotor angle. The structure of the phase-locked loop is shown in Fig. 3.

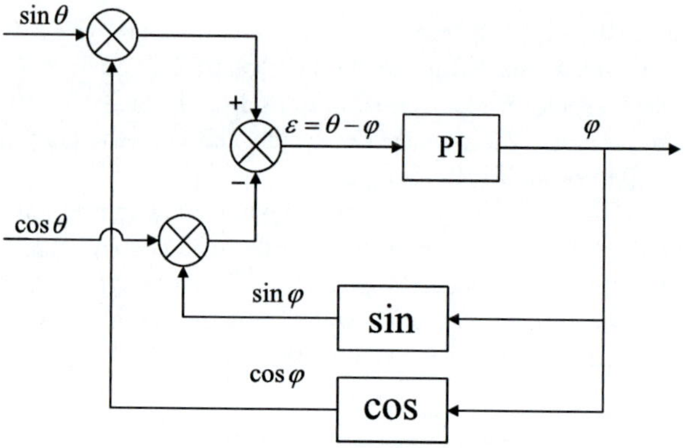

Fig. 3. Phase-Locked Loop.

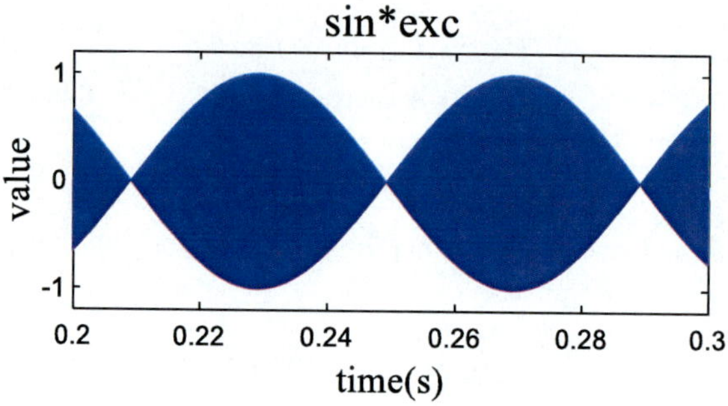

Fig. 4. Resolver's output signal.

To verify the effectiveness of this method, a simulation model was built in MATLAB. Figure 4 shows the original waveform of the resolver.

In the output signal of the resolver, the sine signal is superimposed with the high-frequency excitation signal, and it is input for the demodulation part. The demodulation result is shown in Fig. 5.

After demodulation, the rotor angle sine signal can be obtained with low error. The sine cosine signal is input to the phase-locked loop, and the result is the rotor angle value, the result of the phase-locked loop is shown in Fig. 6.

It can be seen that this method can obtain the rotor angle of the motor, with fast response speed and low error.

3 Sensor Fault Identification

Assuming that an amplitude-phase mismatch fault occurs at this time, but the degree of the fault is small:

$$\begin{aligned} U_{\sin} &= \sin(\theta + \beta) \\ U_{\cos} &= (1 + \alpha)\cos\theta \end{aligned} \tag{5}$$

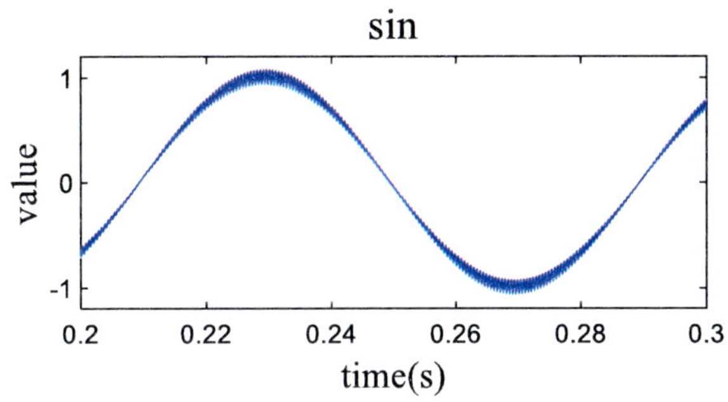

Fig. 5. Demodulation result.

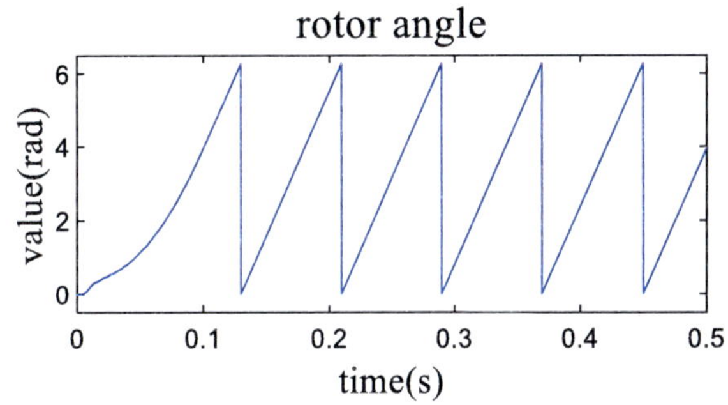

Fig. 6. Rotor angle value from the phase-locked loop.

where, α is the amplitude deviation value, and β is the phase deviation value.

After the signal output from the phase-locked loop, there will be a deviation in the output angle, which is related to the amplitude deviation value and phase deviation value:

$$\theta_{\text{err}} \approx \beta \frac{1 + \cos 2\theta}{2} - \alpha \frac{\sin 2\theta}{2} \tag{6}$$

In the $i_{\text{d}} = 0$ field-oriented control system, the i_{d} and i_{q} in the current loop normally follow the reference value, but when the fault occurs, the current will pulsate due to the error angle:

$$\begin{aligned} i_{\text{d}} &= I_{\text{ref}} \left[\beta \frac{1 + \cos 2\theta}{2} - \alpha \frac{\sin 2\theta}{2} \right] \\ i_q &= I_{\text{nref}} + \alpha I_{\text{nref}} \frac{1 + \cos 2\theta}{2} + \beta I_{\text{n}} \frac{\sin 2\theta}{2} \end{aligned} \tag{7}$$

The normal current is shown in Fig. 7, and under fault conditions, the current is shown in Fig. 8.

Using mean filtering to obtain the current pulsation, and then the deviation value can be extracted by using the sign function, Fig. 9 shows the fault-detection method.

When the error occurs, the angle error is obtained by the identification method in Fig. 9. The result of the system is shown in Fig. 10.

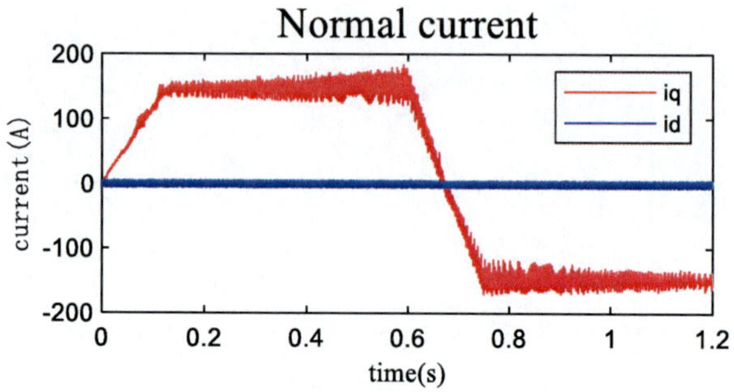

Fig. 7. Normal current waveform.

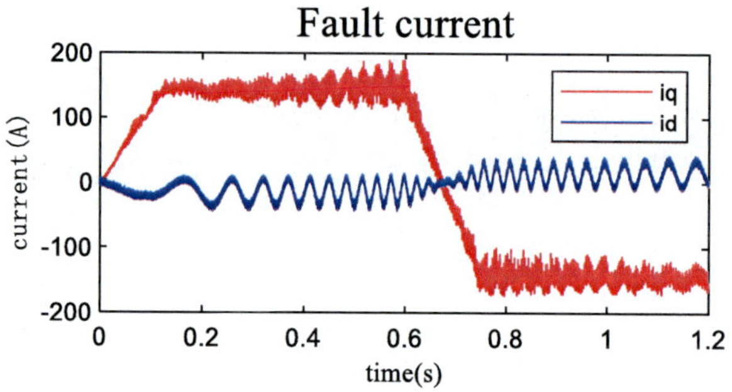

Fig. 8. Fault current waveform.

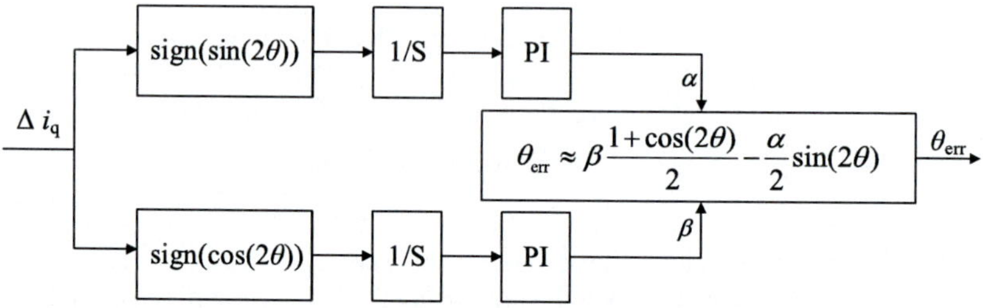

Fig. 9. Fault-detection method

It can be seen that the method can effectively identify the error value, and the result is closer to the real value, so it can be used as an effective sensor fault identification method.

4 Conclusion

This paper proposes an RDC and fault identification method for the resolver. The first part uses phase-sensitive demodulation to extract the fundamental frequency signal from the resolver signal, then uses a phase-locked loop to calculate the rotor's angle. From

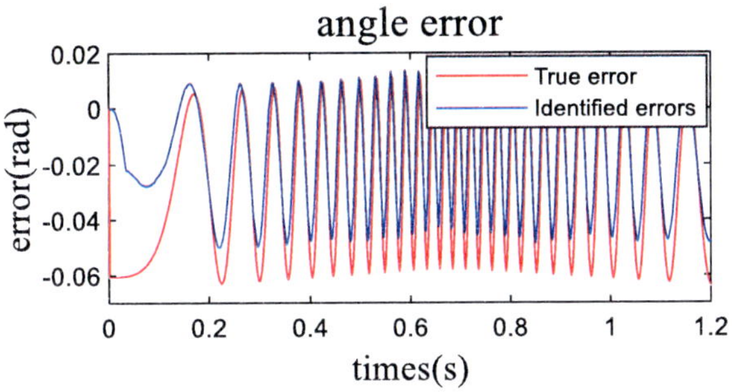

Fig. 10. Angle error comparison

the simulation results, it can be seen that the method can effectively obtain the rotor angle with a low error level. The resolver faults can cause current pulsation, second part builds simulations to verify the Phenomenon of current pulsation. For the case of current pulsation, this paper proposes a method to calculate the fault angle using the pulsation value, and the simulation results show that the method can effectively extract the fault angle.

References

1. Zhang, J., Su, H., Ren, Q.: Review on development and key technologies of permanent magnet synchronous traction system for rail transit. J. Traffic Transp. Eng. **21**(6), 63–77 (2021). (in Chinese)
2. Mo, H., Min, L.: Summary of modern high-performance permanent magnet AC servo system—sensor device and technology. Trans. China Electrotech. Soc. **30**(6), 10–21 (2015). (in Chinese)
3. Alipour, S.: Use of resolvers in limited rotations. IEEE Trans. Energy Convers. **37**(3), 2019–2026 (2022)
4. Joan, B., Coia, F., Gabriel, G., Rodrigo, R., Samuel, G., Joan, R.: High-accuracy all-digital resolver-to-digital conversion. IEEE Trans. Ind. Electron. **59**(1), 326–333 (2012)
5. Jang, P., Lee, T., Hwang, Y., Nam, K.: Quadrature demodulation method for resolver signal processing under different sampling rates. IEEE Access **10**(1), 7016–7024 (2022)
6. Wei, X., Wang, L., Shen, P., Yue, C., Bai, L.: Research on traction motor speed and angle of measurement technique based on resolver. J. China Railw. Soc. **40**(1), 50–56 (2018). (in Chinese)
7. Wang, S., Zhang, H., Bao, Y., Vasyl, V., Prystupa D, Buticchi G, Li J, Zhang X.: A high precision resolver-to-digital conversion method with pre-filtering DDSRF used in electric vehicle. In: 2022 IEEE Transportation Electrification Conference and Expo, Asia-Pacific (ITEC Asia-Pacific), pp. 1–6. IEEE, Haining (2022)
8. Xiong D, Jiang Q, Wang J.: Accuracy measuring system for a resolver and its decoder. In: 2021 IEEE 16th Conference on Industrial Electronics and Applications (ICIEA), pp. 713–717. IEEE, Chengdu (2021)
9. Li, H., Chen, X., Liu, G., Cui, X.: A Disturbance estimation approach to self-calibration of gimbal resolver-to-digital conversion system. IEEE Trans. Ind. Electron. **70**(1), 793–802 (2023)

10. Zhang, H., Fan, T., Guo, J., Liu, Z., Bian, Y., Wen, X.: Analysis and elimination of position signal error of permanent magnet synchronous motor. Proc. CSEE **40**(1), 294–302 (2020). (in Chinese)
11. Li, H., Yao, H., Wang, P.: Position sensor online fault diagnosis and adaptive fault-tolerant control of PMSM drive system. Trans. China Electrotech. Soc. **31**(2), 228–235 (2016). (in Chinese)

Modeling and Control of Long-Stator Linear Synchronous Motors for High-Speed Maglev

Shanhui Zhang[1,2], Chunmei Xu[1,2](✉), Lei Wang[1,2], Hao Zhu[1,2], Zichen Gao[1,2], Qiya Wu[1,2], and Lijun Diao[1,2]

[1] School of Electrical Engineering, Beijing Jiaotong University, Beijing 100044, China
{22121558,chmxu,leiwang,22121565,19117007,20117040, ljdiao}@bjtu.edu.cn

[2] Key Laboratory of Vehicular Multi-Energy Drive Systems (VMEDS), Ministry of Education, Beijing 100044, China

Abstract. Reliable control of high-speed maglev traction motors and power converters is a prerequisite for the stable operation of high-speed maglev trains. Due to the particularity of the high-speed maglev system, a rotor magnetic field directional control strategy with constant excitation and zero direct axis current is usually adopted to achieve independent traction suspension control. Considering the low switching frequency of the traction converter, a three-level topology is adopted. Compared to the two-level, the voltage stress per switch in the three-level topology is only half of the DC bus voltage, and the corresponding DC bus voltage is doubled. With the same switching frequency, the three-level output ripple and harmonics are much smaller. In this paper, traction control simulation models based on two-level and three-level converters are built in MATLAB/Simulink to compare the operation of maglev trains with different inverter topologies and to verify the superiority of the three-level structure.

Keywords: High-speed Maglev · Long Stator Linear Synchronous Motor · Vector control · Three-level SVPWM

1 Introduction

The high-speed magnetic levitation system is a new type of rail transit system. High-speed maglev trains are driven by long-stator linear synchronous motors (LSLSM) for operation, which have many advantages such as contactless operation, high speed, fast start-up, low energy consumption, and low ecological effect, and widely studied in countries around the world [1, 2].

Reliable control of high-speed maglev traction motors and power converters is a prerequisite for the stable operation of high-speed maglev trains. Considering that the LSLSM traction and levitation system of high-speed maglev trains shares a single magnetic pole, a rotor magnetic field orientation control strategy with constant excitation and zero cross-axis currents is adopted to realize independent control of levitation and traction control mutually decoupled [3, 4]. As the high-speed maglev system is a large

L. Jia et al. (Eds.): EITRT 2023, LNEE 1135, pp. 305–312, 2024.
https://doi.org/10.1007/978-981-99-9307-9_33

inertia system, the power converter is subjected to high voltage and high current, and the switching frequency is usually low, so the main circuit needs to consider a multi-level inverter topology to improve traction control performance, and the high-speed maglev usually adopts three-level topology [5–7]. The main modulation methods for inverter speed control based on NPC three-level converter include SPWM [8] and space vector pulse width modulation (SVPWM) algorithms, which have higher voltage utilization compared with the SPWM algorithm [9].

This paper presents a comparative analysis of the output performance of different inverter topologies under directional control of the rotor magnetic field of LSLSM for high-speed maglev and verifies the effectiveness of the three-level topology and modulation strategy through simulation.

2 High-Speed Magnetic Levitation Traction Control Model

2.1 LSLSM Mathematical Model

To determine the traction control strategy applicable to the maglev system, the mathematical model of the LSLSM is first established. The voltage, magnetic chain and torque equations of the LSLSM [10] are

Voltage equation as

$$u_d = ri_d + p\psi_d - \frac{\pi v \psi_q}{\tau_s},\ u_q = ri_q + p\psi_q + \frac{\pi v \psi_d}{\tau_s} \tag{1}$$

where, $U_d, U_q, i_d, i_q, \varphi_d, \varphi_q, r, v$, and τ_s are Stator cross-axis voltage, current, magnetic chain, stator resistance, train running speed, and pole pitch, respectively.

Magnetic chain equation as

$$\psi_d = L_d i_d + M_{afd0} i_{fd},\ \psi_q = L_q i_q \tag{2}$$

where M_{afd0} and i_{fd} are the stator-rotor inductance and excitation current respectively.

Traction force equation as

$$F_e = \frac{3}{2}\frac{\pi}{\tau_s}(\psi_d i_q - \psi_q i_d) = \frac{3}{2}\frac{\pi}{\tau_s} M_{afd0} i_{fd} i_q + \frac{3}{2}\frac{\pi}{\tau_s}(L_d - L_q) i_d i_q \tag{3}$$

Mechanical equations of motion as

$$\frac{dv}{dt} = \frac{(F_e - F_z)}{M},\ S = \int_0^t v dt \tag{4}$$

Among them,

$$F_Z = F_A + F_M + F_B \tag{5}$$

$$F_A = 3.2 * v^2 \cdot (0.6 \cdot n/2 + 0.4) \cdot 10^{-3} \tag{6}$$

$$F_M = 0.5n * ((v/111)^{0.5} + 1.3 * (v/111)^{0.7}) \tag{7}$$

$$F_B = \begin{cases} 0, v = 0 \\ 3.3 * n, 0 < v \leq 41.7 \\ n * (146/v - 0.2), v > 41.7 \end{cases} \tag{8}$$

where, F_Z, F_A, F_M, and F_B are the total running resistance, air resistance, guiding rail vortex resistance, and linear generator resistance of high-speed maglev trains, respectively.

2.2 LSLSM Vector Control

The LSLSM for high-speed magnetic levitation has a common pole for the traction and levitation system, and the excitation current is adjusted according to the levitation system to ensure a constant air gap. It is hoped that the adjustment of the excitation current by the suspension control system will not have a significant impact on the traction performance, i.e., the excitation current is expected to remain constant.

The rotor field orientation control method with constant excitation current and zero axial currents in the d-axis can realize independent control of levitation and traction, which is a suitable traction control strategy for the LSLSM [3, 4]. The control structure of the high-speed maglev motor is shown in Fig. 1.

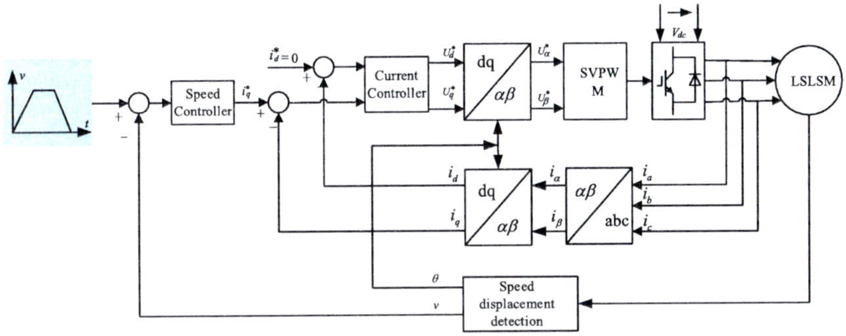

Fig. 1. High-speed magnetic floating LSLSM vector control block diagram

3 NPC Three-Level Topology and SVPWM Modulation

3.1 Three-Level Inverter Topology

The high-speed maglev train adopts a three-level NPC (Neutral Point Clamped) converter with the topology shown in Fig. 2. Where U_{dc} is the DC bus voltage; C1 and C2 are the support capacitors; for phases a, b, and c of the NPC three-level converter, each phase contains four $V_{i1}-V_{i4}$ power switching devices IGBT, four current-continuity diodes $D_{i1}-D_{i4}$ and two clamp diodes $D_{i5}-D_{i6}$.

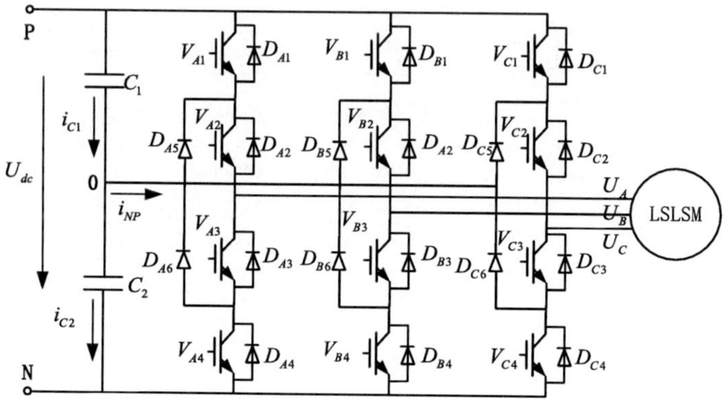

Fig. 2. NPC three-level inverter topology diagram

According to the three-level converter topology, the three-phase bridge arm state function S_i (i = a, b, c) is defined as

$$S_i = \begin{cases} P, \ V_{i1} \text{ and } V_{i2} \text{ conductive} \\ O, \ V_{i2} \text{ and } V_{i3} \text{ conductive} \\ N, \ V_{i3} \text{ and } V_{i4} \text{ conductive} \end{cases} \tag{9}$$

Take A-phase as an example, define the phase voltage of the A-phase bridge arm as u_{an}, then: when S_i = P, $u_{an} = U_{dc}/2$; when S_i = O, $u_{an} = 0$; when S_i = N, $u_{an} = -U_{dc}/2$; so the three-level inverter has $3^3 = 27$ switching states and the corresponding vector distribution as shown in Fig. 3, in which: U_{ref} is the target voltage vector; The spatial voltage vector plane is divided into 6 large sectors at 60° intervals; each large sector is further divided into 6 small sectors.

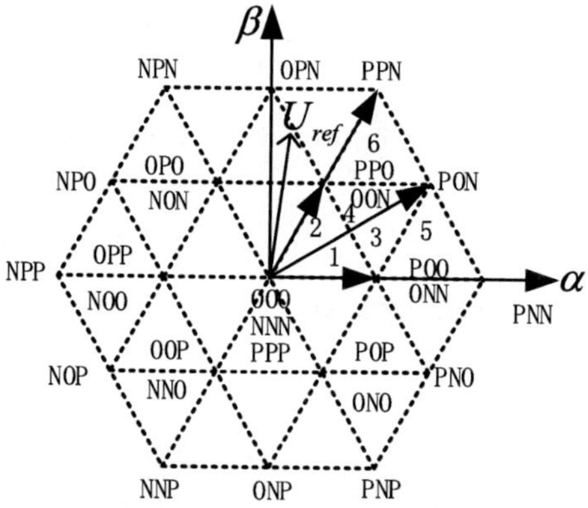

Fig. 3. Three-level inverter voltage space vector distribution diagram

3.2 Modulation Algorithm

The SVPWM method is used to implement the three-level inverter control [5].

Vector selection: Take sector 1 as an example, to ensure that there is no vector mutation during sector switching, each small sector uses (POO or PPO) as the first vector. The vector timing of small sectors 5 and 6 are POO → PON → PNN → ONN → PNN → PON → POO, PPO → PPN → PON → OON → PON → PPN → PPO. The vector timing of the remaining small sectors is shown in Fig. 4. Voltage vector timing diagram. Where, T_x, T_y, and T_z correspond to the action times of the voltage vectors U_x, U_y, and U_z, respectively.

Synthesis time calculation: In one switching period T_s, the voltage vectors U_x, U_y, and U_z are divided into the target voltage vector U_{ref} according to the principle of recent triple vector composite; by the volt-second balance equilibrium, the action time of each voltage vector meets the following formula.

$$U_{ref}T_s = U_xT_x + U_yT_y + U_zT_z \tag{10}$$

Among them, $T_s = T_x + T_y + T_z$.

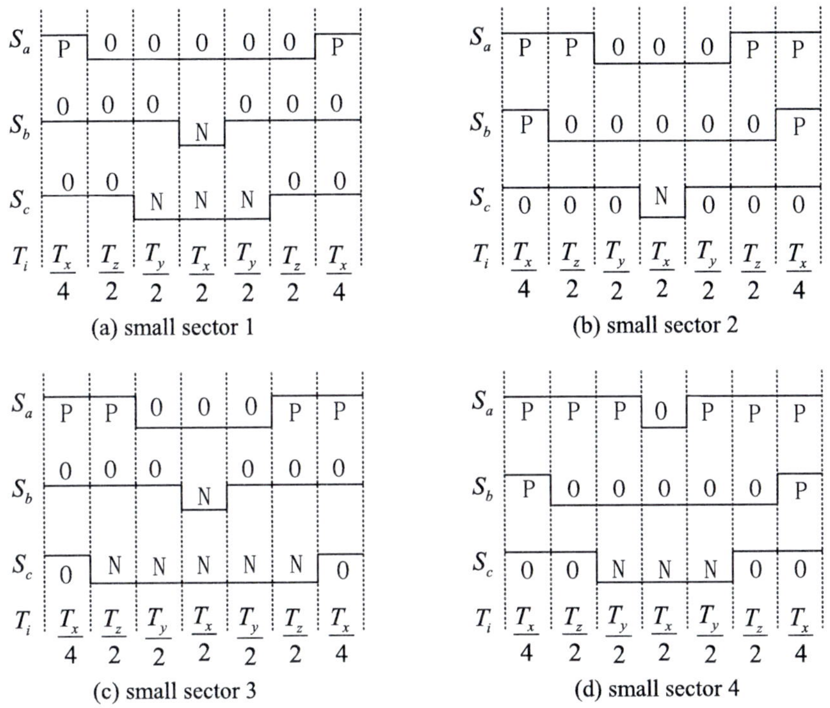

Fig. 4. Voltage vector timing diagram

Define the modulation system m $= \pi U_{ref}/(2U_d)$, and take the timing diagram shown in Fig. 4 as example, the voltage vector action times for small sectors 1 and 2, and small sectors 3 and 4 are shown in Eq. (11).

$$\begin{cases} T_x = \frac{4\sqrt{3}}{\pi} m T_s \sin(\frac{\pi}{3} - \theta) \\ T_y = \frac{4\sqrt{3}}{\pi} m T_s \sin\theta \\ T_z = T_s - \frac{4\sqrt{3}}{\pi} m T_s \sin(\frac{\pi}{3} + \theta) \end{cases}, \begin{cases} T_x = T_s - \frac{4\sqrt{3}}{\pi} m T_s \sin\theta \\ T_y = T_s - \frac{4\sqrt{3}}{\pi} m T_s \sin(\frac{\pi}{3} - \theta) \\ T_z = \frac{4\sqrt{3}}{\pi} m T_s \sin(\frac{\pi}{3} + \theta) + T_s \end{cases} \tag{11}$$

where θ is the phase angle of the target voltage vector U_{ref}.

4 Simulation Verification

A simulation model of a high-speed magnetic levitation traction control system is established in MATLAB/Simulink platform. Two-level and three-level inverter topologies are adopted respectively, SVPWM modulation method, switching frequency of 1000 Hz, and 5000 V for both upper and lower DC bus voltages, and other simulation parameters are shown in Table 1.

Table 1. Simulation parameters for traction control of high-speed maglev trains.

Long Stator Linear Synchronous Motor Parameters	Stator resistance R		0.31 Ω
	Straight axis inductor L_d		3.061 mH
	Cross-axis inductance L_q		2.611 mH
	Stator-rotor mutual inductance M_{afd0}		264 mH
	Excitation current i_{fd}		21A
	Polar distance τ_s		258 mm
Train parameters	Train weight M		155t
	Train grouping n		8
Running speed requirements	Maximum operating speed		600 km/h
	Train acceleration and deceleration	Average acceleration	0–200 km/h: $\geq$ 1.0 m/s^2 0–300 km/h: $\geq$ 0.8 m/s^2 0–600 km/h: $\geq$ 0.7 m/s^2
		Deceleration	600-0 km/h: $\geq$ 1.0 m/s^2 (Electrical brakes); 600-0 km/h: $\geq$ 1.5 m/s^2 (Emergency braking)

To improve the simulation speed, Acceleration becomes 100 times the original speed, and Vehicle weight becomes 1/100 of the original weight, while the thrust and current output are guaranteed, to realize the high-speed magnetic levitation full process traction control simulation. The following is the thrust output of LSLSM under two-level and three-level topologies (Figs. 5, 6 and 7).

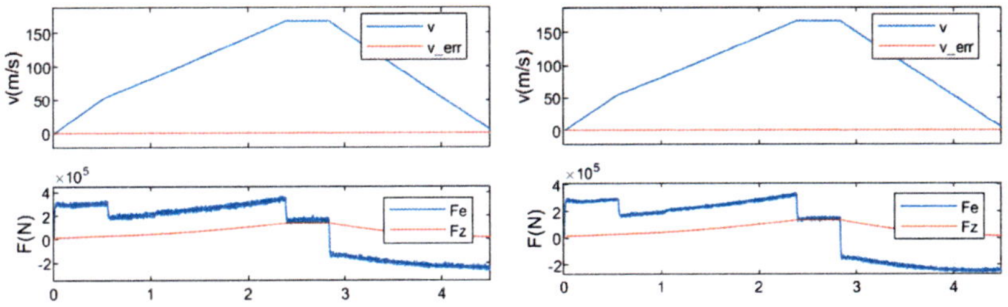

(a)Two-level topology (b)Three-level topology

Fig. 5. Speed and thrust output

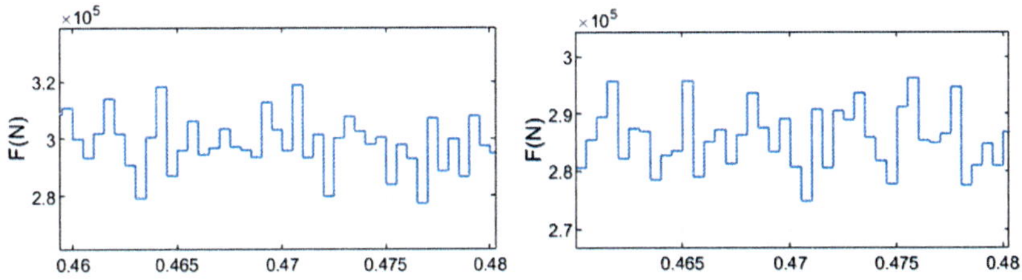

(a)Two-level topology (b)Three-level topology

Fig. 6. Thrust ripple condition

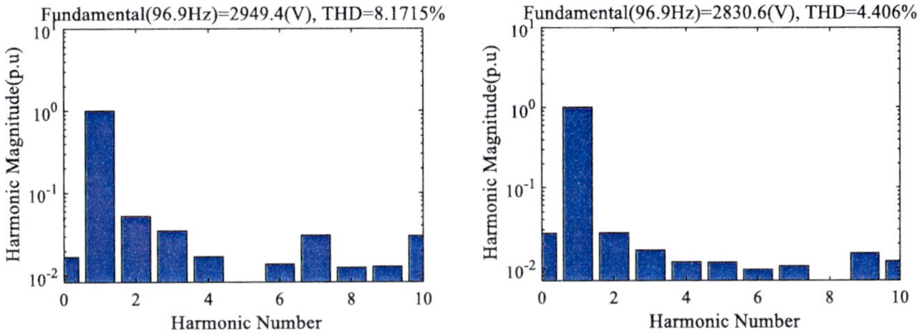

(a)Two-level topology (b)Three-level topology

Fig. 7. Phase current Fourier analysis (Harmonic analysis at 0.5 s and speed of 51 m/s)

From the simulation results, it is obvious that the thrust ripple is reduced in the three-level compared to the two-level, and it is calculated that the thrust fluctuation is 13.3% in the two-level and 6.8% in the three-level under the same operating conditions. The three-level ripple is reduced by 6.5% compared to the two-level ripple. Taking phase an as an example, the harmonic analysis of phase current shows that the THD of the three-level is 4.406% and that of the two-level is 8.1715%, and the harmonics of the three-level are reduced by nearly half compared with the two-level.

5 Conclusion

In this paper, the vector control of LSLSM for a high-speed magnetic levitation train is simulated. The simulation results show that the control method with Constant excitation and zero direct axis current can achieve full-speed operation, fast speed following, and stable thrust output. The output ripple and harmonics of the three-level inverter power waveform are smaller, and it has certain advantages in high-power traction control.

References

1. Xiong, J.Y., Deng, Z.G.: High-speed maglev rail transit research progress. J. Transp. Eng. **21**(1), 177–198 (2021) (in Chinese)
2. Lee, H.W., Kim, K.C., Lee, J.: Review of maglev train technologies. IEEE Trans. Magn. **42**(7), 1917–1925 (2006)
3. Sun, P.K.: Research on Traction Control Strategy of High-speed maglev Train in Double-end Power Supply Mode. University of Chinese Academy of Sciences (2021) (in Chinese)
4. Liu, J.X.: Research on Key Technologies of Multi-zone Traction Control for High-Speed Maglev Traffic. University of Chinese Academy of Sciences (2015) (in Chinese)
5. Liu, J.Q., Jiang, J.G.: Research on the control of permanent magnet synchronous motor based on NPC three-level. Large Motor Technology **290**(5), 25–31+37 (2023) (in Chinese)
6. Kouro, S., Rodriguez, J., Wu, B., et al.: Powering the future of industry: high-power adjustable speed drive topologies. IEEE Ind. Appl. Mag. **18**(4), 26–39 (2012)
7. Rodriguez, J., Franquelo, L.G., Kouro, S., et al.: Multilevel converters: an enabling technology for high-power applications. Proc. IEEE **97**(11), 1786–1817 (2009)
8. Bi, H.D., Liu, Y.H., Liu, J.J., et al.: Research on SPWM carrier modulation strategy for high power three-level inverters. Electrical Drive Automation **43**(04), 1–6 (2021) (in Chinese)
9. Cai, P., Fang, X., Yang, Z., et al.: Research on multiple topology and modulation strategy of large capacity traction inverter for the high-speed maglev train. In: 22nd International Conference on Electrical Machines and Systems (ICEMS), pp. 1–5. IEEE, Harbin, China (2019)
10. Han, J.W.: Analysis of traction characteristics of the high-speed maglev train under different magnetic field oriented control methods. Graduate School of Chinese Academy of Sciences Institute of Electrical Engineering (2006) (in Chinese)

Design and Research of Variable Frequency Direct Feed Power Supply for Urban Rail Transit

Yifei Miao[1](✉) and Xueyan Dong[2]

[1] School of Electrical Engineering, Beijing Engineering Research Center of Electric Rail Transportation, Beijing Jiaotong University, Beijing, China
21121464@bjtu.edu.cn
[2] CRRC Changchun Railway Vehicles Co.,Ltd., Changchun, China
dongxueyan.ck@crrcgc.cc

Abstract. This paper presents a variable frequency direct feed power supply system based on the cascade topology of ZVS three-level Buck+LLC resonant converter. The system takes power directly from the overhead line, and the three-level buck depresses the input voltage with a wide fluctuation range to obtain stable DC power to the LLC converter in the rear stage. This scheme can realize the ZVS of all power switching tubes without adding any auxiliary devices. The key conditions of realizing ZVS of three-level Buck, the working principle of LLC and the control strategy of cascade topology are analyzed and studied. To verify the feasibility and effectiveness of the design scheme, a simulation model is built. It provides theoretical support for the development of a new type of variable frequency direct feed power supply system, so as to optimize the architecture of the auxiliary power supply system of urban rail and improve system efficiency.

Keywords: Variable frequency direct feed · ZVS three-level Buck · LLC converter

1 Introduction

The rapid development of urban rail transit has put forward new development requirements for its auxiliary converters. High frequency, high efficiency and lightweight are the development trends of urban rail auxiliary converters [1]. However, the development of high-power auxiliary power supply system technology has entered a bottleneck period, which is difficult to meet the development needs of green rail transit at the present stage [2].

The capacity of the auxiliary power supply system is determined by the electric equipment on the vehicle. The air conditioner compressor and fan are the main load of the AC380V side. At the same time, with the wide application of variable frequency air conditioning and variable frequency compressor, the input AC380V needs an external frequency converter to be used for the air conditioning compressor, which often causes unnecessary energy loss [3, 4].

L. Jia et al. (Eds.): EITRT 2023, LNEE 1135, pp. 313–321, 2024.
https://doi.org/10.1007/978-981-99-9307-9_34

In order to solve the above problems, this paper improved the power supply link and studied a new type of variable frequency direct feed power supply system. The intermediate auxiliary inverter was cancelled, and the power was directly drawn from the overhead line. The DC1500V was reduced by the power supply to obtain a stable DC600V isolated from the input, which could directly supply power to the inverter at the later stage.

2 Topology Design of Variable Frequency Direct Feed Power Supply

LLC resonant converter, as a series and parallel resonant converter with excellent soft switching characteristics, has been widely used in electric vehicles, new energy power generation and other fields [5, 6]. However, due to the large fluctuation range of input voltage and load of urban rail transit, the design of high-power auxiliary converter based on LLC resonant converter is difficult to design and has few engineering applications [7].

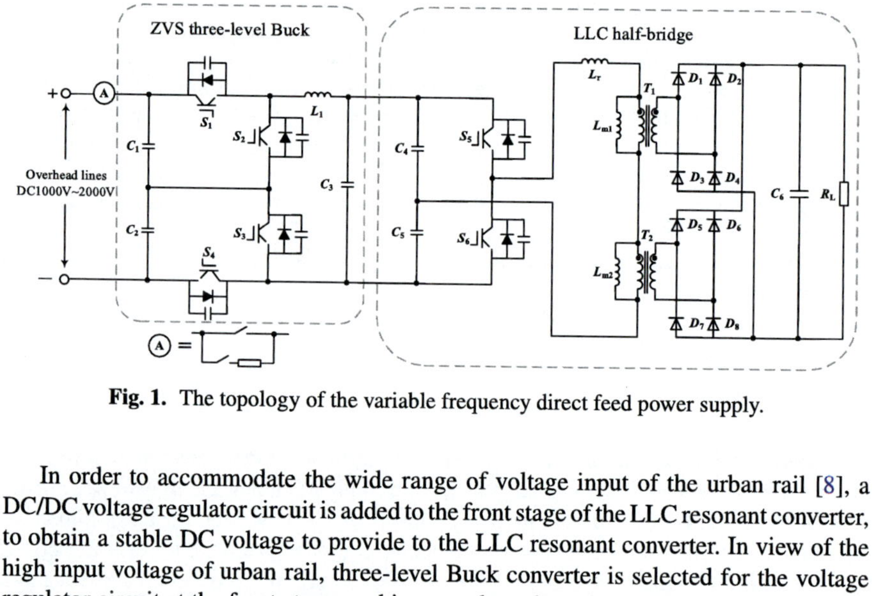

Fig. 1. The topology of the variable frequency direct feed power supply.

In order to accommodate the wide range of voltage input of the urban rail [8], a DC/DC voltage regulator circuit is added to the front stage of the LLC resonant converter, to obtain a stable DC voltage to provide to the LLC resonant converter. In view of the high input voltage of urban rail, three-level Buck converter is selected for the voltage regulator circuit at the front stage, and it can reduce the voltage stress of the switching tube [9–11]. At the same time, in order to realize zero-voltage-switching (ZVS) of three-level Buck, switching tube is used to replace the original diode, so ZVS of switching tube can be realized without adding any auxiliary components. The transformer of the LLC resonant converter of the latter stage adopts the structure of two identical transformers with primary side input in series and secondary side output in parallel. In this way, the input power of the transformer can be evenly divided, the capacity level of the transformer can be reduced, and the magnetic core heating can be dispersed. At the same time, it can reduce the conduction loss of the secondary side rectifier diode and reduce the output current ripple. The topology of the designed variable frequency direct feed power supply is shown in Fig. 1.

3 Working Principle Analysis and Control Scheme Research

3.1 Operation Mode Analysis of Three-Level Buck

When duty cycle D > 0.5, that is, when input voltage is DC1000V ~ DC1800V, the working waveform of ZVS three-level Buck is shown in the Fig. 2.

By analyzing the working mode of ZVS three-level Buck, it can be seen that: in mode 2 and mode 8, the peak value of inductance current for resonance is large enough, so the ZVS of S_2 and S_3 are easy to achieve. However, in mode 5 and mode 11, it is necessary that the reverse inductor current at the beginning of resonance is large enough to discharge the charges in the parasitic capacitors of S_1 and S_4, to achieve the conditions of ZVS. Next, the influence of the amplitude of the reverse inductance current I_R on ZVS during dead time is analyzed.

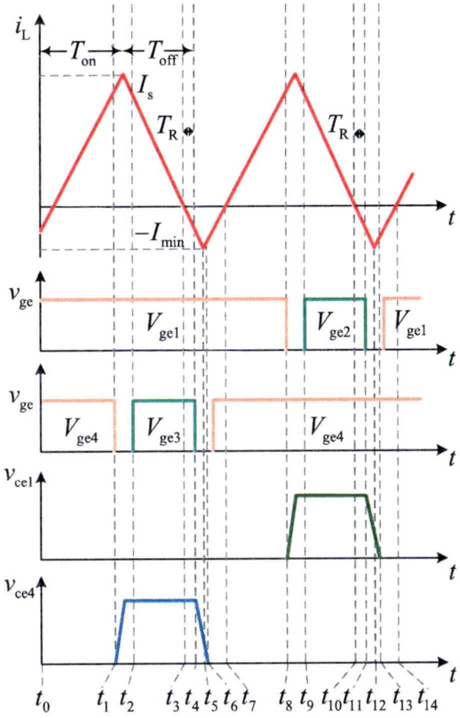

Fig. 2. Operating waveform of ZVS three-level Buck.

As can be seen from the waveform of i_L in the Fig. 2, the change rate of i_L when it decreases meets:

$$\begin{aligned}\frac{I_S}{T_{on}} &= \frac{U_{in} - U_o}{L} \\ \frac{I_S + I_R}{T_{off}} &= \frac{U_o - 0.5U_{in}}{L}\end{aligned} \tag{1}$$

The time for the inductor's current I_L to rise from 0 to the peak value T_{on} is:

$$T_{on} = DT_L \tag{2}$$

where, T_L is the pulsation period of I_L, and D is the duty cycle of S_1 and S_4. The resonant initial current in mode 2 is relatively large, so the resonant time is very short and can be ignored. Therefore, the inductance current pulsation period meets the following relation:

$$\begin{aligned}T_L &= T_{on} + T_{off} + T_{dt} \\ T_{off} &= T_L - T_{on} - T_{dt} \\ &= (1 - D)T_L - T_{dt}\end{aligned} \tag{3}$$

where, T_{dt} is the dead time. Considering that the inductance current pulsation frequency is twice the switching frequency, the expression of critical reverse current I_R can be obtained as follows:

$$I_R = \frac{1}{2Lf_{sw}}\left(U_o - \frac{(1+D)}{2}U_{in}\right) - \frac{1}{L}\left(U_o - \frac{1}{2}U_{in}\right)T_{dt} \tag{4}$$

As can be seen from the above formula, when U_{in}, U_o and D are determined, the amplitude of the critical reverse current can be changed by adjusting f_{sw} and T_{dt}, and then ZVS of S_1 and S_4 can be realized.

3.2 Characteristics of f_{sw} and T_{dt}

When $D > 0.5$, the converter operates in three-level mode. From the Fig. 3:

$$\frac{1}{2}I_{min}T_2 = q_c \tag{5}$$

$$\frac{I_{min}}{T_2} = \frac{U_{in} - U_o}{L} \tag{6}$$

Therefore, the minimum inductance current can be expressed as:

$$I_{min} = \sqrt{\frac{2q_c(U_{in} - U_o)}{L}} \tag{7}$$

And

$$\frac{I_R}{T_R} = \frac{I_{min}}{T_R + T_1} = \frac{U_o - 0.5U_{in}}{L} \tag{8}$$

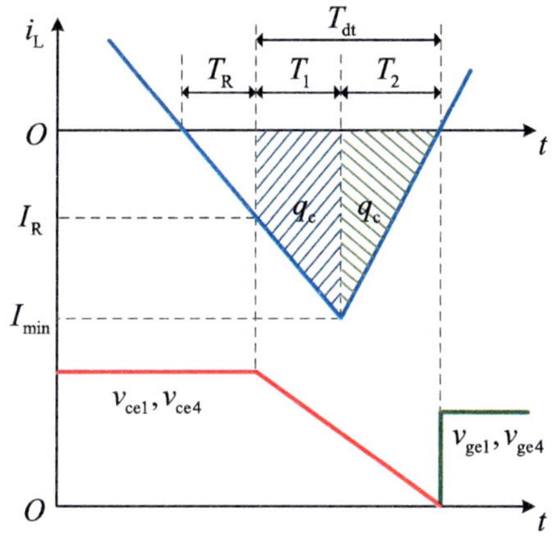

Fig. 3. Waveform diagram of critical ZVS when inductor's current is reversed.

The charge stored in the parasitic capacitor is:

$$\frac{1}{2}I_{\min}(T_R + T_1) - \frac{1}{2}I_R T_R = q_c \tag{9}$$

The expression for I_R is:

$$I_R = \sqrt{\frac{2q_c(1.5U_{in} - 2U_o)}{L}} \tag{10}$$

According to Fig. 1:

$$\begin{aligned} \frac{I_{\min} - I_R}{T_1} &= \frac{U_o - 0.5U_{in}}{L} \\ \frac{I_{\min}}{T_2} &= \frac{U_{in} - U_o}{L} \end{aligned} \tag{11}$$

So

$$T_{dt} = T_1 + T_2 = \frac{\sqrt{2q_c(U_{in} - U_o)L} - \sqrt{2q_c(1.5U_{in} - 2U_o)L}}{U_o - 0.5V_{in}} + \frac{\sqrt{2q_c(U_{in} - U_o)L}}{U_{in} - U_o} \tag{12}$$

In the three-level working mode, ZVS can be realized only when $4/3U_o < U_{in} < 2U_o$ is met.

Also known:

$$\begin{aligned} I_{ave} &= (I_s - I_{\min})/2 \\ \Delta I_L = I_s + I_{\min} &= \frac{U_{in} - U_o}{L} D \frac{T_s}{2} \end{aligned} \tag{13}$$

Therefore, f_{sw} must meet:

$$f_{sw} = \frac{d}{4L} \cdot \frac{U_{in} - U_o}{I_{ave} + I_{\min}} \tag{14}$$

Similarly, I_R under the two-level working mode can be obtained as follows:

$$I_R = \sqrt{\frac{2q_c(0.5U_{in} - 2U_o)}{L}} \tag{15}$$

If the above formula is valid, it must meet $0.5U_{in} > 2U_o$. Therefore, in the two-level working mode, it must meet $U_{in} > 4U_o$ to achieve ZVS. So, the variable frequency direct feed power supply designed in this paper cannot realize ZVS in two-level mode.

A three-level Buck out-of-phase control strategy based on Characteristics of f_{sw} and T_{dt} is proposed for the advanced ZVS three-level Buck converter. The inductor's current is reversed by out-of-phase control to discharge the parasitic capacitor of the power switch tube, and the ZVS of the switch tube can be realized without adding any auxiliary components.

The upper tube and lower tube of the LLC half-bridge resonant converter are complementary on, and the D is 0.5. To prevent misdirection, a certain dead zone is set. The resonant frequency of resonant inductor L_r and resonant capacitor C_r is defined as:

$$f_r = \frac{1}{2\pi\sqrt{L_r C_r}} \tag{16}$$

After the resonant inductor L_r and the excitation inductor L_m are connected in series, the frequency of resonating with the resonant capacitor C_r is:

$$f_m = \frac{1}{2\pi\sqrt{(L_r + L_m)C_r}} \tag{17}$$

When the switching frequency f_s of LLC meets $f_m < f_s < f_r$, the primary side switching tube can achieve ZVS, and the secondary side rectifier diode can achieve ZCS. And as f_s approaches f_r, the more energy transferred from the primary side to the secondary side in unit period, the higher the efficiency. Therefore, $f_s = f_r$ is set in this paper, when the converter has the highest efficiency, works at the optimal operating point, and the soft switching performance is superior.

Based on the above analysis, the LLC fixed-width and fixed-frequency control strategy is proposed. Through the front stage ZVS three-level Buck voltage regulator output, the rear stage LLC works in fixed-width and fixed-frequency mode. At the same time, through reasonable design of resonant parameters, it can work at the resonant frequency point, to achieve the highest efficiency and achieve the optimal working point.

3.3 Control Scheme Research

Based on the above analysis, a cascade topology comprehensive control scheme is proposed as shown in the Fig. 4.

4 Simulation Analysis

Simulation model was built to verify the correctness of the above research content. Set the simulation model parameters as follows: the switching frequency of the three-level Buck is 14 kHz, the filter inductance L_1 is 24 uH, the switching frequency of the LLC is 20 kHz, the resonant inductance L_r is 26.3 uH, the resonant capacitor C_4 and C_5 is 1200 nF, and the excitation inductance L_{m1} and L_{m2} is 66 uH.

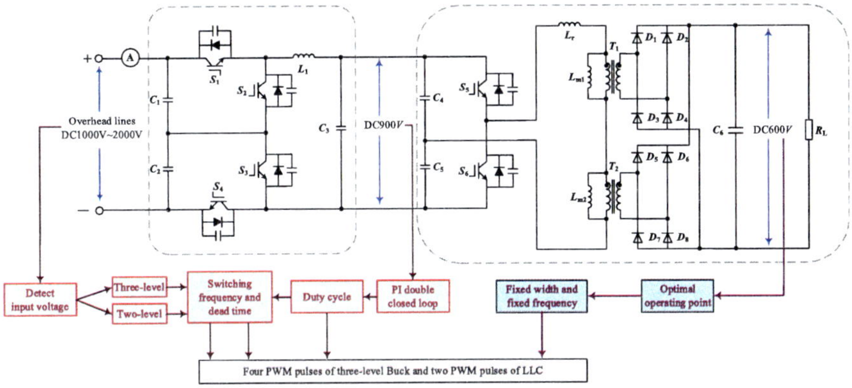

Fig. 4. Cascade topology comprehensive control scheme.

The output voltage and intermediate DC bus voltage waveform are shown in the Fig. 5. Both the output voltage and the intermediate DC bus voltage can be stabilized at the target value, and can be stabilized quickly under closed-loop control even when the input voltage fluctuates.

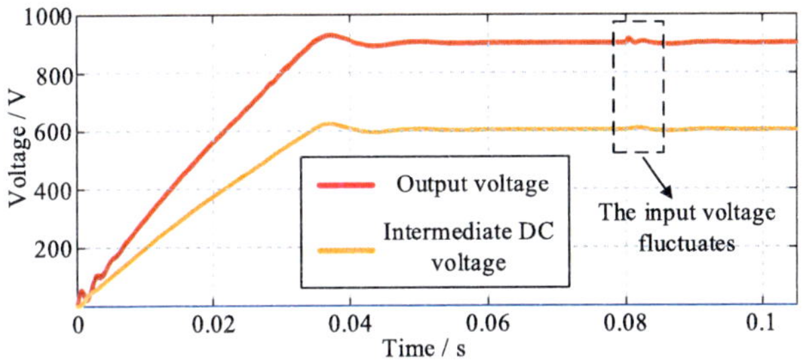

Fig. 5. Waveform of output voltage and intermediate DC voltage.

The ZVS waveform of three-level Buck is shown in the Fig. 6. When the input voltage is 1200V and 1650V respectively, the performance of ZVS is superior.

The Fig. 7 shows the superior soft switch performance of LLC.

When the input voltage is rated at 1500V, the relationship between efficiency and output power is shown in Table 1. It can be seen that the transmission efficiency is positively correlated with the change of output power, and the transmission efficiency increases with the increase of output power. When the output power is 25 kW, the output efficiency is 93.69%, and the energy utilization rate is high, which proves the correctness of the hardware parameters designed and the effectiveness of the control strategy researched in this paper.

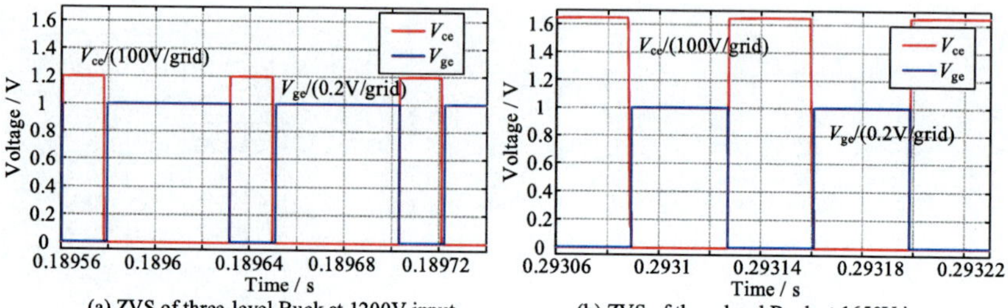

(a) ZVS of three-level Buck at 1200V input (b) ZVS of three-level Buck at 1650V input

Fig. 6. The ZVS waveform of three-level Buck.

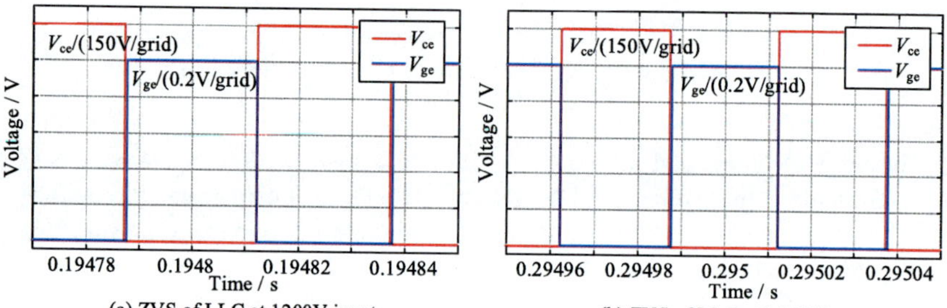

(a) ZVS of LLC at 1200V input (b) ZVS of LLC at 1650V input

Fig. 7. The ZVS waveform of LLC.

Table 1. Table of efficiency changes with output power at rated DC input 1500V.

Output power (kW)	Efficiency (%)
12	88.44
14	90.11
19	91.87
25	93.69
35	95.58

5 Conclusion

In order to optimize the architecture of urban rail auxiliary power supply system and improve energy utilization efficiency and system power density, this paper presents a design scheme of urban rail variable frequency direct feed power supply system based on ZVS three-level Buck+LLC half-bridge resonant converter. The front three-level Buck changes the wide range of input voltage into a stable intermediate DC voltage, allowing the LLC resonant converter to operate at a fixed switching frequency. Through reasonable design parameters and research of control strategy, the ZVS of all power devices can be realized, and the system efficiency is greatly improved. Finally, a simulation model is built to verify the feasibility of the scheme.

References

1. Dong, K., Ma, Y., Yang, E.: Intermediate frequency auxiliary converter based on high power LLC resonant converter. Power Electron. **53**(3), 47–50 (2019) (in Chinese)
2. Jiang, X., Zeng, P., Yu, B.: Optimal design of resonant soft switch converter of rail transit auxiliary powersupply system. Power Electron. **55**(3), 1–4 (2021) (in Chinese)
3. Jeong, Y., Kim, J.K., Lee, J.B.: An Asymmetric half-bridge resonant converter having a reduced conduction loss for DC/DC power applications with a wide range of low input voltage. IEEE Trans. Power Electron. **32**(10), 7795–7804 (2017)
4. Jia, P., Huang, Z., Hao, Y: A Two-stage isolated converter without intermediate capacitor for wide voltage range applications. In: 2019 IEEE Energy Conversion Congress and Exposition, pp. 1315–1320. IEEE, Baltimore, Baltimore, Maryland, USA (2019)
5. Ivensky, G., Bronshtein, S., Abramovitz, A.: Approximate analysis of resonant LLC DC-DC converter. IEEE Trans. Power Electron. **26**(11), 3274–3284 (2011)
6. Hu, H., Fang, X., Chen, F.: A modified high-efficiency LLC converter with two transformers for wide input-voltage range applications. IEEE Trans. Power Electron. **28**(4), 1946–1960 (2013)
7. Yang, Z., Bi, J., Chang, Q.: Design and research of high power and high frequency auxiliary converter for rail vehicles. Power Electron. **56**(3), 38–42 (2022) (in Chinese)
8. Petar, J.: Grbovi´, Philippe Le Moigne: A bidirectional three-level DC-DC converter for the ultracapacitor applications. IEEE Trans. Ind. Electron. **57**(10), 3415–3430 (2010)
9. Liu, B., Ren, R., Wang, F.: A variable frequency ZVS control of a three-level buck without zero crossing detection for wide-range output voltage battery chargers. In: IEEE Applied Power Electronics Conference and Exposition, pp. 2311–2316. Los Angeles CA, USA (2019)
10. Cheng, X. -F.. Liu, C., Wang, D.: A ZVS three-level synchronous buck converter with simple LC circuit. IEEE Trans. Circuits Syst. II: Express Briefs (Early Access) (2023)
11. Baek, S. -W., Kim, H. -W., Kang, J. -W.: Auxiliary power unit for railway vehicles using three-level LLC converter with pulse width modulation control. In: 2019 International Symposium on Electrical and Electronics Engineering, pp. 232–236. IEEE, Ho Chi Minh City, Vietnam (2019)

Research on Power Calculation Method for Auxiliary Converters of Multiple Unit

Bo Zhang(✉), Qingwen Sun, Junbo Zhao, Xuan Su, Xinyu He, Haining Hu, Qiang Li, and Zheng Xie

CRRC Qingdao Sifang Rolling Stock Research Institute Co., Qingdao 266031, China
18806426590@126.com

Abstract. A new power calculation method based on instantaneous power theory is proposed for the special working conditions of the main circuit topology and sensor acquisition position of the auxiliary converter of the high-speed train unit. It can calculate the instantaneous active power and instantaneous reactive power output of the auxiliary converter in real-time. The power calculation method used is simple in calculation, does not require complex calculations, and does not require lag processing on sampling points, resulting in high real-time performance. The positive and negative sequence component extraction technology based on second-order generalized integrator is adopted, and the voltage and current are constructed with orthogonal components while achieving adaptive filtering, which can effectively cope with unbalanced loads. The simulation and experimental results show that this method can accurately calculate instantaneous power and achieve power sharing of parallel auxiliary converters.

Keywords: Auxiliary converter · Parallel control · Second-order generalized integrator · Power calculation

1 Introduction

The auxiliary converter of Multiple unit is an important part of the auxiliary power supply system. It is mainly responsible for providing AC380V power for Multiple unit and supplying power for AC loads such as charger, air conditioner and traction cooling system [1]. In order to improve the reliability and redundancy of the AC power supply system, multiple auxiliary converters are installed in parallel for power supply throughout the train, using a parallel control algorithm without interconnection lines. Each auxiliary converter adjusts the output voltage based on the load power and automatically shares the load current [2]. At present, the auxiliary converter of Multiple unit mostly adopts power frequency isolation topology, and galvanic isolation of input and output is realized through power frequency transformer. In the parallel algorithm without interconnection lines, it is necessary to know the instantaneous active and reactive power output by each auxiliary converter, and adjust the output voltage through droop control to achieve balanced output current of each auxiliary converter [3].

L. Jia et al. (Eds.): EITRT 2023, LNEE 1135, pp. 322–330, 2024.
https://doi.org/10.1007/978-981-99-9307-9_35

According to the "instantaneous reactive power theory" put forward by Japanese scholar Akagi et al. [4] in 1983, the instantaneous reactive power and instantaneous reactive current in the three-phase circuit can be calculated [5, 6], but the voltage and current collected by this calculation method must be at the same position. The voltage and current collected by the auxiliary converter of Multiple unit are located on both sides of the transformer due to isolation of the transformer, and cannot be directly used for calculation. In response to this issue, this article studies the power calculation method in this special case, based on the "instantaneous power theory", and improves the power calculation method. The feasibility and effectiveness of the method have been proven through simulation analysis and experimental verification.

2 Structure of Auxiliary Converter of Multiple Unit

The main circuit topology of the auxiliary converter is mainly composed of a pre charging circuit, an inverter bridge, an isolation transformer, and an AC filtering capacitor [7], as shown in Fig. 1.

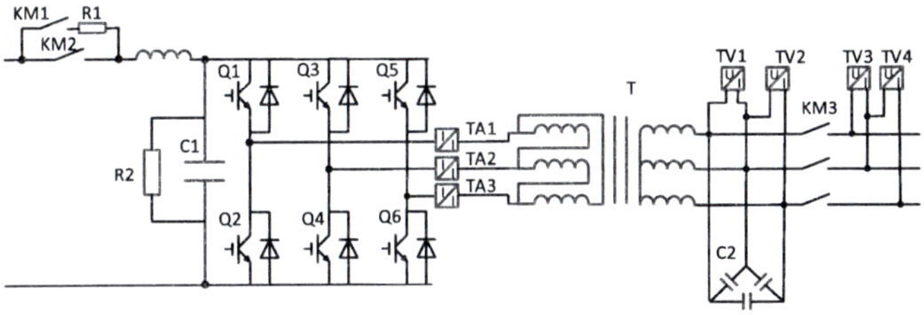

Fig. 1. Main circuit topology of auxiliary converter of Multiple unit

3 Auxiliary Converter Power Calculation

3.1 Power Conversion Calculation

The auxiliary converter uses the power frequency transformer to Galvanic isolation the input and output. The control system collects the output current of the inverter bridge through the current sensors TA1, TA2 and TA3, and the output voltage of the inverter through the voltage sensors TV1 and TV2. The collected current and voltage are located on the primary and secondary sides of the transformer, and cannot be directly calculated using instantaneous reactive power theory.

Set the module output current to I_1, the module side transformer primary voltage to U_1, the transformer secondary voltage to U_2, and the module output current I_1 lag behind the transformer secondary voltage U_2 angle φ. The group of transformers is Dyn, and the transformation ratio is N. According to transformer theory, the secondary voltage of the transformer lags behind the primary voltage by n * 30 degrees, and the ratio of the

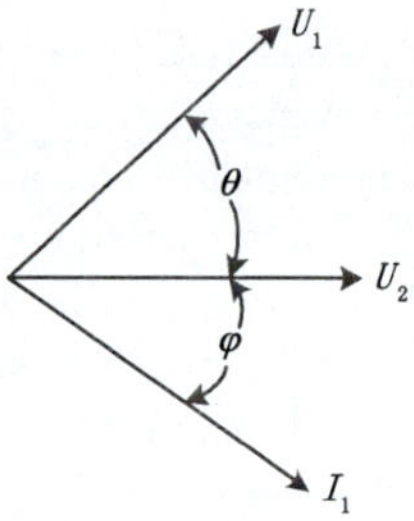

Fig. 2. Voltage and Current Phase Relationship Diagram

primary voltage to the secondary voltage is N, namely $N = U_1/U_2$. If $\theta = n * 30$, the phase relationship between U_1, U_2, and I_1 is shown in Fig. 2.

According to the phase relationship in Fig. 2, the output active and reactive power of the power module are:

$$\begin{cases} P_1 = \sqrt{3}U_1I_1\cos(\phi+\theta) \\ Q_1 = \sqrt{3}U_1I_1\sin(\phi+\theta) \end{cases} \tag{1}$$

According to the transformer ratio relationship and the sum difference product formula, by expanding formula (1), we can obtain:

$$\begin{cases} P_1 = \sqrt{3}NU_2I_1(\cos\phi\cos\theta - \sin\varphi\sin\theta) \\ Q_1 = \sqrt{3}NU_2I_1(\sin\phi\cos\theta + \cos\varphi\sin\theta) \end{cases} \tag{2}$$

Assuming that U_2 and I_1 are the output voltage and current of the auxiliary converter, according to Fig. 2, the simulated output active and reactive power can be calculated as follows:

$$\begin{cases} P' = \sqrt{3}U_2I_1\cos\phi \\ Q' = \sqrt{3}U_2I_1\sin\phi \end{cases} \tag{3}$$

Bringing formula (3) into formula (2) yields:

$$\begin{cases} P_1 = NP'\cos\theta - NQ'\sin\theta \\ Q_1 = NP'\sin\theta + NQ'\cos\theta \end{cases} \tag{4}$$

According to formula (4), as long as the power P' and Q' corresponding to the secondary voltage U_2 and module current I_1 are calculated, the output power P' and Q' on the power module side can be obtained.

The module current collected by the sensor and the output voltage of the inverter are transformed from the abc coordinate system to the $\alpha\beta$ coordinate system using an equal amplitude CLARKE transformation to obtain I_α, I_β, U_α, and U_β. Using the instantaneous power calculation method, it can be concluded that:

$$\begin{cases} P' = 1.5U_\alpha I_\alpha + 1.5U_\beta I_\beta \\ Q' = 1.5U_\alpha I_\beta - 1.5U_\beta I_\alpha \end{cases} \tag{5}$$

By incorporating formula (5) into formula (4), the output power P_1 and Q_1 on the side of the auxiliary converter module can be calculated.

$$\begin{cases} P_1 = N\left(1.5U_\alpha I_\alpha + 1.5U_\beta I_\beta\right)\cos\theta - N\left(1.5U_\alpha I_\beta - 1.5U_\beta I_\alpha\right)\sin\theta \\ P_1 = N\left(1.5U_\alpha I_\alpha + 1.5U_\beta I_\beta\right)\sin\theta + N\left(1.5U_\alpha I_\beta - 1.5U_\beta I_\alpha\right)\cos\theta \end{cases} \quad (6)$$

3.2 Capacitor Reactive Power Compensation

Due to the presence of AC filtering capacitors on the output side of the auxiliary converter, the calculated module output power includes AC filtering capacitor power, which needs to be compensated in the reactive power output of the module.

The capacitance value of the AC current is fixed parameter C, and the output angular velocity of the inverter ω, then the capacitance reactance can be calculated as $X_C = 1/(\omega C)$.

From this, it can be seen that the capacitance power is:

$$Q_c = \frac{U_2^2}{X_c} = 1.5\left(U_\alpha^2 + U_\beta^2\right)\omega C \quad (7)$$

According to the above formulas (1) to (7), it can be seen that the final output active power P and reactive power Q of the auxiliary converter are:

$$\begin{cases} P = P_1 \\ Q = Q_1 + Q_c \end{cases} \quad (8)$$

4 Unbalanced Load Response

The auxiliary converter of Multiple unit supplies power for all AC loads of the train and provides three-phase AC power supply. Due to the existence of single-phase loads, the three-phase power supply cannot be guaranteed to be completely consistent, resulting in three-phase imbalance. Currently, the instantaneous reactive power theory mainly applies to three-phase balanced systems [8], so special treatment methods need to be adopted to reduce the impact of unbalanced loads on power calculation. This article adopts the orthogonal component extraction method based on second-order generalized integrator (SOGI) [9, 10], which reconstructs voltage and current separately in the $\alpha\beta$ coordinate system. By using the orthogonal components of voltage and current, positive sequence voltage and current can be extracted for power calculation.

4.1 Second-Order Generalized Integrator

Extracting the positive and negative sequence components of power grid voltage based on a second-order generalized integrator, with the integrator structure shown in Fig. 3 [11].

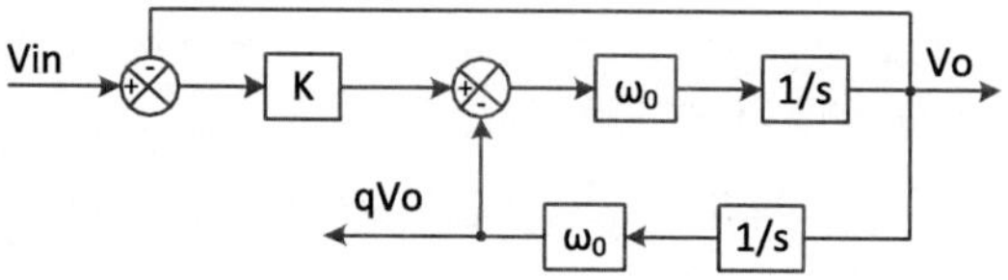

Fig. 3. Principle block diagram of second-order generalized integrator

Vin is the integrator input, V_o and qV_o are two outputs, where V_o and V_{in} have the same phase and amplitude, qV_o and V_{in} have the same amplitude, with a phase lag of 90°. The transfer function between the two outputs and the input is:

$$\begin{cases} D(s) = \frac{v_o(s)}{v_{in}(s)} = \frac{k\omega_0 s}{s^2+k\omega_0 s+\omega_0^2} \\ Q(s) = \frac{qv_o(s)}{v_{in}(s)} = \frac{k\omega_0^2}{s^2+k\omega_0 s+\omega_0^2} \end{cases} \tag{9}$$

4.2 Positive Sequence Component Extraction

The positive and negative sequence calculation method (PNSC) utilizes instantaneous symmetrical components to eliminate the mutual influence between voltage positive and negative sequence components, achieving the separation of positive and negative sequences. To achieve the separation of positive and negative sequences $U_{\alpha\beta}^{P}$ and $U_{\alpha\beta}^{N}$ in the stationary coordinate system $U_{\alpha\beta}$, only the orthogonal quantities of $U_{\alpha\beta}$ and its 90 degree phase lag need to be obtained.

The collected three-phase voltage and current are transformed from ABC coordinate system to $\alpha\beta$ coordinate system through CLARKE transformation, and are respectively converted into orthogonal components through second-order generalized integrator. The positive sequence component can be extracted through response formula. The influence of unbalanced load can be eliminated by calculating active power and reactive power according to the positive sequence component (Fig. 4).

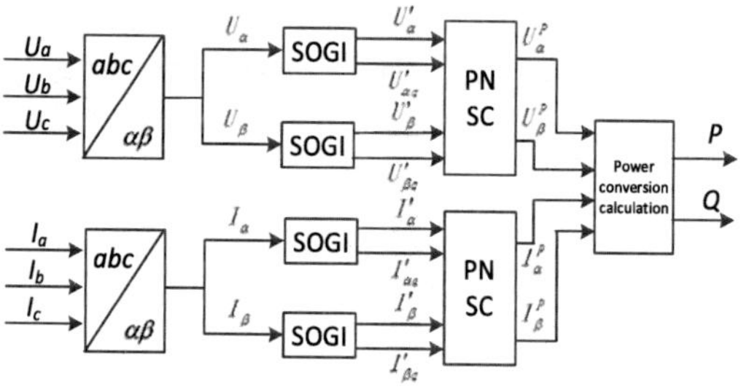

Fig. 4. Unbalanced load power calculation block diagram

5 Simulation Analysis and Experimental Verification

5.1 Simulation Analysis

A simulation model for parallel operation of two auxiliary inverters was built using Matlab simulation software. The feasibility of the power calculation method was verified by observing the power distribution and current sharing of the two inverters under three operating conditions: no load, half load, and full load.

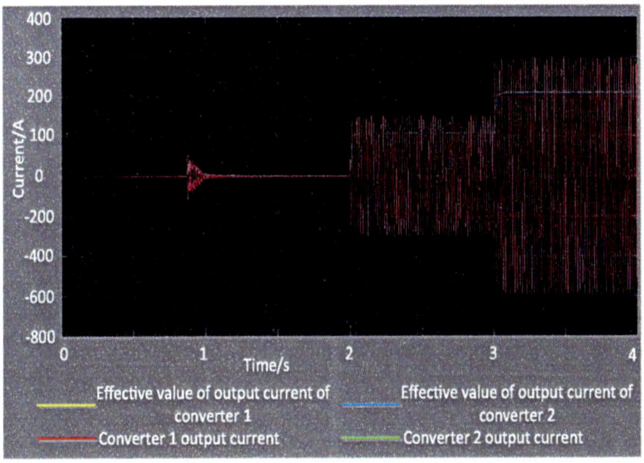

Fig. 5. Output current of parallel auxiliary converter

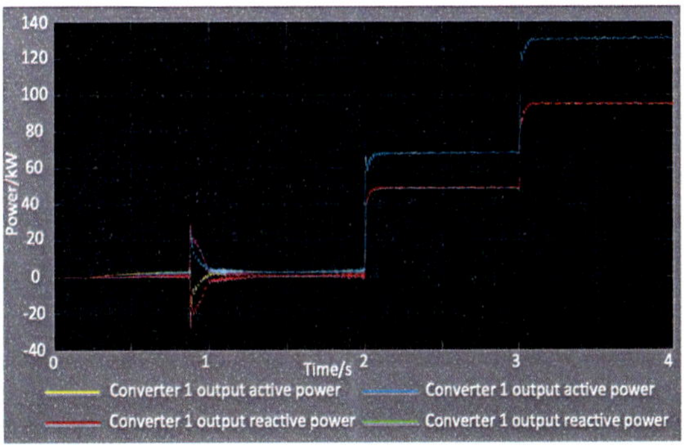

Fig. 6. Output power of parallel auxiliary converter

From the simulation waveforms in Figs. 5 and 6, it can be seen that the two auxiliary converters operate in parallel at around 0.88S, and there are impulse currents and power fluctuations at the moment of parallel connection. After the parallel connection is completed, the current and power tend to stabilize. At 2s and 3s, half load and full load are put into operation respectively, and the current output is consistent. The calculation of active and reactive power is accurate and can be evenly divided, verifying the effectiveness of the above power calculation method.

5.2 Experimentation

Two auxiliary converters of Multiple unit are used to work in parallel, and the power calculation method in this paper is verified by the voltage and current data and waveform under different load conditions during parallel operation.

Parallel light load test.

Figure 7 shows the waveform of the parallel light load test of two auxiliary converters (CH1: output voltage of converter 1, CH2: output voltage of converter 2, CH3: output current of converter 1, CH4: output current of converter 2, the same below). According to the test waveform, it can be seen that the two converters can work stably when connected in parallel with light load, with basically the same output voltage, output current phase and amplitude, and effective current values of 22.1A and 19A.

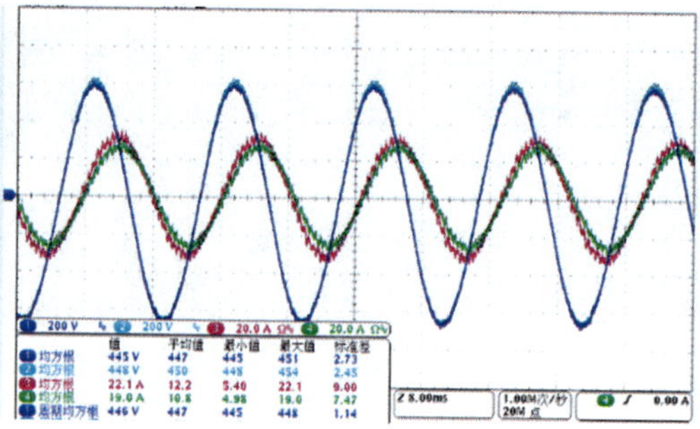

Fig. 7. Parallel light load experimental waveform of auxiliary converter

Parallel half load test.

Figure 8 shows the parallel waveform of two auxiliary converters under half load condition. According to the waveform, the output voltage and current of the two converters are basically the same, with effective values of 102A and 108A respectively. The amplitude and phase of the current are basically the same, achieving equal load sharing.

Parallel full load test.

Figure 9 shows the waveform of the full load test of two auxiliary converters in parallel. The effective output currents of the converters are 219A and 217A, respectively, and the current magnitude and phase are basically the same.

Trial Summary.

From the data of two parallel auxiliary converters under different test conditions such as light load, half load, and full load, it can be seen that using the power calculation method in this paper can meet the requirements for active and reactive power in droop control.

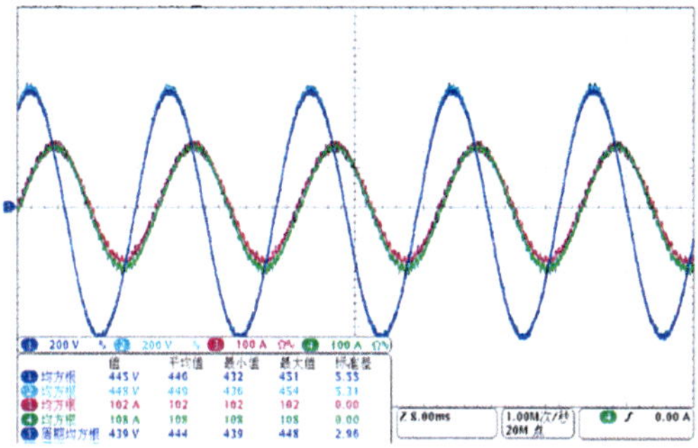

Fig. 8. Auxiliary converter parallel half load experimental waveform

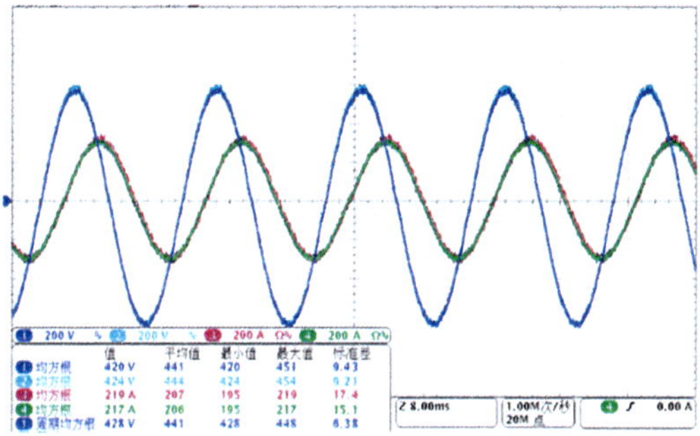

Fig. 9. Parallel full load experimental waveform of auxiliary converter

The calculation method for the imbalance of output current sharing in parallel systems is as follows:

$$I_{avg} = \frac{1}{K}\sum_{x=1}^{K} I_x \tag{10}$$

$$I_{cusx} = \frac{I_x - I_{avg}}{I_N} \times 100\% \tag{11}$$

In the formula, I_x is the output current of the x-th inverter to be evaluated; K is the total number of inverters on the parallel bus; I_{avg} is the average output current of each inverter on the parallel bus; I_N is the rated output current of the inverter. I_{cusx} is the degree of current distribution imbalance.

According to the above formula, the current distribution imbalance under three working conditions of light load, half load, and full load is calculated, which are 0.74%, 1.4%, and 0.48%, respectively, all far better than the technical requirement of 5%.

6 Conclusion

Based on the characteristics of the main circuit topology of the auxiliary converter of Multiple unit, this paper improves the calculation method of instantaneous power, calculates the output active power and reactive power of the load side according to the module output current acquisition signal and the transformer secondary voltage acquisition signal. The calculation method used is concise, does not require lag processing on sampling points, and has high real-time performance. To cope with unbalanced loads, positive and negative sequence component extraction technology has been introduced. The voltage and current reconstruction method used achieves adaptive filtering while constructing orthogonal components, improving the system's anti-interference ability and making the auxiliary power supply system more reliable.

References

1. Bo, L., Bo, Z., Congcong, W., Junbo, Z., Qingwen, S., Xinyu, H., Shen, W., Hongguo, L., Meiyun, L.: Research on network control strategy of parallel auxiliary converters of EMUs. Rolling Stock **61**(2), 30–36 (2023) (in Chinese)
2. Guerrero, J.M., Vicuna, L.G.D., Matas, J., et al.: A wireless controller to enhance dynamic performance of parallel inverters in distributed generation systems. IEEE Trans. Power Electron. **19**(5), 1205–1213 (2004)
3. Xitang, T., Aihua, L., Qinyue, Z., Wei, W., Yaolin, L.: Parallel control of traction auxiliary inverter based on adaptive virtual impedance. Electric Mach. Contr. **26**(10), 98–110 (2022) (in Chinese)
4. Akagi, H., Kanazawa, Y., Nabae, A.: Instantaneous reactive power compensators comprising switching devices without energy storage components. IEEE Trans. Ind. Appl. **20**(3), 625–630 (1983)
5. Menghua, H., Juanjuan, W.: Study and application of instantaneous power calculation method. Proc. CSU-EPSA **31**(4), 17–24 (2019) (in Chinese)
6. Hongliang, W., Xiumei, Y., Xuejun, P., Yong, K.: New power calculation method and realization of parallel inverters. Electric Mach. Contr. **14**(5), 37–43 (2010) (in Chinese)
7. Zhenhuan, Y., Kan, D., Jianan, C.: Application of intermediate frequency auxiliary converter based on silicon carbide MOSFET in high-speed train. Railway Locomotive Car **43**(1), 128–132 (2023) (in Chinese)
8. Li, W., Huijin, L., Chen, W.: Summary of the instantaneous reactive power theory. High Voltage Eng. **32**(2), 98–103 (2006) (in Chinese)
9. Ranjan, A., Kewat, S., Singh, B.: DSOGI-PLL with in-loop filter based solar grid interfaced system for alleviating power quality problems. IEEE Trans. Ind. Appl. **57**(1), 730–740 (2021)
10. Hackl, C.M., Landerer, M.: Modified second-order generalized integrators with modified frequency locked loop for fast har-monics estimation of distorted single-phase signals. IEEE Trans. Power Electron. **35**(3), 3298–3309 (2020)
11. Guihua, L., Xiaojiao, C., Wei, W.: A frequency locked loop grid synchronization method of single-phase grid-connected PV inverter under weak grid. Proc. CSEE **35**(19), 5022–5029 (2015) (in Chinese)

Demagnetization Characteristics Diagnosis of Permanent Magnet Synchronous Motors

Zichen Gao[1], Haiting Wang[1], Qiya Wu[1], Yaoheng Li[1,2], and Lijun Diao[1](✉)

[1] Beijing Jiaotong University, Beijing 100044, China
ljdiao@bjtu.edu.cn

[2] Science and Technology on Vehicle Transmission Laboratory China North Vehicle Research Institute, Beijing, China

Abstract. Permanent magnet synchronous motors (PMSM) are widely used in various fields due to their small size, high efficiency and power density, and simple structure. However, during the application process, it may cause demagnetization of the motor's permanent magnet (PM), resulting in poor or even damaged output performance of the drive system. This article conducts research and comparison on several different demagnetization characteristics diagnosis methods. This includes algebraic identification methods that can identify motor related parameters, voltage model flux estimation methods that use the integration of stator voltage to achieve flux estimation, and sliding mode observers based on motor current signals and rotor speed. The validation results of these three methods for motor flux estimation are presented. Finally, based on the design process and the results obtained, the advantages and disadvantages of three different magnetic flux monitoring methods were summarized.

Keywords: Permanent magnet synchronous motor · Demagnetization mechanism · Demagnetization diagnosis

1 Introduction

The focus of demagnetization state diagnosis for PMSM is to select simple and reliable feature quantities, and different feature quantities correspond to different diagnostic methods. This study based on data-driven analysis, and the analyzed feature signals include two aspects: current and voltage.

Diagnosis based on voltage signals generally involves measuring the back electromotive force of the motor and extracting specific components from the back electromotive force to determine the demagnetization of the PM. Julio Cesar Urresty proposed a method for estimating the back electromotive force by measuring the zero-sequence voltage component (ZSVC) online. By measuring and analyzing the ZSVC at the accessible neutral point in the stator winding, the demagnetization status of the motor can be determined [1]. Yang Feng proposed that a detection device can be added to the stator teeth of the motor, such as embedding a detection coil to detect the induced electromotive

L. Jia et al. (Eds.): EITRT 2023, LNEE 1135, pp. 331–338, 2024.
https://doi.org/10.1007/978-981-99-9307-9_36

force of the stator tooth magnetic flux. The demagnetization situation can be judged by the harmonic component of the decomposed induced electromotive force [2].

Diagnosis based on current signals refers to extracting and analyzing the stator current signal output by the motor, analyzing and processing the current signal to obtain characteristic quantities that can characterize the demagnetization of the PM. Pietrzak Przemyslaw applied the time-frequency domain analysis of short-time Fourier transform to the process of extracting PM fault features from stator phase current signals, achieving reliable state detection and fault diagnosis [3]. Demagnetization faults can affect the current, and Zhang Yecheng et al. established a fault diagnosis module based on the phenomenon of different current fluctuations in the same phase current and out of phase current of the rotor within a certain period [4]. Reference [5] uses the Hilbert Huang transform to extract the instantaneous frequency component from the motor current signal, achieving demagnetization fault diagnosis of PMSM under non-stationary conditions.

1.1 Analysis of Demagnetization Mechanism

The stability of PM materials is greatly affected by factors such as manufacturing defects, environmental corrosion, high temperature, and armature magnetic field, which can easily lead to demagnetization faults of PM [6, 7]. It can be caused by a single factor or by multiple factors coupling with each other. Causing instability in the magnetic properties of PM, resulting in a decrease or distortion in the induced back electromotive force in the winding, ultimately leading to a decrease in the amplitude of the PM flux or the occurrence of non-integer harmonic component in the flux.

Main reasons for demagnetization are electromagnetic and mechanochemical aspects. The electromagnetic aspect includes two types: demagnetizing magnetic field and environmental high temperature [8]. In PMSM, the direction of the PM magnetic field is opposite to the direction of the armature magnetic field caused by the armature reaction. Therefore, the magnetic field generated by the armature current has a natural demagnetization effect, especially under high load and high-speed conditions, the transient value of the armature current increases sharply, and the armature reaction magnetic field strengthens, which can easily cause irreversible demagnetization of the PM and lead to a vicious cycle [9, 10]. In terms of environmental temperature, when the motor drive is overloaded or the heat dissipation is not good enough, the thermal stability of the PM becomes worse when the working environment temperature is high, which can also cause demagnetization of PM, resulting in a similar vicious cycle. The mechanochemical aspect includes several influencing factors such as environmental erosion, external demagnetizing field, vibration, and impact.

2 Algebraic Identification Method

2.1 Magnetic Flux Identification Based on Algebraic Method

Both local demagnetization and uniform demagnetization can lead to a decrease in the magnetic flux value of the permanent magnet in the motor. The method of identifying the rotor flux can be used to diagnose the demagnetization phenomenon. Then use the state variables *id*, *iq*, ω and *t*. Based on the algebraic identification method of permanent magnet rotor flux, parameter identification is transformed into the process of finding the optimal solution of the singular matrix, obtaining the optimal solution of the singular matrix, and identifying the relevant parameters of the motor.

Reorganized q-axis voltage equation:

$$u_q = R_s i_q + L_q \frac{di_q}{dt} + \omega L_d i_d + \omega \psi_f \tag{1}$$

Multiply *t* on both sides of the equation and integrate on [0, *t*] to obtain:

$$\int t u_q = R_s \int t i_q + L_q (t i_q - \int i_q) + L_d \int t \omega i_d + \psi_f \int t \omega \tag{2}$$

Order:

$$P_t = \left[t i_q - \int i_q \int t \omega i_d \int t i_q \int t \omega \right],\ \gamma = \left[L_q\ L_d\ R_s\ \psi_f \right]^T,\ q_t = \int t u_q$$

Then there is:

$$P_t \gamma = q_t \tag{3}$$

The matrix P_t, as a singular matrix, can produce countless solutions. In the identification process, the above equation can be transformed into the solution of the optimization problem, and the error function can be defined respectively:

$$\varepsilon(t) = P_t \gamma - q_t \tag{4}$$

Squared error criterion function:

$$J(\gamma, t) = \frac{1}{2} \int_0^t \varepsilon^2_{(\gamma,\delta)} d\delta = \frac{1}{2} \int_0^t (Pt\gamma - qt)^2 d\delta \tag{5}$$

Differentiate γ in the above equation and obtain:

$$\Delta \gamma J(\gamma, t) = \int_0^t Pt^T (Pt\gamma - qt) d\delta \tag{6}$$

Among them, Pt_T is the transpose matrix of P_t.

Order $\Delta_\gamma J_{(\gamma,t)} = 0$, get:

$$\int_0^t Pt^T (Pt\gamma - qt) d\delta = 0 \tag{7}$$

From the above equation, the expression for the magnetic flux parameter can be obtained as:

$$\hat{\gamma} = \left[\int_0^t Pt^T Pt d\delta \right]^{-1} \int_0^t Pt^T qt d\delta \tag{8}$$

The magnetic flux identification method is a qualitative diagnosis of whether demagnetization occurs or not, and the algebraic identification method has a small overall computational complexity, does not need to provide initial values of the parameters to be identified, does not need to inject signals, and can quickly and accurately identify relevant parameters. The qualitative judgment has high reliability.

2.2 Simulation Verification

The identification results of the algebraic identification method under steady-state conditions are in Fig. 1, which can simultaneously identify the dq-axis inductance, stator resistance, and permanent magnet flux value.

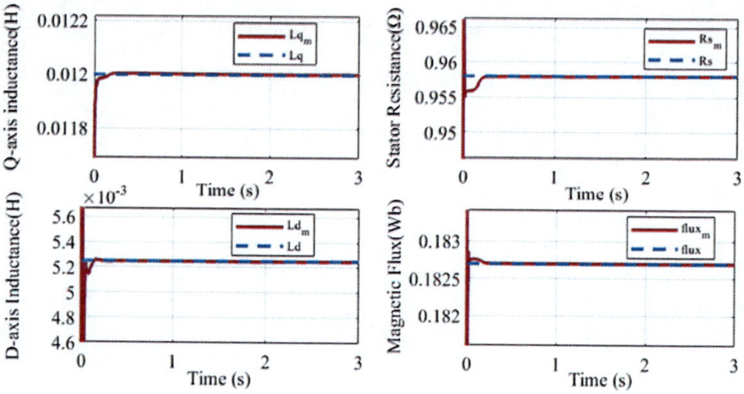

Fig. 1. Algebraic identification results under steady-state conditions.

From the results of the above, it can be seen that under steady-state conditions, the error value of the algebraic identification method for identifying motor parameters is less than 1%, and the identification results are very accurate. The identification time of each parameter of the motor is shown in Table 1, indicating that the algebraic identification method can achieve high-precision online identification of four parameters simultaneously.

Table 1. Evaluation of Algebraic Identification Algorithm.

Identification parameters	Single average time consumption	Average value/Active value
R_s	6.84 ms	0.9577 Ω/0.958 Ω
L_d		5.245 mH/5.25 mH
L_q		11.995 mH/12 mH
ψ_f		0.1827 Wb/0.1827 Wb

3 Magnetic Flux Estimation Method Based on Voltage Model

3.1 Estimation Strategy Based on Voltage Model

The method based on voltage model does not require additional observation of speed, has high robustness, requires fewer motor parameters, and is simple to implement. The observation process is as follows:

The calculation equation for the back electromotive force and magnetic linkage of the motor in two-phase stationary coordinate system is:

$$\begin{cases} e_{\alpha} = u_{\alpha} - R_{s}i_{\alpha} \\ e_{\beta} = u_{\beta} - R_{s}i_{\beta} \end{cases} \tag{9}$$

Among them, u_{α}, u_{β}: the stator voltage of the motor in two-phase stationary coordinate system; e_{α}, e_{β}: the back electromotive force.

$$\begin{cases} \psi_{\alpha} = \int e_{\alpha} \\ \psi_{\beta} = \int e_{\beta} \end{cases} \tag{10}$$

Among them: ψ_{α}, ψ_{β}: the motor flux in two-phase stationary coordinate system.

Directly integrating the back electromotive force can easily lead to DC bias, spikes, and other issues in the obtained magnetic flux waveform. A second-order generalized integrator is used to calculate the magnetic flux value.

To achieve pure integrator function using SOGI, the schematic diagram is shown in Fig. 2:

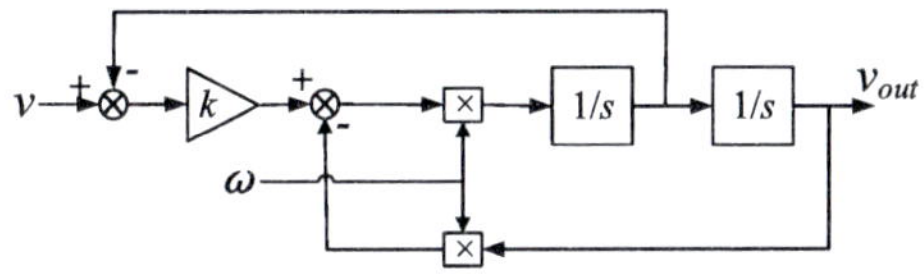

Fig. 2. Integrator based on second-order generalized integrator.

The integrator can achieve integration function, suppress harmonics outside of the fundamental wave, and have strong filtering function.

3.2 Simulation Verification

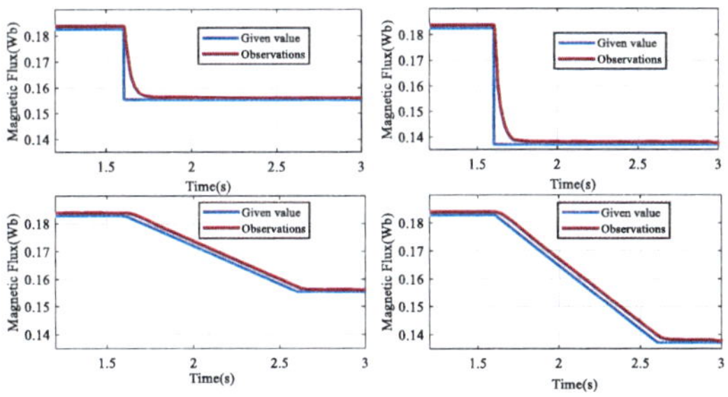

Fig. 3. Results of voltage model flux estimation strategy.

To simulate the extreme conditions during the operation of the motor, the identification effect of the module is verified by setting the motor flux change to sudden changes and slope drops. In Fig. 3, the left side shows the identification result of 15% uniform demagnetization, while the right side shows 25%. When the magnetic flux changes suddenly, the identification module needs 0.15s to track the actual value, and the steady-state error is less than 2%; During the slope descent process, the error is less than 5%. After stabilization, the error decreases and is ultimately less than 2%. The results indicate that the module can identify the degree of uniform demagnetization of the motor accurately and quickly.

4 Effective Magnetic Flux Sliding Mode Observer

To improve the accuracy of the flux observer, a sliding mode observer is used, which takes the stator current and effective magnetic flux value in a two-phase stationary coordinate system as state variables, and calculates the stator flux using the effective magnetic flux observation results.

4.1 Design of Sliding Mode Observer for Effective Magnetic Flux Linkage

Define the effective magnetic flux value of a PMSM to:

$$T_{\mathrm{e}}=\frac{3}{2}p\psi_{ext}i_{\mathrm{q}} \tag{11}$$

Effective magnetic flux value ψ_{ext} is:

$$\psi_{ext}=\psi_{\mathrm{ro}}+(L_{\mathrm{d}}-L_{\mathrm{q}})i_{\mathrm{d}} \tag{12}$$

Construct the state equation of a PMSM in αβ-axis, and design a sliding mode observer for the system. The current error equation can be obtained as:

$$\begin{bmatrix}\dot{\tilde{i}}_{\alpha}\\ \dot{\tilde{i}}_{\beta}\end{bmatrix}=\frac{\omega}{L_{\mathrm{q}}}\begin{bmatrix}0 & 1\\ -1 & 0\end{bmatrix}\begin{bmatrix}\tilde{\psi}_{ext,\alpha}\\ \tilde{\psi}_{ext,\beta}\end{bmatrix}-\begin{bmatrix}k_{11} & 0\\ 0 & k_{22}\end{bmatrix}\begin{bmatrix}\mathrm{sgn}(\tilde{i}_{\alpha})\\ \mathrm{sgn}(\tilde{i}_{\beta})\end{bmatrix} \tag{13}$$

Among them, the quantity labeled with "~" is the error value of the state variable. If appropriate parameters k_{11} and k_{22} are selected, the current observation results will converge, and the error between the current and current observation values can be used as feedback correction for the magnetic flux observation part, achieving closed-loop monitoring of the flux and improving the accuracy of flux monitoring. After the error value of the current state variable enters the sliding mode surface, the observer's flux error equation can be obtained:

$$\begin{bmatrix}\dot{\tilde{\psi}}_{ext,\alpha}\\ \dot{\tilde{\psi}}_{ext,\beta}\end{bmatrix}=\omega\begin{bmatrix}0 & 1\\ -1 & 0\end{bmatrix}\begin{bmatrix}\tilde{\psi}_{ext,\alpha}\\ \tilde{\psi}_{ext,\beta}\end{bmatrix}-\begin{bmatrix}k_{31} & k_{32}\\ k_{41} & k_{42}\end{bmatrix}\begin{bmatrix}\mathrm{sgn}(\tilde{i}_{\alpha})\\ \mathrm{sgn}(\tilde{i}_{\beta})\end{bmatrix} \tag{14}$$

The expression for the current observation part is:

$$\begin{cases}\dot{\hat{i}}_{\alpha}=-\frac{R_{\mathrm{s}}}{L_{\mathrm{q}}}i_{\alpha}+\frac{\omega}{L_{\mathrm{q}}}\hat{\psi}_{ext,\beta}+\frac{u_{\mathrm{q}}}{L_{\mathrm{q}}}+k_{11}\mathrm{sgn}(\tilde{i}_{\alpha})\\ \dot{\hat{i}}_{\beta}=-\frac{R_{\mathrm{s}}}{L_{\mathrm{q}}}i_{\beta}+\frac{\omega}{L_{\mathrm{q}}}\hat{\psi}_{ext,\alpha}+\frac{u_{\mathrm{q}}}{L_{\mathrm{q}}}+k_{22}\mathrm{sgn}(\tilde{i}_{\alpha})\end{cases} \tag{15}$$

The expression for the magnetic flux observation part is:

$$\begin{cases} \dot{\hat{\psi}}_{ext,\alpha} = -\omega\hat{\psi}_{ext,\beta} + k_{31}\mathrm{sgn}(\tilde{i}_\alpha) + k_{32}\mathrm{sgn}(\tilde{i}_\beta) \\ \dot{\hat{\psi}}_{ext,\beta} = \omega\hat{\psi}_{ext,\alpha} + k_{41}\mathrm{sgn}(\tilde{i}_\alpha) + k_{42}\mathrm{sgn}(\tilde{i}_\beta) \end{cases} \tag{16}$$

The expression for the effective magnetic flux value is:

$$\hat{\psi}_{ext} = \sqrt{\hat{\psi}^2_{ext,\alpha} + \hat{\psi}^2_{ext,\beta}} \tag{17}$$

4.2 Simulation Verification

The motor exhibits a uniform demagnetization mutation of 25% at 1s, and the effective magnetic flux observation results in the sliding mode observer are shown in Fig. 4.

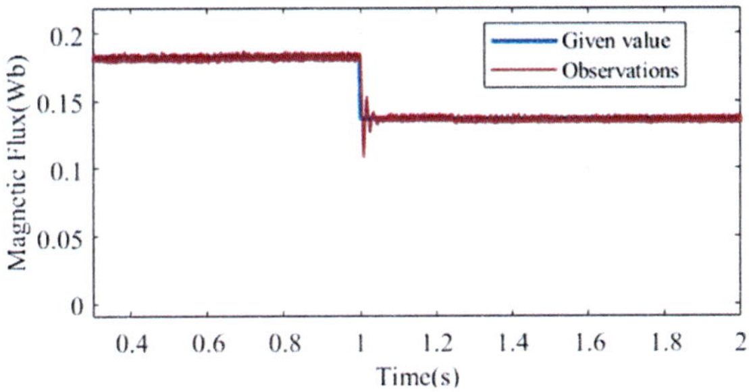

Fig. 4. Result of effective magnetic flux of state observer during PMSM demagnetization.

From the above results, it can be seen that, after the uniform demagnetization phenomenon occurs, the state observer designed can identify the actual effective magnetic flux value of the motor online accurately and quickly. And the observer only involves the stator resistance and q-axis inductance of the motor, eliminating the saliency phenomenon of the motor. It also has strong robustness in suppressing the motor inductance and higher accuracy in flux observation.

5 Conclusion

The algebraic identification method has high accuracy and the fastest identification speed, but its robustness to motor parameters is not high; The flux estimation strategy based on voltage model has the advantages of requiring fewer motor parameters, high robustness, and simple implementation method. It is very sensitive to the degree of motor demagnetization, has good tracking performance, and is suitable for accurate judgment of uniform demagnetization degree. However, this method relies more on the detection of back electromotive force; The sliding mode observer, on the other hand, does not need to rely on the observation of the back electromotive force, and has advantages such as high flux observation accuracy and strong robustness to changes in motor inductance parameters. Its structure is simple, and it is also suitable for salient pole machines and hidden pole machines.

References

1. Julio, U., Jordi, R., Luis, R.: A back-emf based method to detect magnet failures in PMSMs. IEEE Trans. Magn. **49**(13), 591–598 (2013)
2. Yang, F., Yang, Y.: Diagnosis method for permanent magnet synchronous motor loss of excitation fault based on induced electromotive force decomposition. New Industrialization **8**(2), 1–10 (2018) (in Chinese)
3. Przemyslaw, P., Marcin, W.: Demagnetization fault diagnosis of permanent magnet synchronous motors based on stator current signal processing and machine learning algorithms. Sensors **23**(4), 1757–1782 (2023)
4. Zhang, Y., Liu, G., Chen, Q.: Research on the classification and diagnosis of inter turn short circuit and local demagnetization faults in permanent magnet synchronous motors based on current fluctuation characteristics. J. Electr. Technol. **37**(7), 1634–1643, 1653 (2022) (in Chinese)
5. Wang, C., Delgado, M., Romeral, L.: Detection of partial demagnetization fault in PMSMs operating under nonstationary conditions. IEEE Trans. Magn. **52**(7), 1–4 (2016)
6. Hang, J., Zhang, J., Xia, M., et al.: Interturn fault diagnosis for model-predictive-controlled-PMSM based on cost function and wavelet transform. IEEE Trans. Power Electron. **35**(6), 6405–6418 (2020)
7. Chen, H., Cao, C., et al.: A novel method for diagnosing demagnetization fault in PMSM using Toroidal-Yoke-type search coil. IEEE Trans. Instrum. Meas. **71**(4), 1–12 (2022)
8. Faiz, J., Mazaheri, E.: Demagnetization modeling and fault diagnosing techniques in permanent magnet machines under stationary and nonstationary conditions: an overview. IEEE Trans. Ind. Appl. **53**(3), 2772–2785 (2017)
9. Li, H., Chen, T.: PMSM demagnetization fault diagnosis and fault pattern recognition for electric vehicles. J. Electr. Eng. Technol. **32**(5), 1–8 (2017). (in Chinese)
10. Imai, K., Doki, S., Kondo, K., et al.: Current control system with high-frequency signal injection for dual three-phase permanent magnet synchronous motor: paper. IEEJ J. Ind. Appl. **11**(1), 897–903 (2022)

Full Speed Range Field Oriented Control Method for PMSM

Donghui Liu[1], Leiting Zhao[1], Liu Kan[1], Liran Li[1], Chi Ma[1], Zheng Yue[2], Zheming Jin[2], and Lijun Diao[2](✉)

[1] Beijing Zongheng Electromechanical Technology Co., Ltd., Beijing 100094, China
[2] Beijing Jiaotong University, Beijing 100044, China
ljdiao@bjtu.edu.cn

Abstract. Compared with the electric excitation motor, the magnetism and flux linkage of the permanent magnet of the excitation device cannot be changed after magnetization. As the motor speed increases, the back electromotive force of the permanent magnet synchronous motor will gradually increase. When the voltage increases When the output voltage of the inverter reaches the limit, the speed cannot continue to rise. In traditional control strategies, the magnetism of permanent magnet synchronous motors is usually kept at a high level to ensure efficient operation and good control performance of the motor. But in some cases, such as applications such as fast acceleration and variable speed, the motor needs higher acceleration and faster response speed. At this time, it is an effective means to use the flux-weakening speed expansion control strategy. In this paper, the negative direct axis compensation and Maximum Torque-per-Voltage (MTPV) flux-weakening control methods are used to realize the high-speed operation of the permanent magnet synchronous motor, and the corresponding control algorithm is verified by the hardware-in-the-loop test method.

Keywords: PMSM · Flux-weakening Control · MTPV · Negative direct axis compensation

1 PMSM Operating Limits

Due to the limitation of the voltage output on the inverter side of the motor and the influence of the self-heating of the motor winding, the voltage and current of the motor itself cannot increase infinitely during operation [1].

That is, the motor must meet the following conditions when running.

$$\begin{aligned} U_s^2 &= U_d^2 + U_q^2 \le U_{\max}^2 \\ I_s^2 &= i_d^2 + i_q^2 \le I_{\max}^2 \end{aligned} \tag{1}$$

Among them, the value of $U_{\max}$ is determined by the DC bus voltage on the inverter side [2].

$$U_{\max} = \frac{U_{dc}}{\sqrt{3}} \tag{2}$$

L. Jia et al. (Eds.): EITRT 2023, LNEE 1135, pp. 339–346, 2024.
https://doi.org/10.1007/978-981-99-9307-9_37

In order to intuitively describe the running state of the motor under the dual constraints of voltage and current, the above formula is described on the $i_d - i_q$ coordinate plane.

When the motor is running at high speed, ignoring the voltage drop of the motor stator winding, the voltage equation of the motor is

$$\begin{cases} u_d = -\omega_e L_q i_q \\ u_q = \omega_e (L_d i_d + \psi_f) \end{cases} \tag{3}$$

Bringing it into the voltage constraint equation, the voltage limit circle equation when the motor is running can be obtained as [3]:

$$(L_d i_d + \psi_f)^2 + (L_q i_q)^2 \leq (\frac{U_{\max}}{\omega_e})^2 \Rightarrow \frac{i_q^2}{L_d^2} + \frac{(i_d + \frac{\psi_f}{L_d})^2}{L_q^2} \leq (\frac{U_{\max}}{L_d L_q \omega_e})^2 \tag{4}$$

2 Flux-Weakening Control Strategy

According to the change characteristics of parameters such as torque, current and voltage when the motor is in different operating regions, PMSM is divided into three regions when running at full speed: constant torque region, flux-weakening zone I and flux-weakening zone II. As shown in Fig. 1.

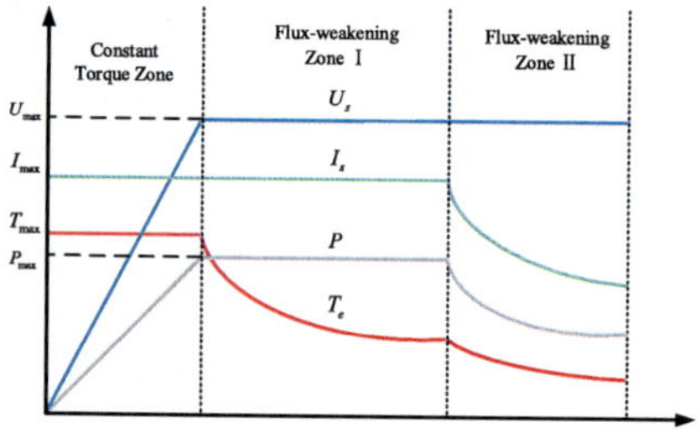

Fig. 1. PMSM Full Speed Range Operation Characteristics Division

2.1 Optimal Running Trajectory in Full Speed Range

According to the characteristics of the motor in different operating regions and the electrical quantity, the optimal operating trajectory of the PMSM in the full speed range is planned, as shown in Fig. 2.

The optimal trajectory is $O \rightarrow A \rightarrow B \rightarrow C$, Among them, the OA section is the constant torque area. In order to make the motor obtain the fastest speed response, MTPA is used for control; the AB section is the flux-weakening Zone I. In order to make the motor have the largest output power, the negative direct axis compensation is used for flux-weakening control; BC The section is the flux-weakening zone II. In order to make the motor obtain the fastest speed response, MTPV is used for control [4].

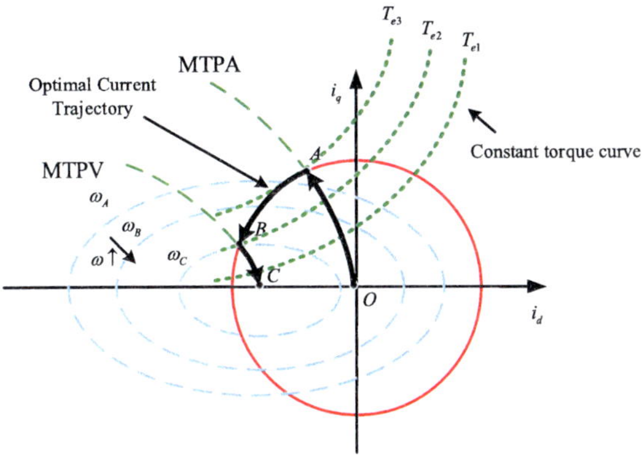

Fig. 2. Optimal Current Trajectory of PMSM in the Full Speed Range

(1) Constant torque area. At this time, the motor is running in the low-speed area (below the base speed ω_{base}), and the current operating point of the motor is inside the current and voltage limit circle, that is, the voltage and current of the motor satisfy $U_s < U_{\max}, I_s < I_{\max}$, and the motor can continue to operate in this area The maximum torque $T_{\max}$ is output, and the output power of the motor is increasing at this time [5].

The torque and current when the motor running in the constant torque area can be calculated by Eq. (5).

$$\begin{cases} T_e = \frac{3}{2} P_n i_q \left[i_d (L_d - L_q) + \psi_f \right] \\ i_d = \dfrac{-\psi_f + \sqrt{\psi_f^2 + 4(L_d - L_q) i_q^2}}{2(L_d - L_q)} \end{cases} \quad (5)$$

(2) Flux-weakening Zone I/Maximum current (MC) zone. When the speed of the motor increases to the critical point of flux-weakening zone I, due to the limitation of the inverter output and the motor itself, the voltage and current of the motor are limited by $U_{\max}$ and $I_{\max}$ and cannot continue to rise. In this case, only rely on MTPA Control strategy, it is no longer possible to increase the speed by increasing the terminal voltage of the motor. It can be seen from Eq. (6) that at this time, the magnetic flux of the motor needs to be weakened through the flux-weakening control, so that the motor can further speed up.

$$\frac{L_d^2 (i_d + \frac{\psi_f}{L_d})^2}{L_q^2} + i_q^2 \le \left(\frac{U_{\max}}{L_q \omega_e}\right)^2 \quad (6)$$

Since the voltage and current of the motor are $U_{\max}$ and $I_{\max}$ respectively and remain constant when the motor is in the flux-weakening zone I, that is, the voltage of the motor satisfies $U_s = U_{\max}, I_s = I_{\max}$, so the output power of the motor is constant, so the motor runs in a constant power state. The torque and current of the motor running in the flux-weakening zone I can be calculated by Eq. (7) [6].

$$
\begin{cases}
T_e = \frac{3}{2}P_n i_q\left[i_d(L_d - L_q) + \psi_f\right] \\
i_d = \frac{-\psi_f + \sqrt{\psi_f^2 + 4(L_d - L_q)i_{q-MTPA}^2}}{2(L_d - L_q)} + i_{d-buchang} \\
i_q = \sqrt{I_{\max}^2 - i_d^2}
\end{cases}
\tag{7}
$$

(3) Flux-weakening Zone II. With the further increase of the motor speed and the further shrinkage of the voltage limit circle, the normal operation of the motor cannot be satisfied by only relying on a single flux-weakening control [7]. Therefore, the BC segment is controlled by MTPV, which can achieve the most effective output while meeting the voltage limit of the motor. The current and torque calculation of the motor in flux-weakening zone II is shown in Eq. (8).

$$
\begin{cases}
i_d = -\frac{\psi_f}{L_d} + \frac{-L_q\psi_f + \sqrt{(L_q\psi_f)^2 + 4L_q^2(L_d - L_q)^2 i_q^2}}{2L_d(L_d - L_q)} \\
(L_d i_d + \psi_f)^2 + (L_q i_q)^2 \le (\frac{U_{\max}}{\omega_e})^2 \\
T_e = \frac{3}{2}P_n i_q\left[i_d(L_d - L_q) + \psi_f\right]
\end{cases}
\tag{8}
$$

2.2 Negative Direct Axis Compensation

The negative direct axis compensation method is one of the voltage feedback regulation methods, and is widely used in various engineering projects at present. When the motor is below the base speed, the MTPA control strategy is adopted. As the speed increases, the negative direct axis compensation method introduces a negative demagnetization current Δi_d to weaken the air gap magnetic field. It can be seen from Eq. (6) that when $U_{\max}$ is a constant value, the motor speed ω_e increases with the decrease of the motor flux ψ_f, so that the purpose of flux-weakening and speed expansion can be achieved. The specific control block diagram is shown in Fig. 3.

It can be seen from the control block diagram that the principle of negative direct axis compensation is based on the traditional double current closed-loop control, using the error of the stator voltage u_s and the voltage limit $U_{\max}$ to generate a feedback closed loop [8]. When the speed of the motor is low, that is, below the base speed ω_{base}, since $u_s < U_{\max}$, the compensation value Δi_d of the current i_d is zero after being limited, and the flux-weakening compensation part does not work at this time.

When the motor speed exceeds the base speed After the speed ω_{base}, the stator voltage u_s of the motor will exceed $U_{\max}$ for a short time. At this time, the input of PI is negative, and the compensation value Δi_d generated after PI is also negative, and then the current i_{d-MTPA} calculated by the MTPA trajectory make compensation; The i_{d-MC} generated after compensation can be calculated through $\sqrt{I_{\max}^2 - i_d^2}$ to calculate

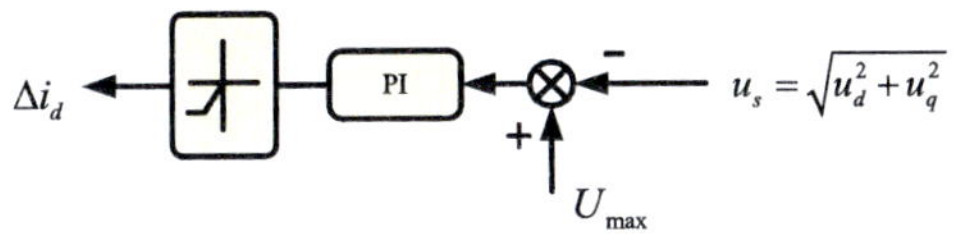

Fig. 3. Negative direct axis compensation

the quadrature axis current i_{q-MC} of the MC segment, Finally, the i_{q-MTPA} output is replaced by the Switch module [9]. In this way, the running trajectory of the motor can be pulled from the MTPA to the MC curve to achieve the purpose of flux-weakening and speed expansion.

2.3 Simulation Model Building

On the basis of the previous discussion, the off-line model of full speed range Field Oriented control of permanent magnet synchronous motor is built as shown in Fig. 4 [10].

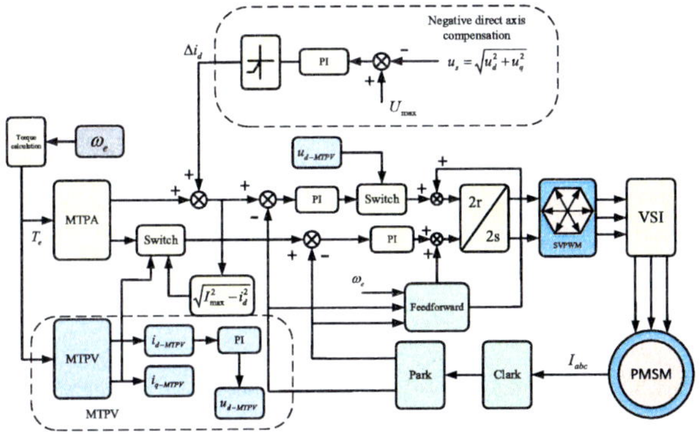

Fig. 4. Composite weak magnetic field control system based on MTPV and negative direct axis compensation

3 Simulation Results

The results shown in the figure below are verified on the built simulink simulation model. The motor starts with a constant torque first, at this time the AC and DC axis currents i_d and i_q remain constant, and the speed N_r and stator voltage U_s rise with a fixed slope.

When the voltage U_s rises to the limit value, the negative direct-axis compensation starts to act, and at this time the compensation value Δi_d and the direct-axis current i_d begin to increase in the opposite direction, and at the same time, under the limitation of the current limit circle, the quadrature-axis current i_q begins to decrease positively, the torque also gradually decreases, and the slope of the speed N_r increase becomes slower (Figs. 5, 6 and 7).

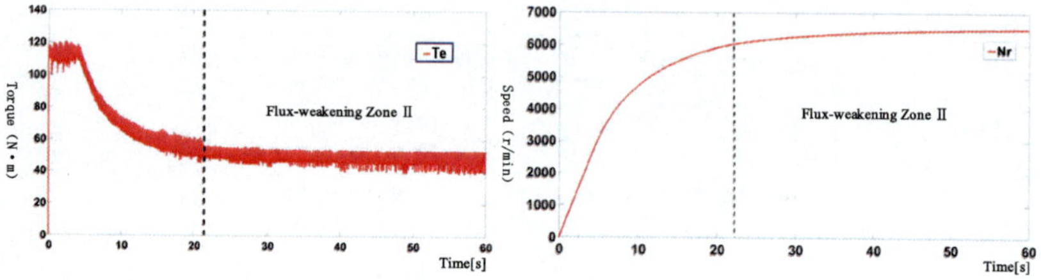

Fig. 5. Torque and speed

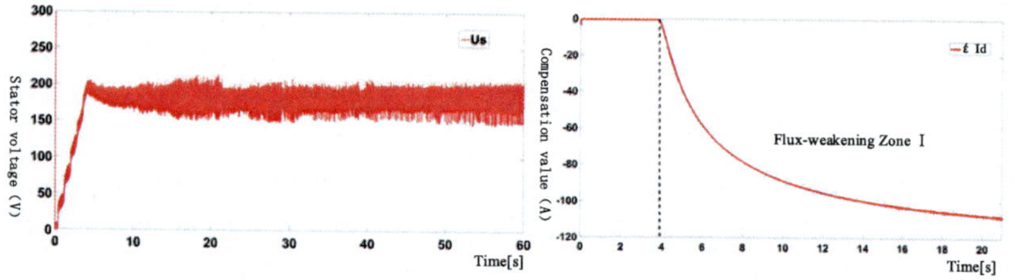

Fig. 6. Stator voltage and straight-axis current compensation values

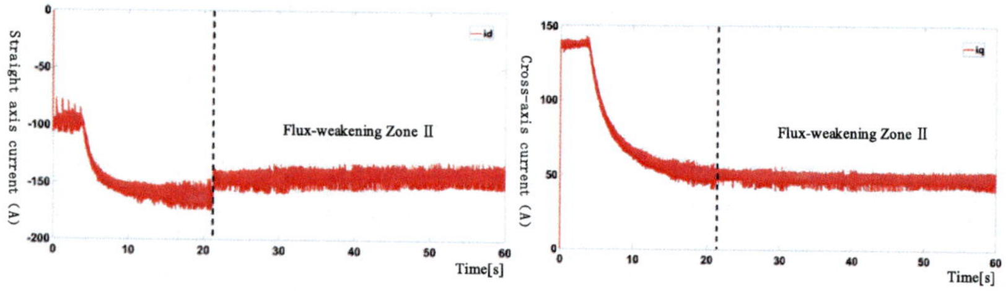

Fig. 7. Cross-axis and straight-axis currents

When the speed reaches 5800 r/min, it enters the flux-weakening zone II. At this time, the quadrature axis current and the direct axis current both decrease positively under the control of MTPV, and the torque T_e also decreases accordingly. At this time, the voltage is still stable near $U_{\max}$, the speed N_r rises more and more slowly.

Build the hardware-in-the-loop verification experiment platform for hardware-in-the-loop simulation as shown in the figure, and carry out the hardware-in-the-loop verification of the above control algorithm, and successfully verify the effectiveness of the algorithm (Figs. 8, 9 and 10).

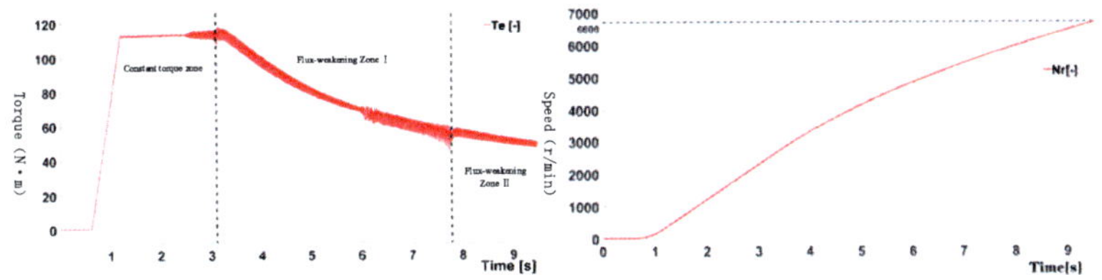

Fig. 8. Torque and speed

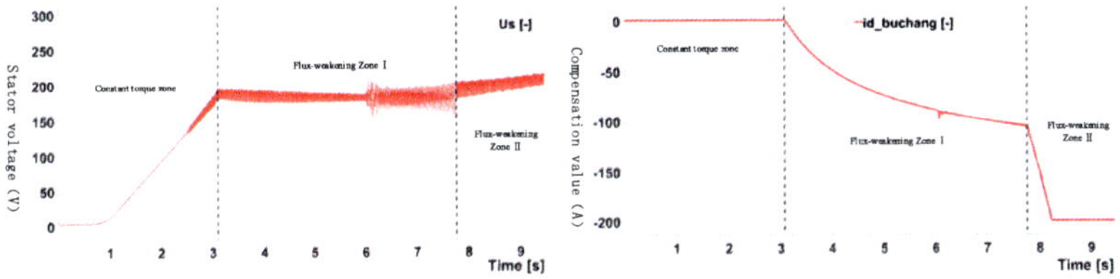

Fig. 9. Stator voltage and straight-axis current compensation values

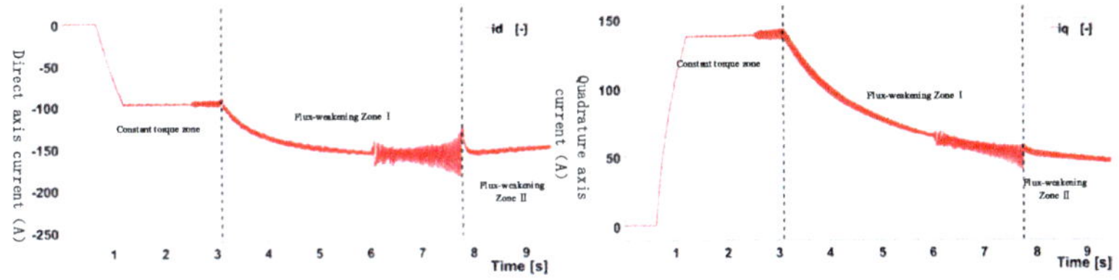

Fig. 10. Stator voltage and straight-axis current compensation values

4 Conclusion

In this paper, combined with the operating limit of the motor and the current trajectory under each control strategy, and according to the characteristics of the electrical quantity of the motor in different speed ranges, a composite flux-weakening control strategy based on negative direct axis compensation and MTPV is proposed, and the speed of the motor is successfully reduced. Increased to 6600r/min, completed the flux-weakening speed expansion of the motor in the deep flux-weakening area, and verified the correctness and effectiveness of the algorithm by hardware-in-the-loop test.

Acknowledgement. This work was supported in part by the Science and Technology Research and Development Plan of China National Railway Group Company Ltd., under Grant EMU and Locomotive Traction and Control State Key Laboratory Fund (2021YJ281).

References

1. Wu, W.: Research on Current Distribution Strategy of Permanent Magnet Synchronous Motor. Hunan University (2020) (in Chinese)
2. Chen, S.: Reliability Analysis of Electric Vehicle Permanent Magnet Synchronous Motor Under Multiple Failure Modes. University of Electronic Science and Technology of China (2022) (in Chinese)
3. Liu, H.: Research on Field-Weakening Control Strategy of Permanent Magnet Synchronous Motor for Electric Vehicles. Wuhan University of Technology (2021) (in Chinese)
4. Jiwon, Y.O.O., Joohyun, L.E.E., Seung-Ki, S.U.L.: Analysis of instability in torque control of sensorless PMSM drives in flux weakening region. IEEE Trans. Power Electron. **36**(9), 10815–10826 (2021)
5. Wang, S., Kang, J., Degano, M., et al.: An accurate wide-speed range control method of IPMSM considering resistive voltage drop and magnetic saturation. IEEE Trans. Ind. Electron. 1–1 (2019)
6. Trancho, E., Ibarra, E., Arias, A., et al.: PM-assisted synchronous reluctance machine flux weakening control for EV and HEV applications. IEEE Trans. Ind. Electron. **65**(4), 2986–2995 (2018)
7. Zhou, Z., Scuiller, F., Charpentier, J.F., et al.: Power control of a nonpitchable PMSG-based marine current turbine at overrated current speed with flux-weakening strategy. IEEE J. Oceanic Eng. **40**(3), 536–545 (2015)
8. Ai, M.: Research on Field Weakening Control Technology of Permanent Magnet Synchronous Motor. Harbin University of Science and Technology (2020) (in Chinese)
9. Zhang, L.: Research on Field Weakening Control Strategy of Permanent Magnet Synchronous Motor. Hunan University (2020) (in Chinese)
10. Zhang, Z.: Research on Key Technologies of Permanent Magnet Synchronous Motor Control in Rail Transit. Beijing Jiaotong University (2019) (in Chinese)

Multi-device Collaborative Anomaly Diagnosis Method for Electric Mining Dump Trucks

Yichao Ma[1], Lei Wang[1](✉), Benchao Zhu[1], and Qubo Xie[2]

[1] Beijing Jiaotong University, Beijing 100044, China
{21121463,leiwang,zhuben}@bjtu.edu.cn
[2] CNR Datong Electric Locomotive Co, Datong, China
xiequbo.dt@crrcgc.cc

Abstract. Electric mining dump trucks (EMDTs) play an essential role in open pit mining. EMDTs are complex electromechanical equipment characterized by messy status information and numerous failure modes. For other devices to also be able to identify new faults, the new fault data should be shared among multiple devices in the poor network environment of the mine. This article proposes a collaborative multi-device fault diagnosis method for EMDTs. At the individual device's level, the powerful multivariate time-series data feature extraction capability of bidirectional long short-term memory (BLSTM) networks is utilized to identify the EMDT's overall fault state. At the multiple devices' level, federal learning is employed to achieve collaborative optimization of numerous devices' diagnostic models to identify new faults effectively. Collect actual operational data on EMDT and simulate faults based on this data to validate the strategy for fault diagnosis.

Keywords: Anomaly Diagnosis · Electric Mining Dump Trucks · Bidirectional Long Short-term Memory Network · Federated Learning

1 Introduction

Electric mining dump trucks (EMDTs) are developing towards large size and high working intensity, and their electric drive system structure is being more and more complex to meet the task requirements. The electric drive system is responsible for driving the huge load of the EMDT and is susceptible to anomalies due to the load shock from complex road conditions. In the daily operation of the EMDTs, if the anomalies cannot be detected in time, it may lead to the failure of the system. Therefore, anomaly diagnosis is crucial to ensure the reliable operation of the EMDTs.

The electric drive system of EMDT is a typical electromechanical composite equipment. Its structure is shown in Fig. 1, which consists of a diesel engine, synchronous generator, rectifier, inverter, asynchronous drive motor, and control device. The signals from the mechanical and electrical parts are coupled with each other, and the transient characteristics of the signals from different parts vary, making it challenging to extract abnormal features directly from them.

L. Jia et al. (Eds.): EITRT 2023, LNEE 1135, pp. 347–355, 2024.
https://doi.org/10.1007/978-981-99-9307-9_38

Methodologies for fault diagnosis can be categorized as model-based, knowledge-based, or data-based [1, 2]. For complex equipment like EMDTs, creating an accurate physical model and comprehensive fault knowledge base is challenging because of the complex structure, and massive state information [3]. Therefore, it is recommended to apply advanced data-based methods for diagnosis.

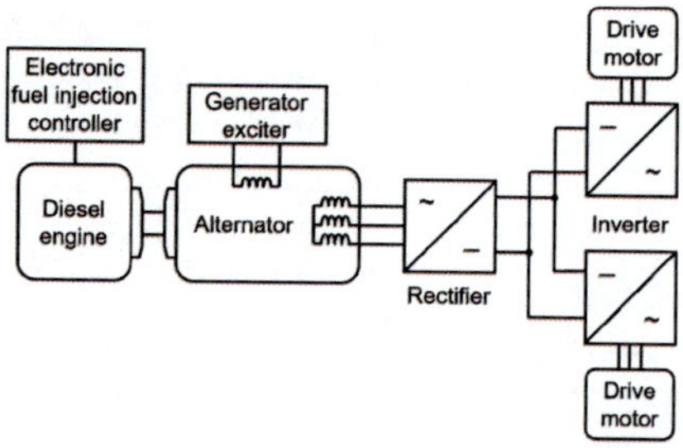

Fig. 1. Structure of the drive system of EMDT.

In recent years, the rapid development of Deep Learning (DL) has revolutionized fault diagnosis for complex equipment [4]. DL-based methods do not require experts to design fault feature representations, achieving good identification accuracy while saving workforce consumption, and are currently a research hotspot. Ince et al. [5] proposed a 1D-CNN model that implements fault feature extraction and classification within a single network and applies it to motor bearings diagnostic. Sun et al. [6] proposed a bearing fault diagnosis approach based on a stacked self-encoder to facilitate the migration of diagnostic models between machines. Ma et al. [7] implemented the automatic extraction of fault features from industrial process data based on Transformer. However, due to the numerous EMDT subsystems and varying probabilities of potential failures in different working environments and tasks, it has become a challenge to deploy and optimize fault diagnosis models based on actual situations.

This article proposes a multi-device collaborative fault diagnosis method for EMDTs. The primary contributions of the article are as follows: first, the challenges of EMDTs' fault diagnosis are analyzed, and an EMDTs' fault diagnosis method is proposed based on the actual operation. Second, we proposed to use the feature extraction capability of the bidirectional LSTM (BLSTM) network at the individual device's level to identify and classify faults; at the multiple devices' level, FL is used to make full use of data from multiple devices through multi-device collaboration to achieve global optimization of the diagnostic model. Finally, the proposed algorithm is validated by developing onboard diagnostic equipment and collecting actual EMDTs operational data.

2 Proposed Methodology

2.1 System Architecture

We propose a fault diagnosis method for EMDTs based on BLSTM and multi-device collaboration, and its overall framework is shown in Fig. 2. The diagnostic strategy contains a server node and multiple onboard nodes. The server node and the onboard node are deployed with global diagnostic models H_{global} and local diagnostic models H_k, respectively, where k is the serial number of the onboard node. The network structures of H_{global} and H_k are the same, both implemented based on BLSTM, but the parameters of w_{global} and w_k of those two networks are different. The onboard node k collects raw data $S_{k\text{-raw}} = \{s_0, s_1, \ldots, s_m\}$ from sensors via Controller Area Network (CAN), where m is the total number of sensors. The onboard node preprocesses S_k to obtain S_k. S_k is then used to train the diagnostic model H_k of this node or is used as an input to the diagnostic model to determine the fault status of this vehicle and output the fault category, denoted as $y_k = H_k(S_k)$. The server node optimizes the model H_{global} by weighing the parameters w_k of H_k and then distributes the parameters of the parameters w_{global} to the onboard nodes so that the H_k of each vehicle node converges to the same distribution.

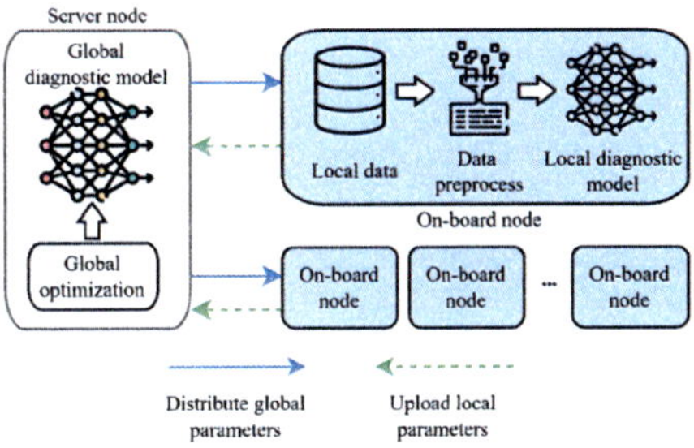

Fig. 2. Framework of fault diagnosis method for EMDTs.

The fault diagnosis method for EMDTs proposed in this article is composed of three parts: data preprocessing, diagnostic models, and multi-device collaborative optimization. These three parts are described in detail below.

2.2 Data Preprocessing

Due to differences in signal scale and sequence length, the raw signals collected from sensors in different subsystems cannot usually be directly used as inputs for diagnostic models. To address these problems, we preprocess the raw data collected by the sensors in order of normalization and segmentation.

Firstly, the raw data are normalized. Due to electromagnetic interference in the working environment, outliers may appear in the data collected by the sensors. 0-mean and max-min normalization are sensitive to outliers. Therefore, robust normalization is used in this article. This method scales signals through a robust statistic as follows:

$$S_k = \frac{S_{k-\text{raw}} - P_{1/2}}{P_{3/4} - P_{1/4}} \tag{1}$$

where $S_{k-\text{raw}}$ denotes the raw sensor signals, $P_{1/2}$ is the median of the signals, $P_{3/4}$ is the 3/4 quantile, and $P_{1/4}$ is the 1/4 quantile. After normalization, the data scales are transformed into a similar range, which facilitates the network to extract the variation characteristics of data at different scales.

Secondly, segment the raw signals from sensors. In order to make the inputs of the diagnostic model have the desired size and strengthen the correlation between adjacent samples, a sliding window with overlap is adopted. The process of segmenting the time-series data is as follows. To begin with, the sample length (the number of points in the sliding window, denoted as l_{segment}) is determined according to the sampling frequency and the duration of the anomaly signal. Then, according to the number of samples to be generated determine the length of the overlap between two consecutive samples (denoted as l_{overlap}). The sliding window moves each time the length of $l_{\text{stride}} = l_{\text{segment}} - l_{\text{overlap}}$ after intercepting from the signal the length of l_{segment} as a sample until it reaches the end of the time series. It is important to note that overlapping sampling by sliding window is only used to process the training samples.

2.3 Diagnostic Model

The LSTM network can extract long- and short-term dependent characteristics from time series data. As an enhancement to the LSTM network, the BLSTM network can simultaneously extract transient characteristics before and after the fault moment from both directions, resulting in enhanced feature extraction capability.

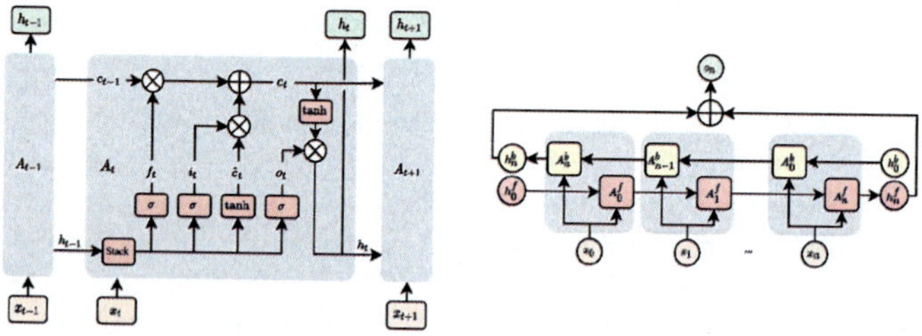

(a) LSTM network structure (b) BLSTM network structure

Fig. 3. LSTM and BLSTM network structure.

A BLSTM-based encoder is applied to extract fault features from preprocessed signals. The structure of LSTM is shown in Fig. 3a, where A represents a cell of LSTM in time step t. LSTM eliminates vanishing/exploding gradients and overcomes RNNs' difficulties in learning long-term dependencies with the aid of forgetting, input, and output gates [8]. The BLSTM stacks two LSTMs with opposite propagation directions, allowing the network to obtain information about the neighboring points before and after a specific point. Its structure is shown in Fig. 3b.

The cell of BLSTM consists of a forward propagation vector *hf t* and a backward propagation vector *hb t*, which are shown as [9]:

$$\begin{cases} h_t^f = \sigma\left(W_t^f X_t + U_t^f h_{t-1}^f + b_t^f\right) \\ h_t^b = \sigma\left(W_t^b X_t + U_t^b h_{t-1}^b + b_t^b\right) \end{cases} \tag{2}$$

where superscripts f and b denote forward and backward propagation, respectively.

The hidden states of these two inverse LSTM layers can be combined as the network's output by summing, averaging, merging, or concatenating. The concatenating is selected to combine the last hidden states of two LSTM layers as output, shown as:

$$O = \left[h_n^f, h_n^b\right] \tag{3}$$

where O is the output of BLSTM, and n is the total number of time steps.

After the encoder consisting of BLSTM extracts the fault features from the signals, a decoder is needed to map the fault features into fault classes. In this article, we use an MLP consisting of a 3-layer fully connected layer as the decoder. The output layer of MLP is a neuron and the output is an anomaly score. A larger anomaly score indicates a higher probability of anomalies in the signal.} For the onboard node k, the diagnosis model can be expressed as:

$$\begin{cases} O_k = f_{\mathrm{BLSTM}}(S_k) \\ y_k = f_{\mathrm{MLP}}(O_k) \end{cases} \tag{4}$$

where S_k is the pre-processed signal, O_k is the output of BLSTM, and y_k is the fault category predicted by the diagnostic model.

2.4 Multi-device Collaborative Optimization

Due to the limited communication network in industrial production environments, we adopt a multi-device collaborative method based on FL to optimize the diagnostic model globally with multiple onboard devices collaboration, sharing the parameters of the diagnostic model between onboard devices instead of raw data. Each of the m edge devices has its own local dataset $\{\mathrm{D}_1, \mathrm{D}_2, \ldots, \mathrm{D}_m\}$ and uses its local dataset to train parameters of the local model $\{\theta_1, \theta_2, \ldots, \theta_m\}$. The cloud server aggregates the parameters of the edge devices and trains a global parameter θ_{fed}, which is then distributed to the edge devices. Eventually, the model at distributed edge devices converges to the same distribution. The objective function of FL is [10]:

$$\min_{\theta} F(\theta) = \sum_{k=1}^{m} p_k F_k(\theta) \tag{5}$$

where m is the number of edge devices, $p_k \geq 0$ is the weighting factor, The N-th cumulative sum of p_k is 1, and $F_k(\theta)$ is the local objective function of edge device E_k. FL not only effectively alleviates the bandwidth constraint in the production environment but also protects the privacy of the local dataset.

We choose the FedAvg algorithm for optimizing the diagnostic model in onboard devices. FedAvg algorithm updates the model as follows [11]:

$$\theta^{t+1} = \sum_{k \in S_t} \frac{n_k}{M} \theta_k^{t+1} + \sum_{k \notin S_t} \frac{n_k}{M} \theta^t \tag{6}$$

where θ^t is the global model parameter at the current step, θ^{t+1} is the global model parameter at the next step, M is the total number of samples in all models, and n_k is the total number of samples of the ith models. FedAvg algorithm first selects some models randomly to form a subset S_t. The models not in the S_t directly use θ^t as their parameters. The models in the S_t employ local data to update the global model parameter to get θ_k^{t+1}. Then the cloud server averages the parameters of all models to generate a new global parameter θ^{t+1}.

3 Experimental Evaluation

We developed an anomaly diagnostic system for the CR240E EMDT and conducted experiments in the Datong coal mine to collect actual operating data of EMDTs. Some of the data collected from the EMDT are shown in Fig. 4a. The collected data in Fig. 4a is a complete cycle with a duration of about 10 min, containing four phases: loading cargo, full-load uphill, unloading cargo, and no-load downhill. We collected data for 10 cycles. After normalizing and segmenting, we obtained a total of 12,300 samples. Due to the risk of damaging the equipment by creating faults artificially, we obtained the abnormal samples by modifying the normal sample data. We have designed three types of anomaly signals, as shown in Fig. 4b. For contextual anomaly [12]: First, referring to the rectifier switching device open-circuit fault, we add a sine wave to the DC voltage. Second, referring to the inverter switching device OC fault, we set the negative half of the motor stator U-phase current to 0. For collection anomaly: referring to the motor inter-turn short-circuit fault, we increase the amplitude of the stator U-phase current and decrease the amplitude of the V-phase.

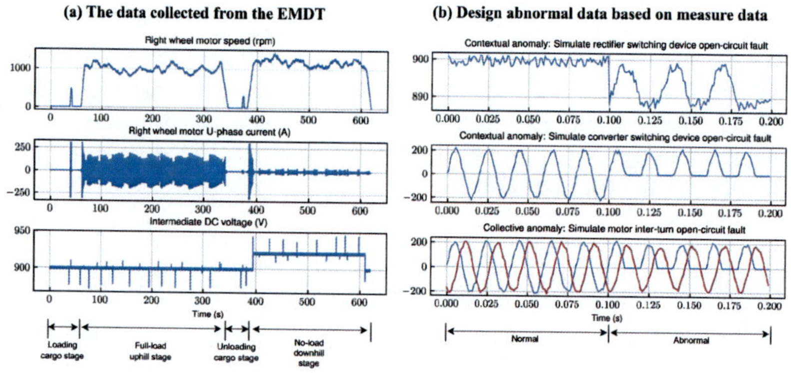

Fig. 4. The collected and simulated abnormal data from EMDT.

The parameters of the diagnostic model are shown in Table 1. Meanwhile, we chose LSTM and gate recurrent unit (GRU) as the encoder of the diagnostic model for comparison. The parameters of these two networks are the same as BLSTM, except that the input dimension of the decoders is changed to 512 to match the output of the encoders.

We first train the model on a single device using the collected data to evaluate the model's performance against anomaly diagnosis for EMDTs. In this paper, we use two metrics, the area under the receiver-operating-characteristic curve (AUC-ROC) and the

Table 1. BLSTM-based diagnostic model parameters.

Parameter	Value	Parameter	Value
Input dimension of encoder	7	Hidden dimension of encoder	256
Number of encoder layer	2	Dropout rate of encoder	0.8
Input dimension of decoder	1024	Hidden dimension of decoder	256
Output dimension of decoder	1	Dropout rate of decoder	0.8
Optimizer	Adam	Learning rate	0.001

area under the precision-recall curve (AUC-PR), to evaluate the performance of the diagnostic model. The X-axis of the ROC curve is the FPR = FP/(FP + TN) and the Y-axis is the TPR = TP/(TP + FN). The X-axis of the PR curve is the precision = TP/(TP + FP) and the Y-axis is the recall = TP/(TP + FN). Where TP, FP, FN, and TN denote true positive, false positive, false negative, and true negative respectively.

The curves of AUC-ROC and AUC-PR of the diagnostic model during the training process are shown in Fig. 5. The results show that the BLSTM-based diagnostic model was able to reach the accuracy of 94.5% in AUC-ROC and 94.3% in AUC-PR. It achieve approximately 1.7% and 2.3% lead in AUC-ROC compared to LSTM-and GRU-based ones, while the lead in AUC-PR was approximately 1.6% and 5.6%, respectively. And the BLSTM-based diagnostic model has a faster convergence speed compared with the other two models. The results show that the BLSTM-based diagnostic model has the diagnostic capability for the anomalies of EMDTs.

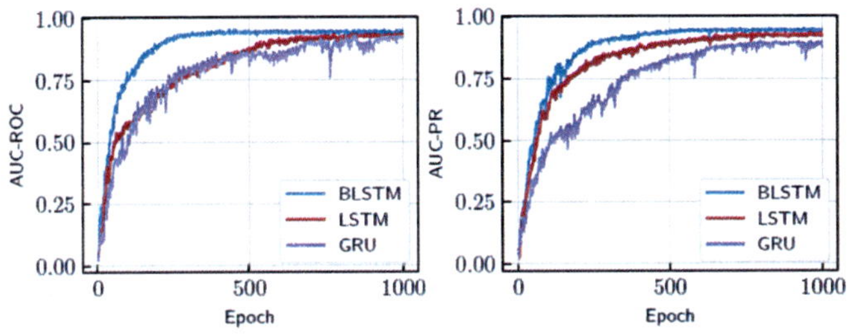

Fig. 5. Performance comparison of diagnostic models during single-device training.

The performance of the BLSTM, LSTM, and GRU-based networks with joint multi-device optimization is shown in Fig. 6. With an accuracy of 92.9% in AUC-ROC and 92.9% in AUC-PR, the BLSTM-based diagnostic model can still accomplish superior anomaly diagnosis results in the scenario of joint training of multiple devices. In the two metrics of AUC-ROC, the BLSTM-based model achieves a lead of 4.81% and 4.77% compared to the LSTM and GRU-based models, respectively. And the leads on AUC-PR were 3% and 5.1%. In addition, the BLSTM-based model also has a faster convergence speed. The results show that the multi-device collaborative anomaly diagnosis method has excellent diagnostic ability for EMDT anomalies.

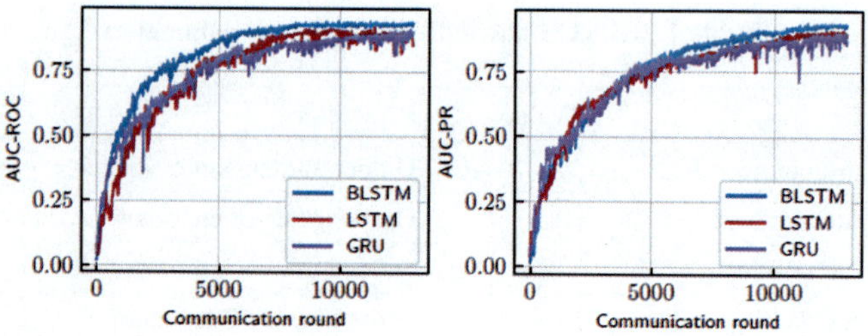

Fig. 6. Performance comparison during multi-device collaborative training.

4 Conclusion

A fault diagnosis method for EMDTs is proposed in this article. For a single EMDT, the fault diagnosis method first preprocesses the EMDT sensor signals. Then an encoder is constructed by BLSTM to autonomously extract fault features from the sensors' signal. For multiple EMDTs, the diagnostic method employs the FL framework, which enables numerous EMDTs to collaborate for the global optimization of the diagnostic model. The diagnostic method improves the model's scalability and optimizes the diagnosis capability of the single device's diagnosis model for unknown faults. This article verifies the proposed fault diagnosis strategy by accumulating actual operation data and simulating faults based on this data.

References

1. Zhiwei, G., Cecati, C., Ding, S.X.: A survey of fault diagnosis and fault-tolerant techniques—Part II: fault diagnosis with knowledge-based and hybrid/active approaches. IEEE Trans. Ind. Electron. **62**, 3768–3774 (2015)
2. Zhiwei, G., Cecati, C., Ding, S.X.: A survey of fault diagnosis and fault-tolerant techniques—Part II: fault diagnosis with model-based and signal-based approaches. IEEE Trans. Ind. Electron. **62**, 3757–3767 (2015)
3. Feng, J., Yao, Y., Lu, S., et al.: Domain knowledge-based deep-broad learning framework for fault diagnosis. IEEE Trans. Ind. Electron. **68**(4), 3454–3464 (2020)
4. Ren, L., Jia, Z., Wang, T., et al.: LM-CNN: a cloud-edge collaborative method for adaptive fault diagnosis with label sampling space enlarging. IEEE Trans. Ind. Inf. **18**(12), 9057–9067 (2022)
5. Ince, T., Kiranyaz, S., Eren, L., et al.: Real-time motor fault detection by 1-D convolutional neural networks. IEEE Trans. Ind. Electron. **63**(11), 7067–7075 (2016)
6. Sun, M., Wang, H., Liu, P., et al.: Stack autoencoder transfer learning algorithm for bearing fault diagnosis based on class separation and domain fusion. IEEE Trans. Ind. Electron. **69**(3), 3047–3058 (2021)
7. Ma, S., Nie, J., Kang, J., et al.: Privacy-preserving anomaly detection in cloud manufacturing via federated transformer. IEEE Trans. Ind. Inf. **18**(12), 8977–8987 (2022)
8. Gers, F.A., Schmidhuber, J., Cummins, F.: Learning to forget: continual prediction with LSTM. Neural Comput. **12**(10), 2451–2471 (2000)
9. Deng, C., Deng, Z., Lu, S., et al.: Fault diagnosis method for imbalanced data based on multi-signal fusion and improved deep convolution generative adversarial network. Sensors **23**(5), 2542 (2023)

10. Li, T., Sahu, A.K., Talwalkar, A., et al.: Federated learning: challenges, methods, and future directions. IEEE Signal Process. Mag. **37**(3), 50–60 (2020)
11. McMahan, B., Moore, E., Ramage, D., et al.: Communication-efficient learning of deep networks from decentralized data. In: Artificial Intelligence and Statistics, pp. 1273–1282. PMLR (2017)
12. Cook, A.A., Mısırlı, G., Fan, Z.: Anomaly detection for IoT time-series data: a survey. IEEE Internet Things J. **7**(7), 6481–6494 (2019)

A New DC Auxiliary Power Supply System Architecture Designed for Electric Multiple Units

Chao Wang[1], Yingtao Ma[2], Zheming Jin[1](✉), and Lijun Diao[1]

[1] Key Laboratory of Vehicular Multi-Energy Drive Systems (VMEDS), Ministry of Education, Beijing 100044, China
zhmjin@bjtu.edu.cn

[2] Locomotive and Car Research Institute, China Academy of Railway Sciences Corporation Ltd., Beijing 100094, China

Abstract. In recent years, the main form of power consumption of electric multiple units (EMU) on-board equipment is undergoing rapid transformation, with the proportion of DC and terminal DC loads gradually increasing (such as inverter air conditioners, speed-controlled fans), and there is a trend to replace the main position of traditional AC loads. At the same time, the third generation of semiconductor devices represented by silicon carbide devices is maturing, the penetration rate is increasing, and the auxiliary power supply for EMU represented by frequency isolation and medium frequency isolation "straight - AC" type auxiliary power supply system has been completely power electronic. In addition, other DC industrial fields are also developing vigorously, fully equipped with the technical basis for further DC conversion of the system as a whole. This paper proposes a new DC auxiliary power supply system architecture for EMU. By referring to the system architecture of the DC distribution network and combining it with the world application cases and related standards, this paper puts forward rationalization and suggestions on the system architecture and voltage level of the new DC auxiliary power supply system for EMU.

Keywords: Electric multiple units · Auxiliary power supply system · Architecture design · DC

1 Introduction

From the introduction of foreign technical standards of the "Harmony" series of EMU to the Chinese standard system dominant "Fuxing", China's high-speed railroads have achieved rapid development [1, 2]. At present, the auxiliary power supply system of EMU is an AC power supply system with an AC busbar, but with the continuous development of information technology and electric energy conversion technology, the main form of power consumption of on-board equipment is undergoing rapid transformation, and its prominent feature is that the proportion of DC and terminal DC loads is gradually increasing (such as inverter air conditioners and speed-controlled fans), and there is a trend to replace the main position of traditional AC loads.

L. Jia et al. (Eds.): EITRT 2023, LNEE 1135, pp. 356–363, 2024.
https://doi.org/10.1007/978-981-99-9307-9_39

The third generation of semiconductor devices represented by silicon carbide power semiconductors is gradually penetrating the power electronics industry. With the advancement of silicon carbide device technology and the upgrading of device grade, Japanese companies represented by Mitsubishi, Hitachi, and Toshiba [3] and European companies represented by Siemens and Alstom [4, 5] are actively developing traction converters and auxiliary converters for silicon carbide devices, and several Chinese companies are also actively following up. At the same time, the auxiliary power supply of the rolling stock represented by the frequency isolation and medium frequency isolation "direct-across" type auxiliary power supply system has realized complete power electronics [6], and it has the technical basis for further DCization through the system as a whole.

In addition, several industrial fields such as power systems [7], communication power [8], naval power [9], and aircraft power [10] are exploring next-generation DC power supply systems, and some breakthroughs have been achieved. Due to the inherent advantages of DC power supply systems, they are flourishing in various application scenarios, which all provide lateral support for onboard power supply systems in rail transportation.

In this context, this paper proposes a DC auxiliary power supply system, whose basic structure is that the auxiliary converter on the train converts the train power supply into stable medium-voltage DC power, and then connects to the DC bus of the whole train for parallel operation. Various inverter air conditioning units, battery chargers, inverter fans, and other loads take power directly from the DC bus. By referring to the system architecture of the DC distribution network and combining it with the world application cases and related standards, this paper puts forward rationalization and suggestions on the system architecture and voltage level of the new DC auxiliary power supply system for EMU.

2 DC Auxiliary Power Supply System Busbar Form Selection

The locomotive vehicle load can be distinguished as power load and control load, on this basis, there are naturally two levels of busbars in the DC auxiliary power supply system, which are medium voltage DC and low voltage DC, as the main change of DC auxiliary power supply system relative to the original AC auxiliary power supply system lies in the medium voltage DC busbar, low voltage DC busbar is not too much, the following two busbar forms are introduced for selection.

2.1 Main Power (Medium Voltage) Busbar Form Selection

Combining domestic and international application experience and relevant application cases, the following key factors should be considered for DC auxiliary power supply systems for rolling stock trains:

(1) Cable laying and load input requirements (unipolar or bipolar):
The bipolar busbar is mostly seen in high-voltage high-power applications. Its main advantage is that the load side can choose among three input forms (PN, PG, NG, P for positive busbar, N for negative busbar, G for neutral) based on demand, while the neutral of the system is generally grounded, making the system have better fault

protection and fault location capability [11]. At the same time, PG and NG busbar voltages can be different, which is an effective multi-voltage system design scheme, and can rely on three busbars to realize two power supply systems of medium voltage and low voltage.

However, the electrical equipment of rolling stock trains generally has the same voltage level, and better electrical isolation is required between the main power bus and the low-voltage DC bus. In this application context, the three access forms of bipolar busbar are not effectively used, but make the system structure complex, the number of cables increases, and the potential grounding problems increase. Therefore, the bipolar busbar is not suitable for the application requirements of the DC auxiliary power supply system of rolling stock, and the single-pole busbar should be selected.

(2) Grounding form (negative line grounding or floating ground):
As the grounding of the rolling stock system needs to rely on the special axle grounding device to achieve, which makes the system grounding resistance relatively large, for the roof high voltage system, the impact of this resistance is relatively small. However, the DC auxiliary power supply system has lower voltage and higher current, and the larger grounding resistance means a poorer grounding effect, and the traditional negative line grounding mode is less applicable in the rolling stock train system. In contrast, the floating ground system has better applicability in this application context. With effective insulation and electrical isolation, the system is more likely to achieve safe, reliable, and stable operation.

Based on the above analysis, the main power bus of the DC auxiliary power supply system should be selected in single-pole floating ground mode, which is shown in Fig. 1.

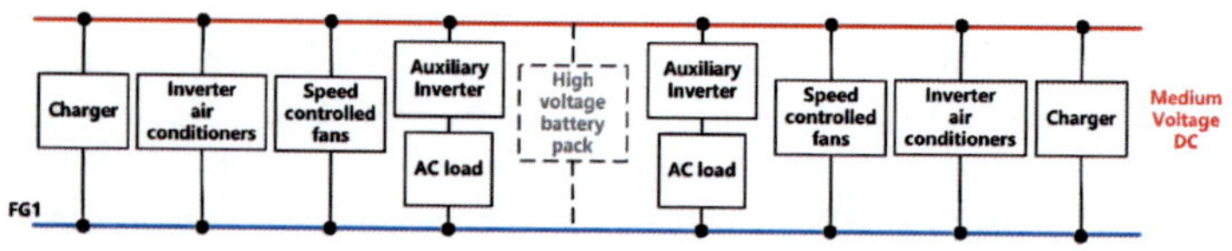

Fig. 1. Schematic diagram of the main power (medium voltage) busbar.

2.2 Lighting/Control (Low-Voltage) Busbar Form Selection

For the low-voltage busbar, its structure selection process can refer to the aforementioned medium-voltage busbar design method. At this time, the low-voltage busbar and medium-voltage busbar constitute a two-stage floating ground system, which can effectively reduce the mutual interference of the two busbars, and the schematic diagram of the form of the medium-voltage busbar and low-voltage busbar is shown in Fig. 2.

2.3 Selection of Segmentation/Expansion Methods for DC Auxiliary Power Supply Systems for Rolling Stock

To ensure the safe and stable operation of the rolling stock, the train power supply system needs to adopt a redundant design. The train is divided into certain units to supply power

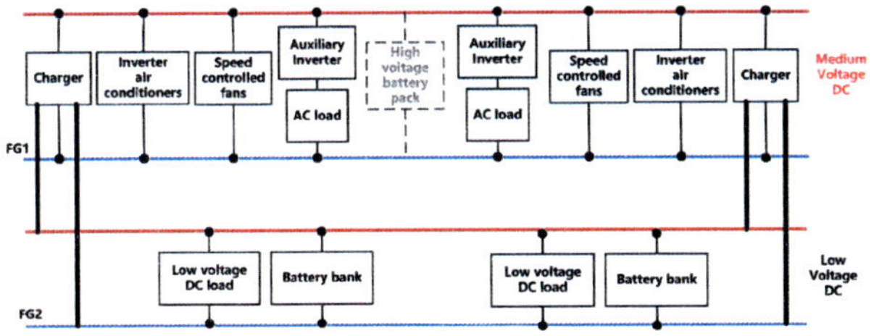

Fig. 2. Schematic diagram of main power (medium voltage) busbar and lighting/control (low voltage) busbar.

separately, and when one unit fails, the DC auxiliary converter of this unit or other units can be used to supply power to ensure the stable operation of the train. There are two options for the segmentation/expansion of this DC auxiliary power supply system: segmentation using busbar circuit breakers and segmentation using auxiliary converters.

Firstly, it is common to use linear segmentation by using circuit breakers on the busbar, and each DC auxiliary converter is connected in parallel to supply power to the medium voltage DC bus. In this case, the DC bus can be set up as a full column through or independent power supply by subunits, and each DC bus is connected by normally open or normally closed circuit breakers. The schematic diagram is shown in Fig. 3.

When the bus is set to be fully connected, all DC auxiliary converters are connected in parallel by normally closed circuit breakers, and when a converter fails, it is isolated from the DC bus and powered by other DC auxiliary converters.

When the busbar is set up for subunit power supply, each DC busbar is electrically isolated by circuit breakers and each converter module supplies power to its assigned load. However, in case of a fault, when a converter fails, this converter is isolated from the DC bus. Instead, the loads connected to it are switched to the DC bus of the adjacent converter output via circuit breaker conversion, respectively.

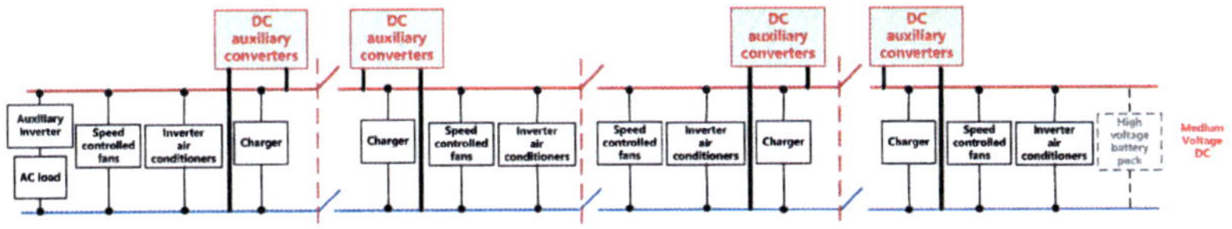

Fig. 3. Busbar circuit breaker segmentation schematic.

In general, the comprehensive cost of busbar circuit breaker segmentation is low, and the similarity with the existing system is high, but the ability of grounding protection is weak in this way, and the cost of DC circuit breaker is high and the maturity is low.

3 DC Auxiliary Power Supply System Bus Voltage Level Selection

3.1 Voltage Levels and Standards of Existing Low- and Medium-Voltage DC Busbars

The study of the DC distribution system is to provide a reference basis for the DC auxiliary power supply system of the rolling stock, and since the DC distribution network is much higher overall compared to the voltage of the rolling stock, we mainly draw on relevant technologies and standards in the field of medium- and low-voltage DC distribution networks.

According to the Chinese standard "GB/T 35727–2017 Guidelines for Medium and Low Voltage DC Distribution Voltage", 110V ~ 1500(±750) V is classified as the low-voltage range, and the nominal voltage of the low-voltage DC distribution system is suitable to be selected from Table 1.

Table 1. The nominal voltage of low voltage DC distribution system.

Preferred value /V	1500 (±750)	–	750 (±375)	–	–	–	–	–	220 (±110)	–
Alternative value /V	–	1000	–	600	440	400	336	240	–	110

3.2 Existing Power Supply Cables, Circuit Breakers, and Other Distribution Equipment Standards

(1) Circuit breaker

The so-called circuit breaker is a mechanical switching apparatus that can turn on, carry and break the current under normal circuit conditions, and also turn on, carry, and break the current for a certain period under the specified non-normal circuit (e.g. short circuit).

DC auxiliary power supply system as a new field of low-voltage circuit breaker application, for the moving train set of medium-voltage DC voltage and complex outdoor use environment, the low-voltage circuit breaker performance, insulation withstand voltage level, and breaking capacity have put forward new requirements. If the technical specifications of the selected low-voltage circuit breakers do not meet the design requirements of the DC auxiliary power supply system, it will affect the reliable operation of the rolling stock, and in serious cases, even lead to train paralysis.

Currently based on the current standard "IEC 60947-2:2019 Low-voltage switchgear and control gear Part 2: Circuit-breakers", in addition to the Chinese national standard "GB/T 14048.2-2020 Low-voltage switchgear and control gear Part 2: Circuit Breakers" is equivalent to or modifies the adopted "IEC 60947-2:2019", the provisions of which apply to DC auxiliary power supply systems for 690Vac-750dc busbar level, the corresponding level in the international well-known companies such as ABB, Sai Xuelong has corresponding batch circuit breaker products can be selected for use.

(2) On-board power distribution cables

The power cable selected for onboard power distribution is generally a metal wire with an insulating layer. Insulated wires and cable lines are usually composed of three parts: conductive core, insulation layer, and protective layer.

The power and lighting wiring in the vehicle starts from the vehicle switchboard, and the three-phase AC or DC power input from the main line under the vehicle is used by the load through the switchboard and the wiring respectively. The under-vehicle wiring includes the main line that carries three-phase AC and DC power, the power connector line, and the load branch line under the vehicle. Under the condition of meeting the cross-sectional area of the wire, the main line under the car is generally connected in parallel in two ways, and they are laid in the steel pipe to increase the strength.

In the Chinese standard "TJ/CL313–2013 Wire and Cable for Locomotives", it is stated that in the DC system, the nominal voltage of this system should be no more than 1.5 times the rated voltage of this cable.

3.3 Main Power (Medium Voltage) Bus Voltage Level Selection

Firstly, in the main power (medium voltage) bus voltage level, the upper and lower limits should be determined first.

About the upper limit: According to the voltage recommendations of the "GB/T 35727-2017 Voltage Guidelines for Medium and Low Voltage DC Distribution" for low-voltage DC distribution systems (Table 1), and the standard "IEC 60947-2:2019 Low-voltage switchgear and control gear Part 2: Circuit-breakers The 690Vac-750dc busbar rating for protection devices, so the medium voltage DC busbar should be no greater than 750 Vdc.

About the lower limit: the lower limit cannot be too low, because it involves the conversion of the former high-voltage 4000 Vdc, of course, cannot be too high, need to consider the conversion of the latter low-voltage DC, due to the calculation in the modulation of three-phase AC 380V, without adding a step-up conversion while the harmonic component is low, the current inverter modulation in the higher DC voltage utilization rate of up to 100%, due to the impact of positive and negative tolerances, according to $\pm$ 10% of the calculation (360Vac ~ 420Vac), as modulation is not possible to achieve the ideal situation, leaving a margin of 1.1 times, according to $420 * \sqrt{2} * 1.1$, at least 650 Vdc DC voltage is required.

Based on this analysis, 690 Vdc (740V–620V, +7.5% ~−10%) is recommended as the medium voltage DC bus voltage, at this time the use of DC circuit breakers rated at 750 Vdc can meet the demand, ABB, Sai Xuelong and other companies have corresponding batch products; about the cable using mature 600Vac cable can meet the standard, in the Chinese standard "TJ/ CL313-201 rolling stock wire and cable" in Table A.1 (Table 2) gives part of the first type of cable rated at 600Vac, at this time applied to the DC system, 690 Vdc does not exceed 600Vac 1.5 times, to meet the requirements.

Table 2. The first type of cable classification table.

Voltage level	Category
600V	NH-WL1 series single-core cable
	NH-C-WL1 Series Lightweight Cable
	NH-WEXO series multi-core shielded cable
	NH-WKK1 series high-temperature resistant cable
	NH-Z-SP0 series of jumper cables

3.4 Lighting/control (Low-Voltage) Busbar Voltage Level Selection

Regarding the lighting/control (low-voltage) bus voltage level of the on-board DC auxiliary power supply system, it is always 110 Vdc (77V–137.5V, +25% −30%) from the current architecture of the rolling stock, and it is already mature and complete, so no change is needed. The use of the original power distribution equipment (circuit breakers, cables, etc.) is sufficient. 110 Vdc is recommended as the lighting/control (low voltage) bus voltage level for DC-loaded auxiliary power systems.

4 Conclusion

In summary, the preliminary design of the main architecture of the new DC auxiliary power supply system for EMU is as follows (Fig. 4).

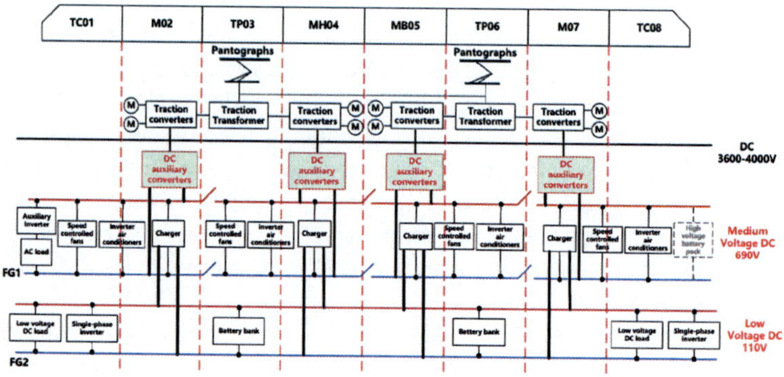

Fig. 4. Preliminary design of the main architecture of the new DC auxiliary power supply system for rolling stock.

The DC auxiliary converter takes power from the high-voltage DC link (3600V ~ 4000V) of the traction converter system and converts it to medium-voltage DC 690V through a high-power high-frequency isolated DC/DC with electrical isolation property (the rear output topology is Buck type), which can supply power to the load containing input rectifier, fan and pump and AC load using The power is supplied to equipment including input rectifier loads, fan and pump loads, and AC loads connected by auxiliary inverters.

Each DC auxiliary converter is connected in parallel to supply power to the medium voltage DC bus, which is connected by busbar circuit breakers. In case of a converter failure, this converter is isolated from the DC bus and powered by other auxiliary converters. Although the parallel operation is more cumbersome in control, the redundancy is better and the capacity design is low, and the high power high-frequency isolated DC/DC at this time can effectively reduce the quality and volume of the converter and utilize the lightweight operation of the rolling stock.

At the same time, medium voltage DC 690V is transformed into 110V low voltage DC through the direct–direct conversion of the charger to power the onboard low voltage DC load.

Acknowledgments. This research has been supported by the Key R&D Projects of China Railway Corporation (No. L2022Z001).

References

1. Yun, Y., Tianyu, L., Wang, H.: Comparative analysis of auxiliary power supply system of Fuxing and Harmony rolling stock. China Strategic Emerging Industries (Theory Edition), **24**, 0235 (2019) (in Chinese)
2. Qi, Y., Zhou, L.: The fuxing: The China standard EMU. Engineering **6**(3), 227–233. ISSN 2095-8099
3. Sato, K., Kato, H., Fukushima, T.: Development of SiC applied traction system for shinkansen high-speed train. In: 2018 International Power Electronics Conference. Niigata, Japan, pp. 3478–3483 (2018)
4. Kestler, T., Damec, V., Bakran, M.: Differences in dimensioning SiC MOSFETs and Si IGBTs for traction inverters. In: 2018 20th European Conference on Power Electronics and Applications (EPE'18 ECCE Europe). Riga, Latvia, pp. 1–9 (2018)
5. Alkayal, F., Saada, J.B.: Compact three-phase inverter in silin arbide technology for auxiliary converter used in railway applications. In: 2013 15th European Conference on Power Electronics and Applications (EPE). Lille, France, pp. 1–10 (2013)
6. Jiahui, R.: Research on Medium-Frequency Isolated DC-DC Converters with All-Silicon Carbide Devices. Beijing Jiaotong University (2022) (in Chinese)
7. Peng, L.I., Chengyi, C.A.O., Fangying, C.A.O., et al.: A review on the application technology of customer-side DC power distribution. Build. Electricity **40**(10), 43–48 (2021)
8. Shineng, L., Wenbo, S., Fuping, H.: Telecom 240 V DC power supply scheme study and development strategy. Guangdong Commun. Technol. **33**(11), 2–5 (2013) (in Chinese)
9. Zhao, Y., Jiang, N., Zhu, L., et al.: Numerical study on the arc-firing pressure rise characteristics of the cable chamber of a ship's medium-voltage DC switchboard. Ship Electricity Technol. **42**(11), 6–10 (2022) (in Chinese)
10. Brombach, J., Jordan, M., Grumm, F., et al.: Converter topology analysis for aircraft application. In: International Symposium on Power Electronics, Electrical Drives, Automation and Motion (SPEEDAM). Sorrento, Italy, pp. 446–451 (2012)
11. Cuzner, R.M., Esmaili, D.A.: Fault-tolerant shipboard MVDC architectures. In: 2015 International Conference on Electrical Systems for Aircraft, Railway, Ship Propulsion and Road Vehicles (ESARS), Aachen, Germany, pp. 1–6 (2015)

Topology Design of High-Frequency DC Auxiliary Converter for Electric Multiple Units

Kan Dong[1,2], Yingtao Ma[1,2], Long Cheng[1,2](✉), Chao Wang[3], Zheming Jin[3], and Lijun Diao[3]

[1] State Key Laboratory for Traction and Control System of EMU and Locomotive, China Academy of Railway Sciences Corporation Limited, Beijing 100081, China
707282247@qq.com

[2] Locomotive and Car Research Institute, China Academy of Railway Sciences Corporation Limited, Beijing 100081, China

[3] Key Laboratory of Vehicular Multi-Energy Drive Systems (VMEDS), Ministry of Education, Beijing 100044, China

Abstract. With the rapid development of power electronic technology and the commercialization of high-voltage and high-power semiconductor materials, rail transit field has experienced significant advancements. Auxiliary power supply system plays a crucial role in rail vehicles due to the prospect of high-frequency, lightweight and miniaturization. In recent years, DC power system has gained considerable attention and development due to its inherent advantages such as high efficiency, the absence of reactive power and harmonic issues. With the development of DC power distribution and DC electrical equipment, it is expected that DC power will become increasingly prevalent in rail transit field. This paper proposes a forward-looking topology for a DC auxiliary power supply system for rolling stock and conducts research on topology design which focus on aspects of lightweight design, miniaturization, and reliability.

Keywords: Rail transit · Auxiliary converters · Architecture design · DC system

1 Introduction

The new demand for rail transportation needs on-board devices to be more efficient and lightweight, with the gradual commercialization of high-voltage and high-power silicon carbide (SiC) devices, it is anticipated that the power density and efficiency of auxiliary converters will experience further enhancements [1].

Meanwhile, the development of inverter technology has brought onboard equipment of rail vehicles significant changes. One notable characteristic is the increasing proportion of variable frequency air conditioners and speed-controlled fans, which plays a role of the vehicle's auxiliary power supply system [2]. During the last 20 years,

State Key Laboratory for Traction and Control System of EMU and Locomotive (Contract No. 2022YJ268), China Academy of Railway Sciences Corporation Limited.

L. Jia et al. (Eds.): EITRT 2023, LNEE 1135, pp. 364–370, 2024.
https://doi.org/10.1007/978-981-99-9307-9_40

the continuous progress of power electronic technology, particularly the emergence of new devices and topologies, has improved DC systems' performance and offer the following advantages: high efficiency, high current-carrying capacity, etc. As a conseque, DC transmission and distribution technology has gained significant attention and witnessed substantial development in the last decade [3, 4]. In [5], an integrated DC system designed for both traction and auxiliary power system is proposed, and a set of DC-oriented solution to improve rail vehicles' performance is presented.

In this paper, a DC auxiliary power system is proposed, and focusing on lightweight design, miniaturization and reliability, the topology design for DC converters is carried out.

2 DC Auxiliary Power Supply System

To improve the performance of auxiliary power supply system, this paper presents a future-oriented DC power system for rolling stock, as illustrated in Fig. 1. The key equipment is on-board DC auxiliary converter, which converts high DC voltage to medium level voltage. This converted power is subsequently connected to the medium level DC bus that spans the entire train, enabling parallel operation of multiple components.

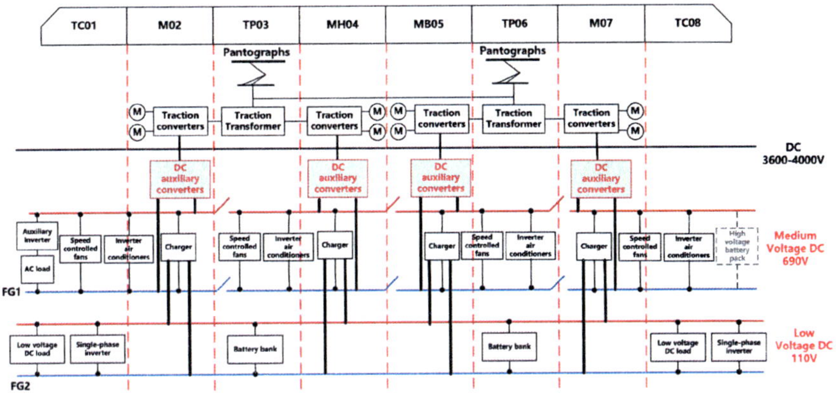

Fig. 1. DC auxiliary power supply system.

3 DC Converter Topology Design

The design of DC auxiliary converter for rolling stock should consider several key structural features based on the specific application scenario: wide range input voltage, high reliability, high voltage ratio, lightweight and high frequency. To leverage the advantages of certain topologies and ensure high-efficiency operation while meeting various operational requirements, multi-stage DC-DC converters have gained increasing attention [6]. Based on the multi-stage concept, DC auxiliary converter is designed as a three-stage structure, including pre-stage chopper regulator circuit, intermediate-stage high-frequency isolation and rear-stage output protection circuit, as shown in Fig. 2.

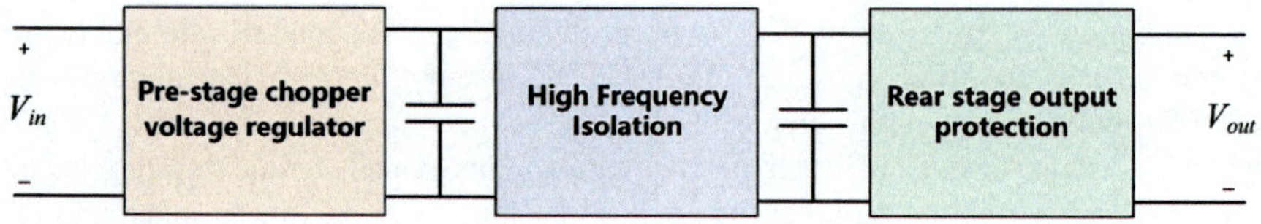

Fig. 2. DC auxiliary converter schematic.

3.1 Chopper Regulator

In pre-stage chopper regulator circuit, conventional DC/DC converters are typically employed. The main objective of pre-stage is to ensure the accurate voltage conversion, low ripple and good dynamic performance required for the normal operation of the converter [7]. In this subsection, two circuits are analyzed: a three-level buck circuit and a three-level boost circuit. The schematic diagrams for these circuits are depicted in Fig. 3.

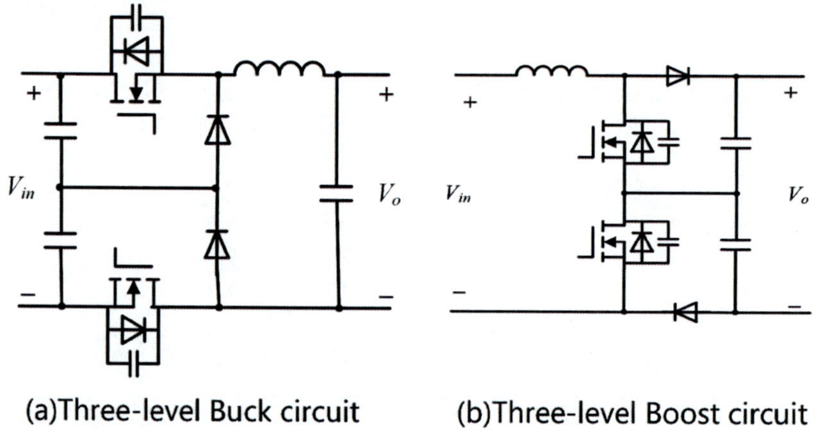

Fig. 3. Chopper regulator circuit.

Three-level buck circuit offers several advantages. It has a simple structure, requires small number of power devices based on a mature and reliable circuit design. Additionally, buck-type circuit inherently provides voltage reduction, which is easy for device selection. However, three-level buck circuit also has some limitations. It requires a capacitive voltage divider at the input, as well as an independent capacitor at the output, which means it has slightly higher losses compared to other topologies.

Three-level boost employs interleaved control, which can effectively reduce current ripple, enabling a reduction in inductance, which is beneficial for lightweight design. Moreover, three-level boost circuit also enables the realization of a split capacitor branch for voltage balance control. This is suitable for high-frequency bridge, leading to cost and power loss reduction.

From the above analysis, three-level boost circuit emerges as a better choice due to its efficiency improvements and potential for enhancing the performance of post-stage circuit in DC auxiliary converter.

3.2 High-Frequency Isolation Circuit

In order to improve the operating frequency of DC/DC converters and improve system efficiency, it becomes essential to employ the topologies that can achieve soft switching. Among such topologies, LLC-type resonant converters offer significant advantages due to their superior performance and simple structure [8]. In this subsection, three types of circuits are analyzed: half-bridge LLC circuit, full-bridge LLC circuit, and differential three-level half-bridge LLC circuit, as shown in Fig. 4.

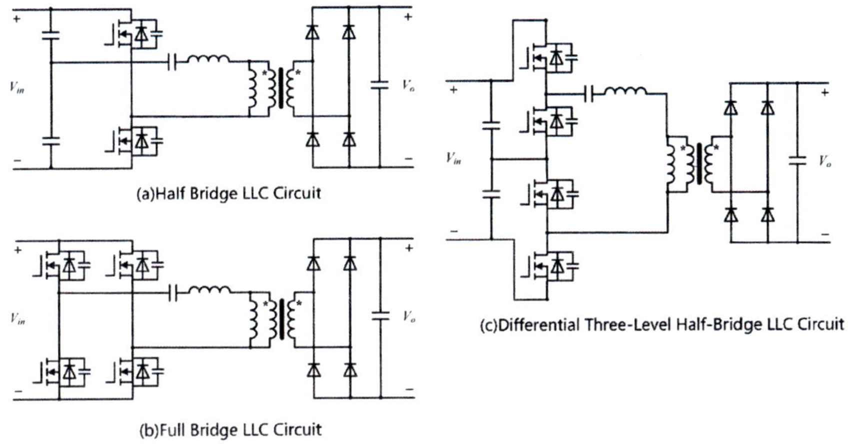

Fig. 4. High-frequency isolation circuit.

The number of power devices in half-bridge converter is only half that of the full-bridge circuit. As a result, the corresponding switching losses are also reduced by half, which contributes to improve overall efficiency. Furthermore, the interaction between the capacitor bridge arm and high-frequency transformer in half-bridge circuit is beneficial for the design of high-frequency resonant parameters. Additionally, the use of multiple parallel structures in half-bridge LLC circuit allows for smaller capacity requirements for transformer. However, it's important to note that in half-bridge LLC circuit, the input of the half-bridge capacitor needs to be controlled and regulated by the front stage.

Full-bridge inverter provides a wide range of output voltage, which is twice that of the half-bridge structure. Additionally, the input voltage for full-bridge circuit can be lower, providing further flexibility in power supply options. However, it is important to consider that full-bridge LLC circuit requires more power devices. The increased devices can be disadvantageous in terms of system size and weight reduction.

Differential three-level half-bridge LLC circuit presents unique characteristics that differentiate it from the full-bridge and half-bridge topologies [9]. Differential three-level half-bridge LLC circuit offers a higher voltage withstand capability than both the full-bridge and half-bridge topologies. Similar to the other LLC circuit topologies, the interaction between the capacitor bridge and the high-frequency transformer contributes to favorable high-frequency parameter design.

As a comprehensive evaluation, the differential three-level half-bridge LLC circuit offers a more favorable solution. It combines the benefits of a three-level operation, improved voltage withstand capability and redundancy, making it a better choice for DC auxiliary power supply system.

3.3 Output Circuit

To mitigate the potential issues caused by short circuits in the rear stage, it becomes necessary to utilize a buck-type topology at the output side of DC converter. Half-bridge buck converter features a reduced number of switching devices, and has simple structure and high reliability, can be a solution for output circuit, as shown in Fig. 5.

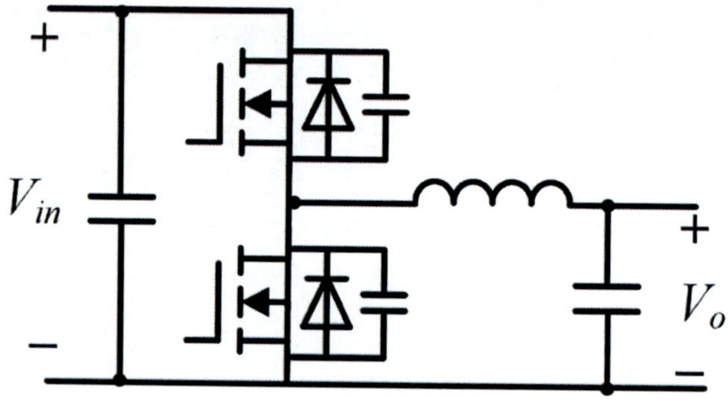

Fig. 5. Output circuit.

3.4 Number of Cascades

As previously mentioned, high voltage and high power create a challenge in terms of the voltage withstand capability of the devices and the desired switching frequency [10]. One effective approach is to utilize an ISOP (Input Series Output Parallel) structure, which helps overcome the voltage and frequency limitations of individual devices. The scheme of two-stage cascades and four-stage cascades is shown in Fig. 6, which is more competitive depends on voltage and power level.

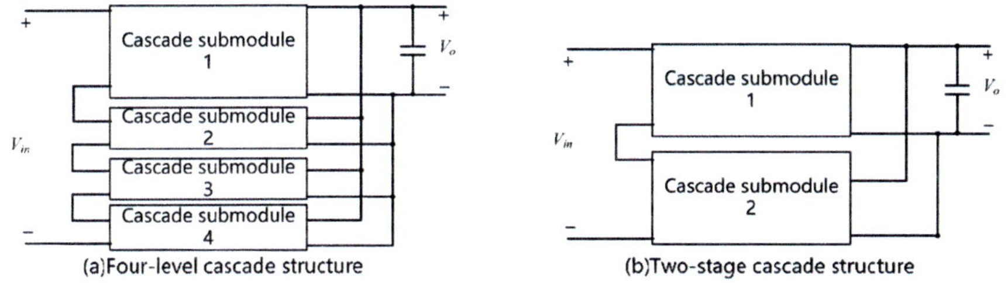

Fig. 6. Number of cascades selection.

4 Conclusion

Based on the above analysis, we propose a structure for DC auxiliary converter, which consists of a two-stage ISOP system, as shown in Fig. 7. Power devices of this topology can be SiC MOSFETs with a voltage rating of 1.2 kV or 1.7 kV, enabling an improvement in operating frequency while maintaining a high level of efficiency. Moreover, it presents the design of integration and symmetry, promotes structural lightweight design, and can be used as an implementation solution for DC system.

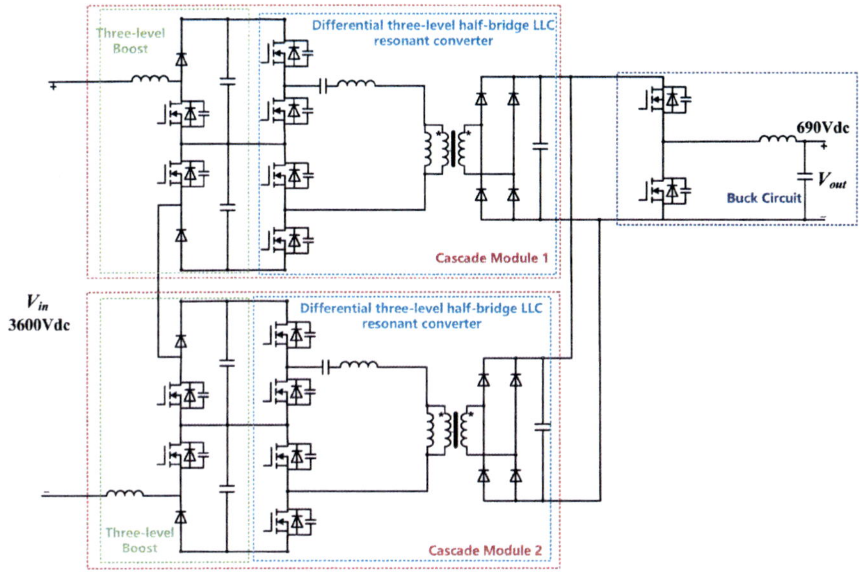

Fig. 7. DC auxiliary converter topology for rolling stock.

Acknowledgments. This work is supported by a grant from the State Key Laboratory for Traction and Control System of EMU and Locomotive (2022YJ268).

References

1. Ocklenburg, M., Döhmen, M., et al.: Next generation DC-DC converters for auxiliary power supplies with SiC MOSFETs. In: 2018 IEEE International Conference on Electrical Systems for Aircraft, Railway, Ship Propulsion and Road Vehicles & International Transportation Electrification Conference (ESARS-ITEC). Nottingham, UK, pp. 1–6 (2018)
2. Yangyu, Y.: Research on High-Performance Passenger Train Auxiliary Power Supply System. China University of Mining and Technology (2020) (in Chinese)
3. Fan, S.Y., Wang, C., Xiang, X., et al.: A review of medium and high voltage DC transformer topologies. J. Northeastern Electric Power Univ. **42**(5), 15–24 (2022). (in Chinese)
4. Jin, Z., Sulligoi, G., Cuzner, R., Meng, L., Vasquez, J.C., Guerrero, J.M.: Next-generation shipboard DC power system: introduction smart grid and dc microgrid technologies into maritime electrical networks. IEEE Electrific. Mag. **4**(2), 45–57 (2016)
5. Yingtao, M.: An integrated DC system for traction and auxiliary power supply of locomotives. J. Power Supply **19**(4), 112–117 (2021) (in Chinese)
6. Liu, F., Ruan, X., Huang, X., et al.: Second harmonic current reduction for two-stage inverter With DCX-LLC resonant converter in front-end DC–DC converter: modeling and control. IEEE Trans. Power Electron. **36**(4), 4597–4609 (2021)
7. Lee, M., Lai, J.-S.: Spread-spectrum frequency modulation with adaptive three-level current scheme to improve EMI and efficiency of three-level boost DCM PFC. IEEE Trans. Power Electron. **36**(3), 2476–2480 (2021)
8. Xu, J., Ma, S.: A review of LLC resonant converter topology. J. Xi'an College Arts Sci. (Natural Science Edition) **25**(1), 10–18 (2022). (in Chinese)
9. Lu, C., Hu, W., Lee, F.C.: Neutral-point voltage balancing methods of series-half-bridge LLC converter for solid state transformer. IEEE Trans. Power Electron. **36**(6), 7060–7073 (2021)
10. Sun, D.N.: Study on the main circuit of high-frequency auxiliary variable current system for locomotives and electric locomotives. New Technol. Electric Power **38**(3), 74–81 (2019). (in Chinese)

Residual Energy Estimation of Battery Packs for Energy Storage Based on Working Condition Prediction and the Representative Cell

Yong Chen[1], Tao Wen[1], Lintao Hou[2](✉), Chunhua Hou[1], and Xizheng He[2]

[1] ZTE Corporation, Shenzhen 518057, China
{chen.yong2,wen.tao1,hou.chunhua}@zte.com.cn
[2] School of Electrical Engineering, Beijing Jiaotong University, Beijing 100044, China
20126138@bjtu.edu.cn

Abstract. Residual energy is a direct description of the energy supply capacity of batteries and its accurate estimation is a key issue in current research. However, the residual energy of batteries is significantly influenced by the future operating conditions, and for an energy storage system consisting of several cells, the inconsistency of the cells will lead to differences in the residual energy among them, and it is difficult to estimate the residual energy of all cells with limited hardware resources. Therefore, this paper proposes a method for estimating the residual energy of battery packs in energy storage based on the prediction of operating conditions and the representative cell. Firstly, the definition of residual energy and its influencing factors are analyzed; secondly, the residual energy of batteries is estimated based on the prediction of future operating conditions using Markov chains; finally, a simulation model of battery packs is established to take into account the inconsistency, and the residual energy of battery packs is accurately estimated based on the representative cell with a relative error of less than 0.6%, which is of some guidance for the residual energy assessment of energy storage battery packs in practical applications.

Keywords: Lithium-ion battery · Energy storage battery packs · Markov chain · Residual energy

1 Introduction

A fast and accurate estimation of the battery state is the primary task of battery management in electrochemical energy storage systems [1]. However, the battery energy rating varies with battery ageing and a complex estimation of the maximum remaining available energy of the battery is required to obtain an accurate energy state.

Therefore, for energy storage systems, it would be more intuitive and convenient to estimate the remaining energy of the battery directly. However, in the current research on battery residual energy, it still corresponds directly to the state of charge, and the battery residual energy is only estimated based on current and past battery information, lacking consideration of future battery operating conditions affecting the battery residual

L. Jia et al. (Eds.): EITRT 2023, LNEE 1135, pp. 371–378, 2024.
https://doi.org/10.1007/978-981-99-9307-9_41

energy, which leads to inaccurate estimation of the pool residual energy, resulting in large calculation errors [2–5]. In addition, due to the limited voltage and energy of a single cell, the energy storage system is necessarily composed of several single cells connected in series and parallel, and the inconsistency of the single cell parameters will inevitably lead to the differentiation of the single cell state, and it is difficult to achieve the estimation of the residual energy of all single cells with limited hardware resources [6, 7]. In summary, there is an urgent need to explore a method for estimating the residual energy of a battery pack that is both accurate and feasible, taking into account the future operating conditions of the battery.

In order to solve the above problems, this paper takes into account the influence of the battery operating conditions on its residual energy and uses Markov chains to predict the future operating conditions of the battery, and on this basis, the residual energy of the battery is estimated. Finally, the relationship between the inconsistency of the cells in the battery pack and the residual energy of the battery pack is analyzed. From both theoretical and practical aspects, the cells with average voltage in the battery pack are selected as representative cells and their residual energy is estimated as the residual energy of the battery pack at the current moment.

The rest of the paper is arranged as follows: In Chap. 2, the definition of residual battery energy will be briefly introduced; in Chap. 3, the Markov chain prediction method is used to predict the future battery current of the energy storage system, and the residual battery energy is estimated on the basis of the working condition prediction; in Chap. 4, the single cell with the voltage at the mean value within the battery pack is selected as a representative single cell for the estimation of the residual battery energy; in Chap. 5, the conclusions of this paper will be explained.

2 Definition of Residual Energy and the Factors Affecting It

For energy storage systems, the residual energy of the battery is the cumulative energy charged or discharged from the current moment until the battery reaches the charge/discharge cut-off voltage when the energy storage battery is charged or discharged at a certain operating condition. This section discusses the remaining discharge energy of the battery. The remaining discharge energy of the battery can be expressed as the area enclosed by the cumulative discharge capacity and terminal voltage of the battery and the coordinate axis, as shown in Fig. 1.

The current moment is t, the corresponding battery terminal voltage is $U(t)$, the battery charge state and the accumulated discharge capacity are $SOC(t)$ and $Q(t)$ respectively; the moment corresponding to the discharge cut-off voltage is recorded as t_{lim}, at which time SOC is 0 and the accumulated discharge capacity is $Q(t_{lim})$; then the remaining discharge capacity of the battery at the current moment is shown in the shaded part of Fig. 1, and the calculation process is as shown in Eq. (1) is shown. It can be seen from Eq. (1) that the integration process of calculating the remaining energy of the battery involves the SOC of the battery and its remaining capacity at the future moment, so when estimating the remaining discharge energy of the battery at the current moment,

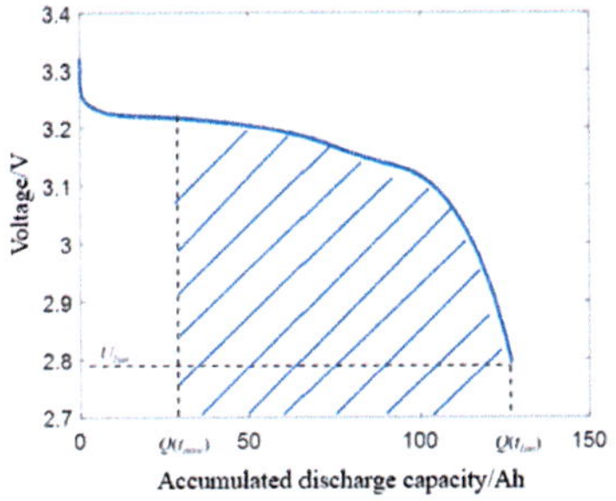

Fig. 1. Schematic diagram of residual discharge energy of battery

the current state of the battery and its future state must be combined [8].

$$E(t_{now}) = \int_{t_{now}}^{t_{\lim}} U(t)I(t)dt = \int_{SOC(t_{now})}^{SOC(t_{\lim})} U(t)Q_{\max}dSOC = \int_{Q(t_{now})}^{Q(t_{\lim})} U(t)dQ \quad (1)$$

Figure 2 shows a comparison of the discharge curves of a lithium iron phosphate monomer discharged at two different multipliers for constant current, where blue is the large multiplier discharge curve and orange is the small multiplier discharge curve. It is clear from the graphs that a significant drop in battery voltage occurs at the initial stage of high current discharge, mainly due to the large voltage drop across the ohmic and polarized internal resistance of the battery, and the fact that the battery reaches the cut-off voltage earlier at high discharge rates, which together leads to a rapid decay in the remaining dischargeable energy of the battery. In summary, there are significant differences in the remaining energy of the battery when it is charged and discharged at different rates, which means that the operating conditions of the battery have a significant impact on the remaining energy of the battery.

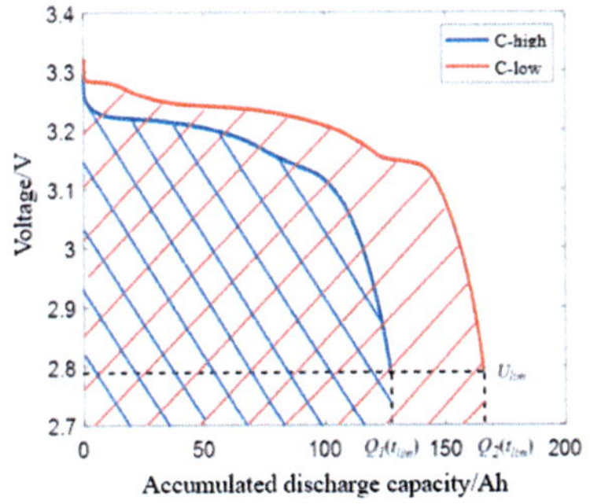

Fig. 2. Characterization diagram of battery residual energy under different discharge rates

3 Estimation of Remaining Battery Energy Based on Working Condition Prediction

3.1 Markov Chain-Based Prediction of Battery Operating Conditions

Markov chain is a stochastic process of transition from one state to another in the state space, in which the probability distribution of the next state is only determined by the current state, independent of the historical state [9, 10]. Similarly, in the energy storage system, the battery operating conditions at each moment are independent of each other, and the battery operating conditions at different moments do not affect each other. Therefore, this paper predicts the operating conditions of the energy storage battery based on Markov chain, as shown in Fig. 3. Firstly, the current battery history is clustered, and the predicted battery condition at the next moment is obtained by calculating the cluster transfer probability matrix between neighboring samples, and the predicted value will form a new sample space with the historical conditions at that moment.

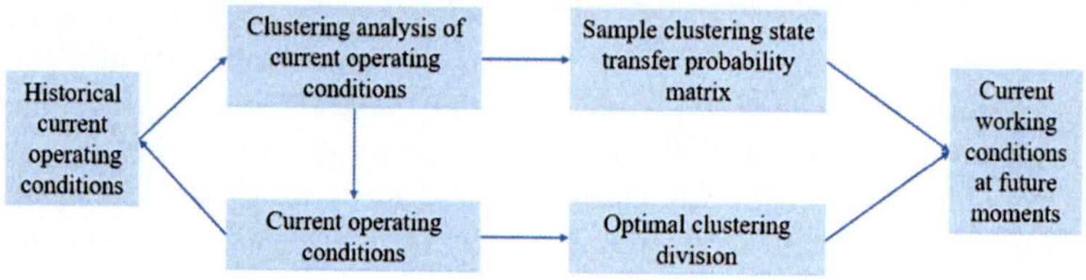

Fig. 3. Flow chart of current condition prediction based on Markov chain

3.2 Residual Energy Estimation of Battery

On the premise of accurate prediction of the working condition based on Markov chain, the residual energy of the energy storage battery is estimated, and the estimation process is shown in Fig. 4. The core idea is to make coupled predictions of the battery state, battery model parameters and terminal voltage output for the future charging and discharging process based on the established battery model based on the estimation of the current state of the battery and the prediction of the working condition current, thus realizing the real-time estimation of the remaining energy of the battery.

Figure 5 shows a comparison of the estimated residual energy of the energy storage batteries for a particular day in spring using both the condition-based prediction and the extended Kalman filter-based methods. As can be seen from the figure, the method of estimating the remaining energy of the battery based on the working condition prediction always has a high accuracy, with a maximum error of no more than 2%, which provides a good basis for the following estimation of the remaining energy of the battery pack of the energy storage system.

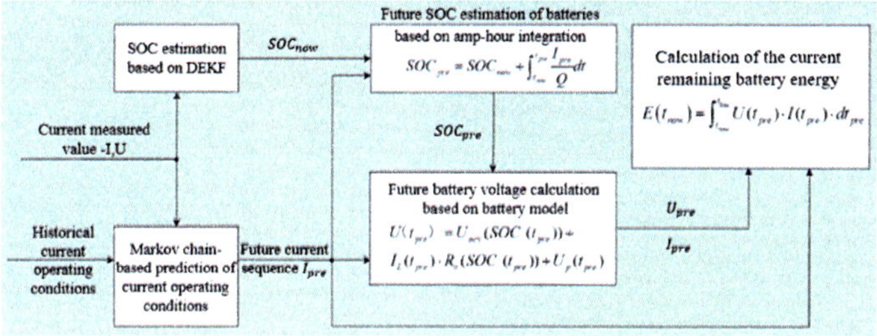

Fig. 4. Flow chart of battery residual energy estimation

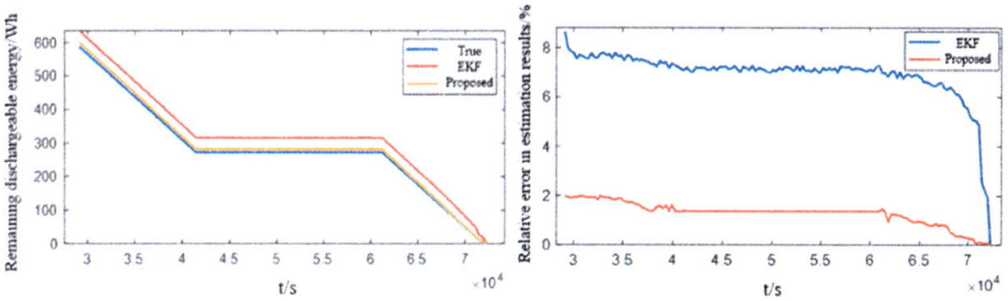

Fig. 5. Estimation results and errors of battery residual energy on a certain day in spring

4 Estimation of Residual Energy in Energy Storage Battery Packs Based on the Representative Cell

Energy storage systems consist of a number of single cells connected in series and parallel. Inconsistencies in the parameters of the cells lead to differences between the states of the cells, yet the method of estimating the residual energy of each cell is poorly feasible on limited hardware resources. Therefore, in this paper, the residual energy of a battery pack is estimated based on representative cells within the battery pack by analyzing the inconsistency of the cells within the pack.

The inconsistency of the cells within the battery pack is mainly manifested in the inconsistency of capacity, internal resistance and SOC between cells. Based on the consideration of the inconsistency differences of the cells, a battery pack simulation model was built in Matlab/Simulink to simulate the series-connected battery pack in the energy storage system as shown in Fig. 6. The SOC calculation module, the model parameter identification module and the external cell voltage calculation module are included in each cell module.

Assuming that the battery pack is fully charged at the initial moment, and that there are differences in capacity, internal resistance and SOC of individual cells in the battery pack and they approximately obey the Weber distribution, the battery pack is discharged at a constant current close to the actual energy storage operating condition of 0.15 C, and when a single cell in the battery pack reaches the cut-off voltage or SOC becomes 0 first, the battery pack is considered to be discharged to investigate the capacity, internal resistance and This is used to investigate the trend of single cell capacity, internal resistance and SOC inconsistency and the remaining dischargeable energy. From

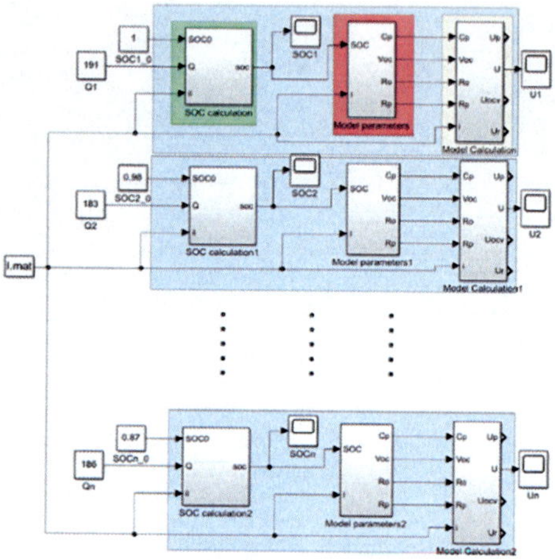

Fig. 6. Battery pack simulation model

the simulation analysis, it can be seen that the trend between the remaining available energy and the remaining available capacity of all the cells in the battery pack is shown in Fig. 7, which shows that the remaining available energy and the remaining available capacity of a single cell maintain a high consistency.

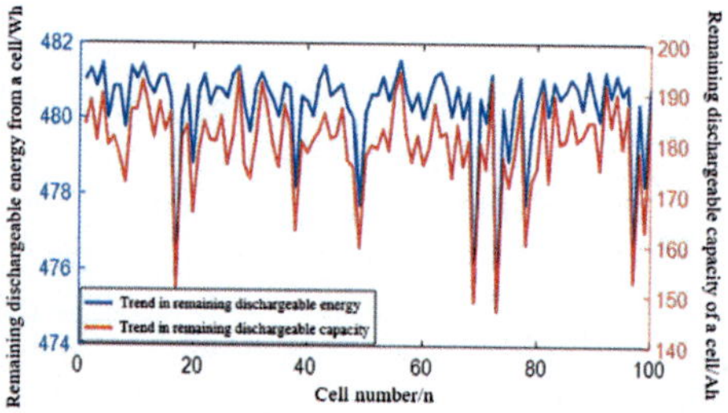

Fig. 7. Variation trend of remaining dischargeable energy and remaining dischargeable capacity

However, it is difficult to accurately estimate the remaining available capacity of each individual cell in real time because of the high demands on algorithms and hardware resources, so it is necessary to explore a method that is both simple and guarantees a certain level of accuracy in estimating the remaining energy of a battery pack. During operation, voltage, current and temperature are directly accessible battery parameters, with battery voltage as an output quantity coupled with a variety of factors affecting the battery intrinsically and externally, so it is most feasible to select a battery pack's characterization unit through voltage. One hundred and sixty strings of battery data from an energy storage system are used as the basis for exploring the relationship between cell voltage and residual energy, as shown in Fig. 8, which shows the relationship between the remaining available energy and voltage for each cell at different times.

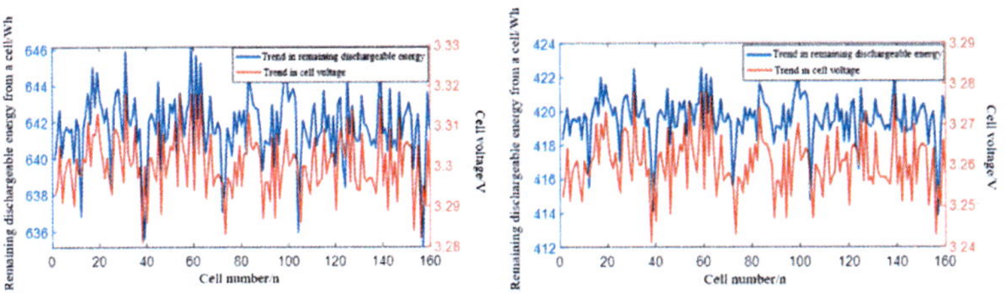

Fig. 8. Variation trend of residual energy and voltage of cells

The energy storage system was selected for 10 cycles, and the single cell with the average voltage was selected at each time to estimate the remaining energy according to the watt-hour integration method. The relative error is shown in Fig. 9. This illustrates the feasibility of selecting a single cell with a voltage at the mean value as the characterizing single cell, and lays a good foundation for estimating the remaining available energy of the battery pack of an energy storage system in practical applications.

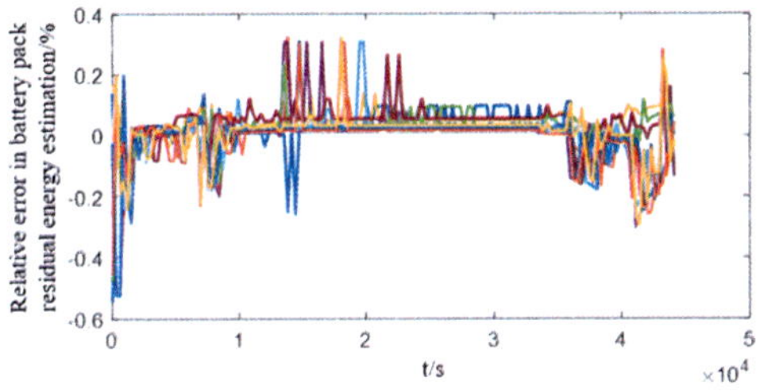

Fig. 9. Residual energy estimation error of battery pack based on the characterized cell

5 Conclusion

This paper presents a method for estimating the residual energy of an energy storage battery pack based on the prediction of working conditions and representative single cells. Firstly, on the basis of the definition of battery residual energy, the necessity of adopting the method of battery residual energy estimation based on work condition prediction is analyzed; secondly, the prediction method of Markov chain and the process of battery residual energy estimation based on work condition prediction are introduced; then, a battery pack simulation model is established on Matlab/Simulink according to the inconsistency of the battery, and the relationship between the inconsistency of the battery singles and the residual energy is explored; finally. And the relationship between the residual energy; finally. The method of selecting representative cells based on the single cell voltage is proposed from practical applications, and the estimation error is less than 0.6%, which verifies the accuracy and feasibility of estimating the remaining energy of the battery pack based on this characterized single cell, and lays a good foundation for estimating the remaining available energy of the battery pack of energy storage system in practical applications.

Acknowledgments. This research was partially funded by the ZTE Industry-University-Institute Cooperation Funds.

References

1. Zheng, L., Zhu, J., Wang, G., et al.: Novel methods for estimating lithium-ion battery state of energy and maximum available energy. Appl. Energy **178**, 761–791 (2016)
2. Ren, D., Languang, L., Shen, P., et al.: Battery remaining discharge energy estimation based on prediction of future operating conditions. J. Energy Storage **25**, 100836 (2019)
3. Seongyun, P., Jeongho, A., Taewoo, K., et al.: Review of state-of-the-art battery state estimation technologies for battery management systems of stationary energy storage systems. J. Power Electron. **20**(6), 1526–1540 (2020)
4. Liu, G., Ouyang, M., Languang, L., et al.: A highly accurate predictive-adaptive method for lithium-ion battery remaining discharge energy prediction in electric vehicle applications. Appl. Energy **149**, 297–314 (2015)
5. Zheng, L., Zhu, J., Wang, G., et al.: Novel methods for estimating lithium-ion battery state of energy and maximum available energy. Appl. Energy **178**, 1–8 (2016)
6. Lai, X., Huang, Y., Han, X., et al.: A novel method for state of energy estimation of lithium-ion batteries using particle filter and extended Kalman filter. J. Energy Storage **43**, 103269 (2021)
7. Li, X., Jianhua, X., Hong, J., et al.: State of energy estimation for a series-connected lithium-ion battery pack based on an adaptive weighted strategy. Energy **214**, 118858 (2021)
8. Zhang, S., Zhang, X.: Joint estimation method for maximum available energy and state-of-energy of lithium-ion battery under various temperatures. J. Power. Sources **506**, 230132 (2021)
9. Qiao Zhang, X., Chen, S.L.: Energy management control strategy for hybrid energy storage systems in electric vehicles. Int. J. Electrochem. Sci. Electrochem. Sci. **17**(1), 220121 (2022)
10. Zhu, M., Quan, O., Wang, Y., et al.: Remaining useful life prediction of lithium-ion batteries: a hybrid approach of Grey-Markov chain model and improved gaussian process. IEEE J. Emerg. Sel. Top. Power Electron. **11**(1), 143–153 (2023)

Research on the Design Scheme of Routine Conductive High-Speed Maglev Integrated Grounding System

Guosheng Li(✉), Di Su, and Ruichao Chen

China Railway Design Corporation, Tianjin 300308, China
{liguosheng01,sudi,chenruichao}@crdc.com

Abstract. By analyzing and researching the main components and grounding requirements of routine conductive high-speed maglev transportation facilities, using natural grounding bodies of structural facilities such as line stations, bridges, tunnels as the main grounding network, propose specific grounding design plans for line track facilities, propulsion and power supply facilities, operation control facilities, and maglev vehicles, and then design a unified, high and low voltage compatible, strong and weak current integrated grounding scheme. It can meet the requirements for working grounding, safety grounding, lightning protection grounding, and shielding grounding of various facilities of routine conductive high-speed maglev, and ensure personal and equipment safety.

Keywords: High-speed maglev · Integrated grounding · Propulsion and power supply · Operation control

1 Introduction

High-Speed maglev, as a novel form of rail transportation, is gradually advancing towards large-scale implementation in trunk-line projects. In response to the routine conductive high-speed maglev transportation, the National Railway Administration and the Ministry of Housing and Urban-Rural Development have successively issued the "Technical Standards for Maglev Railways" (TB10630–2019) and the "Design Standards for High-Speed Maglev Transportation" (CJJT310–2021) to regulate and guide the construction of high-speed maglev transportation projects. These standards outline specific requirements for lightning protection and grounding of various facilities associated with high-speed maglev transportation.

Integrated grounding technology, falling under the purview of interdisciplinary science and technology, is one of the critical aspects in the high-speed rail domain. It constitutes an essential measure to ensure the normal operation of equipment along the rail lines and the safety of passengers [1]. A standardized approach for integrated grounding in high-speed railways has been established, but as of now, there is no standardized integrated grounding scheme specifically tailored for routine conductive high-speed maglev systems. While high-speed maglev and high-speed railways share many similarities, they

L. Jia et al. (Eds.): EITRT 2023, LNEE 1135, pp. 379–386, 2024.
https://doi.org/10.1007/978-981-99-9307-9_42

also exhibit distinct differences [2]. By analyzing the primary components of routine conductive high-speed maglev transportation facilities and the grounding requirements of various systems, and drawing from the design experiences of the Shanghai Maglev Demonstration Line, Tongji University's "Three Ones" Maglev Test Line, and the Qingdao High-Speed Maglev Test Line, this research endeavors to explore and develop a integrated grounding design proposal for routine conductive high-speed maglev transportation. The findings can serve as a reference for future implementations of high-speed maglev transportation projects.

2 Analysis of the Main Components and Grounding Requirements of High-Speed Maglev Transportation Facilities

High-Speed maglev transportation mainly includes track facilities, propulsion and power facilities, operation control facilities, and maglev vehicles [3]. These numerous facilities, forming a dense grounding system, are distributed along the high-speed maglev transportation route. If the grounding is carried out separately according to function or system, on the one hand, in order to meet the respective requirements for grounding resistance, a large number of grounding electrodes need to be installed. On the other hand, it is necessary to isolate the different grounding systems separately, which poses challenges and can easily result in unequal potential between different grounding bodies, causing damage to the equipment [4]. Therefore, drawing lessons from the experience of high-speed railways, a integrated grounding technology should be adopted for the grounding of high-speed maglev lines, creating a unified grounding system that is compatible with high and low voltages, and combines strong and weak electric grounds. This system will meet the working, safety, and lightning protection grounding requirements of various facilities and equipment [5].

2.1 Line Facilities

Line facilities mainly refer to various equipment attached to the high-speed maglev track beam, including track beam, track functional components, stator cores and coil windings, power rails, steel turnouts, positioning markers, and steel components of fixed equipment along the track. Line facilities is shown in Fig. 1 below:

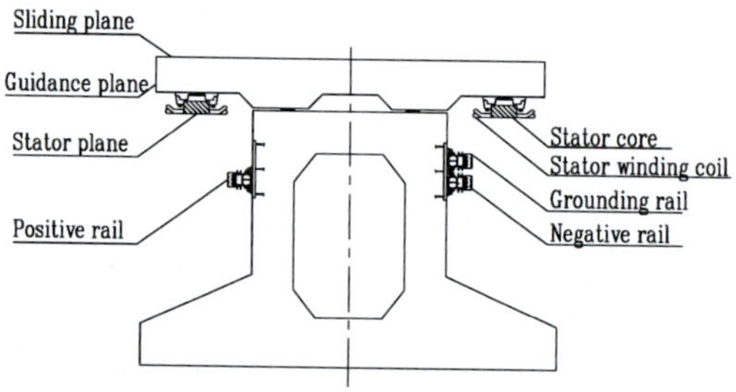

Fig. 1. Diagram of line facilities.

Track functional components generally refer to the three functional planes of the high-speed maglev track, including the stator plane for suspension and propulsion, guidance plane for guidance, and the sliding plane for supporting the train, as well as their associated fixtures. The design of track functional components needs to consider the intrusion of lightning currents during the operation of the maglev train and ensure the timely discharge of lightning currents into the ground through effective lightning protection grounding design [6].

The stator core and coil windings generate currents that vary with the frequency of the applied voltage, providing the necessary suspension force and propulsion for the maglev train. The stator core is made by laminating and punching electrical steel sheets into a toothed shape in the factory. The long stator coil windings are firmly embedded in the tooth slots of the stator core. Generally, lightweight aluminum core cables with a cross-section of 300 mm^2 are used without armor for the stator coil windings. To prevent personal injury caused by induced voltages and equipment damage caused by fault currents, it is necessary to ground the outer surface of the stator core and coil windings. Furthermore, the high-speed maglev propulsion and power supply system belongs to a neutral ungrounded system. In the event of a single-phase grounding fault in the stator windings, the fault can be identified by detecting the zero-sequence current in the detection system [7]. Reliable grounding helps to accurately identify faults.

The power rails laid along the track beam are used to supply power to the train during low-speed travel or when the train is stopped. The power rails consist of a positive rail, a negative rail, and a grounding rail (PE rail). The rails are installed with insulation, and the fixed base of the power rails is generally installed in the form of reserved metal channels. They are fixed approximately every 3 m along the direction of the track. All metal bases need to be properly grounded to ensure personal safety and equipment safety. The ground rail is electrically connected to the train body through collector shoes during low-speed operation, ensuring good grounding when the maglev vehicle operates at low speed or comes to a stop.

The turnouts in high-speed maglev transportation generally use integral steel beams. The turnout equipment includes high-powered mechanical or hydraulic switch mechanisms and control mechanisms for controlling the turnout. Grounding is necessary to meet the requirements of lightning protection grounding and the operational needs of control equipment.

In the operation control system, positioning sign boards and communication equipment are installed at regular intervals along the line to determine the train's position during operation and facilitate control. The metal fixed bases of positioning signboards, communication equipment, and other steel components installed along the line need to be properly grounded for safety purposes.

2.2 Propulsion and Power Supply Facilities

The propulsion and power supply facilities primarily consist of main substations, propulsion substations, track-side substation, and stator switch stations. The main substation imports power from the external power system and supplies it to the propulsion substations and track-side substations along the line after being stepped down by the main transformer. Unlike high-speed railways, the high-speed maglev propulsion and power supply system is an internal system with complex equipment that combines AC, variable frequency, and DC power supply. During normal operation, the circuit does not pass through the ground; only when there is an insulation fault in the line, current passes through the ground. In order to ensure the quick identification and removal of faults by the automatic protection devices of the power supply system, prevent damage from lightning strikes, ensure personal and equipment safety, as well as meet electromagnetic compatibility requirements, grounding protection is necessary for the high-speed maglev power supply facilities.

2.3 Operation Control Facilities

The operation control system mainly consists of the central control subsystem, partition control subsystem, onboard control subsystem, and communication transmission subsystem. The hardware components include monitors, sensors, communication lines, microwave transmitters and receivers, microcomputers, and control rooms. To ensure the reliable operation of the high-speed maglev transportation operation control system, meet electromagnetic compatibility requirements, prevent lightning damage, and ensure the safety of personnel and equipment, surge protection, shielding, and grounding protection are necessary. In addition, effective grounding is required for communication towers, disaster and intrusion monitoring devices, emergency evacuation equipment, and switch station equipment along the line [8].

2.4 Vehicles

There is no lightning protection system similar to the overhead contact system on high-speed maglev tracks. In addition, the train does not have direct contact with the track during high-speed operation. When the train is subjected to lightning strikes in open areas, it will experience overvoltage on the body, which affects the safety of train operation. It is necessary to reduce the body-to-ground voltage by utilizing the sharp tip effect between the train and the track through the structure of guiding electromagnetic magnets [9]. Static electricity can also be generated by high-speed maglev vehicles during operation, or potential differences can be formed with fixed grounding bodies such as the track and platforms due to induction and conduction. The train body is reliably grounded when operating at low speeds or during stops to ensure the safety of personnel, maglev trains, and their equipment [10].

3 Design Scheme for Integrated Grounding of High-Speed Maglev Systems

The high-speed maglev integrated grounding system fully utilizes the grounding devices within the infrastructure of bridges, tunnels, and station sections along the line to form a low-resistance equipotential integrated grounding platform. The grounding devices should prioritize the use of non-prestressed structural steel bars in the structures as natural grounding bodies. When there are no structural steel bars, dedicated grounding bars can be added. If the grounding resistance of the natural grounding body does not meet the requirements, additional artificial grounding bodies should be installed, and the integrated grounding resistance should not exceed 1 Ω. The specific implementation plans for grounding in each facility are as follows.

3.1 Grounding Scheme for Line Facilities

Taking the bridge section as an example, each bridge pier serves as a reliable natural grounding electrode. Longitudinal grounding steel bars are embedded in each beam hole, and grounding terminals are pre-embedded at the beam ends. By connecting grounding jumpers reliably through grounding steel bars, which are embedded longitudinally inside the track beam, a continuous ground line along the longitudinal direction of the track is objectively formed. Diagram of grounding for bridge piers is shown in Fig. 2 below.

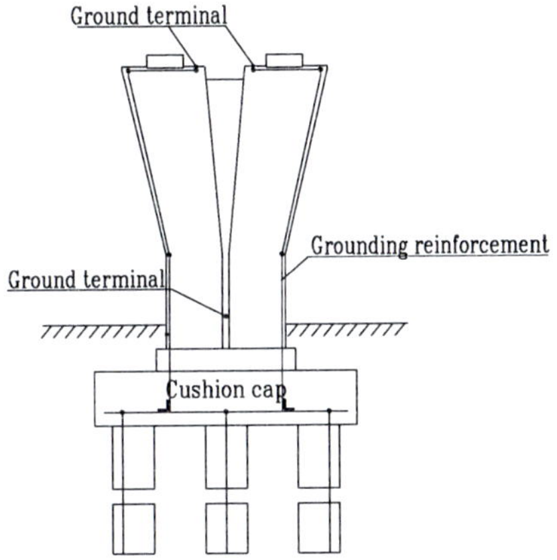

Fig. 2. Diagram of grounding for bridge piers.

At the same time, all cable supports, including other non-energized metallic bodies in the cable trench, are connected along the route direction using grounding flat steel passing through the cable trench, forming a longitudinal underground grounding connection. The cable trench is generally laid near the bridge pier along the cable line. At the bridge pier, the grounding cable is led out from the cable trench and horizontally connected to the grounding terminal reserved at the lower end of the bridge pier, entering the ground grid through the grounding penetration inside the pier.

The metal components such as the guide plates, sliding plates, and stator cores on the track beam are connected to the embedded installation sleeves and the grounding reinforcements inside the track beam to form a natural grounding.

During the installation of the stator winding cables into the tooth slot, grounding sleeves are simultaneously installed. The grounding sleeves extend out of the core groove and bend upward, forming grounding wire clamps. They are then connected to special grounding cables and the reserved grounding terminals on the track beam to achieve grounding of the stator winding coils. Diagram of grounding for stator winding coils is shown in Fig. 3 below:

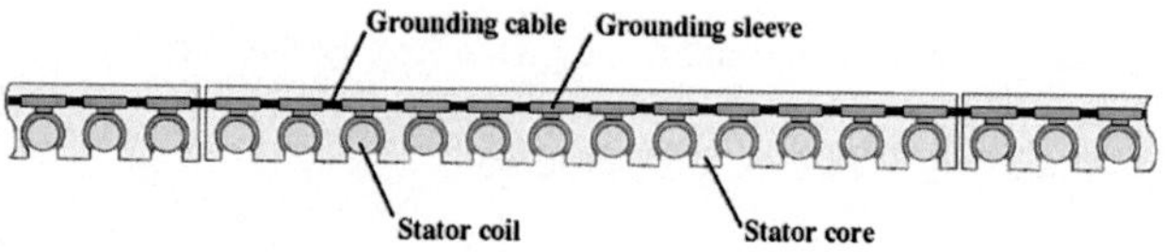

Fig. 3. Diagram of grounding for stator winding coils.

The power rails are installed in a form of pre-embedded channel in track beam. The embedded channels are connected to the grounding reinforcements inside the track beam, achieving natural grounding for the installation base of the power rails. The grounding rail (PE rail) in the power rails is connected to the reserved grounding terminals on the track beam through grounding cables, and there are no fewer than two grounding downleads within each grounding section.

3.2 Grounding Scheme for Propulsion and Power Supply Facilities

The main substations, propulsion substations, wayside converter stations, and fixed switching stations of the propulsion and power system should all be equipped with a primary grounding grid. The grounding resistance of various propulsion facilities' grounding grids is determined based on system grounding configuration and step voltages, among other factors. The grounding grid of the propulsion facilities is connected to natural grounding grids, such as bridge piers, tunnels, and stations, to access to the integrated grounding system. Interfering equipment within the power supply facilities and equipment requiring protection against interference are shielded by metal shields installed in buildings. These shields are reliably connected to the grounding system of the propulsion and power supply facilities to ensure compliance with electromagnetic compatibility requirements.

3.3 Grounding Scheme for Operation Control System Facilities

The lightning protection grounding, safety grounding, and shielding grounding of the power supply and facilities in operation control system are integrated into the integrated grounding system of the building in which the facilities are located. The integrated grounding system of the building is connected to the high-speed maglev integrated grounding grid.

In the equipment rooms of the supervisory control system, a grounding busbar is securely fastened against the wall beneath the anti-static floor. Various grounding wires are routed and fixed to bolt holes. The grounding busbar is connected to the overall grounding structure of the building through grounding conductors. Internal equipment grounding is connected to the cabinet chassis, and the ground wire of the cabinet chassis is routed to the indoor grounding busbar for connection.

3.4 Grounding Scheme for Maglev Vehicles

When high-speed maglev vehicles encounter lightning during operation, the lightning current is conducted to the side sliding plate of the track beam through discharge gaps installed below the vehicle body, and then dissipated into the ground. During low-speed operation of maglev vehicles, such as at stations, auxiliary parking areas, and maintenance bases, a power rail system is installed alongside each track. The vehicles are connected to the power rail through collector shoes, which are in contact with the grounding rail (PE rail) in the power rail, achieving natural grounding for the maglev vehicle body.

In summary, the integrated grounding system of high-speed maglev utilizes natural grounding structures such as bridges, tunnels, and stations as the primary grounding grid. Grounding connections are established both on the line side of the track beam and in the underground cable trenches. Propulsion, supervisory control equipment, and buildings are connected to the integrated grounding system in proximity. To prevent backlashes, weak electrical access points, such as supervisory control facilities, are located at least 15 m away from high-voltage access points, such as propulsion and power supply. The overall design of the high-speed maglev integrated grounding system is presented in Fig. 4 below.

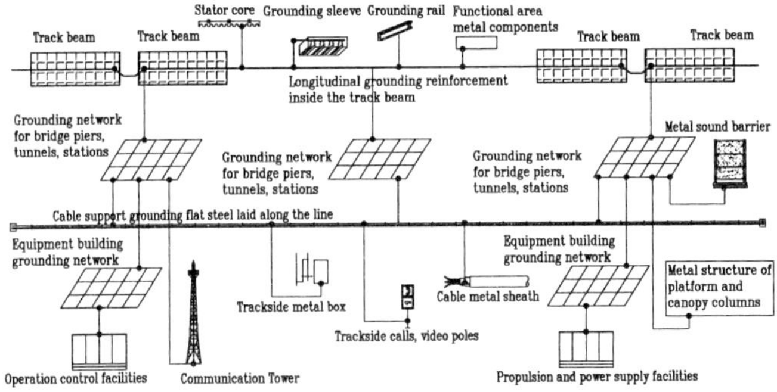

Fig. 4. Integrated grounding design plan of high-speed maglev.

4 Conclusion and Recommendations

To ensure the safety of personnel and equipment, the grounding of high-speed maglev transportation systems should follow the design of a integrated grounding system, creating a unified, high- and low-voltage compatible, and integrated grounding system throughout the entire line. And the system should meet the functional requirements of working grounding, safety grounding, lightning protection grounding, and shielding grounding for various equipment within the stations.

Specific engineering designs should adhere to the principles of human-centeredness, system optimization, and comprehensive protection, taking into account comprehensive planning, overall coordination, and holistic considerations. Considering factors such as maglev line classification, operational mode, equipment system composition, and civil engineering structures, location-specific measures should be adopted in a scientific and reasonable manner to meet the requirements of protecting personnel and ensuring equipment safety.

Acknowledgments. This research is supported by Science and Technology Development Project of China Railway Design Group Co., Ltd, China (No. 2021A240810).

References

1. Xie, B.J., Yang, G., Liu, L.F.: Research and engineering application of high-speed railway integrated grounding technology. Railw. Signal. Commun. **56**(12), 1–5 (2020) (in Chinese)
2. Xiao, S., Xing, L., Zhang, D.: Lightning protection design of high-speed maglev system. In: 6th Global Electromagnetic Compatibility Conference (GEMCCON), pp. 1–4. IEEE, Xi'an (2020)
3. Liang, Y.: Design scheme of high-speed maglev power rail system. Electr. Railw. **31**(sl), 97–99+104 (2020) (in Chinese)
4. Wu, G., Gao, G., Dong, A., et al.: Study on the performance of integrated grounding line in high-speed railway. IEEE Trans. Power Deliv. **26**(3), 1803–1810 (2011)
5. Yang, G.: Research on integrated earthing system scheme of passenger dedicated line journal of railway engineering society. Railw. Signal. Commun. **07**, 76–80 (2006). (in Chinese)
6. Wu, D., Xiao, S., Xu, Y., et al.: Study on grounding and lightning protection for high-speed EMS Maglev. In: International Conference on Electromagnetics in Advanced Applications (ICEAA), pp. 661–666. IEEE (2018)
7. Du, P.C., Xu, X.D.: Research of electromotive force in stator of high speed maglev vehicle. Power Syst. Prot. Control **40**(6), 109–112 (2012) (in Chinese)
8. Chen, R.C.: Engineering design analysis of high-speed maglev operation control system. Constr. Des. Proj. **16**, 84–85 (2020). (in Chinese)
9. Li, K.L.: Research on overvoltage characteristics of maglev train and suppression method. J. Railw. Sci. Eng. **17**(09), 2191–2197 (2020) (in Chinese)
10. Wang, D.F.: Key Points of lightning protection and grounding design for changsha maglev express project. Urban Mass Transit **12**(20), 113–115 (2017) (in Chinese)

Design and Research of Propulsion Substation for High-Speed Maglev Railway

Di Su(✉) and Guosheng Li

China Railway Design Corporation, Tianjin 300308, China
{sudi,liguosheng01}@crdc.com

Abstract. At present, The country is committed to the research of high-speed maglev technology for long routes line. Due to the significant increase in speed and length, it is necessary to study the propulsion and power supply system for high-speed maglev railway. The research and engineering design of propulsion substation, track-side substation, stator switch station, propulsion control system are important parts of the research of propulsion and power supply system for high-speed maglev railway. Combining with the design of Qingdao 600 km high-speed maglev test line, this paper analyzes the main circuit diagram,general layout, propulsion control system of high-speed maglev railway propulsion substation, track-side substation and stator switch station, and introduced the interface design scheme between propulsion substation specialty and specialties such as overhead contact system, station yard,geology, HVAC, water supply and drainage, geophysical exploration, architecture, distribution, communication and operation control, It will be helpful to accumulate experience in the degisn of high-speed maglev railway propulsion substation.

Keywords: High-speed maglev railway · Propulsion substation · Track-side substation · Stator switch station · Design

1 Introduction

The propulsion and power supply system of high-speed maglev railway can be divided into two parts: propulsion system and power supply system. The propulsion system adopts the technology of routine conductive long stator power supply, which provides the required electric energy for the operation of the train by supplying power to the long stator coils through propulsion substations and stator switch stations. The power supply system supplies power to the stations, along-line power loads, and power rails after voltage reduction or rectification through the track-side substation. This paper analyzes and studies the design proposal of the propulsion substation for the Qingdao 600 km high-speed maglev test line.

L. Jia et al. (Eds.): EITRT 2023, LNEE 1135, pp. 387–395, 2024.
https://doi.org/10.1007/978-981-99-9307-9_43

2 Overview of Propulsion and Power Supply System for High-Speed Maglev Test Line

The construction of the high-speed maglev test line includes one propulsion substation, one track-side substation, and four stator switch stations. A single 35 kV power supply line is used to provide power to the propulsion substation. The track-side substation is connected to the power supply from the propulsion substation through a 35 kV medium-voltage looped cable.

The power supply system is divided into two propulsion segment, which are supplied in single-end and double-end modes. Switching between single-end and double-end power supply modes can be accomplished by using a switch at stator switch station 3. The stator section power supply adopts a two-step method (with provisions for three-step conditions), and power rails are installed throughout the line. The main circuit diagram of propulsion and power supply system is shown in Fig. 1.

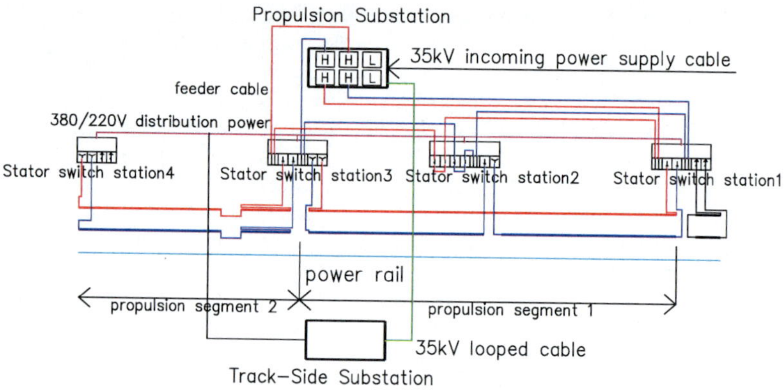

Fig. 1. The main circuit diagram of the propulsion and power supply system.

3 Design Proposal for Propulsion Substation

3.1 Main Circuit Diagram

Propulsion Substation and Stator Switch Station

Main Connection of Incoming Power Supply System

The 35 kV busbar of the propulsion substation adopts a single bus connection wiring mode. The incoming power supply is a 35 kV line, which provides power for the propulsion substation. It is equipped with regional power supply introduction cabinet, metering cabinet, incoming cabinet, outgoing cabinet and four feeder cabinets. On the 35 kV busbar, a set of voltage transformer and lightning arrester are installed for busbar voltage measurement, relay protection and overvoltage protection. The main circuit diagram of the incoming power supply system is basically consistent with that of the circuit diagram of the incoming power supply system under the decentralized power supply mode. Similar projects can be referred to for detailed design.

Main Connection of Propulsion System

The propulsion system mainly consists of high-power converter units,low-power converter units, input cabinet, output cabinet, input transformer, output transformer, propulsion control system and stator switch station. The stator switch station is a box-type substation, which includes input cabinet, output cabinet, feeder cabinet, star connection cabinet and short-circuit cabinet in different stator switch stations. The main circuit diagram of the propulsion system is shown in Fig. 2.

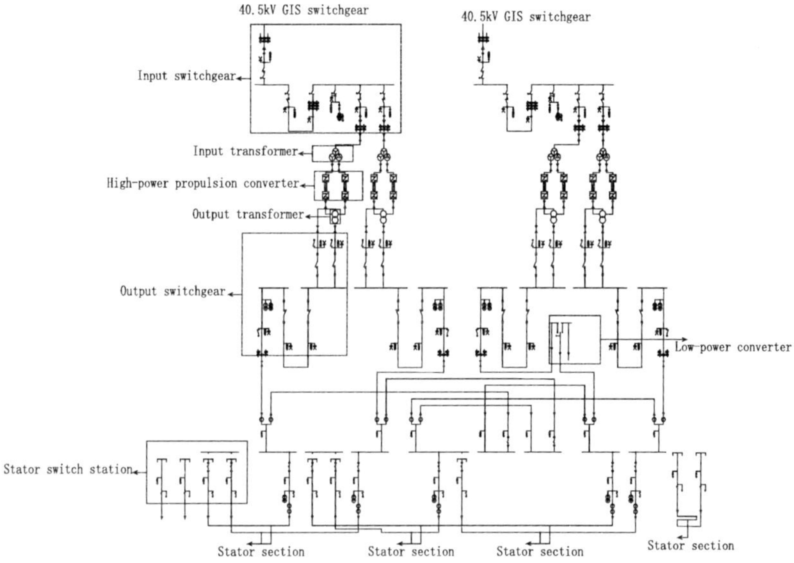

Fig. 2. Main circuit diagram of propulsion system.

Track-Side Substation

The 35 kV busbar of the track-side substation adopts a single bus connection wiring mode. It consists of incoming cabinet, PT cabinet and three feeder cabinets. The design scheme of track-side substation is basically the same as that of traction and step-down hybrid substation in urban rail transit. The detailed design can be referred to similar projects.

4 General Layout

Propulsion Substation

The propulsion substation is designed with a indoor configuration, featuring independent housing and single-story layout, along with the incorporation of cable interlayers. The equipment rooms encompass input and output transformer room, high-power propulsion converter room, high-power input and output cabinet room, propulsion segment control room, power monitoring room, distribution room, low-power cabinet room, reactive compensation room and filtering room. Particular attention needs to be given to the distinctive design aspects that distinguish it from conventional electrified railway and urban rail transit:

(1) High-power propulsion converter generates significant amounts of heat, making it impractical to rely on traditional air-cooling methods to meet their temperature requirements. Consequently, water-cooling systems are required for the high-power propulsion converter units. Within the cable interlayer of this project, water-cooling equipment for the high-power propulsion converter is deployed.
(2) To accommodate the dissipation of braking energy from vehicles, brake resistors must be installed. However, due to the substantial amount of heat they generate, it is unsuitable to house them within the buildings of the propulsion substation. As a solution, a dedicated outdoor area adjacent to the propulsion substation premises referred to as the "brake resistor chamber" is established, enclosed by fencing.

The general layout of the propulsion substation is shown in Fig. 3.

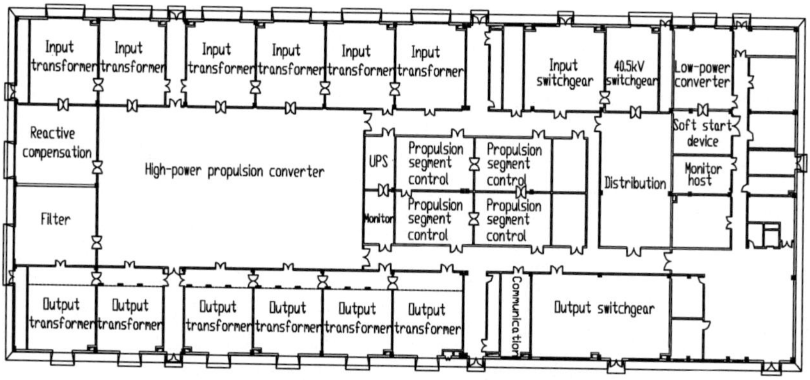

Fig. 3. Propulsion substation general layout.

Track-Side Substation

The track-side substation is designed with a indoor configuration, featuring independent housing and single-story layout, along with the incorporation of cable interlayers. The equipment rooms consist of the high-voltage cabinet room, power transformer, low-voltage cabinet room, rectifier transformer room, DC cabinet room and control room.

Stator Switch Station

Stator switch station adopts a box-type substation design, with its typical layout is shown in Fig. 4.

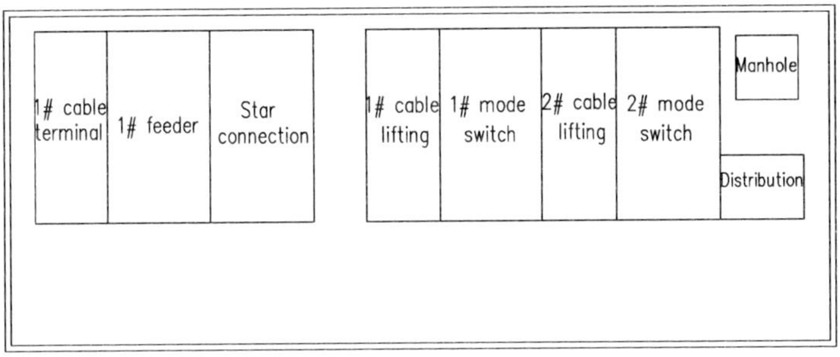

Fig. 4. Typical plane general layout of stator switch station.

4.1 Propulsion Control System

The propulsion control system is the central controlling element of the propulsion and power supply system. Its primary function is to regulate the amplitude, frequency, and phase of the output voltage and current of the high-power power electronic converter in real-time, based on instructions from the train operation control system. This control ensures the provision of propulsion power to the stator sections of the maglev train through feeder cables and stator switch stations, facilitating operations such as propulsion, braking, real-time monitoring, and protection.

The propulsion control system consists of multiple components, including the motor control unit (MCU),converter control unit (CCU), and its subsidiary devices, data transmission system and human-machine interface. The topological diagram representing the system configuration is shown in Fig. 5.

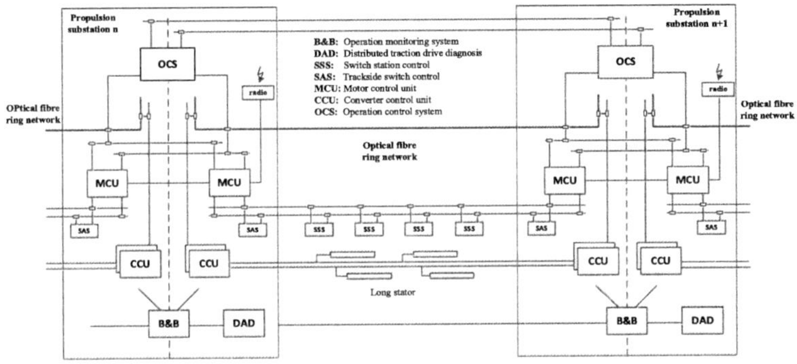

Fig. 5. Propulsion control system network topology diagram.

4.2 Interface Design

The interface between the propulsion substation and other disciplines is divided into internal interfaces and external interfaces. Internal interfaces involve interactions among electrical engineering disciplines, while external interfaces involve interactions with non-electrical engineering disciplines.

Interface design holds significant importance within the design process, as it enables a holistic approach to the design of the propulsion substation and facilitates the realization of its optimized design outcomes. Clarifying the interface design is essential to consider the entire process of designing the propulsion substation comprehensively. Interface is shown in Tables 1 and 2.

Table 1. Internal interface.

Interface professional	Technical Requirements for Interfaces	Interface professional
Propulsion substation	(a) Voltage levels of the power supply system (b) Proposals for the propulsion and power supply system (c) Operation modes of various types of substations under different working conditions (d) Capacities of various types of inverter units (e) Rated currents of incoming lines and feeder lines (f) Short-circuit currents of 35 kV incoming lines and feeder lines in substations during maximum and minimum operating conditions of the power system (g) Requirements for reactive power compensation and harmonics	Propulsion and power supply
Propulsion substation	(a) The installation position of feeder cables and DC cables inside the substation (b) The position of the grounding busbar inside the substation	Overhead contact system

5 Conclusions

At present, there are limited case studies available in domestic literature regarding the design of propulsion substations for high-speed maglev railways. This paper builds upon the Qingdao 600 km/h high-speed maglev test line project and conducts a comprehensive design analysis and investigation into key design aspects such as main circuit diagram,general layout, propulsion control system and interface design for high-speed maglev propulsion substation. The Qingdao 600 km/h high-speed maglev test line project was formally completed in 2021, with the propulsion substation running smoothly. The

Table 2. External interface.

Interface professional	Technical Requirements for Interfaces	Interfacen professional
Propulsion substation	Propulsion substation specialty provides the preliminary location of the site of the station range to the station yard specialty. After the confirmation of each specialty, the station yard specialty produces drawings,including elevation, earthwork statistics, drainage slope, access road, etc.	Station yard
Propulsion substation	Propulsion substation specialty provides the location of propulsion substation,the geology specialty is responsible for providing geological survey data	Geology
Propulsion substation	Propulsion substation specialty provides propulsion substation temperature, ventilation, gas fire protection and other requirements	HVAC
Propulsion substation	Propulsion substation specialty provides water supply and drainage requirements for propulsion substation facilities. Water supply and drainage specialty is responsible for water source introduction and drainage	Water Supply and drainage
Propulsion substation	Propulsion substation specialty provides the location of the propulsion substation to the geophysical exploration specialty, and the geophysical exploration specialty is responsible for providing the soil resistivity of each layer of the site	Geophysical exploration
Propulsion substation	Propulsion substation specialty provides housing plane and stress, hole embedded parts requirements, equipment foundation stress requirements, cable trench and drainage pipe, oil collection pit, drainage pipe requirements, field site, road and wall requirements, Architecture specialty is responsible for providing building drawing	Architecture
Propulsion substation	Propulsion substation specialty provides the technical requirements of various substations,Distribution specialty designs the power distribution according to the requirements	Distribution

(continued)

Table 2. (*continued*)

Interface professional	Technical Requirements for Interfaces	Interfacen professional
Propulsion substation	(a) Propulsion substation equipment is interconnected with the transmission system equipment through the network port, and the interface is located at the wiring frame (b) Propulsion substation specialty provides dispatching telephone technical requirements, communication specialty is responsible for providing configuration dispatching telephone	Communication
Propulsion substation	Propulsion substation specialty provides data transmission, and propulsion shutdown requirements of propulsion and power supply system. Operation control specialty carries out relevant design according to the requirements	Operation control

design experience gained from these high-speed maglev propulsion substations can provide valuable references for similar engineering projects and contribute to the accumulation of knowledge for future high-speed maglev railway propulsion substation designs along longer routes.

Acknowledgments. This research is supported by Science and Technology Development Project of China Railway Design Group Co., Ltd, China (No. 2021A240810).

References

1. National Railway Administration: Standard for technology of maglev railway. China railway publishing house co., Ltd., Beijing (2019)
2. Ministry of Housing and Urban-Rural Development of the People's Republic of China: Code for design of metro. China Architecture and Building Press, Beijing (2013)
3. National Railway Administration: Code for Design of Railway Traction Power Supply. China railway publishing house co., Ltd., Beijing (2016)
4. Wang, L.T., Li, L.P.: High-Speed Magiev Traction Power Supply System. Beijing Jiaotong University Press, Beijing (2014)
5. Liu, H.T., Feng, Y.X., Xu, Y.J. et al.: Optimal layout design of high-speed maglev traction substation. Railw. Tech. Stand. **5**(01), 9–14 (2023) (in Chinese); Chen, C.R.: Engineering design analysis of high-speed maglev operation control system. Constr. Des. Proj. **438**(16), 84–85 (2020) (in Chinese)
6. Wu, W., Lu, X., Liu, G., et al.: Operation control system of 600 km/h high-speed maglev levitated transport system in China. In: 2019 IEEE Vehicle Power and Propulsion Conference (VPPC), pp. 1–5. IEEE (2019)

7. Zhang, Y., Qing, G., Zhang, Z., et al.: Calculation and simulation of propulsion power supply system for high-speed maglev train. In: 2020 IEEE Vehicle Power and Propulsion Conference (VPPC), pp. 1–5. IEEE (2020)
8. Warder, S.B.: Traction substations: their general layout, operation and maintenance. Proc. IEE-Part IA: Electr. Railw. Tract. **97**(1), 140–151 (1950)
9. Chen, C.R.: Engineering design analysis of high-speed maglev operation control system. Constr. Des. Proj. **438**(16), 84–85 (2020) (in Chinese)
10. Phaenkongngam, T., Chinnawong, K., Patumasuit, N., et al.: Reviewing propulsion and levitation system for magnetic levitation train. In: 2021 9th International Electrical Engineering Congress (iEECON), pp. 185–188. IEEE (2021)

Design of IGBT Drive and Protection Based on M57962AL

Jiaxi Gou(✉), Zhankai Fang, Guo Chen, Qiang Lai, and Shuang Liu

Chongqing Chuanyi Automation Co., Ltd., Chongqing 401121, China
319087982@qq.com

Abstract. Aiming at the anti-interference performance requirements of high-power IGBT drive circuit in active power filter, this paper designs IGBT drive and protection circuit based on M57962AL chip. The voltage change of VCE is detected by two diode series resistors to the chip. Combined with the output characteristic curve of IGBT, the threshold of IGBT short circuit protection current can be set by adjusting the resistance value. The operation experiment of active power filter shows that the design can effectively maintain the stability of the fault output signal, so it has stronger anti-interference performance and improves the reliability of IGBT short-circuit protection.

Keywords: IGBT drive · Short-circuit protection · Reliability

1 Introduction

Active power filter is used to compensate the harmonic current of each order in the power grid and prevent the interference and even damage of the high order harmonic to the electrical equipment and power generation equipment. It is widely used in factories, hospitals, shopping malls, rail transit and other places [1–3]. IGBT is the most important switching device in active power filter, and its safe and stable operation is crucial to the reliability of the equipment. IGBT overcurrent is a common cause of its damage. Overcurrent can be divided into two types: one is the overcurrent caused by overload, which can reach 2 times of the rated value, and is generally allowed to last for about 1ms. Therefore, the current sampling circuit can be connected to the blocking circuit for protection; The other is overcurrent caused by short circuit, which can reach more than 4 times of the rated value, and can only last for 10 μs, so higher requirements are put forward for the response time of the protection circuit [4, 5].

At present, the VCE threshold detection and protection circuit built in the IGBT driver chip is commonly used for short circuit protection of IGBT [6–9]. There are many IGBT driver chips, such as 2SD315AI, FOD8332, 2ED300C17-S, etc. [10–14], of which the M57962AL chip produced by Mitsubishi Corporation of Japan has fast protection response time and low cost, and is often used in high-power IGBT driver circuits. In literature [15], the Schottky diode series resistance is used to adjust the threshold of short circuit protection current. However, in the active power filter, the IGBT switching frequency is as high as tens of kHz, which puts forward higher requirements for the

L. Jia et al. (Eds.): EITRT 2023, LNEE 1135, pp. 396–403, 2024.
https://doi.org/10.1007/978-981-99-9307-9_44

anti-interference performance of the short-circuit protection circuit. Therefore, based on the M57962AL chip, this paper designs a drive and short circuit protection circuit with stronger anti-interference performance. Schottky diode series resistor and then series Schottky diode are used to adjust the short circuit protection current threshold to prevent misoperation caused by the fluctuation of fault output signal.

2 Working Principle of M57962AL

The internal principle of M57962AL is shown in Fig. 1. When 10 mA current flows through pins 13 and 14, the optocoupler is turned on, the potential of point A is pulled down to VEE, the output of point B is high, Q1 and Q2 are turned on, Q3 and Q4 are turned off, and VCC is connected to IGBT gate. During normal operation, D5 is on, and the potential at point F is R18 lower than that of VCC, so Q6 is connected, and VCC is connected to R10 and D1. $\frac{R11}{R10+R11} > \frac{R9}{R8+R9}$, so the output of IC1D is pulled down and Q7 is cut off. $\frac{R7}{R6+R7} > \frac{R9}{R8+R9}$, so IC1B outputs high level and IC1C outputs high level, making Q5 cut off. VCC is connected to pin 8 through R19, D3 and R15, so pin 8 is high.

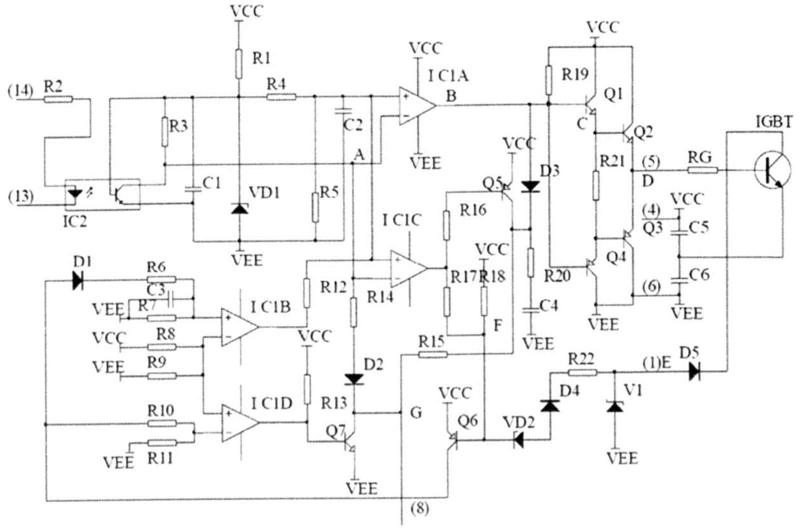

Fig. 1. M57962AL internal schematic diagram.

When there is no current in pin 13 and 14, the optical coupling is off, the potential of point A is VCC, and the IC1A in-phase input terminal potential is $\frac{(VCC-VEE)R5}{R1+R4+R5}$, so the point B output is low, the Q1 and Q2 are off, and Q3 and Q4 are on, so the gating voltage is −VEE. The IC1C output is low-level, Q5 is on, and the VCC passes Q5, R15 to pin 8. Q6 is on because the F-point potential is $\frac{VCC \bullet R17}{R17+R18} < VCC$. Similarly, the partial voltage of R11 is greater than that of R9, so the IC1D output is low and Q7 is off.

When a short circuit fault occurs, the IGBT collector voltage increases, the F point potential increases, and Q6 is cut off, so the IC1D reverse input is pulled down, the IC1D output is high, Q7 is on, and the pin 8 output is low. At the same time, the voltage of point B is discharged through D3, R15, and Q7 and gradually decreases, so that Q1 and Q2 are off and soft shutdown is realized.

3 Driver Circuit Design

The driving circuit based on M57962AL is shown in Fig. 2. In order to reduce the power consumption of the main control chip, the PNP triode amplifier circuit is used. When Vin is low voltage, Q3 is on, pin 14 is high, so the optical coupling inside the chip is on, pin 5 outputs high level, so that Q1 is on, IGBT gate-emitter voltage is the voltage of regulator diode D4. The chip power supply voltage is VCC = 24 V, and the voltage of regulator diode is 15 V, so the IGBT is on. Similarly, when the Vin is high voltage, the IGBT gate-emitter voltage is $V\_GND - V\_E = -9V$, so the IGBT is off.

The input end R7 and R9 constitute a voltage divider circuit, which can adjust the critical level of Q3 switching state to prevent the interference signal of Vin from making Q3 misoperate. C3 is a filter capacitor, which filters out the high-frequency interference of Vin. R10 is a bleed resistor that quickly drains the junction capacitance charge of the light-emitting diode when the optical coupling inside the chip is turned off, so that the optical coupling is turned off quickly and reliably.

The output VCC generates +15 V and −9 V switching voltages through the voltage regulator diode D4, and the parallel aluminum electrolytic capacitors CE1 and CE2 stabilize the voltage output. Q1 and Q2 constitute a power push-pull amplifier circuit to increase the driving capacity. R3 and R4 are gate closing and opening resistors of IGBT. The LED D5 is used to indicate the on status of the IGBT. Pin 1, pin 2, pin 8 of the chip are short-circuit protection related pins.

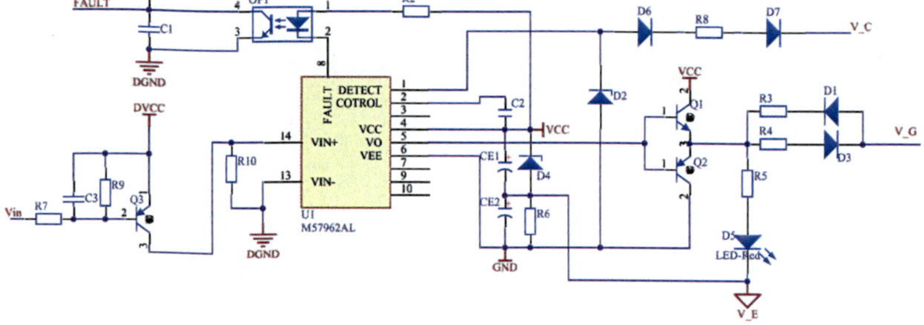

Fig. 2. Driver circuit diagram based on M57962AL.

4 Short Circuit Protection Design

4.1 Short Circuit Protection Mechanism

When the IGBT is working normally, the collector voltage VCE is low, and with the increase of the collector current IC, the VCE also gradually increases, as shown in Fig. 3 (refer to Infineon FF225R12ME4_B11 IGBT datasheet). Therefore, the short circuit current can be protected by detecting the size of VCE and setting an appropriate VCE protection threshold. When VCE is greater than the set threshold, Q1 and Q2 inside the chip are cut off, and the chip output level is low to turn off the IGBT. The desaturation recovery time of M57962AL is several μs, which is less than the IGBT short circuit withstand time, so the IGBT short circuit current can be turned off in time.

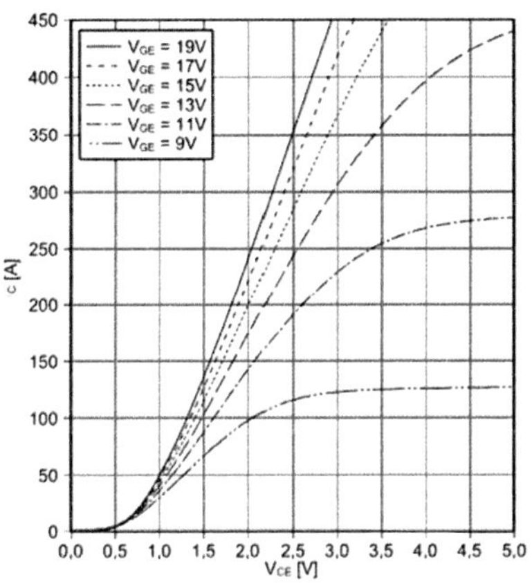

Fig. 3. IGBT output characteristic curve (FF225R12ME4_B11).

4.2 Short Circuit Protection Circuit

Pins 1, 2 and 8 of the M57962AL chip are short-circuit protection related pins. Pin 2 is connected to VCC by capacitor C2. Adjusting C2 can control the desaturation recovery time and adjust the protection sensitivity. The larger the C2, the greater the desaturation recovery time, pin 2 suspended, desaturation recovery time is 2.6 μs. Pin 8 is the fault signal output pin, which is high in normal operation and low in short circuit fault. Pin 8 outputs the FAULT signal to the MCU through the optocoupler OP1. During normal operation, the optocoupler does not conduct, so the FAULT is high, and the FAULT becomes low when the short-circuit fault occurs. Pin 1 is a short-circuit current detection pin, which is connected to the IGBT collector through the series resistance R8 of Schottky diode D6 and then the series resistance of Schottky diode D7. Therefore, the detection voltage of pin 1 increases with the increase of the IGBT current. Voltage regulator diode D2 is used to clamp pin 1 voltage to prevent chip overvoltage damage.

When the resistance R8 is increased, the lower the protection threshold voltage of VCE, the smaller the short circuit current protected. Therefore, the short circuit protection current value can be set by adjusting R8 and combining the IGBT output characteristics. The VCE threshold voltage and short circuit protection current ISCP (refer to Infineon FF225R12ME4_B11, VGE = 15V) under different resistance values of R8 can be measured using the signal generator and adjustable power supply. The test results are shown in Table 1. In this paper, R8 = 18 kΩ is used, and the short-circuit protection current is 650A.

Table 1. VCE and ISCP under different R8 resistance values

R8/kΩ	VCE/V	ISCP/A
0	8.6	900
10	6.6	850
18	5	650
20	4.6	550
30	2.2	230
40	0.1	<10

Reference [15] uses a Schottky diode string resistor for detection, but in the operation experiment of active power filter, it is found that the 8-pin fault signal of this circuit (Fig. 2 with D6 removed) will be disturbed and reduced to a minimum of 3.137 V, as shown in Fig. 4a. This situation is easy to lead to short-circuit protection misoperation, especially when the electrical fast transient pulse group and surge interference. The use of two Schottky diode intermediate series resistance can significantly reduce the interference of 8-pin fault signal, the lowest is 26.35 V, as shown in Fig. 4b.

The above phenomenon may be caused by the reverse recovery time of the diode. Due to the presence of junction capacitance, the current flowing through the diode does not immediately turn to 0, but first increases in the reverse direction and then decreases to 0. This period is called the reverse recovery time trr, as shown in Fig. 5. Trr is generally at the ns level. When the IGBT is on, the two diodes D6 and D7 on pin 1 of M57926AL are conducting forward. When it is turned off, D6 and D7 should have turned off, but due to the reverse recovery time, they are conducting backward, causing pin 1 to detect a high level, so the fault signal of pin 8 appears a low level pulse. Adding two diodes at both ends of resistor R8 can suppress the reverse recovery current flowing through R8, so the fault detection effect is better than that of a diode string resistor.

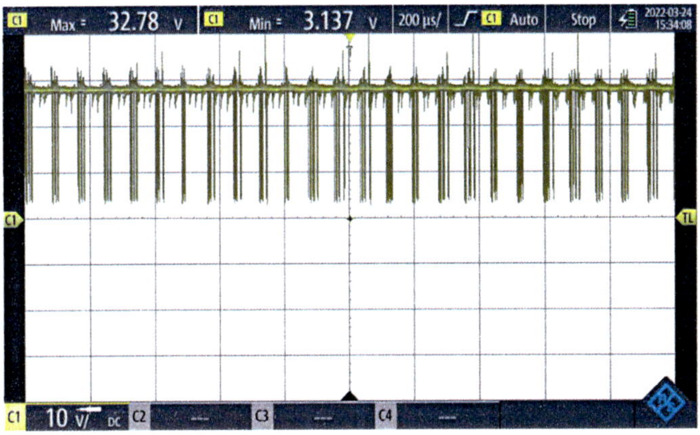

(a) A Schottky diode string resistor

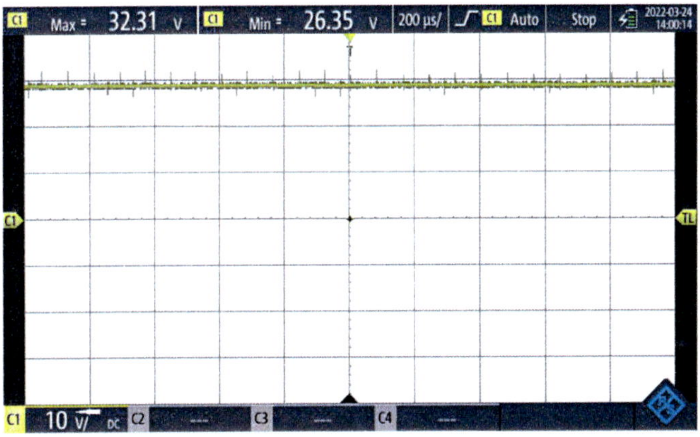

(b) Two Schottky diodes intermediate series resistor

Fig. 4. Voltage of fault signal on pin 8.

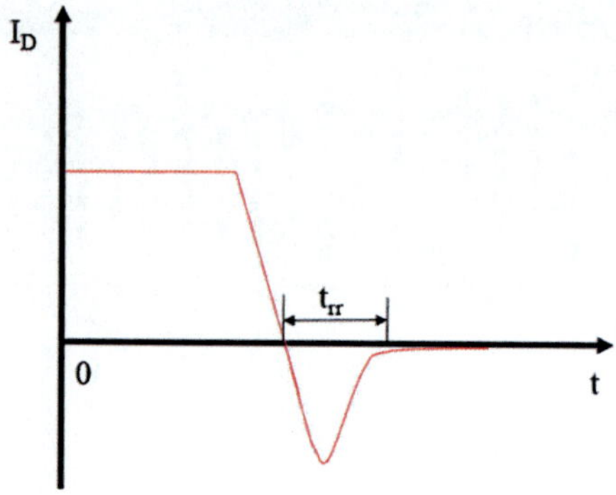

Fig. 5. Schematic diagram of diode reverse recovery time.

5 Conclusion

Based on M57962AL chip and IGBT output characteristics, the IGBT driving circuit for active power filter is designed in this paper. The change of VCE voltage is detected by using the intermediate series resistance of two diodes. The short circuit current protection threshold can be set by adjusting the resistance based on the curve of VCE changing with IC. The experimental results show that the circuit design in this paper enhances the anti-interference performance of short-circuit fault detection and improves the reliability of IGBT short-circuit protection.

References

1. Gong, J., Li, D., Wang, T., et al.: A comprehensive review of improving power quality using active power filters. Electr. Power Syst. Res. **199**(3), 107389 (2021)
2. Dong, X., et al.: Morphological characteristics and technical prospects of new distribution systems. High Volt. Technol. **47**(09), 3021–3035 (2021). (in Chinese)
3. Feng, Y.: Analysis of measures to improve power quality under filters. Commun. Power Technol. **38**(03), 229–231 (2021). (in Chinese)
4. Li, Y.: Design of threshold adjustable IGBT overcurrent protection based on VCE detection. Autom. Instrum. (02), 169–171+175 (2021) (in Chinese)
5. Wang, L., Yuan, P., Xie, M., Wang, W.: Analysis of the causes of IGBT damage in electromagnetic stirring power supply and solutions. Power Electron. Technol. **54**(08), 132–135 (2020). (in Chinese)
6. Li, Y.: An IGBT short-circuit protection method using variable VCE detection threshold. J. Phys. Conf. Ser. **16**(01), 022004 (2020)
7. Lei, P., Huang, Y., Yu, Z.: Application research of IGBT gate drive circuit based on 1ED020I12-FA. Micromotor **44**(04), 91–94 (2011). (in Chinese)
8. Li, K.: High stability IGBT driver protection circuit. Electr. Transm. **50**(06), 98–102 (2020). (in Chinese)
9. Huang, X., Li, X., Liu, Y., Wang, F., Gao, G.: IGBT state monitoring and protection circuit based on quantitative voltage parallel comparison. J. Electr. Eng. Technol. **36**(12), 2535–2547 (2021). (in Chinese)

10. Hua, A., Li, W., Hua, Z., et al.: Research on a IGBT drive circuit based on 2ED300C17-S. Electr. Weld. Mach. (2009)
11. Sun, Q., Tao, J.: Study on a new type hybrid IC (2SD315A) for IGBT drive and protection. Mod. Electr. Power (2003)
12. Cong, X., Shang, L., Lu, Y.: Design of IGBT driver circuit for PFN charging high-voltage power supply. High Volt. Technol. **02**, 78–80 (2006). (in Chinese)
13. Huang, Q., Kang, M., Bian, H.: Design and implementation of multiphase motor drive control system. Ind. Instrum. Autom. Equip. **04**, 58–63 (2021). (in Chinese)
14. Wang, L., Zou, Z., Liu, Y., Yuan, P., Chen, L., Hu, L., Wang, F., Xu, D., Fu, H.: Design and application of class II short circuit detection circuit for electromagnetic stirring power supply. Electr. Technol. (12), 7–12+15 (2022) (in Chinese)
15. Chen, Q., Lu, Y., Yu, H., Huang, X.: Design of high-power IGBT drive and protection based on M57962AL chip. Power Electron. Technol. **50**(07), 104–108 (2016). (in Chinese)

Speed Balance Control of the Dual-Motor System Based on Three-Vector Model Prediction

Yuqi Liu[1], Yuanfeng Huang[1(✉)], Bin Xiong[1], and Haifeng Wang[2]

[1] Institute of Electrical Engineering, Chinese Academy of Sciences, Beijing 100190, China
huangyuanfeng2023@163.com

[2] Ganjiang Innovation Academy, Chinese Academy of Sciences, Jiangxi 341000, China

Abstract. In recent years, dual-motor drive systems have been widely applied in various fields, including rail transportation and industrial production. The control of dual-motor drive systems has become a key research topic, with the focus on determining the most effective control method. In the past, magnetic field orientation control or direct torque control methods were commonly used for driving dual-motor systems. However, with the significant improvement in microprocessor chip performance, more complex and precise algorithms have emerged, such as model predictive control. Therefore, this paper compared the Three-Vector Model Predictive Current Control (TV-MPCC) with a similar framework as the Field-Oriented Control (FOC) algorithm but with different principles. The simulation experiments were conducted to verify the feasibility and effectiveness of TV-MPCC in improving the control accuracy and coordination of the dual-motor drive system.

Keywords: TV-MPCC · PMSM · Dual-motor drive system

1 Introduction

Currently, the three main control methods for controlling Permanent Magnet Synchronous Machines (PMSMs) using voltage source inverters are Field Oriented Control (FOC), Direct Torque Control (DTC), and Model Predictive Control [1, 2]. Unlike DTC, FOC manipulates the stator current vector's orthogonal projection on the rotating coordinate system axes to indirectly control the required torque and magnetic flux. Therefore, the control effectiveness of FOC primarily depends on the performance of the current loop controller. The main current control schemes include PI control, current hysteresis control, and zero-difference control. Compared to the first two control methods, Model Predictive Control has faster dynamic response, stronger nonlinear control performance, and is more easily digitized. However, this control method also has some drawbacks, such as higher computational complexity compared to the previous two control methods and weaker robustness due to system parameter uncertainties. In recent years, with the rapid development of microprocessors, there has been increasing attention on Model Predictive Current Control [3, 4].

L. Jia et al. (Eds.): EITRT 2023, LNEE 1135, pp. 404–412, 2024.
https://doi.org/10.1007/978-981-99-9307-9_45

Dual-motor control systems are widely used in various aspects of people's daily lives, such as transportation, industrial production, and underwater robotics. Leveraging the advantage of fast dynamic response offered by model predictive control, this paper applies this control algorithm to the dual-motor control system. The aim is to reduce torque and speed fluctuations in the dual-motor system through model predictive current control. The objective is to enhance the dynamic response speed of the motors, achieve coordinated control of their speeds, and improve the control system's disturbance rejection capability while ensuring high steady-state control accuracy.

2 Model Predictive Current Control Strategy for SPMSM

2.1 Model Predictive Current Control (MPCC)

Model Predictive Control (MPC) can be divided into two categories: Continuous Control Set Model Predictive Control (CCS-MPC) and Finite Control Set Model Predictive Control (FCS-MPC). In CCS-MPC, within one switching period, any desired voltage vector can be outputted through modulation, but it requires solving complex voltage prediction equations and requires a modulator. In typical three-phase inverters, FCS-MPC only has eight basic voltage vectors, matching the discrete characteristics of the inverter itself, and does not require a modulator, resulting in better dynamic performance. However, compared to CCS-MPC, general FCS-MPC algorithms cannot output arbitrary desired voltage vectors and can only generate eight basic voltage vectors, leading to larger torque ripple [5].

Three-vector model predictive current control is a type of FCS-MPC. As shown in Fig. 1, it uses two adjacent basic voltage vectors and one zero voltage vector to predict the corresponding axis current. The objective is to select the optimal desired voltage vector using the cost function (Eq. 1) to expand the candidate voltage vectors [6], thereby achieving smoother torque output and reducing torque ripple. The core idea is to synthesize voltage vectors with different directions and magnitudes by applying different durations of basic voltage vectors. This improves current control accuracy and reduces torque ripple in the motor.

$$J(S_k) = |i_d^* - i_d(n+1)| + \lambda |i_q^* - i_q(n+1)| \tag{1}$$

$$S_{opt} = \arg\min J(S_k) \tag{2}$$

In the equation, $S_k(k = 1, 2..., 6)$ represents the kth sector, $J(S_k)$ represents the prediction error of the voltage vector in the kth sector, i_d^* and i_q^* represent the reference currents in the dq-axis, $i_d(n+1)$ and $i_q(n+1)$ respectively represent the predicted current value at the next sampling moment, λ represents the weighting factor, and S_{opt} represents the sector where the optimal voltage vector is located.

In the synchronous rotating dq coordinate system, the current state equation of a Surface Permanent Magnet Synchronous Machine (SPMSM) can be expressed as:

$$\begin{cases} L_s \dfrac{di_d}{dt} = u_d - R_s i_d + \omega_e L_s i_q \\ L_s \dfrac{di_q}{dt} = u_q - R_s i_q - \omega_e(\psi_f + L_s i_d) \end{cases} \tag{3}$$

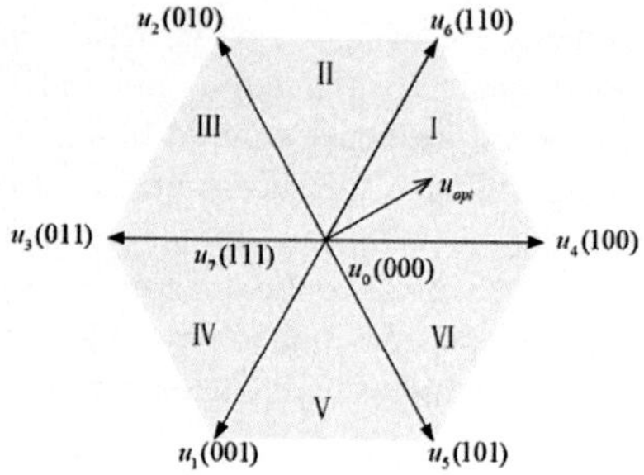

Fig. 1. Voltage vector synthesis diagram

where, i_d and i_q represent the direct and quadrature components of the stator current, respectively. L_s represents the stator inductance. u_d and u_q represent the direct and quadrature components of the stator voltage, respectively. R_s represents the stator resistance. ω_e is the rotor electrical angular velocity. ψ_f represents the magnetic flux of the rotor permanent magnet.

Firstly, based on the mathematical model of the Surface Permanent Magnet Synchronous Motor (SPMSM), a dual-loop control strategy for the motor is established as shown in Fig. 2. The speed regulator (ASR) utilizes a traditional PI controller to enhance the overall robustness of the system. The current regulator (ACR) employs Model Predictive Control (MPC) with three voltage vectors (two active voltage vectors and one zero voltage vector) to improve the dynamic response speed of the SPMSM.

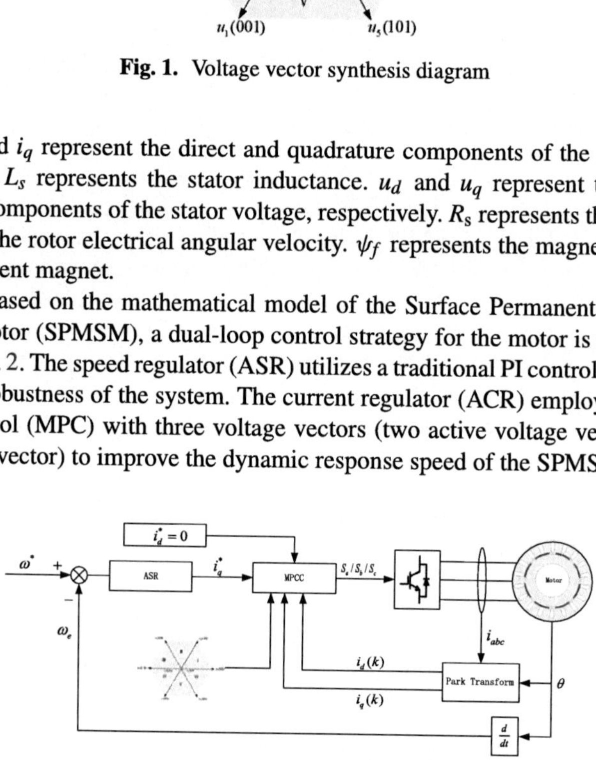

Fig. 2. Model predictive dual-loop control strategy

When the zero voltage vector and the two adjacent active voltage vectors u_i and u_j are applied, the rate of change of the dq-axis currents can be calculated as follows:

$$k_{d0} = \left.\frac{di_d}{dt}\right|_{u_d=0} = \frac{1}{L_s}[-R_s i_d + \omega_e L_s i_q] \tag{4}$$

$$k_{q0} = \left.\frac{di_q}{dt}\right|_{u_q=0} = \frac{1}{L_s}[-R_s i_q - \omega_e L_s i_d - \omega_e \psi_f] \tag{5}$$

$$k_{di} = \left.\frac{di_d}{dt}\right|_{u_d=u_{di}} = k_{d0} + \frac{u_{di}}{L_s} \tag{6}$$

$$k_{qi} = \left.\frac{di_q}{dt}\right|_{u_q=u_{qi}} = k_{q0} + \frac{u_{qi}}{L_s} \tag{7}$$

$$k_{dj} = \left.\frac{di_d}{dt}\right|_{u_d=u_{dj}} = k_{d0} + \frac{u_{dj}}{L_s} \tag{8}$$

$$k_{qj} = \left.\frac{di_q}{dt}\right|_{u_q=u_{qj}} = k_{q0} + \frac{u_{qj}}{L_s} \tag{9}$$

By applying the forward Euler method, the current state equation (Eq. 1) can be discretized to obtain the predicted current values at the next sampling instant [7].

$$i_d(n+1) = i_d(n) + \frac{T_s}{L_s}[u_d(n) - R_s i_d(n) + \omega_e(n)L_s i_q(n)] \tag{10}$$

$$i_q(n+1) = i_q(n) + \frac{T_s}{L_s}[u_q(n) - R_s i_q(n) - \omega_e(n)L_s i_d(n) - \omega_e(n)\psi_f] \tag{11}$$

When using current hysteresis control [8], the $dq-$ axis current reference value is given by:

$$i_d^* = i_d(n+1) = i_d(n) + k_{di}t_i + k_{dj}t_j + k_{d0}t_0 \tag{12}$$

$$i_q^* = i_q(n+1) = i_q(n) + k_{qi}t_i + k_{qj}t_j + k_{q0}t_0 \tag{13}$$

$$t_i + t_j + t_0 = T_s \tag{14}$$

In the equation, t_i and t_j represent the durations of u_i and u_j, respectively. t_0 represents the duration of the zero vector, and the sum of the durations of the three voltage vectors is the sampling period T_s.

By combining Eqs. (6)–(14).

$$t_i = \frac{(i_d^* - i_d(n))(k_{qj} - k_{q0}) + (i_q^* - i_q(n))(k_{d0} - k_{dj})}{Den} + \frac{T_s(k_{q0}k_{dj} - k_{qj}k_{d0})}{Den} \tag{15}$$

$$t_j = \frac{(i_d^* - i_d(n))(s_{q0} - s_{qi}) + (i_q^* - i_q(n))(s_{di} - s_{d0})}{Den} + \frac{T_s(s_{qi}s_{d0} - s_{q0}s_{di})}{Den} \tag{16}$$

$$Den = k_{q0}k_{dj} + k_{qi}k_{d0} + k_{qj}k_{di} - k_{qi}k_{dj} - k_{qj}k_{d0} - k_{q0}k_{di} \tag{17}$$

By calculating the durations of the two active voltage vectors based on Eqs. (12)–(14), we can then substitute them into Eqs. (15) and (17) to solve for the optimal synthesized voltage vector components u_{opt} in the dq-axis.

$$u_d = \frac{t_i}{T_s}u_{di} + \frac{t_j}{T_s}u_{dj} + \frac{t_0}{T_s}u_{d0} \tag{18}$$

$$u_q = \frac{t_i}{T_s} u_{qi} + \frac{t_j}{T_s} u_{qj} + \frac{t_0}{T_s} u_{q0} \tag{19}$$

Based on the current state equation for PMSM (Eq. 3) and the Model Predictive Control equations (Eqs. 1, 2 and 4–19), we can build a corresponding Simulink model (Fig. 3). Following the basic principles of vector control, we can use the sector K, the durations of the three voltage vectors, and their combinations from the Three-Vector Model Predictive Current Control (TV-MPCC) as the modulation waveform for the PWM generation module. After comparing the modulation waveform with a sawtooth carrier waveform, we can generate the PWM drive signals for the three-phase full-bridge circuit.

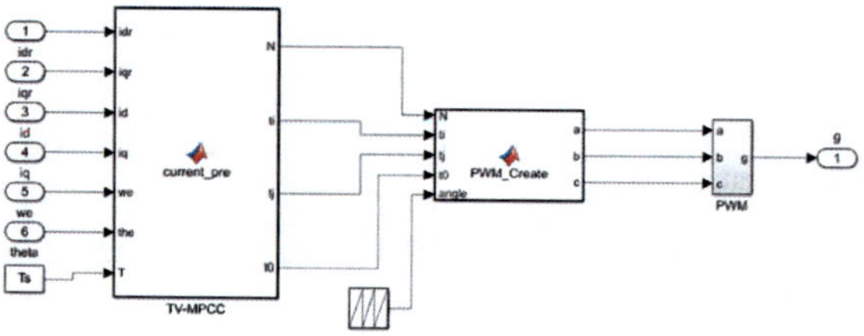

Fig. 3. The TV-MPCC model

3 The Dual-Motor Control System

In a conventional dual-motor coordinated control system, a dual-loop strategy is commonly used for speed and current control. Some approaches aim to achieve consistent torque by equalizing the current feedback values of both motors, while others aim to achieve coordinated control by ensuring equal speed feedback values. In this simulation, a 90 kW PMSM with a rated speed of 2400 rpm and rated torque of 200 Nm is considered. The model predictive current control algorithm, as discussed earlier, is applied using a dual-loop strategy to regulate the speed of a single motor. To achieve coordinated control of the dual-motor system and ensure speed synchronization, a speed difference feedback loop is introduced. This ensures the synchronization of speeds in the dual-motor control system.

The control methods for dual-body torque motors are generally classified into three categories: master-slave control, parallel control, and cross-coupled control [9], as shown in Fig. 4.

Master-slave control involves using the speed output of the first motor as the speed reference for the second motor to achieve speed synchronization between the two motors. Its advantage lies in its simple structure and ease of control. However, it has clear drawbacks. The speed control of the two motors has a time delay, and it takes a long time to achieve synchronization. When starting up or during speed changes, the synchronization effect is poor.

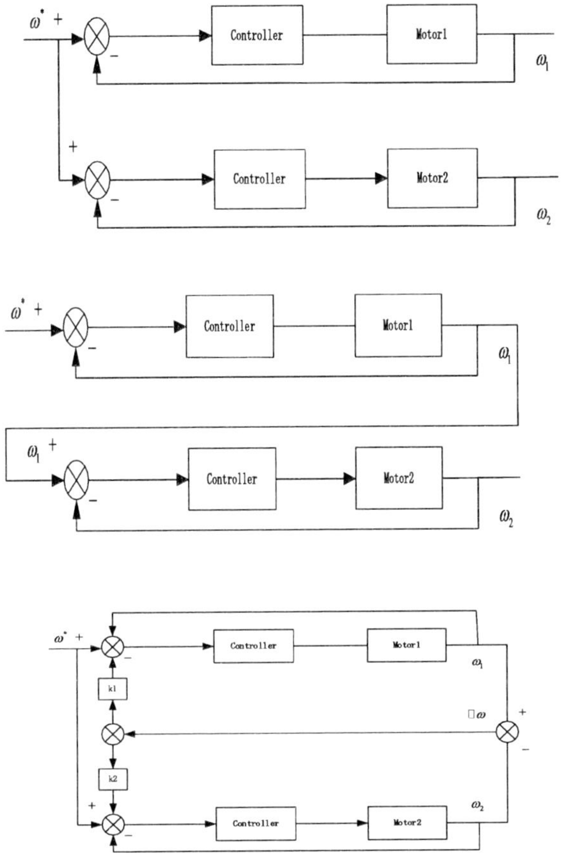

Fig. 4. **a** Master-slave control. **b** Parallel control. **c** Cross-coupled control

Parallel control connects two motors in parallel, receiving the same control signal for synchronized operation, offering a simple and easy-to-control structure. However, it lacks mutual interaction between the motors, resulting in poor synchronization and disturbance rejection.

Cross-coupled control is a closed-loop control method that combines the features of both master-slave and parallel control. It incorporates speed and speed difference feedback signals. Its advantages lie in the closed-loop control, strong disturbance rejection capability, and good dynamic performance. However, it has the disadvantage of complex calculations due to the presence of cross-coupling feedback.

Considering overall performance, this study selects the cross-coupled control method.

4 Simulation and Result Analysis

The dual-motor control system model built using MATLAB is shown in Fig. 5.

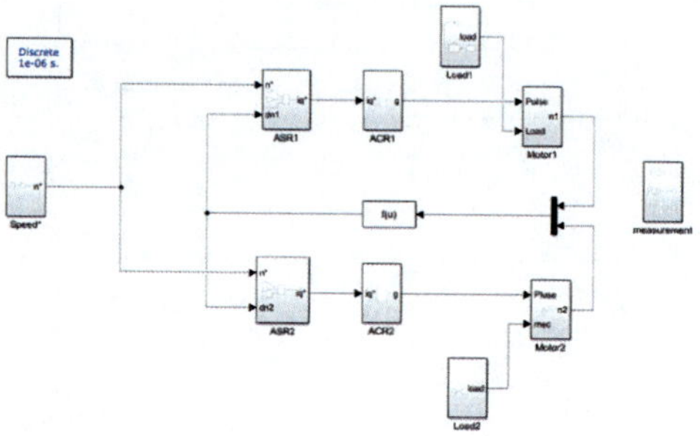

Fig. 5. The dual-motor control system model

From 0 to 0.5 s: The motor starts and reaches the rated speed steadily. At 0.6 s: The load torque of Motor 1 suddenly decreases by 20 Nm, while the load torque of Motor 2 remains unchanged. Due to the similarity in core principles between TV-MPCC and FOC [10], both using three voltage vectors for PMSM control, this paper compares their simulation results to validate the superiority of the TV-MPCC algorithm.

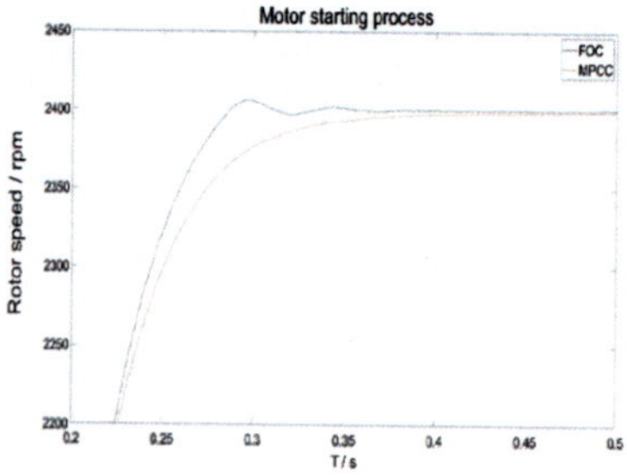

(a) S

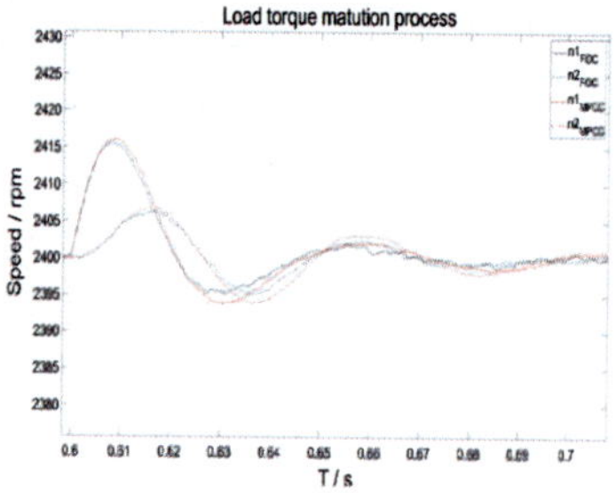

(b) Load mutation process

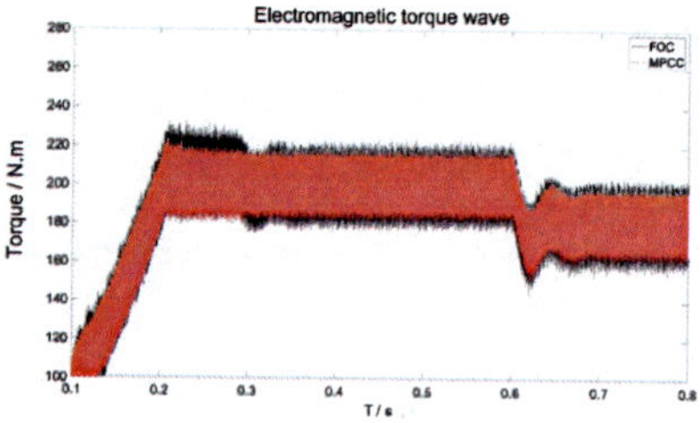

(c) Torque ripple

Fig. 6. Dual motor system speed and torque curve

Starting from Fig. 6a, both control methods show similar rise times, meeting dynamic response speed requirements. Using FOC, the motor reaches 2400 rpm at 0.28 s with a 0.75% overshoot and settling time of 0.37 s. On the other hand, MPCC achieves 2400 rpm at 0.38 s with minimal overshoot and a settling time of 0.4 s. TV-MPCC outperforms FOC, displaying smaller overshoot and settling time, indicating superior steady-state performance.

Figure 6b demonstrates that the dual-motor control system utilizing the TV-MPCC algorithm achieves faster speed synchronization and smoother speed response. In Fig. 6c, during the transition phase from motor startup to stable speed (0.28–0.37s), the torque difference with FOC is 60 Nm, whereas with TV-MPCC, it is 40 Nm. During the stable speed phase (0.4–0.6s), the torque difference with FOC is 46 Nm, whereas with TV-MPCC, it is 35 Nm.

From the above data comparison, it can be concluded that compared to FOC, TV-MPCC provides a smoother motor startup process with smaller speed overshoot and torque ripple. Moreover, when the speed reaches steady-state, TV-MPCC exhibits reduced torque ripple. Therefore, using TV-MPCC as the current regulator improves the overall speed control performance of the motor.

5 Conclusion

Due to the nonlinearity, strong coupling, and multivariable nature of the PMSM, the dual-motor drive system becomes even more complex. The traditional combination of PI control and FOC lacks robustness and has limited dynamic performance. This paper proposes a new approach to control the dual-motor drive system using TV-MPCC. TV-MPCC utilizes the mathematical model of PMSM to construct constraints and employs a flexible design of the cost function to select the optimal voltage vector. Simulation results demonstrate that compared to FOC, the adoption of TV-MPCC yields superior dynamic performance and improved speed control.

References

1. Feng, N., Kui, L., Yao, W.: Model predictive direct torque control for permanent magnet synchronous machines based on duty ratio modulation. Trans. China Electrotech. Soc. **29**(11), 20–29 (2014). (in Chinese)
2. Saeed, M.S.R., Song, W., Huang, L., et al.: Double-vector-based finite control set model predictive control for five-phase PMSMs with high tracking accuracy and DC-link voltage utilization. IEEE Trans. Power Electron. **37**(12), 15234–15244 (2022)
3. Preindl, M., Bolognani, S.: Model predictive direct speed control with finite control set of PMSM drive systems. IEEE Trans. Power Electron. **28**(2), 1007–1015 (2013)
4. Petkar, S.G., Eshwar, K., Thippiripati, V.K.: A Modified model predictive current control of permanent magnet synchronous motor drive. IEEE Trans. Indus. Electron. **68**(2), 1025–1034 (2021)
5. Lin, H., Niu, S., Xue, Z., et al.: A simplified virtual-vector-based model predictive control technique with a control factor for three-phase SPMSM drives. IEEE Trans. Power Electron. **38**(6), 7546–7557 (2023)
6. Nguyen, H.T., Jung, J.W.: Finite control set model predictive control to guarantee stability and robustness for surface-mounted PM synchronous motors. IEEE Trans. Indus. Electron. **65**(11), 8510–8519 (2018)
7. Stellato, B., Geyer, T., Goulart, P.J.: High-speed finite control set model predictive control for power electronics. IEEE Trans. Power Electron. **32**(5), 4007–4020 (2017)

8. Saeed, M.S.R., Song, W., Yu, B., et al.: Generalized deadbeat solution for model predictive control of five-phase PMSM drives. IEEE Trans. Power Electron. **38**(4), 5178–5191 (2023)
9. Yuhao, Y., Fei, P., Yunkai, H.: Overview of multi-motor synchronous motion control technology. Trans. China Electrotech. Soc. **36**(14), 2922–2935 (2021)
10. Abu-Ali, M., Berkel, F., Manderla, M., et al.: Deep learning-based long-horizon MPC: robust, high performing, and computationally efficient control for PMSM drives. IEEE Trans. Power Electron. **37**(10), 12486–12501 (2022)

Initial Rotor Position Detection Based on Peak Dc-Link Current Comparison for SPMSM

Peng Chen(✉) and Ruiqing Ma

School of Automation, Northwestern Polytechnical University, Xi'an 710129, China
cooper_chenpeng@mail.nwpu.edu.cn, marq@nwpu.edu.cn

Abstract. Initial rotor position detection is important for the control of PMSM (Permanent Magnet Synchronous Motor) in rail transit. One method is proposed to detect initial rotor position at standstill based on peak dc-link current comparison in this paper. Initial rotor position detection with resolution of 30° is achieved by injecting four pulse voltages into the motor windings. This method is suitable for the application that rotor keeps at standstill during the starting process. One SPMSM is used to validate the proposed method.

Keywords: Initial rotor position detection · Peak dc-link current comparison · Pulse voltage injection · SPMSM

1 Introduction

Initial Rotor position detection is of great significance for the successful starting and high-performance control of PMSM. Obtaining accurate initial position can prevent the rotor from reversing or even failing to start during the starting process, especially for the application in the electric vehicle and the rail transit [1].

The saliency tracking based method is an important mean to estimate initial rotor position at standstill for PMSM. According to the injection signal is continuous or not, this kind of method can be divided into two categories: high-frequency (HF) voltage injection method and pulse voltage injection method.

Although methods based on HF voltage injection can estimate rotor position precisely while the rotor is standstill [2, 3]. The process that HF voltage injection, HF current demodulation, filter and observer design make the methods complex, and the widely application of the methods is limited.

Different from the HF voltage injection methods, the pulse voltage injection methods inject discrete voltage signals into the windings of the motor, initial rotor position is detected based on voltage comparison [4, 5], inductance calculation [6], current response [7–11]. The resolution may be lower than the methods based on HF voltage injection, but pulse voltage injection methods are more robust and simpler, which can be a desirable option for low-cost systems. For the methods based on voltage com-parison, in [4], two pulse voltages are injected into the AB and AC phases respectively to estimate the possible sectors of rotor position by comparing the terminal voltages measured, the

L. Jia et al. (Eds.): EITRT 2023, LNEE 1135, pp. 413–421, 2024.
https://doi.org/10.1007/978-981-99-9307-9_46

third pulse voltage is used to identify magnetic polarity, the resolution of this method is 30°. In [5], only one dc-link voltage sensor is needed for initial rotor position estimation, the maximum error is ±15°, But the method is sensitive to the dc-link voltage measurement errors, dc-link capacitance, dc-link resistance, and dc-link voltage ripples etc. For the methods based on inductance calculation [6], fundamental and second harmonic components of the line inductance are calculated and extracted by injecting six voltage pulses into the windings. The resolution of initial rotor position is high, but the procedure is complex. For methods based on current response [7–11], the *d*-axis current response-based method is used in [7], as the number of subdivisions increases, the identification accuracy becomes higher. However, when the rotor angle approaches the actual position, the difference of the current before and after subdivisions is small, and angle identification is easily confused. dc-link current response-based method is used in [8, 9]. In [8], five voltage pulses are needed at worst condition and resolution is 30°. In [9], an iterative sequence of voltage pulses with a fuzzy logic processing of the current responses and phase currents derivation based on the dc-link current measurements is combined. Phase current response-based method is used in [10, 11]. In [10], three short time voltage pulses are used for sectors of rotor position detection at unsaturated condition and another three long time pulses are used for polarity identification. The average of the estimation initial rotor position error is 1.14 electrical degrees, and the maximum estimation error is 7.4°. In [11], only three pulses are needed and 15° estimation resolution is achieved by boundary detection strategy. However, this method requires the motor to have a saliency ratio of 10% or above, which imposes a certain challenge for SPMSM.

Aiming at the problems mentioned above, one method is proposed in this paper to detect initial rotor position based on peak dc-link current comparison for SPMSM. Three pulse voltages are injected into the windings of the motor, the possible sectors of the rotor position are determined by comparing the peak dc-link current corresponded. Then, the fourth pulse voltage is injected for magnet polarity identification. Ultimately the initial rotor position is achieved with resolution of 30°. One SPMSM was used to validate the proposed method.

2 Detect Initial Rotor Position Bases on Peak Dc-Link Current Comparison

2.1 Line Inductances Vary Against Rotor Position

For a symmetrical three-phase PM machine, Self-inductance and mutual line inductances of PMSM are shown in (1) and (2), the components above the second harmonic are ignored for simplification. Where, L_{AA}, L_{BB}, L_{CC} are self-line inductances, M_{AB}, M_{BC}, M_{AC} are the mutual-line inductances. *Ls0* is the constant component representing the self-inductance of the winding and *Ls2* is the second harmonic component of the winding, θ is the electrical angle of the rotor.

$$\begin{cases} L_{AA} = L_{s0} + L_{s2}\cos(2\theta) \\ L_{BB} = L_{s0} + L_{s2}\cos(2(\theta - \frac{2}{3}\pi)) \\ L_{CC} = L_{s0} + L_{s2}\cos(2(\theta + \frac{2}{3}\pi)) \end{cases} \tag{1}$$

$$\begin{cases} M_{AB} = -\frac{1}{2}L_{s0} - L_{s2}\cos(2(\theta + \frac{\pi}{6})) \\ M_{BC} = -\frac{1}{2}L_{s0} - L_{s2}\cos(2(\theta + \frac{\pi}{6} - \frac{2}{3}\pi)) \\ M_{AC} = -\frac{1}{2}L_{s0} - L_{s2}\cos(2(\theta + \frac{\pi}{6} + \frac{2}{3}\pi)) \end{cases} \tag{2}$$

Line inductance can be calculated by self-inductances and mutual inductances, take L_{AB} as example:

$$L_{AB} = L_{AA} + L_{BB} - M_{AB} - M_{BA} \tag{3}$$

The line-inductances are calculated by (1), (2), (3), shown as (4).

$$\begin{cases} L_{AB} = 2(\frac{3}{2}L_{s0} + \frac{3}{2}L_{s2}\cos(2\theta + \frac{\pi}{3})) \\ L_{BC} = 2(\frac{3}{2}L_{s0} + \frac{3}{2}L_{s2}\cos(2\theta + \pi)) \\ L_{AC} = 2(\frac{3}{2}L_{s0} + \frac{3}{2}L_{s2}\cos(2\theta - \frac{\pi}{3})) \end{cases} \tag{4}$$

It is shown in Fig. 1 that the relationship between the line inductances varies every 30° electrical angle. So, Rotor position can be determined by the line inductances comparison.

2.2 Initial Rotor Position Detection by Peak Dc-Link Current Comparison

By injecting pulse voltages into phase A and B, voltage equation is shown as (5),

$$U_{dc} = 2iR + L_{AB}\frac{di}{dt} \tag{5}$$

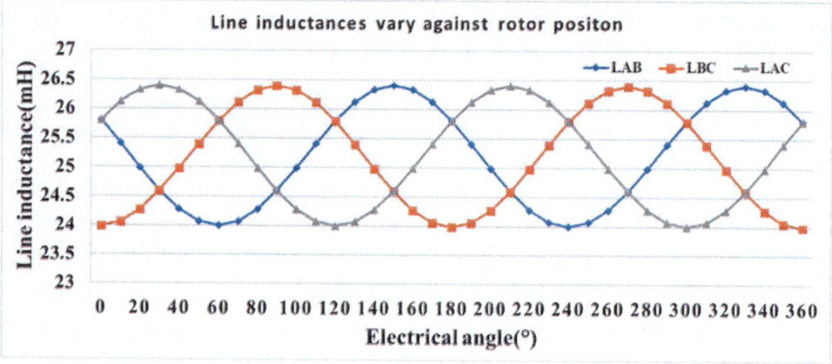

Fig. 1. Line inductances vary against rotor positions

where U_{dc} is the bus voltage, R is the phase resistance, L_{AB} is the line inductance between phase A and phase B, and i is the line current.

Considering that the pulse voltage width typically lasts only a few hundred microseconds, so, the voltage drop on the resistor can be ignored. In addition, during the pulse voltage injection process, (5) can be simplified as (6) when the average change rate is used instead of differential calculation.

$$U_{dc} = L_{AB}\frac{i_{peak}}{\Delta T} \tag{6}$$

It is shown that peak dc-link current is inversely proportional to the line inductance when the bus voltage and injection time are constant. So, initial rotor position can be detected by comparing peak dc-link current corresponded.

3 Proposed Method to Detect Initial Rotor Position

There are two steps consisted of the proposed method to detect initial rotor position.

Step 1: Determine the possible sectors the rotor is in. When pulse voltages of BHCL, AHCL, and AHBL are injected into the motor winding, I1, I2, and I3 are the peak dc-link currents corresponded. Take the pulse voltage AHBL as an example, AHBL represents upper transistor of phase A and the lower transistor of phase B are switched on, as shown in Fig. 2a. Sectors of rotor position are detected by comparing the peak values of bus current after three pulse voltages are injected.

Step 2: Magnetic polarity identification. The fourth pulse voltage is selected for polarity identification based on the magnetization or demagnetization effect of positive and negative pulse voltage injection. The selection of the fourth pulse voltage is shown in Fig. 2b.

Assuming that rotor position is determined at sector 4 or sector 10 by the first step. Then, pulse voltage CHAL is selected as the fourth pulse voltage for magnetic polarity identification. For example, by comparing the peak value of current responded to pulse voltage of CHAL and AHCL, when I2 > I4, it indicates that magnetization is increased by the second pulse voltage, as a result, magnetic field saturation is increased and inductance is reduced. Ultimately the rotor initial position is determined at sector 4 by the two steps. The detailed rules of the proposed method are shown in Table 1.

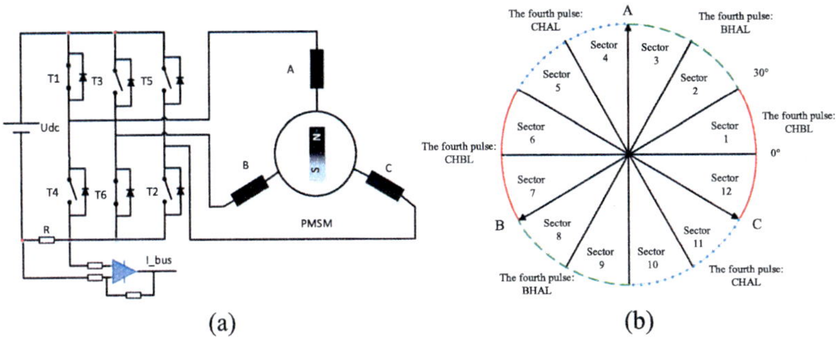

Fig. 2. Proposed method to detect initial rotor position. **a** hardware for the proposed method, **b** the fourth pulse voltage injected for polarity identification.

Table 1. Rotor initial position detection method proposed in this paper

Comparison of dc-link current	Comparison of line inductances	Possible sectors	The fourth pulse	Current comparison	Rotor position
I2 < I3 < I1	$L_{AC} > L_{AB} > L_{BC}$	1,7	CHBL	I4 > I1	0–30°
				I4 < I1	180–210°
I2 < I1 < I3	$L_{AC} > L_{BC} > L_{AB}$	2,8	BHAL	I4 < I3	30–60°
				I4 > I3	210–240°
I1 < I2 < I3	$L_{BC} > L_{AC} > L_{AB}$	3,9	BHAL	I4 < I3	60–90°
				I4 > I3	240–270°
I1 < I3 < I2	$L_{BC} > L_{AB} > L_{AC}$	4,10	CHAL	I4 < I2	90–120°
				I4 > I2	270–300°
I3 < I1 < I2	$L_{AB} > L_{BC} > L_{AC}$	5,11	CHAL	I4 < I2	120–150°
				I4 > I2	300–330°
I3 < I2 < I1	$L_{AB} > L_{AC} > L_{BC}$	6,12	CHBL	I4 < I1	150–180°
				I4 > I1	330–360°

4 Experiment Validation

4.1 Experiment Platform

Parameters of the SPMSM used are shown in Table 2.

During the experiment, two drivers are connected to the power supply through common DC bus. Motor I is the tested SPMSM, Motor II is used to pulls the tested motor to the accurate position, then pulse voltages are injected into the tested SPMSM for initial rotor position detection.

Table 2. Parameters of the test motor

Parameter	Value	Parameter	Value
Rated voltage	110 VAC	Pole pairs	5
Rated current	1.5 A	Resistance	2.8 Ω
Rated power	200 W	*Lq*	13 mH
Rated torque	0.64 Nm	*Ld*	13 mH

4.2 Initial Rotor Position Detection

Figure 3 shows the detection result when rotor position is in the range of 90–120°. Firstly, three pulse voltages BHCL, AHCL, AHBL are injected into the windings of the motor, by comparing of the peak dc-link currents responded, it is obtained that I1 < I3 < I2. The rotor position is determined within the range of 90–120° or 270–300° by Table 1. Then, CHAL is selected as the fourth pulse voltage injection, I4 < I2 indicates that the rotor position is between 90 and 120°.

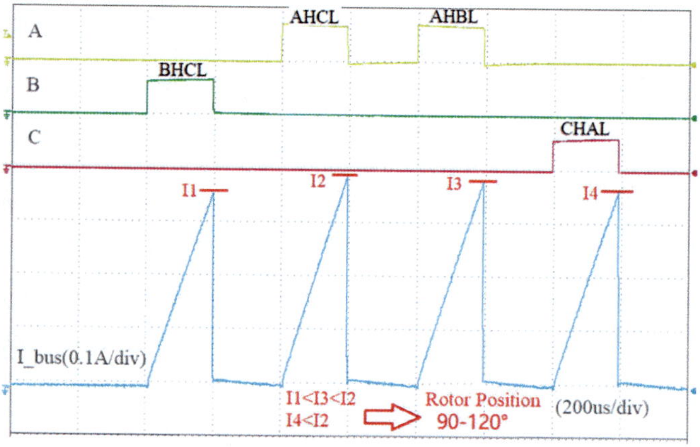

Fig. 3. Rotor position detection at 90–120°

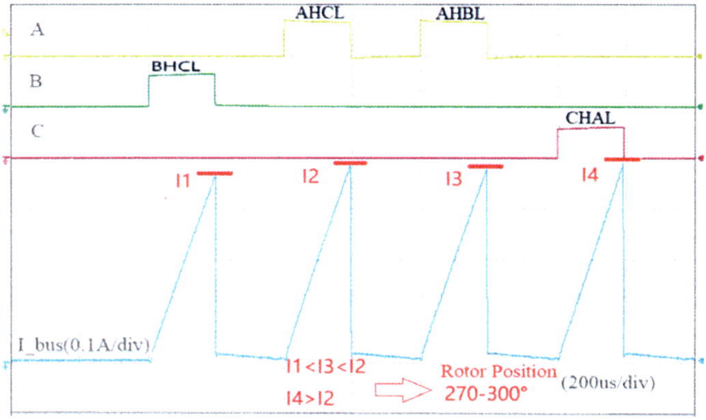

Fig. 4. Rotor position detection at 270–300°

Figure 4 is the detection results when the rotor position is in the range of 270–300°. Different from the result in Fig. 3, by the injection of fourth pulse voltage of CHAL voltage, the corresponding bus peak current I4 > I2. The rotor position is determined in the range of 270–300° according to Table 1.

Figure 5 shows the initial rotor position detection result within the 360°. The horizontal axis is the actual position from encoder, and the vertical axis is the detected rotor position. It can be seen that the proposed method is capable to detect rotor initial position of SPMSM within 360°, the resolution is 30°.

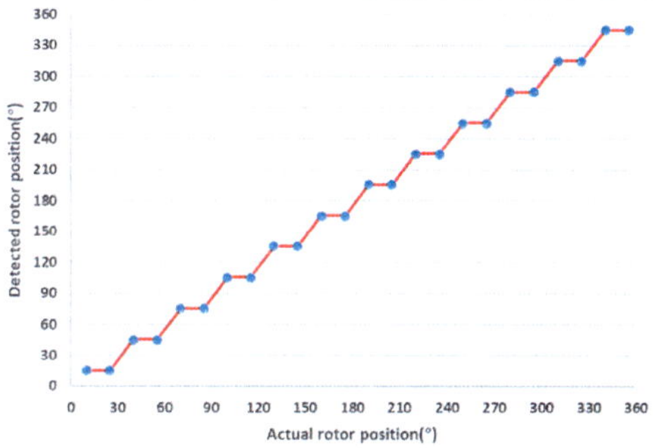

Fig. 5. Initial rotor position detection from 0° to 360°

Some features of the proposed method and the methods in [4, 6, 7, 10] are compared in Table 3.

Table 3. Comparison of features in [4, 6, 7, 10] and the proposed method

Method	[4]	[6]	[7]	[10]	Proposed method
Features	Voltage comparison	Inductance calculation	Current analysis	Current analysis	Current analysis
Numbers of pulses	3	6	21	6	4
Average estimation error	15°	7.94°	3.8°	1.14°	15°
Polarity identification	√	√	x	√	√
Voltage detection circuit	2	x	x	x	x
Phase current sensors	x	2	2	3	x
Dc-link current sensors	1	x	x	x	1
Complexity	Simple	Complex	Complex	Complex	Simple

5 Conclusion

One method is proposed to detect initial rotor position for SPMSM based on peak dc-link current comparison in this paper. By utilizing the magnetic saturation saliency effect, four pulse voltages are injected into the windings of the motor, and the initial rotor position with resolution of 30° is achieved by comparing the peak dc-link current responded. The advantage of the method proposed is that it does not require additional phase voltage detection circuits, nor does it require complex inductance parameter identification or current calculation processes. This method is suitable for the application that rotor keeps at standstill during the starting process.

References

1. Anayi, F.J., Al Ibraheemi, M.M.A.: Estimation of rotor position for permanent magnet synchronous motor at standstill using sensorless voltage control scheme. IEEE/ASME Trans. Mechatron. 25(3), 1612–1621 (2020)
2. Xu, P., Zhu, Z.Q.: Initial rotor position estimation using zero-sequence carrier voltage for permanent-magnet synchronous machines. IEEE Trans. Ind. Electron. 64(1), 149–158 (2017)
3. Shuang, B., Zhu, Z.-Q.: A novel sensorless initial position estimation and startup method. IEEE Trans. Ind. Electron. **68**(4), 2964–2975 (2021)
4. Champa, P., Somsiri, P., Wipasuramonton, P., Nakmahachalasint, P.: Initial rotor position estimation for sensorless brushless DC drives. IEEE Trans. Ind. Appl. 45(4), 1318–1324 (2009)

5. Wu, X., Zhu, Z.Q., Wu, Z.: A novel rotor initial position detection method Utilizing DC-link voltage sensor. IEEE Trans. Ind. Appl. 56(6), 6486–6495 (2020)
6. Zhang, S., Kang, J., Siyuan, M.: Initial rotor position detection for permanent magnet synchronous motor based on identical width voltage pulse injection. Proc. CSEE **40**(19), 6085–6092 (2020). (in Chinese)
7. Nakashima, S., Inagaki, Y., Miki, I.: Sensorless initial rotor position estimation of surface permanent-magnet synchronous motor. IEEE Trans. Ind. Appl. **36**(6), 1598–1603 (2000)
8. Lee, W.-J., Sul, S.-K.: A new starting method of BLDC motors without position sensor. IEEE Trans. Ind. Appl. **42**(6), 1532–1538 (2006)
9. Boussak, M.: Implementation and experimental investigation of sensorless speed control with initial rotor position estimation for interior permanent magnet synchronous motor drive. IEEE Trans. Power Electron. **20**(6), 1413–1422 (2005)
10. Tursini, M., Petrella, R., Parasiliti, F.: Initial rotor position estimation method for PM motors. IEEE Trans. Ind. Appl. **39**(6), 1630–1640 (2003)
11. Wu, X., Zhu, Z., Wu, Z.: A rotor initial position estimation method for surface-mounted permanent magnet synchronous machine. IEEE Trans. Energy Convers. **36**(3), 2012–2024 (2021)

Research on Electromagnetic Force Characteristics Based on Asymmetric 120° Pitch Figure-8 Coils in EDS System for Maglev Train

Zhigang Xue[1,2], Rong Wei[1](✉), Yukang Gao[1,2], and Qiongxuan Ge[1]

[1] Key Laboratory of Power Electronics and Electric Drive, Institute of Electrical Engineering, Chinese Academy of Sciences, Beijing 100190, China
{xuezhigang,weirong,gaoyukang,gqx}@mail.iee.ac.cn
[2] University of Chinese Academy of Sciences, Beijing 100190, China

Abstract. Insufficient suspension force of superconducting electrodynamic suspension (EDS) system at low speed is an important problem in the construction of maglev trains. The structure of the figure-8 ground coil plays a crucial role in the performance of suspension and guidance. An asymmetric 120° pitch figure-8 coil used in the suspension system of maglev train is proposed. Compared with the symmetrical structure, the levitation force generated by the figure-8 coil of the asymmetric structure is greatly improved and the levitation force can reach twice as much as that of the symmetrical structure. Then the corresponding relationship between the train offset and the asymmetry of the figure-8 coil is proposed. Under this corresponding relationship, the structural change of the figure-8 coil reaches about 10%. Finally, the simulation model of suspension system of superconducting linear motor is established by using 3D finite element software to verify the validity of the calculated results.

Keywords: EDS · Figure-8 coil · Spatial harmonic method · 3D-FEM

1 Introduction

With the development of transportation, it is difficult for traditional wheel-rail transportation to reach a higher speed due to factors such as wheel-rail relationship, arch network relationship, aerodynamics and energy consumption. Continuing to develop wheel-rail transport would be more cost and less effective [1–3]. Maglev transportation benefits from the absence of direct mechanical contact between maglev train and track. It is not limited by the adhesion limit of traditional wheel-rail system, which is the only passenger transportation vehicle on the ground that can reach 500 km/h. At present, the mainstream magnetic levitation technology is divided into electro magnetic suspension (EMS), electro dynamics suspension (EDS), high-temperature superconducting magnetic levitation (HTS Maglev) using magnetic flux pinning technology. At present, the technology of EMS and EDS is mastered by Germany and Japan, respectively [4]. And the HTS Maglev is in the stage of research and development and testing, its commercial operation has not yet been realized.

L. Jia et al. (Eds.): EITRT 2023, LNEE 1135, pp. 422–430, 2024.
https://doi.org/10.1007/978-981-99-9307-9_47

However, maglev trains cannot generate enough levitation force at low speed and need auxiliary wheels to support the vehicle. As a result, there is a lot of energy consumption and wheel-rail wear [5]. How to reduce the float speed is an important topic in the development of maglev train.

The suspension and guidance system of EDS maglev consists of figure-8 ground coils (abbreviated as figure-8 coils). The structure of the figure-8 coil affects the levitating and guiding performance of maglev. It also has a great influence on the floating speed of the maglev. At present, the research on figure-8 coils mainly focuses on coils with 60° pitch. In [6], Yao Cai and others calculated the electromagnetic force of symmetrical coil by dynamic circuit combined with semi-analytical method. In [7], Guo and others proposed a vacuum tube EDS maglev train architecture and conducts electromagnetic force analysis. In [8], H. Huang and others studied the dynamic electromagnetic characteristics of EDS trains through 3D finite element simulation and analyzed the dynamic performance of its suspension force and magnetic resistance. In [9], T. Murai and others optimized propulsion and suspension steering system by optimizing algorithm to improve suspension force and motor efficiency. At present, the 60° pitch figure-8 coil studied by above scholars cannot provide enough suspension force when the train is running at low speed. If a larger pitch coil is used and the asymmetry of the coil is changed, the suspension ability of the train will be improved at medium and low speed.

In this paper, a mathematical model of suspension system of superconducting linear motor is established. The spatial magnetic field distribution generated by the superconducting magnet is derived by the spatial harmonic method. Based on the fixed parameters of the superconducting magnet and the 120° pitch figure-8 coils, the magnetic flux, voltage and current induced by the space magnetic field of the figure-8 coil are deduced. Then the electromagnetic force is solved by energy method. Finally, three-dimensional finite element simulation model (3D-FEM) is used to verify the validity of the analytical results.

2 Analytical Model of EDS System

The EDS system cross-section is shown in Fig. 1a. The whole system is composed of maglev train, on-board superconducting magnet, U-shaped ground track and ground coil. Ground coils include propulsion coil and figure-8 (suspension) coil. Superconducting magnets are placed on both sides of the vehicle. And the magnetic fields of the left and right superconducting magnets face out in opposite directions. The propulsion coil and superconducting magnet form the long-stator linear motor, and the figure-8 coil and SCM form the suspension steering system. As shown in Fig. 1b.

The upper and lower loops of the figure-8 coil are connected in reverse. This structure ensures that the magnetic poles generated by the upper and lower loops of the coil are always opposite.

The principle of generating suspension force is shown in Fig. 1b. When the SCM runs under the center line of the figure-8 coil, (that is saying the downward offset $\Delta z > 0$, shown in Fig. 1a) the magnetic flux of the upper and lower loop of the figure-8 coil is different. Therefore, the total flux is not zero, and the induced current is generated. Due to the reverse connection of the upper and lower loops of the figure-8 coil, magnetic

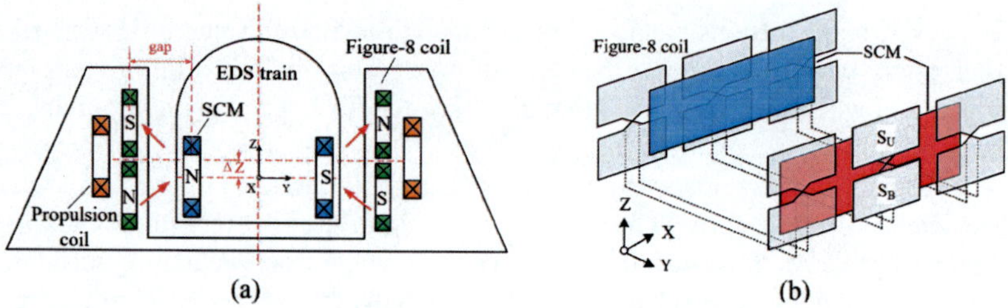

Fig. 1. Cross-section (**a**) and side view (**b**) of the levitation system of the EDS maglev train

fields of different polarity are generated. The lower loop creates a repulsive force and the upper loop creates an attractive force, which causes the train to levitate to balance the weight of the train. Therefore, as long as the SCM movement, there will be a suspension force.

The magnetic induction intensity generated by SCM in the y direction in space can be obtained by the spatial harmonic method [10],

$$B_y = \sum_{m=1}^{\infty}\sum_{n=1}^{\infty} \frac{8\mu_0 N_s I_s}{mn\pi^2}\lambda_{mn}\sin(k_{xm}\frac{a_0}{2})\sin(k_{zn}\frac{b_0}{2})\cos(k_{xm}x)\cos[k_{zn}(z-z_0)e^{-\lambda_{mn}y}] \tag{1}$$

where μ_0 is the vacuum permeability, N_sI_s is the current of the superconducting magnet, a_0 and b_0 are the length and width of the SCM respectively, $k_{xm} = m\pi/\tau$, $k_{zn} = n\pi/W_z$, $\lambda_{mn} = \sqrt{k_{xm}^2 + k_{zn}^2}$, τ and W_z are the polar distance of the superconducting magnet in the x and z directions respectively, and y is the distance from the superconducting magnet in the y direction.

As shown in Fig. 1b, the net magnetic flux of the v-th figure-8 coil can be obtained by integrating the magnetic induction intensity over the area where a single figure-8 coil is located

$$\Phi_{1v} = N_l \iint_{S=S_B-S_U} B_y\big|_{y=g}\, dS \tag{2}$$

where N_l is the number of turns in the figure-8 coil, S_U and S_B are the magnetic flux of the upper and lower loops respectively, g is the distance between the SCM and the figure-8 coil.

Let the resistance and inductance of the figure-8 coil be R and L, by

$$Ri_{1v} + L\frac{di_{1v}}{dt} = -N_1\frac{d\Phi_{1v}}{dt} \tag{3}$$

The current of the vth figure-8 coil can be obtained.

Finally, the energy method is used to obtain the force in a direction by taking the derivative in that direction

$$\vec{f_r} = \frac{\Phi_{1v} i_{1v}}{\partial \vec{r}} \tag{4}$$

3 Electromagnetic Force Characteristics

The model was calculated according to the parameters in Table 1 [11].

Table 1. Parameters of EDS maglev

Parameter	Symbol	Value
SCM		
Width (m)	a_0	1.07
Height (m)	b_0	0.50
Pitch (m)	τ_s	1.35
Turns	N_s	1400
Current (A)	I_s	500
Running velocity (km/h)	v	500
120° pitch figure-8 coil		
Width (m)	c	0.35
Height (m)	h	In Table 2
Pitch (m)	W_z	0.90
Turns	N_l	12
Gap between SCM and figure-8 coil (m)	g	0.185

Table 2. Parameter of 120° pitch figure-8 coil

Figure-8 coil	Case 1	Case2	Case3	Case4
h_1 (mm)	205	240	275	310
h_2 (mm)	415	380	345	310
R1 (Ω)	0.0120	0.0124	0.0129	0.0133
R2 (Ω)	0.0146	0.0142	0.0137	0.0133
L_1 (uH)	347	388.2	426.3	457.5
L_2 (uH)	561.25	529.5	496.35	457.5
M_{12} (uH)	42.6	44.5	45.6	44.86

In order to measure the degree of asymmetry of the figure-8 coil, the asymmetry coefficient K is defined to represent the ratio between the upper coil and the total coil height.

$$K = \frac{h_1}{h_1 + h_2} \tag{5}$$

For case 1–4, the K value fraction is 0.331, 0.387, 0.444, 0.500. Among them, $K = 0.5$ represents a symmetrical coil, and the smaller the value of the asymmetry coefficient K, the greater the degree of asymmetry.

3.1 Effect of K on the Characteristics Levitation System

In the case of case 1–4, the effect of the asymmetry of 120° pitch figure-8 coil on the force of the train during operation can be clearly seen.

Figure 2a shows the trend diagram of the levitation force increasing with the speed under the case of case 1–4. Generally, the levitation force curve rises rapidly with the increase of the speed within the range of 0–200 km/h. And the levitation force reaches its maximum value at 200 km/h, and then tends to be stable with the increase of the speed, basically no longer rising. The figure-8 coil with a higher asymmetry has a greater levitation force in the high-speed phase (200–600 km/h). The levitation force in the case of case 1 is twice that in the case of case 4, which means that the use of the levitation coil with a higher asymmetry can obtain a greater levitation force and the train can withstand a larger load. In addition, the figure-8 coil with higher asymmetry increases faster with the speed at the low speed stage (0–200 km/h). And the levitation force equivalent to the train can be obtained earlier. So that the fluctuating speed can be reached earlier. Figure 2b shows the trend of magnetic resistance increasing with speed in case 1–4. It can be seen that the figure-8 coil with higher asymmetry has greater magnetic resistance at each speed. This is also expected because the higher the current passing through the levitation coil, the greater the magnetic resistance to the superconducting magnet. At low speed, the magnetic resistance difference is obvious, and at high speed, the magnetic resistance of the suspended coil with different symmetry tends to be the same. Figure 2c shows the levitation/resistance ratio in case of case 1–4. It can be seen that the levitation/resistance ratio of the levitation coil with higher asymmetry is larger and the levitation/resistance ratio is better.

3.2 Relationship Between the ΔZ and K

Figure 3 shows the levitation force of the train under the same speed and different symmetry. There are three ways for EDS maglev to achieve maximum levitation. First, the speed must be large, but the higher the speed, the less obvious the growth of the levitation force. Second, the train offset ΔZ should be large, but the offset setting is related to the size of the levitation system and is not easy to change. Third, the asymmetry of the 8-word coil is high, but the asymmetry of the speed will be an inflection point. That is saying, a certain speed corresponds to a specific figure-8 coil asymmetry at high speed.

Figure 4 shows the relationship between the offset and the optimal asymmetry at high speed.

4 Model Validation

ANSYS Maxwell software was used for 3D simulation to verify the results of analytical calculation. The simulation model is shown in Figs. 5 and 6. The solution region is a static region. The Moving region where the superconducting coil is located is set to

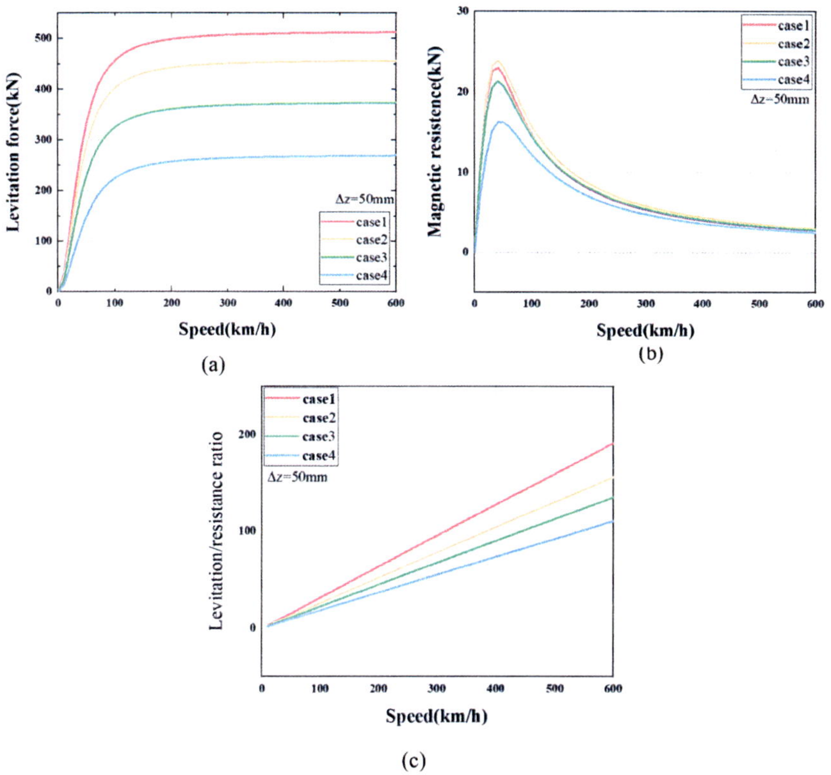

Fig. 2. The levitation force (**a**), magnetic resistance (**b**) and levitation/resistance ratio (**c**) of the vehicle based on 120° pitch figure-8 coil at different symmetries

the moving region. The fixed region where the levitation coil is located is set to the stationary region. The two faces in the x direction of the solution domain boundary are set respectively as the independent-dependent boundary. The SCM drives out of the independent boundary and in from the depend boundary, forming a loop to simulate the motion of the SCM in the infinite figure-8 coils.

It can be seen from Fig. 7 that the trend of analytical results and simulation results is the same, and the value is similar. The correctness of the analytic results can be verified.

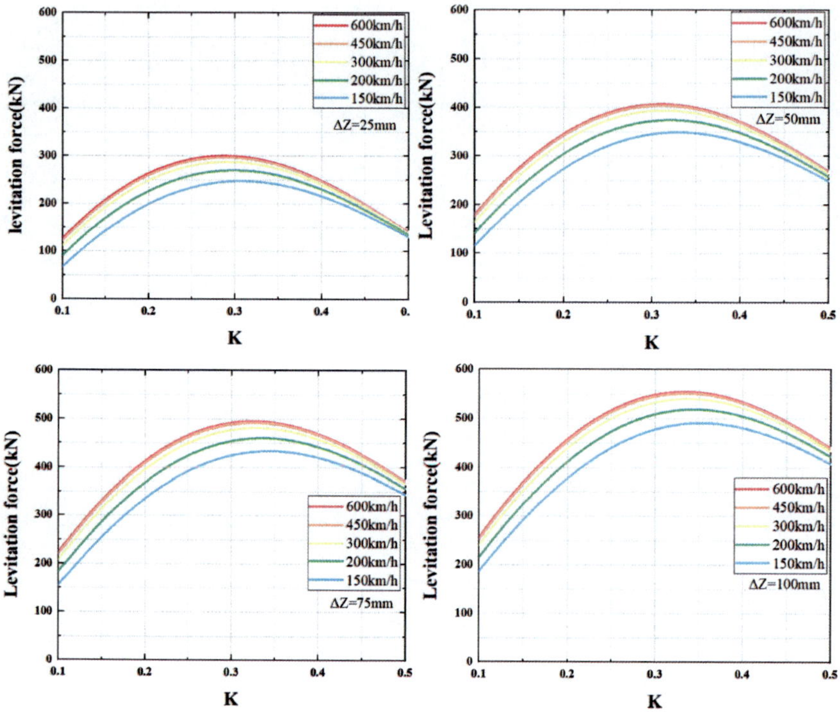

Fig. 3. Levitation force of the train under the different asymmetry and offset

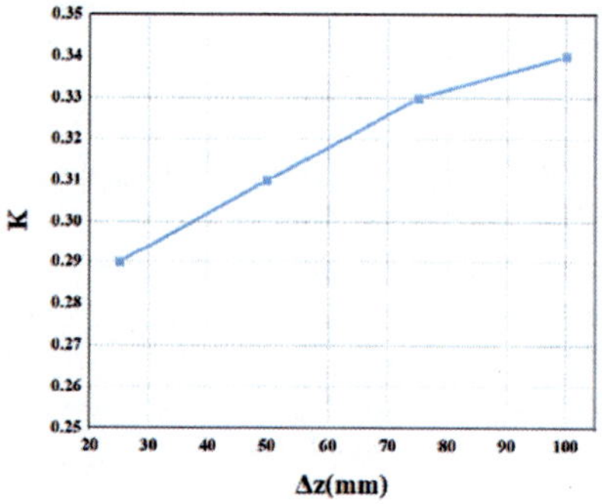

Fig. 4. Relationship between the offset ΔZ and the optimal asymmetry K

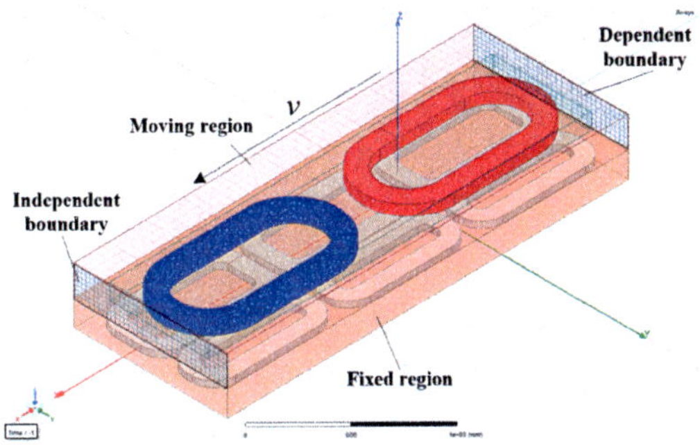

Fig. 5. 3D-FEM simulation model of EDS system.

5 Conclusion

In this paper, the mathematical analytical model of EDS levitation system is established by using spatial harmonic method. The influence of the 120° pitch figure-8 coil with different asymmetry on the running performance of EDS system is studied.

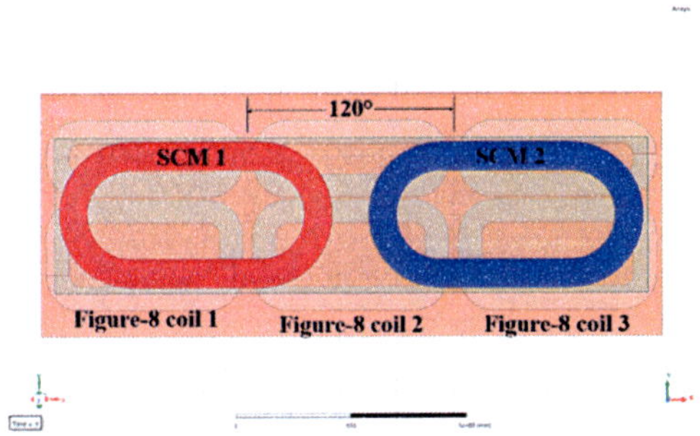

Fig. 6. Side view of simulation model of EDS system

And a three-dimensional finite element simulation model is established. Compared with the calculated results, the calculation error is 7.13%, which can prove the validity of the calculated results. The conclusions are as follows:

1. The levitation system using a 120° levitation coil with a higher asymmetry has a higher levitation force at all speeds. At the same time, the magnetic resistance is lower when the speed is higher. The levitation/resistance ratio is also correspondingly larger. In terms of levitation force and levitation/resistance ratio, the performance of asymmetric structure is more than twice that of symmetrical structure.
2. At high speeds, the specific offset has a unique correspondence with the most suitable asymmetry. With the change of offset, the size optimization of the figure-8 coil can reach 30 mm, which is close to 10% for the figure-8 coil with a width of 310 mm.

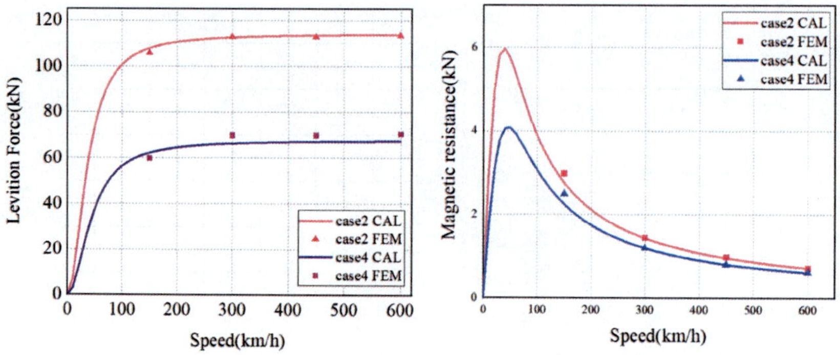

Fig. 7. Comparison of levitation and magnetic resistance between 3D-FEM simulation and analytical calculation

References

1. Thornton, R., Thompson, M.T., Perreault, B.M.: Linear motor powered transportation. Proc. IEEE **97**(11), 1754–1757 (2009)
2. Lv, G., Zhang, Z., Liu, Y., Zhou, T.: Characteristics analysis of linear synchronous motor integrated with propulsion, levitation, and guidance in high-speed maglev system. IEEE Trans. Transp. Electrification 7(4), 3185–3193 (2021)
3. Wang, J., Wang, Y., Peng, Y., Lu, J.J.: Examining partial proportional odds model in analyzing severity of high-speed railway accident. Smart Resilient Transp. 3(1), 12–24 (2021)
4. Tsuchiya, Y., Sakai, I., Mizuno, K., Kohama, Y., Yoshida, Y., Awaji, S.: Critical current measurements of HTS tapes using pulsed current in high fields at low temperatures. IEEE Trans. Appl. Supercond. 33(5), 1–5 (2023)
5. Murai, T., Fujiwara, S.: Characteristics of combined propulsion, levitation and guidance system with asymmetric figure between upper and lower coils in EDS. IEEJ Tran. Ind. Appl. 1289–1296 (1996)
6. Cai, Y., Ma, G., Wang, Y., et al.: Semianalytical calculation of superconducting electrodynamic suspension train using figure eight-shaped ground coil. IEEE Trans. Appl. Supercond. **30**(5), 1–9 (2020)
7. Guo, Z., Li, J., Zhou, D.: Study of a null-flux coil electrodynamic suspension structure for evacuated tube transportation. Symmetry **11**, 1239 (2019)
8. Huang, H., Deng, Z., Li, H., Zhu, H., Shi, H., Zheng, J.: Numerical simulation of dynamic electromagnetic characteristics of superconducting electrodynamic suspension (EDS) train. In: IEEE Trans. Appl. Supercond. **31**(5), 1–5 (2021)
9. Murai, T., Fujiwara, S.: Design of coil specifications in EDS Maglev using an optimization program. In: Proceedings of 2nd International Symposium on Linear Drives for Industry Applications (LDIA), pp. 343–346 (1998)
10. Nonaka, S., et al.: Analysis of characteristics of repulsive magnetic levitated train using a space harmonic technique. Electr. Eng. Jpn. **100**, 80–88 (1980)
11. Fujimoto, T., et al.: Characteristics of electromagnetic force of ground coil for levitation and guidance at the Yamanashi maglev test line. Quart. Rep. RTRI **41**(2), 63–67 (2000)

Performance Evaluation on Levitation Control System of High-Speed Maglev Train

Mingda Zhai(✉), Pinglang Yi, Xiaolong Li, and Zhiqiang Long

College of Intelligence Science and Technology, National University of Defense Technology, Changsha 410000, China

zhaimd@126.com

Abstract. After the completion of the levitation control system design for high-speed maglev train, the performance of the control system under complex working conditions needs to be further analyzed. In this chapter, the nominal performance of the levitation system is firstly analyzed and evaluated using the nominal performance index. Secondly, according to the working characteristics of the levitation system under dynamic operating conditions, the dynamic performance index and evaluation criteria suitable for the levitation system of the high-speed maglev train are proposed. Finally, the performance of the levitation system under non-deterministic disturbance is quantitatively analyzed.

Keywords: Levitation control system · High-speed maglev train · Performance evaluation · Non-deterministic disturbance

1 Introduction

The high-speed maglev train is an entirely new mode of transport without wheels. It takes full advantage of electromagnetic force to achieve active levitation and runs the train by the linear motor. Compared with the conventional high-speed rail, it is generally believed that the high-speed maglev train have not only less resistance, less noise, lower cost, but also higher speed [1–3]. The levitation control system of high-speed maglev train plays a significant role in suspending the maglev train above the rail against the gravity [4, 5]. The safety and reliability issues on levitation control system are actually the most critical factors for the high-speed maglev train. However, the performance evaluation on levitation control system of high-speed maglev train is often overlooked after the completion of the levitation control system design for high-speed maglev train. The performance of the control system under complex working conditions needs to be further analysed. This subject is concerned chiefly with the study of performance evaluation on levitation control system of high-speed maglev train. Firstly, the nominal performance of the levitation system is analysed and evaluated using the nominal performance index. Secondly, according to the working characteristics of the levitation system under dynamic operating conditions, the dynamic performance index and evaluation criteria suitable for the levitation system of the high-speed maglev train are proposed. Finally, the performance of the levitation system under non-deterministic disturbance is quantitatively analysed.

L. Jia et al. (Eds.): EITRT 2023, LNEE 1135, pp. 431–438, 2024.
https://doi.org/10.1007/978-981-99-9307-9_48

2 Nominal Performance of the Levitation Control System

To analyze the control performance of the levitation system with the implemented nominal controller, it is essential to characterize its corresponding evaluation criteria and performance indicators. The nominal performance of the system is analyzed and evaluated in the following.

The initial levitation gap is set to 20 mm, while the desired gap is set to 12 mm. When the levitation command is sent, the reference input signal of the levitation system can be represented by:

$$\omega(t) = \begin{cases} 0.02\, t < 0 \\ \int_{0.02}^{0.012} -0.003\mathrm{d}t\; t \geq 0 \end{cases} \tag{1}$$

When the levitation system receives the levitation command, the dynamic responses of the levitation electromagnets on the left and right sides of the joint structure are shown in Figs. 1 and 2.

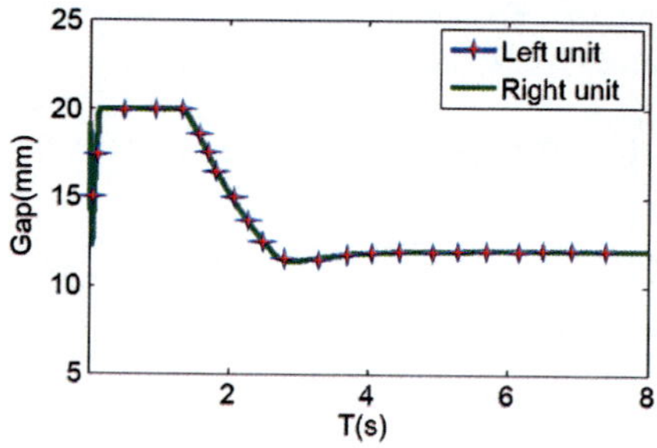

Fig. 1. Dynamic responses of the levitation gap

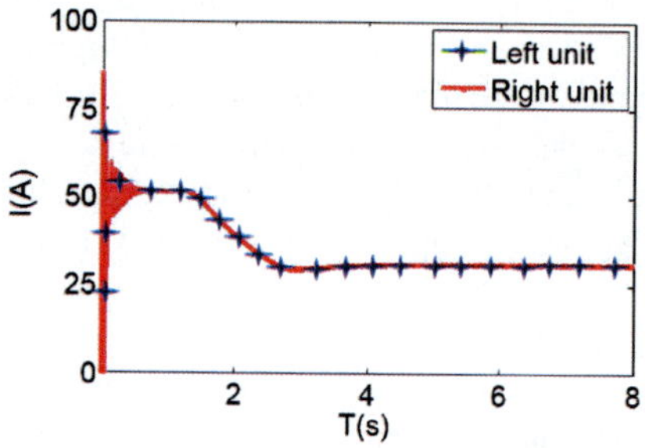

Fig. 2. Dynamic responses of the levitation current

The levitation electromagnets on both sides can finally be stabilized at the target gap, which indicates that the levitation control system with the nominal controller is stable. When the given reference signal of the levitation system experiences a sudden

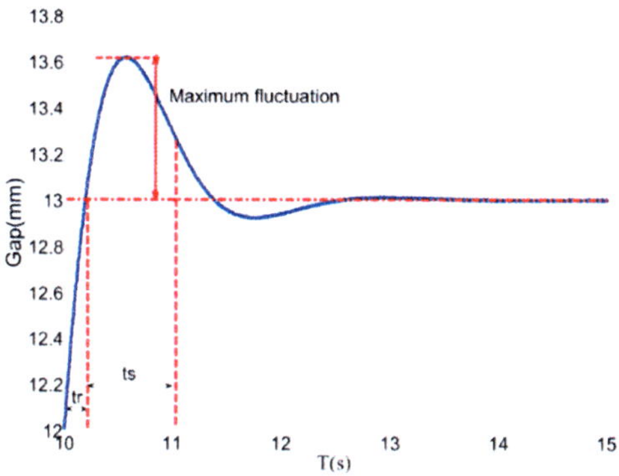

Fig. 3. Transient response of the levitation control system for high-speed maglev train

step change with an amplitude of 13 mm, the control system must be able to adapt to it immediately. The transient response of the levitation gap is shown in Fig. 3.

By analyzing the transient response given in Fig. 1, the transient performance of the levitation system for the high-speed maglev train under the action of the nominal controller is summarized in Table 1.

Table 1. Transient performance index of levitation control system for high-speed maglev train

Transient performance index	Maximum overshoot σ %	Rise time tr	Settling time ts	Cycles of oscillation μ
Value	4.7%	0.2	1.03	1

3 Performance Evaluation Method of Levitation Control System Under Dynamic Operating Conditions

After the design of the nominal controller for the levitation system is completed, the nominal performance index has been applied to analyze and evaluate the nominal performance of the levitation system. However, during the operation phase of the high-speed maglev train, the levitation system is always in a dynamic balance state, in which case using the nominal performance index to evaluate the performance of the levitation system is deficient. Therefore, a dynamic performance evaluation index that applies to the levitation control system is proposed.

3.1 Deficiencies in Using Nominal Performance Index

The levitation system of high-speed maglev train is a tracking control system, and the control objective is to achieve reference tracking and disturbance rejection [6]. Therefore, the levitation control algorithm is always regulating the system in real time so that a dynamic balance state can be maintained between the output and input [7, 8]. The levitation gap fluctuation curve is shown in Fig. 4.

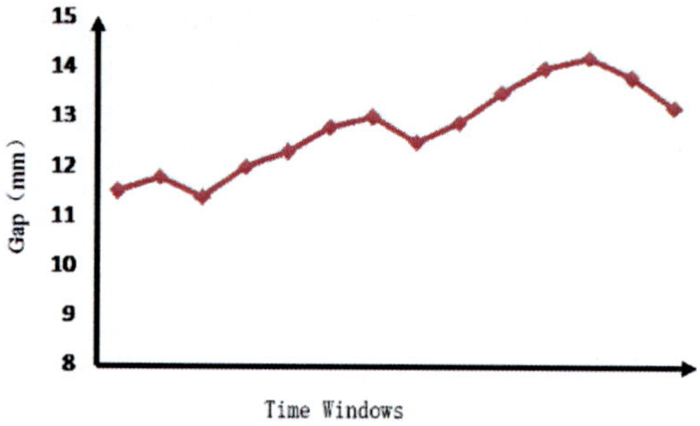

Fig. 4. Levitation gap fluctuation curve

3.2 Performance Evaluation of Levitation Control System Under Dynamic Operating Conditions

Generally, a smaller fluctuation range of the levitation gap indicates better control performance of the levitation system. The fluctuation range of the levitation gap can be directly leveraged to evaluate the control performance of the levitation system for high-speed maglev trains during the operation phase. The fluctuation interval of the levitation gap and the corresponding control performance level are defined in Fig. 5.

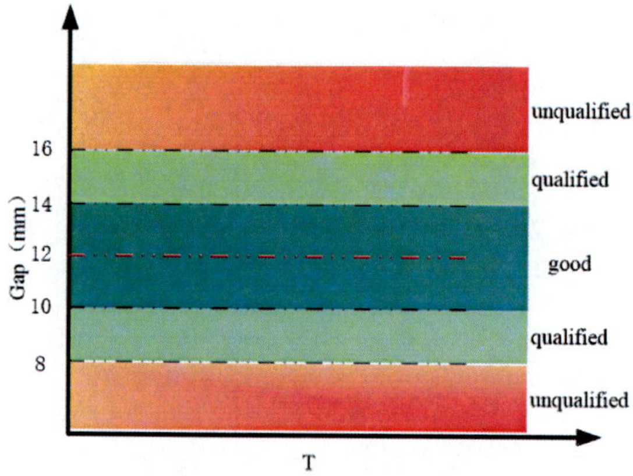

Fig. 5. Levitation gap fluctuation range and control performance level

However, when the fluctuation range of the levitation gap is within the range of 10–14 mm, and the levitation electromagnet vibrates violently at this time, it will also affect the stability and comfort of the high-speed maglev train. It is found that it is not enough to rely on the fluctuation range of the levitation gap as the basis for evaluating the control performance only.

The sperling stability index reflecting the vibration impulse and vibration kinetic energy is calculated and used as the basis for judging the stability of the train [9, 10]. Considering the different structural and functional characteristics, the calculation equation of the sperling stability index of the levitation electromagnet as shown in Eq. (2) is given. The sperling stability index of the levitation electromagnet as the dynamic performance index of the levitation system of the high-speed maglev train is defined.

$$w = \begin{cases} 0.896\sqrt[10]{0.325a^3 f} & f \in [0.5, 5.9] \\ 0.896\sqrt[10]{\dfrac{400a^3}{f^3}} & f \in [5.9, 20] \\ 0.896\sqrt[10]{\dfrac{a^3}{f}} & f \in [20, 20] \end{cases} \tag{2}$$

where, a is the vibration acceleration of the levitation electromagnet. f is the vibration frequency of the levitation electromagnet.

During the actual operation of the high-speed maglev train, the vibration of the levitation electromagnet is random. Therefore, the vibration acceleration of the levitation electromagnet can be decomposed according to the frequency, and the vibration amplitude of the levitation electromagnet in each frequency range can be obtained. The dynamic performance indicators of the levitation electromagnet in the entire frequency range are:

$$W = \left(\mathbf{w}_1^{10} + \mathbf{w}_2^{10} + \ldots + \mathbf{w}_n^{10}\right)^{0.1} \tag{3}$$

According to the actual operation of the high-speed maglev train and the on-site riding experience, the dynamic performance index and control performance level of the levitation electromagnet are given in Table 2.

Table 2. Dynamic performance index and control performance level

Control performance level	Assessment	Dynamic performance indicators
1	Excellent	<2.1
2	Good	2.1–2.3
3	Qualified	2.3–2.5
4	Failed	>2.5

4 Performance Analysis of Levitation System Under Non-Deterministic Disturbance

As indicated in Fig. 6, the high-speed maglev train usually uses an elevated track structure. However, changes in the external environment such as foundation settlement and changes in ambient temperature will result in the deformation of the track beam, and the tangential torsion angle of the beam end will also change accordingly. The larger the tangential torsion angle of the beam end is, the larger the deflection-span ratio is.

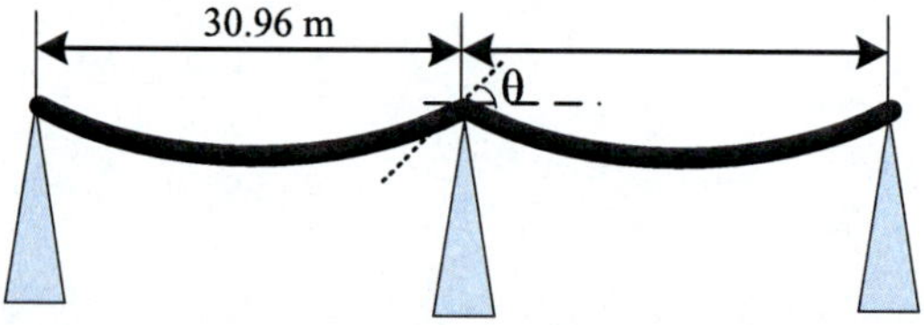

Fig. 6. Schematic diagram of the tangential torsion angle of the beam end of the high-speed maglev line

When the high-speed maglev train passes through the above-mentioned road section at a speed of 600 km/h, the dynamic characteristic curves of the high-speed levitation joint structure are obtained as shown in Fig. 7.

The dynamic characteristics of the levitation system passing through the track beams with the tangential torsion angles of the beam ends of 0.0004, 0.0006 and 0.0008 rad are simulated respectively and results are obtained as shown in Fig. 8.

According to the classification standard of the control performance, the control performance of the levitation system under different beam end tangential torsion angles is shown in Table 3.

To sum up, the increase of the tangential torsion angle of the beam end leads to the degradation of the control performance of the levitation system of the high-speed maglev train. The specific performance is that with the increase of the tangential torsion angle of the beam end, the fluctuation range of the levitation gap becomes larger and larger, and the dynamic performance index becomes larger and larger. When the tangential torsion angle of the beam end increases to 0.0008 rad, the control performance of the levitation system has been seriously degraded, which seriously endangers the stability and comfort of the high-speed maglev train.

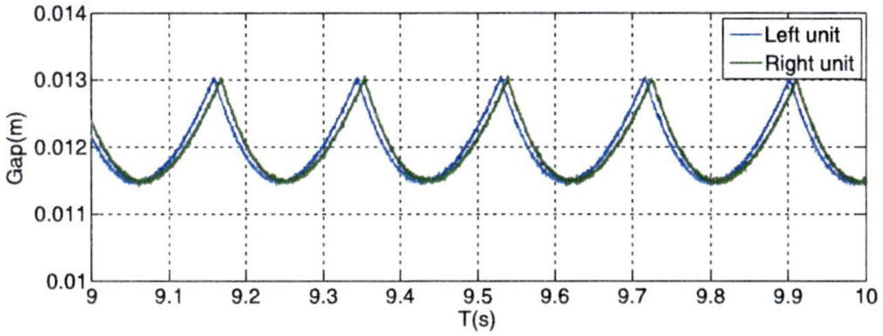

(a) Levitation gap fluctuation curve

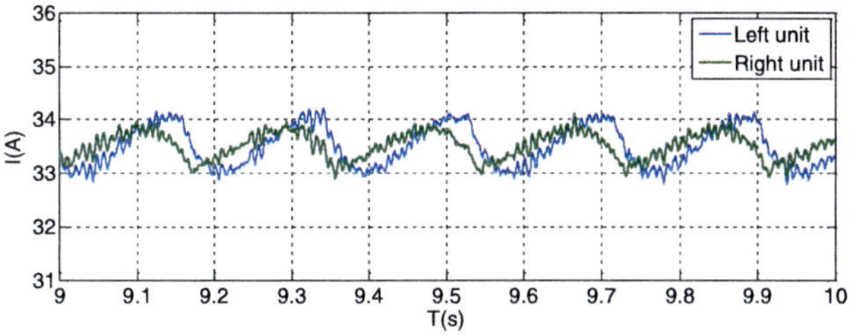

(b) Levitation current fluctuation curve

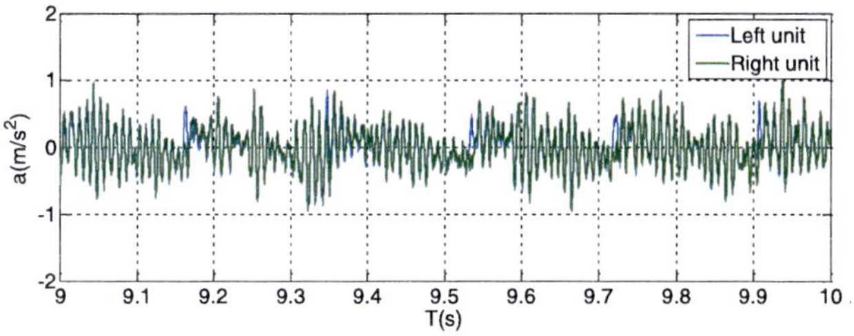

(c) Acceleration fluctuation curve

Fig. 7. The dynamic characteristic curves of the levitation system when the angle is 0.00021 rad

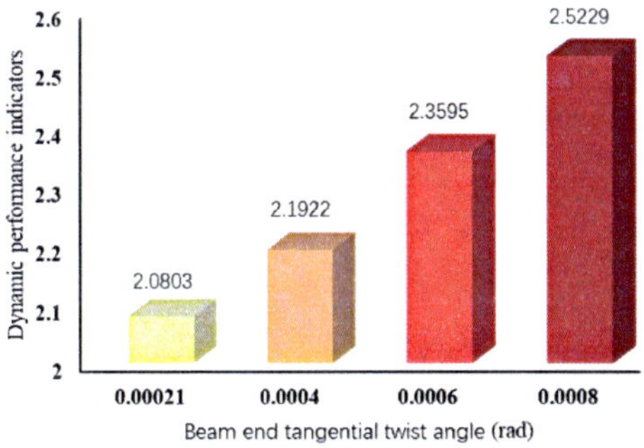

Fig. 8. Dynamic performance index of the levitation system corresponding to the change of the tangential torsion angle of the beam end

Table 3. Dynamic performance of levitation system under the disturbance of beam end tangential torsion angle

Beam end tangential twist angle	0.0002 rad	0.0004 rad	0.0006 rad	0.0008 rad
Levitation gap fluctuation range	11.4–13	10.9–14	10.4–15.1	9.8–16.1
Dynamic performance indicators	2.0803	2.1922	2.3595	2.5229
Performance evaluation	Excellent	Good	Acceptable	Severely degraded

5 Conclusion

In this paper, the nominal performance of the levitation system by using the nominal performance index is firstly analyzed and evaluated. Secondly, according to the working characteristics of the levitation system under dynamic operating conditions, the dynamic performance index and evaluation criteria suitable for the levitation system of the high-speed maglev train are proposed. Finally, the performance of the levitation system under non-deterministic disturbances is quantitatively analyzed.

References

1. Janic, M.: Multicriteria evaluation of the high speed rail, transrapid maglev and hyperloop systems. Transp. Syst. Technol. **4**(4), 5–31 (2018)
2. Wang, R., Yang, B., Gao, H.: Nonlinear feedback control of the inductrack system based on a transient model. J. Dyn. Syst. Meas. Contr. **143**(8), 1–13 (2021)
3. An, F., Sun, H., Li, X.: Adaptive active control of periodic vibration using maglev actuators. J. Sound Vib. **331**(9), 1971–1984 (2021)
4. Yang, J., Su, J., Li, S.: High-order mismatched disturbance compensation for motion control systems via a continuous dynamic sliding-mode approach. IEEE Trans. Ind. Inf. **10**(1), 604–614 (2013)
5. Xu, Z., Xu, S., Shi, L.: Study on direct fuzzy adaptive control for hybrid magnets used on ems maglev systems. Proc.-Chin. Soc. Electr. Eng. **25**(18), 157 (2005)
6. Zhai, M., Long, Z., Li, X.: Fault-tolerant control of magnetic levitation system based on state observer in high speed maglev train. IEEE Access **7**, 31624–31633 (2019)
7. Yan, Z., Chan, C.L., Yao, Y.: Multivariate control performance assessment and control system monitoring via hypothesis test on output covariance matrices. Ind. Eng. Chem. Res. **54**(19), 5261–5272 (2015)
8. Yamamoto, T., Kinoshita, T., Ohnishi, Y.: Design and experimental evaluation of a performance-driven PID controller. J. Robot. Mechatron. **28**(5), 616–624 (2016)
9. Zhang, Z., Liu, X., Hu, H.: Passenger rail station safety improvement and analysis of end-of-track collisions based on systems-theoretic accident modeling and processes (STAMP). Smart Resilient Transp. **3**(2), 94–117 (2021)
10. Liao, Z., Qin, Y., Wang, M.: Research on high-speed railway operation adjustment model based on priority. Smart Resilient Transp. **4**(1), 12–21 (2022)

Design of a Three-Level APF Controller Based on Dual DSP

Fang Zhankai(✉), Gou Jiaxi, Liu Shuang, Huang Yunbiao, Chen Guo, and Yang Ziqing

Chongqing Chuanyi Automation Co., Ltd, Chongqing 401121, China
1597132881@qq.com

Abstract. To improve the compensation performance of Active Power Filter (APF), this paper proposes a three-level APF controller based on dual DSP, which improves resource allocation. The main DSP is responsible for load current sampling, compensation current sampling, DC side voltage sampling, harmonic and reactive compensation current extraction, current tracking control calculation, PWM wave modulation, and other tasks, while the auxiliary DSP is responsible for sharing some operations and signal sampling, Including harmonic calculation, grid voltage sampling, touch screen display, power outage data storage, etc., communication between the two DSP is achieved through dual port random access memory RAM; More reasonable allocation of software tasks has improved the parallel computing ability of dual DSP. And it has been verified in a three-level APF prototype that it can achieve a switching frequency of 20 kHz, reduce the grid current distortion rate from 15 to 2%, and achieve a harmonic filtering rate of 86.7%, demonstrating good harmonic filtering ability.

Keywords: Dual DSP · Three level · Active power filter · Harmonic filtering

1 Introduction

Active Power Filter (APF) is a power electronic device used for dynamic suppression of harmonics. Due to nonlinear loads such as frequency converters, intermediate frequency furnaces, welding machines, and rolling mills, a large amount of harmonics are introduced into the power grid, thereby affecting the power quality of the power grid [1, 2]. APF first detects the voltage and current values of the object to be compensated through voltage and current transformers, and then conducts harmonic current analysis on DSP to obtain a reference instruction current signal. The tracking control circuit then outputs a PWM drive signal based on this reference signal, thereby controlling the inverter circuit to output compensation current and achieving the purpose of harmonic compensation [3, 4].

The compensation performance of APF mainly depends on its controller. Currently, APF controllers mostly use single DSP [5, 6], single FPGA [7], DSP + CPLD [8, 9], DSP + FPGA [10, 11] or dual DSP [12–15]. The performance of a single DSP is far inferior to that of a combination of two processors; Although a single FPGA can leverage its parallel processing advantages by observing the internal running status online,

L. Jia et al. (Eds.): EITRT 2023, LNEE 1135, pp. 439–447, 2024.
https://doi.org/10.1007/978-981-99-9307-9_49

it cannot be debugged or modified online. Observing the internal running status of the FPGA through SignlaTapII will add observation points to the designed circuit, which sometimes affect the timing of the design, leading to inaccurate observation results and occupying additional resources. The controller is still at the theoretical research and simulation level [7]. The development time of APF controllers with DSP + CPLD and DSP + FPGA is relatively long, and the software's portability is poor; The current dual DSP APF controllers are mostly limited to the main DSP taking on the majority of calculations in terms of functional allocation, while the auxiliary DSP has limited application of computing power, making it difficult to fully utilize the dual DSP computing performance. Moreover, most of them only apply dual DSP control to two-level APF, which has significant limitations on harmonic compensation performance and is difficult to meet the growing demand for power quality in the current power grid.

Therefore, this article proposes a three-level APF control system design based on dual DSP (F28335 + F28335). The main DSP is responsible for AD sampling, communication, relay control, protection, and pulse width modulation PWM generation. The two DSP communicate with each other through dual port random access memory (RAM); Improved hardware resource allocation, more reasonable allocation of software tasks, fully leveraging the parallel computing capabilities of dual DSP, and verified in a three-level APF physical.

2 Controller Design

The hardware design of the APF controller mainly includes the main processor module, signal conditioning circuit, AD sampling module, PWM level conversion module, communication module, and power module.

2.1 Main Processor Module Design

The hardware design of the main processor module mainly considers two TMS320F28335 chips (hereinafter referred to as DSP1 and DSP2), fully utilizing the computing power of the two TMS320F28335 chips, achieving parallel operations as much as possible, and making reasonable use of their peripheral resources. The system adopts a dual DSP design, with DSP1 as the main controller, responsible for load current sampling, compensation current sampling, DC side voltage sampling, harmonic and reactive compensation current extraction, current tracking control calculation, PWM wave modulation, etc. As an auxiliary controller, DSP2 is responsible for some calculations and signal sampling, including harmonic calculation, grid voltage sampling, power outage data storage, etc. Two DSPs perform parallel computing to improve computing speed and leave a lot of space for on-chip resources, increasing the scalability of the system.The block diagram of the main processor module is shown in Fig. 1.

Two DSP exchange data through a dual port SRAM, using Renesa IDT70V24L20PFGI as shown in Fig. 2. The working level of this SRAM is 3.3 V, with 16 data bits and a read/write speed of 15ns, enabling high-speed data exchange between dual DSP.

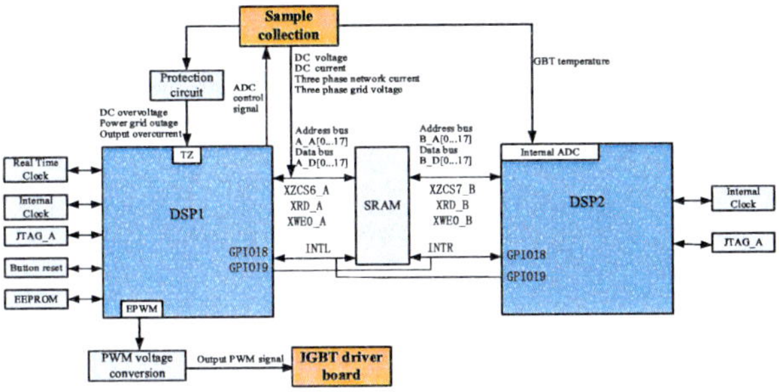

Fig. 1. Main processor module block diagram

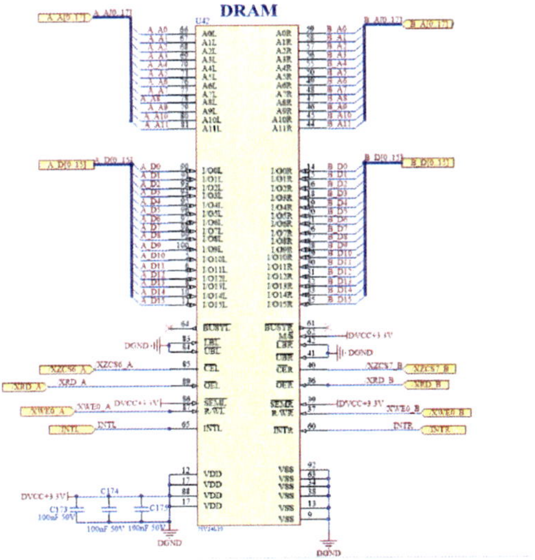

Fig. 2. Dual port SRAM circuit diagram

2.2 Signal Conditioning Circuit

For signal conditioning circuits, it mainly includes signal conditioning of load current, output current, grid voltage, and DC side voltage. For load current and output current, Hall sensors are used to convert large current signals into small current signals. Taking load current sampling as an example, A-phase load current sampling is taken here. The A-phase load current sampling circuit is shown in Fig. 3.

The load current sampling needs to go through two levels of current transformers. After passing through the first external transformer, the current is converted into 5A. The internal second level current transformer uses TA0913-2M with a transformation ratio of 2000:1, so the input current is 5A. The output current of the transformer is ±2.5 mA. After passing through a 0.1% precision sampling resistor R5 of 1 k Ω, the current signal is converted into a voltage signal with a range of ±2.5 V. The first operational amplifier and the second motion are both voltage following effects. The passive RC filter

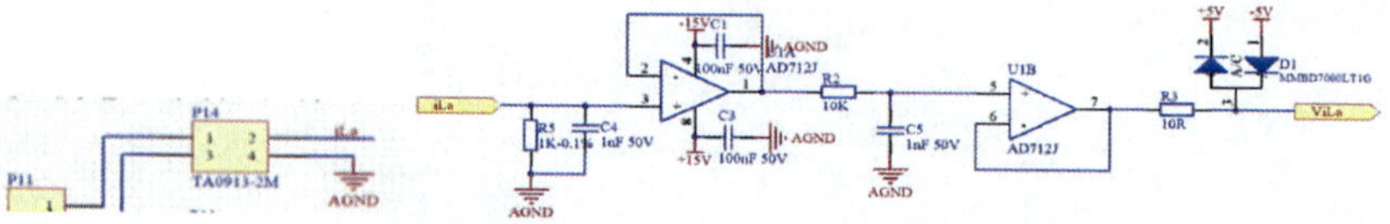

Fig. 3. Load current sampling circuit diagram

used, with R2 = 10 k Ω and C5 = 1nF, has a cut-off frequency of 16 kHz to ensure compensation for harmonics up to 50 times (2.5 kHz).

According to the characteristics of ViLa entering external AD (MAX1308), in order to reduce gain error, it is necessary to ensure that the R3 resistance value is less than 100 Ω. After testing, selecting a R3 resistance value of 10 Ω can achieve the function; Finally, add a limiting diode to limit the output signal to ±5V.

2.3 AD Sampling Module Design

While ensuring the sampling accuracy of load current and output current, APF should also ensure the real-time and synchronization of sampling. The AD of TMS320F28335 can only maintain 10 bit accuracy when sampling at high speed, and can not achieve synchronous sampling of three-phase current. Therefore, external AD is selected for current sampling, and MAX1308 (with 8 HOLDs) with 8-way parallel sampling capability is used. MAX1308 is an AD chip produced by MAXSIM company with 8-channel parallel adoption capability, with a sampling accuracy of 12 bits; The minimum sampling time for 8 channels is 1.98us; Capable of achieving bipolar input with a single power supply; Data exchange with DSP through parallel port; Isolation between digital power supply and analog power supply. The circuit schematic of the external AD is shown in Fig. 4. In the design, a 5V power supply is used as the analog power supply, with a bipolar input connection method. Therefore, the sampling input voltage range is ±5V; The digital power supply is 3.3 V, ensuring that the MAX1308 data cable can be directly connected to the DSP data cable.

2.4 Design of PWM Level Conversion Module

Due to the same PWM control circuit for the three phases, the PWM control circuit for the C-phase upper and lower tubes is taken as an example. The input signals are EPWMC1, EPWMC3, EPWMC2, EPWMC4, and the output signals are C1, C3, C2, C4. The PWM control circuit of the four switch tubes in phase C is shown in Fig. 5.

After PWM is output from the DSP pin, it is at a 3.3 V voltage level. First, it undergoes a first level voltage conversion to convert the 3.3 V PWM wave into a 5 V voltage level PWM wave. Then, it is output to the drive board to drive the IGBT through a three-level hardware interlocking circuit, mainly including the non gate MC14049UBDG and the AND gate SN74LS08DR, to prevent IGBT short circuits. Finally, a two input NAND gate is used, with one input being a PWM wave and the other being a fault blocking signal, IGBT driver input low level effective. The three-level bridge arm consists of four switches, V1 to V4, for each phase. The control logic of the four switch tubes from V1

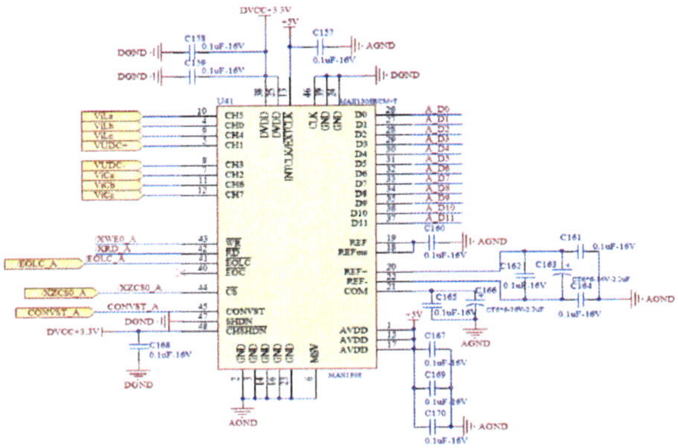

Fig. 4. External AD circuit diagram

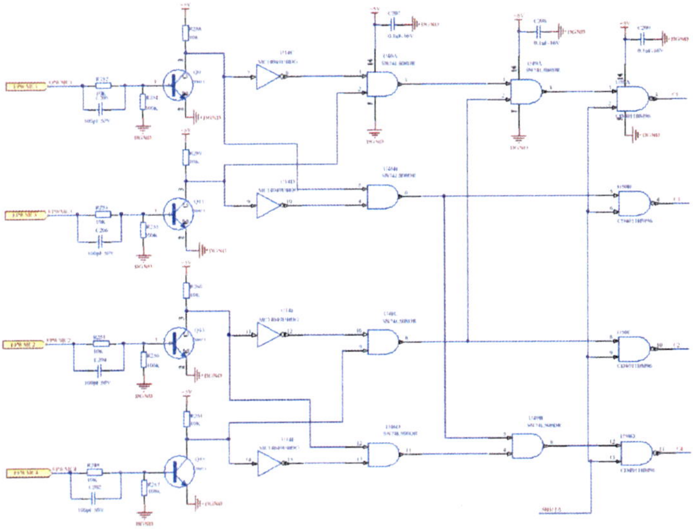

Fig. 5. PWM level conversion circuit diagram

to V4 requires that the driving signal of switch tube V1 and the driving signal of switch tube V3 are mutually reversed, and the driving signal of switch tube V2 and the driving signal of switch tube V4 are mutually reversed; The driving signal of switch tube V1 must be at a high level when the driving signal of switch tube V2 is low, and the driving signal of switch tube V4 must be at a high level when the driving signal of switch tube V3 is high [16].

Due to the use of a transistor for the flip of the 3.3–5 V level, there is a certain delay. Therefore, through debugging, 100 pF capacitors C202, C204, C205, and C206 are added to the DSP output for time compensation.

2.5 Communication Module Design

Both touch screen communication and upper computer communication are implemented using RS485, using the TD501M485 communication module from Jin Shengyang. This module is equipped with efficient isolation power supply, automatic data transmission and transmission function, and two end isolation (2.5 kV DC). The specific circuit is shown in Fig. 6. In addition, the terminating resistor R247 is used for impedance matching of transmission lines, with a typical value designed as 120 ohms; The bus uses 12 V transient suppression diodes for lightning protection.

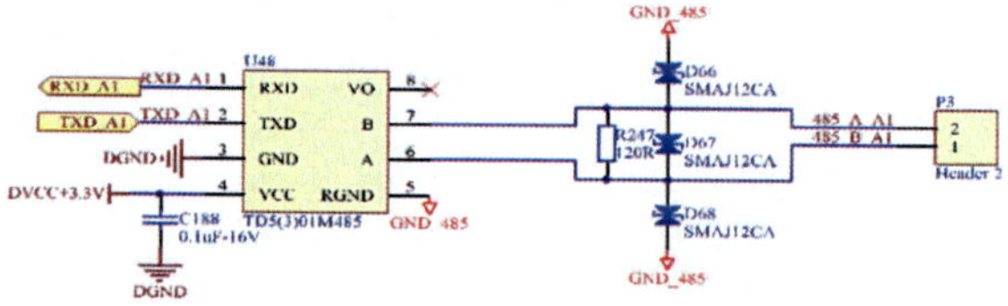

Fig. 6. Communication module circuit diagram

2.6 Power Supply Design

Due to the dual voltage supply of 3.3 and 1.8 V required for TMS320F28335, an LDO voltage conversion chip is used here. The input power supply is 5V, and the 3.3 V voltage is generated by the LDO chip AMS117-3.3. As shown in Fig. 7. The 1.8 V voltage is generated by the combination of LDO chips AMS117-3.3 and AMS117-1.8, as shown in Fig. 8. First, 5 V is converted to 3.3 V, and then 3.3 V is converted to 1.8 V, in order to reduce the power consumption of the chip AMS117-1.8 and prevent overheating damage.

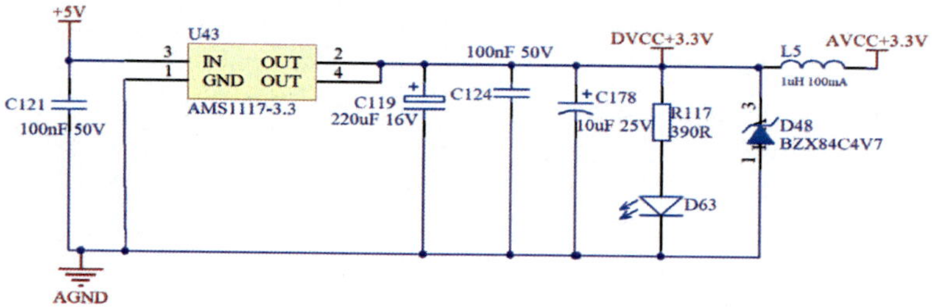

Fig. 7. Circuit diagram from 5 to 3.3 V

Combining the above hardware design, a three-level APF main control board based on dual DSP is obtained, as shown in Fig. 9.

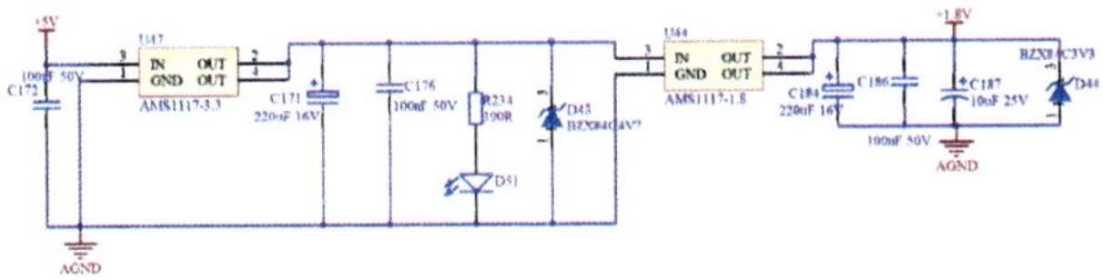

Fig. 8. Circuit diagram of 5 to 1.8 V

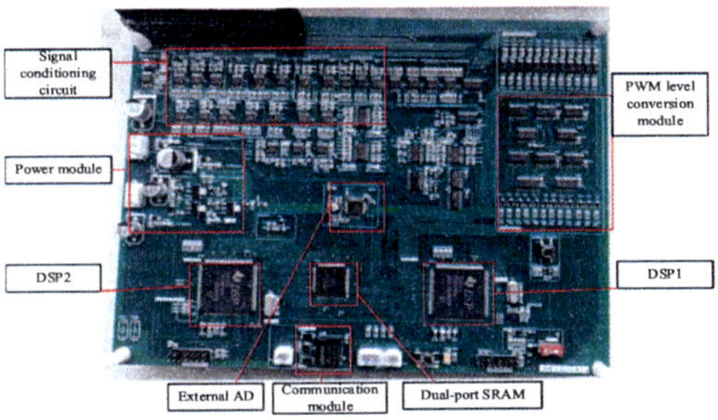

Fig. 9. Physical diagram of APF main control circuit board

3 Experimentation

By designing a dual DSP three-level APF controller and verifying it in actual hardware circuits, the running time of the control program was measured, the DSP1 control interrupt program has a running time of 38 us. For this, a switching frequency of 20 kHz and a cycle of 50 us can be designed, with a margin of 24% reserved. Tested in the actual three-level APF product, the Dead time is 2 us. Establish a non-linear load, and the current distortion rate of the power grid before governance is about 15%, as shown in Fig. 10; After treatment, the current distortion rate decreased to 2%, as shown in Fig. 11. The harmonic filtering rate reached 86.7%, and the treatment effect was good. It can be seen that the design of a dual DSP three-level APF controller can improve the compensation performance of the APF.

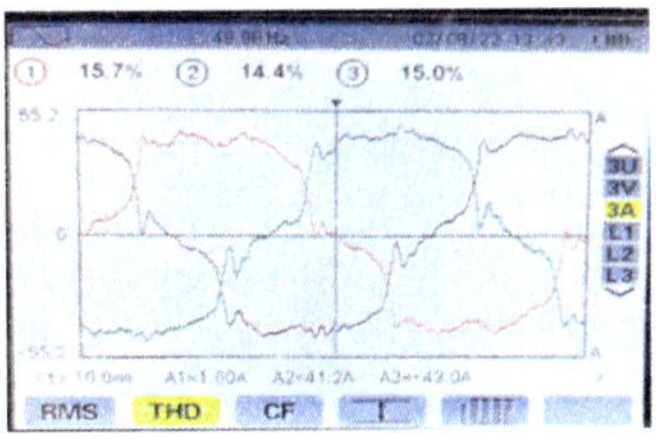

Fig. 10. Current waveform of the power grid before governance

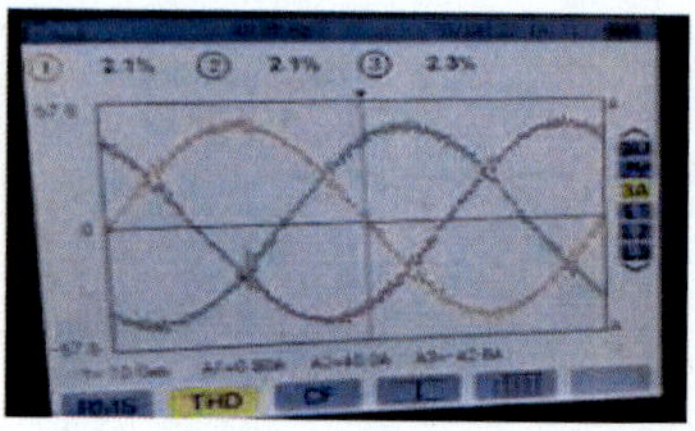

Fig. 11. Current waveform of the power grid after governance

4 Conclusion

This article is based on the design of a three-level APF controller with dual DSP. The hardware and software of the control system have been designed and analyzed. Through experiments, it has been verified that the system can achieve a switching frequency of up to 20 kHz. It has been verified in practical three-level APF products, and the experimental results have shown good harmonic suppression performance.

References

1. Huabei, X., Lei, J., Pengjun, S., et al.: Research review of three-phase three-wire active power filters. J. Electr. Eng.Electr. Eng. **1**, 51–54 (2023). (in Chinese)
2. Erickson, R.W., Maksimovi, D.: Fundamentals of power electronics. Kluwer Academic/Plenum Publishers, New York (2020)
3. Akagi, H.: Active harmonic filters. Proc IEEE **93**(12), 2128–2141 (2006)
4. Ameran, I.A.: Harmonic reduction in thvristor converters by harmonic current injection. IEEE Trans. Power Appar. Syst. **95, 96**(2), 441–449 (1976)
5. Jiazheng, R.: Design and Implementation of Active Power Filter Design Based on DSP. Hangzhou Dianzi University (2022) (in Chinese)
6. Ruilong, Y., Wang Gang, X., Yuan.: Research on three-level APF based on DSP. Electr. Appl. **38**(03), 21–25 (2019). (in Chinese)
7. Xinlei, Z.: Development of T-Type Three-Level Active Power Filter Device Based on FPGA. North China University of Technology (2019) (in Chinese)
8. Teichmann, R., Bernet, S.: A comparison of three-level converters versus two-level converters for low-voltage drives, traction, and utility applications. IEEE Trans. Ind. Appl. **41**(3), 855–865 (2005)
9. Lisa Research on Active Power Comprehensive Treatment Device of Distribution Network [D]. Harbin University of Science and Technology (2021) (in Chinese)
10. Lianqing, S., Zhuo, W., Xinlei, W., et al.: Hardware design of APF based on DSP FPGA. Electron. Technol. Appl. **46**(5), 5 (2020) (in Chinese)
11. Liping, Z.: Research and Application of Harmonic Control Technology in Hospital Power Supply System. Lanzhou University of Technology (2022) (in Chinese)
12. Zheng, Z., Jiwei, L., Guopeng, Z.: Control system design of dual DSP active power filter. J. Power Syst. Autom. **31**(2), 7 (2019) (in Chinese)
13. Sensarma, P.S., Padiyar, K.R., Ramanarayanan, V.: A comparative study of harmonic filtering strategies for a shunt active filter. In: Thirty-Fifth IAS Annual Meeting and World Conference on Industrial Applications of Electrical Energy, pp. 2509–2516. Rome, Italy (2002)

14. Qian Yejuan, X., Weibing, S.G., et al.: Research on active power filters controlled by dual DSP. Autom. Instrum. **1**, 3 (2017). (in Chinese)
15. Dongshan, Z., Ning, G., Yuan, D.: Power quality monitoring and governance based on dual DSP structure. J. Lanzhou Inst. Technol. **28**(01), 81–85 (2021). (in Chinese)
16. Wenqiang, C., Yuan, G., Yiyong, Z., et al.: A PWM interlock protection circuit suitable for three-level IGBT bridge arms: 201910568919 [2023–06–21] (in Chinese)

Study on Dynamic Characteristics and Electromagnetic Thrust of Unilateral Linear Induction Motor for Maglev Train

Shuncai Yao[1], Zhiping Cheng[1](✉), Huixing Chen[2], and Bailin Li[1]

[1] School of Electrical and Control Engineering, North University of China, Shanxi 030051, China
1274873556@qq.com, false863471835@qq.com, 719180089@qq.com

[2] The Third Academy of China Aerospace Science and Industry Corporation, Beijing 100110, China
chx0808@sina.com

Abstract. This paper analyzes the structure and working principle of unilateral linear induction motor in order to study the dynamic characteristics and electromagnetic thrust. Then, the electromagnetic model has been established. It has been deduced that the end effect of unilateral linear induction motor air gap magnetic density, electromagnetic thrust related analytical calculation expression. The FEM model has been created to get unilateral linear induction motor specific time air gap magnetic density distribution map and electromagnetic thrust, by using the transient solver. Large thrust has to been provided to unilateral linear induction motor for maglev train by adopting composite secondary. This paper adopts the method of parameterization of the thickness of the secondary plate to improve the efficiency of the motor and maximum thrust. The electromagnetic thrust curve has been got by using ANSYS optimization module. Meanwhile, optimized the structure and performance of the motor have been achieved by optimizing the secondary pole thickness curve.

Keywords: Maglev train · Unilateral linear induction motor · Dynamic characteristics · Electromagnetic thrust

1 Introduction

With the rapid development of transportation, many countries have developed maglev trains by using linear motors as the driving force of trains and combining maglev technology. Because it cancels the wheel and axle, breaks the traditional mode of propulsion, making the train has the advantages of high speed, comfort and safety, and the linear motor has the characteristics of non-adhesive drive, simple structure and stable performance, making it the best choice for maglev trains [1].

For the high-power unilateral linear induction motor used in maglev train, it is necessary to know the electromagnetic thrust in the design of motor structure and maglev system. Reference [2–8] is the study of electromagnetic thrust of linear induction motor.

L. Jia et al. (Eds.): EITRT 2023, LNEE 1135, pp. 448–455, 2024.
https://doi.org/10.1007/978-981-99-9307-9_50

In this paper, through the derivation of the analytical calculation expression of the air gap flux density and electromagnetic thrust of the unilateral linear induction motor used in maglev train, the finite element simulation is used to intuitively and accurately analyze the air gap flux density distribution and electromagnetic thrust characteristic curve of the unilateral linear induction motor, and the parameterized motor thrust curve is analyzed through the parameterized secondary plate-thickness. Thus, the structure and performance of unilateral linear induction motor can be optimized.

2 Structure and Working Principle of Unilateral Linear Induction Motor for Maglev Train

The unilateral linear induction motor is composed of primary core, winding and composite secondary, and its structure diagram is shown in Fig. 1.

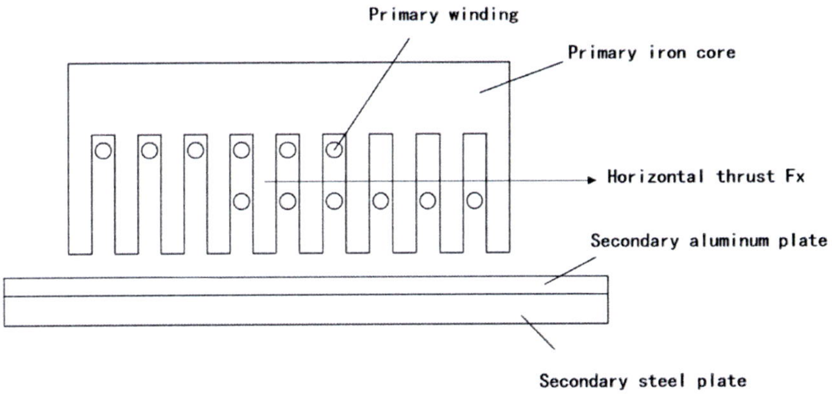

Fig. 1. Structure diagram of unilateral linear induction motor.

The unilateral linear induction motor used in maglev trains will be mounted on the bogie, and the reaction plate will be mounted on the back of the track and its surface will be attached with aluminum to provide eddy current path. The suspension magnet is used to form a closed magnetic circuit with the track, and the air gap sensor and suspension controller are used to ensure stable suspension. At the same time, the closed magnetic circuit composed of the track and suspension magnet follows the law that the magnetic flux always closes along the path of the least reluctance, so that the lateral stability of the vehicle can be guaranteed [9].

3 Electromagnetic Model and Characteristic Analysis of Unilateral Linear Induction Motor for Maglev Train

The electromagnetic analysis model of unilateral linear induction motor is shown in Fig. 2.

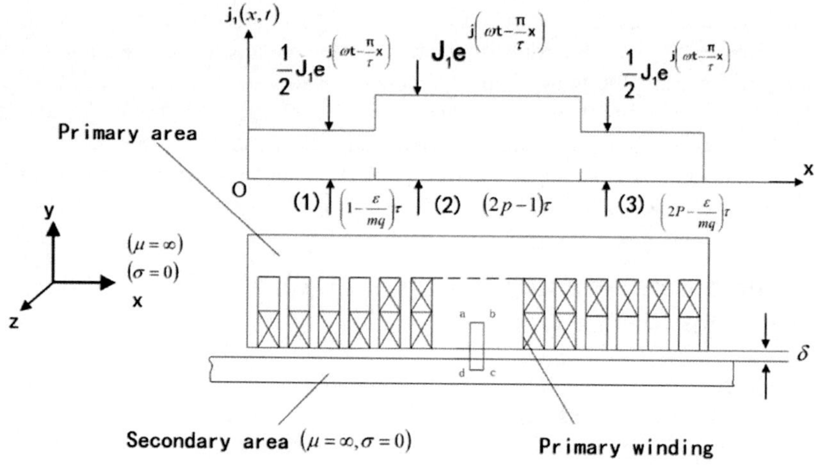

Fig. 2. Analysis model of unilateral linear induction motor.

The following equation is the motor air gap magnetic field equation, and its analytical solution is determined by the boundary conditions.

$$\frac{\delta'}{\mu_0}\frac{\partial^2 B_y}{\partial x^2} - \sigma_s V_x \frac{\partial B_y}{\partial x} - \sigma_s \frac{\partial B_y}{\partial t} = \frac{\partial j_1}{\partial x} \tag{1}$$

where δ' is the effective electromagnetic air gap of the motor; μ_0 is the air permeability; B_y is the air gap magnetic flux density; σ_s is the surface conductivity of the secondary conductor; V_x is the secondary moving speed and j_1 is the primary current.

Now assume that the primary current layer of the three sections in Fig. 1: primary, secondary aluminum plate and secondary steel plate is:

$$j_1(x,t) = \begin{cases} 0.5J_1 e^{j(\omega t - kx)}; & 0 < x < \left(1 - \frac{\varepsilon}{m_1 q_1}\right)\tau \\ J_1 e^{j(\omega t - kx)}; & \left(1 - \frac{\varepsilon}{m_1 q_1}\right)\tau < x < (2p-1)\tau \\ 0.5J_1 e^{j(\omega t - kx)} & (2p-1)\tau < x < \left(2p - \frac{\varepsilon}{m_1 q_1}\right)\tau \end{cases} \tag{2}$$

where J_1 is the amplitude of the traveling wave current layer; ω is the angular frequency of the power supply; ε is the dielectric constant; m_1 is the number of phases of the primary winding; q_1 is the actual number of slots per pole per phase in the primary; τ is the pole distance and p is the primary winding pole number.

When the longitudinal end effect is ignored, the motor air gap magnetic field has only a normal traveling wave magnetic field component, so it can be assumed to be:

$$B_y = B(x,t) = B_m e^{j(\omega t - kx)} \tag{3}$$

where B_m is the amplitude of the air gap magnetic field.

By substituting Eqs. (2) and (3) into Eqs. (1), the air gap magnetic flux density of the three intervals is obtained as follows:

$$B_1(x,t) = \frac{j0.5\pi J_1 e^{j(\omega t - kx)}}{\tau\left(\frac{\delta'\pi^2}{\mu_0\tau^2} + js\omega\sigma_s\right)} \quad 0 < x < \left(1 - \frac{\varepsilon}{m_1 q_1}\right)\tau \tag{4}$$

$$B_2(x,t) = \frac{j\pi J_1 e^{j(\omega t - kx)}}{\tau\left(\frac{\delta'\pi^2}{\mu_0\tau^2 + js\omega\sigma_s}\right)} \quad \left(1 - \frac{\varepsilon}{m_1 q_1}\right)\tau < x < (2p-1)\tau \tag{5}$$

$$B_3(x,t) = \frac{j0.5\pi J_1 e^{j(\omega t - kx)}}{\tau\left(\frac{\delta'\pi^2}{\mu_0\tau^2} + js\omega\sigma_s\right)} \quad (2p-1)\tau < x < \left(2p - \frac{\varepsilon}{m_1 q_1}\right)\tau \tag{6}$$

In order to estimate the electromagnetic force F, Maxwell stress tensor is an effective method. Equation (7) estimates F by surface integration of a closed surface containing a primary winding:

$$F = \int_S \left[\frac{1}{\mu_0}(n \cdot B)B - \frac{1}{2\mu_0}B^2 n\right] dS \tag{7}$$

where n is the unit normal vector on the closed surface; B is the magnetic induction intensity vector and S is the area of the closed surface.

The thrust generated by the magnetic field is the x direction components of force F, which can be expressed as:

$$F_x = \int_l \frac{\omega}{2\mu_0}\left[\left(B_x^2 - B_y^2\right)n_x + 2B_x B_y n_y\right] dl \tag{8}$$

where ω is the width of the primary core; μ_0 is the vacuum permeability; B_x is the x-oriented component of magnetic induction intensity; B_y is the y-oriented component of magnetic induction intensity; n_x is the unit orthogonal vector of x direction on a closed surface; n_y is the unit orthogonal vector of y direction on a closed surface and l is a closed line around the primary winding region in two-dimensional analysis [10].

4 Finite Element Model and Characteristic Analysis of Unilateral Linear Induction Motor for Maglev Train

4.1 Finite Element Analysis Model of Unilateral Linear Induction Motor

Ignoring the lateral side-end effect of the unilateral linear induction motor, the model of unilateral linear induction motor for maglev train is presented. The main parameters of the motor are shown in Table 1 [11].

Table 1. Main dimensional parameters of linear induction motors.

Parameter	Value
Motor size $a \times b \times c/$mm	$1800 \times 200 \times 100$
Pole number	6
Phase number	3
Tooth width b/mm	10
Slot width d/mm	10
Slot depth h/mm	85
Mechanical air gap g/mm	13
Pole distance τ/mm	300
Aluminum plate thickness D_{Al}/mm	15
Steel plate thickness D_{st}/mm	30

4.2 Analysis of Dynamic Characteristic of Unilateral Linear Induction Motor

Figure 3 shows the longitudinal distribution of the air gap magnetic field density of the unilateral linear induction motor at $t = 0.1$ s. When the distance between the non-model line and the secondary plate is 0.06 m, it is the primary core edge, and when it is 0.0625 m, it is the primary winding region. After that, the magnetic flux density begins to fluctuate up and down, reaching a peak value of about 31 mTesla at 0.68 m, and falling almost to 0 at 1.92 m.

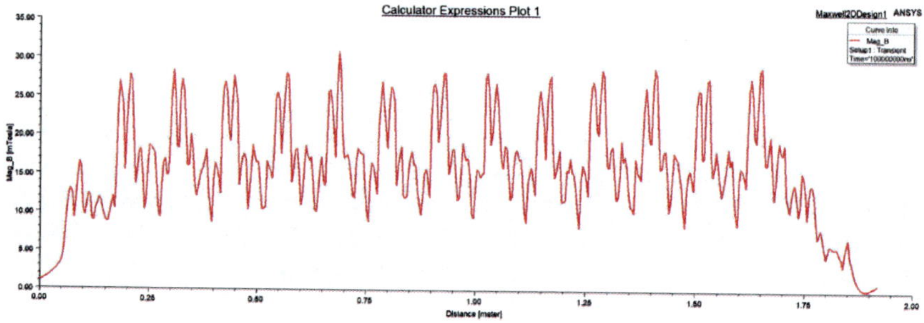

Fig. 3. Longitudinal distribution of air gap flux density of unilateral linear induction motor ($t = 0.1$ s).

The curve of thrust change is shown in Fig. 4 in case of over time for unilateral linear induction motor. When the motor is in no-load operation, the thrust is in a state of oscillation at the initial stage, and tends to steady state after 0.155 s. The actual thrust direction is the same as the thrust direction defined in the software, so the thrust is positive.

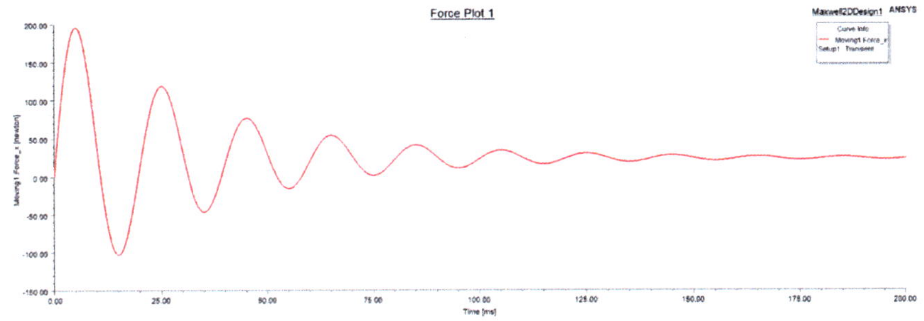

Fig. 4. Thrust curve of unilateral linear induction motor with time.

5 Application of Parametric Processing of Unilateral Linear Induction Motor for Maglev Train in Structural Optimization

5.1 Structure Optimization of Unilateral Linear Induction Motor for Maglev Train

Structural optimization design is to seek the best design sequence of the structure from many design sequences. On the premise of ensuring that the geometric structure of the structure meets the requirements of the specification, by changing the geometric size of the structure, material type and other design variables, the design resources are reasonably allocated in the engineering system, which saves the material consumption on the one hand and reduces the cost on the other hand. ANSYS optimization module adopts three optimization variables to describe the optimization process [12].

5.2 Influence of Secondary Plate Thickness Parameterization of Unilateral Linear Induction Motor for Maglev Train on Motor Structure and Performance

Set the thickness of the secondary board as variable d in the drawing area of ANSYS Maxwell 2D, select *Optimetrics* optimizer to add parameters, scan variable d, set the starting point of the linear step as 29mm, the end point as 30.4mm and the step size as 0.2 mm. The thrust curve with different d values as shown in Fig. 5 and the average thrust curve with d in the time range of 0.075–0.125 s as shown in Fig. 6 are obtained.

From Figs. 5 and 6, it can be seen that with the gradual increase of the thickness of the secondary plate, the thrust of the unilateral linear induction motor becomes smaller and smaller when it reaches the steady state. When the thickness of the secondary plate increases by 0.2 mm, the thrust decreases by about 0.15 N on average. When the thickness of the secondary plate is 29 mm, the thrust of the unilateral linear induction motor reaches the maximum. Therefore, when designing the motor, the structure of the motor can be more reasonable and the performance is more prominent by properly reducing the thickness of the secondary plate, and it can also save materials and save costs.

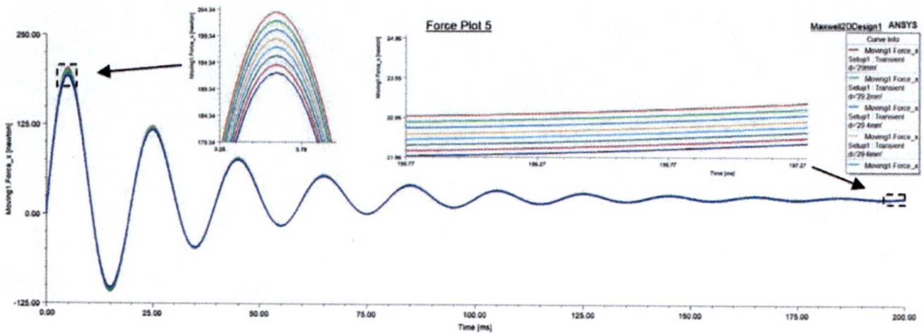

Fig. 5. Thrust curves with different d values.

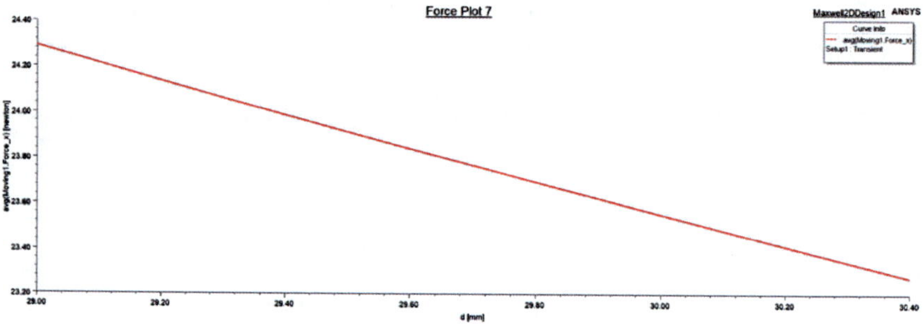

Fig. 6. Curve of the average thrust force with thickness d.

6 Conclusion

In this paper, the structure, working principle, electromagnetic model and characteristics of unilateral linear induction motor for maglev train were analyzed first, and then the finite element model was generated by using the transient solver of ANSYS Maxwell 2D software. The characteristic curves of air gap magnetic density, thrust of unilateral linear induction motor were obtained and analyzed by using the finite element method. The simulation results reflect the transient change law of unilateral linear induction motor, and then the thickness of the secondary plate is parameterized, and the optimization module of ANSYS software is used to optimize the structure. The influence of the secondary plate thickness on the thrust of the unilateral linear induction motor is obtained, which provides an important basis for the improvement of the structure and performance of the motor design and the saving of materials and costs.

References

1. Long, X.: Theoretical and Electromagnetic Design Method of the Linear Induction Motor, 1st edn. Science Press, Beijing (2006). (in Chinese)
2. Zhang, S., Huang, S., Cao, G.: Analysis of electromagnetic force characteristics of linear motor for rail transit. Micro Electric Mach. **49**(12), 23–28 (2021). (in Chinese)

3. Liu, X., Cui, Z., Gao, L.: Electromagnetic thrust analysis of linear induction motor used for UAV electromagnetic ejection. Comput. Simul. **39**(8), 16–19 (2022). (in Chinese)
4. Wang, Y., Liu, X., Liu, F.: Influence of medium and low speed maglev train speed on the traction force of linear induction motor. J. Tongji Univ. (Natural Science Edition) **49**(12), 1671–1677 (2021). (in Chinese)
5. Han, Y., Nie, Z., Xu, J.: Control strategy for optimizing the thrust of a high-speed six-phase linear induction motor. IET Power Electron. **13**(11), 2260–2268 (2020)
6. Jeong, J., Choi, J., Sung, S.: A study on the thrust characteristic analysis of linear induction motor according to secondary reaction plate using the container scanner vehicle. Trans. Korean Inst. Electr. Eng.Electr. Eng. **65**(1), 65–72 (2016)
7. Liu, H., Wang, B.: Research on traction calculation simulation system of mid-low speed maglev train. In: 2021 International Conference on Computer Application in Transportation Engineering (CATE 2021), pp. 1–6. IOP Publishing, England (2021)
8. Ma, J., Zhu, Y., Jiang, J.: A novel method to calculate the thrust of linear induction motor based on instantaneous current value. J. Modern Transp. **20**(2), 76–81 (2012)
9. Lv, G.: Linear Induction Motor and Control of Rail Transit, 1st edn. Machinery Press, Beijing (2021). (in Chinese)
10. Li, X., Tao, C.: Finite element analysis of linear-induction motors for traction based on ANSOFT software. Res. Urban Rail Transit **17**(3), 79–82 (2014). (in Chinese)
11. Liu, H., Zhang, Z., Song, T.: ANSYS Maxwell + Workbench 2021 Motor Multi-Physics Coupling Finite Element Analysis from the Introduction to Engineering Practice, 1st edn. Chemical Industry Press, Beijing (2022). (in Chinese)
12. Yan, Y.: Structure optimization design based on the ANSYS parametric design language. J. East China Jiao Tong Univ. **21**(4), 52–55 (2004). (in Chinese)

Research on Magnetic Levitation Bearing Three-Level Switching Power Amplifier

Yong Yu and Lei Mei(✉)

Nanjing Tech University, Nanjing, China
meilei@njtech.edu.cn

Abstract. Magnetic levitation bearings are widely used in flywheel energy storage because of the advantages of frictionless and low mechanical loss. Its performance directly affects the control effect of the whole system. In order to reduce the switching frequency of the switching tubes while reducing the current ripple, this paper proposes a new three-level control, firstly, the error and direction signals are obtained by processing the control feedback signals through operational amplifiers and comparators, while the full-bridge topology is used to realize the on and off of the switching tubes by using digital logic devices. And When the direction signal remains unchanged (from left to right, for example), the lower right tube is kept in the on state, and the current rise and fall is controlled by adjusting the on and off of the left bridge arm switch tube. The simulation results show that the proposed switching amplifier can meet the requirement of following control signal.

Keywords: Magnetic bearing · Switching power amplifier · Three-level control

1 Preface

As a typical mechatronic device, the high-speed flywheel rotor support technology [1] included in flywheel energy storage technology has been the focus of research. And the use of magnetic bearing technology is the best choice in order to realise the advantages of flywheel energy storage device such as high energy storage density, long service life and high conversion efficiency.

The magnetic bearing technology is widely used in flywheel energy storage because there is no mechanical technology, so there is no need for lubrication, small mechanical loss, and also has the advantages of small noise [2], long service life, etc. It is generally composed of four parts: controller, power amplifier, sensor and magnetic bearing coil [3], where the sensor mainly detects the rotor displacement change and transmits the displacement signal to the controller, the controller mainly processes the displacement signal through the DSP to generate the corresponding control signal, and the power amplifier amplifies the control signal so that the magnetic bearing coil current can follow the control signal without distortion. The power amplifier amplifies the control signal so

L. Jia et al. (Eds.): EITRT 2023, LNEE 1135, pp. 456–462, 2024.
https://doi.org/10.1007/978-981-99-9307-9_51

that the magnetic bearing coil current can follow the control signal without distortion, and the magnetic bearing coil generates the corresponding electromagnetic force to balance the rotor.

As an important component of magnetic bearing [4], the performance of power amplifier directly affects the control performance and stability of the whole magnetic bearing system. Switching amplifier is widely used in magnetic bearing instead of linear amplifier because of its low power consumption, and it has been a hot spot for research scholars at home and abroad. The main research objectives of switching amplifiers are to minimize current ripple, improve the dynamic performance of power topology circuits, reduce switching losses, accurately track control signals, and reduce device costs as much as possible. There are many related researches in ancient and modern times.

In 1990, T. Bardas in Canada and F.J. Keith in the United States first proposed the application of switching power amplifiers to magnetic bearing systems, the disadvantage of two-level switching amplifiers is that the current ripple is large, and the current ripple is proportional to the DC bus voltage, and reducing the DC bus voltage will affect the dynamic performance of the system. 1995, Jing et al. [5] proposed a three-level PWM modulation technique can reduce the current ripple, and the current ripple is independent of the DC bus voltage, which greatly reduces the switching losses compared with the two-level switching amplifier. Zhou et al. [6] proposed a new single-period control method, established the current ripple model of unipolar and bipolar single-period control, and deduced that the current ripple of unipolar single-period control is smaller than that of bipolar single-period control, and is independent of DC bus voltage. Aiming at the phase delay problem of the power topology of magnetic bearing power amplifier, Fang and Ren [7] proposed an adaptive phase compensation method based on asymmetric current sampling resistance network, which can effectively realize the advanced phase compensation.

The switching amplifier studied in this paper is based on a full-bridge power topology that processes the error and control signals, and then converts them into control signals for the switching tubes through digital logic devices to achieve precise following of the control signals for the magnetic bearing coil currents. The main advantage is the ability to reduce the switching frequency and current ripple, while the control structure can be achieved with only two non-gates, which greatly reduces the device cost.

2 Amplifier Main Circuit Topology Working Principle

There are many topologies of amplifiers, including unipolar, push-pull, half-bridge, full-bridge, etc. [8], of which half-bridge is commonly used for active magnetic bearing, while full-bridge is mainly used for hybrid switching amplifiers. This paper focuses on hybrid magnetic bearing switching amplifiers with full-bridge power topology, the schematic diagram of which is as follows.

As shown in the figure, assuming that the current is from left to right, the full-bridge power topology generally has three operating states:

1. When the switch tube S1, S4 is on, the voltage at both ends of the coil +VDC, the coil current increases rapidly.

2. When the switch tube S2, S3 is on, the voltage across the coil −VDC, the coil current rapidly decreases.
3. When the switch S1, S3 or S2, S4 is on, the voltage across the coil is zero, and the coil current decreases slowly.

3 New Control Structure Fundamentals

The three-level switching power amplifier designed in this paper is divided into three main parts, the direction and error signal generation circuit, the digital logic circuit and the power topology circuit. Among them, the power topology circuit mainly adopts the full-bridge power topology, whose operating states are introduced in Chap. 2.

3.1 Direction and Error Signal Generation Circuit

As shown in the figure, the voltage source V3 is the control signal given by the magnetic bearing controller, FED is the feedback signal detected by the current sensor, firstly, the two signals are amplified by AD8072jn/AD, U1, U4 is the voltage comparator. The control signal and the ground signal are compared by U1 to produce the direction signal, while the feedback signal and the control signal are compared by U4 to produce the error signal, adding AD8072jn/AD is mainly to adjust the power amplifier input-output relationship, the input of this paper amplifier is −3 to 3V, the output is −3 to 3A, so the control and feedback signals can be amplified by the same multiple. Suppose the output is −6 to 6A, then the control signal can be compared with the feedback signal twice to generate the error signal.

In practice, since the feedback signal is measured by the current sensor, and the current sensor when the input signal is zero, the output is not zero, so the zero adjustment circuit must be added, that is, the feedback signal is added to the subtraction circuit consisting of a Sliding rheostat.

3.2 Digital Logic Circuits

The main purpose of this design is to realise that the magnetic bearing coil current can follow the control signal accurately and at the same time reduce the switching frequency, so as to reduce the current ripple [9]. The main working principle is as follows:

(1) When the direction signal is 1, the current goes from left to right, keeping the switching tube S4 on all the time, as can be seen from the above figure, the error signal is the feedback signal minus the given signal. When the error signal ERR is 0, that is, the given signal is greater than the feedback signal, the switch tube S1 conducts, the coil current rises rapidly, when ERR is 1, that is, the given signal is smaller than the feedback signal, the switch tube S2 conducts, the full bridge circuit works in the renewal state, and the coil current decreases.
(2) Similarly when the direction signal is 0, the current goes from right to left, keeping the switch tube S2 always on, if the coil current needs to increase in the reverse direction, the switch S3 is on, if the coil current needs to decrease in the reverse direction, the switch S4 is on, making it work in the continuous current state.

Table 1. Logical relation truth table

DIR	ERR	S1	S2	S3	S4
0	0	1	0	1	0
0	1	0	1	1	0
1	0	1	0	0	1
1	1	0	1	0	1

The truth table is as follows (Table 1).

The relationship between the switching signals S1, S2, S3, S4 and DIR, ERR is introduced from the truth table as follows (let DIR = A, ERR = B):

$$S1 = \overline{B}$$

$$S2 = B$$

$$S3 = \overline{A}$$

$$S4 = A$$

The equation shows that only two non-gates are needed to control the four switching tubes, which greatly reduces the control cost [10].

4 Simulation Results

In order to verify the correctness of the designed amplifier theory, this paper uses cadence simulation software to simulate it with magnetic bearing coil resistance $R = 2.7\ \Omega$ and inductance $L = 7$ mH, DC bus voltage is 100 V. The simulation in this chapter is divided into two parts:

(1) Assuming that the control signal of the magnetic levitation bearing is a step signal, the main purpose is to verify the dynamic performance of the switching power amplifier designed in this paper when the control signal has a step change. The simulation waveform is as follows:

As can be seen from the simulation Fig. 1, when the step signal changes abruptly, the switching amplifier designed in this paper only needs 0.08ms to follow the control signal, and as can be seen from the simulation Fig. 2 when the control signal is unchanged, the current ripple is very small. It is almost zero (Figs. 3, 4 and 5).

(2) Assume that the magnetic bearing control signal is a sinusoidal signal with amplitude of 3 and period of 0.02s. The simulation results are as follows:

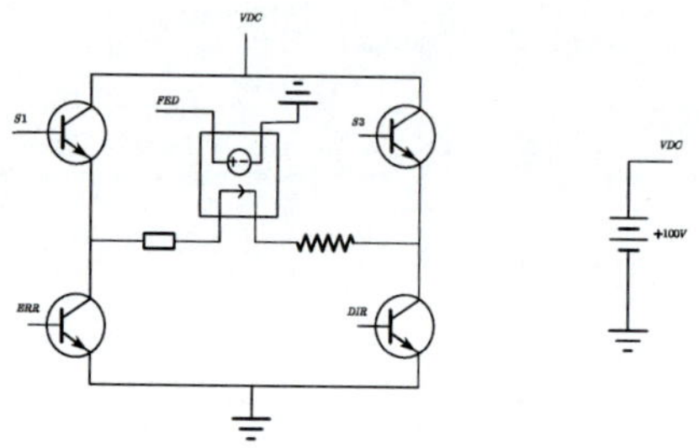

Fig. 1. Bridge power topology

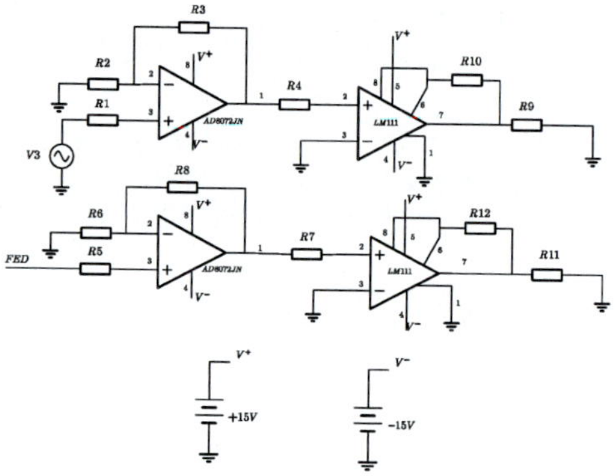

Fig. 2. Direction and error signal generation circuit

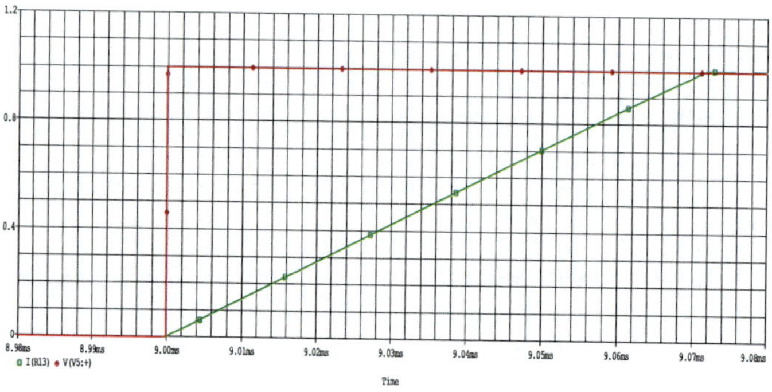

Fig. 3. Step signal simulation 1

As shown in the figure, the red waveform is the control signal, and the green waveform is the magnetic bearing coil current, which shows that the switching amplifier designed in this paper can follow the control signal more accurately. At the same time, the switching frequency is reduced, thus reducing the switching loss.

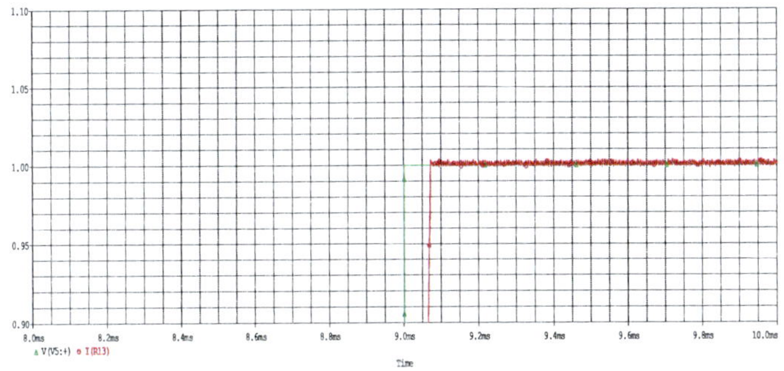

Fig. 4. Step signal simulation 2

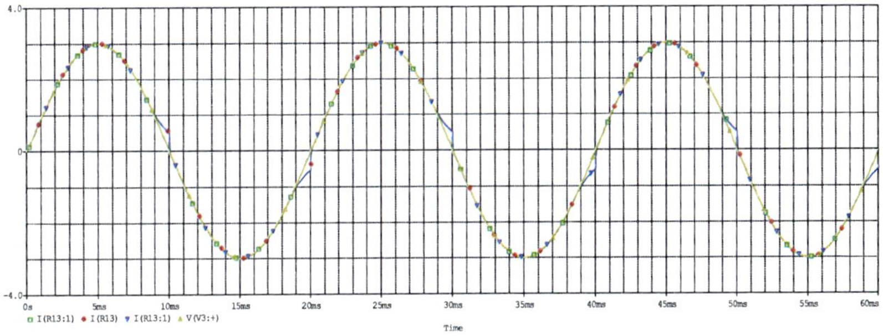

Fig. 5. Sinusoidal signal simulation 1

Suppose the input of our designed switching amplifier is −3 to 3 V and the output is −6 to 6 A, we only need to adjust the resistor R3, or R8 in the direction and error signal generation circuit to make the amplification of the control signal twice as much as the switching signal, assuming the input signal is still sinusoidal. The simulation waveform is as follows (Fig. 6).

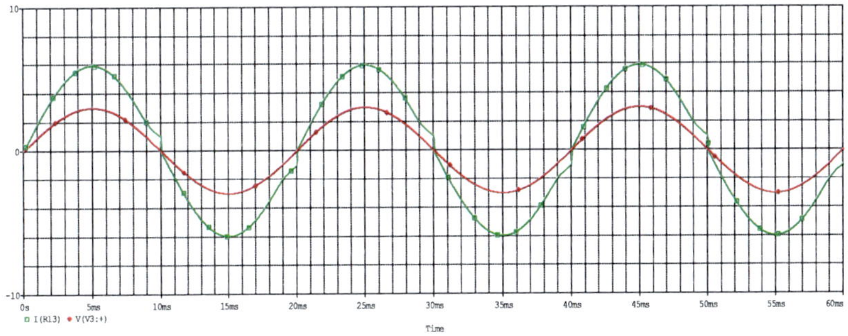

Fig. 6. Sinusoidal signal simulation 2

The magnetic bearing coil current approximates a sinusoidal signal of amplitude 6 and period 0.02. From the simulation waveform, it can be seen that the switching amplifier designed in this paper can not only follow the input and output accurately, but also adjust the input-output ratio by adjusting the resistor as needed.

5 Conclusion

The switching amplifier designed in this paper firstly processes the control and feedback signals so that the amplifier input and output can be adjusted by adjusting the resistor resistance value, and then converts them into the control signals of the switching tubes using only two digital logic devices. Moreover, the designed switching amplifier has a reduced switching frequency and low switching losses.

References

1. Bamisile, O., Zheng, Z., Adun, H., et al.: Development and prospect of flywheel energy storage technology: a citespace-based visual analysis. Energy Reports **9**, 494–505 (2023)
2. Wang, J., Xu, L.: Analysis and modeling of a switching power amplifier for magnetic bearing. In: 2009 4th IEEE Conference on Industrial Electronics and Applications, pp. 2257–2261. IEEE, Xian, China (2009)
3. Cao, G., Sun, H., Yang, G., et al.: Active magnetic bearing amplifier design based on SiC devices. In: 2021 IEEE Workshop on Wide Bandgap Power Devices and Applications in Asia (WiPDA Asia), pp. 382–386. IEEE, Wuhan, China (2021)
4. Yu, W., Fan, Y., Liu, S., et al.: Research on MATLAB simulation of three level power amplifier for magnetic bearing. In: 2010 Asia-Pacific Power and Energy Engineering Conference, pp. 1–4. IEEE, Chengdu, China (2010)
5. Jiang, Y., Wang, K., Zhang, H.: Rotor displacement self-sensing of permanent magnet bias magnetic bearings using three-level switching amplifier power demodulation. In: 2017 20th International Conference on Electrical Machines and Systems (ICEMS), pp. 1–6. IEEE, Sydney, Australia (2017)
6. Zhou, J., Deng, Z., Liu, C., et al.: Current ripple analysis of five-phase six-leg switching power amplifiers for magnetic bearing with one-cycle control
7. Fang, J., Ren, Y.: Self-adaptive phase-lead compensation based on unsymmetrical current sampling resistance network for magnetic bearing switching power amplifiers. IEEE Trans. Ind. Electron. **59**(2), 1218–1227 (2012)
8. Gupta, S., Biswas, P.K., Laldingliana, J., et al.: Comparative analysis among different types of power amplifier for active magnetic bearing system. In: 2022 IEEE International Power and Renewable Energy Conference (IPRECON), pp. 1–4. IEEE, Kollam, India (2022)
9. Liu, C., Zhan, J., Wang, J., et al.: An improved one-cycle control algorithm for a five-phase six-leg switching power amplifier in active magnetic bearings. IEEE Trans. Ind. Electron. **69**(12), 12564–12574 (2022)
10. Hu, H.-J., Liu, K., Wang, H., et al.: A wide bandwidth GaN switching power amplifier of active magnetic bearing for a flywheel energy storage system. IEEE Trans. Power Electron. **38**(2), 2589–2605 (2023)

Research on Damping System of High Speed Superconducting Maglev Train

Haoyun Wang and Ruikun Mai(✉)

School of Electrical Engineering, Southwest Jiaotong University, Chengdu 611756, China
wanghaoyun_swjtu@163.com, mairk@swjtu.cn

Abstract. When the high-speed superconducting maglev train is running at high speeds, it generates mechanical vibrations due to air resistance and track structure errors. These vibrations can be classified into high-frequency vibrations and low-frequency vibrations based on their frequencies. Mechanical dampers (hydraulic or pneumatic) between the primary suspension and the secondary suspension can effectively suppress the low-frequency vibrations of the train, but they are difficult to effectively suppress the high-frequency vibrations. High-frequency vibrations can affect passenger experience and pose a threat to the safe operation of the train. To suppress high-frequency vibrations, this paper proposes a method that utilizes the damping force generated by the collector coil, and the core of this method is to use a PWM rectifier to change the power factor angle of the collector current in the collector coil to generate the desired damping force for the train.

Keywords: Mechanical vibration · Damping suppression · PWM rectifier · Power factor

1 Introduction

Maglev trains have the advantages of non-contact operation, high speed, low noise, and small turning radius, making them one of the ideal solutions for future high-capacity high-speed transportation systems. Compared to traditional wheel-rail trains, maglev trains have higher speeds, which require them to adopt a non-contact power supply method. Therefore, the use of non-contact linear generators for high-speed maglev trains has become an important research direction [1, 2]. Non-contact linear generators for high-speed maglev trains are primarily used in both conventional conductive maglev systems and superconducting maglev systems [3–5].

The superconducting Maglev system suspended by Electro-Dynamic Suspension (EDS) intrinsically has negative magnetic damping [6–8]. This factor degrades the ride comfort of the Maglev system. In view of this defect, some scholars put forward the method of using damping coil to generate damping force to know the train vibration. However, due to the limited installation space of the train, there is no position where the damping coil can be installed, so this method is not feasible. Some scholars have proposed a method to control the current in the superconducting coil to generate damping force, but because the superconducting coil is installed in the Dewar container, the control

L. Jia et al. (Eds.): EITRT 2023, LNEE 1135, pp. 463–476, 2024.
https://doi.org/10.1007/978-981-99-9307-9_52

process is complicated, so this method is not feasible. Therefore, a method to improve EDS negative damping by multiplexing collector coils is proposed [9]. This method is to control the phase (power factor) between the induced voltage and the currents in the linear generator coils in order to generate a damping force [10]. This method not only does not require the installation of additional coils, but also is relatively easy to control.

This paper takes the Japanese MLX01 maglev train as the research object and conducts in-depth research on the damping system of high-speed superconducting maglev trains. It provides precise mathematical models, simulation models, and control strategies.

2 MLX01 Train Key Components and Working Principle

Figure 1 shows a cross-sectional view of the direction of movement of the high-speed superconducting maglev train. The diagram depicts key components of the MLX01 maglev train, including the superconducting coils and collector coils installed on both sides of the bogie, as well as the propulsion coils and suspension coils installed on both sides of the U-shaped track.

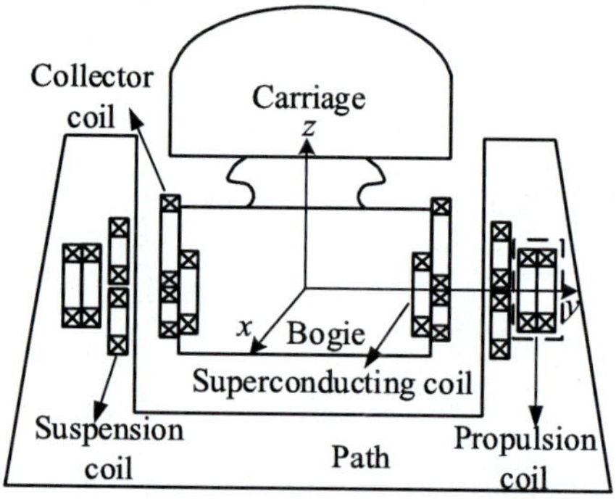

Fig. 1. MLX01 train profile

The suspension coils periodically cut through the superconducting coils' generated superconducting magnetic field, thereby inducing suspension currents. These suspension currents generate a suspension magnetic field in the air gap. The upper end of the suspension magnetic field exerts an upward attractive force on the superconducting coils, while the lower end exerts an upward repulsive force. The combination of these forces is the suspension force exerted on the train. The collector coil is a closed asymmetric fig-eight coil, and the fifth harmonic of the suspended magnetic field periodically passes through the closed surface of the collector coil to generate the collector current in the collector coil. The collector current is rectified by a PWM rectifier to supply power to the electrical equipment on the train. The collector coil is a closed asymmetric figure-eight-shaped coil. It periodically cuts through the fifth harmonic of the suspension magnetic field, thereby inducing collector current. After rectification by a PWM rectifier, the collector current supplies power to the electrical equipment on the train.

The PWM rectifier can change the amplitude and phase of the collector current. When the power factor angle of the collector current is 0, no damping force is generated. The damping force increases as the power factor angle increases. Details are as follows.

3 Working Principle of PWM Rectifier

The waveform of single-phase collector voltage is shown in Fig. 2.

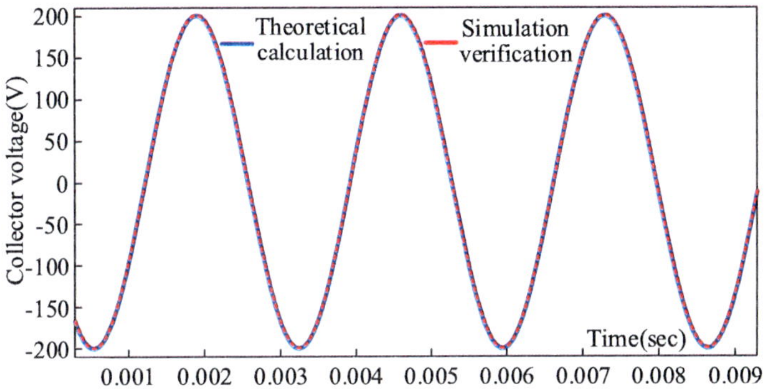

Fig. 2. Collector voltage waveform

The collector current needs to be rectified by a three-phase PWM rectifier before it can be supplied to the electrical equipment on the vehicle. The equivalent circuit model of the collector coil and the three-phase PWM rectifier is shown in Fig. 3.

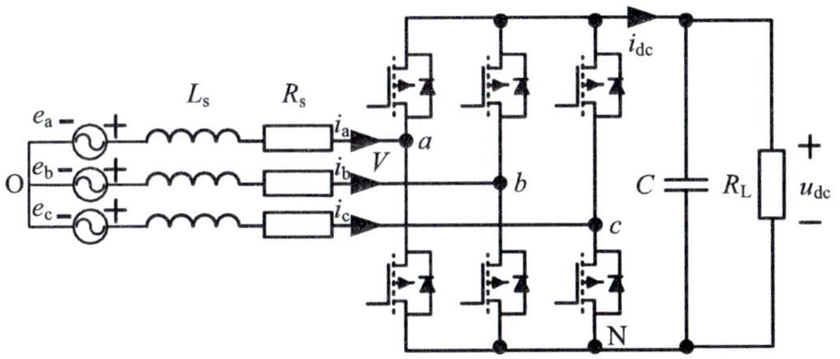

Fig. 3. Topology of main circuit of three-phase PWM rectifier

For the convenience of analysis, the unipolar binary logic switching function S_k is first defined as:

$$S_k = \begin{cases} 1 : \text{Upper bridge arm conduction} \\ 0 : \text{Lower bridge arm conduction} \end{cases} \quad (k = a, b, c) \tag{1}$$

According to Kirchhoff's voltage law, the AC side loop equation is established:

$$\begin{cases} -e_a + R_s i_a + L_s \dfrac{di_a}{dt} + u_{aN} + u_{NO} = 0 \quad ① \\ -e_b + R_s i_b + L_s \dfrac{di_b}{dt} + u_{bN} + u_{NO} = 0 \quad ② \\ -e_c + R_s i_c + L_s \dfrac{di_c}{dt} + u_{cN} + u_{NO} = 0 \quad ③ \end{cases} \tag{2}$$

where, L_s, R_s are respectively the inductance and resistance of a phase collector coil.

From the main circuit:

$$\begin{cases} u_{aO} = u_{aN} + u_{NO} = S_a u_{dc} + u_{NO} \\ u_{bO} = u_{bN} + u_{NO} = S_b u_{dc} + u_{NO} \\ u_{cO} = u_{cN} + u_{NO} = S_c u_{dc} + u_{NO} \end{cases} \tag{3}$$

Because the system is a three-phase symmetric system, so:

$$\begin{cases} e_a + e_b + e_c = 0 \\ i_a + i_b + i_c = 0 \end{cases} \tag{4}$$

From Formulas (1)–(4), it can be obtained:

$$\begin{cases} u_{aO} = S_a u_{dc} - \dfrac{1}{3}(S_a + S_b + S_c)u_{dc} = [S_a - \dfrac{1}{3}(S_a + S_b + S_c)]u_{dc} \\ u_{bO} = S_b u_{dc} - \dfrac{1}{3}(S_a + S_b + S_c)u_{dc} = [S_b - \dfrac{1}{3}(S_a + S_b + S_c)]u_{dc} \\ u_{cO} = S_c u_{dc} - \dfrac{1}{3}(S_a + S_b + S_c)u_{dc} = [S_c - \dfrac{1}{3}(S_a + S_b + S_c)]u_{dc} \end{cases} \tag{5}$$

Applying Kirchhoff's current law to the positive capacitor node on the DC side:

$$\begin{cases} i_{dc} = i_a S_a + i_b S_b + i_c S_c \\ C\dfrac{du_C}{dt} = i_{dc} - i_L \end{cases} \tag{6}$$

$$\begin{cases} L_s \dfrac{di_a}{dt} = -R_s i_a + 0 \times i_b + 0 \times i_c - (S_a - \dfrac{1}{3}\sum\limits_{k=a,b,c} S_k)u_{dc} + e_a \\ L_s \dfrac{di_b}{dt} = 0 \times i_a - R_s i_b + 0 \times i_c - (S_b - \dfrac{1}{3}\sum\limits_{k=a,b,c} S_k)u_{dc} + e_b \\ L_s \dfrac{di_c}{dt} = 0 \times i_a - 0 \times i_b - R_s i_c - (S_c - \dfrac{1}{3}\sum\limits_{k=a,b,c} S_k)u_{dc} + e_c \\ C\dfrac{du_{dc}}{dt} = i_a S_a + i_b S_b + i_c S_c - \dfrac{u_{dc}}{R_L} + 0 \end{cases} \tag{7}$$

Formula (7) is expressed in matrix form:

$$\begin{pmatrix} L_s & 0 & 0 & 0 \\ 0 & L_s & 0 & 0 \\ 0 & 0 & L_s & 0 \\ 0 & 0 & 0 & C \end{pmatrix}\begin{pmatrix} \dot{i}_a \\ \dot{i}_b \\ \dot{i}_c \\ \dot{u}_{dc} \end{pmatrix} = \begin{pmatrix} -R_s & 0 & 0 & -(S_a - \frac{1}{3}\sum\limits_{k=a,b,c} S_k) \\ 0 & -R_s & 0 & -(S_b - \frac{1}{3}\sum\limits_{k=a,b,c} S_k) \\ 0 & 0 & -R_s & -(S_c - \frac{1}{3}\sum\limits_{k=a,b,c} S_k) \\ S_a & S_b & S_c & -\frac{1}{R_L} \end{pmatrix}\begin{pmatrix} i_a \\ i_b \\ i_c \\ u_{dc} \end{pmatrix} + \begin{pmatrix} e_a \\ e_b \\ e_c \\ 0 \end{pmatrix} \tag{8}$$

In order to simplify the design of the control system, the three-phase symmetrical stationary coordinate system (a, b, c) is transformed into a (d, q) coordinate system that rotates synchronously with the fundamental frequency of the grid. This paper adopts the "equivalence amplitude transformation". Let:

$$C = \frac{2}{3} \times \begin{pmatrix} \sin(\omega t) & \sin(\omega t - \frac{2}{3}\pi) & \sin(\omega t + \frac{2}{3}\pi) \\ \cos(\omega t) & \cos(\omega t - \frac{2}{3}\pi) & \cos(\omega t + \frac{2}{3}\pi) \end{pmatrix} \tag{9}$$

$$\begin{cases} Ce_j = \frac{2}{3}\begin{pmatrix} e_d \\ e_q \end{pmatrix} \\ CR_s i_j = \frac{2}{3}R_s\begin{pmatrix} i_d \\ i_q \end{pmatrix} \\ CL_s\frac{di_j}{dt} = \frac{2}{3}\begin{pmatrix} 0 & -\omega L_s \\ -\omega L_s & 0 \end{pmatrix}\begin{pmatrix} i_d \\ i_q \end{pmatrix} + \frac{2}{3}L_s\frac{d}{dt}\begin{pmatrix} i_d \\ i_q \end{pmatrix} \\ Cu_{jO} = \frac{2}{3}\begin{pmatrix} u_d \\ u_q \end{pmatrix} \end{cases} \quad (j = a, b, c) \tag{10}$$

Therefore, the original KVL expression can be expressed as:

$$\begin{pmatrix} e_d \\ e_q \end{pmatrix} = R_s\begin{pmatrix} i_d \\ i_q \end{pmatrix} + \begin{pmatrix} 0 & -\omega L_s \\ \omega L_s & 0 \end{pmatrix}\begin{pmatrix} i_d \\ i_q \end{pmatrix} + L_s\begin{pmatrix} \frac{di_d}{dt} \\ \frac{di_q}{dt} \end{pmatrix} + \begin{pmatrix} u_d \\ u_q \end{pmatrix} \tag{11}$$

The Laplace transform of the above equation gives:

$$\begin{cases} e_d = (R_s + sL_s)i_d - \omega L_s i_q + u_d \\ e_q = (R_s + sL_s)i_q + \omega L_s i_d + u_q \end{cases} \tag{12}$$

i_d, i_q as output, u_d, u_q as input:

$$\begin{cases} i_d = \dfrac{1}{R_s + sL_s}(e_d + \omega L_s i_q - u_d) = \dfrac{1}{R_s + sL_s} V_d \\ i_q = \dfrac{1}{R_s + sL_s}(e_q - \omega L_s i_d - u_q) = \dfrac{1}{R_s + sL_s} V_q \\ \dfrac{i_d}{V_d} = \dfrac{i_q}{V_q} = \dfrac{1}{R_s + sL_s} = G(s) \end{cases} \tag{13}$$

Therefore, the input and output model of PWM rectifier is shown in Fig. 4:

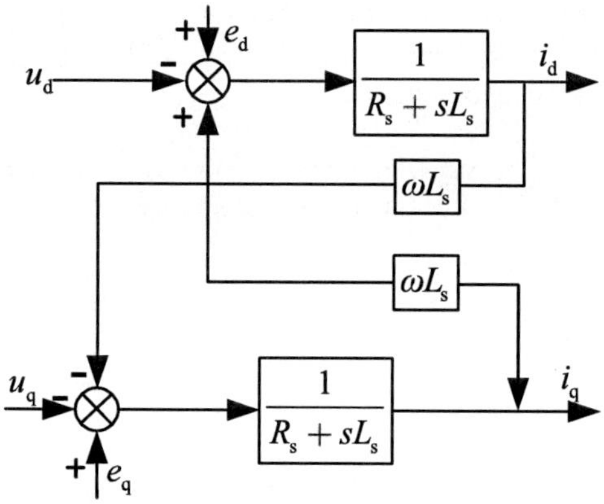

Fig. 4. Input and output model of PWM rectifier

On the d and q axes, a controller $G_c(s)$ is designed to control $G(s)$. The schematic block diagram is shown in Fig. 5.

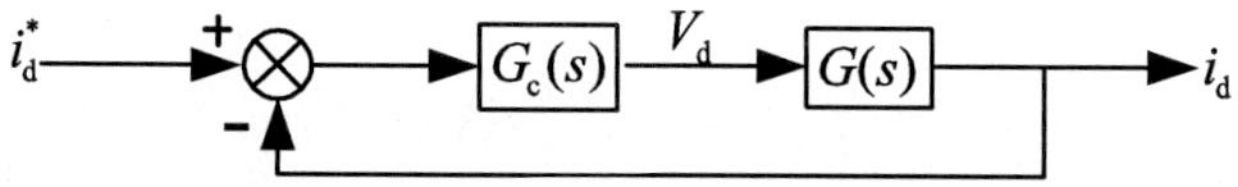

Fig. 5. Controller schematic block diagram

Let the expression for $G_c(s)$ be:

$$G_c(s) = K_{P1} + \frac{K_{I1}}{s} \tag{14}$$

Then:

$$\begin{cases} V_d = (i_d^* - i_d)(K_{P1} + \dfrac{K_{I1}}{s}) \\ V_q = (i_q^* - i_q)(K_{P1} + \dfrac{K_{I1}}{s}) \end{cases} \tag{15}$$

And because:

$$\begin{cases} V_d = e_d + \omega L_s i_q - u_d \\ V_q = e_q - \omega L_s i_d - u_q \end{cases} \tag{16}$$

The combination of vertical (15) and formula (16) gives:

$$\begin{cases} u_d^* \to u_d = -(i_d^* - i_d)(K_{P1} + \frac{K_{I1}}{s}) + e_d + \omega L_s i_q & ① \\ u_q^* \to u_q = -(i_q^* - i_q)(K_{P1} + \frac{K_{I1}}{s}) + e_q - \omega L_s i_d & ② \end{cases} \tag{17}$$

The output of ① and ② is used as the output command signal of the controller to make the inner loop controller model of the PWM rectifier. Similarly, the external loop controller model of PWM rectifier can be obtained. The PWM rectifier control model is shown as follows (Fig. 6).

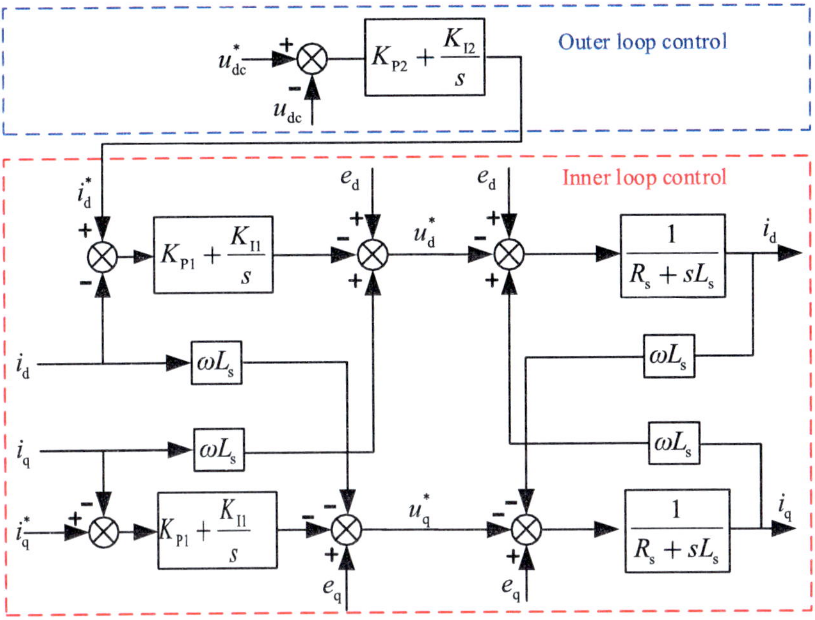

Fig. 6. The controller model of PWM rectifier

According to experience, the PI parameters of the inner current loop and the outer voltage loop can be obtained, as shown in Table 1.

By integrating the above models and parameters with simulink, the simulation model of PWM rectifier can be obtained, as shown in Fig. 7.

By adjusting i_q^*, the proportion of reactive current in AC side current of PWM rectifier can be changed, so as to control the power factor Angle of AC side. Taking phase a as an example, the voltage and current of the AC side of the PWM rectifier under different power factor angles are shown in (Fig. 8).

Table 1. PI parameter

	KP	KI
Current inner loop	$K_{P1} = 20$	$K_{I1} = 10$
Voltage outer loop	$K_{P2} = 0.1$	$K_{I2} = 10$

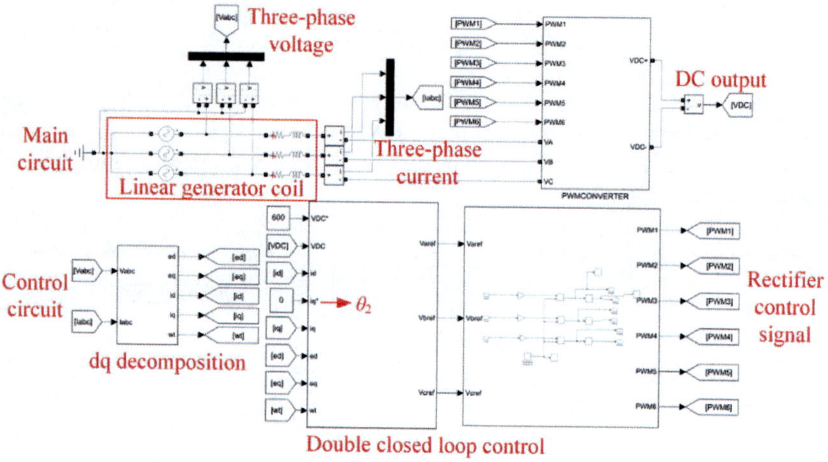

Fig. 7. PWM rectifier simulation model

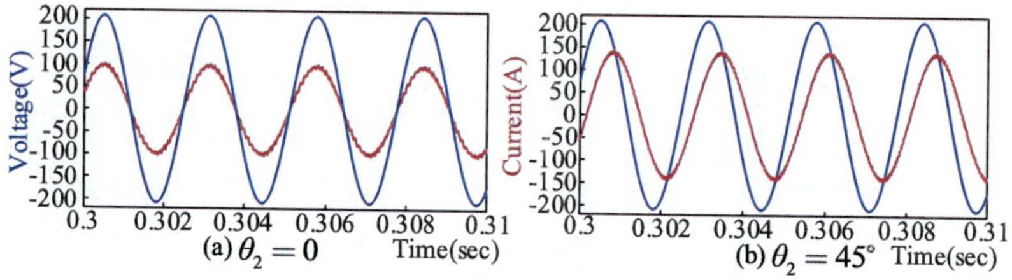

Fig. 8. Voltage and current waveform on AC side of PWM rectifier

4 Analytical Calculation of Set Electromagnetic Field

After obtaining the AC side current (i.e., the current on the collector coil), the magnetomotive force of the collector coil can be solved. By using the spatial harmonic method, the distribution of the magnetomotive force of the collector coil along the y-axis direction can be obtained. With the change of the x-coordinate, the magnetomotive force distribution of the upper part of the collector coil is as follows (Fig. 9).

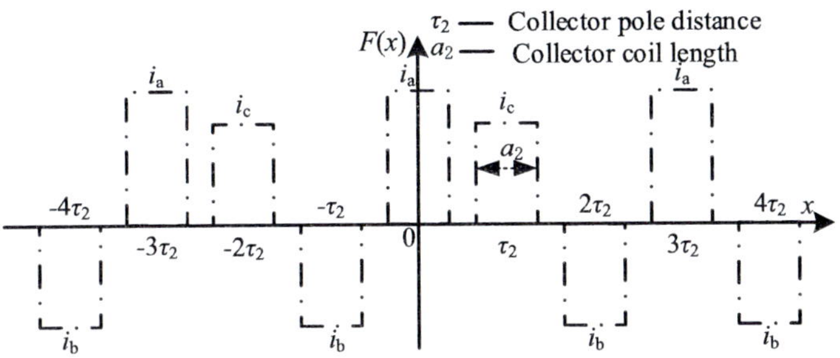

Fig. 9. Distribution of the magnetomotive force of a collector coil on the x-axis

The fourier series coefficients of the magnetomotive force of the upper half of the phase a collector coil in the x-axis direction are:

$$\begin{aligned} a_{axm} &= \frac{2}{3\tau_2}\int_0^{3\tau_2} N_2 i_a \cos(\frac{m\pi x}{3\tau_2})dx \\ &= \frac{2}{3\tau_2}\left[\int_0^{a_2/2} N_2 i_a \cos(\frac{m\pi x}{3\tau_2})dx + \int_{3\tau_2 - a_2/2}^{3\tau_2} N_2 i_a \cos(\frac{m\pi x}{3\tau_2})dx\right] \\ &= \frac{4N_2 i_a}{\pi m}\sin(\frac{m\pi a_2}{6\tau_2}) \end{aligned} \tag{18}$$

where $m = 2, 4, 6, 8\ldots$ (For ease of calculation, $m = 0$ can be ignored).

The magnetomotive force of the upper half of the a-phase collector coil at any point on the z-axis is:

$$F_{za}(x) = \sum_{m=2}^{\infty} a_{axm}\cos(\frac{m\pi x}{3\tau_2}) \tag{19}$$

On the z-axis, the collector coils are also arranged in reverse polarity, and the pole distance is W_z. It should be noted that W_z is assumed for convenient calculation, and its value should meet the requirement that W_z is much higher than the height of the collector coil. In this paper, W_z is taken as 6m. The distribution of $F_2(z)$ on the z-axis is as follows (Fig. 10).

In the same way as finding the Fourier series coefficients on the x-axis, the Fourier coefficients on the z-axis are:

$$a_{azn} = \frac{4F_{za}(x)}{n\pi}\sin(\frac{n\pi b_{2U}}{6W_z}) \tag{20}$$

where, b_{2U} is the height of the collector coil, $n = 2, 4, 6, 8\ldots$

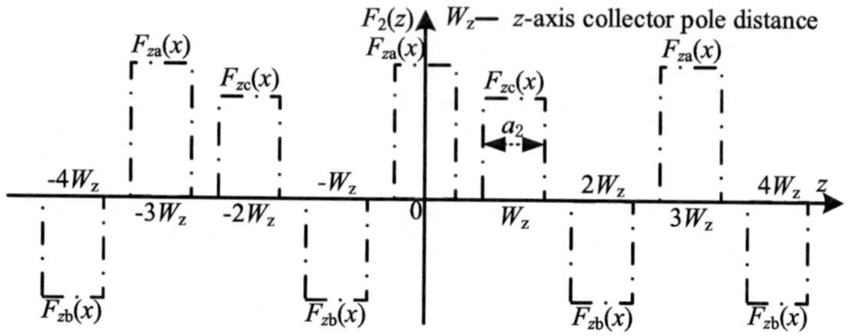

Fig. 10. Distribution of the magnetomotive force of a collector coil on the *x*-axis

The distribution function of the magnetomotive force of the upper half of the collector coil on the *xoz* plane can be obtained as follows:

$$F_{\mathrm{aU}}(x,z) = \frac{16N_2 i_{\mathrm{a}}}{9\tau_2 W_z}\sum_{m=1}^{\infty}\sum_{n=1}^{\infty} f_{2\mathrm{U}}(m,n)\cos(k_{2xm}x)\cos[k_{2zn}(z - z_{2\mathrm{U}})] \tag{21}$$

where $\sum$ is the sum of non-zero even numbers, $k_{2xm} = \frac{m\pi}{3\tau_2}$, $k_{2zn} = \frac{n\pi}{3W_z}$, $z_{2\mathrm{U}}$ is the *z*-axis coordinate of the middle point of the upper half of the collector coil, $f_{2\mathrm{U}}(m,n)$ is a function related to the size of the coil on the collector, whose expression is:

$$f_{2\mathrm{U}}(m,n) = \frac{\sin(k_{2xm}\frac{a_2}{2})}{k_{2xm}}\frac{\sin(k_{2zn}\frac{b_{2\mathrm{U}}}{2})}{k_{2zn}} \tag{22}$$

The same method can be used to obtain the magnetomotive force of the upper half of the b-phase and c-phase collector coils, whose expressions are as follows:

$$F_{\mathrm{bU}}(x,z) = \frac{16N_2 i_{\mathrm{b}}}{9\tau_2 W_z}\sum_{m=1}^{\infty}\sum_{n=1}^{\infty} f_{2\mathrm{U}}(m,n)\cos(k_{2xm}(x-\tau_2))\cos[k_{2zn}(z - z_{2\mathrm{U}})] \tag{23}$$

$$F_{\mathrm{cU}}(x,z) = \frac{16N_2 i_{\mathrm{c}}}{9\tau_2 W_z}\sum_{m=1}^{\infty}\sum_{n=1}^{\infty} f_{2\mathrm{U}}(m,n)\cos(k_{2xm}(x-2\tau_2))\cos[k_{2zn}(z - z_{2\mathrm{U}})] \tag{24}$$

The magnetomotive force of the upper half of all collector coils is:

$$F_{2\mathrm{U}}(x,z) = F_{\mathrm{aU}}(x,z) + F_{\mathrm{bU}}(x,z) + F_{\mathrm{cU}}(x,z) \tag{25}$$

Similarly, the expression of the magnetomotive force of the lower half of the phase a, b, and c collector coil is:

$$F_{\mathrm{aB}}(x,z) = -\frac{16N_2 i_{\mathrm{a}}}{9\tau_2 W_z}\sum_{m=1}^{\infty}\sum_{n=1}^{\infty} f_{2\mathrm{B}}(m,n)\cos(k_{2xm}x)\cos[k_{2zn}(z - z_{2\mathrm{B}})] \tag{26}$$

$$F_{\mathrm{bB}}(x,z) = -\frac{16N_2 i_{\mathrm{b}}}{9\tau_2 W_z}\sum_{m=1}^{\infty}\sum_{n=1}^{\infty} f_{2\mathrm{B}}(m,n)\cos(k_{2xm}(x-\tau_2))\cos[k_{2zn}(z - z_{2\mathrm{B}})] \tag{27}$$

$$F_{cB}(x, z) = -\frac{16N_2 i_c}{9\tau_2 W_z} \sum_{m=1}^{\infty} \sum_{n=1}^{\infty} f_{2B}(m, n) \cos(k_{2xm}(x - 2\tau_2)) \cos[k_{2zn}(z - z_{2B})] \quad (28)$$

where, $f_{2B}(m, n)$ is a function related to the size of the lower half of the collector coil, whose expression is:

$$f_{2B}(m, n) = \frac{\sin(k_{2xm}\frac{a_2}{2})}{k_{2xm}} \frac{\sin(k_{2zn}\frac{b_{2B}}{2})}{k_{2zn}} \quad (29)$$

The magnetomotive force of the lower half of all collector coils is:

$$F_{2B}(x, z) = F_{aB}(x, z) + F_{bB}(x, z) + F_{cB}(x, z) \quad (30)$$

The collector coil magnetomotive force is:

$$F_2(x, z) = F_{2U}(x, z) + F_{2B}(x, z) \quad (31)$$

According to the boundary conditions of scalar magnetic potential and Laplace's equation:

$$\begin{cases} \psi_2(x, y_2, z) = F_2(x, z)/2 \\ \psi_2(x, \infty, z) = 0 \\ \nabla^2 \psi_2(x, y, z) = 0 \end{cases} \quad (32)$$

The component of the magnetic induction intensity generated by the collector current on the y-axis can be obtained:

$$B_{2y} = -\mu_0 \frac{\partial \psi_2(x, y, z)}{\partial y} = \frac{1}{2}\mu_0 \lambda_{2mn} F_2(x, z) e^{-|\lambda_{2mn}(y - y_2)|} \quad (33)$$

Let $t = 0$, $y = 0.185$, $z = 0.21$, and the distribution of y-axis components of the set electromagnetic field with x-coordinate can be obtained from Eq. (33), as shown in Fig. 11.

5 Analytical Calculation of Damping Force

As the train operates, the suspension coil will periodically cut through the collector magnetic field. According to the law of electromagnetic induction, an induced voltage will be generated in the suspension coil, known as the collector-suspension voltage. Its waveform is shown in Fig. 12.

The equivalent circuit of the suspension coil is shown in Fig. 13. L_1 represents the equivalent inductance, R_1 represents the equivalent resistance, u_{21} represents the collector-suspension voltage, and i_{21} represents the collector-suspension current.

According to Kirchhoff's law, the equivalent circuit equation of the suspended coil can be written:

$$L_1 \frac{di_{21}}{dt} + R_1 i_{21} = u_{21} \quad (34)$$

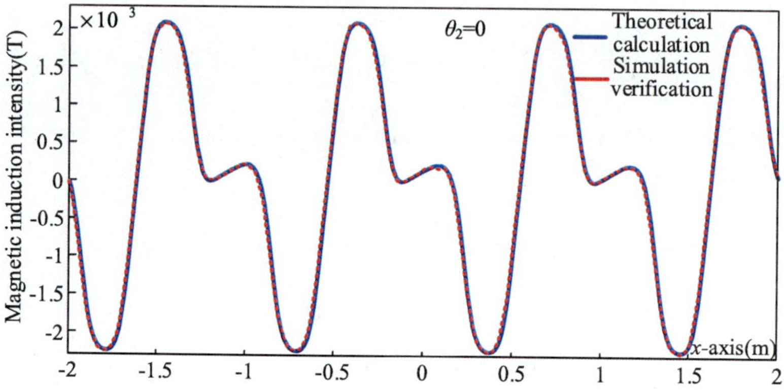

Fig. 11. *y*-axis component of the set electromagnetic field

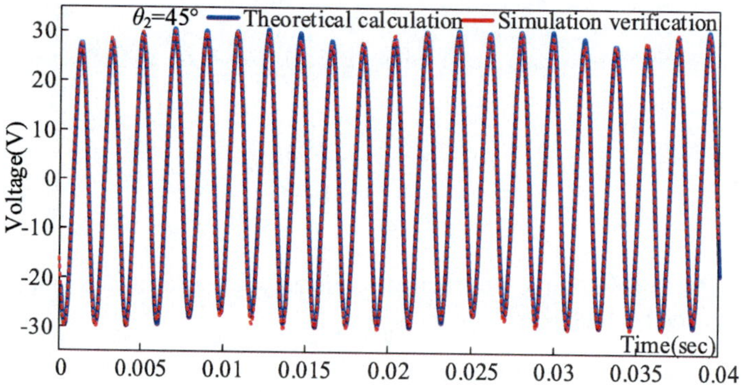

Fig. 12. Collector-suspension voltage waveform

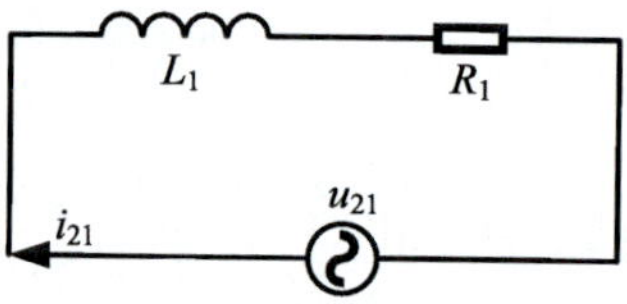

Fig. 13. Levitating coil equivalent circuit mode

The waveform of collector-suspension current can be obtained by solving this differential equation, as shown in (Fig. 14).

Fourier decomposition is performed on the collector-suspension current, and the result is shown in Fig. 15. It can be seen that the collector-suspension current mainly contains fundamental waves and a higher harmonic.

In this paper, only the damping force generated by fundamental current is considered. The relationship between the damping force generated by fundamental current and the power factor Angle of collector current is shown in Fig. 16.

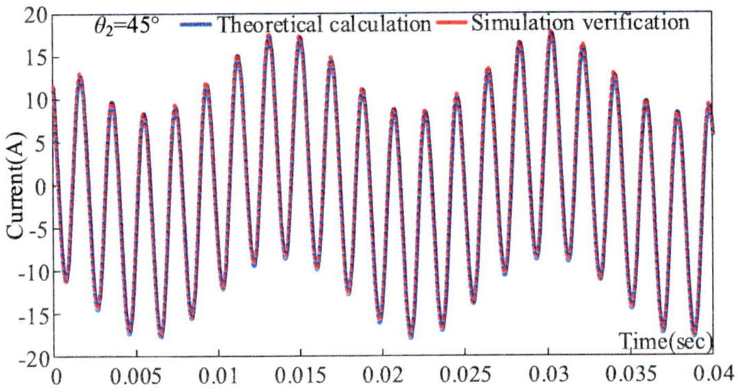

Fig. 14. Collector—suspension current waveform

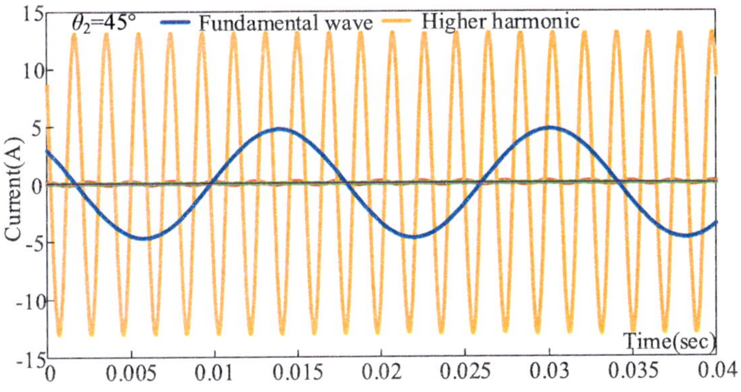

Fig. 15. Fourier decomposition results

The desired damping force can be obtained by adjusting the collector current power factor Angle according to the train operating condition.

6 Conclusion

This article validates the feasibility of using the multiplexed collector coil to generate damping force. Based on the characteristics of the MLX01 linear generator, a model of a three-phase PWM rectifier is established. By adjusting the reactive power setting of the three-phase PWM rectifier to change the power factor angle on the AC side of the rectifier, the magnitude and direction of the damping force can be controlled. Based on the train's operating state, the power factor angle is adjusted in a timely manner to maximize the effectiveness of the damping system, thereby improving the stability of the train.

The research findings in this article have guiding significance for the study of the damping system in high-speed superconducting maglev trains, laying a foundation for further research on high-speed superconducting maglev trains.

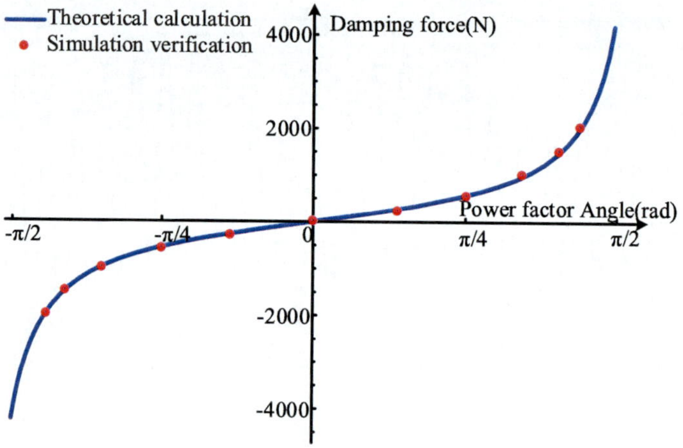

Fig. 16. Relation between damping force and power factor Angle

Acknowledgments. This work was supported by the Natural Science Foundation of Hebe Province under Grant (E202105047), and by the Key Laboratory of Railway industry of Maglev Technology (TJU), National Railway Administration of P.R.C (202207), and by the Fundamental Research Funds for the Central Universities (2682023ZTPY041).

References

1. Xu, F., Luo, S., Deng, Z.: Study on key technologies and whole speed range application of Maglev Rail transport. J. China Railway Soc. **41**(3), 40–49 (2019) (in Chinese)
2. Wang, Y., Cai, Y., Song, X., Li, G., Ma, G.: Characteristic analysis and experiment of the equivalent simulation system for null-flux electrodynamic suspension. Trans. China Electrotech. Soc. **36**(8), 1628–1635 (2021) (in Chinese)
3. Hu, D., Feng, X., Zhang, Z.: Study on the damping characteristics of superconducting electrodynamic suspension system. Proc. Chinese Soc. Electr. Eng. **41**(13), 4679–4687 (2021) (in Chinese)
4. Wang, Z., Cai, Y., Gong, T., Liu, K., Li, J., Ma, G.: Characteristic studies of the superconducting electrodynamic suspension train with a field-circuit-motion coupled model. Proc. Chinese Soc. Electr. Eng. **39**(4), 1162–1170 (2019) (in Chinese)
5. Lee, H.W., Kim, K.C., Ju, L.: Review of maglev train technologies. IEEE Trans. Magn. **42**(7), 1917–1925 (2006)
6. Yamada, T., Iwamoto, M., Ito, T.: Magnetic damping force in inductive magnetic levitation system for high speed trains (in Japanese). Trans. IEE Jpn. **94-B**(11), 49–54 (1974)
7. Fujiwara, S., Yamaguchi, H.: Magnetic damping of electro-dynamic levitation system (in Japanese). Linear Drive Trans. IEE Jpn. **LD-92-55**, 47–54 (1992)
8. Higashi, K., Ohashi, S., Osaki, H., Masada, E.: Magnetic damping of the electrodynamic suspension type superconducting levitation system (in Japanese). Trans. IEE Jpn. **117-D**(8), 1015–1023 (1997)
9. Murai, T., Hasegawa, H., Fujiwara, S.: Improvement of inductive power collection in null-flux EDS Maglev (in Japanese). Trans. IEE Jpn. **117D**(1), 81–90 (1997)
10. Fujiwara, S., Murai, T., Hasegawa, H.: Magnetic damping method of EDS system using power collection coil (in Japanese). Trans. IEE Jpn. **119D**(2), 254–259 (1999)

A High and Low Voltage DC Distribution System for Special Vehicles

Fengpu Qiao(✉), Yanming Li, Zhiyuan Wang, and Lu Xu

China North Vehicle Research Institute, Beijing 100072, China
neuqfp@126.com

Abstract. As an important part of special vehicles, the power distribution system is responsible for the power supply and distribution of all loads of the whole vehicle and the protection of related abnormal power consumption. Its performance and intelligence directly affect whether the key power consumption loads of special vehicles can work normally. In order to improve the reliability and intelligent level of the power supply and distribution system of special vehicles, this paper applies the secondary power distribution device and solid-state power switch to the intelligent power distribution system of special vehicles, which greatly improves the intelligent power distribution level and system integration level of special vehicles.

Keywords: Distribution · Reliability · Intelligence

1 Introduction

With the in-depth development of military, police and other special vehicles, multi-source power grids are increasingly applied to these special vehicles. Especially, the medium-voltage and high-voltage power supply systems will bring many problems such as reliability, stability and anti-interference of the power supply and distribution system, and put forward higher requirements for the automation, intelligence, modularization, integration and other aspects of the power distribution system. The traditional power distribution system based on fuse or air circuit breaker and contactor or relay for distribution line short circuit and overload protection cannot better meet the power supply demand of special vehicles. It is urgent to develop modular, intelligent and highly integrated power distribution system, power distribution device and equipment to ensure the safety, efficiency and reliability of power supply for special vehicles.

With the rapid development of modern power electronic technology, sensor technology, communication technology, computer and network technology and automatic control technology, the degree of distribution automation and intelligence is becoming higher and higher, and the distributed intelligent distribution system based on secondary power distribution device is also becoming more mature in weapons, aviation, aerospace, shipbuilding, metallurgy and other industries [1]. It uses advanced secondary power distribution device to monitor and control the power load in real time, realizing the intelligence, visualization and high integration of the power distribution system.

L. Jia et al. (Eds.): EITRT 2023, LNEE 1135, pp. 477–488, 2024.
https://doi.org/10.1007/978-981-99-9307-9_53

The research content and solution of this paper is based on combing high-voltage and low-voltage electrical loads and electrical stress curves of special vehicles, and has developed a high-voltage distributed intelligent power distribution system, which can meet the power demand of various types of special vehicles. In terms of power distribution informatization, automatic control, real-time monitoring of operation status, load power abnormal protection, and intelligent management, it is more centralized and safe. The fusible and irrecoverable power distribution mode has much improved.

2 Research Status

2.1 Research Status

At present, the distribution system of most special vehicles adopts the regional distribution system of main distribution box and distribution box, and the protection of most electrical loads adopts relay protection components such as fuse or air circuit breaker to protect the overload, overheating and short circuit of distribution lines. When the power load is not working to find the cause, it needs to check multiple links to locate the cause, and the degree of informatization is low; This distribution mode still has the following disadvantages:

(1) The distribution line is long, resulting in excessive voltage drop and loss. The distribution line must be laid from the main distribution box to each electrical load, and the control line of relay protection elements and components involving safety must also be laid to each electrical load of the whole vehicle, so the distribution line is long and heavy, and the line voltage drop and power loss are too large.
(2) There are few distribution branches, resulting in a large number of distribution boxes. A power supply branch is protected by a relay protection element. Due to the limitation of volume and weight, the number of power supply branches in the distribution box is limited.
(3) The setting value of relay protection components of distribution box often needs to be set in advance during the design of distribution system, and it is not easy to change after setting. When the working current of the power supply branch of the distribution box changes, it may be necessary to remanufacture the fuse base or re-select the air circuit breaker, which may increase the work quantity and work cycle.
(4) The distribution box has no power supply branch current, voltage, instant power and other electrical parameters and power quality signal output. Although the distribution box is equipped with relay protection components such as fuse or air circuit breaker, it cannot provide the power measurement signal of each power supply branch, and cannot realize the visualization of distribution information.

With the rapid development of power distribution technology and the field of electronic components, the research of automatic power distribution system has been carried out in foreign countries since the 1960s, and the high automation and intelligence of the power distribution system has been realized by adopting multiplex transmission technology, which automatically manages the electrical load through bus controller, load management center, electrical remote terminal, solid state power controller, and large current electromechanical hybrid power controller, The operation and control are simple, the weight of the power grid is greatly reduced, and the maintenance is also very convenient. After the 1980s, the United States has made full use of distributed power distribution and automatic load management technology in F-16, AH-64 and Navy LAMPS shipborne helicopters, which has greatly improved aircraft reliability, survivability, maintainability and flexibility [2–4]. Representatives of foreign new aircraft models, such as F-22, F-35, Airbus A380 and Boeing 787, whose power supply system consists of three major parts: 270V voltage DC power generation system, solid-state electrical logic system and general multiplex transmission system. Programmable solid-state switches, microprocessors and automatic load management, self detection (HIT) and fault isolation are used to improve the ability to control and protect the power supply system [5].

2.2 Development Needs

In recent years, with the introduction and promotion of the concept of all-electric special vehicles, the demand for electrical equipment and power consumption has increased sharply, and higher requirements have been put forward for power supply quality, intelligent power distribution and safety protection technology. The conventional power distribution system cannot meet the needs of the development of all-electric special vehicles in terms of maintainability, scalability, reliability and automation, so it is urgent to adopt advanced distributed intelligent power distribution system to meet the performance requirements of the new generation of special vehicles [6]. The intelligent power distribution system can be embodied in the following aspects: remote control, intelligent protection, fault recovery, terminal viewing, and whole process recording [7, 8].

The CAN bus, LIN bus and FLEXRAY bus are used for the interactive transmission of data, and modular power distribution executive components are designed to replace relays, power switching devices, self-protection switches, etc.; Real-time detection of output current and voltage changes, automatic protection of the load and itself according to the changes, and transmission of the detected information, fault and alarm information to the upper computer through the bus [9, 10]. The upper computer then interacts with the bus network of the whole vehicle through the bus to meet the requirements of intelligent, reliable and real-time of the whole vehicle power distribution system [11].

Through the optimization design of the distributed distribution of the whole vehicle, reduce the wiring, reduce the cost, and realize the lightweight design of the distribution system. With the improvement of the power demand and quality of special vehicles, the intelligent power distribution system needs to monitor and display the current, voltage, operation status and other information of each load in the vehicle in real time, and provide real-time alarm and fault protection [12, 13]; Fault information can be directly displayed or indicated at the terminal, further improving intelligence and reliability.

3 System Composition and Working Principle

The distributed intelligent power distribution system for special vehicles is divided into management layer (core processor), control layer (control power distribution unit, information interaction unit and power distribution protection circuit) and equipment layer (power load) on the system architecture, and realizes the interaction of control command and operation, load distribution status feedback and other information through field bus (Fig. 1).

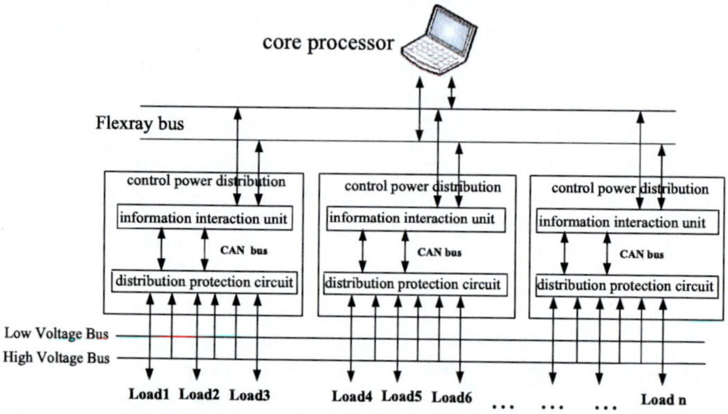

Fig. 1. Composition of special vehicle distributed power distribution system

The management layer is responsible for the control and management of the power consumption of special vehicles and the monitoring and control of the distributed intelligent power distribution system. The control layer is distributed in the power cabin, cockpit, passenger cabin and other parts, and is composed of control electric power supply unit, information interaction unit, power distribution and protection circuit. It carries out power supply control and power distribution protection for the field load. At the same time, the information interaction module uploads the electric energy parameter information to the core processor through the dual-redundant field bus. The information interaction module also receives the instructions of the core processor, and carries out direct power supply and power off control for the load. The equipment layer load is arranged at each specific position of special vehicles.

The working principle and process of distributed intelligent power distribution system for special vehicles are divided into four parts, as shown in Fig. 2:

(1) System parameter configuration

The management layer and control layer of special vehicles set the current setting value for the information interaction module of each control layer on the whole vehicle power distribution console according to the short-circuit current calculation results of the whole vehicle power distribution system and the overload, overheat and short-circuit protection characteristics of the electrical load. When the electrical load is abnormal, the control layer can report the abnormal information to the management layer.

(2) Power supply control and power distribution protection

The power supply management layer of the whole vehicle carries out power supply control on each power supply branch of the site electrical load according to the set control protocol, and the control layer carries out automatic power supply control and power distribution protection on each power supply branch of the site electrical load according to the working condition demand and current setting value.

(3) Parameter monitoring

The display of the power distribution management layer of the whole vehicle displays the voltage, current, power and other electric energy parameters of each load branch uploaded to the dual redundant Flexray bus by the control layer.

(4) Alarm and fault handling

When the distribution protection of a distribution branch is broken, such as overload, overheat, overcurrent or short circuit, the distribution control layer of the whole vehicle reports the alarm information and prompts the alarm through the driver display and control terminal. At the same time, the whole train information recording device records the fault branch for later fault analysis and processing [14].

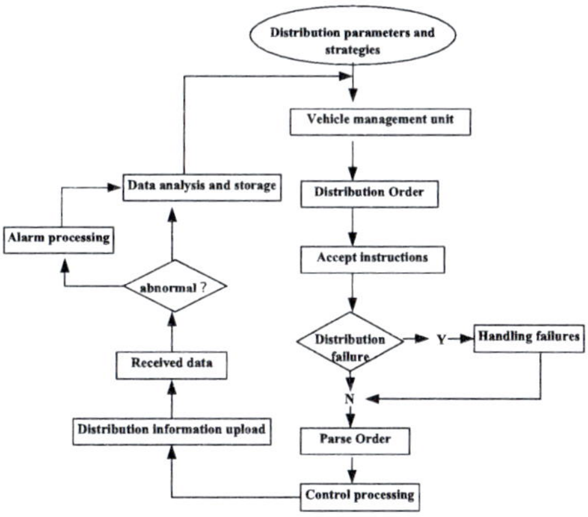

Fig. 2. Principle flow of special vehicle distributed power distribution system

4 Power Distribution System Design

4.1 Management (Core Processor)

The management layer is the monitoring and command center of the distributed intelligent power distribution system, which monitors, controls and alarms all electrical loads in the power cabin, cockpit and passenger cabin of the distributed intelligent power distribution system of special vehicles, and realizes the power supply control and power distribution protection of each power supply branch of the on-site electrical load.

Multi-function standard display and control console is adopted, which is composed of display, computer, mouse, keyboard and other structures. The main configurations are as follows:

(1) Display

The LCD that can rotate up and down, left and right can meet the display requirements of different angles of view.

(2) Computer

Run the foreground human-machine interface software and background task processing program. The human-machine interface software provides the display interface of electric energy parameter information of each power supply branch of the site electrical load, the power supply control interface, and the current setting interface. The task processing program is used to complete the real-time calculation and fault recording of the electric energy parameter information of each power supply branch of the site electrical load.

4.2 Control Layer

The control layer is composed of control distribution unit, information interaction unit, distribution and protection circuit, and is the actual implementer, executor and information reporter of distributed intelligent distribution system [15].

The control power supply unit provides DC28V and high-precision DC12V, DC5V, DC3.3V and other voltage levels of power required for each control and acquisition circuit.

The information interaction unit is mainly composed of a microcontroller, a CAN bus communication module, a Flexray communication module, and an information transfer circuit. The information interaction unit and the body network are realized through the Flexray communication module. The microcontroller completes the control and information exchange with the power distribution and protection current through the CAN bus communication module. The microcontroller completes the mutual conversion of CAN and Flexray information through the information transfer circuit.

Low-voltage bus and high-voltage bus provide power input for distribution circuit in the form of busbar, and realize intelligent distribution, protection and status monitoring of load through distribution and protection circuit.

According to the layout of electrical loads in the power cabin, cockpit, passenger cabin and other cabins, a single on-site distribution board is composed of multiple distribution and protection circuits to ensure regional power supply, and together with the core computer and electrical loads, it forms a distributed intelligent distribution system for special vehicles.

In order to ensure that the intelligent power distribution system can work normally under the complex electromagnetic interference of high and low voltage, the low-voltage power distribution and protection circuits are fixed in the intelligent power distribution box through circuit boards and aviation plugs, and external power supply is provided; At the same time, some low-voltage power distribution and protection circuits are packaged into integrated modules (2, 4, and 10 power distribution and protection circuits with different power levels can be packaged) for the expansion of power distribution functions, so that the intelligent distribution box can adapt to more types of special vehicles; The high-voltage distribution and protection circuits are all packaged into integrated modules to facilitate installation. At the same time, installation slots are reserved inside the distribution box, which can be dynamically adjusted according to the number of load paths. Low-voltage and high-voltage intelligent distribution and protection circuits are isolated in the intelligent distribution box to reduce electromagnetic coupling paths.

4.3 Low-Voltage Secondary Power Distribution and Protection Module

The low-voltage secondary power distribution and protection module adopts a digital control system with power MOS tube as the core to realize the transmission, distribution, control and protection of electric energy, and has the fault-tolerant power supply function and automatic load management capability. The circuit composition diagram is shown in Fig. 3.

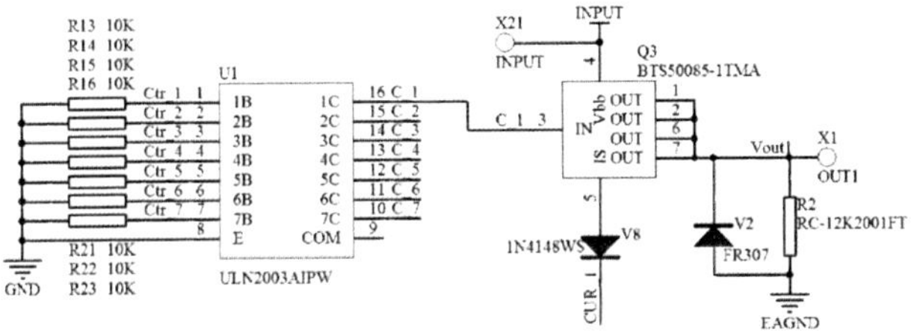

Fig. 3. Low voltage secondary distribution schematic

The power distribution circuit uses Darlington tube ULN2003AIPW control power chip BTS50085 to realize power distribution. BTS50085 has current output signal and short circuit protection function, and current sampling can be realized through current output signal, as shown in Fig. 4. However, when the load power is abnormal, the output can be cut off in time to achieve abnormal protection. The microcontroller uploads the abnormal information to the core computer through the bus.

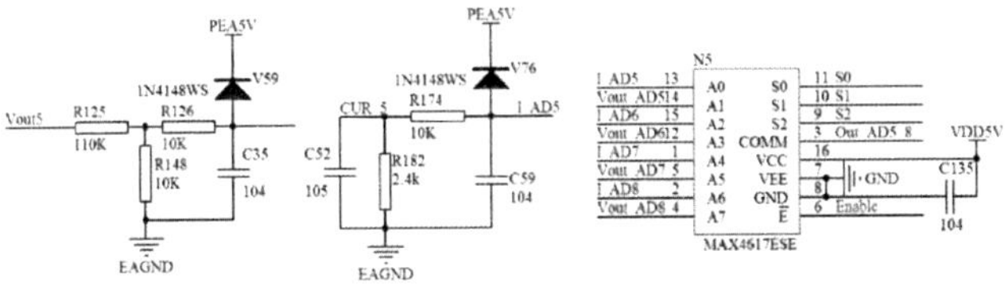

Fig. 4. Protection circuit

4.4 High Voltage Secondary Power Distribution and Protection Module

The high-voltage power distribution and protection circuit is integrated into a module through encapsulation. The power supply branch can be switched on and off by using programmable MOSFET or IGBT and other power electronic devices as solid state switches. The power supply control can be carried out according to the control command of the whole vehicle power distribution console, and the power distribution protection can be carried out according to the set value to realize the breaking of the power supply branch in the case of overload, overheat and short circuit faults, and cut off the overload, overheat and short circuit current. The module also has the functions of power distribution status indication and bus communication indication, displaying the on-off status of the module, and uploading the power parameter information of each distribution branch to the bus.

The function is shown in Fig. 5. The circuit receives the CAN bus on command, controls the on signal to make the drive signal high, and the drive chip outputs the MOSFET drive signal Vout, which can make the Power MOSFET on. The circuit receives the off command, controls the on signal to make the control signal low, and the drive chip turns off the MOSFET drive signal, which can turn off the MOSFET.

After the driving function circuit is connected, a circuit is formed through circuit R1 and diodes V2, V3, V4, and MOSFET. Monitor the voltage drop between the DESAT pin and VE pin of the driver chip. When the voltage drop exceeds the set threshold, the fault pin outputs a low level interrupt signal and turns off the MOSFET driver signal. When a short circuit occurs in the functional circuit, the short-circuit current increases, which can increase the above threshold. When the voltage between the DESAT pin and the VE pin exceeds the set threshold, the drive signal is turned off, and the MOSFET output is turned off to achieve short-circuit protection.

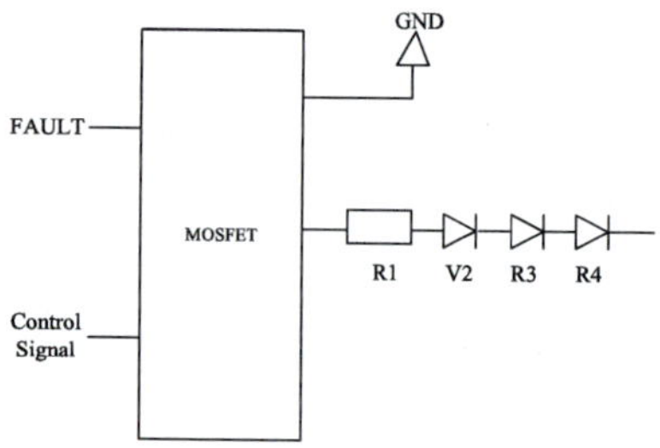

Fig. 5. Schematic diagram of high voltage drive circuit

4.5 System Integration Design

The outline of the high and low voltage intelligent power distribution system is shown in Fig. 6, which is mainly composed of control and information interaction unit, low voltage distribution board, busbar, low voltage secondary power distribution and protection module, high voltage secondary power distribution and protection module, low voltage aviation plug, high voltage aviation plug, locking device, etc., as shown in Fig. 6.

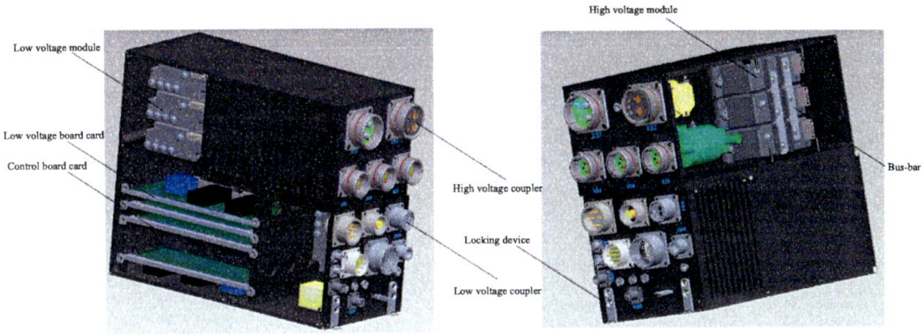

Fig. 6. Intelligent power distribution system integration

5 Function Test

5.1 Real Vehicle Test

The distributed intelligent power distribution test system for special vehicles takes the specific load of a certain type of military vehicle as the test object to test the low-voltage power distribution and protection functions, the hign-voltage power distribution and protection functions of the distributed intelligent power distribution test system. After the real load test of the real vehicle, the short circuit is tested in the way of fault injection, which fully meets the use requirements of the real load of the whole vehicle; The distributed intelligent distribution test system can also upload the fault information to the bus and display it on the display screen of the core computer, achieving the design purpose.

In case of overcurrent of a low-voltage load, the distributed intelligent distribution system can timely protect and turn off the output (the protection time can be set by software, and the protection time of this distribution system is 2s), and upload the fault information to the bus, and the core computer display will display it, as shown in Fig. 7. The blue curve is the voltage change curve, and the purple curve is the current change curve.

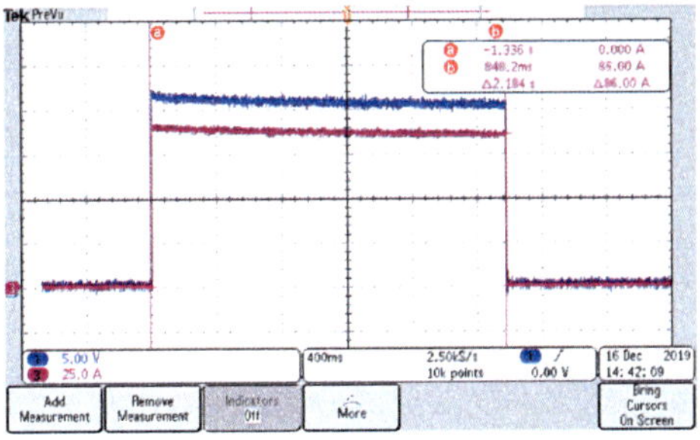

Fig. 7. Low voltage overcurrent protection curve

When a low voltage load has a short circuit (generated by fault injection), the distributed intelligent distribution system can timely protect and turn off the output. This protection is hardware circuit protection, and upload the fault information to the bus, and the core computer display will display it, as shown in Fig. 8. The blue curve is the voltage change curve, and the purple curve is the current change curve.

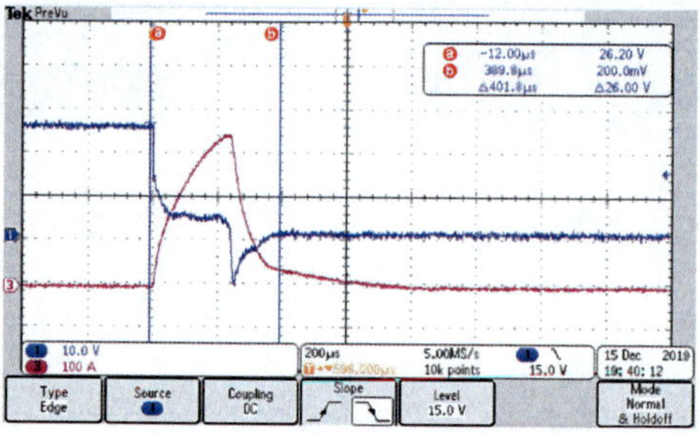

Fig. 8. Low voltage short current protection curve

In case of short circuit of a high-voltage load (generated by fault injection), the distributed intelligent distribution system can timely protect and turn off the output. This protection is hardware circuit protection, and upload the fault information to the bus, and the core computer display will display it, as shown in Fig. 9. The yellow curve is the voltage change curve, and the green curve is the current change curve.

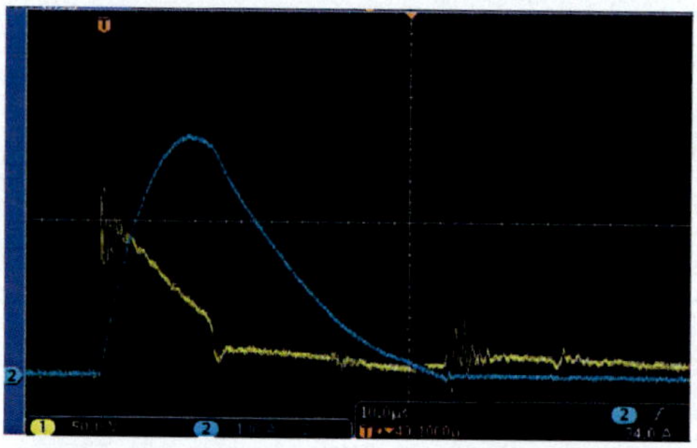

Fig. 9. High voltage short current protection curve

5.2 Test Conclusion

From the above physical tests, it can be seen that the distributed intelligent power distribution system can fully meet the use of low-voltage and high-voltage loads of special vehicles under various extreme conditions, and can realize real-time monitoring of load distribution, and upload the monitoring information. When the load has abnormal power consumption, it can timely cut off the power output, protect the load, and isolate the fault.

6 Summary

The distributed intelligent power distribution system of special vehicles based on the secondary power distribution device realizes the real-time power distribution monitoring and control of the electrical load, and provides over-current, over-voltage and over-heat protection, which greatly improves the reliability, efficiency and safety of power distribution; It is beneficial to control and manage the electric load of the whole vehicle and save energy; It improves the distributed automation degree of the distribution system and promotes the effective improvement of the informatization level of special vehicles; The distributed and regional intelligent power distribution system design can be used to meet the requirements of the optimized design of the whole vehicle power distribution system of special vehicles, which can greatly reduce the length of wiring, reduce the weight of cables, further reduce the cost, play an active role in the lightweight design of the power distribution system of special vehicles, and effectively reduce the power loss of the line.

Through the test and test of real vehicle load, it can fully meet the use requirements of special vehicles. The application results of real vehicle show that the distributed intelligent power distribution system of special vehicles based on secondary power distribution device has far-reaching significance to ensure the electrification, intelligence, modularization and integration of the whole vehicle, and has high practical value.

References

1. Zhao, W.: Design of Electrical Load Management Center for Large Aircraft. Northwestern Polytechnical University, Xi'an (2016) (in Chinese)
2. Hua, S.: Design of intelligent power supply system for special vehicles. J. Military Commun. Technol. **37**(2), 54–55 (2016) (in Chinese)
3. Wu, X.: Research on Distributed Distribution System in Avionics Laboratory. Nanjing University of Aeronautics and Astronautics, Nanjing (2014) (in Chinese)
4. Wen, Y.: Comprehensive protection design and countermeasures for ship power distribution system. Ship Electr. Technol. **24**(7), 1–2 (2011) (in Chinese)
5. Xie, K.: Research on ship energy flow analysis and its comprehensive utilization technology. Ship Sci. Technol. **39**(5), 65–70 (2017) (in Chinese)
6. Renkuan, Y.: Design of fault monitoring and automatic control system for high voltage distribution cabinet. Modern Manuf. Technol. Equip. **57**(9), 30–31 (2021) (in Chinese)
7. Li, Q.-l.: Development of a high voltage distribution cabinet for heat dissipation and dust prevention. Electr. Switch **59**(3), 53–54 (2021) (in Chinese)
8. Li, Z.-H.: Research and design of high voltage dc power supply and distribution system for special vehicles. Northeastern University, Shenyang (2017) (in Chinese)
9. Yin, F.: Random vibration simulation analysis and experimental verification of high-voltage distribution box FOR electric vehicles. Bus. Coach Technol. Res. **44**(3), 12–14 (2022) (in Chinese)
10. Avery, C.R., Burrow, S.G., Mellor, P.H.: Electrical generation and distribution for the more electric aircraft, pp. 1007–1009. UPEC, England (2007)
11. Song, S.: Reliability improvement method for high voltage dc power supply and distribution system of special vehicles. Meas. Control Technol. **40**(12), 47–49 (2021) (in Chinese)
12. Qian, Y.: Control technology of DC solid state power controller. Modern Electron. Technol. **23**(17), 25–27 (2012) (in Chinese)

13. Xie, S.: Design and implementation of an aviation DC solid state power controller. Comput. Meas. Control **18**(7), 1553–1554 (2010) (in Chinese)
14. Abbasi, V., Shayestch, M.G.: A method for improving power distribution characteristics of space time block codes. Communications **17**(6), 223–226 (2020)
15. Norman, P.J., Galloway, S.J., Bun, G.M.: Trasient analysis of the more electric engine electrical power distribution. In: 4th IET Conference on Machines and Drives, pp. 681–683 (2018)

IGBT Open-Circuit Fault Diagnosis for MMC Based on OCDWPT

Qinyue Zhu, Yichen Yu(✉), Yanwen Zhan, Jie Li, and Yanbiao Yang

School of Electronic Information and Engineering, Tongji University, Shanghai 201804, People's Republic of China
2130679@tongji.edu.cn

Abstract. Considering the issue of high sensor demand and susceptibility to load changes in existing IGBT fault diagnosis for MMC sub-module (SM), a fault diagnosis method based on over-complete discrete wavelet packet decomposition (OCDWPT) is proposed in this paper. Firstly, based on the research on the fault mechanism of MMC, the bridge arm voltage is used as the diagnostic parameter, eliminating the need for collecting capacitor voltage of each SM, thereby reducing the dimension of the fault diagnostic signal. Secondly, the diagnostic parameter is decomposed by OCDWPT and converted to time-frequency domain information. Additionally, the iterations of down-sampling are determined based on practical requirements to minimize computational load and storage space while preserving signal phase characteristics as much as possible. Finally, fault diagnosis is performed using an optimized deep neural network.

Keywords: Fault diagnosis · OCDWPT · MMC · Deep neural network

1 Introduction

Modular multilevel converter (MMC) adopts a topology with sub-modules (SMs) series connection and has widely used in traction inverters and other high-voltage applications due to its lower voltage steps, switching frequency, and losses [1, 2]. However, as the number of SMs increases, the potential faulty points increase as well. Any fault that occurs can affect the output quality and may even lead to further damage to the devices.

The IGBT faults are the most common faults in MMC and can be divided into open circuit fault and short circuit fault. The fault characteristics and damages of short circuit fault are much more significant, therefore current MMC driver circuits are integrated with short circuit protection measures. However, the fault characteristics of open circuit fault are inconspicuous and more difficult to detect. Persistent open circuit faults can also lead to severe consequences [3]. Diagnosis methods for IGBT faults can be divided into hardware-based and algorithm-based approaches. Hardware-based methods incur additional hardware costs and are gradually being replaced by algorithm-based methods, especially with the development of artificial intelligence, machine learning-based methods have been widely applied in MMC fault diagnosis [4]. Refs. [5–7] utilize several methods to estimate system parameters and compare them with actual parameters

L. Jia et al. (Eds.): EITRT 2023, LNEE 1135, pp. 489–497, 2024.
https://doi.org/10.1007/978-981-99-9307-9_54

for diagnosis. However, these approaches are unable to diagnose specific fault type. As for diagnosis methods using machine learning, Ref. [8] extracts fault features from output current and circulating current and utilizes the capsule network for fault diagnosis, but it only locates the faulty bridge arm. Ref. [9] achieves similar diagnostic results using OS-ELM. Ref. [10, 11] extract fault features by Markov distance and short-time wavelet entropy respectively, then employ machine learning methods for fault diagnosis. However, both approaches rely on numerous SM voltage sensor data.

This paper proposes an IGBT open-circuit fault diagnosis method based on bridge arm voltage, eliminating the need for numerous SM voltage data. The time-frequency information in the bridge arm voltage is deeply explored through OCDWPT. By setting the iterations of down-sampling, the computational load and storage space of the algorithm are reduced without compromising the diagnosis accuracy.

2 Analysis of IGBT Open Circuit Fault

The three-phase MMC and SM structure are shown in Fig. 1. Each phase consists of upper and lower bridge arms, with each bridge arm consisting of N SMs connected in series with buffer inductors. Each SM includes a capacitor C and two complementary IGBT devices T_1 and T_2 with anti-parallel diodes connected in series.

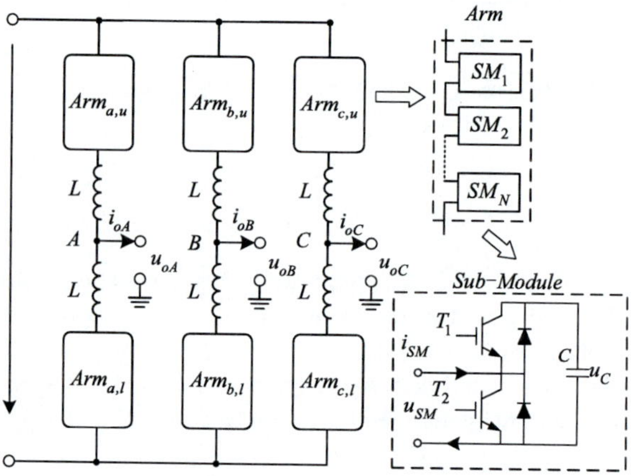

Fig. 1. Main circuit and SM topology of the MMC

For single SM, the drive signals of IGBTs are complementary. When T_1 is turned on, T_2 is turned off, and the output voltage of SM is equal to the voltage of capacitor. On the contrary, the output terminal is shorted, and the output voltage of SM is zero. The switching function of single SM can be expressed as:

$$S_{i,j,k} = \begin{cases} 1, (T_{1|(i,j,k)} = ON, T_{2|(i,j,k)} = OFF) \\ 0, (T_{1|(i,j,k)} = OFF, T_{2|(i,j,k)} = ON) \end{cases} \tag{1}$$

where $S_{i,j,k}$ is switching function of the kth SM($k = 1,2,\ldots N$) on the jth bridge arm($j = u,l$) of the i phase($i = a,b,c$). The output voltage of the SM is given by:

$$u_{SM|(i,j,k)} = S_{i,j,k} \cdot u_{c|(i,j,k)} \tag{2}$$

where $u_{c|(i,j,k)}$ represents the capacitor voltage of corresponding SM.

When IGBT open-circuit fault occurs, the current path is shown in Fig. 2, the summarized output voltage of SM under different scenarios are presented in Table 1.

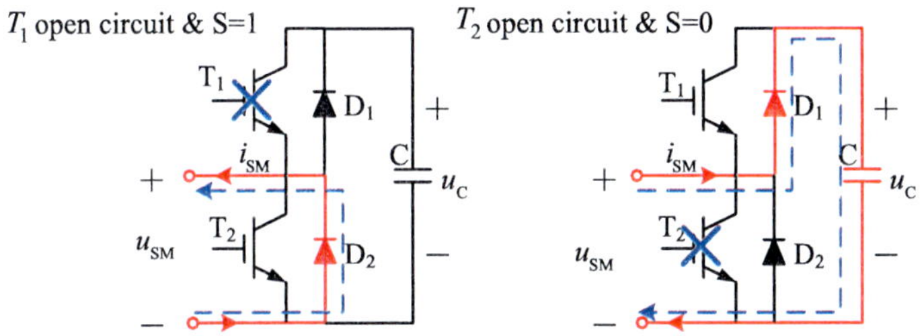

Fig. 2. Fault path for IGBT open circuit

Table 1. The output voltage of SM under different conditions

iSM	*S*	Output voltage		
		Normal	T_1 fault	T_2 fault
>0	1	uc	u_C	u_C
>0	0	0	0	u_C
<0	1	u_C	0	u_C
<0	0	0	0	0

Table 1 indicates that abnormal output voltage can be observed externally regardless of the fault type. Therefore, fault diagnosis can be performed by collecting electrical parameters including the output voltages of SMs such as output voltage, output current, and bridge arm voltage. Since current-based parameters may vary with load changes, voltage-based parameters should be prioritized. For a single phase of the MMC, neglecting the voltage on the buffer inductor, the voltages of the bridge arm and output voltage can be respectively given by:

$$\begin{cases} u_{uSM} = \sum_{k=1}^{N} S_{u,k} \cdot u_{c|u,k} \\ u_{lSM} = \sum_{k=1}^{N} S_{l,k} \cdot u_{c|l,k} \end{cases} \tag{3}$$

$$u_{output} = \frac{u_{lSM} - u_{uSM}}{2} \tag{4}$$

Above equations indicate that the bridge arm voltage is a combination of the voltage at all SM within the bridge arm, while the output voltage is half of the difference between upper and lower bridge arm voltages. Obviously, when specific types of faults occur simultaneously in the upper and lower bridge arms, the fault characteristics in the output voltage might be masked. Therefore, this paper selects the bridge arm voltage as the electrical parameter for fault diagnosis.

3 Methodology

3.1 Signal Processing Based on OCDWPT

Wavelet transform (WT) is a signal analysis technique which decomposes signal into different scales. Discrete wavelet transform (DWT) is implemented by Mallat algorithm, which involves repeatedly decomposing signal using high-pass and low-pass filters based on wavelet basis. In DWT, down-sampling is performed after each filtering operation to accelerate the decomposition process so as conserve storage space.

However, real-world filters do not have ideal brick-wall response, leads to the shift variance in the signal after DWT. Therefore, the stationary wavelet transform (SWT), also known as Atrous algorithm was proposed. SWT does not perform down-sampling during decomposition and preserves shift invariance. The decomposition processes of DWT and SWT are illustrated in Fig. 3. Where X(n) is original signal, H(z) and G(z) are low-pass and high-pass filters. a^i and d^i ($i = 1, 2, \ldots$)are approximate coefficients and detailed coefficients obtained from the ith level decomposition.

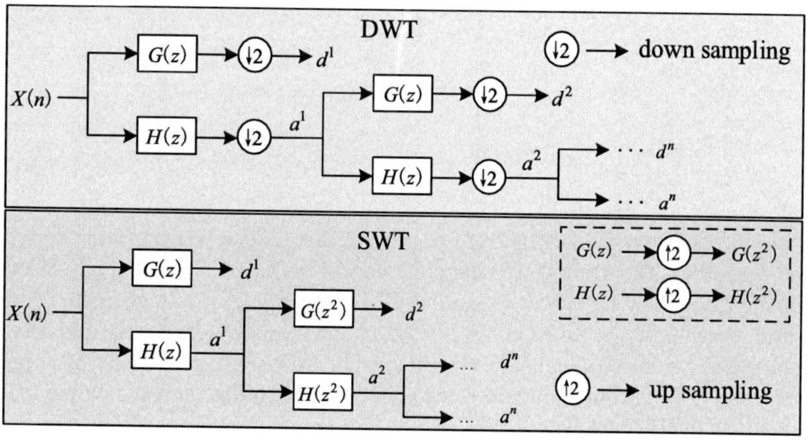

Fig. 3. Process diagrams for DWT and SWT decomposition

Although SWT preserves shift invariance, it also increases computational load and storage space. To balance those demands, one can employ the over complete discrete wavelet transform (OCDWT), where DWT is applied in the first L layers of a N-level decomposition, while the remaining N-L layers use SWT. This approach combines the advantages of both DWT and SWT. When applied in scenarios where strict shift invariance is not required, increasing L can accelerate computation speed. Conversely, reducing L can achieve acceptable shift invariance while obtaining faster computation speed. Furthermore, to obtain more detailed information, one can simultaneous iterative decompose high-frequency and low-frequency coefficients. This transform is called over complete discrete wavelet packet transform (OCDWPT). The structure of OCDWPT is illustrated in Fig. 4, where $c^{i,j}$ is coefficients of jth sub-band of ith decomposition ($i = 1, 2, \ldots, N; j = 1, 2, \ldots, 2^i$).

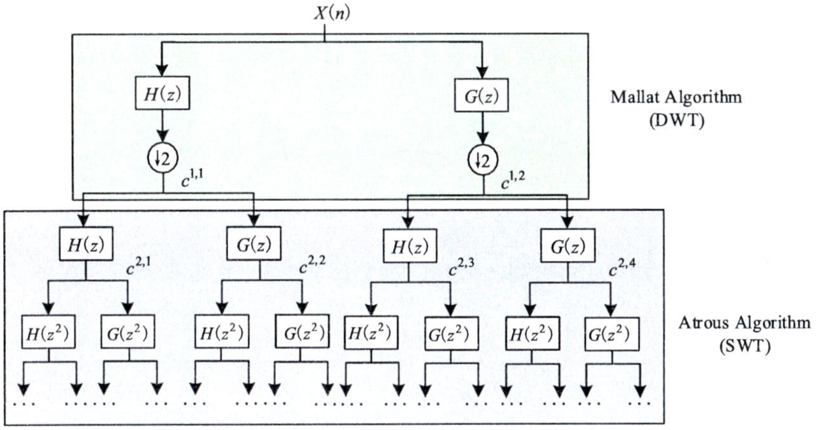

Fig. 4. The schematic diagram of OCDWPT

3.2 Establishment of Diagnosis Model Based on DNN

The bridge arm voltage is sampled based on power frequency period. After OCDWPT, the wavelet coefficients of the last decomposition are collected. The length of each sub-band depends on the length of original signal and the value of L. Obtained data is then organized into a 2-D matrix with number of sub-bands as length and the length of each sub-band as width. This matrix serves as fault diagnosis sample for the model. The fault diagnosis is improved based on ResNet50. By modifying the convolution channels, the model is able to read the sample matrix. Additionally, dropout layer is added after the convolution layer to randomly deactivate part neurons during the training process, thereby enhancing the model's generalization performance. The model structure is illustrated in Fig. 5.

The fault diagnosis process based on Deep Neural Network (DNN) is as follows:

Step 1: Collect the bridge arm voltage for each phase and each bridge arm.
Step 2: Convert the voltage values in the time domain into time-frequency domain data using OCDWPT and construct a 2-D matrix as the input sample.
Step 3: Randomly divide original data into training, test and validation set. Train the DNN with training and validation set, then evaluate the model with test set.
Step 4: Input constructed samples into the trained network and obtain the fault codes.

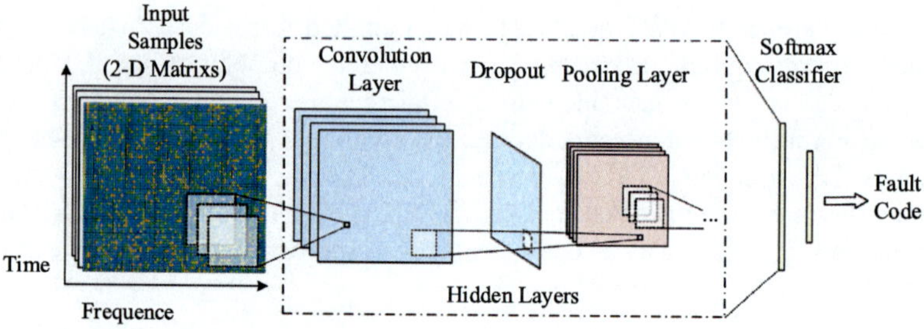

Fig. 5. The structure of the fault diagnosis model based on Deep Neural Network (DNN)

4 Simulation Verification

A 65-level MMC system is established based on the MATLAB/Simulink. Faults are simulated by changing drive signals of corresponding IGBTs. System parameters are as follows: DC bus voltage is 32 kV, N = 64, load resistance is 20 Ω, load inductance is 10 mH, buffer inductance is 15 mH, capacitance is 22 mF, carrier frequency is 1250 Hz, rated frequency is 50 Hz, sampling rate is 100 kHz and modulation index is 0.95.

Take *a* phase upper arm as example, waveforms of the bridge arm voltage are collected under normal condition and various faulty scenarios, including single SM faults and dual SMs faults at different positions. Partial fault codes are shown in Table 2.

Table 2. Fault codes table (partial)

Fault type	Fault code	Fault type	Fault code
Normal condition	0	$SM_{u,1}T_1$ & $SM_{u,2}T_1$ *fault*	11
$SM_{u,1}T_1$ *fault*	1	$SM_{u,1}T_1$& $SM_{u,4}T_1$ *fault*	12
$SMu,1T2$ *fault*	2	$SMu,1T1$& $SMu,2T2$ *fault*	13
$SM_{u,2}T_1$ *fault*	3	$SM_{u,1}T_1$& $SM_{u,4}T_2$ *fault*	14
...	...	...	...

Assuming the T_1 fault of $SM_{u,1}$ occurs at 0.5 s, the bridge arm voltage is decomposed by OCDWPT with N = 7 and L = 5. Partial wavelet coefficients of first cycle after fault occurred and normal working condition are shown in Fig. 6.

Compared to the normal condition, there is significant difference in bridge arm voltage with T_1 fault when the fault module is switched in and $i_{sm} < 0$, consistent with the analysis in Table 1. After decomposing by OCDWPT, the information of waveform can be converted into multiple sub-bands. The length of each sub-band is the same, but the representative frequency increases sequentially. By this way, the fault characteristics are obviously enlarged. The decomposed wavelet packet coefficients can be used as inputs for network to achieve fault diagnosis. Partial constructed input samples and their differences are shown in Fig. 7.

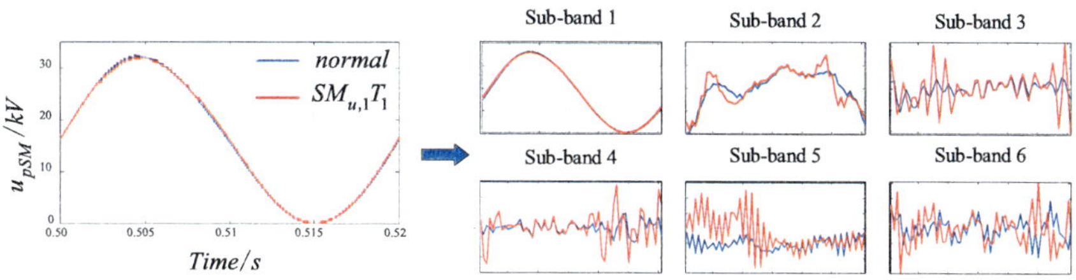

Fig. 6. Partial sub-band coefficients in normal working condition and faulty condition

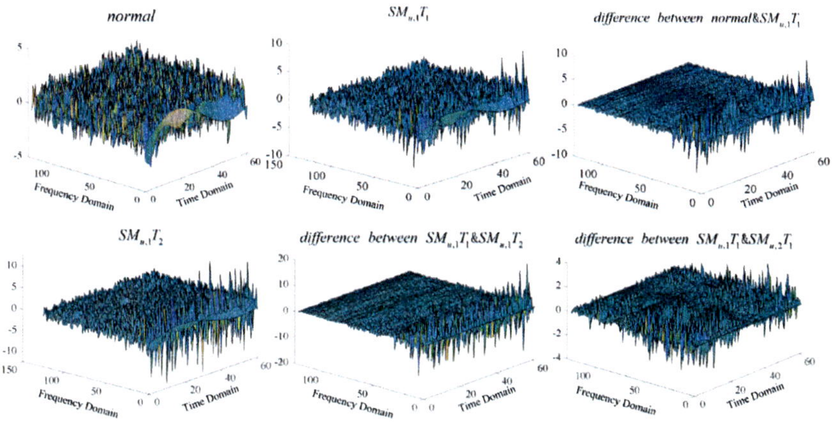

Fig. 7. The input samples and differences between different faults

Figure 7 demonstrates that after OCDWPT, the differences between T_1 fault and T_2 fault exist in partial frequency domain across the entire time domain, whereas the differences between the same fault at different positions exist in the entire frequency domain across partial time domain, providing the basis for classification.

The training process is set as follows: batch size = 32, epoch = 50, dropout rate = 0.1. The dataset is divided into training, test and validation sets in 8:1:1 ratio. Then save the best model according to the minimum of val_loss.

Training is conducted for different values of L and samples are collected by randomly setting the starting point for sampling to eliminate correlation. The total of 17 scenarios and 12,580 samples are collected. The diagnostic results of the original network and modified network are shown in Table 3.

It can be seen from tables that, modified network generally achieved better result than original one. Additionally, as the down-sampling iterations decrease, the shift invariance of DWPT gradually increases, improving the accuracy of the diagnosis network that utilized signal phase for fault localization. However, as L increases, the size of the sample also doubles, leading to longer diagnosis time and increased storage space. In this scenario, L = 5 can be chosen, where both test and validation sets achieve an accuracy above 99%, and the average diagnosis time remains around 1ms.

Table 3. The accuracy of different networks for different values of L

L	Predict time (ms)		Train set accuracy (%)		Valid set accuracy (%)		Test set accuracy (%)		Sample size
	Orig	Modif	Orig	Modif	Orig	Modif	Orig	Modif	
7	0.881	0.734	92.73	98.66	83.38	93.00	84.58	92.13	15 × 128
6	1.124	0.780	98.22	99.03	95.15	95.39	94.03	95.71	31 × 128
5	1.628	1.218	98.85	99.99	97.29	99.21	95.86	99.20	62 × 128
4	2.737	2.112	99.74	99.54	98.96	99.05	99.04	99.21	125 × 128

5 Conclusion

This paper proposes an effective IGBT open-circuit fault diagnosis method for MMC. Firstly, by analyzing the MMC circuit and identifying the use of bridge arm voltage as diagnostic parameter, eliminating the need for numerous SM capacitor voltage sensor. Subsequently, a fault feature extraction method using OCDWPT is proposed based on the analysis of SM fault mechanism. Time-frequency domain information is extracted from the original signals and transformed into a 2-D sample matrix. Then a deep learning network is constructed for fault diagnosis, with the ResNet50 network structure optimized for matrix inputs and improved generalization capability. Finally, various faulty data are collected through simulations to validate the effectiveness of proposed fault diagnosis method.

References

1. Ronanki, D., Williamson, S.S.: Modular multilevel converters for transportation electrification: challenges and opportunities. IEEE Trans. Transport. Electrif. **4**(2), 399–407 (2018)
2. Gontijo, G.F., Wang, S., Kerekes, T., et al.: Novel converter topology with reduced cost, size and weight for high-power medium-voltage machine drives: 3 × 3 modular multilevel series converter. IEEE Access **9**, 49082–49097 (2021)
3. Jia, H., Deng, Y., Hu, X., et al.: A concurrent diagnosis method of IGBT open-circuit faults in modular multilevel converters. IEEE J. Emerg. Sel. Top. Power Electron. **11**(1), 1021–1034 (2023)
4. Ke, L., Liu, Y., Yang, Y.: Compound fault diagnosis method of modular multilevel converter based on improved capsule network. IEEE Access **10**, 41201–41214 (2022)
5. Chen, X., Liu, J., Deng, Z., et al.: A diagnosis strategy for multiple IGBT open-circuit faults of modular multilevel converters. IEEE Trans. Power Electron. **36**(1), 191–203 (2021)
6. Liu, M., Li, Z.T.: Parameter estimation of sub-module capacitors of modular multi-level converter based on adaptive observer. In: 2021 China Automation Congress (CAC), pp. 4803–4807 (2021)
7. Wu, J., Liu, C.: An expert system for fault diagnosis in internal combustion engines using wavelet packet transform and neural network. Expert Syst. Appl. **36**(3), 4278–4286 (2009)
8. Zhang, B., Hong, X., et al.: Diagnosis method for sub-module open-circuit fault in modular multilevel converter based on unsupervised learning. Power Syst. Protect. Control **49**(12), 98–105 (2021). (in Chinese)

9. An, Y., Sun, X., Yuan, Q.: IGBT open circuit fault diagnosis method for a modular multilevel converter based on PNN-MD. In: 2020 4th International Conference on HVDC (HVDC), pp. 911–916 (2020)
10. Han, Y., Qi, W., Ding, N., et al.: Short-time wavelet entropy integrating improved LSTM for fault diagnosis of modular multilevel converter. IEEE Trans. Cybern. **52**(8), 7504–7512 (2022)
11. Wang, Z., Peng, L.: Grouping capacitor voltage estimation and fault diagnosis with capacitance self-updating in modular multilevel converters. IEEE Trans. Power Electron. **36**(2), 1532–1543 (2021)

Sensorless Control with Sliding-Mode Observer for Superconducting Linear Motors

Shaobo Yin(✉), Xin Jiang, and Kekang Wei

China Aerospace Science and Industry Co., Ltd, Beijing 100143, China
bxdlysb@163.com, 1027255625@qq.com, weikekang@163.com

Abstract. In the application of the ultra-high-speed low-vacuum tube maglev transportation system, the superconducting linear motors are utilized for propulsion. The cross-inductive wire which is used to obtain position and speed information needs to be installed along the line. The cost of the hardware and maintenance is expensive. These limitations greatly promote the sensorless control of the superconducting linear motors, especially in high speed region. In this paper, one sensorless control strategy based on sliding-mode observer is proposed for superconducting linear motors, which achieves good performance in medium to high speed region and maintains stable during transient procedure. The control strategy under power supply fed at two ends is also discussed. The effectiveness of the proposed method is demonstrated by simulation studies.

Keywords: Linear motor · Sensorless control · Sliding-mode observer

1 Introduction

The ultra-high-speed low-vacuum tube maglev transportation system adopts the superconducting electric suspension propulsion and segmented long stator superconducting linear motors, which is one of the technical directions of the new generation of high-speed trains.

High performance vector control requires position information of the motor and the inductive radio system is utilized for position detecting. The position and speed sensors along the track greatly increase the construction and maintenance cost of the system because the cross-inductive wire needs to be installed along the line precisely. Sensorless control has great advantages in reducing the hardware complexity and improving system's reliability.

Various sensorless control methods for drive systems have been proposed in the literature, including model reference adaptive systems(MRAS) [1, 2], extend Kalman Filter method [3, 4], and sliding-mode observers [5–10].

Among these methods, the sliding-mode observer has the advantages of simple implementation and strong robustness which attracts a lot of attention [6]. To weaken the chatting problem of the sliding-mode control, the high-order fast nonsingular terminal sliding mode control is proposed in [7] to connect sliding mode surface and the current

L. Jia et al. (Eds.): EITRT 2023, LNEE 1135, pp. 498–505, 2024.
https://doi.org/10.1007/978-981-99-9307-9_55

which can achieve high tracking accuracy. In [8], a piecewise square root switching function is utilized to suppress sliding mode chattering. An improved super-twisting sliding mode control strategy is proposed in [9], which can accurately observe the unknown disturbance part. Another method [11] presents a new sliding mode current observer in which the sliding mode coefficient is adjusted dynamically.

In this paper, one sensorless control strategy by using of sliding-mode observer for superconducting linear motors is proposed, which maintains stable during medium to high speed region and achieves good performance in transient procedure when torque command changes. The simulation results are presented based on the actual propulsion system for validation.

2 Sliding-Mode Observer

The block diagram of sliding-mode observer for superconducting linear motors can be shown in Fig. 1, where the motor control is based on field oriented control. The angle and speed information is obtained from the proposed sliding-mode observer.

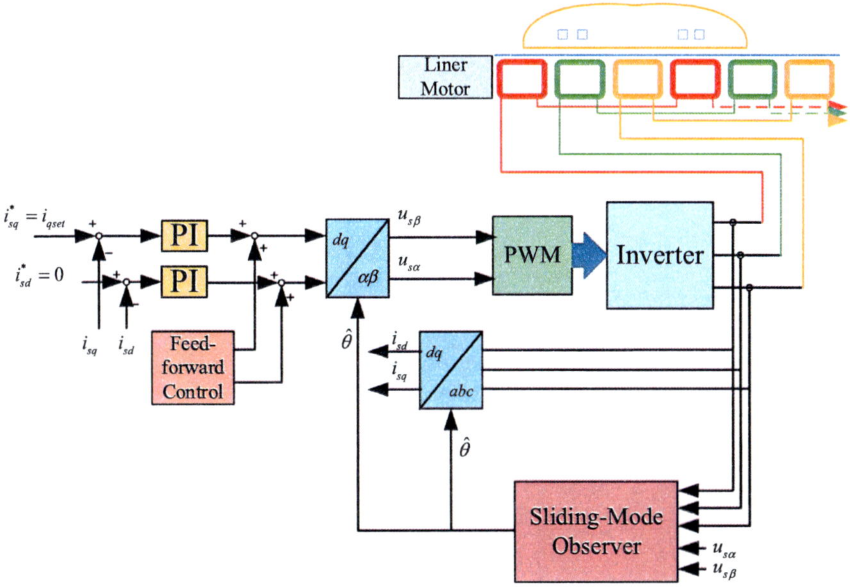

Fig. 1. Block diagram of sliding-mode observer for superconducting linear motors.

The mathematical model of the superconducting linear motor in synchronous reference frame can be expressed as

$$\begin{cases} \dfrac{di_d}{dt} = \dfrac{1}{L_d}(-R_s i_d + u_d + L_q \omega_e i_q - e_d) \\ \dfrac{di_q}{dt} = \dfrac{1}{L_q}(-R_s i_q + u_q - L_d \omega_e i_d - e_q) \end{cases} \tag{1}$$

where, i_d, i_q, u_d, u_q are the d and q axis components of stator current and voltage, respectively, R_s is the stator resistance, ω_e is the stator electrical frequency, L_d, L_q are the d and q axis components of stator inductance, e_d, e_q are the d and q axis components of electromotive force.

To obtain the angle and speed, a sliding-mode observer in the continuous time domain can be derived as

$$\begin{cases} \frac{d\hat{i}_d}{dt} = \frac{1}{L_d}(-R_s\hat{i}_d + u_d + L_q\hat{\omega}_e\hat{i}_q - Z_d) \\ \frac{d\hat{i}_q}{dt} = \frac{1}{L_q}(-R_s\hat{i}_q + u_q - L_d\hat{\omega}_e\hat{i}_d - Z_q) \end{cases} \\ \begin{cases} Z_d = k \cdot sat(\hat{i}_d - i_d) \\ Z_q = k \cdot sat(\hat{i}_q - i_q) \end{cases} \tag{2}$$

where the superscript ^ represents the estimated value, k is gain which is set to a reasonable value to achieve satisfactory convergence speed and stable performance of the sliding-mode observer. The saturation function is expressed as

$$sat(S) = \begin{cases} 1 & S > \Delta \\ S/\Delta & -\Delta < S < \Delta \\ -1 & S < -\Delta \end{cases} \tag{3}$$

where Δ is the adjustable parameter which can decrease chattering phenomenon. According to the Lyapunov stability theorem, the sliding-mode observer can reach a stable state and Z_d approaches e_d which actual value is zero.

In the beginning, the estimation error S is relatively large, the sliding-mode control law will be sign function. The estimation error S decreases quickly under the sliding-mode control and the sliding-mode control law changes from sign function to linear function. The estimation error S converges to zero gradually. Therefore, the utilization of the saturation function can not only estimate the speed information, but also suppress the chattering phenomenon of the traditional sliding-mode observer.

To extract angle and speed information, the estimated value Z_d can be utilized. If the estimated value $\hat{\theta}$ is not equal to actual value θ, the estimated value Z_d will not be zero either. Therefore, the phase-locked loop is usually used for angle and speed estimation. The block diagram of the angle and speed estimation is shown in Fig. 2.

3 Sensorless Control Strategy

To realize high speed region up to 1000 km/h, the propulsion system adopts power supply fed at two ends to achieve this goal in the ultra-high-speed low-vacuum tube maglev transportation system. Two inverters which are located at two substations far apart output current in parallel to drive the high speed trains. In this way, the capacity of the inverters can be relatively small which can reduce construction costs. On the other hand, power supply fed at two ends increases the redundancy of the propulsion system. The topology of the power supply system is illustrated in Fig. 3.

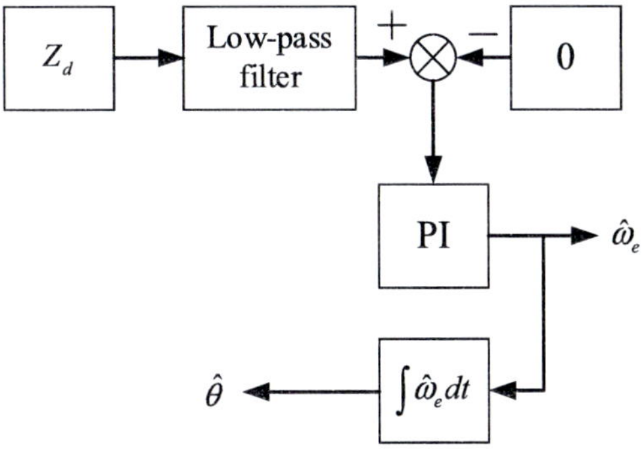

Fig. 2. Block diagram of angle and speed estimation based on phase-locked loop.

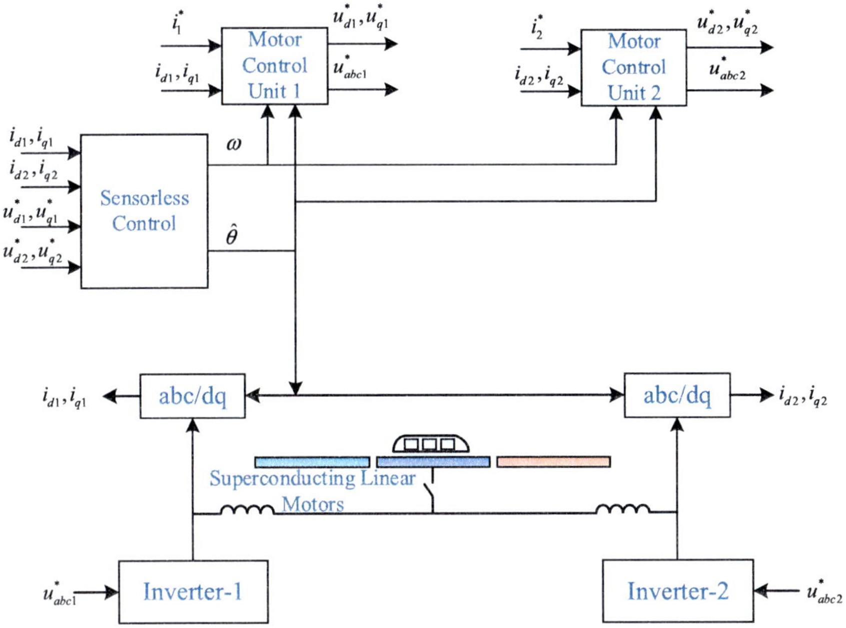

Fig. 3. Sensorless control strategy under power supply fed at two ends.

When the high speed train drives from inverter-1 to inverter-2, the distance between the high speed train and inverter-1 is getting farther and farther, and the distance between the high speed train and inverter-2 is getting closer. The output voltages of the two inverters will be different due to voltage drop in long-distance power supply cables.

To realize sensorless control under power supply fed at two ends, the control strategy needs to be adjusted accordingly. The main point is to find the terminal voltage and the total current at the junction of the superconducting linear motors. In real application, the values of the output currents and voltages in two motor control units are transmitted to sensorless control unit. Based on the distance between high speed train and the inverter, the cable voltage drop can be calculated. The terminal voltage in the position of the high speed train can be obtained by u_{d1}^*, u_{q1}^*, u_{d2}^*, u_{q2}^* and the cable voltage drop. The terminal current at the junction can be obtained by i_{d1}, i_{q1}, i_{d2}, i_{q2}. At last, the terminal voltage and current are brought into sliding-mode observer and position information extraction is realized.

The phase of the power supply fed at two ends needs to be consistent in order to reduce the circulating current. The observer angle and speed information by sensorless control unit will be transmitted to two motor control units at the same time. Due to the long transmission distance, the communication delay needs to be taken into account.

4 Simulation Results

To verify the proposed methodology, the detailed simulation of the superconducting linear motors under power supply fed at two ends has been established using MATLAB/Simulink. The switching frequency of the inverter and the controller sampling rate are selected as 3 kHz and 10 kHz, respectively.

To demonstrate the sensorless control performance over a wide range of speeds, the simulation results of the estimated angle and actual angle, the estimated speed and actual speed, the estimated error of angle from 100 to 1000 km/h are illustrated in Fig. 4. As it can be observed in Fig. 4, the sliding-mode observer gives position and speed estimation results that perfectly match with the actual value. This validates the proposed sliding-mode observer achieves good sensorless control performance for the superconducting linear motor from medium to high speed region.

To give more insight into the stability of the proposed sliding-mode observer, the simulation results under constant speed when the torque command changes is shown in Fig. 5 and the simulation performance of sensorless control under the same conditions are illustrated in Fig. 6.

As it can be observed in Figs. 5 and 6, the speed of the superconducting linear motor is stabilized at around 108 km/h during the process when torque command changes and the estimated speed does not show significant fluctuations during the transient procedure. The error between estimated angle and actual angle is almost zero.

Therefore, from the angle and speed observation in steady-state and dynamic conditions, it is validated that the proposed sliding-mode observer can meet the requirements of sensorless control for superconducting linear motors.

5 Conclusion

In this paper, one sensorless control strategy by using of sliding-mode observer is proposed for superconducting linear motors. The sensorless control strategy under power supply fed at two ends is discussed for actual application. The performance of the proposed sensorless control method is validated using simulation studies. It has been verified that presented algorithm is very effective during medium to high speed region.

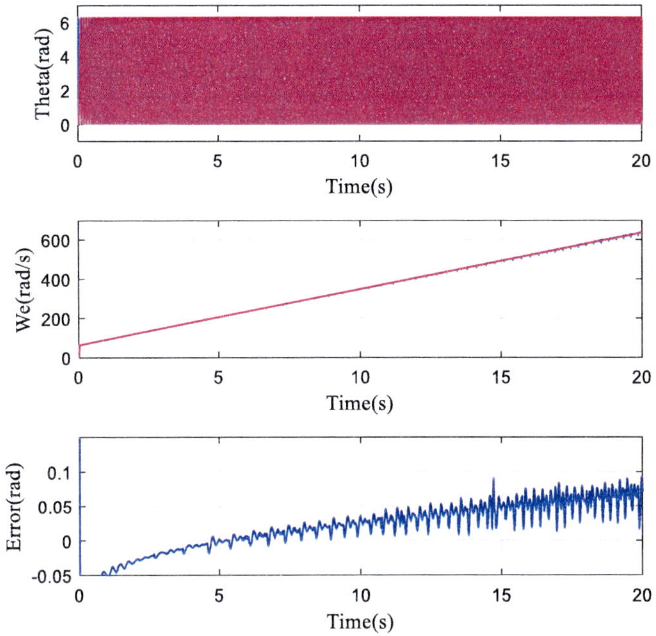

Fig. 4. Simulation results of speed estimation from 100 to 1000 km/h.

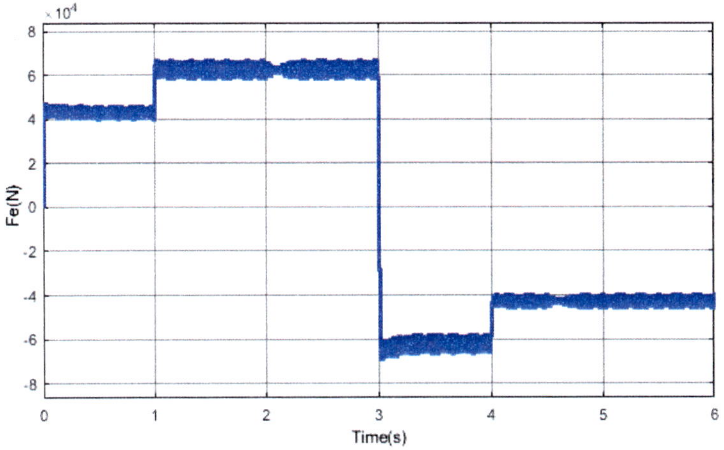

Fig. 5. Simulation results under 108 km/h when the torque command changes.

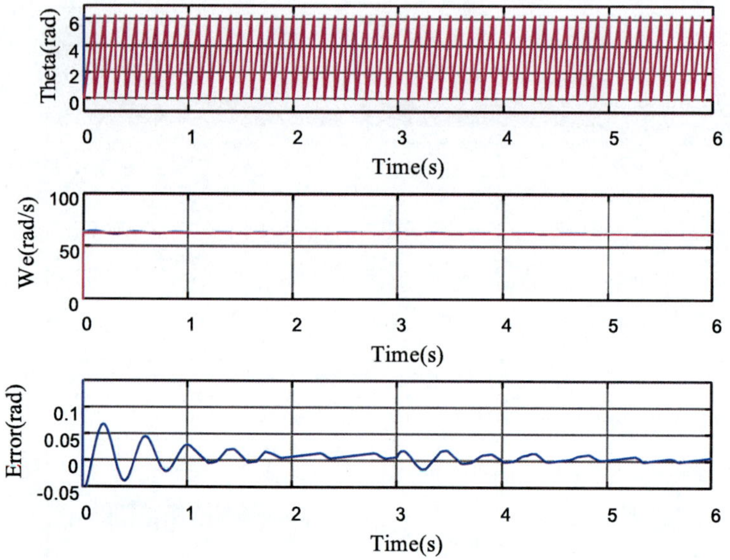

Fig. 6. Simulation performance of sensorless control when the torque command changes.

Acknowledgments. This work is supported by major project in Shanxi Province: Full-scale experimental line of ultra-high-speed low-vacuum pipeline maglev transportation system (Phase I).

References

1. Teja, A.V.R., Chakraborty, C., Maiti, S., et al.: A new model reference adaptive controller for four quadrant vector controlled induction motor drive. IEEE Trans. Ind. Electron. **59**(10), 3757–3767 (2012)
2. Benlaloui, I., Drid, S., Chrifi-Alaoui, L., et al.: Implementation of a new MRAS speed sensorless vector control of induction machine. IEEE Trans. Energy Convers. **30**(2), 588–595 (2015)
3. Yin, Z., Li, G., Zhang, Y., Liu, J., Sun, X., Zhong, Y.: A speed and flux observer of induction motor based on extended Kalman filter and Markov chain. IEEE Trans. Power Electron. **32**(9), 7096–7117 (2017)
4. Zerdali, E., Barut, M.: The comparisons of optimized extended Kalman filters for speed-sensorless control of induction motors. IEEE Trans. Ind. Electron. **64**(6), 4340–4351 (2017)
5. Wang, C., Gou, L., Dong, S., Zhou, M., You, X.: Sensorless control of IPMSM based on super-twisting sliding mode observer with CVGI considering flying start. IEEE Trans. Transport. Electrif. **8**(2), 2106–2117 (2022)
6. Wang, T., Wang, B., Yu, Y., Xu, D.: Discrete sliding-mode-based MRAS for speed-sensorless induction motor drives in the high-speed range. IEEE Trans. Power Electron. **38**(5), 5777–5790 (2023)
7. Zhang, K., Wang, L., Fang, X.: High-order fast nonsingular terminal sliding mode control of permanent magnet linear motor based on double disturbance observer. IEEE Trans. Ind. Appl. **58**(3), 3696–3705 (2022)

8. Sun, Q., Zhu, X., Niu, F.: Sensorless control of permanent magnet synchronous motor based on new sliding mode observer with single resistor current reconstruction. CES Trans. Electr. Mach. Syst. **6**(4), 378–383 (2022)
9. Li, X., Liu, J., Yin, Y., Zhao, K.: Improved super-twisting non-singular fast terminal sliding mode control of interior permanent magnet synchronous motor considering time-varying disturbance of the system. IEEE Access **11**, 17485–17496 (2023)
10. Jukić, F., Pravica, L., Sumina, D., Erceg, I.: Framework for sensorless control and flying start of a permanent magnet generator based on a sliding mode observer. IEEE Trans. Ind. Electron. **71**(1), 294–304 (2023)
11. Xu, W., Qu, S., Zhao, L., Zhang, H.: An improved adaptive sliding mode observer for middle- and high-speed rotor tracking. IEEE Trans. Power Electron. **36**(1), 1043–1053 (2021)

Review on Energy Management Strategies of On-Board Hybrid Energy Storage Systems for Urban Rail Vehicles

Xinni Peng[1], Haiquan Liang[2](✉), and Jingtai Hu[2]

[1] The Key Laboratory of Road and Traffic Engineering, Ministry of Education, Tongji University, Shanghai 201804, China
2133355@tongji.edu.cn

[2] Institute of Railway Transit, Tongji University, Shanghai 201804, China
lhq@tongji.edu.cn

Abstract. With the increasing energy consumption of urban rail transportation, the on-board hybrid energy storage system, which integrates various energy storage technologies, can effectively recycle the regenerative braking energy. It not only solves the problems of voltage increase, temperature rise, and energy consumption increase caused by the energy inversion feedback and braking resistor heat dissipation, but also compensates for the inability of a single energy storage technology to meet the energy and density needs of urban rail vehicles at the same time. This paper first illustrates the composition, topologies and applications of the hybrid energy storage system. Then various energy management strategies of the on-board hybrid energy storage system for urban rail transit are introduced in detail. The characteristics, advantages and disadvantages of different control strategies are analyzed and summarized. Finally, the conclusions make an overall comb and prospect about the paper which lay a certain foundation for the future research of the on-board hybrid energy storage system for urban rail transit.

Keywords: Urban rail transit · On-board hybrid energy storage system · Energy management strategy

1 Introduction

As a globally recognized green transportation mode, urban rail transit has brought great convenience to resident travel and alleviated urban traffic pressure because of its fast speed, large capacity, low energy consumption, high efficiency and low pollution. In recent years, although rail transit has solved some of the traffic problems faced by large and medium-sized cities in China, due to the expansion of urban road network scale and the growth of passenger traffic, the total electric energy consumption and traction energy consumption of urban rail transit have been increasing year by year, with traction energy consumption accounting for nearly 50%. The excessive energy consumption has become the main problem faced by urban rail transit. According to the statistical data from China Association of Metros, the trend of total electric energy consumption and traction energy consumption in the past five years is shown in Fig. 1.

L. Jia et al. (Eds.): EITRT 2023, LNEE 1135, pp. 506–517, 2024.
https://doi.org/10.1007/978-981-99-9307-9_56

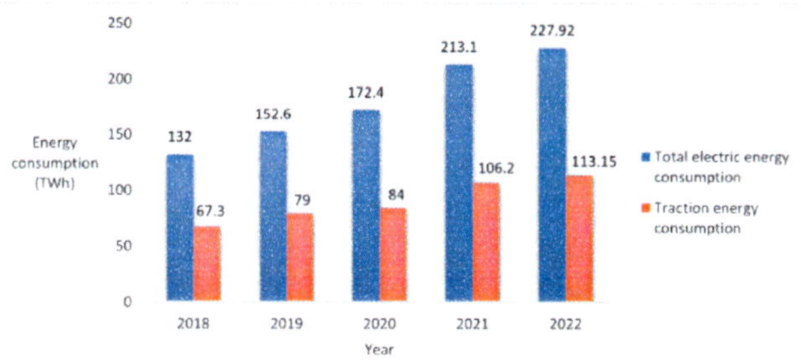

Fig. 1. The trend of energy consumption in the past five years in China.

At present, previous studies have shown that regenerative braking energy of urban rail transit trains can reach 30–40% of traction energy consumption [1]. If the energy storage system equipped on the train can recycle the braking energy, the economical and environmental protection of urban rail transit systems will be greatly improved.

This paper focuses on the energy management of on-board hybrid energy storage system (HESS) for urban rail transit. Section 2 introduces the composition, topologies and applications of on-board HESS. In Sect. 3, it provides an overview of energy management strategies for on-board HESS, analyzing and summarizing their characteristics, advantages and disadvantages. Finally, a few conclusions about energy management strategies of on-board HESS are made to lay a foundation for the future research in Sect. 4.

2 On-Board Hybrid Energy Storage System

2.1 Energy Storage Technologies and On-Board HESS

According to the different principles, there are mainly three types of energy storage technologies: flywheel, battery and ultracapacitor (UC) [2]. Flywheels have been favored by researchers in the field of urban rail transit due to the advantages of environmental friendliness, long working life and so on. However, the disadvantages of high-rate discharge ability and high cost block off their development and application [3]. Batteries have high energy density, small size, low-rate discharge ability and stable performance, making them suitable for on-board energy storage systems. But they need greater security and they're limited by some application conditions. Compared with flywheels and batteries, UCs have higher power density, longer working life and are applicable to both stationary and on-board energy storage systems. They are able to hold a reasonable amount of energy and can take it in or release it quickly. Yet it's worth noting that UCs may cause large voltage fluctuations during charging and discharging, meanwhile, the energy density is lower than batteries (Table 1).

To sum up, the energy storage system with a single technology cannot simultaneously meet the energy and power requirements of urban rail vehicles, so it's necessary to use hybrid energy storage system composed of multiple technologies which can give full play to their complementarity of different energy storage technologies. In recent years,

Table 1. Comparison of energy storage technologies [4–6].

Energy storage technology	Energy density (W h/kg)	Power density (W/kg)	Cycle lifetime (cycles)	Cost
Ultracapacitor (UC)	0.2–20	100–10000	10^5–10^6	High
Lithium-ion battery (LIB)	120–250	150–315	1500–2500	Low
Flywheel	5–130	400–1600	10^4–10^7	High

the HESS composed of batteries and UCs has become the key research object in the field of urban rail transit, which can maximize the advantages of high energy density of batteries and high power density of UCs. This type of HESS not only meets the needs of train's operation, but also has small size, long working life, low cost, and are able to recycle the most braking energy.

2.2 Topologies of HESS

The topology of HESS affects the control effectiveness of energy management strategies to a certain extent. It's known that the research on the topologies begun early and has now tended to ripe. Taking UC-LIB HESS as an example, the topologies can be mainly divided into three types based on the number of bidirectional DC/DC converters or the connection scheme between energy storage elements and DC bus (Figs. 2, 3 and 4): passive parallel topology, semi active topology and fully active topology [7]. The diagrams and merits of the three topologies are shown in the Table 2.

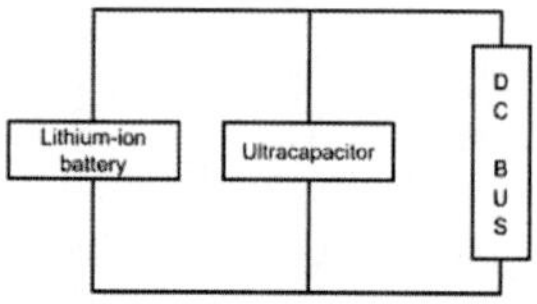

Fig. 2. Passive parallel topology.

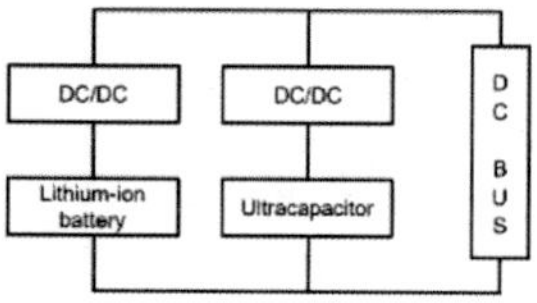

Fig. 3. Fully active topology.

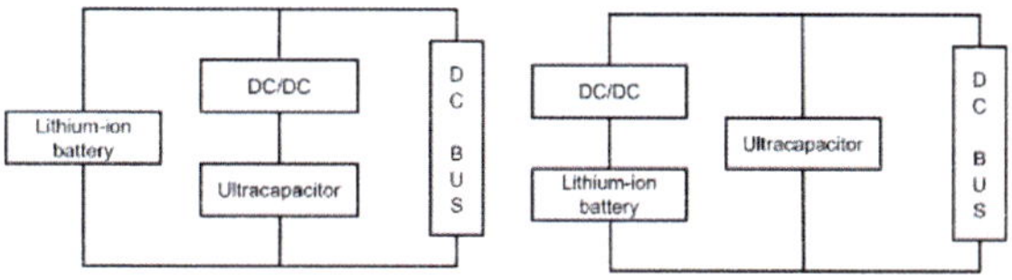

(a) Semi active topology I. (b) Semi active topology II.

Fig. 4. Semi active topology.

Table 2. Advantages and disadvantages of topologies.

Topology	Advantage	Disadvantage
Passive parallel topology	• Easy to control • Lowest cost	Unable to control HESS effectively
Semi active topology	• Capable of controlling energy storage elements • A good balance between control complexity and cost	Unable to fully utilize elements' performance
Fully active topology	• Complete decoupling of energy sources • Flexible distribution • Stable voltage	• Difficult to control • Increase energy loss, system size and cost

2.3 The Domestic and International Research Application

The Sitras HES technology developed by Siemens has been successfully applied to the Combino Plus Tram in Alma, Portugal. The Sitras HES system, which is installed on the roof of the tram, consists of double-layer capacitors and Ni-MH batteries [8]. When the tram is under optimal operating conditions, it's expected to reduce energy demand by 30% through recycling the braking energy [9].

In 2018, the first Delingha Modern Tram came off the assembly line at the CRRC Qingdao Sifang Locomotive & Rolling Stock Co., Ltd. The train adopts a HESS composed of UCs and lithium titanate batteries, which has significant advantages such as charging quickly, operating reliably and environmentally [10].

3 Energy Management Strategy

At present, compared with the relevant studies in the fields of electric vehicles (EVs) and microgrids, there are relatively few researches about urban rail transit. However in recent years, many remarkable researches have been achieved in capacity configuration optimization, charging management method and energy management strategy of HESS. Among them, the energy management strategy, as the brain of HESS which decides the power distribution of the system, not only helps to achieve efficient management and optimization of energy, but also improves the economy of HESS under the premise of energy saving and environmental protection. Therefore, this section provides an overview of energy management strategies of on-board HESS for urban rail vehicles.

A energy management strategy controls the power distribution of HESS based on the energy demand of the urban rail vehicle during its operation so as to obtain the best power and economic characteristics. It plays an important role in extending the working life of energy storage elements and reducing internal energy loss of the system. According to the distribution principles and algorithms, this paper divides the energy management strategies of HESS into three main categories: rule-based energy management strategy, optimization-based energy management strategy and intelligent algorithm-based energy management strategy.

3.1 Rule-Based Energy Management Strategy

The rule-based energy management strategy is based on the empiric experience. It makes corresponding power distribution by judging the demand of trains and the status of HESS [11]. This type of strategies has low computational complexity, strong stability and fast real-time response, making it particularly suitable for rail lines with relatively fixed operating conditions, such as trams. The rule-based energy management strategies mainly include fuzzy logic control (FLC), threshold method, filter method, ratio method, etc.

As shown in Fig. 5, the fuzzy logic control-based energy management strategy takes the real-time information as the input variables, which are expressed in the form of fuzzy set through the membership functions at first. Then the output variables are obtained through the fuzzy inference, and finally the results of power distribution can be gained by defuzzification. FLC does not establish an exact mathematical model for the system, so it's highly adaptable and robust. In [12], an adaptive FLC-based energy management strategy is introduced to minimize the daily operating costs of trams. In [13], the author designed 72 fuzzy rules based on line conditions and characteristics of on-board HESS to get the power distribution results.

Fig. 5. The logical structure of FLC.

Although FLC can ensure the real-time performance of nonlinear systems, its high dependence on the fuzzification rules and membership functions also makes it difficult to achieve optimal control. Therefore, it is often used in conjunction with other algorithms. The author in [14] used a two-level method of FLC and particle swarm optimization to adjust the weight coefficients in the multi-objective optimization equation. On the other hand, by considering the decoupling of the demand power and the requirement of rapid power distribution, a fuzzy control strategy based on real-time wavelet transform is proposed in [15], which effectively improves the load capacity of the on-board HESS.

The threshold method usually sets the output power thresholds of the energy storage element based on its maximum output capacity. By comparing the needed power of the train with these thresholds, the working state and the power distribution of HESS are determined. This method is simple in principle and can fully utilize the characteristics of UCs while protecting the batteries, so it has been widely used in engineering. For example in [16], the authors set a discharge power threshold of LIB to a condition for distinguishing system operating states and determining power distribution rules. Many researchers often choose a fixed threshold to distribute power on the whole line, which may lead to a decrease in energy efficiency of HESS between different stations and cannot fully utilize the capabilities of energy storage elements. It's also not conducive to the recycling of braking energy. To address the above issues, researchers have proposed a dynamic optimization threshold method by considering the differences of the demand power between different stations. The optimal power thresholds obtained from model simulation, algorithm optimization, and other methods are used to distribute power in the corresponding interval [17].

The filter method refers to dividing the train's required power into high-frequency and low-frequency parts based on the energy storage characteristics of batteries and UCs. The high-frequency power is provided by the UCs and the low-frequency power is provided by the batteries. This method utilizes the advantages of the two elements, which is easy to operate and apply. In [18], the authors extracted the high-frequency part of the power generated by the urban rail vehicle with a first-order high-pass filter. Also, the decomposition of the demanded power can be realized by using wavelet transform [19]. In order to give full play to the characteristics of each energy storage element and ensure that the SOC of each element is always in the safe range, it is feasible to use the SOC of UC for correcting the filtering time on the basis of the low-pass filter method [20].

The basic principle of the ratio method is shown as follows:

$$\begin{cases} P_{\text{bat}} = P \cdot \alpha \\ P_{\text{sc}} = P \cdot (1 - \alpha) \end{cases} \tag{1}$$

With a reasonable choice of the scale factor α, the strategy enables the on-board HESS to satisfy the power demand of the train operation. The fixed-ratio method means that the HESS always distributes power according to the same scale factor no matter how the power demand of the train changes [21]. Obviously, the fixed-ratio method cannot meet the operational requirements of the whole line. It not only fails to fully utilize the charging and discharging characteristics of batteries, but also fails to function as normal when the train encounters sudden situations such as charging spots failures. Therefore, two solutions have been proposed in existing researches. Firstly, by defining multiple operating modes for HESS, different fixed-ratio power distribution methods can be set [22]. Secondly, by introducing a working state of energy exchange between energy storage elements, the battery can charge the UC when the SOC of UC reaches the minimum limit [23].

3.2 Optimization-Based Energy Management Strategy

The optimization-based energy management strategy can transform the objective equations into an optimization problem by setting optimization objectives and constraints according to the actual operating conditions. This type of strategies is complicated in calculation and highly dependent on the model, so it is more difficult to achieve online real-time control. The optimization-based energy management strategies are mainly divided into two categories: global optimization and local optimization.

The global optimization energy management strategy refers to conducting global search optimization given train operating conditions and power requirements. It can reduce energy consumption and improve system efficiency by finding the extremum of objective functions under multiple constraints. The commonly used global optimization strategies include genetic algorithm (GA) [24], particle swarm optimization (PSO) [16], Pontryagin's minimization principle (PMP) [25] and dynamic programming (DP). Among them, DP is one of the most widely used global optimization strategies. This algorithm transforms multi-stage problems into a series of interconnected single stage problems, and then solves them one by one (see Fig. 6). It is a mathematical method for solving the optimization of multi-stage decision-making processes. In the current study, DP is generally used to optimize the power distribution in HESS [26, 27]. The optimization effect of DP is great, but its computational complexity is closely related to the model selected, and the stability is relatively poor, making it suitable for offline control of the system.

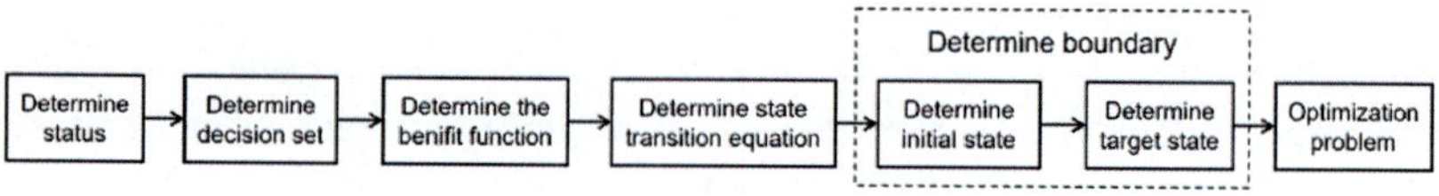

Fig. 6. The steps of DP.

In response to the problem that global optimization cannot be transformed into an online control strategy, an adaptive multi-mode dynamic proportional control strategy based on pseudo spectral method (PSM) is proposed in [28], which uses PSM to solve the optimization model and then extracts online control rules from the optimal solution. Compared with DP, PSM not only achieves global optimization online control, but also improves the convergence speed and accuracy of model calculations, making the optimization results more accurate (Fig. 7).

The local optimization energy management strategy usually targets real-time energy consumption as the optimization objective. It can generally achieve online control of the system. Model predictive control (MPC) is a typical local optimization method, which consists of three main steps: establishment of state predictive model, online optimization and feedback correction (see Fig. 8). As a nonlinear control method, MPC has the advantages of good control effect and strong robustness. It can compensate for the shortcomings of global optimization and real-time control strategies. The energy management strategies based on MPC for single-step prediction and multi-step prediction are studied in [29] with the optimization goal of minimizing energy loss, which further

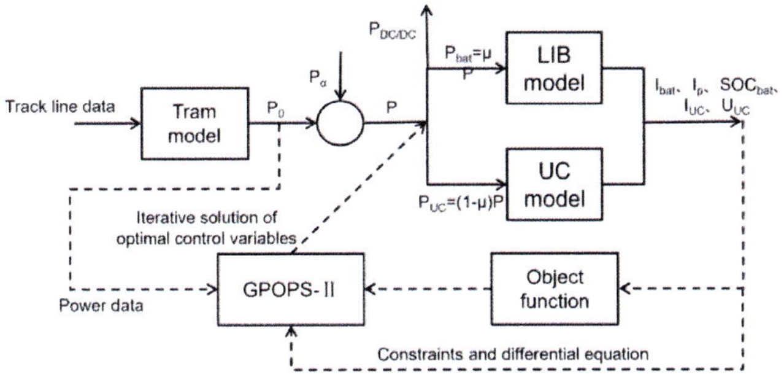

Fig. 7. Pseudo-spectral optimization model of HESS [28].

ensure the global optimal control in traction stage. In addition to the SOC of energy storage elements, the length of the peak power duration can also have an impact on the performance of HESS, so in [30], the author distributed power more reasonably by adding the prediction window of the acceleration time between different stations, which not only makes full use of the characteristics of UCs, but also slows down the degradation of batteries.

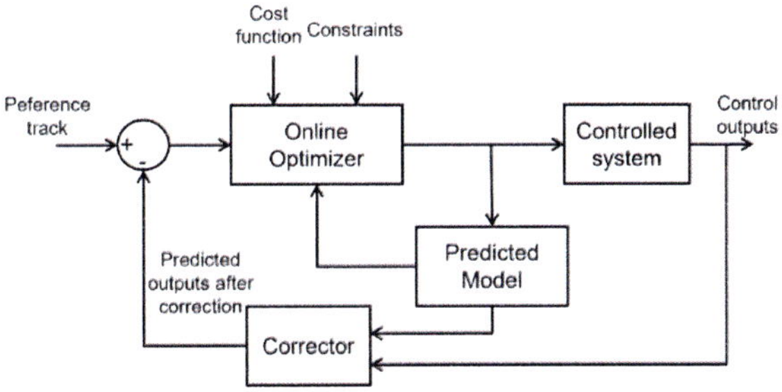

Fig. 8. The logical structure of MPC.

3.3 Intelligent Algorithm-Based Energy Management Strategy

With the development of artificial intelligence technology, many scholars have found that compared with traditional energy management strategies, the intelligent Algorithm-based Energy Management Strategy can not only achieve real-time control, but also effectively reduce energy consumption and have good control effects on power distribution. Therefore, the intelligent Algorithm-based Energy Management Strategy is becoming a hotspot. The energy management strategies of HESS for EVs, which combine with artificial intelligence algorithms such as neural networks [31], reinforcement learning [32, 33] and machine learning [34], have made some achievements. In recent years, many researchers in the field of urban rail transit have also conducted relevant studies on this. For example in [13], the author obtained power distribution strategies under different working conditions through reinforcement learning.

3.4 Other Energy Management Strategies

Due to the various advantages and disadvantages of each energy management strategy, researchers turn their attention to the integrated energy management strategies which have benefits of multiple methods. In [35], the authors used a filter to optimize the power distribution results derived from the fixed-ratio method. Among the integrated control strategies, the integration of rule-based and optimization-based energy management strategies not only achieves the simplification of the strategy, but also gets close to global optimization, making it less difficult and more valuable for actual engineering application. That's why it receives more attention. A few integrated control strategies have been proposed in previous studies, such as the strategy based on FLC and PSO [14], the strategy based on ratio method and DP [26], the strategy based on MPC and threshold method [36], etc. (Fig. 9).

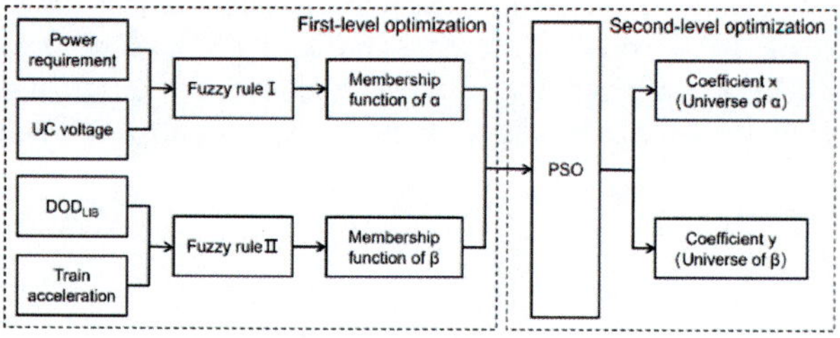

Fig. 9. The logical structure of a two-level optimization strategy based on FLC and PSO [14].

The operating conditions of urban rail vehicles are always changing because of the environment changes, which leads to huge differences in the power requirements of vehicles under different operating conditions. For this reason, the online condition recognition of urban rail vehicles is added to enhance the adaptability of the control strategy [19], which can optimize the control effect under a variety of operating conditions (Fig. 10).

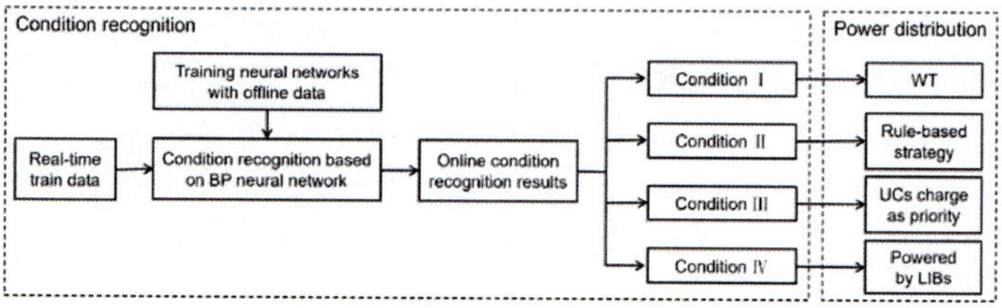

Fig. 10. Adaptive energy management strategy based on real-time condition recognition [19].

4 Conclusions

In summary, energy management strategies of on-board HESS for urban rail vehicles can be mainly divided into three types: rule-based, optimization-based, and intelligent algorithm-based energy management strategy. The rule-based control strategy has the advantages of small computation, easy implementation, good robustness and real-time performance. However, its control effect is limited because of the rules based on the experience of experts or engineers, which makes this type of strategies often used in conjunction with other algorithms. The optimization-based control strategy can achieve multi-objective optimization, while also extending battery life and reducing energy loss. But the complexity of the control and the computation makes online control of some global optimization strategies more difficult. Currently, the strategies based on PSM and PMP have achieved real-time online control of HESS. The intelligent algorithm-based control strategy is capable of high-speed, efficient and high-precision online control for nonlinear systems, but it requires a large amount of training data and has poor control stability. In addition to the three major types, strategies integrated multiple control methods or combined with the train condition recognition can make up for deficiencies in a single strategy and improve the performance of on-board HESS.

Based on existing studies, several key development directions of energy management strategies for on-board HESS in urban rail transit are proposed.

(1) Multi-objective optimization. Most existing energy management researches focus on a single objective. In addition to the common goals such as energy consumption minimization and cost minimization, battery life maximization and train mileage maximization can also be considered.
(2) Integration of different energy management strategies. It's still difficult to eliminate the drawbacks of a single strategy in a short time, so integration of different energy Management Strategies must be the focus of future research.
(3) Intelligent algorithm-based energy management strategy. In the field of EVs, the intelligent Algorithm-based power distribution strategy has gained satisfactory results, which indicates that it also has great development and application potential in the field of urban rail transit.
(4) Global optimization strategy considering HESS capacity configuration, charging spots configuration and energy management strategy. Studies have shown that there exist interactions among HESS capacity configuration, charging spots configuration and energy management strategy. Therefore, the global optimization strategy that comprehensively considers the three should be studied deeply.

Acknowledgments. This work was funded by "National Key R&D Program of China (No. 2021YFB3703805)".

References

1. Hu, J.X., Lin, S.L., Song, W.J.: Energy storage for urban rail transportation. Energy Storage Sci. Technol. **3**(2), 106–116 (2014). (in Chinese)
2. Siang, F.T., Chee, W.T.: A review of energy sources and energy management system in electric vehicle. Renew. Sustain. Energy Rev. **20**(Apr.), 82–102 (2013)
3. Yu, Q.: Application of flywheel energy storage technology in urban rail transit. Electr. Railw. **31**(2), 53–57 (2020). (in Chinese)
4. Wu, J., Zhang, Y.R., Zheng, X.J.: Study on the applicability of lithium titanate battery in urban rail transit. Urban Rapid Rail Transit **34**(6), 39–46 (2021). (in Chinese)
5. Dong, Z., Zhang, Z.B., Li, Z., et al.: A survey of battery-supercapacitor hybrid energy storage systems: concept, topology. Control Appl. Symmetry **14**(6), 1085 (2022)
6. Xiong, R., Chen, H., Wang, C., et al.: Towards a smarter hybrid energy storage system based on battery and ultracapacitor—A critical review on topology and energy management. J. Clean. Prod. **202**(Nov.20), 1228–1240 (2018)
7. Lencwe, M.J., Chowdhury, S.P., Olwal, T.O.: Hybrid energy storage system topology approaches for use in transport vehicles: a review. Energy Sci. Eng. **10**(4), 1449–1477 (2022)
8. Simms, M.: Hybrid energy storage system: high-tech traction battery meets tram's hybrid energy storage system requirements. Ind. Technol. **2010**(APR/MAY), 20 (2010)
9. Meinert, M.: Experiences of the hybrid energy storage system Sitras HES based on a NiMH-battery and double layer capacitors in tram operation. In: International Symposium on Large EC Capacitor Technology and Application (ECCAP 2009), pp. 134–144. Long Beach, California, USA (2009)
10. New hybrid energy storage tram rolls off the line at CSR Sifang. Zhejiang Chem. Ind. **49**(5), 54 (2018) (in Chinese)
11. Zheng, C.H., Wei, S.Y., Wang, Z.G., et al.: Research and implementation on hybrid energy storage system experimental platform of rail transit. Energy Procedia **105**, 3603–3608 (2017)
12. Herrera, V., Milo, A., Gaztanaga, H., et al.: Adaptive energy management strategy and optimal sizing applied on a battery-supercapacitor based tramway. Appl. Energy **169**(C), 831–845 (2016)
13. Mo, H.N.: Reinforcement learning energy management strategy of tram based on condition identification. Trans. China Electrotech. Soc. **36**(19), 4170–4182 (2021). (in Chinese)
14. Lin, H.T.: Multi-objective model predictive control for hybrid energy storage system with adaptive adjustment of weight coefficients. Proc. CSEE **38**(18), 5538–5547 (2018). (in Chinese)
15. Wang, S.Q., Guo, Y.M.: Research on an energy management strategy for composite energy storage urban rail vehicles. Mech. Electr. Inf. **6**, 55–60 (2022). (in Chinese)
16. Chen, W.R., Shi, F.L., Dai, C.H., et al.: Energy management strategy of hybrid tram based ondynamic degree of hybrid. J. Southwest Jiaotong Univ. **55**(02), 409–416 (2020). (in Chinese)
17. Lu, Y.: Study on power parameter matching based on tram operation characteristics. Modern Urban Transit **5**, 25–29 (2018). (in Chinese)
18. Yang, F.P., Zheng, W.Q., Liu, F., et al.: Research on vehicle-mounted hybrid energy storage system for urban rail transit. J. East China Jiaotong Univ. **36**(1), 33–40 (2019). (in Chinese)
19. An, X.K., Yang, Z.P., Wang, Y., et al.: Pareto solution set of tram hybrid energy storage system capacity allocation based on improved convex optimization. Trans. China Electrotech. Soc. **35**(14), 3116–3125 (2020). (in Chinese)
20. Zhang, B.G., Lin, P., Zhang, Z., et al.: Energy management strategy of hybrid energy storage system for urban rail trains. Energy Storage Sci. Technol. **9**(1), 204–210 (2020). (in Chinese)

21. Wang, Y., Yang, Z.P., Lin, F.: Collaborative optimization of management strategy with energy interaction and sizing for the tram's hybrid storage system. WIT Trans. Built Environ. **181**, 121–132 (2018)
22. Yang, J.B., Zhang, J.Y., Song, P.Y.: Multi-objective optimization of energy management strategy for a tramway with onboard energy storage system. J. Power Supply **15**(5), 137–143 (2017). (in Chinese)
23. Wang, Y., Yang, Z.P., Li, F., et al.: Energy management strategy with energy interaction and configuration optimization for the tram's hybrid storage system. Trans. China Electrotech. Soc. **34**(08), 1780–1788 (2019). (in Chinese)
24. Herrera, V.I., Gaztanaga, H., Milo, A., et al.: Optimal energy management of a battery-supercapacitor based light rail vehicle using genetic algorithms. Appl. Energy **169**, 831–845 (2016)
25. Li, F., Yang, Z.P., Wang, Y., et al.: Energy management strategy of tram with hybrid energy storage system based on Pontryagin's minimum Principle. Trans. China Electrotech. Soc. **34**(S2), 752–759 (2019). (in Chinese)
26. Wang, Y., Yang, Z.P., Lin, F., et al.: Dynamic ratio distribution strategy for hybrid storage system of tram. Trans. China Electrotech. Soc. **34**(S1), 405–413 (2019). (in Chinese)
27. Zhang, C.: Optimization of energy management strategy of hybrid energy storage device for streetcars based on V2I communication. In: The 22nd China Power Supply Society Conference (CPSSC2017). Shanghai, China (2017)
28. An, X.K., Yang, Z.P., Wang, Y., et al.: Multi-mode dynamic proportional energy management strategy for battery-supercapacitor hybrid energy storage system of tram based on pseudo-spectrum method. In: 2019 IEEE Transportation Electrification Conference and Expo, Asia-Pacific (ITEC Asia-Pacific), pp. 1–8. IEEE, Seogwipo, Korea (South) (2019)
29. Jia, Z.D.: A real-time MPC-based energy management of hybrid energy storage system in urban rail vehicles. Energy Procedia **152**(1), 526–531 (2018). (in Chinese)
30. Zhu, F.Q.: Research on acceleration-time-prediction-based energy management and optimal sizing of onboard energy storage system for modern trams. Trans. China Electrotech. Soc. **32**(23), 158–166 (2017). (in Chinese)
31. Zgheib, R., Al-Haddad, K.: Neural network controller to manage the power flow of a hybrid source for electric vehicles. In: 2015 IEEE Vehicle Power and Propulsion Conference (VPPC), pp. 1–5. Montréal, Canada (2015)
32. Liu, T., Zou, Y., Liu, D.X., et al.: Reinforcement learning of adaptive energy management with transition probability for a hybrid electric tracked vehicle. IEEE Trans. Ind. Electron. **62**(12), 7837–7846 (2015)
33. Xiong, R., Cao, J.Y., Yu, Q.Q.: Reinforcement learning-based real-time power management for hybrid energy storage system in the plug-in hybrid electric vehicle. Appl. Energy **211**(1–2), 538 (2018)
34. Murphey, Y.L., Chen, Z.H., Kiliaris, L., et al.: Intelligent power management in a vehicular system with multiple power sources. J. Power Sour. **196**(2), 835–846 (2011)
35. Chen, H.X., Wang, Y.S.: Research on new control scheme and strategy of on-board hybrid energy storage device for urban rail transportation. Electr. Railw. **29**(1), 65–71 (2018). (in Chinese)
36. Cheng, L., Pable, A., Ricardo, A.P., et al.: Model predictive control for energy management of a hybrid energy storage system in light rail vehicles. In: 2017 11th IEEE International Conference on Compatibility, Power Electronics and Power Engineering (CPE-POWERENG), vol. 345, p. 10017. IEEE, New York, NY USA, Cadiz, Spain (2017)

Energy-Efficient Train Control Based on Improved Dynamic Programming Algorithm for Online Applications

Fuwang Chen, Yang Peng, and Shaofeng Lu(✉)

Shien-Ming Wu School of Intelligent Engineering, South China University of Technology Guangzhou, Guangzhou 511442, China
chenfuwang88@163.com, lushaofeng@scut.edu.cn

Abstract. The optimization algorithm for Energy Efficient Train Control (EETC) must meet train operation's safety, punctuality, and energy-saving characteristics under complex engineering constraints, which poses significant challenges theoretically and practically. This paper proposes an improved Dynamic programming (DP) algorithm to solve the EETC problem. This algorithm creates a state space table, layers the states according to the location of the road section, solves the objective function value of the state space table from the back to the front, and then obtains the optimized speed track. This algorithm optimizes the state space twice before and after offline computation, greatly reducing the number of state spaces and achieving memory storage of only about 400 MB (or even less). Based on this state storage, the optimized speed trajectory can be obtained within 20 ms. The proposed algorithm is compared with mixed integer Linear programming (MILP) and Convex optimization (CO). This method's real-time energy-saving optimization performance is similar to that of MILP and CO, and the real-time track acquisition speed is at the millisecond level. Moreover, it has no requirements for the high performance of the computer CPU and the allocation of high-quality stable solvers and has more advantages in security and practicality.

Keywords: Energy efficient train control · Dynamic programming · State space table

1 Introduction

Energy-efficient train control (EETC) has attracted the attention of many scholars in recent decades, and the methods used have achieved good energy-saving effects. In early research, based on the PMP method, researchers indirectly derive the optimal control strategy that trains operation needs to meet: a combination of maximum acceleration, cruise, coasting, and maximum deceleration braking stages [1–3]. With the continuous improvement of computer performance, many scholars have gradually applied heuristic algorithms [4, 5] and mathematical programming methods [6, 7] to EETC problems. Although these methods have achieved good results, they still have some shortcomings. Heuristic algorithms are prone to premature convergence and falling into local

L. Jia et al. (Eds.): EITRT 2023, LNEE 1135, pp. 518–529, 2024.
https://doi.org/10.1007/978-981-99-9307-9_57

optima. Mathematical algorithms require high-performance computers and use commercial solvers to solve optimal trajectories. In addition, due to the use of relaxation techniques and piecewise linearization and other simplified methods, the accuracy of the solution has decreased.

Dynamic programming (DP) applied in EETC has achieved excellent results recently. The traditional DP method performs well on EETC problems [8, 9]. The optimization results of DP are compared with those of genetic algorithm and Ant colony optimization algorithms, and the results show that the optimization effect of DP is significantly better [4]. But the traditional DP method is ineffective in the real-time performance of train operations. Intelligent methods based on DP have also made significant progress in studying EETC problems. Yin et al. [10] used the Q-learning method and obtained good high-quality speed trajectories through iterative learning of driver data. Ning et al. [11] used the Deep deterministic policy gradient (DDPG) method to propose an algorithm to speed up the network's training and generate high-quality train speed tracks through offline training. Although these intelligent methods have achieved good results in real-time and high-quality resolution, it is challenging to ensure and interpret the security of the results.

This paper proposes an improved DP method for backward solving applied to EETC problems. Compared to commonly used mathematical programming methods, this method has significant advantages in real-time algorithm operation, and has little demand for computer performance and software solvers.

The main contributions of this paper are as follows:

1. The improved DP method proposed constructs the optimal substructure relationship for train operation and obtain a static lookup table that can be searched for. This searchable table can quickly search for high-quality train speed trajectories in practical use, without requiring high requirements for computer processors and solvers.
2. This paper analyzes the characteristics of train operation and conducts state reduction operations before and after offline DP calculation. In practical use, only a few hundred MB of memory is needed to store searchable tables, which can quickly obtain high-quality energy-saving operation trajectories that can resist train rescheduling.

The rest of this paper is organized as follows. In the Sect. 2, the EETC optimization model is proposed and discretized to prepare for the DP process. In the Sect. 3, the principle and implementation process of the proposed DP method are described. In the Sect. 4, numerical experiments were conducted to compare the algorithm with MILP and CO methods, analyzing its characteristics and proving its superiority.

2 EETC Model

This section constructs an discretized EETC model. This model divides the entire journey into N segments in terms of distance, and the motion of the train in each segment is considered as a process of uniform acceleration or deceleration. According to this model, this paper calculates the energy consumption and time consumption of each segment (v_i, v_{i+1}), and forms an energy-time matrix, which provides a basis for the solution of improved DP.

2.1 Train Dynamics Model

The train dynamics model is as follows:

$$\frac{dv(t)}{dt} = \frac{F(v,x) - D(v) - G(x)}{M} \tag{1}$$

Among them, t is the time when the train has already run, $v(t)$ is the speed of the train at time t, and $F(v,x)$ is the traction or braking force applied by the motor to the train.

$D(v)$ represents the basic resistance a train encounters during operation.

$$D(v) = D_0 + D_1 v + D_2 v^2 \tag{2}$$

where D_0, D_1, D_2 represent non-negative running resistance coefficients for train running resistance.

G(x) represents the impact of train track geometry on train dynamics.

$$G(x) = Mg\sin(\alpha) \tag{3}$$

The following formula can further express the traction/braking force:

$$F(v,x) = k_a(x)F_a(v,x) - k_b(x)F_b(v,x) \tag{4}$$

$$0 \le F_a(v,x) \le F_{a\max}(x) \tag{5}$$

$$0 \le F_b(v,x) \le F_{b\max}(x) \tag{6}$$

where $F_a(v,x), F_b(v,x)$ represents the traction and braking force of the train. When $k_a(x) = 1, k_b(x) = 0$, the train is in traction state; When $k_a(x) = 0, k_b(x) = 1$, the train is in a braking state.

2.2 Motor Traction/Braking Characteristics

During train operation, the motor's maximum traction/braking force dynamically changes with speed. The formulation is as follows:

$$F_{a\max}(v) = \begin{cases} F_{t\max} & 0 \le v \le v_T \\ \frac{P_{a\max}}{v} & v_T \le v \end{cases} \tag{7}$$

$$F_{b\max}(v) = \begin{cases} F_{b\max} & 0 \le v \le v_T \\ \frac{P_{b\max}}{v} & v_B \le v \end{cases} \tag{8}$$

In the motor traction stage, the electric motor has a fixed maximum power $P_{a\max}$ and the maximum traction force that can be applied $F_{t\max}$. When the train is in the low-speed stage, and the motor is in the constant torque zone, the motor power is less than the maximum power, and the maximum traction force is constant at $F_{t\max}$. When the speed exceeds v_T, and the motor is not sufficient to provide greater power, the motor enters the constant power zone $F_{a\max}(v) = F_{a\max}(v)$.

In the motor braking stage, the motor is in the generator stage, which recycles the train's kinetic energy into electrical energy and stores it. During the braking phase of the train, the motor also has a constant torque zone and a constant power zone.

The energy conversion relationship between traction/braking is as follows:

$$\Delta e = \begin{cases} \frac{f \cdot v \cdot \Delta t}{\eta_a} & f \geq 0 \\ f \cdot v \cdot \Delta t \cdot \eta_b f & < 0 \end{cases} \tag{9}$$

2.3 Model Discretization

The distance of each segment is Δd, and i represents the index of the discrete distance segment. The formula is as follows:

$$\sum_{i=1}^{M} \Delta d_i = D \tag{10}$$

Trains have different maximum allowable speed constraints at different sections.

$$0 \leq v_i \leq v_{lim}(i) \tag{11}$$

If there are m maximum allowable speed limit switching points along the road, the total distance is divided into $N = M + m$ segments. Modified Formulas (10) and (11) as follows:

$$\sum_{i=1}^{N} \Delta d_i = D \tag{12}$$

$$\sum_{i}^{N} \Delta t_i \approx T_r \tag{13}$$

T_r is the set time when the train runs from the starting point to the midpoint. Excluding the endpoint, a total of N position points, including the starting point, are assigned along the total route.

$$v_{i+1} - v_i = a_i \cdot \Delta t_i \tag{14}$$

The acceleration a_i also has certain limitations as follows:

$$a_{\min} \leq a_i \leq a_{\max} \tag{15}$$

After the speed (v_i, v_{i+1}) is determined, the energy consumption of section i can also be determined according to Formulation (9).

$$e_{i,i+1} = \int \Delta e(f, v, x) \tag{16}$$

$$\Delta t_{i,i+1} = \frac{2 \cdot \Delta d_i}{v_i + v_{i+1}} \tag{17}$$

According to Formulas (16) and (17), the energy and time consumption under different speed changes on all road sections can be calculated, and an energy time consumption data table can be formed.

3 DP Process

3.1 Objective Function

The objective optimization function can be obtained after the discretization of the model.

$$\max J(x, t, v, f) = -\sum_{i=1}^{N} e_{i,i+1} - \omega \cdot |t_N| \tag{18}$$

Subject to: (1)–(9), (13), (15)–(17).

ω is a weight coefficient. The larger the ω, the higher the punctuality of the train reaching the destination. This objective function combines time and energy considerations, and the value of this function evaluates the quality of the train's speed trajectory.

3.2 DP Status

Considering a train as an intelligent agent, its motion process is continuous with infinite states. According to the discrete DP model in the last section, the entire operation process can be represented in a discrete finite state.

$$S_i = [i, v_i, t_i] \tag{19}$$

Among them, i represents the position point number, v_i represents the speed of the train at position point i, and $t_i = T_r - T_i$ represents the remaining planned running time of the train at position point i. T_i represents the time it has taken for the train to run from the starting point to the position point i.

$$i \in \{0, 1, 2, \ldots, N\} \tag{20}$$

Multiple velocity points can be taken on a continuous velocity region to represent the velocity region discretely.

$$v_i \in \{0, 0.1, 0.2, 0.3, \ldots, v_{lim}(i)\} \tag{21}$$

The average speed of the remaining distance when the train is at position point i is expressed as follows:

$$v_{aver}(i) = \frac{(X_{end} - X_i)}{t_i} \tag{22}$$

When the train's average speed is less than $v_{\min}$, the energy-saving control of the train is no longer switched on. At this time, it can be directly switched to the low-speed control to meet the requirement of punctuality.

Transform the inequality:

$$t_i \leq \frac{(X_{end} - X_i)}{v_{\min}} \tag{23}$$

$$t_{\max}(i) = \frac{(X_{end} - X_i)}{v_{\min}} \tag{24}$$

Similarly, multiple time points can be taken on a continuous time region to represent the time region discretely.

$$t_i \in 0, 0.1, 0.2, 0.3, \ldots, t_{\max}(i) \tag{25}$$

3.3 State Transition

The state transition is an orderly process, starting from position point 0, selecting the speed of position point 1 under constraints, and so on.

$$S(0, v_0, t_0) \xrightarrow{v_1} S(1, v_1, t_1) \xrightarrow{v_2} S(2, v_2, t_2) \xrightarrow{v_3} \cdots \xrightarrow{v_N} S(N, v_N, t_N) \tag{26}$$

By selecting the velocity v of the next state from the previous state and searching for the energy time consumption matrix obtained in Sect. 2, the energy consumption and time information in the next state can be obtained.

$$t_{i+1} = t_i - \frac{2 \cdot \Delta d_i}{v_i + v_{i+1}} \tag{27}$$

3.4 Optimal Substructure and Backward Solving

The overall problem is to solve the velocity trajectory that maximizes the objective function J according to (18), which is the optimal state switching trajectory from the starting state $(0, 0, t_0)$ to the ending state $(N, 0, t_N)$. The subproblem can be defined as solving the objective function J from a certain position point (i, v_i, t_i) to the endpoint state $(N, 0, t_N)$.

The optimal substructure of the DP problem is shown in Fig. 1.

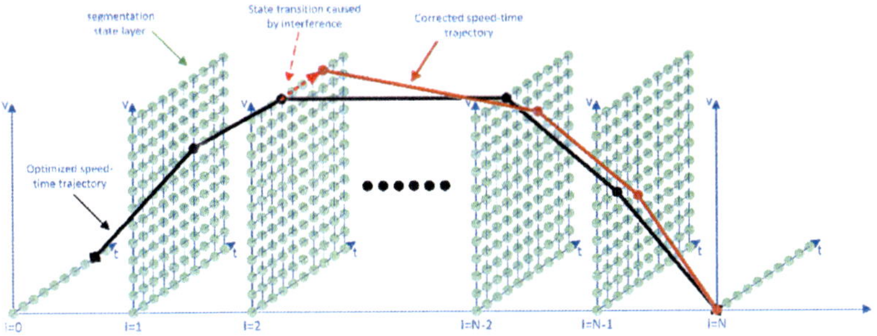

Fig. 1. The state space diagram of the improved DP method, where the green plane is a state plane composed of states with the same distance component, and the black and orange curves are optimized train speed-time remaining trajectories.

The total distance is divided into N small segments from left to right, with different boundary distances for each segment. $i = 0$ represents the starting point of the platform, and $i = N$ represents the ending point of the platform. Based on this, an $N + 1$ layer state plane is created. According to the above definition of optimal structure, each state corresponds to a problem, and for adjacent two-layer states, the problem corresponding to the right state is a sub problem of the problem corresponding to the left state. The objective function of the layer i state can be defined by analogy with Formula (18).

$$\max J_i(x, t, v, f) = -\sum_{k=i}^{N} e_{k,k+1} - \omega \cdot |t_N| \tag{28}$$

Its optimal substructure can be described as follows:

$$\max J_0 = max(-e_1 - e_2 - \ldots - e_i) + \max J_i \tag{29}$$

The backward-to-forward solve mode is similar to the value function representation of Reinforcement learning. This solve method starts from the sub-problem corresponding to the right-layer state and gradually updates to the left layer state. Figure 1 shows that the layer $i = 0$ state is a one-dimensional vector about time, which means that this table can solve the running situations of different expected times. In addition, if an uncontrollable event occurs that affects train operation, it is necessary to change the expected time of the train arriving at the destination to obtain an optimized speed trajectory. The state space storage of this method can solve this problem in real time. Figure 1 shows an example where a train receives a command at $i = 2$ and needs to be delayed for a period of time t_{delay} to reach the destination.

4 Numerical Experiments

In this section, the results of improved DP are compared with the results of MILP and CO to verify the superiority of the DP method for the EETC problem. The optimization object of this paper takes the line data from Xierqi Station to Life Science Park Station of Beijing Metro Changping Line as an example, and the trains and modeling parameters are shown in Table 1. The software optimizer for this paper runs on a 12th Gen Intel(R) Core(TM)i7-12700F 2.10GHz processor and python3.8 interpreter on a 16GB RAM computer (Table 1).

4.1 MTTC and EETC

Firstly, compare the proposed improved DP method's optimization trajectory performance under the minimum running time. From the left side of Fig. 2, it can be seen that the velocity trajectory curves of MILP and CO are almost perfectly fitted. In contrast, the DP curve deviates slightly from the MILP and CO curves. The DP method tends to make train operations more energy-efficient in the case of small time delays. The energy consumption of DP is 35.27 kWh, MILP is 38.24 kWh, and CO is 38.43 kWh. The energy consumption of the DP method is significantly lower than that of the other two methods, which indicates this characteristic.

The energy-saving performance of the three methods in the case of predefined arrival time $T_r = 308$s is compared. The result is shown on the right side of Fig. 3. The three curves all reach the end within a certain error time range, and all meet the punctuality requirements of train operation. In this experiment, the expected arrival time of the train was 308 s, with DP's actual arrival time being 308.01 s, MILP's being 307.15 s, and CO's being 310.76 s. The error of the final running time of MILP and CO methods after calculation correction and set expected time is much larger than that of DP methods. The energy consumption of DP is 19.04 kWh, MILP is 19.65 kWh, and CO is 18.74 kWh. The DP method proposed in terms of energy efficiency is comparable to mathematical programming methods.

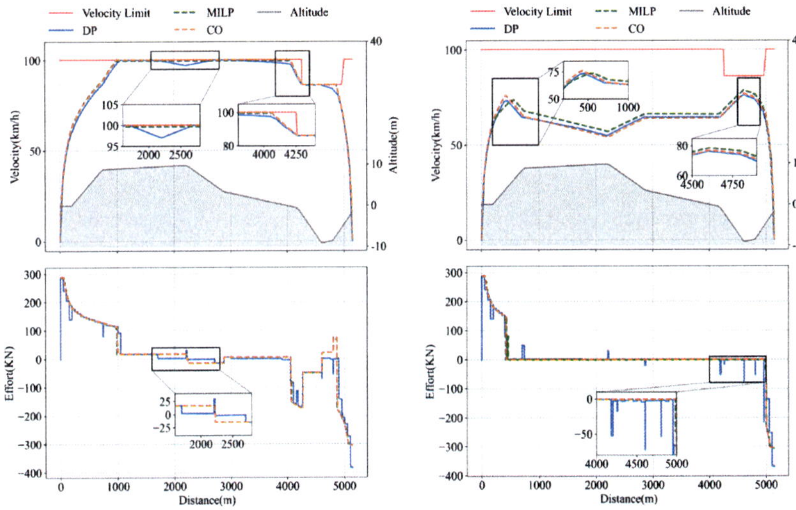

Fig. 2. Comparison diagram of MTTC (left) and EETC (right) speed trajectory and effort trajectory between the improved DP method and MILP, CO methods.

4.2 Algorithm Self-adaptability

During train operation, there are various interferences that can occur. To maintain energy-saving operation of trains while meeting the requirement of delayed arrival at the destination, algorithms need to be able to quickly and accurately generate subsequent train operation trajectories.

As shown in Fig. 4, when the train reaches the 1500 m position, it receives a signal that the train needs to delay reaching the destination, and the DP algorithm regenerates the energy-saving operation trajectory of the remaining sections. As shown by the red dashed line in Fig. 1, after the train receives the delay signal, the current state of the train in the DP algorithm moves from the black state point to the brown state point, and the next position's state table is searched again to obtain the optimal speed trajectory. This DP algorithm can obtain real-time information about the true state of trains and readjust the subsequent train operation trajectory.

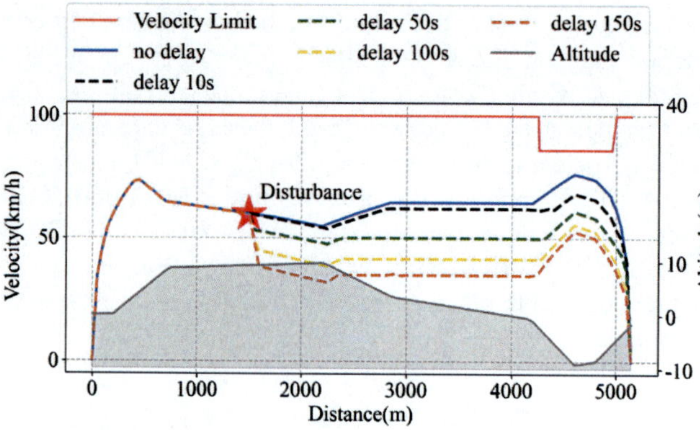

Fig. 3. Comparison diagram of train operation trajectories under different delay times.

4.3 Reduction of State Space

This subsection discusses the characteristics of the DP method and how to maximize the overall performance of DP. The improved DP method is essentially a space-for-time, off-line slow running to online fast running method, and this section discusses how to reduce the scope of memory space and improve the speed of DP running.

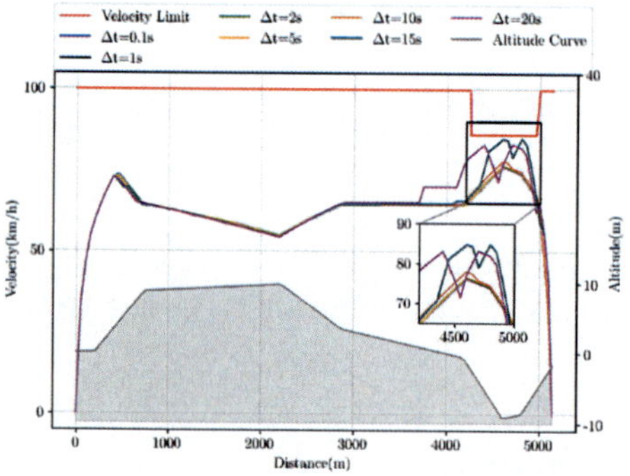

Fig. 4. Comparison diagram of energy-saving operation control speed trajectory at different scales in the time component of offline training state data.

- **Reduce state during offline calculation**: At the expected arrival time $T_r = 5145$ s, the average speed of the total journey is 1m/s. At this time, the energy optimization effect of the train is close to convergence, and the feasibility of optimization is small. If the average minimum speed of the total sections are 3 and 5 m/s, the optimization effect of the energy-saving operation of the train has little change. The space storage size and calculation time of their state target values can be greatly reduced. This section also considers reducing the number of states by expanding the speed scale. From the comparison of the energy optimization of the curves from 0.1 to 1 km/h, it can be found that the difference in energy optimization between the other two curves is not significant. This proves that the storage of states can be reduced by increasing the speed scale.

- **Reduce state space after offline computation**: The above two methods are both operations to reduce the state before offline calculation. In the offline calculation of state information, as the target value of the state has been determined, it becomes very feasible to extend the time scale to obtain critical states, as shown in Fig. 4. The graph shows that the extension of the time scale from $\Delta t = 0.1$ s to $\Delta t = 10$ s has a minimal impact on the optimized velocity trajectory. If the original scale $\Delta t = 0.1$s is extended to $\Delta t = 5$s, the state space can be reduced to the original $\frac{1}{50}$, which is an astonishing and significant result.
- **Comprehensive DP Effect**: The use of different parameter values in the improved DP algorithm will have an impact on the usage conditions and effectiveness. This section considers the impact of speed scale, time scale, and lower limit of average speed on the effectiveness of the algorithm, considering the estimated running time of the train being 308 s. The results are shown in Table 2. From the table, it can be seen that regardless of the parameters, the algorithm's solving speed is very fast. By changing the parameters of the algorithm, it can have a significant impact on the memory space, with the maximum memory space reaching around 20 GB and the minimum only needing around 20 MB. This proves that the DP method does not require high-performance computers and commercial solvers, and can quickly obtain optimized energy-saving operation curves of trains with minimal storage space.

Table 1. Modelling parameter of urban rail transit

Train mass (t)	Rotary allowance (%)	Maximum traction/braking force (KN)	Maximum traction/braking power (KW)	Motor traction/braking efficiency	Davis resistance (KN)
287	8	289/−390	3144/4243	0.9/0.6	$3.48 + 0.14v + 0.01\,v^2$

Table 2. Performance comparison of improved DP algorithm under different parameters with expected arrival time of 308 s

Δv (km/h)	Δt (s)	$v_{\min}$ (m/s)	State storage	Energy-time data storage	Energy consumption E (kWh)	Actual arrival time (s)	Solving time (ms)
0.1	0.1	1	19.5 GB	1.5 GB	19.05	308	21
0.1	1	1	1.95 GB	1.5 GB	19.06	308	18
0.1	1	3	668 MB	1.5 GB	19.19	308.1	15
0.1	5	3	134 MB	1.5 GB	19.21	308.5	11
0.2	5	4	51 MB	378 MB	19.17	308.4	8
0.5	5	4	20 MB	61 MB	19.81	309.7	6
1	5	4	5 MB	15 MB	20.31	307.5	6

5 Conclusion

This paper proposes a static lookup table DP method for offline calculation from back to front. This paper constructs a Discretization DP model, creates a time space state table, and defines the optimization problem for each state, and derives the optimal substructure relationship between the states mathematically. In this paper, the numerical experiments are compared with MILP and CO. The first experiment shows that this method strictly adheres to various constraints and has excellent energy-saving effects for train operation. The second experiment demonstrated that this DP algorithm can obtain real-time information about the true state of trains and readjust the subsequent train operation trajectory. Finally, the spatiotemporal state space of improved DP is optimized. Experiments show that only 400 MB (or even less) of memory is needed to obtain the speed trajectory. In addition, its real time solving speed is extremely fast, and the optimized velocity trajectory can be solved within 20 ms.

Funding Information. This work was supported in part by the National Science Foundation of Guangdong Province, China (Grant No.2023A1515012949), in part by the 2022 Fundamental and Applied Fundamental Research Project of Guangzhou Municipal Basic Research Program (Grant No.202201010716) and the Featured Creative Project of Higher Education Institution of Guangdong Province, China (No. 2021KTSCX001).

References

1. Ichikawa, K.: Application of optimization theory for bounded state variable problems to the operation of train. Bull. JSME **11**(47), 857–865 (1968)
2. Howlett, P.: An optimal strategy for the control of a train. ANZIAM J. **31**(4), 454–471 (1990)
3. Khmelnitsky, E.: On an optimal control problem of train operation. IEEE Trans. Autom. Control **45**(7), 1257–1266 (2000)
4. Lu, S., Hillmansen, S., Ho, T.K., Roberts, C.: Single-train trajectory optimization. IEEE Trans. Intell. Transp. Syst. **14**(2), 743–750 (2013)

5. Yildiz, A., Arikan, O., Keskin, K.: Traction energy optimization considering comfort parameter: a case study in Istanbul metro line. Electr. Power Syst. Res. **218**, 109196 (2023)
6. Cao, Y., Zhang, Z., Cheng, F., Su, S.: Trajectory optimization for high-speed trains via a mixed integer linear programming approach. IEEE Trans. Intell. Transp. Syst. **23**(10), 17 666–17 676 (2022)
7. Wu, C., Lu, S., Xue, F., Jiang, L., Chen, M., Yang, J.: A two-step method for energy-efficient train operation, timetabling, and onboard energy storage device management. IEEE Trans. Transp. Electrif. **7**(3), 1822–1833 (2021)
8. Wu, Z., Gao, C., Tang, T.: An optimal train speed profile planning method for induction motor traction system. Energies **14**(16), 5153 (2021)
9. Gao, H., Zhang, Y., Guo, J.: A novel dynamic programming approach for optimizing driving strategy of subway trains. In: MATEC Web of Conferences, vol. 325, p. 01002. EDP Sciences (2020)
10. Yin, J., Chen, D., Li, L.: Intelligent train operation algorithms for sub-way by expert system and reinforcement learning. IEEE Trans. Intell. Transp. Syst. **15**(6), 2561–2571 (2014)
11. Ning, L., Zhou, M., Hou, Z., Goverde, R.M., Wang, F.-Y., Dong, H.: Deep deterministic policy gradient for high-speed train trajectory optimization. IEEE Trans. Intell. Transp. Syst. **23**(8), 11 562–11 574 (2021)

Check for updates

Long Distance Two Terminal Power Supply Simulation Technology Based on RTLAB Hardware-in-The-Loop Simulation Platform

Xin Jiang(✉), Shaobo Yin, Yongyong Guo, and Kekang Wei

China Aerospace Science and Industry Co., Ltd, Beijing, China
1027255625@qq.com, bxdlysb@163.com, guoyong10@126.com, weikekang@163.com

Abstract. In order to accelerate the research and development process of the two terminal power supply control system for ultra-high-speed linear motors, a two-terminal power supply simulation system has been built by using the RTLAB Hardware-in-the-loop (HIL) simulation platform. Based on the operating characteristics of each equipment in the traction system, build a master-slave simulation platform by using the "CPU + FPGA" architecture and set different operating simulation time steps. The RTLAB simulation model has been compared with the Simpower model in the discrete environment to verify the accuracy of the RTLAB model. On the basis, the RTLAB platform is connected to a real traction control unit to complete normal operating conditions tests. The results indicate that the two terminal converters can stably output electrical energy according to the current distribution instructions issued by the control system. The simulation platform is important to the test and verification of the two terminal power supply control system.

Keywords: Two terminal power supply · Long distance · RTLAB · Simulation technology

1 Introduction

When the aircraft run on the line, converters are required to supply power. The aircraft is drove according to the target planning curve. The types of power supply include single terminal power supply and two terminal power supply. The single terminal power supply mode is relatively simple in connection and requires less power from the substation. However, limited by the capacity of the electrical substation, the aircraft cannot reach a very high operating speed in the single terminal power supply mode. Therefore, the double terminal power supply mode is adopted, that is, two electrical stations at both ends of the line are used to supply power to the aircraft line in parallel at the same time.

At present, long distance two terminal power supply electronic simulation is mostly offline. Although it has high accuracy and good convergence, its timeliness and speed are poor, and it cannot truly reflect the dynamic changes of the system. Therefore,

L. Jia et al. (Eds.): EITRT 2023, LNEE 1135, pp. 530–537, 2024.
https://doi.org/10.1007/978-981-99-9307-9_58

semi physical simulation is used to solve the above problems. Semi physical simulation refers to the integration of real system devices into a simulation platform. The RTLAB software utilizes real-time Linux and can be used for a wide range of experiments [1]. In addition, RTLAB can be configured with analog input/output (AI/AO) cards and digital input/output cards, which can send signals when we need them [2]. The RTLAB simulation platform has been widely used in wind power generation, DC distribution and other systems for preliminary verification of system control design [3–5].

The long-distance two terminal power supply model is constructed in the simulation system, and the real control system is tested to obtain the same experimental effect as the real scenes as much as possible. The RTLAB platform has the advantages of safety, economy, controllability, easy observation, non-destructive, and repeatable experiments. It can assist in researching and analyzing the operating characteristics of parallel inverters [6]. Meanwhile, it can simulate normal operating conditions and some extreme special conditions, providing sufficient test cases for designers to verify the correctness and effectiveness of the design. This method can shorten the equipment development cycle and release technical risks in advance.

2 Architecture and Energy Distribution Strategy of Long Distance Two Terminal Power Supply System

The structure of the two terminal power supply system for the aircraft is shown in the following figure. The distance between dual end traction converters is usually between 20 and 30 km [7]. The components of dual terminal power supply system include: traction converter, traction power cable, trackside switch station, and linear motor. It has many advantages, such as high utilization rate of DC voltage, strong redundancy and fault tolerance [8]. The traction converter outputs electrical energy according to the instructions of the traction control system. At the same time, according to the position of the aircraft, the trackside switch stations along the line perform opening and closing actions. The output electrical energy of the traction converter is transmitted to the linear motor through the traction power supply cable and trackside switch station, driving the aircraft according to the specified curve. The substations at both ends of the line supply power to the linear motor on the line. The linear motor adopts sectional power supply to reduce the motor energy loss. In future development, photovoltaic power generation can be combined to further reduce power generation losses [9] (Fig. 1).

The long distance two terminal power supply energy distribution strategy is mainly applied in the aircraft testing stage. In this stage, the motor controller receives traction cut-off release and departure instructions. Then, the aircraft starts running. Based on the current position of the aircraft and the output status of the converters, the output current is distributed to the two converters, the energy distribution control algorithm is executed, and the command current is given. The motor controller performs closed-loop control, three step control algorithm operation. The inverters output current, and aircraft is drove. When the operation is completed, the motor controller blocks the pulse, receives the traction cancellation command issued by the operation control system, and the system test ends.

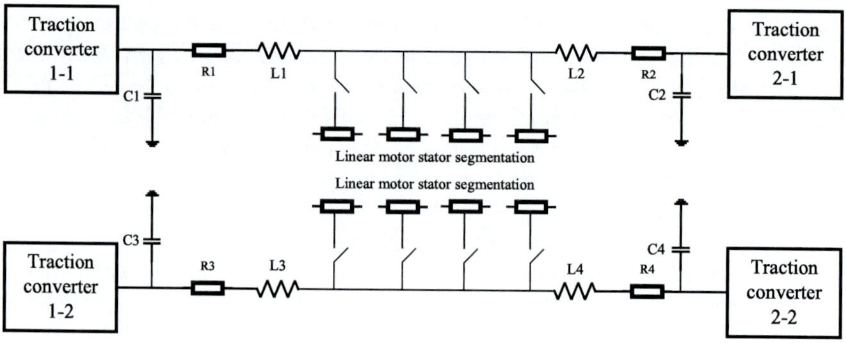

Fig. 1. Architecture of long distance two terminal power supply system

3 Composition of RTLAB Hardware-in-The-Loop Simulation Platform

The semi physical simulation platform for traction is divided into three parts: the actual traction control system device, interface equipment, and traction simulation system. The system architecture diagram is shown in the following figure. The actual device is used to achieve control functions related to the traction control system. The simulation system is used to simulate the power devices, mainly including traction converters, traction power cables, trackside switch stations, and linear motor systems. The interface device mainly realizes interface conversion and data exchange between the actual control devices and the traction simulation system (Fig. 2).

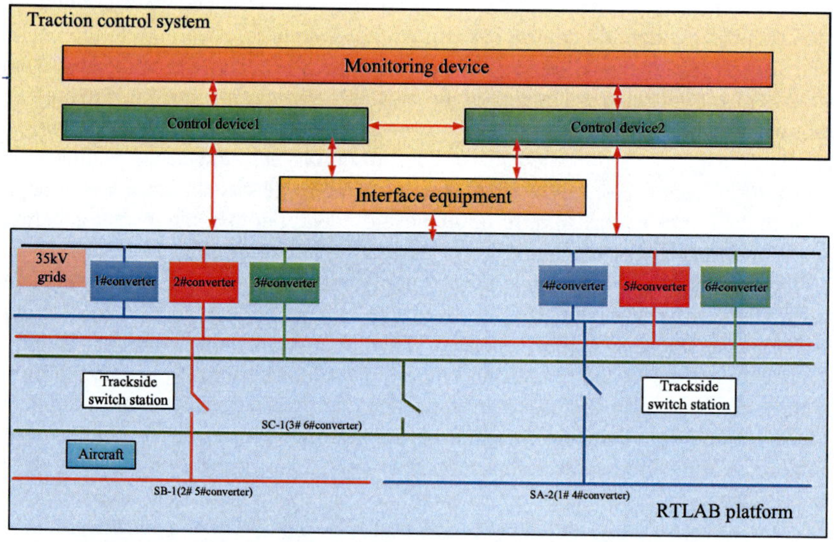

Fig. 2. Composition of RTLAB Hardware-in-the-loop simulation platform

The traction control system is responsible for implementing core control strategies, including calculation of operating curves, three-step method for changing steps, motor vector control, fault handling algorithms, etc. At the same time, based on system instructions and aircraft position information, the control system can switch and control the feeder switches in the trackside switching station, enabling the aircraft to operate stably according to the planned curve. The monitoring device is responsible for monitoring the status of various equipment during the testing process, collecting, processing, and storing operational data of each controller and equipment.

The traction simulation system is mainly used to simulate the operation of traction power equipment, including power grid system, traction converter system, traction power supply system, linear motor system, etc. For grids and aircrafts, their dynamic response time scale is in milliseconds. The discrete nature of power electronics and motor systems enables their dynamic response time scale to reach microseconds. Therefore, it is necessary to build a CPU + FPGA master-slave control architecture semi physical simulation platform, and construct large step models and small step models in the CPU and FPGA cores respectively. The composition of the RTLAB simulation machine is shown in the following Fig. 3.

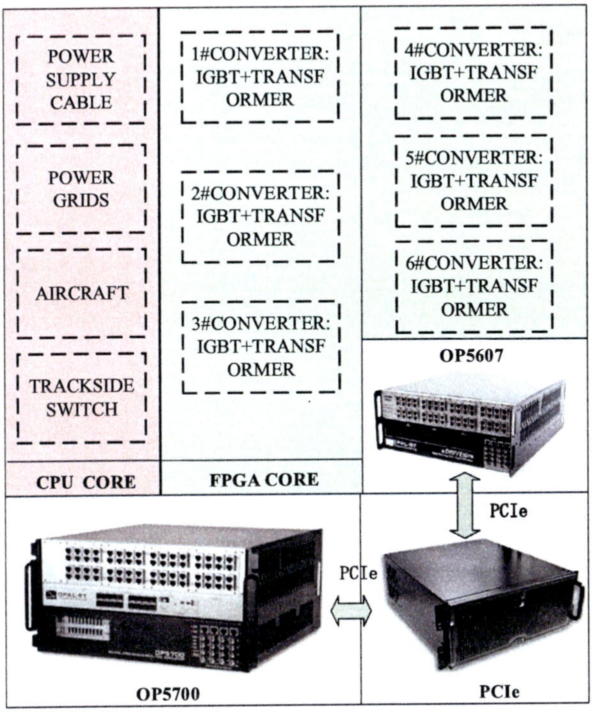

Fig. 3. The composition of the RTLAB simulation machine

RTLAB adopts a distributed platform architecture, which can achieve high-precision real-time simulation of large-scale complex models [10]. The RTLAB simulation platform includes the power grid system, traction converter system, traction power supply system, and linear motor system. The functions of each simulation model are as follows:

- Power grid system: It mainly inputs the grid side voltage into the electrical energy of the traction converter system input transformer, which is the energy input of the system.
- Traction converter system: According to the instructions issued by the traction control system, the IGBT is driven to output electrical energy, generate variable amplitude and variable frequency voltage, provide electrical energy for different segmented motor modules, and generate current.
- Traction power supply system: The electrical energy output by the traction converter is transmitted to the linear motor module, and the trackside switch station executes opening and closing commands based on the position to supply power to different linear motor modules.
- Linear motor system: Calculate the real-time current of the linear motor modules based on the electrical energy transmitted by the traction converter system and traction power supply system, and calculate the back electromotive force and output electromagnetic thrust of the linear motor based on the information such as current, aircraft speed and position.

4 System Simulation

After completing the construction of the long-distance two terminal power supply system model, the model is tested and compared with the Simpower model. The simulation results of the two models are basically consistent. On this basis, the semi physical simulation test is conducted with the controller for traction. The total output of the two terminal converter is 1000 A. The two ended converter performs current closed-loop control based on 5:5 and 4:6 current distribution ratios. The simulation results are shown in the following figure. It can be seen that the converter can output current according to the command, verifying the correctness and effectiveness of the control program (Figs. 4, 5, 6 and 7).

5 Conclusion

To accelerate the development process of long-distance two terminal power supply traction control system, a long-distance power supply traction simulation platform based on RTLAB is designed. This article elaborates on the architecture and components of the RTLAB platform, with a focus on describing the modeling and simulation of the traction simulation system. On the basis of offline model simulation and comparison verification, the effectiveness and correctness of the current control program are verified through joint test between the real controller and the simulation platform system. This simulation technology provides a reliable and efficient means for the research of traction control strategies in long-distance two terminal power supply, and is of great significance.

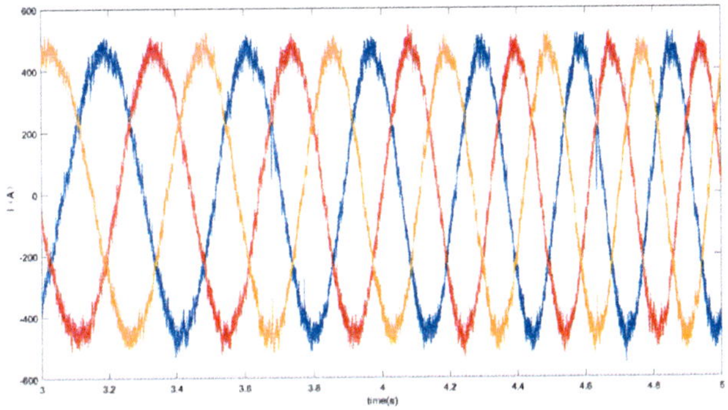

Fig. 4. Output current of one converter (5:5 current distribution ratios)

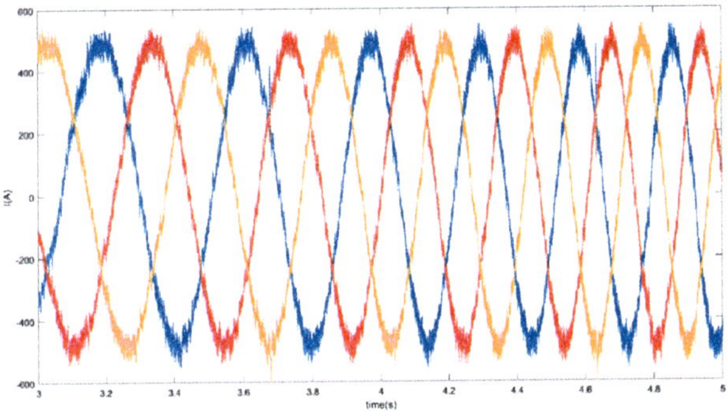

Fig. 5. Output current of another converter (5:5 current distribution ratios)

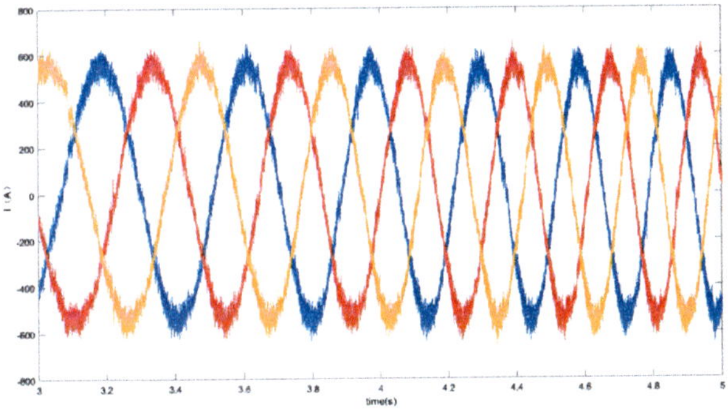

Fig. 6. Output current of one converter (4:6 current distribution ratios)

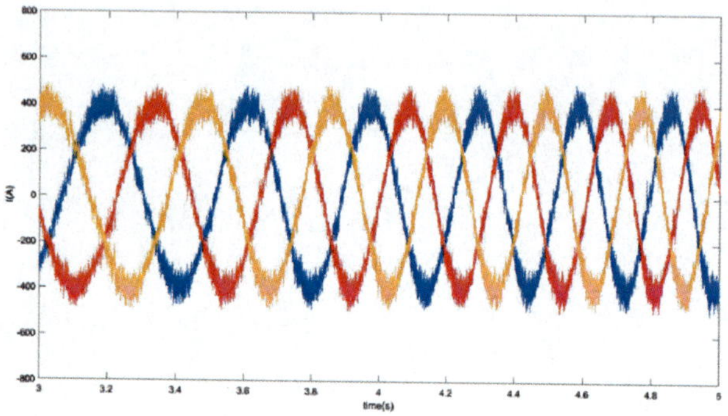

Fig. 7. Output current of another converter (4:6 current distribution ratios)

Acknowledgments. This work is supported by major project in Shanxi Province: Full-scale experimental line of ultra-high-speed low-vacuum pipeline maglev transportation system (Phase I).

References

1. Culianu, C.A., Christini, D.J.: Real-time Linux experiment interface system: RTLab. In: 2003 IEEE 29th Annual Proceedings of Bioengineering Conference, pp. 51–52, Newark, NJ, USA (2003)
2. Dong, W., Zhang, X., Jing, L., Wu, W.: Design of rapid-control-prototype platform for modular multilevel converter based on RT-lab. In: 2019 IEEE 10th International Symposium on Power Electronics for Distributed Generation Systems (PEDG), pp. 94–98, Xi'an, China (2019)
3. Liang, C., Mengmeng, J., Qiang, H., Xiaodong, Y., Fei, L.: Engineering simulation analysis and demonstration application of multi-terminal DC distribution system. In: 2018 International Conference on Power System Technology (POWERCON), pp. 2343–2349, Guangzhou, China (2018)
4. Dou, C., Zhang, Z., Yue, D., Zheng, Y.: MAS-based hierarchical distributed coordinate control strategy of virtual power source voltage in low-voltage microgrid. IEEE Access **5**, 11381–11390 (2017)
5. Zhang, X., et al.: Electromagnetic transient modeling and verification of DFIG unit and wind farm based on RTLAB. In: 2022 IEEE 5th International Electrical and Energy Conference (CIEEC), pp. 302–307, Nangjing, China (2022)
6. Gao, S., Zhao, H., Wang, P., Gui, Y., Terzija, V., Blaabjerg, F.: Comparative study of symmetrical controlled grid-connected inverters. IEEE Trans. Power Electron. **37**(4), 3954–3968 (2022)
7. Dai, N., Lao, K., Wong, M.: A hybrid railway power conditioner for traction power supply system. In: 2013 Twenty-Eighth Annual IEEE Applied Power Electronics Conference and Exposition (APEC), pp. 1326–1331, Long Beach, CA, USA (2013)
8. Deng, C., Zhou, L., Yan, M., Han, J., Zhong, Y., Li, W.: Modeling and force fluctuation analysis of open winding PMLSM driven by dual-terminal. In: 2021 13th International Symposium on Linear Drives for Industry Applications (LDIA), pp. 1–4, Wuhan, China (2021)

9. Ji, L., Zhang, Y., Shi, R., Jia, L., Zhang, X.: A scheduling strategy for a new energy highway integrated network with clean green energy synergy. Smart Resil. Transp. **4**(2), 173–191 (2022)
10. Zhang, X., He, G., Wang, J., Chen, D.: New energy cluster and large-scale power grid co-simulation application based on RTLAB and CloudPSS. In: 2023 IEEE 6th International Electrical and Energy Conference (CIEEC), pp. 699–703, Hefei, China (2023)

Research on Maglev Pipeline Trains Speed and Positioning Based on Improved Kalman Filtering Algorithm

Jusong Jiang[1,2], Jie Yang[1,2](✉), Chuanshu Meng[3], and Zhixin Li[1,2]

[1] Jiangxi University of Science and Technology, Ganzhou 341000, China
yangjie@jxust.edu.cn
[2] Key Laboratory of Maglev Technology of Jiangxi Province, Ganzhou 341000, China
[3] China Academy of Railway Sciences Co., Ltd., Beijing 100081, China

Abstract. This study presents the A novel Adaptive Kalman Filter Algorithm (APRKF) to address system noise, measurement noise, and sensor anomalies arising from train vibrations in multi-sensor fusion for maglev pipeline trains velocity and positioning. The algorithm focuses on rapid train vibration changes, estimating actual innovation sequence covariance using a constrained limited-memory approach. An adaptive factor is derived from the ratio of actual to theoretical innovation covariance. Further, an anomaly detection threshold restricts adaptive factor variations, constraining the scaled covariance matrix size and eliminating abnormal data. The adaptive factor dynamically corrects measurement noise and predicted state covariance in real-time, enhancing noise adaptability and mitigating filter divergence risk. This approach effectively mitigates the impact of sensor data distortion and noise variations on system state estimation. Simulation reveals that APRKF outperforms traditional Kalman Filter (KF) and Sage-Husa Adaptive Filter (SHKF) in velocity estimation, closely approximating true values, reducing oscillations, and efficiently filtering abnormal data to converge rapidly. For position estimation, APRKF excels by effectively reducing integration-induced errors. Comparative analysis demonstrates the superior accuracy and stability of APRKF in velocity and position estimation through standard deviation, mean squared error, and mean value metrics.

Keywords: Pipeline trains · Multi-sensor fusion · Adaptive Kalman filtering

1 Introduction

The permanent magnet maglev pipeline logistics system [1] employs permanent magnet levitation for zero-power suspension within underground comprehensive pipe corridors. This eco-efficient and secure system addresses urban logistics, industrial parks, and pharmaceutical transportation, minimizing surface space usage, relieving traffic congestion, and lowering operational expenses.

Train speed and positioning technology is a key element of train control systems [2]. With increasing demands for rail transit safety and efficiency, relying solely on a single sensor for train speed and positioning falls short. Multi-sensor fusion [3] speed and

L. Jia et al. (Eds.): EITRT 2023, LNEE 1135, pp. 538–547, 2024.
https://doi.org/10.1007/978-981-99-9307-9_59

positioning technology, achieved through a judicious sensor application architecture, can compensate for the insufficient speed measurement accuracy of a single sensor. Zhang et al. [4] employed multi-sensor fusion technology to address wheel slip in pipeline positioning systems. However, during train operation, intermittent impact vibrations arising from uneven tracks, pipeline joints, and vehicle structures lead to sporadic shock vibrations, resulting in system noise, measurement noise, and sensor data anomalies. Consequently, a noise-adaptive-capable filter is required. Kalman filtering updates and optimally estimates system states by analyzing input data with noise and disturbances [5]. Literature [6] classifies adaptive Kalman filtering algorithms into proportional scaling state covariance matrix, multi-model adaptive estimation, and adaptive stochastic model modeling. Among them, the proportional scaling system noise covariance matrix algorithm exhibits robustness with fewer tuning parameters, but faces the issue of being unable to adjust the variance of a single noise. However, direct use of the Sage-Husa algorithm often leads to filter divergence [7]. Zhou et al. [8] introduced an adaptive fading Sage-Husa extended Kalman filter algorithm, adaptively adjusting measurement noise covariance, but overlooked the influence of process noise on the system. In practical applications, to avoid filter divergence, a simplified Sage-Husa filter is typically employed, assuming process noise stability and no longer updating its mean and variance [9]. However, the inevitable impact vibrations encountered by maglev pipeline trains also affect accelerometers, causing abnormal measurements in both accelerometers and speed sensors. Adopting a simplified Sage-Husa filter by merely assuming constant process noise and handling contradicts reality. Therefore, a novel noise-adaptive filtering method needs to be explored.

In summary, this paper addresses the issues of system noise, measurement noise, and sensor data anomalies caused by train vibration in the context of multi-sensor fusion speed and positioning for maglev pipeline trains. Building upon the combination of Doppler radar and accelerometers for positioning, a novel Adaptive Kalman Filter (APRKF) algorithm is proposed. Firstly, the algorithm addresses the rapid temporal variations of train vibrations by employing a limited-memory weighted method to estimate the actual innovation sequence covariance. The ratio of the actual innovation covariance to its theoretical value is used as the adaptive factor. Subsequently, a threshold for abnormal data detection is established to constrain the adaptive factor's variation range and eliminate outliers. Finally, the adaptive factor is utilized to real-time modify the measurement noise covariance and predicted state covariance, thereby indirectly adjusting system noise and measurement noise. Through comparison with traditional Kalman filtering, Sage-Husa adaptive filtering, and the proposed APRKF using simulation data, the advantages of the APRKF algorithm in enhancing filtering reliability, accuracy, and adaptability are validated.

2 State-Space Model

This paper employs Doppler radar and accelerometers to measure the speed and acceleration of the train. Doppler radar, based on the Doppler effect [10], enables contactless speed measurement [11], unaffected by the unique environment of underground pipelines. Accelerometers are electronic devices capable of measuring acceleration forces [12]. Based on adjacent sampled data, the maglev pipeline trains can be treated as an inertial body under constant velocity or acceleration [13]. The Doppler radar is mounted on the top of the train body, while the accelerometers are installed inside the train body. as shown in Fig. 1.

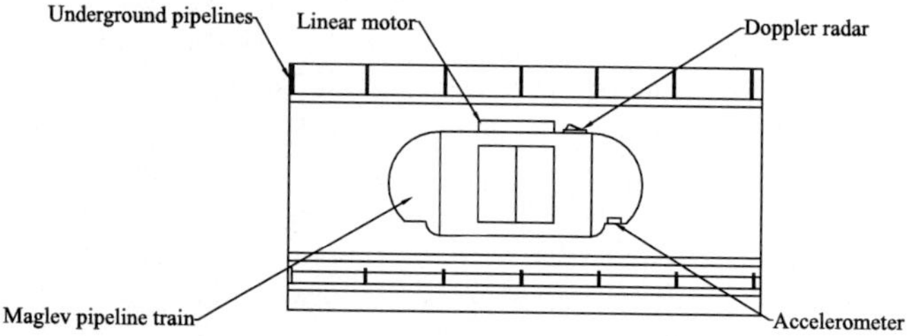

Fig. 1. Schematic diagram of speed and position measurement for Maglev pipeline train

Given an initial velocity of $v_0 = 0$ m/s, acceleration of a, and a sampling time of Δt, the kinematic equation at time instant k is expressed as follows:

$$\begin{cases} V_k = V_{k-1} + a\Delta t \\ S_k = S_{k-1} + \frac{1}{2}a\Delta t^2(2k-1) \\ t_k = \Delta t \cdot k \end{cases} \tag{1}$$

Constructing the state equation with train velocity v and displacement s as system state variables, using acceleration measurements from the accelerometer as the input, the state equation is formulated as follows:

$$X_k = AX_{k-1} + BU + W_{k-1} \tag{2}$$

In the equation, X_k represents the two-dimensional state vector, $X_k = \begin{bmatrix} v_k & s_k \end{bmatrix}^T$, with A being the state transition matrix and B the control matrix, $A = \begin{bmatrix} 1 & 0 \\ \Delta t & 1 \end{bmatrix}$, $B = \begin{bmatrix} \Delta t & \frac{1}{2}\Delta t^2 \end{bmatrix}^T$. U denotes the control input, specifically the accelerometer measurement, while W_{k-1} signifies the process noise.

Construct the measurement equation using Doppler radar velocity information:

$$Z_k = HX_k + V_k \tag{3}$$

H is the measurement matrix, $H = \begin{bmatrix} 1 & 0 \\ 0 & 1 \end{bmatrix}$, and V_k is the measurement noise. Where W_{k-1} and V_k are uncorrelated white noises with mean values and variances: $E\{W_k\} = 0$, $E\left\{W_k W_j^T\right\} = Q_k$, $E\{V_k\} = 0$, $E\left\{V_k V_j^T\right\} = R_k$, $\text{cov}\{W_i V_j\} = 0$.

3 Method for Speed and Position Estimation System

3.1 Kalman Filter

The Kalman filtering algorithm consists of two steps: prediction and update. Prediction involves forecasting the state using prior information, while update combines measurement data and prior estimates to compute the posterior state estimate. The formulas are given below:

$$\hat{X}_{k,k-1} = A_k\hat{X}_{k-1} + B_k U_k \tag{5}$$

$$P_{k,k-1} = A_k P_{k-1} A_k^T + Q_k \tag{6}$$

$$K_k = P_{k,k-1} H_k^T \Big[H_k P_{k,k-1} H_k^T + R_k \Big]^{-1} \tag{7}$$

$$v_k = Z_k - H_k \hat{X}_{k,k-1} \tag{8}$$

$$\hat{X}_k = \hat{X}_{k,k-1} + K_k \upsilon_k \tag{9}$$

$$P_k = (I - K_k H_k) P_{k,k-1} \tag{10}$$

wherein, $\hat{X}_{k,k-1}$ represents the predicted state; $P_{k,k-1}$ is the predicted error covariance; K_k stands for the Kalman gain; v_k represents the innovation sequence; $\hat{X}_k$ is the updated state; P_k denotes the updated error covariance.

3.2 Adaptive Kalman Filtering Algorithm with Novel Approach

The theoretical variance of the innovation sequence is:

$$C_{\upsilon_k} = E\Big[\upsilon_k \upsilon_k^T\Big] = H_k\Big(A_k P_{k-1} A_k^T + Q_{k-1}\Big) H_k^T + R_k \tag{11}$$

Commonly used fading memory exponential weighting method is employed to estimate the actual innovation sequence variance, as follows:

$$\hat{C}_{\upsilon_k} = (1 - d_k)\hat{C}_{\upsilon_{k-1}} + d_k \upsilon_k \upsilon_k^T \tag{12}$$

When noise statistics change rapidly, outdated data provide little significant help to the current noise statistical estimation [14]. However, the variation in sensor measurement noise caused by the vibration of trains exhibits a fast time-varying characteristic, and recent historical data have a greater impact on the current noise statistical truth. Therefore, the limited-memory weighted method is introduced, controlling the historical data used for the current noise statistical estimation through a sliding window, and applying weighted processing only to a fixed preceding historical data segment. The estimated variance of the actual innovation sequence is obtained as follows:

$$\hat{C}_{\upsilon_k}=\frac{1}{N}\sum_{j=j_0}^{k}\upsilon_j\upsilon_j^T \quad (13)$$

In the equation, N represents the sliding window length, and $j_0 = k - N + 1$ denotes the first index of the correlated innovation sequence within the window.

In theory, C_{υ_k} and $\hat{C}_{\upsilon_k}$ should be consistent, approximated as follows:

$$\hat{C}_{\upsilon_k} \approx H_k\left(A_kP_{k-1}A_k^T + Q_{k-1}\right)H_k^T + R_k \quad (14)$$

For practical systems, if inaccurate modeling leads to deviations in system structural parameters $A_k,\ H_k,\ B_k$ or noise parameters $Q_k,\ R_k$, it may disrupt the balance of the equation above to some extent. In such cases, it is considered that the measurement covariance matrix R is improperly set, and its size should be dynamically adjusted in real-time to maintain the validity of the equation, as follows:

$$\hat{C}_{\upsilon_k} - H_k\left(A_kP_{k-1}A_k^T + Q_{k-1}\right)H_k^T = \lambda_k^R R_k \quad (15)$$

Let $N_k = H_k(A_kP_{k-1}A_k^T + Q_{k-1})H_k^T$, and define the time-varying scaling factor λ_k^R:

$$\lambda_k^R = \frac{tr\left(\hat{C}_{\upsilon_k} - N_k\right)}{tr(R_k)} \quad (16)$$

Replace the modified $\lambda_k^R R_k$ with R_k for filtering update, achieving adaptive adjustment of measurement noise.

Sequential processing can also be employed for a more targeted adaptation of each individual measurement noise component. The formula is as follows:

$$\left(\lambda_k^R\right)_j = \frac{\left(\hat{C}_{\upsilon_k} - H_k(A_kP_{k-1}A_k^T + Q_{k-1})H_k^T\right)_{jj}}{(R_k)_{jj}}, j = 1, 2..., n \quad (17)$$

To enhance the stability of the filtering process and mitigate the risk of filter divergence resulting from simultaneous adaptive adjustments of process noise Q and measurement noise R, a more robust approach is adopted in this study. Specifically, instead of independently adapting process noise Q, a time-varying scaling factor λ_k^P is introduced to indirectly adjust process noise Q by modifying the covariance of the predicted state P (as process noise Q is related to P). This approach enhances the adaptive capability of the filter to handle noise while simultaneously reducing the risk of filter divergence. The formula for calculating the time-varying scaling factor λ_k^P is as follows:

$$\left(\lambda_k^P\right)_i = \frac{tr\left(\widehat{C}_{V_k}\right)}{tr(C_{V_k})} \quad (18)$$

The modified $\lambda_k^P P_{k,k-1}$ is substituted for $P_{k,k-1}$ in the filtering update process to achieve adaptive adjustment of the predicted information.

Maglev pipeline trains often encounter the issue of anomalous sensor measurement data in complex conditions such as pipeline joints, vehicle turning, and track water accumulation. A threshold value F is set for anomaly data determination. When the absolute difference f between the measured value and the predicted system state value exceeds the threshold value F, the measured data is considered anomalous. The magnitudes of the proportional factors A and B are then constrained, with each component of λ_k^P and λ_k^R restricted within the bounds of $[\lambda_{\min}, \lambda_{\max}]$, $\lambda_{min} \in [0, 1]$, and $\lambda_{\min} \cdot \lambda_{\max} = 1$ (Fig. 2).

$$f = \left| Z_k - \hat{X}_{k,k-1} \right| \tag{19}$$

$$\lambda_k^{(j)} = \begin{cases} \lambda_{\min}, \lambda_k^{(j)} < \lambda_{\min} \\ \lambda_k^{(j)}, \lambda_{\min} < \lambda_k^{(j)} < \lambda_{\max} \\ \lambda_{\max}, \lambda_k^{(j)} > \lambda_{\max} \end{cases} \tag{20}$$

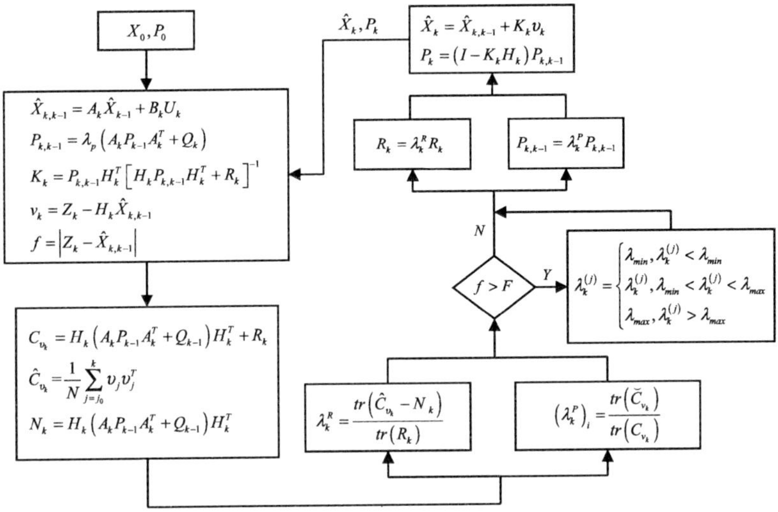

Fig. 2. Flowchart of the novel adaptive Kalman filtering algorithm

4 Simulation Verification

4.1 Simulation Parameter Configuration

To verify the efficacy of the algorithm, this study employs the PSINS integrated navigation toolbox in Matlab to design the operational states of the maglev pipeline trains. The entire process is categorized into uniform acceleration, crossing seam, and turning stages. The Doppler radar and accelerometer output frequencies are set at 20 Hz and 100 Hz, respectively. The measurement parameters during the train's operation are presented in Table 1.

Table 1. Sensor noise parameters

State	Sensor type	Noise values (m/s)
Constant variable acceleration	Doppler radar	0.025
	Accelerometer	0.500
Passing Pipe Seams	Doppler radar	30 × 0.025
	Accelerometer	30 × 0.500
Turning	Doppler radar	20 × 0.025
	Accelerometer	0.500

4.2 Simulation Results and Analysis

Based on the simulation parameter settings in Sect. 4.1, the performance of the KF, SHKF, and APRKF algorithms in velocity estimation is illustrated in Fig. 3.

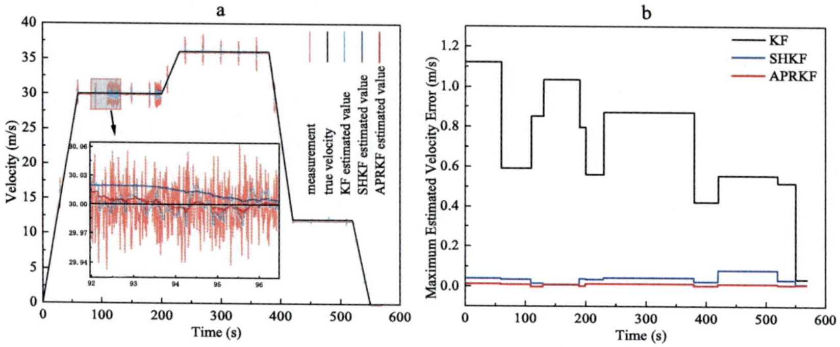

Fig. 3. Comparison of velocity estimation

Figure 3 illustrates a comparison of velocity estimates among three algorithms. The APRKF algorithm closely approximates the true velocity value throughout the entire motion, exhibiting reduced curve amplitudes and avoiding pronounced fluctuations. The continuous outliers at 110 and 190 s in the graph represent simulated measurement anomalies arising from the train's turning process, while other outliers simulate measurement anomalies caused by the train passing over pipe joints and encountering water accumulation on the tracks. The APRKF algorithm demonstrates smaller errors when confronted with measurement anomalies, implying enhanced adaptability in complex environments and improved accuracy in velocity estimation. Additionally, Fig. 3b quantitatively analyzes the maximum estimation error during different motion phases. The results reveal that the APRKF algorithm consistently yields the smallest maximum errors across various stages, further substantiating its superior performance in velocity estimation.

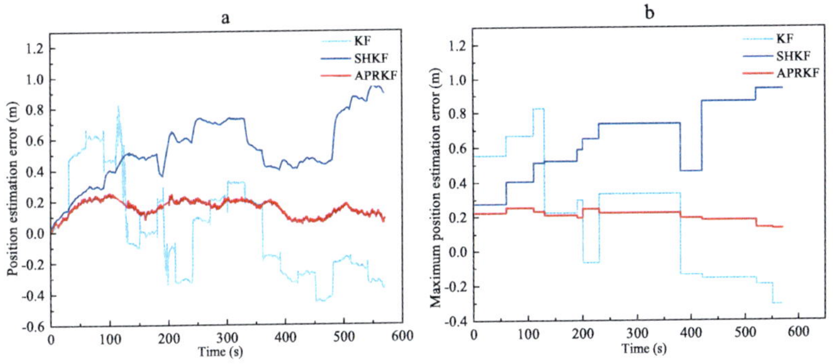

Fig. 4. Position estimation error

In Fig. 4, we analyze the performance of the APRKF algorithm in position estimation. KF and SHKF algorithms exhibit rising position estimation errors due to error accumulation during velocity integration, compounded by initial errors, noise, and disturbances. In contrast, the APRKF algorithm incorporates an adaptive factor to efficiently manage system and measurement noise, effectively suppressing error accumulation. This leads to substantially reduced position estimation errors, enabling more accurate and precise position estimation.

Table 2 provides a comparative analysis of velocity and position estimation errors among the KF, SHKF, and APRKF algorithms. The APRKF algorithm demonstrates the lowest values for standard deviation, mean squared error, and mean, underscoring its enhanced accuracy and stability in both velocity and position estimation. Notably, the APRKF algorithm exhibits a pronounced advantage over the SHKF algorithm, particularly in the context of position estimation.

Table 2. Comparison of velocity and position estimation error statistics

Type	Algorithm	Standard deviation	Mean squared error	Average
Velocity estimation error	KF	0.0881	0.0881	−0.0006
	SHKF	0.0146	0.0149	0.0028
	APRKF	0.0044	0.0044	0.0001
Position estimation error	KF	0.3134	0.3138	0.0168
	SHKF	0.2078	0.5712	0.5320
	APRKF	0.0534	0.1634	0.1545

5 Conclusion

In this paper, we address the challenges posed by train vibration-induced system noise, measurement noise, and sensor data anomalies in the multi-sensor fusion speed and position estimation of maglev pipeline trains systems. We delve into the performance optimization and adaptive adjustment mechanisms of the Kalman filter algorithm. Through comparative simulations, we conduct an in-depth analysis of the performance of three algorithms: KF, SHKF, and APRKF, in terms of velocity and position estimation. Simulation results demonstrate that the APRKF algorithm provides velocity estimates that are closer to the true values, particularly excelling in handling measurement anomalies. Throughout various stages of train operation, APRKF exhibits the smallest maximum velocity estimation errors. Moreover, APRKF yields significantly lower position estimation errors compared to the KF and SHKF algorithms. This improvement is attributed to APRKF's adaptive factor that dynamically adjusts system and measurement noise, effectively mitigating error accumulation and achieving more precise position estimation. These outcomes underscore the adaptability and robustness of the APRKF algorithm. Considering comprehensive statistics of velocity and position estimation errors, the APRKF algorithm displays the lowest standard deviation, mean squared error, and mean values, reflecting heightened accuracy and stability. This highlights the algorithm's capacity to accurately capture the train's motion state, making it suitable for diverse operational environments and complex scenarios. In conclusion, the results of this study establish the significant advantages of the APRKF algorithm in the multi-sensor fusion speed and position estimation of maglev pipeline trains systems. Its adaptive adjustment mechanism enhances estimation accuracy and stability, providing substantial support for practical engineering applications. Future research may further explore the applicability of the algorithm under various operating conditions while optimizing real-time performance and computational efficiency.

Acknowledgments. This research is supported by Research and Development Fund Key Project of China Academy of Railway Sciences Co., Ltd. (2022YJ077).

References

1. Cheng, L., Yang, J., Feng, F.: Research on intelligent velocity measurement algorithm of maglev pipeline logistics system. J. Railw. Sci. Eng. 1–11 (2023). https://doi.org/10.19713/j.cnki.43-1423/u.T20222288. (in Chinese)
2. Otegui, J., Bahillo, A., Lopetegi, I., Diez, L.E.: A survey of train positioning solutions. IEEE Sens. J. **17**(20), 6788–6797 (2017). https://doi.org/10.1109/JSEN.2017.2747137
3. Li, W., Wei, S.G., Cai, B.G., Wang, J.: Signal simulation of the train integrated positioning system. In: 2013 International Conference on Materials, Mechatronics and Automation, ICMMA, vol. 740, pp. 86–91. Trans Tech Publications Ltd., Singapore (2013)
4. Zhang, Q., Wu, L., Li, Q.: Multi-sensor pipeline location method based on federated filter. Navig. Position. Timing. **9**(01), 111–118 (2022). (in Chinese)
5. Deng, C.Y.: Research on Key Technologies of Precise Location for BDS/INS Integrated Navigation. Hefei Polytechnic University, Master (2019). (in Chinese)
6. Ding, W.D., Wang, J.L., Rizos, C., Kinlyside, D.: Improving adaptive Kalman estimation in GPS/INS integration. J. Navig. **60**(3), 517–529 (2007)
7. Zhou, M.M., Zhang, B., Zhao, Q., Pan, M.T., Zuo, S.Y.: Cooperative localization algorithm based on adaptive fadingSage-Husa extended Kalman filter. Chin. J. Ship Res. **17**(04), 92–97 (2022). (in Chinese)
8. Sun, X., Zhang, J.L., Zhang, Y.S., Zhou, Y.B.: An improved Sage-Husa adaptive tracking filter. Electron. Inf. Warfare Technol. **35**(06), 37–41 (2020). (in Chinese)
9. Wu, A.P., Lu, Y.F., Guo, Z., Hou, Z.X.: Three-dimensional localization for moving target using modified Sage-Husa adaptive filter. J. Natl. Univ. Defense Technol. **45**(02), 146–154 (2023). (in Chinese)
10. Jung, J., Lim, S., Lee, E., Heo, M.O., Kim, S.C.: Localization based on distributed heterogeneous radar sensor network. Measurement **204** (2022)
11. Shariff, K. K. M., Hoare, E., Daniel, L., Antoniou, M., Cherniakov, M.: Comparison of adaptive spectral estimation for vehicle speed measurement with radar sensors. Sensors **17**(4) (2017)
12. Cai, X., Wang, C.L., Lin, Y.: Train idling/sliding detection and adjust method based on multiple sensors. Urban Mass Transit. **18**(01), 22–27 (2015). (in Chinese)
13. Cheng, S., Liu, C., Song, L.W.: Maglev train integrated positioning and speed measuring method based on multi-source information fusion. Urban Mass Transit. **25**(08), 136–140 (2022). (in Chinese)
14. Zhao, L., Wang, X.X., Sun, M., Ding, J.C., Yan, C.: Adaptive UKF filtering algorithm based on maximum a posterior estimation and exponential weighting. Acta Autom. Sinical **36**(07), 1007–1019 (2010). (in Chinese)

Dynamic Trajectory Planning Method on ARM Platform for the Vacuum Pipelines High-Speed Maglev Train

Yongyong Guo(✉), Zhihong Yang, Kekang Wei, and Zhiyong Zhao

China Aerospace Science and Industry Co., Ltd., Beijing, China
guoyong10@126.com, yzhm1516@126.com, weikekang@163.com, zhaozhiyong_bj@163.com

Abstract. Dynamic trajectory planning is the key function of Automatic Train Operation System for the vacuum pipelines high-speed maglev train. In order to solve the problem of accelerating process control, the train's running state is analyzed. During the train's operation, the train's dynamical model is necessary to be established based on the traction, braking force, and resistance. Under the constraints of protection curve, jump, and optimal time, the speed, acceleration and distance curve is adjusted. The traction control system operates based on acceleration feedforward and current speed control error, ultimately achieving precise control of trajectory operation for the vacuum pipeline high-speed maglev trains.

Keywords: The vacuum pipeline high-speed maglev trains · Online planning · Dynamic trajectory planning · Automatic train operation system

1 Introduction

The vacuum pipeline high-speed maglev fills the speed gap between aircraft and high-speed rail, and plays an indispensable role in building a comprehensive transportation system in the full speed domain. In September 2019, the Central Committee of the Communist Party of China and the State Council issued the "Outline for the Construction of a Strong Transportation Country", which clearly proposed the "reasonable coordination and arrangement of technical reserve research and development for high-speed maglev systems with a speed of 600 km per hour, and high-speed wheel rail (including variable gauge) passenger train systems with a speed of 400 km per hour".

The Automatic Train Operation System is one of the important functions of the vacuum pipeline high-speed maglev traction control system, which can ensure the safe and stable operation of trains, and is also a necessary condition for achieving high-speed operation. Among them, dynamic planning of operating curves is an important segment [1–4], therefore, it is necessary to study the dynamic trajectory planning of The Automatic Train Operation System [5, 6].

L. Jia et al. (Eds.): EITRT 2023, LNEE 1135, pp. 548–555, 2024.
https://doi.org/10.1007/978-981-99-9307-9_60

The Automatic Train Control system (ATC) is the core part of Automatic Train Operation System for operation control. The ATC for the vacuum pipelines high-speed maglev trains includes three functions:

(1) The Automatic Train Protection system (ATP) sets the speed position with velocity limit boundary for train operation, and in order to ensure train safety, the train automatically takes emergency protection measures when detecting train overspeed or overtaking;
(2) The Automatic Train Supervision system (ATP) supervises the line and train status, and dynamically adjusts the train's operation and stopping time to ensure the timely arrival of the train;
(3) The Automatic Train Operation system (ATO) dynamically generates a running trajectory curve based on the ATP speed limit curve, and achieves the train operate control according to the trajectory curve.

The power system of the vacuum pipeline high-speed maglev is located on the ground, which determines its natural suitability for using ATO system for control. When the traction control system receives the protective speed limit curve issued by ATP, the ATO system generates a speed position curve, and the motor control system adjusts the train speed based on the speed position curve. The ATO system is an important component of the traction control subsystem, and its relationship diagram is shown in Fig. 1.

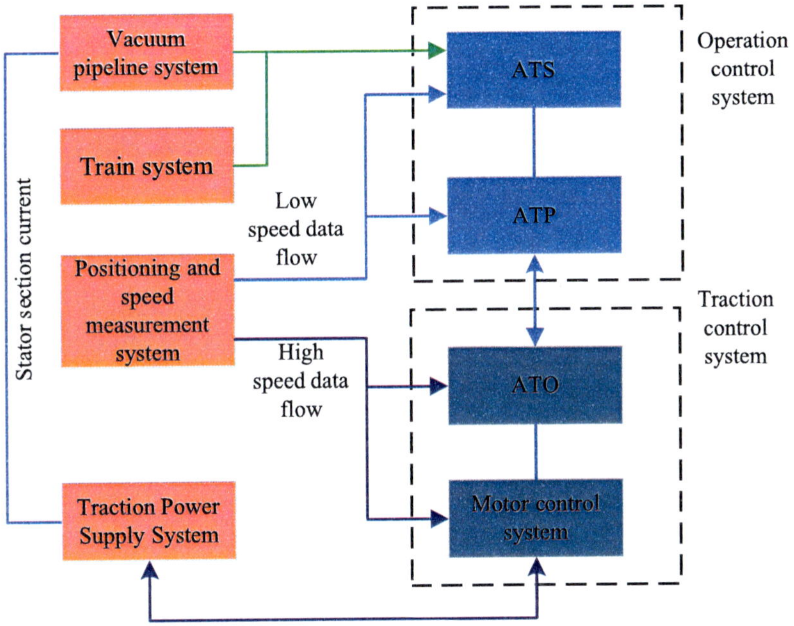

Fig. 1. The relationship diagram of high-speed maglev train ATO system

The speed of the vacuum pipeline high-speed maglev trains is fast, and precise control of the operation trajectory has a significant impact on the comfort, safety, and parking accuracy of trains. In response to this requirement, this article proposes a high-speed maglev train dynamic trajectory planning method based on the ARM platform. Firstly, a dynamical model of the train was established based on the traction force and resistance for operation process. Next, a 7-segment S-shaped curve planning method was introduced, and an efficient method for dynamically generating trajectory planning curves was proposed based on constraints such as optimal time, train performance data, line magnetic resistance, wind resistance, jump, etc. According to the position of the train, real-time acceleration and speed of the next position are issued. The motor controller uses acceleration as a closed-loop control feedforward parameter, and speed as the control target to adjust the speed of the train. Finally, a semi-physical simulation test and validation of the experimental line were conducted, demonstrating the efficiency and reliability of the algorithm.

2 Train Operation Model

Using finite element simulation and curve fitting, the motion model of the train is obtained, which includes three parts: the wind resistance, the magnetic resistance, and the traction force [7, 8]:

(1) The wind resistance is related to the atmospheric pressure inside the pipeline and the train speed, and its mathematical model is: $F_{air} = K_1 V^2$;
(2) The magnetic resistance is related to the maglev track and train speed,, and its mathematical model is: $F_{track} = K_2 \ln(V + 20) + K_4 e^{K_5 V}$;
(3) The traction force is related to the output current, train and track, and its mathematical model is: $F_{thrust} = K_6 I$;

Among them, V is the train running speed, $K_1, K_2, K_3, K_4, K_5, K_6$ is the correlation coefficient of curve fitting, and I is the output current of the traction converter

3 Trajectory Planning Method

3.1 Division of Speed Limit Intervals

According to the protective speed limit curve issued by ATP, it is divided into several speed limit sections according to different speed limit sections. The data contained in the speed limit section is shown in the Table 1.

3.2 Reverse Solution of Interval Maximum Limited Speed

Calculate from the last speed limit interval, that is, reverse estimating from the Terminal position issued by ATP.

Find interval appropriate v_e and $v_{\max}$, Based on the constraint of Jerk, $a_{\max}, v_{lim}, p_s, p_e$, use the S-curve programming method to solve the interval velocity, acceleration, position, etc. and parameters are as shown in Table 2.

Among them, the constraint conditions are as follows:

Table 1. The data contained in the speed limit section

Parameter	Describe	Datatype	Unit
p_s	Starting position	Uint32_t	cm
p_e	Terminal position	Uint32_t	cm
v_s	Starting speed	Uint16_t	cm/s
v_e	Terminal speed	Uint16_t	cm/s
v_{lim}	Maximum limited speed in the interval	Uint16_t	cm/s

Among them, e_pos in the previous interval e_pos is equal to the current interval, same as e_vel

Table 2. Obtained data in the speed limited intervals

Serial number	Intervals	Acceleration	Velocity	Position
1	Acceleration increase stage	Jt_1	$v_s + \frac{1}{2}Jt_1^2$	$v_s t_1 + \frac{1}{6}Jt_1^3$
2	Constant acceleration stage	$a_{\max}$	$v_1 + a_{\max}t_2$	$s_1 + v_1 t_2 + \frac{1}{2}Jt_1 t_2^2$
3	Acceleration reduction stage	$a_{max} - Jt_3$	$v_2 + a_{\max}t_3 - \frac{1}{2}Jt_3^2$	$s_2 + v_2 t_3 + \frac{1}{2}Jt_1 t_3^2 + \frac{1}{6}Jt_3^3$
4	Constant velocity stage	0	v_3	$s_3 + v_3 t_4$
5	Deceleration increase stage	$-Jt_5$	$v_4 - \frac{1}{2}Jt_5^2$	$s_4 + v_4 t_5 - \frac{1}{2}Jt_5^3$
6	Constant deceleration stage	$-a_{\max}$	$v_5 + a_{\max}t_6$	$s_5 + v_5 t_6 + \frac{1}{2}Jt_5 t_6^2$
7	Deceleration reduction stage	$-a_{\max} + Jt_7$	$v_2 + a_{\max}t_3 - \frac{1}{2}Jt_3^2$	$s_6 + v_6 t_7 - \frac{1}{2}Jt_5 t_7^2 + \frac{1}{6}Jt_7^3$

① $J < Jerk$;
② $a_{\max} < (F_{thrust} - F_{track-}F_{air})/M_{train}$;
③ $v_{\max} < v_{lim}$;
④ $s_1 + s_2 + s_3 + s_4 + s_5 + s_6 + s_7) < (p_e - p_s)$.

When $v_s < v_e$, The speed limit section is generally in the acceleration stage, which can be divided into two situations [9, 10]:

(1) When $v_e < v_{lim}$, explains that the train accelerates first and then decelerates. The distance is required to be calculated for the entire journey of the train. If the distance does not meet the requirements, within the interval of v_e and v_{lim}, Using the dichotomy method to solve the appropriate interval maximum velocity.

(2) When $v_e = v_{lim}$, explains that the train accelerates first and then travels at a constant speed. The distance is required to be calculated for the entire journey of the train. If the distance does not meet the requirements, within the interval of v_s and v_e, Using the dichotomy method to solve the appropriate interval maximum velocity.

When $v_s > v_e$, The speed limit section is generally in the deceleration stage, which can be divided into two situations:

(1) When $v_s < v_{lim}$, explains that the train accelerates first and then decelerates. The distance is required to be calculated for the entire journey of the train. If the distance does not meet the requirements, within the interval of v_s and v_{lim}, Using the dichotomy method to solve the appropriate interval maximum velocity.
(2) When $v_s = v_{lim}$, explains that the train accelerates first and then travels at a constant speed. The distance is required to be calculated for the entire journey of the train. If the distance does not meet the requirements, within the interval of v_s and v_e, Using the dichotomy method to solve the appropriate interval maximum velocity.

The ending speed of the current interval is equal to the starting speed of the next interval, and the starting speed, ending speed, and maximum operating speed of each interval are calculated in sequence.

3.3 Forward Calculation Verification

From the starting position of the train, calculating and verifying the starting speed, ending speed, and maximum operating speed of each speed limited interval in the forward direction. When the verification fails, recalculate the interval data using the method in Sect. 3.2.

3.4 Generate Trajectory Curve Sequence

According to the interval calculation data and the calculation method in Table 2, the acceleration, velocity, and position data sequences are calculated in chronological order with a time interval of 100 ms.

4 Simulation Test

4.1 Test Parameters

In the semi-physical simulation platform, the input parameters of ATP protection data are shown in Table 3. The traction acceleration limit value is set to 200 cm/s^2, the braking acceleration limit value is set to 200 cm/s^2, the starting speed of the train is 0 cm/s, the starting position is 0 cm, and Jerk is set to 0.7.

Table 3. Speed limit data from ATP

ID	Maximum speed (cm/s)	Line offset (cm)	Line ID	Absolute starting point position of the line (cm)
1	4000	0	0 × 690108	0
2	7000	80000	0 × 690109	200000
3	5000	40000	0 × 690110	400000
4	7000	70000	0 × 690111	600000
5	0	50000	0 × 690112	800000

4.2 Test Results

From the semi-physical simulation results is shown in Fig. 2, it can be seen that the acceleration increases or decreases linearly according to the value of Jerk, and the velocity presents a 7-segment S-shaped curve. This process takes 8ms. After the curve data sequence is generated, target acceleration, speed, and position data are sent to the traction system every 10 ms. The traction system utilizes a speed closed-loop strategy of acceleration feedforward, and the speed tracking curve is shown in Fig. 3. The results indicate that this method has low impact on train operation and precise speed control.

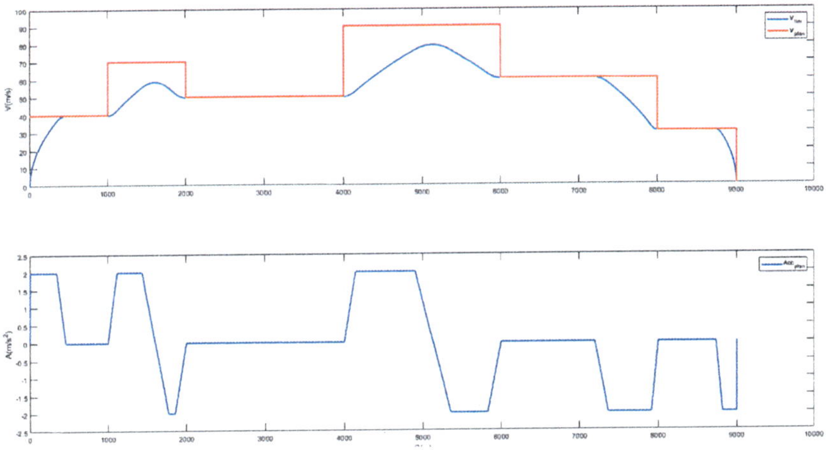

Fig. 2. Curve planning results

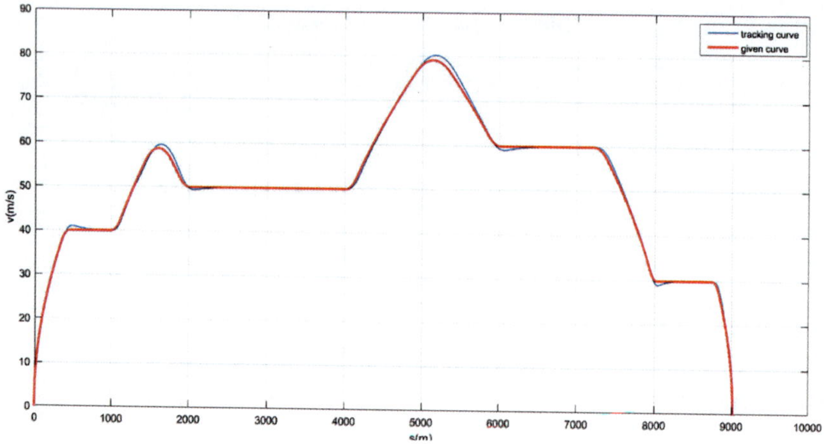

Fig. 3. Curve tracking process

5 Conclusion

In this paper, a dynamic trajectory planning method is proposed with Jerk constraints and the train's dynamical model constraints, and targeting for the Automatic Train Operation System precise control of the speed of the vacuum pipeline high-speed maglev trains. The experimental results show that the algorithm can complete curve trajectory calculation within milliseconds, and the trajectory control is precisely, under the constraints of protection curve, jump, and optimal time.

Acknowledgments. This work is supported by major project in Shanxi Province: Full-scale experimental line of ultra-high-speed low-vacuum pipeline maglev transportation system (Phase I).

References

1. Gao, B., Zheng, K., Zhang, F., Su, R., Zhang, J., Wu, Y.: Research on multi-target tracking method based on multi-sensor fusion. Smart Resil. Transp. **4**(2), 46–65 (2022)
2. Abudakka, F.J., Rubio, F., Valero, F., et al.: Evolutionary indirect approach to solving trajectory planning problem for industrial robots operating in workspaces with obstacles. Eur. J. Mech. A/Soilds **42**(6), 210–218 (2013)
3. Huang, J.S., Hu, P.F., Wu, K.Y., et al.: Optimal time-jerk trajectory planning for industrial robots. Mech. Mach. Theory **121**, 530–544 (2018)
4. Liu, S.: An on-line reference-trajectory generator for smooth motion of impulse-controlled industrial manipulators. In: 7th International Workshop on Advanced Motion Control, Proceedings. IEEE (2002)
5. Badr, E., Almotairi, S., Chamry, A.E.A.: Comparative study among new hybrid root finding algorithms and tradition-Al methods. Mathematics **9**(11), 1306 (2021)
6. Etesami, R., Madadi, M., Mashinchi, M., et al.: A new method for rooting nonlinear equations based on the bisection method. MethodsX **8**, 101502 (2021)

7. Hashemi, L., Mahmoodi, A., Jasemi, M., Millar, R.C., Laliberté, J.: Modeling a robust multi-objective locating-routing problem with bounded delivery time using meta-heuristic algorithms. Smart Resil. Transp. **3**(3), 283–303 (2021)
8. Yang, Y., Pan, J., Wan, W.: Survey of optimal motion planning. IET Cyber-Syst. Robot. **1**(1), 13–19 (2019)
9. An, Y.J., Chen, X.H., Li, Y.H., et al.: An improved non-dominated sorting biogeography-based optimization algorithm for the (hybrid) multi-objective flexible job-shop scheduling problem. Appl. Soft Comput. **99**, 106869 (2021)
10. Liu, M., Yao, X.F., Li, Y.X.: Hybrid whale optimization algo-rithm enhanced with Levy flight and differential evolution for job shop scheduling problems. Appl. Soft Comput. **87**, 105954 (2020)

The Top-Level Scheme Analysis of Medium Speed Maglev Operation Control System

Kuanxin Li(✉), Xinghu Chen, Yuefeng Liu, and Yonggang Wang

CRRC Tangshan Co., Ltd, Tangshan 064000, China
likuanxin@tangche.com

Abstract. The Operation control system plays a key role in automatic train operation control and safety protection in the whole traffic system. Its external interface is numerous. The top-level scheme of the operation control system, such as the top-level architecture, top-level system interface, and system functions, is directly related to the speed index of the traffic system and some key systems. Starting from the maglev traffic system, this paper expounds the application scenario of the medium speed maglev operation control system. And it analyzes the characteristics and distribution of the external system which is closely related to the operation control system by means of comparison. Finally, a top-level scheme of the operation control system suitable for the medium speed maglev traffic system is proposed, which can provide technical input for the design of the operation control system subsystem.

Keywords: Maglev transportation · Medium speed maglev system · Operation control system · System architecture

1 Introduction

The medium speed maglev transportation system is based on the "13th Five-Year Plan" national project "Maglev system Key Technology research" research system, the system adopts the "permanent magnet synchronous intermediate drive +F rail maglev" technical route. As the core system, the operation control system needs to cooperate with other systems in the traffic system to complete the train operation control, safety protection and scheduling management. Maglev traffic system analysis is directly related to whether the input materials of the overall technical scheme of the operation control system are sufficient and feasible. Therefore, it is necessary to make a comparative analysis of the maglev traffic system, and comprehensively consider the characteristics of the key systems, and then put forward a top-level scheme suitable for medium-speed maglev. Compared with the low-speed maglev, the increase in speed will expand the operation scenario of the medium-speed maglev traffic system. From the low-speed maglev urban rail transit to intercity rail transit, compared with the high-speed maglev technology route. In order to propose the operation control system architecture which is suitable for the medium speed maglev traffic system, it is necessary to make a comparative analysis of the maglev traffic system. The differences are in technical routes and speed

L. Jia et al. (Eds.): EITRT 2023, LNEE 1135, pp. 556–563, 2024.
https://doi.org/10.1007/978-981-99-9307-9_61

indicators, key systems and key technologies used in maglev traffic systems, that will lead to differences in the top-level structure, composition, interface and functional constraints, early warning Settings and control strategies of the operation control system, so these problems are analyzed.

2 Maglev Traffic System Problem Analysis

2.1 Key System Analysis

The interface and function design of the operation control system mainly depends on the index, key system layout and key technology that have functional coupling relationship with the operation control in the traffic system. The system related to the operation control system includes vehicle, traction system, brake system, positioning speed measurement system, switch line system and wireless communication system.

(1) Traction system
In the low-speed maglev traffic system, the linear short stator linear motor is used to achieve traction, and the induction plate is located on the ground. The traction system in this way is located on the vehicle, and the interface of traction control is located on the vehicle. High-speed maglev adopts long stator linear motor, and long stator cable with iron core is laid on the ground track. So traction system traction/power supply is considered by the ground system, and traction control/power supply interface is located on the ground. The long stator hollow coil [1] used in the medium speed maglev system is laid on the ground track, which means that the traction control and power supply are provided by the ground system. So the interface between the operation control system and traction control/power supply is located on the ground.

(2) Braking system
The braking system of low-speed maglev and high-speed maglev traffic are located in the vehicle, But the braking technology used are different. The common braking system of medium-speed maglev is borne by the ground traction system. And the emergency braking system is borne by the vehicle braking system. So the operation control system has emergency braking control in the vehicle, And the common braking control is performed on the ground.

(3) Positioning speed measurement system
Although the low speed maglev and high speed maglev [2] adopt different positioning speed measurement technology, the positioning speed measurement system is arranged on the vehicle. The positioning speed measurement technology used by the medium speed maglev train is different from them, and the positioning speed measurement system is located next to the track and the ground rail. So the control interface of the medium speed maglev operation control system and the positioning speed measurement system are located on the ground, and the interface form and data protocol are the focus of the system design.

(4) Switch system
The switch system is related to the track laying equipment and its performance requirements, but the switch system is located on the ground. The interfaces between the switch system and the operation control system are also located on the ground.

Low speed maglev adopts side wall lifting switch or multi-joint guide rail crossing switch. High-speed maglev adopts elastic bending switches with steel beams. Considering the characteristics of the medium speed maglev line, in order to avoid the engineering problems of the sidewall lifting type switch and the multi-joint guide type switch in dealing with the track traction cable and speed measurement coil, the medium speed maglev system chooses the elastic bending type switch of steel beam, and the interface scheme can be improved on the basis of the original high speed maglev.

The main differences between medium-speed maglev and low-speed maglev are as follows:

(1) The improvement of the operating speed
(2) The traction system is transferred from the vehicle to the ground
(3) The positioning speed measurement is transferred from the vehicle to the ground
(4) The common braking is performed by the ground traction system
(5) The technology of key systems is different

The main differences between medium-speed maglev and high-speed maglev are as follows:

(1) The operating speed is reduced
(2) The central type hollow long stator traction
(3) The positioning speed measurement is transferred from the vehicle to the ground
(4) The technology of key systems is different
(5) Commonly used brake transfer to the ground

According to the above analysis, the key system technology adopted by the medium-speed maglev traffic are different from that of the low-speed maglev and high-speed maglev, but the system structure beam is closer to the high-speed maglev, and the external interface of the operation control system is closer to the high-speed maglev. The analysis of the plastic ducking with the high-speed maglev will be more beneficial to the formation of the top degree scheme.

2.2 Functional Constraint Analysis

Due to the strict delay requirement of traction control system for train positioning data, the delay requirement of high speed maglev based on vehicle positioning to vehicle-ground communication is relatively high. When the positioning system of medium speed maglev system is moved to the ground, the positioning data does not need to be transmitted by vehicle-ground communication, but is directly transmitted to the traction control system and the ground operation control subsystem. The ground operation control subsystem can directly control the traction, thus reducing the delay requirement for vehicle-ground communication.

The traction system of high-speed maglev has an electromagnetic coupling relationship with the suspension system, and the vehicle can only control the opening and closing of the door in the state of stopping and dropping off. The medium-speed maglev can realize the electromagnetic decoupling of traction and suspension [3], which allows the vehicle to control the door in the two states of suspension and dropping off.

The operation control system should consider the fault handling capability of the key system with new technology and the setting of auxiliary parking area, which directly affects the train operation control and safety protection area controls.

2.3 Data Transmission Flow and Control Process Analysis

Because the train positioning speed measurement system is moved from the vehicle to the ground, the direction of the positioning data will be changed. Originally, the vehicle transport control system receives the speed and position information of the positioning system, and transmits the speed and position information to the partition operation control system through the vehicle-ground communication system. The medium speed maglev system will receive the speed and position information of the positioning system by the divisional operation control system, and transmit it to the vehicle transport control system through the vehicle-ground communication system.

Changes in data flow will result in changes in some control processes. When the positioning speed measurement system is located in the vehicle, the vehicle transport control system receives the positioning data before the partition operation control system. The vehicle transport control system will find the speeding situation before the partition operation control system and first requests the partition operation control system to cut off the traction power supply. Then the vehicle transport control system start the vehicle brake after confirming the traction is cut off. The partition operation control system of medium speed maglev system will find speeding before the vehicle control system. It is the main undertaker of the train speed curve monitoring and it will control the traction system to implement electric braking. When the traction fault is not controlled, It will cut off the traction power supply and require the vehicle control system to start emergency braking.

2.4 Parameters Related to Maglev Transportation System

In order to give a relatively specific top-level scheme of the operation control system, the following parameters should also be provided, such as the actual operating environment of the train, the natural environment of the line, the total number of trains, the departure interval, the maximum speed, the redundancy requirement, the line parameters such as the length of the line, the number of substations, the number of single or double lines, the number of stations, the number of line ramps, the number of switches, etc.

3 Aarchitecture of the Operation Control System

3.1 Top-Level Architecture

Analyzed from the architecture of the entire traffic system, the operation control system includes the central operation control system located in the control center, the partition operation control system located in the traction substation or switch room, and the vehicle operation control system located on the train. Reliable data transmission between the maglev train and the ground system is realized through the wireless vehicle-ground communication system, and the architecture is shown in Fig. 1. Considering the different number of switch systems contained in the station and the interval, the divisional operation control is divided into three cases including single group of switches, multiple groups of switches and no switches. The interface architecture of the operation control system and the external system is sorted out according to the analysis of the above chapters, and the interface architecture is shown in Fig. 2.

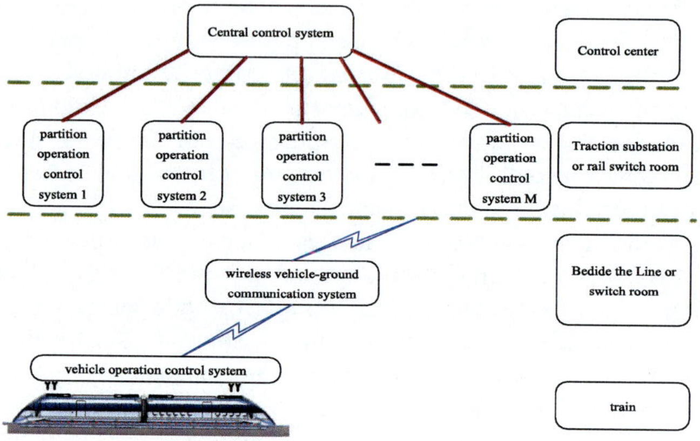

Fig. 1. The top-level architecture of operation control system

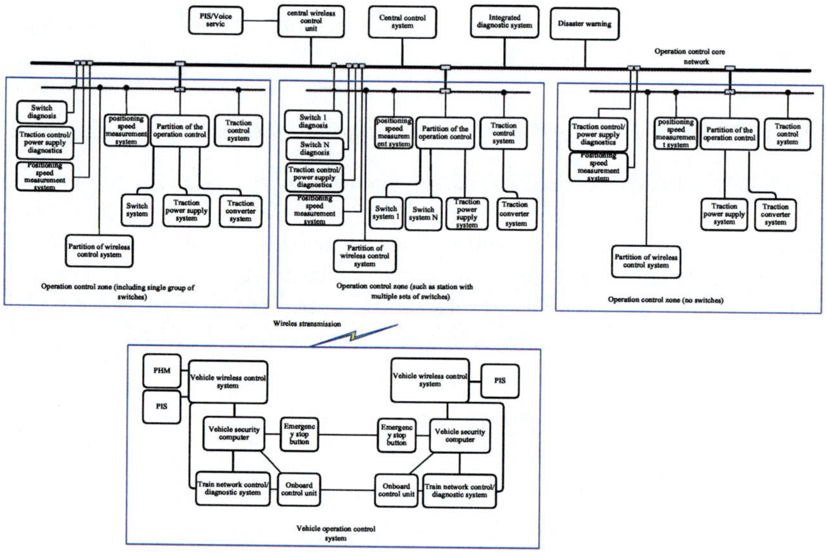

Fig. 2. External interface architecture of the operation control system

3.2 Division of Operation Control Area

Through the discussion of the above chapters, we know that the traction power supply system and the positioning speed measurement system of the medium speed maglev system are all on the ground. The traction power supply system has the division of the stator section and the division of the traction zone. The partition division of the positioning speed measurement system is basically consistent with the traction partition. The partition of the operation control system can be divided into two kinds. The first way is to learn from the architecture of high-speed maglev, that is, the operation control partition corresponds to the traction partition one by one. The other way refers to the corresponding operation control zone and traction zone in the line section. The operation control zone in the station can be managed and monitored by increasing the number of operation control zones according to the actual number of switches in the station, controllable track sections and upstream and downstream lines.

3.3 Functional Requirement Analysis

According to the external interface architecture of the medium-speed maglev operation control system and learning from the functional ports and relevant standards of the high-speed maglev operation control system [4–6], we summarize the functional requirements of the operation control system. The central operation control system needs to display the operating status of the lines and trains of the maglev system; The central operation control system can complete the maglev train operation plan, train scheduling, route presetting, switch control and other functions by automatic or manual setting. The central operation control system can record the operation data, process the train schedule or manual instructions, and transmit the generated operation parameters to the regional operation control system. Through the above functions, the central operation control system can control the train operation of the whole line, and summarize and display the online diagnosis information of each functional component of the operation control system within its jurisdiction.

The integrated diagnostic system is located in the control center and can display the online diagnostic information from the ground traction control/power supply diagnosis, turnout diagnosis, positioning diagnosis and other systems. The central operation control system can obtain the diagnostic information of the key systems within its jurisdiction in real time, which is convenient for train scheduling.

The partition operation control system needs to realize reliable data communication with the central operation control system, traction power supply system, traction control system, positioning speed measurement system, line switch system and ground wireless communication system to realize the operation control of the train. It needs to proofread, analyze, execute and distribute the central instructions to the vehicle. It controls the operation of vehicles within its jurisdiction, implements switch control and protection, and is the core system of train traction brake control and safety protection. It needs to cooperate with other systems to complete the operation mode control, automatic scheduling, driving sequence control, train safety positioning and train login, route protection, switch protection, train protection, train speed curve monitoring, traction safety cutting and other functions.

The vehicle operation control system needs to realize reliable data communication with the central operation control system and the partition operation control system. It receives instructions and parameters from the the central operation control system or the partition operation control system, and cooperates with the vehicle system to complete operation mode switching, speed curve monitoring, train status monitoring, vehicle control device monitoring such as receiver monitoring, train door monitoring, forced parking management, vehicle suspension and landing, etc.

4 Closing Remarks

The operation control system has many external interfaces, and the top-level system architecture, top-level system interface, and system functions contained in its top-level scheme are directly related to some factors including the speed index, traction system, braking system, positioning speed measurement system, line switches and their system layout, and key characteristics/capabilities in the traffic system. The research [6–10] on the operation control of medium speed maglev system should be based on the top-level scheme. That is to say the top-level scheme is the premise of the subsequent operation control and safety protection algorithm. Starting from the forward design, this paper first sorts out the above problems from the perspective of the traffic system, forms the architecture of the operation control system, completes the basic function assignment, and provides accurate input for the design of the subsystem technical scheme of the operation control system.

Acknowledgments. This work was supported by National Key R&D Program of China (Grant No. 2016YFB1200601).

References

1. Wang, Y., Yang, J., Zhang, L.: Analysis of force angle characteristics of hollow Halbah permanent magnet linear synchronous motor for maglev train. Railw. Roll. Stock. **41**(1), 26–32 (2021). (in Chinese)
2. Wang, F., Zhao, D., Wu, D., Xiao, S.: Research on reliability of positioning and speed detection system in high-speed maglev train. J. Beijing Jiaotong Univ. **46**(1), 147–154 (2022). (in Chinese)
3. Wang, Y., Yang, J., Cui, Y.: Characteristics analysis of permanent magnet electromagnetic hybrid suspension guide system for maglev train. Locomot. Electr. Drive **6**, 2–8 (2021). (in Chinese)
4. Qiu, Z., Deng, Z., Liu, X.: Cooperative control scheme for operation control system of 600 km/h high speed maglev train. Railw. Comput. Appl. **31**(1), 75–80 (2022). (in Chinese)
5. Zhang, W., Wei, W., Yang, Y., Nan, N.: An operation control strategy for the connected maglev trains based on vehicle-borne battery condition monitoring. Wirel. Commun. Mob. Comput. (2018)
6. Cao, B., Gao, Y., Li, K., Zhang, J., Zhang, Z., Zhang, W.: A fractional order medium-speed maglev train speed control method based on particle swarm optimization. In: 2020 IEEE 9th Data Driven Control and Learning Systems Conference (DDCLS), pp. 1446–1451. IEEE, Guang Xi (2020)

7. Yang, Y., Nan, N., Wei, W., Zhang, W., Liu, Y.: A section decentralization operation strategy for medium-speed maglev train based on battery states. In: 2018 37th Chinese Control Conference (CCC), pp. 7804–7809. IEEE, Wu Han (2018)
8. Wei, W., Zhang, W., Nan, N., Yang, Y., Liu, Y.: An operation control strategy for medium-speed maglev train in station based on on-board battery condition monitoring. In: 2018 Chinese Control and Decision Conference (CCDC), pp. 6541–6546. IEEE, Shen Yang (2018)
9. Gao, Y., Wang, Y., Zhang, J., Li, M., Yue, Q., & Zhang, W.: A sliding mode periodic adaptive learning operation control method for medium-speed maglev trains. In: 2020 39th Chinese control conference (CCC), pp. 5548–5553. IEEE, Bei Jing (2020)
10. Li, M., Yang, J., Zhang, J., Cao, B., Yue, Q., Zhang, W.: A periodic adaptive learning control method for the medium-speed maglev train with input saturation. In: 2020 39th Chinese Control Conference (CCC), pp. 5596–5601. IEEE, Bei Jing (2020)

Research on Key Technologies of Hydrogen Energy Powered Commuter Trains

Hang Yin[1(✉)], Caihui Zheng[1], Zhonghai Wang[1,2], and Xuekun Li[1,2]

[1] CRRC Changchun Railway Vehicles Co., Ltd., JiLin 130062, China
yinhang_cnr@126.com
[2] Beijing Jiaotong University, Beijing 100044, China

Abstract. Hydrogen energy trains have the characteristics of energy conservation, environmental protection, and operating without overhead lines, which are in accord with the development trend of green rail transit and provide ideal solutions for non-electrified lines. The hydrogen energy system inputs hydrogen and oxygen in a certain ratio into the hydrogen fuel cell, which is converted into electrical energy through chemical reactions and discharged into water. The hydrogen energy system, combined with energy storage system, not only rapid response to vehicle power demand but also absorb regenerative braking energy reliably and efficiently. This article elaborates on the key technologies of hydrogen powered commuter trains from three aspects: vehicle integration technology, energy management and control, fault diagnosis and safety protection, providing effective reference for the design of hydrogen powered rail vehicles in the future.

Keywords: Hydrogen powered commuter trains · Vehicle integration technology · Energy management and control

1 Introduction

Hydrogen technology innovation and its practical application is considered to be the most promising in the 21st century, the most from the root to solve the environmental crisis and energy crisis strategy. Currently in the field of urban rail transit has begun to explore the hydrogen power system, hydrogen power system can replace the original network power supply, energy by hydrogen and oxygen in hydrogen fuel cell electrochemical reaction, emissions only for water, thus have the characteristics of clean, environmental protection, zero carbon.

There have been many successful cases of applying hydrogen fuel cells in the field of rail transit at home and abroad. In 2007, the world's first hydrogen fuel cell hybrid light rail train with lithium battery developed by East Japan Railway Company was officially put into operation, In 2009, BNSF Railway developed a fuel cell shunting locomotive with a maximum power of 1.5 MW, In 2013, China's first hydrogen fuel cell motor vehicle "Blue Sky", Successfully developed by Southwest Jiaotong University; In 2016, China released the world's first commercial hydrogen fuel hybrid tram—Tangxu Railway tourist line, In 2018, the CoradiaiLint train developed by The French company

L. Jia et al. (Eds.): EITRT 2023, LNEE 1135, pp. 564–571, 2024.
https://doi.org/10.1007/978-981-99-9307-9_62

Alstom was put into commercial operation in Germany, In 2019, the hydrogen energy tram developed and designed by Qingdao Sifang was officially operated in Foshan, In the future, the Mireo Plus H train developed by Siemens of Germany and the FV-E991 train developed by Japan's Hitachi, It is scheduled to be officially put into operation in 2024.

2 Hydrogen Energy Power of Urban Train Vehicle Integration Technology

2.1 Overall Technical Parameters

According to the current market analysis, urban A trains are mainly divided into 4 marshalling forms and 8 marshalling forms. Therefore, urban A trains based on hydrogen energy power are also planned according to the two marshalling forms. The detailed parameters are shown in the following Table 1.

Table 1. Technical parameters of vehicles

Item	Technical parameter	
Group form	Four groups	Eight groups
Dynamic drag ratio	3:1 or 1:1	3:1
Train length	95,760 mm	185,600 mm
Maximum width of train	3000 mm	
Gross rail load on axle	≤17 t	
Operation speed	140–160 km/h	
Average acceleration	≥0.65–0.8 m/s^2(0–50 km/h)	≥1.0 m/s^2(0–50 km/h)
	≥0.35 m/s^2(0–140 or 160 km/h)	≥0.4–0.5 m/s^2(0–140 or 160 km/h)
Service braking deceleration	≥1.0 m/s^2	
Hydrogen fuel cell power	1200 kW	2400 kW
Energy storage system power	3200 kW	6400 kW
Motor power	280 kW	

2.2 High Voltage Main Circuit and Principle

Hydrogen energy city trains adopt a hybrid power system that combines hydrogen energy system and energy storage system. The output electric energy of Hydrogen fuel cell is connected to the high-voltage bus of the vehicle through DC/DC boost, and together with the energy storage system, it provides energy for the load of the whole vehicle. As shown in the figure below, the high-voltage main circuit of the vehicle is as follows.

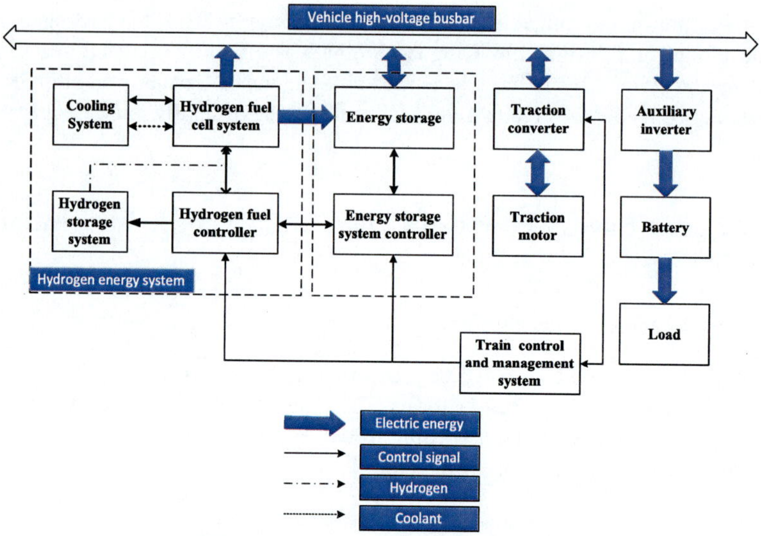

Fig. 1. Vehicle high voltage main circuit

Hydrogen energy system includes Hydrogen fuel cell system, heat dissipation system, hydrogen storage system and control system, of which Hydrogen fuel cell system is the core of the whole hydrogen power system. The Hydrogen fuel system converts the hydrogen in the hydrogen storage system into electrical energy through electrochemical reaction for the use of vehicle traction and auxiliary systems. The Hydrogen fuel cell system is mainly composed of fuel cell stack module, air supply subsystem, hydrogen supply subsystem, cooling subsystem, electrical subsystem and control subsystem.

When the vehicle starts, the energy storage system first operates to provide AC 380 V power to the entire vehicle. The hydrogen energy system takes AC 380 V power from the vehicle and converts it into DC 24 V to supply power to the radiator. At the same time, it converts it into DC 600 V to supply power to the air supply subsystem and hydrogen supply subsystem. The hydrogen storage system supplies hydrogen and oxygen to the anode and cathode of the fuel cell respectively through the hydrogen supply subsystem, Hydrogen and oxygen undergo electrochemical reactions under the action of catalysts inside the fuel cell stack, generating electrical energy. The output of the fuel cell is boosted to DC 1450–1950 V through a DC–DC circuit and connected to the vehicle's DC bus. The reaction product water is discharged to the external environment through the vehicle structure or used as water for other systems. The heat generated during the reaction process is discharged to the external environment through the cooling circuit and fan of the heat dissipation system.

2.3 Train Traction Calculation and System Configuration Analysis

According to the operation condition of the target line, the project adopts the combination of Hydrogen fuel cell and super capacitor, with Hydrogen fuel cell system as the main power source and super capacitor as the auxiliary power source. In order to ensure that the total energy consumption (excluding regenerative energy consumption) of the vehicle during operation is provided by the Hydrogen fuel cell, the power configured by the Hydrogen fuel cell is at least the average power of the vehicle operation. According to the calculation of the traction of the target line, the energy consumption of a round trip vehicle operation is about 1200 kWh, and the operation time is about 80 min, so the average power is $1200 \div (80 \div 60) = 900$ kW. In addition, the power setting of Hydrogen fuel cell should also consider the following two aspects: first, when the energy storage system is completely faulted, the hydrogen energy system can still support AW3 loaded vehicles to run at a speed of 25 km/h on a 30‰ ramp, at which time the output power of a single motor is at least 150 kW, so the total demand power is $150 \times 8 = 1200$ kW. The second is that during the braking and stopping intervals of the vehicle, the energy consumed by the energy storage system can be replenished. The time for the vehicle to accelerate from start to 160 km/h is $160 \div 3.6 \div 0.35 \approx 126s$, Assuming that all vehicles output their maximum power of 2755 kW during this process, the energy consumption of the supercapacitor during the traction process is $(2755 - 900) \times (126 \div 3600) \approx 64.9$ kWh and the braking time is approximately $160 \div 3.6 \div 1.0 \approx 44$ s, the charging power required is $(64.9 - 29) \div [(44 + 60) \div 3600] \approx 1196$kW, so the total power of the hydrogen energy system is set to 1200 kW.

Supercapacitors provide vehicle traction and peak braking power changes, and can maximize the absorption of energy feedback from vehicle braking. Taking a 4-car urban vehicle as an example, as shown in Fig. 1 of the traction characteristic curve, under AW3 working conditions, the maximum traction power is about $310 \times 8 \div 0.9 = 2755kW$, the maximum braking power is about $330 \times 8 \times 0.9 = 2376kW$, and the maximum regenerative energy consumption in the section is about $2376 \times (44 \div 3600) \approx 29kWh$. Considering that even if there are 2 sets of supercapacitors in fault, the vehicle can still operate normally, the maximum output power of each set of supercapacitors should be 500 kW, and the maximum absorbed energy can be reached $29 \div 6 = 4.84kWh$ each time. Therefore, the entire train is equipped with 8 sets of supercapacitors, The rated output power is 400 kW, the maximum output is 600 kW, and the maximum stored energy is about 20 kWh. Therefore, the capacity of the braking recovery section of each group of capacitors accounts for 20% of the total capacity (Figs. 2 and 3).

3 Energy Management and Control Technology of Hybrid Power Systems

The instantaneous power control strategy of hybrid power system is used for power distribution of supercapacitor and Hydrogen fuel cell. The power distribution of supercapacitor and Hydrogen fuel cell is mainly based on the train power demand, supercapacitor status, residual hydrogen gas volume, Hydrogen fuel cell status, etc. When the SOC (remaining power) of a supercapacitor is too high or too low, the output voltage will be unstable, and the output power will also be greatly reduced. In order to ensure the application performance of the supercapacitor, it is necessary to specify the limit value for the application of the supercapacitor. If the SOC is above the upper limit, the supercapacitor will stop charging, and if the SOC is below the lower limit, the supercapacitor will stop discharging. The SOC of supercapacitor is characterized by voltage.

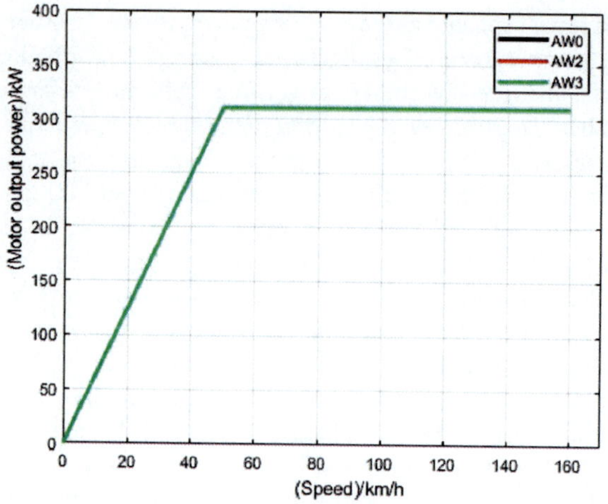

Fig. 2. Motor output power speed curve under AW3 load under traction Condition

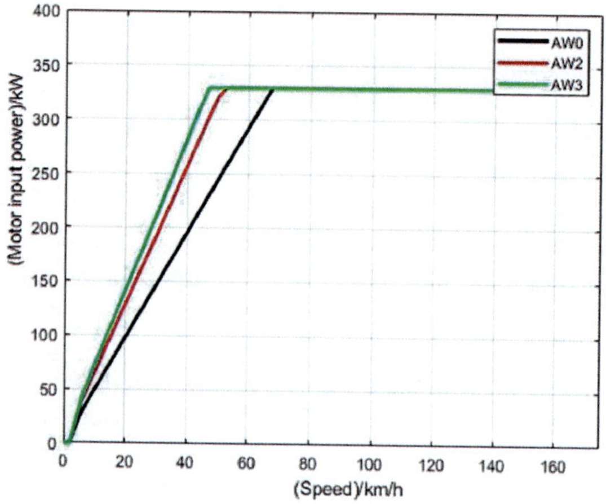

Fig. 3. Motor output power speed curve under AW3 load under braking conditions

As shown in the figure below, the upper and lower voltage limits of the supercapacitor are set to V_{chg_Lim} and V_{dis_Lim}, the full traction condition of a hybrid train requires approximately 20% of the energy consumption of the supercapacitor. After the traction is completed, the voltage of the supercapacitor cannot be lower than V_{dis_Lim}, so the voltage of the supercapacitor during train startup cannot be lower than V_{tra_Lim}; When the supercapacitor voltage $V < V_{brk\_Lim}$, the Hydrogen fuel cell will charge the supercapacitor on the premise of meeting the vehicle power. Under the braking condition, the supercapacitor will recover the braking energy, and the capacitance reaches the upper limit value V_{chg_Lim} (Fig. 4).

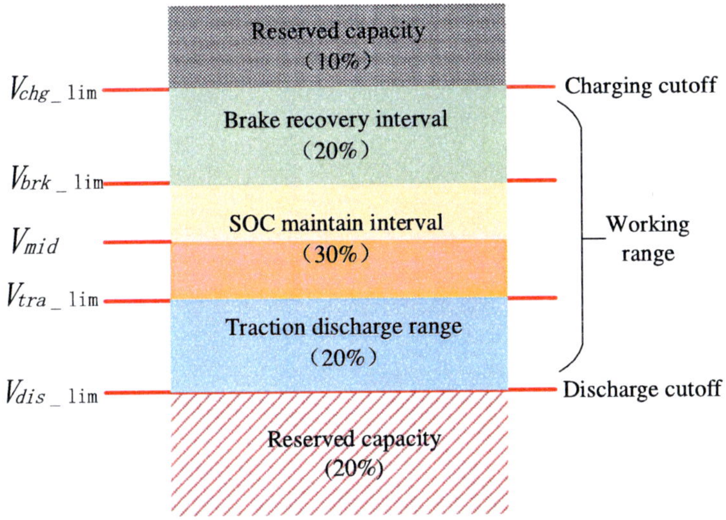

Fig. 4. Supercapacitor SOC working area division

The following is an introduction to the rule-based instantaneous power control of hybrid systems under different operating conditions:

(1) When the vehicle brakes, the demand power of the whole vehicle is less than 0: the Hydrogen fuel cell system is in standby state, and when the voltage value of the energy storage system is $V < V_{chg\_\lim}$, the braking energy fed back by the vehicle is normally absorbed; When the voltage value of the energy storage system is reached $V \geq V_{chg\_\lim}$, the energy storage system stops absorbing energy, the vehicle cuts off the electric braking, and applies Railway air brake to stop.

(2) When the vehicle is being towed, if the required power of the entire vehicle is between 0 and 1200 kW, there are two situations that need to be explained:

a. If the voltage of the energy storage system is $V \geq V_{chg\_\lim}$, the demand power of the Hydrogen fuel cell is the demand power of the whole vehicle. Since the output of the Hydrogen fuel cell is slow, the energy storage system will supplement the output power. When the output power of the Hydrogen fuel cell meets the demand power of the vehicle, the energy storage system will exit;

b. If the voltage value of the energy storage system is $V_{chg\_\lim} > V > V_{dis\_\lim}$, The demand power of Hydrogen fuel cell is 1200 kW, the output power of the Hydrogen fuel cell is still supplemented by the energy storage system. When the output power of the Hydrogen fuel cell meets the vehicle demand power, the energy storage system exits, and the Hydrogen fuel cell charges the energy storage system at the same time;

c. If the voltage value of the energy storage system is $V \leq V_{dis\_\lim}$, the demand power of the Hydrogen fuel cell is 1200 kW. At this time, the energy storage system cannot output power to supplement due to insufficient capacity, so it can only wait for the output of the Hydrogen fuel cell. When the output power of the

Hydrogen fuel cell meets the demand power of the vehicle, use the remaining power to charge the energy storage system.

(3) When the required power of the whole vehicle is greater than 1200 kW, regardless of the voltage of the energy storage system, the requested power of the Hydrogen fuel cell power System request is 1200 kW. When the voltage of the energy storage system is $V \leq V_{dis\_\mathrm{lim}}$, the energy storage system discharges to supplement the output power of the Hydrogen fuel power cell. When the voltage of the energy storage system is reached $V \leq V_{dis\_\mathrm{lim}}$, the discharge stops, and the vehicle alarms, indicating that the power of the whole vehicle cannot meet the demand.

4 Conclusion

At present, a hydrogen energy market A-type vehicle has been developed to test and verify the above research results, filling the gap in relevant vehicle models. However, the application of hydrogen energy in the field of rail transit in China is still in its infancy, lacking mature application cases and standard systems. During the research process, it has also been found that there are some technical difficulties that need to be solved:

(1) The hydrogen storage capacity of vehicles is a problem. The more hydrogen storage capacity a vehicle has, the stronger its endurance. However, due to limitations in vehicle space, clearance, safety, and other aspects, the hydrogen storage capacity has reached a certain bottleneck. If the vehicle's endurance or passenger carrying capacity wants to be further improved, it is necessary to research new hydrogen storage solutions;

(2) The safety assessment of vehicles and even the entire line is currently in the exploratory stage for hydrogen energy vehicles and lines, and can mainly refer to foreign cases. However, safety is particularly important for hydrogen energy lines, including the construction of vehicles, lines, and depots, vehicle operation organization models, maintenance regulations, etc. Therefore, if hydrogen energy is to be promoted and applied in the field of rail transit in the future, Safety assessment is the primary issue to be addressed;

(3) The development of relevant standard systems and the design of hydrogen energy rail transit lines and vehicles do not yet have relevant regulatory guidance and standard constraints. Therefore, the current design is not unified, and product diversity also brings great safety hazards and cost risks.

Hydrogen technology innovation and its practical application are considered as promising, clean, efficient and safe secondary energy in the 21st century. With the continuous development and improvement of Hydrogen fuel cell technology, the above bottlenecks will be broken one by one, and the hydrogen market vehicles will become more and more competitive in the market.

References

1. Chen, W., Bu, Q., Liu, Z., et al.: Design of fuel cell hybrid electric tram power system. J. Southwest Jiaotong Univ. **51**(3), 430–436 (2016) (in Chinese)

2. Xiao, L.: Modeling and Simulation of a Fuel Cell/Supercapacitor Hybrid Power System for Trams. Southwest Jiaotong University, Chengdu (2015). (in Chinese)
3. Chen, W., Zhang, G., Meng, X.: Power performance analysis and design of fuel cell hybrid electric trams. J. Southwest Jiaotong Univ. **52**(1), 1–8 (2017) (in Chinese)
4. Chen, Y., Wang, X.: Application of hydrogen fuel cell in modern tram. J. Railw. Veh. **55**(4), 14–17, (2017) (in Chinese)
5. Qingjun, H., Zhengbin, Y.: Research on structural design and control method of Hydrogen fuel cell tram. J. Locomot. Electr. Drive **2**, 48–51 (2016). (in Chinese)
6. Garcia, P., Fernandez, L.M., Garcia, C.A., et al.: Energy management system of fuel-cell-battery hybrid tramway[J]. IEEE Trans. Ind. Electron. **57**(12), 4013–4023 (2010)
7. Chen, Y., Wang, X.: Application of hydrogen fuel cells in modern trams [J]. Railw. Veh. **55**(4): 14–17 (2017) 4 (in Chinese)
8. Gao, L., Douga, R.A., Liu, S.: Power enhancement of an actively controlled battery/ultracapacitor hybrid[J]. IEEE Trans. Power Electron. **20**(1), 236–243 (2005)
9. Heam, C.S., Flynn, M.M., Lewis, M.C., et al.: Low cost flywheel energy storage for a fuel cell powered transit bus[J]. 2007 IEEE Veh. Power Propuls. Conf., 829–836 (2007)
10. Bo, C.A.I., Kunpeng, L.I.: On contactless traction power supply for modern trams [J]. Urban Mass Transit **1**, 72–77 (2015). (in Chinese)

Optimal Sizing of Hybrid Power System Based on Typical Driving Cycle of the Shunting Locomotive

Mingming Kang[1(✉)], Mingjie Jiang[2], Chenguang Fan[2], Bo Yao[2], Xin Bi[2], and Xiaoxu Wang[2]

[1] CRRC Zhuzhou Locomotive Co., Ltd, Zhuzhou, Hunan 412001, China
long_kamin@yahoo.com.cn

[2] National Active Distribution Network Technology Research Center, Beijing Jiaotong University, Beijing 100044, China

Abstract. With the advancement of new energy technologies and the promotion of carbon peaking and carbon neutral double carbon targets, energy saving and emission reduction have become an important research direction in the field of rail transportation. The optimal sizing of the hybrid power system is important for saving energy and improving economy, while the typical driving cycle cover the operation of shunting locomotives and is the basis for capacity configuration, so this paper studies and analyses the typical driving cycle of oil–electric hybrid shunting locomotives and the capacity configuration of the hybrid power system. Firstly, the typical driving cycle of the shunting locomotive is extracted using genetic algorithm, and the deviation of the unit energy consumption is controlled within 0.2%. Finally, using the typical driving cycle as input conditions, a capacity configuration model based on the convex optimisation method is established, and finally the optimised hybrid system configuration parameters are obtained, and its total life-cycle cost economy is improved by 19.26% compared to the original configuration.

Keywords: Hybrid power system · Sizing optimization · Driving cycle · Genetic algorithm · Convex optimization

1 Introduction

According to the "14th Five-Year Plan" Modern Comprehensive Transportation System Development Plan, during the "14th Five-Year Plan" period, China will add 19,000 km of railroad operation, and it is expected that by the end of 2025, the total operation mileage of the railroad will reach about 165,000 km. From the information released by the State Railway Administration, the electrification degree of railroads reached 71.90% by 2022. However, there are still some non-electrified railroads exist due to the limitation of terrain, environment, and workplace [1, 2]. While traditional internal combustion locomotives are capable of working in non-electrified railroads, their shortcomings of low fuel efficiency, serious pollution and high noise are constantly being magnified, so green and economical hybrid shunting locomotives are beginning to be put into research and development [3–5].

L. Jia et al. (Eds.): EITRT 2023, LNEE 1135, pp. 572–586, 2024.
https://doi.org/10.1007/978-981-99-9307-9_63

At present, the more common energy storage media are supercapacitors and batteries, both of which are widely used in urban rail transit, electric vehicles and smart microgrids [6, 7]. Energy storage media are generally used with internal combustion engines or fuel cells to form a hybrid power system, and both of them can reach the goal of energy saving and emission reduction compared with the single power supply mode of internal combustion engines [8, 9]. At the same time, the configuration parameters of the hybrid power system and the power distribution strategy between power sources need to be further improved to save as much cost as possible under the premise of meeting the demanded power of the locomotive. In the research of hybrid power system configuration, it is mainly divided into two directions: capacity configuration based on vehicle power performance; parameter matching based on optimization of capacity configuration and energy management strategy.

Capacity configuration based on vehicle power performance is mostly used in the early stage of vehicle design, because at this time there is no readily available data on working conditions, and the parameters can only be calculated according to the design requirements such as maximum speed, maximum traction, maximum output power, etc. [10]. Although this method is simple to calculate, it is difficult to ensure the economy of the locomotive in the whole life cycle process.

Another approach is to optimize the power allocation strategy and configuration parameters to obtain the optimal parameters under the corresponding objective (highest fuel savings or lowest whole life cycle cost). For this optimization method, there are generally two configuration schemes, one is to optimize the energy management strategy and the capacity allocation parameters at the same time [11], and the other is to optimize the matching parameters and their corresponding optimal energy management strategies separately by using two-layer optimization [12–14].

In [15] and [16], a regular energy management strategy was chosen, and genetic algorithms were used to simultaneously optimize the thresholds and the adaptation parameters. Research in [17] optimized the power allocation strategy of a supercapacitor-battery hybrid power system with the aim of minimizing the battery energy consumption by using a convex optimization method, and compared it with a regular energy management strategy, which reflects the superiority of the convex optimization method.

The above capacity configurations have one thing in common: the conditions used for the configurations are either derived from traction calculations or are randomized, making it difficult to obtain configuration parameters with high applicability.

The working characteristics of shunting locomotives determine the randomness of their working lines, so the working conditions of a specific line obtained only by traction calculation are not universal. Based on the research on hybrid power system optimization and matching in the field of electric vehicles, first typical driving conditions that can represent the historical operating state of the vehicle are extracted from the historical working condition data, and then the driving conditions are used to obtain the required power and required energy consumption as inputs, and the appropriate energy management strategy is selected, and finally the optimal configuration of the power system is obtained.

In this paper, the hybrid power system topology of the shunting locomotive is first analyzed, and the diesel engine fuel consumption model, circuit model and power battery life model are established. Then, based on the historical operating condition data, typical operating conditions are proposed. Finally, the convex optimal configuration problem model is established to obtain the configuration parameters with the minimum total cost under the whole life cycle.

2 Hybrid Power System Modeling

2.1 Hybrid Powertrain Topology

As shown in Fig. 1 for the hybrid power system topology of the shunting locomotive, the energy is provided by the diesel generator and the power battery dual power source, which can be supplied by a single power source or by dual power sources in the actual operation process, and the specific energy flow mode is related to the power allocation strategy and the working condition demand. It is worth stating that the power battery can provide energy for the locomotive as well as recover braking energy, so the power relationship at the DC bus can be obtained:

$$P_{de} = \begin{cases} P_{DG}\eta_{re} + P_{BAT}\eta_{dc/dc}, P_{BAT} \geq 0 \\ P_{DG}\eta_{re} + \dfrac{P_{BAT}}{\eta_{dc/dc}}, P_{BAT} < 0 \end{cases} \tag{1}$$

where P_{DG}, P_{BAT} and P_{de} denote the power of the diesel generator set, the power of the battery pack, and the demanded power (kW) of the shunting locomotive, respectively; η_{re} and $\eta_{dc/dc}$ represent the efficiency of the rectifier and the efficiency of the bidirectional DC/DC, respectively.

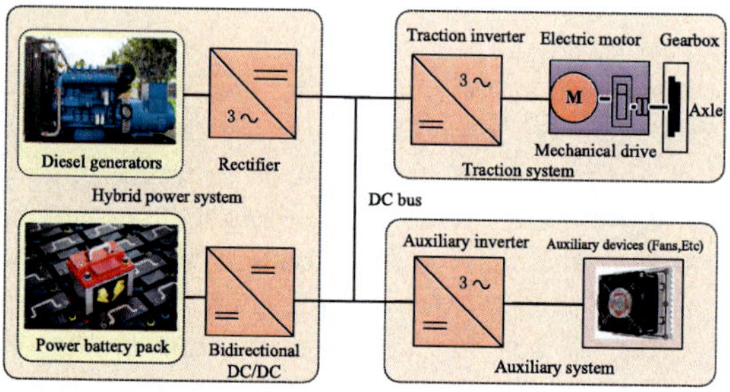

Fig. 1. The topology diagram of hybrid power system

2.2 Diesel Genset Fuel Consumption Model

Since this paper focuses on considering the fuel consumption of diesel generator sets, this section ignores the complex physicochemical processes inside the diesel generator sets and utilizes the experimentally measured fuel consumption rates at different power for modeling. In order to increase the power suitability of the diesel genset, the fuel consumption rate curves at four different power levels are normalized and fitted (the maximum output power of the four diesel gensets are 375 kW, 880 kW, 1324 kW, and 1864 kW, respectively), and ultimately the fitted curves of the relative output power-fuel consumption rate of the diesel genset are obtained as shown in Fig. 2.

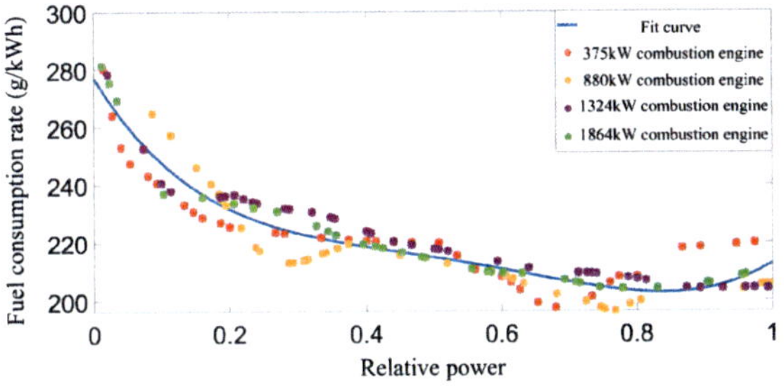

Fig. 2. Relative power-fuel consumption rate of diesel generator set

Based on this fitting relationship, the fuel consumption of diesel generator sets at different power levels can be obtained:

$$\begin{cases} P_{\text{rate}} = \dfrac{P_{\text{DG}}}{P_{\text{DG, max}}} \\ \dot{m}\left(P_{\text{rate}}, P_{\text{DG, max}}\right) = \dfrac{\dot{m}_{\text{g}}(P_{\text{rate}})P_{\text{rate}}P_{\text{DG, max}}}{3600} \end{cases} \tag{2}$$

where P_{rate} denotes the relative power of the diesel generator set; $P_{\text{DG\_max}}$ denotes the maximum output power of the model diesel generator set (kW); P_{DG} denotes the output power of the diesel generator set (kW); $\dot{m}_{\text{ g}}$ denotes the fuel consumption rate of the diesel generator set (g/kWh).

2.3 Power Battery Model

The circuit model of the power battery is shown in Fig. 3, and the Rint model is chosen in this paper.

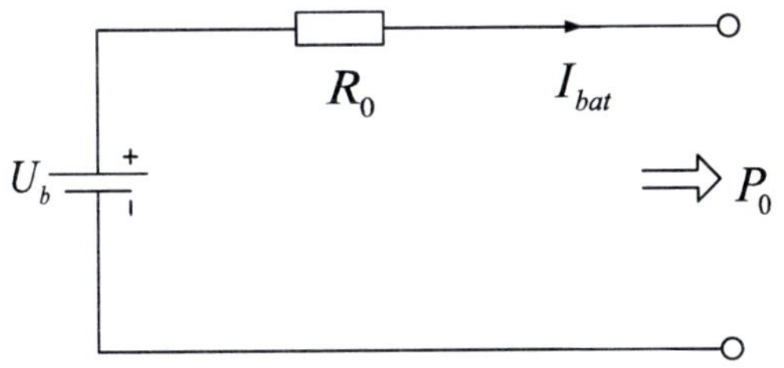

Fig. 3. Rint model of battery

Where U_b denotes the open-circuit voltage of the power battery pack; R_0 denotes the equivalent ohmic internal resistance of the power battery; I_{bat} denotes the output current of the power battery; and P_0 denotes the output power of the battery. According to the Rint model of the power battery and the ampere-time integration method, the state of charge (SOC_{bat}) of the battery pack can be found, so the battery output can be expressed as:

$$\begin{cases} I_{bat} = \dfrac{U_b - \sqrt{U_b^2 - 4R_0P_{DG}}}{2R_0} \\ SOC_{bat} = SOC_0 - \dfrac{1}{3600Q_{bat}} \displaystyle\int I_{bat}dt \end{cases} \tag{3}$$

where SOC_0 represents the charging state of the power battery at the initial moment and Q_{bat} represents the rated capacity (Ah) of the power battery.

The battery pack ignores the inconsistency between monomers and the effects brought by different series and parallel connection methods, and calculates other relevant parameters of the power battery pack, as shown in Eq. (4):

$$\begin{cases} Q_{bat} = N_p Q_{bat\_cell} \\ U_b = N_s U_{b\_cell} \\ R_0 = N_s R_{0\_cell}/N_p \end{cases} \tag{4}$$

where N_s and N_p denote the number of series and parallel connections of the power battery pack, respectively, and Q_{bat_cell}, U_{b_cell} and R_{0_cell} denote the single rated capacity, open-circuit voltage and ohmic internal resistance of the power battery, respectively.

2.4 Power Battery Life Model

The objective function of this paper is to minimize the whole life cycle cost, so it is necessary to calculate the life loss of the battery in the operation process, and this paper chooses the semi-empirical aging model of the battery to estimate the battery life loss.

$$Q_{loss} = Me^{-\left(\frac{E_a + B\times C}{RT}\right)}(A_h)^z \tag{5}$$

where Q_{loss} is the capacity decline rate of the battery; M is the ideal gas constant; E_a is the activation energy; C is the charging and discharging multiplication rate of the battery; T is the temperature of the battery, which is taken as 298.15 K here; and A_h denotes the throughput of the battery.

In this paper, it is considered that the battery reaches the end of its life when its capacity declines to 20%, so the total throughput A_{h_life} of the battery during its life time can be found as:

$$A_{h\_life} = \left[\frac{20}{A_0 e^{-\left(\frac{E_a + B\times C}{RT}\right)}}\right]^{\frac{1}{z}} \tag{6}$$

Similarly, the inconsistency between monomers is ignored when calculations are performed using the lifetime model, so that the total throughput power A_{h_toalt} of the battery pack over its full life cycle is:

$$A_{h\_total} = N_p \left[\frac{20}{A_0 e^{-\left(\frac{E_a + B \times C}{RT}\right)}} \right]^{\frac{1}{z}} \tag{7}$$

3 Shunting Locomotive Typical Working Condition Construction

With regard to the working lines of shunting locomotives and the strong randomness of working conditions, in order to improve the applicability of capacity configuration, this paper extracts a representative driving condition from the original working condition data.

3.1 Data Processing for Actual Working Conditions

The data used in this paper is collected from a certain type of shunting locomotive, the specific operation situation for the marshalling, push-and-pull, humping operation. In this paper, the marshalling operation is selected to introduce its data processing and analyzing process in detail, and the other two groups of working conditions also adopt similar methods. In the raw data, in addition to the running speed, the gear data of the locomotive is collected at the same time, and then the traction force data is derived from the traction characteristic curve of the locomotive (Fig. 4).

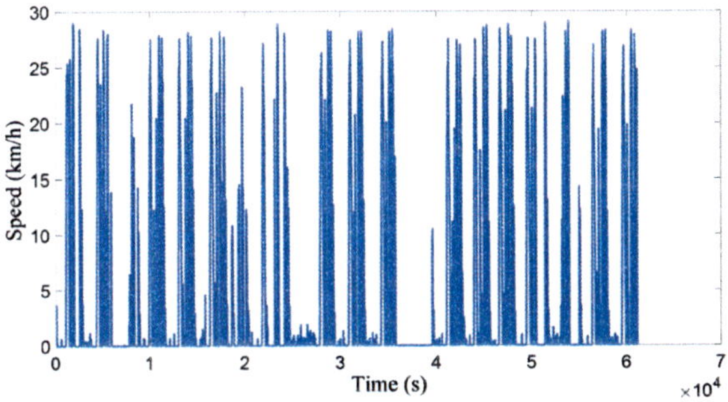

Fig. 4. Measured data on one day

Pre-Processing of Raw Operating Data. In the process of actual working condition data acquisition, due to the influence of locomotive vibration, data acquisition equipment error and other external environment factors, there will be certain abnormal segments of the original data acquired. In order to improve the reliability and validity of the data, this paper firstly screens the abnormal segments of the original working condition data, and the screening rules are as follows:

(1) If the total running mileage of a single short travel segment does not exceed 30 m, the segment is considered to be a very short start and stop, which does not affect the construction of the overall working condition.
(2) The maximum speed of a single short-trip segment does not exceed 5 km/h, which is an error if the shunting locomotive is in operation; if it is in transportation, it is also considered as a short start–stop; if it is a hooked state, it has no effect on the construction of working conditions.

Division of Short Trips. A complete kinematic segment is the movement of a shunting locomotive from one standby starting point to the next standby starting point, and has four working modes: standby, uniform speed, deceleration, and acceleration, and a schematic diagram of a kinematic segment in an actual working mode is shown in Fig. 5. As defined above for short travel segments, the raw data was divided into 1580 kinematic segments.

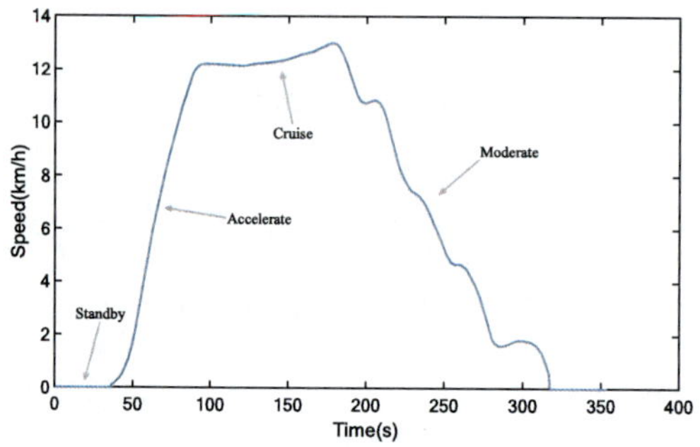

Fig. 5. Schematic diagram of the kinematic segment

Calculation of Characteristic Parameters. After dividing the short trips, it is also necessary to calculate the corresponding feature parameters to further express the information in different kinematic segments, so in this paper, the kinematic features such as maximum speed, average speed, standard deviation of speed, maximum acceleration, maximum power, average power, mileage run, energy consumption per unit of mileage, standby ratio, acceleration ratio, deceleration ratio, and uniform speed ratio, and the statistical feature parameters are selected for each segment.

3.2 Analysis of Working Condition Data

The kinematic segmentation and related feature parameter calculation result in a multi-dimensional feature parameter matrix. In order to reduce the amount of computation in the subsequent cluster analysis and improve the computational efficiency, it is necessary to simplify and downscale the original feature parameter matrix. As correlation of different feature parameters, principal component analysis can be chosen to realize the dimensionality reduction of the feature parameter matrix.

Principal component analysis is often used to deal with high-dimensional data, through the calculation of correlation coefficients between different feature parameters and eigenvector values, such as linear transformation, and finally use the resulting comprehensive indicators to replace the original data for subsequent analysis. As shown in Table 1, the cumulative contribution rate of principal components is shown in the statistical graph, when the cumulative contribution rate reaches 85%, it can be considered that the selected principal components can cover most of the information in the original data.

Table 1. Contribution of principal components.

Work mode	Principal component	Contribution rate (%)	Cumulative contribution rate (%)
Marshalling	Q_1	52.86	52.86
	Q_2	21.41	74.27
	Q_3	11.84	86.11

Cluster Analysis. In this paper, K_means clustering method is used to categorize the kinematic segments into six classes of working conditions, and the results are shown in Fig. 6, which shows that the boundaries between different classes are obvious and the sample distances between the same classes are similar, indicating that the clustering results are good.

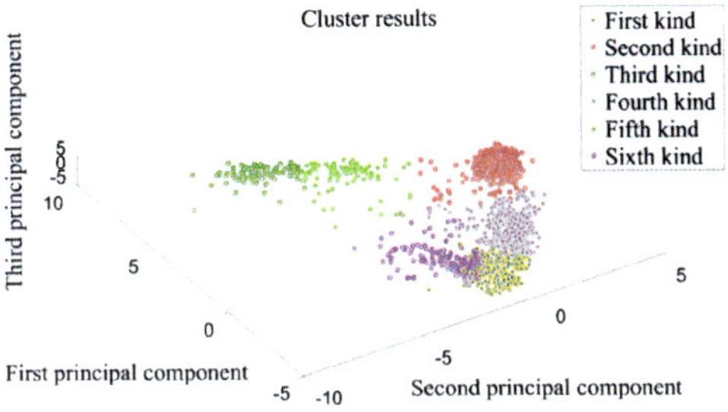

Fig. 6. Graph of clustering results

3.3 Driving Condition Construction and Validation

Typical Working Conditions Splicing. The kinematic segments closest to the clustering center of each class are selected from each class of working conditions and spliced to get the complete driving conditions. The typical working condition extraction process of push-and-pull and humping operations are the same as the construction of driving conditions for marshalling operation, so the driving conditions of the other two operations

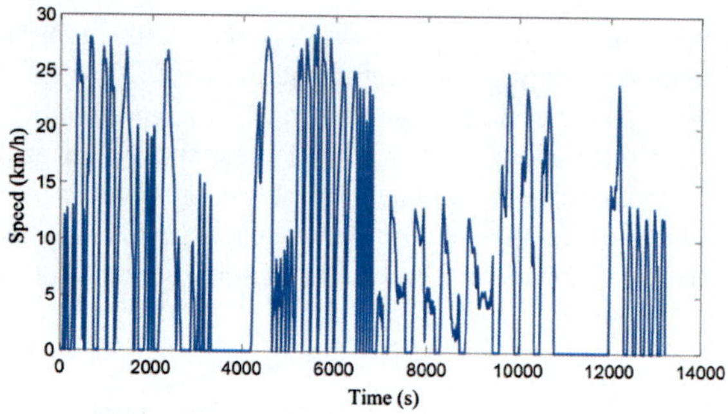

Fig. 7. Result graph of the typical driving cycle

can be obtained, and finally the three conditions are spliced in equal proportions, and the final result is shown in Fig. 7.

Typical Working Condition Verification. After obtaining the typical driving working conditions in the three operation modes, the validity of the working conditions is verified. Considering that the constructed conditions will be utilized for capacity allocation in the future, the focus is on the deviation of the energy consumption per unit mile. For the energy consumption per unit mile, the deviation of the data from the original data for the three operation modes is within 3%, so the constructed driving conditions are valid.

4 Hybrid Power System Capacity Allocation

4.1 Convex Optimization Problem Processing

(1) Processing of the diesel engine model

In Fig. 2, since the fuel consumption rate-relative power relationship curve in the diesel engine model does not constitute a functional relationship and naturally fails to satisfy the convex optimization condition, the following transformation is performed:

$$P_{out}(s) = aP_{DG} + b/(2500/P_{DG,max}) \tag{8}$$

where P_{out} is the fuel output power and a and b are the fitting coefficients, respectively.

(2) Processing of the Power battery model

According to the experimental data of the power battery, the relationship between the open-circuit voltage U_b and SOC can be obtained. From Fig. 8, it can be seen that when the SOC of the battery is maintained at 0.2 ~ 0.9, U_b and SOC are approximately linear, and the following fitting equation can be obtained:

$$U_b = kSOC_{bat} + U_{b0} \tag{9}$$

where k and U_{b0} were obtained by fitting the experimental data, respectively.

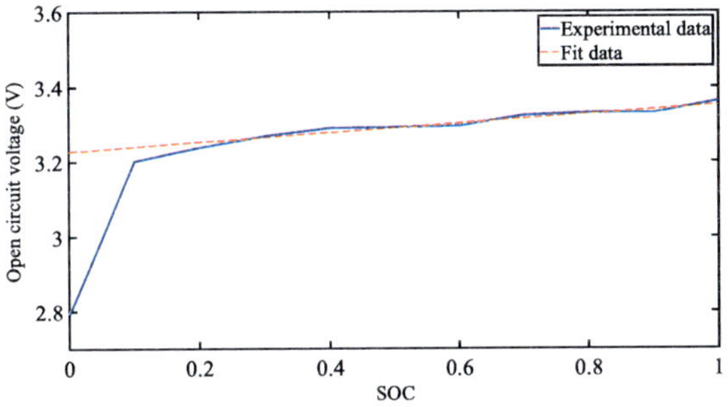

Fig. 8. Open circuit voltage versus SOC curve

The equivalent circuit shown in Fig. 3 and the processing described above results in a relational equation for the output power P_0 of the battery, the internal power P_b of the battery, the internal loss power P_{loss}, and the battery current I_{bat}:

$$\begin{cases} P_0 = P_b - P_{loss} = N_b(I_{bat}U_b - I_{bat}^2 R_b) \\ P_b = U_b I_{bat} \end{cases} \tag{10}$$

Using the battery energy E_{bat} instead of SOC calculation, the relationship between E_{bat} and P_b is as follows:

$$\frac{dE_{bat}(t)}{dt} = -P_b(t) \tag{11}$$

Thus, the constraint on the maximum charging and discharging current of the battery can be transformed into a constraint on the discharging power:

$$\begin{cases} P_b \le I_{bat,max}\sqrt{N_b(\frac{2E_{bat}}{C_{bat}} + N_b U_{b0}^2)} \\ P_b \ge I_{bat,min}\sqrt{N_b(\frac{2E_{bat}}{C_{bat}} + N_b U_{b0}^2)} \end{cases} \tag{12}$$

Likewise, the constraints on the upper and lower SOC thresholds of the battery can be translated into constraints on the battery power:

$$\begin{cases} E_{bat} \le \frac{N_b C_{bat}}{2}(U_b(SOC_{max})^2 - U_{b0}^2) \\ E_{bat} \ge \frac{N_b C_{bat}}{2}(U_b(SOC_{min})^2 - U_{b0}^2) \end{cases} \tag{13}$$

(3) Processing of power battery life model

According to the semi-empirical life model in the first chapter and in the literature [18], the loss of battery life can be described in terms of the capacity decline rate of the battery, dSOH_{bat}/dt. Also, in order for the optimization process to satisfy the conditions of convex optimization, the decline rate is linearized and fitted:

$$\frac{dsoh_{bat}(t)}{dt} = \begin{cases} a_1 P_{b\_cell}(t)^2 + b_1, \left|P_{b\_cell}(t)\right| \le Z_1 \\ a_2 P_{b\_cell}(t)^2 + b_2, Z_1 < \left|P_{b\_cell}(t)\right| \le Z_2 \\ a_3 P_{b\_cell}(t)^2 + b_3, \left|P_{b\_cell}(t)\right| > Z_2 \end{cases} \quad (14)$$

The significance of adding a life model is that it ensures that the life of the battery can meet the full life cycle of the service life, which is more reasonable than directly specifying the service life of the battery without considering the battery life model.

4.2 Convex Optimized Capacity Allocation

(1) Whole life cycle objective function

$$J_{op} = \min(C_{bat}E_{bat} + N \int_0^{t_f} C_{fuel}\dot{m}(t)dt) \quad (15)$$

where C_{fuel} is the unit price of diesel fuel, C_{bat} is the unit price of power battery, E_{bat} is the amount of power battery power distribution, t_f is the length of operation of the shunting locomotive in typical conditions, and N is the number of times that the locomotive operates in typical conditions during the whole life cycle.

(2) Constraints

In addition to the constraints in the processing of the convex optimization problem, constraints such as the output power of the diesel generator are also imposed to ensure the validity of the configuration optimization results.

Constraints on diesel engine output power:

$$\begin{cases} P_{DG} \ge 0 \\ P_{DG} \le P_{DG,max} \end{cases} \quad (16)$$

To facilitate the calculation of energy consumption, the initial and end SOC of the power cell are made consistent:

$$\begin{cases} E_{bat}(t_0) = E_{bat}(t_f) \\ E_{bat}(t_0) = \dfrac{N_b C_{bat}}{2}(U_b(t_0)^2 - U_{b0}^2) \end{cases} \quad (17)$$

Equations (1), (10), (11), (12), (13), (14), (16), and (17) together form the constraints of the convex optimal capacity allocation problem.

4.3 Analysis of Simulation Results

Based on the typical driving conditions of shunting locomotives constructed in Chapter 2, a convex optimization method is adopted to optimize the configuration of the hybrid power system, and a comparative analysis is conducted between the optimization results and the original configuration scheme.

(1) Input parameters

The input bus demand power curve is shown in Fig. 9.

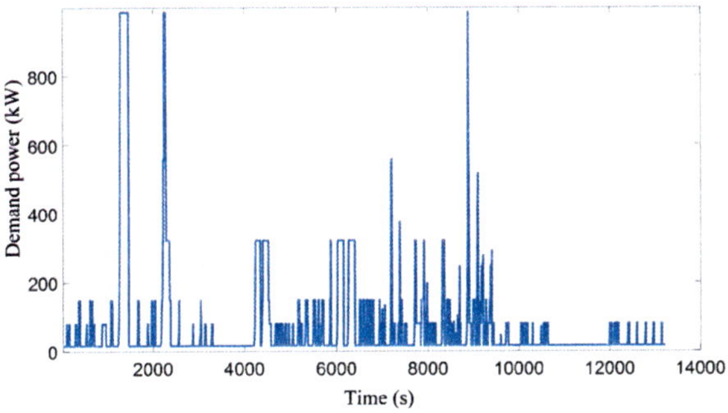

Fig. 9. Curve of busbar demand power

Power battery related parameters are shown below (Table 2):

Table 2. Parameters of power batteries

Battery maximum charge and discharge rate	Power battery unit price (10,000 yuan/kWh)	Initial SOC (%)
1.25C/1.5C	0.3	60

(2) Optimization results

The optimal configuration parameters based on the typical working conditions of shunting locomotives are obtained through optimization calculations, as shown in the following table.

Calculate the fuel consumption and battery acquisition cost in the original configuration and compare it with the optimized cost.

As shown in Table 4, the original configuration has a lower battery acquisition cost because it is a light hybrid vehicle with a lower rated power battery. As for the fuel consumption, it is obviously more fuel-efficient in the optimized configuration.

According to the optimized configuration parameters in Table 3 and the power allocation strategy shown in Fig. 10, the rated power of the diesel engine is smaller in the

Table 3. Optimal parameters of hybrid power system

	Parameter	Value
Power battery	Maximum charging power (kW)	868
	Maximum discharge power (kW)	724
	Power distribution (kWh)	579
Diesel generator sets	Rated output power (kW)	415

Table 4. Cost comparison with different configuration options

Configuration method	Fuel consumption cost (10,000 yuan)	Battery purchase cost (10,000 yuan)	Total cost (10,000 yuan)
Original configuration	1922	37	1959
Optimal configuration	1521	116	1637

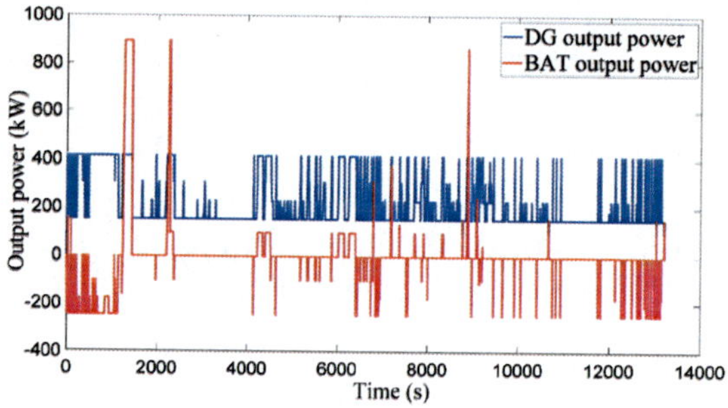

Fig. 10. Strategy of power allocation

optimized configuration, which is close to the average demand power, so the diesel generator can be kept on for a long time, and the fuel power is larger at this time. In contrast, the rated output power of the diesel engine in the original configuration scheme is 2500 kW, which is much higher than the demand power, resulting in low fuel efficiency and high fuel consumption when the power is turned on.

Compared with the original configuration scheme, the optimized scheme achieves a fuel saving of 20.86% and an economy of 16.44%.

5 Conclusion

This paper takes the shunting locomotive as the research object, establishes a typical driving condition based on the actual operation data, and proposes a convex optimized capacity allocation method based on the typical condition, carries out simulation calculations, and obtains the following conclusions:

(1) Relative to the traditional internal combustion locomotive, the hybrid shunting locomotive is able to achieve the fuel saving effect;
(2) Compared with the original configuration scheme, the optimized scheme can achieve a fuel saving rate of 20.86% and a total life cycle cost reduction of 16.44%, even though the battery acquisition cost is increased.

Acknowledgments. This research is funded by the Open Project Fund of the State Key Laboratory of Heavy-duty and Express High-power Electric Locomotive.

References

1. Department of Transportation: National development and reform commission: Introduction to the national "medium and long-term railway network planning." Railw. Knowl. **5**(4), 18–21 (2005). (in Chinese)
2. Zhou, R.: Thirteenth five-year development plan for railways" released. Urban Rail Transit Res. **2017**(12), 37 (2017). (in Chinese)
3. Cipek, M., Pavkovic, D., Kljaic, Z., et al.: Assessment of battery-hybrid diesel-electric locomotive fuel savings and emission reduction potentials based on a realistic mountainous rail route. Energy **173**(APR.15), 1154–1171 (2019)
4. Mayet, C., Pouget, J., Bouscayrol, A., et al.: Influence of an energy storage system on the energy consumption of a diesel-electric locomotive. IEEE Trans. Veh. Technol. **63**(3), 1032–1040 (2014)
5. Li, Q., Huang, W., Chen, W., et al.: Regenerative braking energy recovery strategy based on Pontryagin's minimum principle for fuel cell/supercapacitor hybrid locomotive. Int. J. Hydrogen Energy **44**(11), 5454–5461 (2018)
6. Xilei, Li Yudan, Chen Jianfeng, et al.: Intelligent power generation control based on virtual wolf pack control strategy. Chin. J. Electr. Eng. **38**(10), 2966–2979+3147 (2018) (in Chinese)
7. Sánchez, M., Delprat, S., Hofman, T.: Energy management of hybrid vehicles with state constraints: A penalty and implicit Hamiltonian minimization approach. Appl. Energy **260**, 114149 (2020)
8. Yang, Y., Zhang, Y., Tian, J., et al.: Adaptive real-time optimal energy management strategy for extender range electric vehicle. Energy **197**, 117237 (2020)
9. Peng, H., Li, J., Lars Löwenstein, et al.: A scalable, causal, adaptive energy management strategy based on optimal control theory for a fuel cell hybrid railway vehicle. Appl. Energy **267** (2020)
10. Jian, W., Chi, Z., Weige, Z., et al.: Comparison of different simulation models and simulation methods of energy storage trams. Proc. CSEE **40**(16), 5148–5162 (2020). (in Chinese)
11. Jibin, Y., Pengyun, S., Jiye, Z., et al.: Research on hybrid modern tram simulation system. J. Mech. Eng. **53**(18), 161–168 (2017). (in Chinese)
12. Haezah, M.N., et al.: "A conceptual design of main components sizing for UMT PHEV powertrain." IOP Conf. Ser.: Mater. Sci. Eng. **100**(1) (2015), IOP Publishing
13. Colzi, F., Nigris, M.D., Corti, A., et al.: Extended range electric vehicles components preliminary sizing based on real mission profiles 2013 World Electric Vehicle Symposium and Exhibition (EVS27). IEEE (2014)
14. Yu, H.F., Wang, W., Liu, Z.Q.: Parameter sizing of hybrid energy storage system for hybrid electric vehicle. Adv. Mater. Res. **608–609**, 1643–1646 (2013)
15. Peng, J.: Plug-in hybrid bus working condition construction and variable time domain predictive energy management. (2016) (in Chinese)

16. Wu, X., Cao, B., Li, X., et al.: Component sizing optimization of plug-in hybrid electric vehicles. Appl. Energy **88**(3), 799–804 (2011)
17. Suh, B., Frank, A., Chung, Y.J., et al.: Powertrain system optimization for a heavy-duty hybrid electric bus. Int. J. Automot. Technol. **12**(1), 131–139 (2011)
18. Zhang, L., Xiaosong, H., Wang, Z., et al.: Multiobjective optimal sizing of hybrid energy storage system for electric vehicles. IEEE Trans. Veh. Technol. **67**(2), 1027–1035 (2017)

Application Research of New Contactless Power Supply System for Medium Speed Maglev Train

Kuanxin Li[1](✉), Yonggang Wang[1], Xinghu Chen[1], Zhenggang Yin[2], and Liming Shi[2]

[1] CRRC Tangshan Co., Ltd, Tangshan 064000, China
likuanxin@tangche.com

[2] Institute of Electrical Engineering, Chinese Academy of Sciences, Beijing 100190, China

Abstract. The current power supply system is an indispensable part of maglev train, which mainly provides the electric energy needed by the on-board equipment of the train. The traction system of the medium speed maglev train with permanent magnet synchronous intermediate drive is transferred to the ground, and the electric energy required for traction is provided by the ground equipment. The current power supply system of maglev train should consider the actual operating environment of the train to provide the current service for the train. This article introduces a new contactless power supply system for medium speed maglev train And it describes the composition, control principle, performance parameters of main components and typical test. The system combines redundancy technology, mobile inductive coupling power transmission technology, and adopts modular design. The system has been developed for the first time on the maglev train and test line.

Keywords: Medium speed maglev · Contactless power supply control · High frequency inverter · Electromagnetic coupling

1 Introduction

The medium-speed maglev train is developed based on "Research on Key Technologies of Medium-speed Maglev Transportation System" for the National Key R&D Program of China. In order to meet the requirements of running speed, traction efficiency and suspension, the train adopts the technical material scheme of "permanent magnet excitation and no iron core winding". This scheme determines the medium-speed maglev train should solve the power supply problem of above 120 km/h train in combination with the application environment.

There is a relatively mature contactless power supply mode in maglev train power supply, which uses linear generator power supply. Shanghai high-speed maglev train adopts the mode of the linear generator power supply. Wang et al. [1] suggests a method for figuring out the maximum output power of the EDS maglev's linear generator. Wang et al. [1] proposes a new control structure to achieve the loose coupling control between the stator side and the mover side, where the scalar control is used for the stator side controller (SSC), and the field-oriented control (FOC) is adopted for the mover side

L. Jia et al. (Eds.): EITRT 2023, LNEE 1135, pp. 587–595, 2024.
https://doi.org/10.1007/978-981-99-9307-9_64

controller (MSC). However, Halboch permanent magnet linear synchronous motor [2] adopting for the medium speed maglev train, it means that the medium speed maglev train cannot use the linear generator power supply mode when it exceeds 120 km/h.

Another contactless power supply method is studied in the article [3], which is initially used for rail trains. It has the advantages of no friction, easy maintenance, free from the speed limit of train movement, and no noise pollution. This provides a new choice for the power supply receiving scheme of the medium speed maglev train at high speed.

2 Design of Contactless Power Supply System

2.1 System Working Mechanism

The contactless power supply system is developed based on the principle of mobile induction-coupled power transmission technology [4]. The schematic diagram of the mobile inductively coupled power transmission system is shown in Fig. 1. The system includes ground side converter [5], electromagnetic coupling mechanism (ground transmitting coil structure and vehicle receiving coil structure), vehicle converter and so on. The ground side high-frequency converter converts the mains or other standard external power supply into high-frequency current to supply power to the ground coil. The vehicle receiving coil senses the voltage through the high-frequency resonance of the ground coil. The vehicle induced output voltage is rectified by the vehicle converter to obtain direct current, and then connected to the vehicle power supply trunk, so as to supply power to the vehicle load.

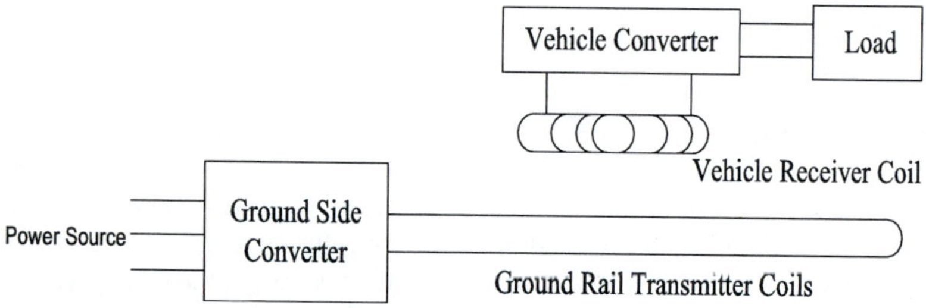

Fig. 1. Schematic diagram of mobile inductive coupling power transmission principle

2.2 Arrangement of Contactless Power Supply System for Medium Speed Maglev

The medium speed maglev contactless power supply system is mainly composed of ground high frequency inverter cabinet, electromagnetic coupling mechanism [6] (transmitting coil structure on the laying line, vehicle receiving coil structure) and vehicle rectifier. The medium-speed maglev adopts the linear synchronous motor drive mode of "permanent magnet excitation installed at the bottom of the vehicle + no iron core winding laid in the center of the track", that is, the stator winding with no iron core length is laid in the center of the track, and the excitation permanent magnet corresponding to the ground length stator winding is installed at the bottom of the vehicle, as shown in Fig. 2.

In order to adapt to the layout of line equipment and vehicle on-board equipment, and consider as far away from the laying equipment on the track as possible to reduce the impact on traction and suspension, the high-frequency cable launching coil of the contactless power supply system is fixed on the track cement beam under the vehicle.

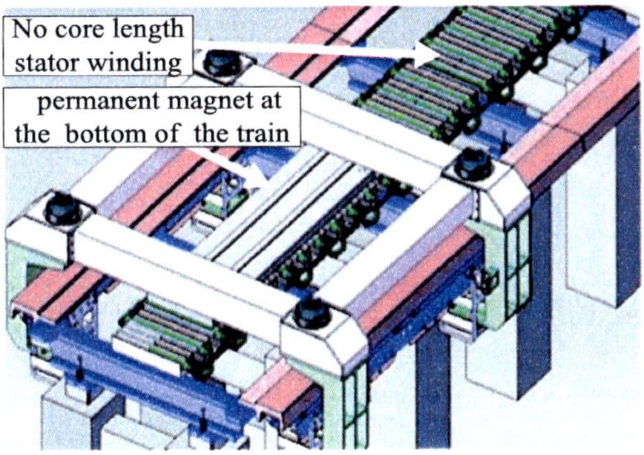

Fig. 2. Layout of linear synchronous motor

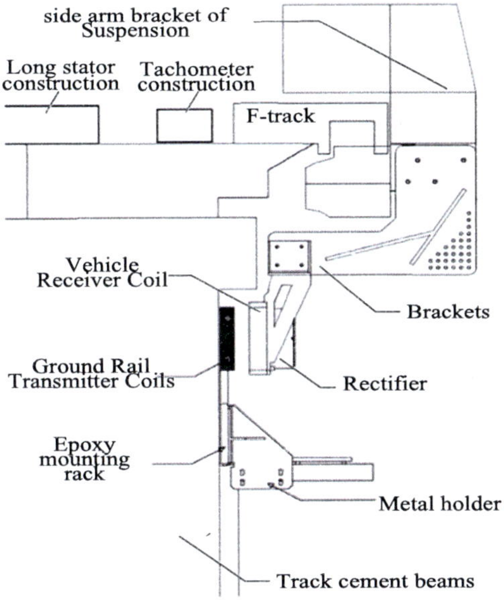

Fig. 3. Layout of linear synchronous motor Installation scheme of electromagnetic coupling mechanism

As shown in Fig. 3, the vehicle receiving device is fixed on the vehicle suspension frame [7] by mounting bolts through the side arm bracket. Due to the need to match the adjustment of the transverse and longitudinal height of the bend, the installation holes are set as long round holes in the design, and the adjustment installation is carried out for different bends.

In order to ensure that the side arm bracket does not have problems in the operation of the vehicle, such as bump, scratch, etc., and reduce the cost of trial and error, customized wooden bracket is adopted in the actual project implementation for vehicle operation test. The structure is corrected through the round-trip operation of the vehicle, and then the final structure of the side arm bracket is determined. In this way, the rework of the real part is avoided, and the overall test efficiency of the project is improved.

2.3 Maglev System Power Supply Scheme

The power supply of the contactless system should not only consider the power supply conditions of the external system to itself, but also take into account the line conditions, power supply efficiency, field operability, cable laying scheme, etc. In comprehensive consideration of project cost, site power location and in view of the stability of the converter main circuit and the power supply efficiency of the whole system, We choose DC 750 V as the input. The power supply voltage required by the vehicle is DC 450 V, and the voltage fluctuation is $\pm 10\%$. The on-board modular design of contactless power supply considers two types of single car 90 kW and single car 60 kw to meet the different power supply requirements of the train. The high frequency inverter converts the direct current DC 750 into high frequency alternating current, which passes into the primary transmitting coil. The high-frequency inverter adopts the partition power supply mode, Considering the factors including line loss caused by long-distance power supply, the electromagnetic radiation of the vehicle without coverage, power supply efficiency, power supply stability and verification indicators.

3 Design Principle and Related Parameters

3.1 High Frequency Inverter

The high frequency inverter is placed in the high frequency inverter cabinet and alternately supplies power to the transmitting coil on the track through a control switch. In order to facilitate maintenance and parts replacement, the high frequency inverter in the cabinet adopts a modular design structure. Figure 4 shows the main circuit structure of the high frequency inverter.

The performance parameters of the high-frequency inverter are as follows:

- Input voltage: DC 750 V;
- Current frequency: 30 kHz;
- Bicycle output power: 60 kW;
- Transfer efficiency: $\geq 80\%$.

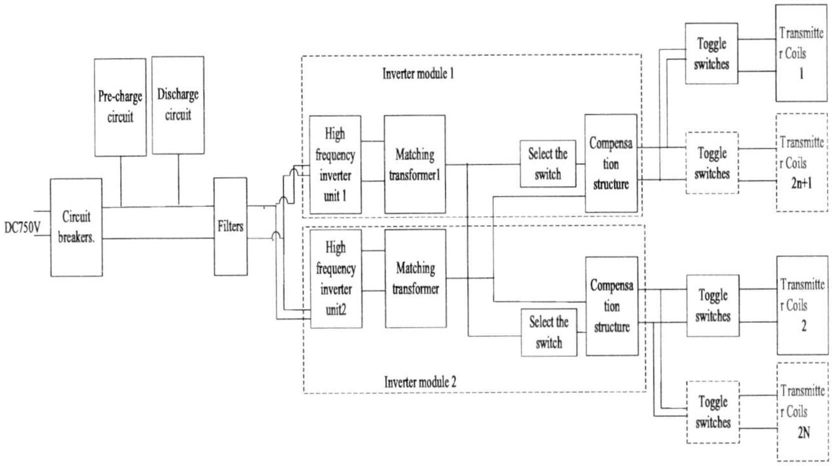

Fig. 4. Main circuit structure diagram

3.2 Electromagnetic Coupling Mechanism and Vehicle Rectification

The electromagnetic coupling mechanism is divided into a transmitting coil structure installed along the track cement beam and a vehicle receiving structure installed on the side arm of the vehicle suspension frame. According to the principle of mobile inductive coupling power transmission technology, the length of the transmitting coil is different and the compensation capacitance value is not allowed. So it means that the type of transmitting coil length directly determines the type of compensation structure in the high-frequency inverter. Through field investigation, we finally determined four transmitting coil lengths. The vehicle suspension frame meets the requirements of the minimum curve of the vehicle.

In principle, the length of the vehicle receiving structure installed on the side arm of the vehicle suspension frame can be less than the longitudinal direction of the side arm length of the suspension frame. In order to make the weight of the vehicle receiving structure evenly distributed relative to the vehicle, the vehicle receiving structure adopts a distributed arrangement. Considering many factors such as vehicle boundary limit, under-vehicle installation space, maximum left and right lateral displacement during vehicle operation, vehicle suspension fluctuation, under-vehicle wiring, convenient component maintenance/replacement, etc., the vehicle receiving structure adopts the form of modular design and grouping installation. In view of the form of 120 kW or 60 kW for a single vehicle and taking into account the installation space and ease of installation, the single vehicle receiving unit is divided into 30 kW. Each vehicle receiving unit consists of 4 vehicle receiving modules, as shown in Fig. 5. The space on one side of a suspension frame can accommodate up to one vehicle receiving unit.

The each vehicle receiving module consists of coil plate and high-frequency rectifier. The output power of each DC/DC rectifier box is made by the current equalization control strategy 7.5 kW, whose maximum output is 10 kW. The main performance parameters of the vehicle receiving structure are as follows:

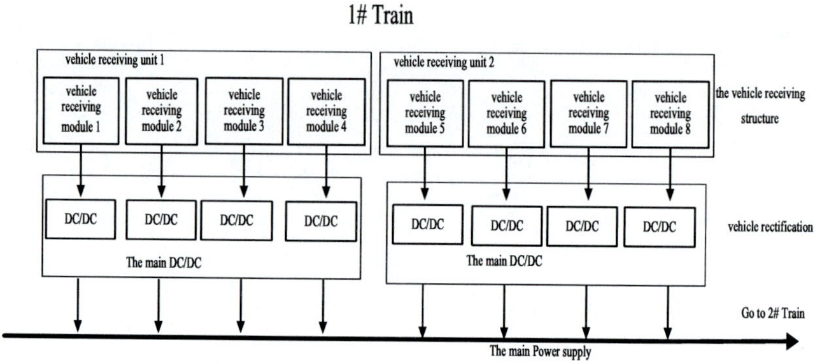

Fig. 5. Modular diagram of vehicle receiving structure

- Transmitting coil current: 200 A
- Resonant frequency: 30 kHz
- Output rated voltage: DC 450 V
- Output power: 60 kW
- Rated output power of single receiving module: 7.5 kW

4 Typical Test and Experiment of System

The whole set of contactless power supply system is installed in the field. On the ground we test the cabinet control logic, the maximum input voltage and the rated temperature rise of the ground system. On board we Carry out air gap test, receiving voltage test, vehicle DC/DC lift voltage test and so on. This article describes only a few typical tests: ground maximum input voltage, air gap and receiving voltage test, and light load efficiency test.

4.1 Ground Maximum Input Voltage Test

As can be seen from the structure diagram of the main circuit in Fig. 4, the DC 750 V input power is output in parallel to the two inverters through the pre-charging circuit and filtering circuit, and the maximum input voltage test is carried out on the two inverters, as shown in Fig. 6.

4.2 Air Gap and Receiving Voltage Test

The magnetic field generated by the ground transmitting coil is captured by the vehicle receiving coil through the air gap. The difference between the ground transmitting coil and the air gap of the vehicle receiving plate will cause the induced voltage of the vehicle receiving plate to be different. The induced voltage of a certain vehicle receiver plate under different air gaps is measured, as shown in Table 1.

It can be seen from the table that when the current of the primary transmitting coil is basically constant, the induced voltage of the vehicle receiver board will decrease with the increase of the air gap, and the corresponding voltage will decrease by about 1/3 as the air gap increases.

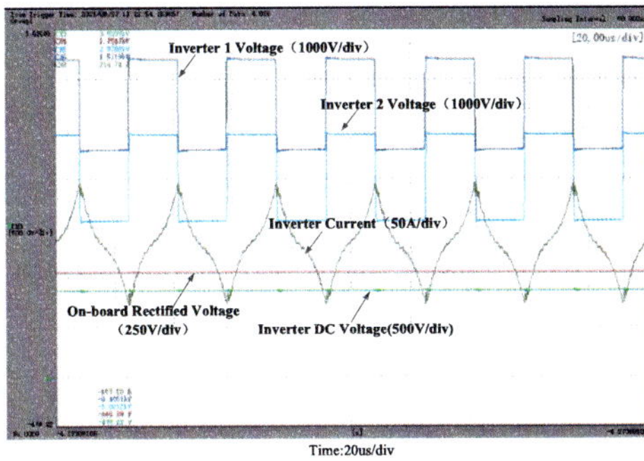

Fig. 6. Voltage and current waveform when the maximum input voltage is 900 V

Table 1. Senses the voltage of the vehicle receiver board and air gap

Serial number	Air gap/mm	Primary side current/A	Secondary induced voltage/V
1	27	190	466.4
2	36.5	190.35	424.3
3	40	196.25	400
4	56	196.25	330

4.3 Efficiency Test

The efficiency test of the system under rated transmission power is carried out. The input end of the test is the DC side of the ground high-frequency inverter, and the output end is the output side of the vehicle DC/DC. In order to get the load resistance power to 60 kW, We connect the DC/DC output to the 4Ω resistance box and set the DC/DC output voltage to 500 V. The voltage and current of the ground input terminal are shown in Fig. 7. Other test [7–10] of subsequent devices still needs to be tested.

Figure 7 shows that input voltage is 782.09 V, input current is 89.06 A, calculated input power is 69.65 kW. The on-board display displays an output voltage of 498.90 V and an output current of 122.40 A. By calculation, the output power is 61.07 kW. The efficiency of the system at rated power is the output power than the input power, and the static efficiency of the system is 87.7% by bringing the above power value into the system.

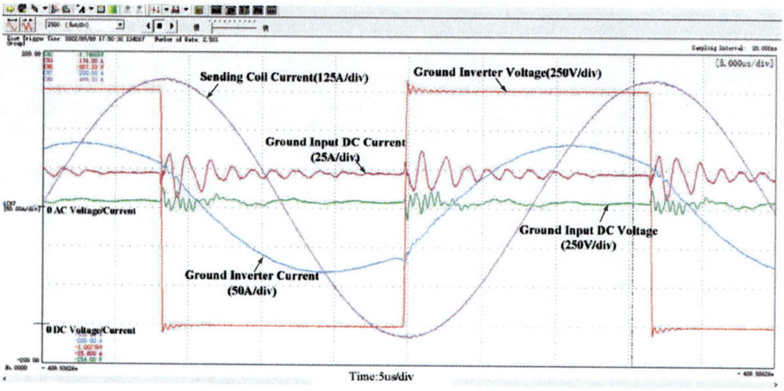

Fig. 7. System input voltage/current

5 Conclusion

At present, the whole contactless power supply system has been installed. Contactless power supply system can operate reliably at DC-900 V input voltage and adapt to the field voltage fluctuation range. Carrying out the oven test under the current of the transmitting coil 240 A, The temperature rise of the ground system and the vehicle system are controllable, which ensures the long-term reliability and stability of the system. At present, it has completed the trial installation and some static test on the Maglev Test Line. and other static tests and dynamic tests are being carried out, which will provide a new solution for the flow of medium-speed maglev train.

Acknowledgments. This work was supported by National Key R&D Program of China (Grant No. 2016YFB1200601).

References

1. Wang, Y., Shao, Z., Zhong, Z., et al.: Quasi-synchronous operation of a long-stator doubly fed linear motor with accurate control of both propulsion and contactless power transfer. IEEE Trans. Transp. Electrification **9**(1), 689–698 (2022)
2. Yonggang, W., Jun, Y., Li, Z.: Analysis of force angle characteristics of hollow Halbach permanent magnet linear synchronous motor for maglev train. Railw. Roll. Stock. **41**(1), 26–32 (2021). (in Chinese)
3. Shi, L., Yin, Z., Jiang, L., Li Y.: Advances in inductively coupled power transfer technology for rail transit. CES Trans. Electr. Mach. Syst. **1**(4), 383–396 (2017)
4. Fin, Z., Shi, L., Yin, Z., Jiang, L., Zhang, F.: Mutual inductance compensation method and output power balance strategy of multi-pick coil in inductive power transmission system. J. Trans. China Electrotech. Soc. **3**(17), 3998–4006 (2018). (in Chinese)
5. Zhang, F., Shi, L., Yin, Z., Fan, M., Jiang, L.: A Parallel current control strategy for receiving modules in inductively coupled power transmission system. J. Proc. CSEE **38**(20), 6091–6098 (2018) (in Chinese)

6. Yonggang, W., Jun, Y., Yumeng, C.: Characteristics analysis of permanent magnet electromagnetic hybrid suspension guide system for maglev train. Locomot. Electr. Drive **6**, 2–8 (2021). (in Chinese)
7. Zhang, F., Shi, L., Yin, Z., Fan, M., Jiang, L.: Current balance control strategy for inductively coupled power transfer system with parallel pickup modules. In: 2017 IEEE Transportation Electrification Conference and Expo, pp. 1–6. IEEE, Asia-Pacific (2017)
8. Zhang, F., Shi, L., Yin, Z., Fan, M., Jiang, L.: Efficiency analysis of inductively coupled power transfer with multi-pickups. In: 20th International Conference on Electrical Machines and Systems (ICEMS), pp. 1–5. IEEE, Sydney (2017)
9. Yin, Z., Shi, L., Fan, M.: Optimal setting of primary coil current for efficiency in electromagnetic inductive wireless power transfer system. Trans. China Electrotech. Soc. **36**(s1), 46–53 (2021)
10. Yin, Z., Shi, L., Yang, J., et al.: The power control of a multiple pick-up bidirectional wireless power transfer system. In: International Conference on Wireless Power Transfer, pp. 444–454. Springer Nature Singapore, Singapore (2022)

An Enhanced Feedforward Compensation Speed Closed Loop Control Strategy for the Vacuum Pipelines High-Speed Maglev Train

Zhihong Yang(✉), Yongyong Guo, and Kekang Wei

China Aerospace Science and Industry Co., Ltd., Beijing, China
yzhm1516@126.com

Abstract. Speed closed-loop control is a key operational control strategy for trains to achieve speed tracking, controllable position, and on-time parking. The vacuum pipeline maglev train aims at the high-speed operation, with the principle of safe, reliable. It should fully consider the precise control of traction operation. This article proposes an enhanced feedforward compensation control strategy. Firstly, the operation status of the train was analyzed, and established a motion model based on the traction and resistance. Under the constraints of train operation status, such as speed, acceleration, and position, and achieves high-precision speed closed-loop control through traditional PID control and enhanced feedforward compensation. Adjust and control according to the control strategy within the six adjustment intervals given for operating speed control error. Meanwhile, the strong robustness of the control strategy was verified based on artificial acceleration disturbance input.

Keywords: The vacuum pipeline high-speed maglev trains · Dynamic constraints · Enhanced feedforward compensation · Speed closed-loop

1 Introduction

For high-speed operating systems, whether it is urban rail transit, high-speed rail, aircraft systems or other transportation tools, high-precision speed closed-loop control systems are the necessary, and it is the guarantee for transportation systems. The power system of the vacuum pipeline high-speed maglev is located on the ground, which determines its natural suitability for using high-precision speed closed-loop control.

The traditional closed-loop control algorithm adopts PID control [1, 2]. With the development of computer technology, various algorithms have been added to PID control to achieve precise tracking of operating curves. Professor She's team conducted research on train operation control based on the Badaling maglev line and the line projected of the university [3]. They established a train master PLC system, collected train data, and combined fuzzy control algorithms to complete speed control on a DSP hardware platform. Professor Wei's team mainly focuses on tracking the operation curve

L. Jia et al. (Eds.): EITRT 2023, LNEE 1135, pp. 596–604, 2024.
https://doi.org/10.1007/978-981-99-9307-9_65

of maglev trains during braking and emergency braking processes, with the auxiliary parking positions, potentially dangerous points are considered between the upper and lower speed protection curves [4]. Through mathematical models, factors such as line slope that affect the dynamic characteristics of vehicle movement are analyzed. In resistance analysis, speed segmentation is considered, with the goal of precise parking. At the same time, combining the principle of simple fuzzy control, the running curve tracking is achieved. Professor Xu's team proposed an adaptive dynamic surface control algorithm for trains, which can achieve high-precision tracking of target speed and displacement while eliminating the problem of computational explosion [5]. The adaptive dynamic surface control law is designed to treat uncertain running resistance as the total disturbance of the system, and an extended state observer is introduced for online estimation.

Foreign research teams have proposed a nonlinear predictive control method based on feedback linearization and model predictive control for high-speed aircraft systems [6, 7]. A control system was designed considering constraints such as thrusters and starting angle, and through simulation analysis of the re-entry flight process of a lift body aircraft, it was verified that this method has stronger robustness.

Therefore, this article proposes a high-precision speed closed-loop control algorithm that combines traditional PID control and enhanced feedforward compensation.

Firstly, based on the train dynamics model, the algorithm flow for high robustness speed closed-loop control is analyzed. According to the position of the train, real-time acceleration and speed of the next position are issued. The speed closed-loop control strategy uses acceleration as the closed-loop control feedforward parameter, and speed as the control target to adjust the speed of the train. Based on setting parameters and specific operating scenarios, simulation tests were conducted to verify the effectiveness and reliability of the algorithm. At the same time, random disturbances were added to verify the robustness of the control algorithm.

2 Train Dynamics Model

The basic dynamic operation problems study the motion laws and operating states of trains under constant traction force [8–10]. The basic control problem of trains is to study the initial and terminal states of known train operation, and to solve the problem of how much traction or braking force is required to control train operation (Tables 1 and 2).

How to solve the traction and braking forces during operation is the goal of closed-loop speed control. According to the state equation, the acceleration during operation directly represents the state of the train's acceleration and deceleration processes. There is the critical correlation between force and acceleration.

3 Running Speed Closed-Loop Control Algorithm

3.1 Function and Composition of Closed-Loop Control of Operating Speed

The speed closed-loop control algorithm mainly refers to the speed closed-loop control that enables trains to operate within the limits of the V–S curve. Based on the deviation between real-time speed and given speed, a control algorithm is adopted to achieve traction and braking force calculation.

Table 1. Constraint condition

State equation		$v = \frac{ds}{dt}, a = \frac{F-W(v)}{M}$
Constraint condition	Point constraint	start time $t_0 = 0$, terminal time t_{final} start constraint $s(0) = 0$, $v(0) = 0$ terminal constraint $s(t_{final}) = s_{final}, v(t_{final}) = 0$
	Control constraints	$-F_{zm} \leq F \leq F_{qm}$
	Speed constraint	$v_{min} \leq v \leq v_{max}$, $\lvert\dot{v}\rvert \leq a_{max}, \lvert\dot{a}\rvert \leq b_{max}$

Table 2. Parameters table

ID	Parameter	Definition	ID	Parameter	Definition
1	S	Train position	8	W(v)	Resistance function
2	V	Train velocity	9	F	Traction force
3	A	Acceleration	10	v_{max}	Maximum velocity
4	$\dot{v}$	Acceleration	11	v_{min}	Minimum velocity
5	$\dot{a}$	Train acceleration change rate	12	a_{max}	Permitted maximum acceleration
6	F_{qm}	Maximum traction force	13	b_{max}	Permitted maximum train acceleration change rate
7	F_{zm}	Maximum braking force			

The enhanced compensation speed control algorithm selects different command information base values and enhanced feedforward coefficients based on feedforward within different speed deviation intervals. Decompose the tracking process of the running curve, and calculate it in small time periods or short distance periods. The final output command information is used for the next step of operation control, achieving stable control within the upper and lower limits of the operation curve.

3.2 Logic Explanation of Enhanced Feedforward Compensation Speed Closed-Loop Control Algorithm

Due to the two round-trip processes of train operation, the acceleration is also different during the acceleration, constant speed, and deceleration stages, it is divided into two major stages based on upward and downward, and then subdivided into three smaller stages under each major stage. The enhanced feedforward compensation speed closed-loop control block diagram is shown in the following Fig. 1.

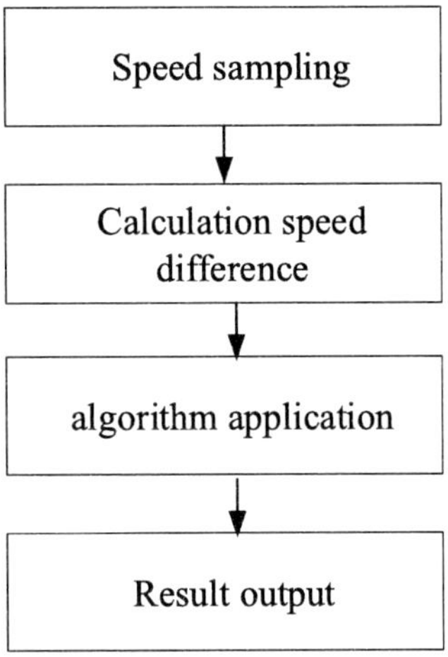

Fig. 1. The enhanced feedforward compensation speed closed-loop control block diagram

(1) Upward process

Calculate speed deviation Δv is the difference between the given speed and the actual speed Obtained from sampling.

Multiple control intervals are selected according to different working conditions.

1. Acceleration phase:

   Divide multiple control intervals based on different tracking speed requirements.

   The Output is the total of General output value and the compensation value.

   Within the closest community, the Output is the total of General output value and the compensation value and the multiple value of speed.

2. Constant velocity phase:

   Divide multiple control intervals based on different tracking speed requirements.

   The Output is the total of General output value and the compensation value and the multiple value of speed;

3. Decelerating phase:

   Divide multiple control intervals based on different tracking speed requirements.

   The Output is the total of General output value and the compensation value.

   Within the closest community, the Output is the total of General output value and the compensation value and the multiple value of speed.

(2) Downward process

Calculate speed deviation Δv is the difference between the given speed absolute value and the actual speed absolute value.

The rest are the same as the upward process.

4 Simulation Test

4.1 MATLAB Simulation Platform

The MATLAB simulation platform includes a traction operation curve generation module, a speed tracking control module, and a current control module.

The traction operation curve generation module uses existing operation curve data input to construct structural variables, with time variables as reference. Each time point corresponds to the corresponding numerical value, and sequentially generates sequence numbers, acceleration, displacement, and velocity curves.

The speed tracking control module includes a speed closed-loop control module, an acceleration (adjustable) feedforward module, a resistance generation module, and a dynamics operation simulation module.

The current control module includes a current closed-loop module, ABC-DQ module, DC-ABC module, angular velocity calculation module, output current module, etc.

4.2 Test Parameters

In the MATLAB simulation platform, the traction acceleration limited value is set to 4 m/s^2, and the braking acceleration limited value is set to 5 m/s^2. The starting velocity of the train is 0 m/s, and the starting position is 0 m. Kp is 600, Ki is 15. The acceleration curve is generated by the upper software, like as Fig. 2. The speed zones are divided into six intervals (Fig. 3).

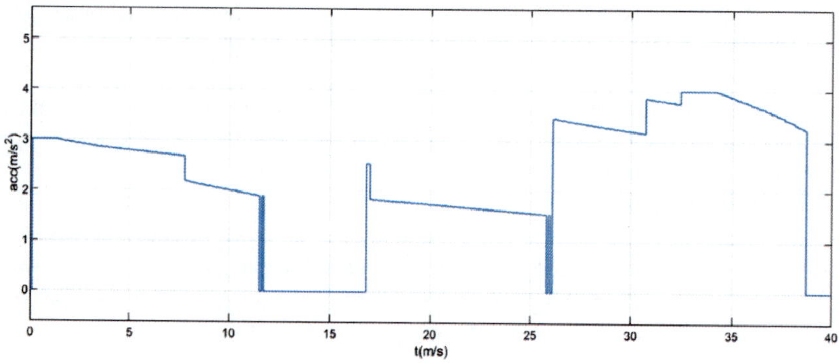

Fig. 2. Input acceleration acc curve

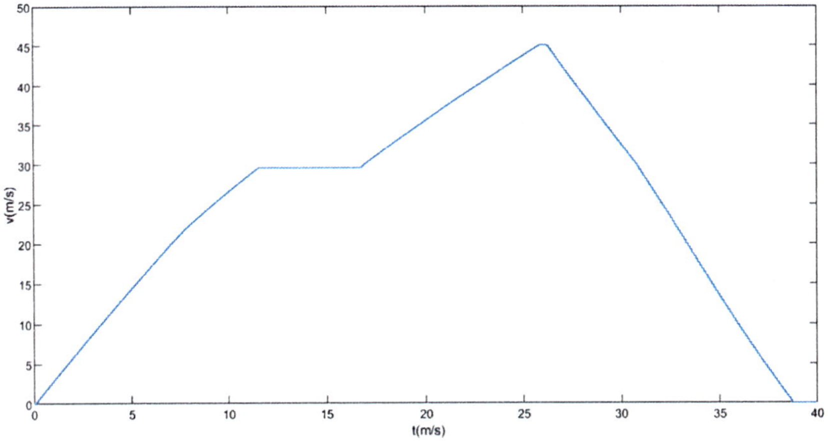

Fig. 3. Input velocity curve

4.3 Test Results

The traction system utilizes the received acceleration as a parameter to implement a feedforward compensation speed closed-loop strategy. From the simulation results, it can be seen that the control strategy with feedforward is significantly more accurate in achieving speed closed-loop control than the results without feedforward compensate, and the tracking curve runs closer to the given curve. The results show that this control method reduces the impact of train operation, provides precise speed control, and the control algorithm can achieve a speed error tracking of ±3 km/h (Figs. 4 and 5).

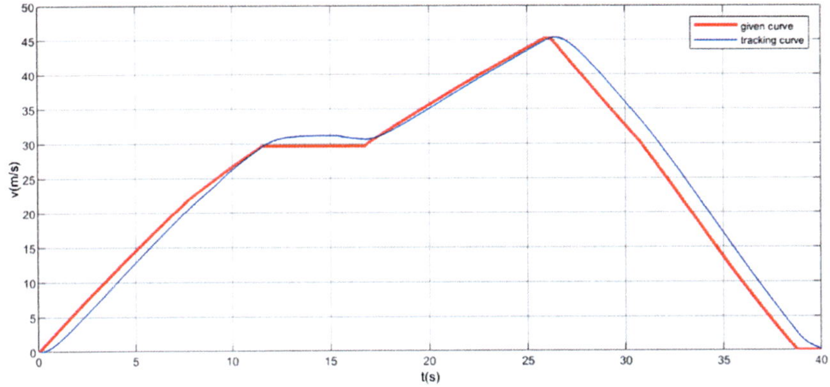

Fig. 4. Test results of control experiment without feedforward compensation

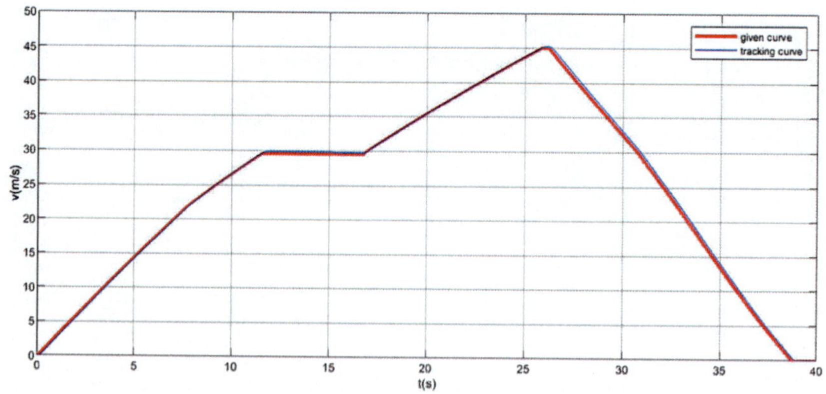

Fig. 5. Test results of control experiment with 0.8 ratio feedforward compensation

4.4 Robustness Verification Test of Control Algorithm

Due to the use of an enhanced feedforward compensation control algorithm in the control algorithm, feedforward plays a significant role in the strategy. Therefore, a fluctuation of 30–80% is randomly added to the input acceleration parameters, as shown in the Fig. 9. Using another working condition with a relatively constant acceleration for simulation testing. Comparing the experimental results of Figs. 6 and 8, even with the large fluctuation acceleration input like Fig. 9, this control algorithm can still achieve stable tracking. Then the strong robustness of the control algorithm has been verified (Fig. 7).

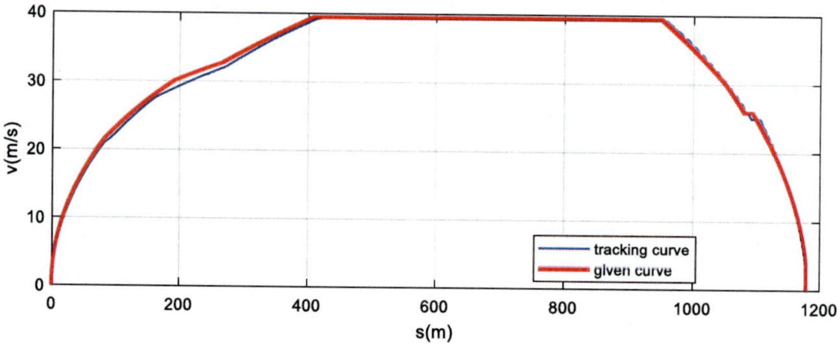

Fig. 6. Test results of velocity tracking test with no random fluctuations in acceleration input

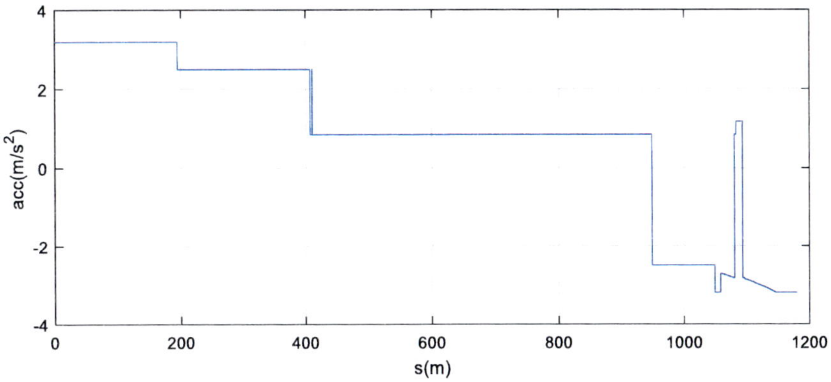

Fig. 7. Acceleration input curve

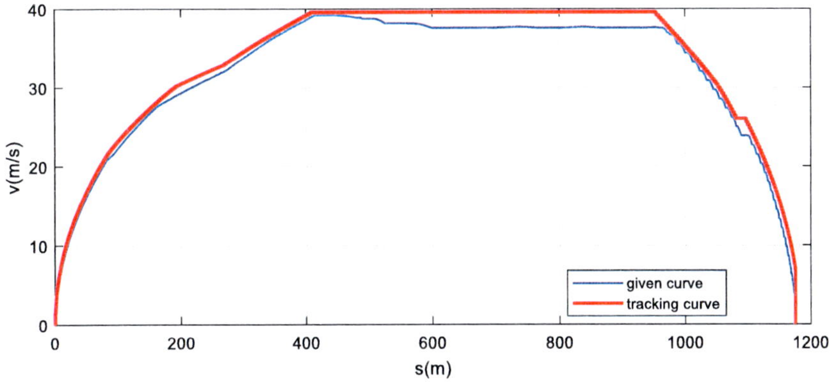

Fig. 8. Test results of velocity tracking test with random fluctuations in acceleration input

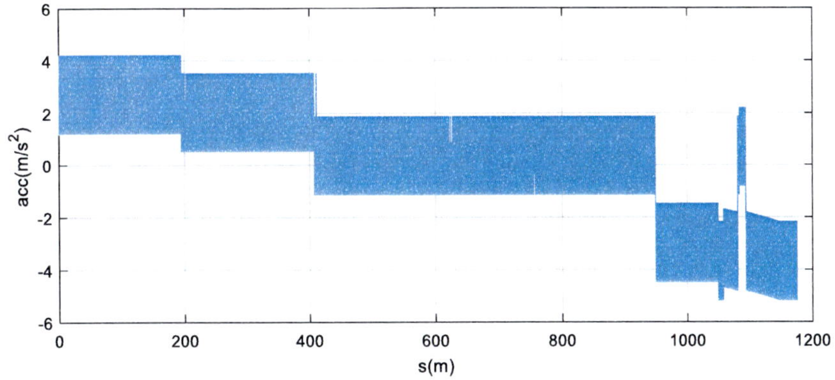

Fig. 9. Acceleration input curve with 30 ~ 80% fluctuation added

5 Conclusion

This article proposes an enhanced feedforward compensation speed closed-loop control strategy based on traditional PID control. Through testing on the simulation platform, the experimental results were compared between the algorithm that only considers PID control and the control algorithm proposed in this paper. The data shows that the control algorithm proposed in this article can achieve speed error tracking of ± 3 km/h. At the same time, in order to eliminate the dependence on the stability of the feedforward input parameters, random fluctuations were added to the input variables, and the experimental results verified the strong robustness of the control algorithm.

Acknowledgments. This work is supported by major project in Shanxi Province: Full-scale experimental line of ultra-high-speed low-vacuum pipeline maglev transportation system (Phase I).

References

1. Han, S.H., Byen, Y.S., Baek, J.H., et al.: An optimal automatic train operation (ATO) control using genetic algorithms (GA) [C]. In: Proceedings of the IEEE Region 10 Conference. IEEE, vol. 1, pp. 360–362 (1999)
2. Duarte, M.A., Sotomayor, P.X.: Minimum energy trajectories for subway systems [J]. Optim. Control. Appl. Methods **20**(6), 283–296 (1999)
3. Wang, H., Yin, L., et al.: Research on Speed Control and auto drive system of Maglev Train. Master of Engineering Thesis (2002) (in Chinese)
4. Liu, Z., Wei, W., et al.: Research on safety speed curve algorithm and emergency braking control of high speed maglev trains. Master's Thesis (2004) (in Chinese)
5. Xu, C., Chen, X., Zheng, X., et al.: Adaptive dynamic surface velocity displacement tracking control for high-speed trains [J]. J. Electr. Mach. Control. **23**(12), 8 (2019) (in Chinese)
6. Chang, C.S., Sim, S.S.: Optimising train movements through coast control using genetic algorithms[C]. Electr. Power Appl., IEE Proc.—IET **144**(1), 65–73 (1997)
7. Wong, K.K., Ho, T.K.: Coast control for mass rapid transit railways with searching methods[C]. Electr. Power Appl., IEE Proc.—IET **151**(3), 365–376 (2004)
8. Wong, K.K., Ho, T.K.: Dynamic coast control of train movement with genetic algorithm [J]. Int. J. Syst. Sci. **35**(13–14), 835–846 (2004)
9. Gao, B., Zheng, K., Zhang, F., Su, R., Zhang, J., Wu, Y.: Research on multi-target tracking method based on multi-sensor fusion. Smart Resilient Transp. **4**(2), 46–65 (2022)
10. Abudakka, F.J., Rubio, F., Valero, F., et al.: Evolutionary indirect approach to solving trajectory planning problem for industrial robots operating in workspaces with obstacles [J]. Eur. J. Mech. A/Soilds **42**(6), 210–218 (2013)

Study on Variable Frequency Power Supply Cable and Leakage Current Characteristics of High-Speed Maglev Train

Qinglai Li(✉), Pengcheng Du, Yongyong Guo, and Kekang Wei

China Aerospace Science and Industry Co., Ltd., Beijing 100143, China
714375417@qq.com

Abstract. The high-speed maglev train uses long distance variable frequency power supply in operation, so the distance between the traction substation and the stator section of the train is usually thousands of meters, which causes a large leakage current in the power supply cable. Aiming at the leakage current of power supply cables, this paper builds a cable model in PSCAD, and applies voltages of different amplitudes and frequencies and different length parameters to power supply cables according to the actual working environment and structural parameters, so as to obtain the relationship between leakage current and these parameters. According to the actual structure of XLPE power supply cable used in the power supply system of high-speed maglev train, a power supply cable model considering ground access parameters is proposed on the basis of the traditional power supply cable model. The proposed model is simulated in Simulink and compared with the simulation results of PSCAD to verify the rationality of the model. It provides a theoretical basis for analyzing the leakage current of variable frequency power supply of high-speed maglev train.

Keywords: High-speed maglev train · Leakage current · Supply cable · PSCAD

1 Introduction

As an important part of high-speed maglev train, traction power supply system has become a hot research topic. Its schematic diagram is shown in Fig. 1. The high-speed maglev train uses double-ended long-distance power supply, and the electric energy comes out from the traction substation at both ends and reaches the long stator section where the train is located through the long-distance power supply cable, so as to carry out traction power supply to the train [1].

Usually, the power supply cable can reach thousands of meters, because the output voltage is in the form of frequency conversion PWM, so it is different from DC power supply, the distribution parameters need to be considered, so that the input current of the power supply cable is not equal to the output current, and the difference current will flow into the ground through the distributed conductance and distributed capacitance, which is currently called leakage current by scholars [2]. Due to the existence of leakage current,

L. Jia et al. (Eds.): EITRT 2023, LNEE 1135, pp. 605–612, 2024.
https://doi.org/10.1007/978-981-99-9307-9_66

it must be considered in the traction power supply system, if the leakage flow is large, it will damage the equipment near the power supply cable and endanger the safety of the train. At the same time, the leakage current is also closely related to the characteristics and parameters of the power supply cable itself, so that the leakage current can reflect the state of the power supply cable, that is, insulation and aging [3]. So, the analysis of leakage current is very important.

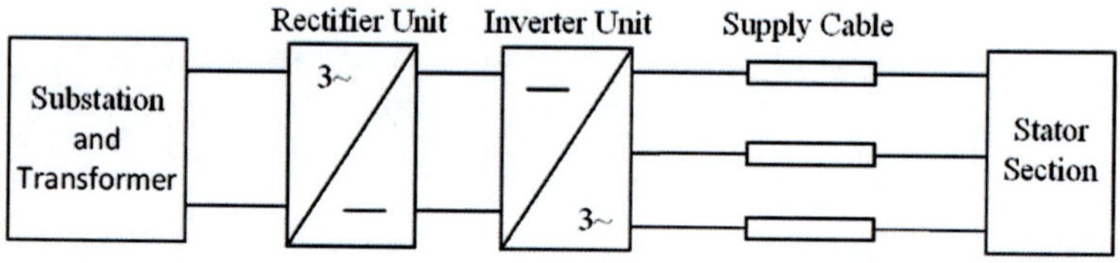

Fig. 1. Schematic diagram of traction power supply structure of high-speed maglev train.

For the problem of long-distance variable frequency power supply, most of the studies in the existing literature focus on the analysis and suppression of overvoltage caused by long-distance power supply cables and how to reduce the influence of long-distance power supply cables on control performance, etc., and there are few studies on the characteristics of long-distance power supply cables leakage current [4–6]. Therefore, this paper builds a power supply cable model in PSCAD simulation software, applies voltage excitation with different amplitude and frequency, and different length parameters to it, so as to obtain the regular relationship between leakage current and these parameters. As the power supply cable of high-speed maglev train is usually grounded in the metal sheath shielding layer, there is a large error in the analysis of the traditional power supply cable model. Therefore, this paper puts forward a power supply cable model considering the parameters of the grounding line. Simulink software simulates the two models, and the simulation results are compared with those in PSCAD software. The rationality of the model is verified by the comparison results.

2 Power Supply Cable Leakage Current Characteristics

2.1 XLPE Power Supply Cable

Cross-Linked Polyethylene (XLPE) cables are used in the traction power supply system of high-speed maglev train because of their good electrical characteristics and good environmental adaptability [7]. XLPE cables usually have single-core and three-core cables, as shown in Fig. 2, in which Fig. 2(a) is the schematic diagram of single-core cable structure, and Fig. 2(b) is the schematic diagram of three-core cable structure. It can be found from the figure that the structure of XLPE cable is from inside to outside, in turn, the core conductor, the inner semiconductor shield layer, the main insulation layer, the outer semiconductor shield layer, the metal sheath, the wrapping tape and the outer sheath [8].

For the traction power supply system of high-speed maglev train in this paper, the single-core XLPE cable is selected for traction power supply considering the carrying capacity, heat dissipation performance and installation conditions.

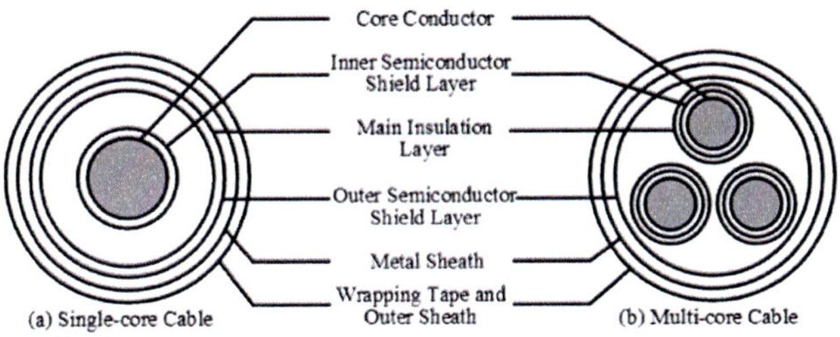

Fig. 2. XLPE single-core cable and multi-core cable structure.

2.2 Simulation Analysis of Leakage Current Characteristics

The single-core XLPE cable model as shown in Fig. 2(a) is established in PSCAD software, and the structural parameters of the power supply cable are shown in Table 1.

Table 1. Structure parameters of the power supply cable.

Parameter	Value	Material
Cable core conductor diameter	17.76 mm	Electrical resistivity 0.028264 Ω mm^2/m
Thickness of the internal semiconductor shield	0.65 mm	Volume resistivity 500 Ω m
Thickness of main insulation layer	5.5 mm	Volume resistivity 3×10^{14} Ω m
Thickness of outer semiconductor shield	0.65 mm	Volume resistivity 300 Ω m
Metal sheath thickness	0.85 mm	Electrical resistivity 0.01724 Ω mm^2/m
Thickness of outer sheath	2.5 mm	Volume resistivity 3×10^{14} Ω m

For the cable model built above, after setting the structural parameters of the power supply cable, simulation is carried out by changing one of the conditions. The simulation and comparison results are shown in Table 2.

It can be seen from the simulation results that the leakage current is very small compared with the supply current. At the same time, it can be found from Table 2 that the leakage current is approximately proportional to the voltage source excitation voltage amplitude, the voltage source excitation voltage frequency, and the cable length. Therefore, it can be shown that under the condition of determining the cable structure, the leakage current is closely related to the output voltage of the converter and the cable length, which is very important for the relay protection of high-speed maglev traction power supply system.

Table 2. Simulation results of different voltage amplitudes.

Voltage/V	Leakage current/A	Frequency/Hz	Leakage current/A	Cable length/km	Leakage current/A
250	0.023	50	0.044	0.5	0.044
500	0.045	100	0.09	1	0.09
1000	0.09	200	0.18	2	0.18
2000	0.18	400	0.35	4	0.35

3 Power Supply Cable Model

For the system with complex control logic and circuit such as high-speed magnetic levitation, there is usually no distributed traveling wave model for the power supply cable in the current widely used simulation, and only a simplified equivalent model can be used for analysis. Because the high-speed magnetic levitation adopts the single-core XLPE power supply cable which is grounded by the metal sheath shielding layer, there will be a large error if the traditional cable model is adopted. Therefore, this paper will improve on the traditional simplified equivalent T-model, and propose a new cable model considering the ground path, which makes the simulation results more accurate.

3.1 Distribution Parameter Model of Power Supply Cable

On the basis of the simple R–L lumped parameter model, because the power supply cable usually works at high frequency or contains high frequency signals, the characteristics of the cable itself, such as ground capacitance and skin effect, make the model lack accuracy [9, 10]. At the same time, the use of this model needs to meet the size of the power supply cable is much smaller than the wavelength of the electromagnetic wave contained in the circuit. Therefore, in long-distance cable power supply, the traditional equivalent model is usually to use multiple ideal components, including resistance, inductance, capacitance, and admittance, while setting up a different number of components according to the length of the supply cable, as shown in Fig. 3 [11].

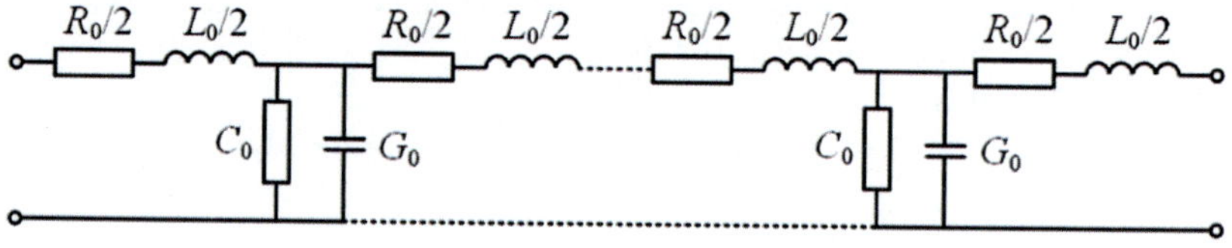

Fig. 3. Traditional T-type equivalent model.

R_0 is the distributed resistance of the cable, L_0 is the distributed inductance of the cable, C_0 is the distributed capacitance of the cable, and G_0 is the distributed insulation resistance of the cable.

3.2 Power Supply Cable Model with Ground Branch

XLPE single-core power supply cable is selected in the traction power supply system of high-speed maglev train. Usually, the metal sheath of the inner layer of the cable will be grounded to make the leakage current have a path to the ground. Therefore, if the traditional equivalent model is used in the simulation, there will be simulation errors. Since the metal sheath in the power supply cable is continuously and evenly distributed, it will be grounded at both ends. From the perspective of uniform distribution parameters, the flow path of the leakage current is as follows: Cable core conductor-distributed capacitance and distributed conductance-metal sheath-grounding. Therefore, the uniform distribution parameter model as shown in Fig. 4 is proposed and established, which is closer to the actual power supply cable when considering the grounding treatment of metal sheath.

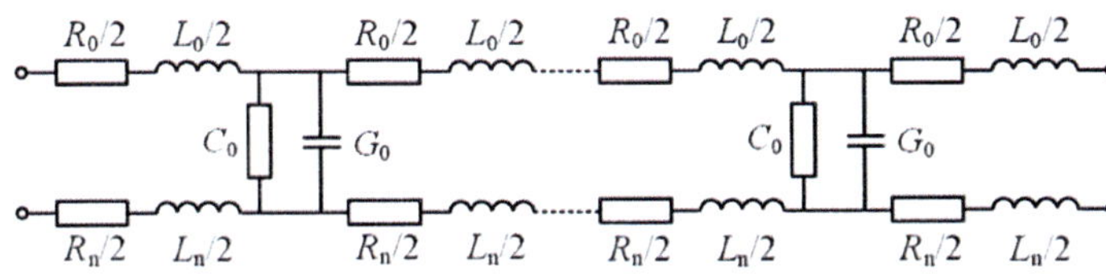

Fig. 4. Improved T-type equivalent model.

According to the power supply cable model shown in Fig. 4, for the resistance parameters R_0 and R_n and the inductance parameters L_0 and L_n, the low-frequency AC voltage excitation can be applied to the cable core conductor and metal sheath respectively to solve the problem. Since the insulation resistance is much greater than the conductor resistance, it can be considered that the insulation resistance and the distributed capacitance branch are open states. By collecting the current flowing through the core conductor of the cable and the voltage at both ends, Ohm's law is used to solve the problem. For the calculation of insulation resistance G_0, the current flowing through the insulation resistance can be obtained by applying DC voltage excitation, collecting the current difference between the conductor and the conductor, and then calculating it according to Ohm's law. For the distributed capacitance C_0 of the cable, it can be calculated by the leakage current under high frequency excitation. At the same time, due to the resistance and inductance of the sheath, the resistance and inductance in the resistance value need to be corrected in the calculation, and then calculated.

4 Power Supply Cable Parameter Calculation and Model Comparison Verification

In this paper, the cable model shown in Table 1 is built in PSCAD software, and all parameters of the power supply cable model proposed in Fig. 4 are calculated according to the above calculation methods. Then the traditional T-type cable model and the cable model proposed in this paper are built in Simulink software, and the calculated parameters are brought into the simulation. Finally, the simulation results in PSCAD software are compared under the same conditions to verify the accuracy of the cable model proposed in this paper.

4.1 Simulation Results of Power Supply Cable Parameters

Figure 5 shows the voltage and current waveforms at both ends of the cable core conductor under low-frequency 1 Hz excitation.

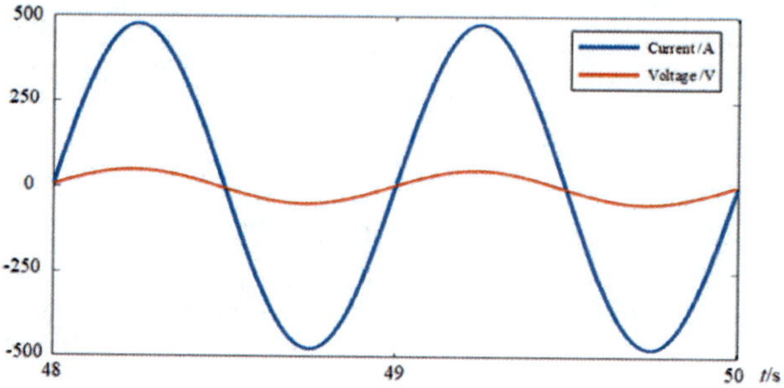

Fig. 5. Voltage and current waveforms at both end of the cable core conductor under 1 Hz excitation.

Under 1 Hz voltage excitation, the voltage amplitude, current amplitude and phase difference at both ends of the cable core conductor are 48.8 V, 475.6 A and 5.544° respectively, and the resistance and inductance of the cable core conductor an be calculated. By calculating the resistance and inductance under different frequency excitation, R_0 is 0.0938 Ω and L_0 is 2.395 mH. According to the above calculation method, it can be obtained that the resistance R_n and inductor L_n of the metal sheath, the insulation resistance G_0 and the modified distributed capacitance C_0 are respectively 0.1162 Ω, 2.379 mH, 29 MΩ, 0.28 μF.

4.2 Comparison of Leakage Current Simulation Results

After all the parameters of the power supply cable are obtained, the cable model shown in Fig. 4 is built in Simulink software, where the simulation conditions are the same as PSCAD. The leakage current simulation results are compared as follows, and the leakage current of the traditional T-type equivalent cable model is compared as follows.

Under 1000 Hz voltage excitation, the simulation errors between the proposed cable model and PSCAD are as follows: amplitude error percentage 0.06%, Angle error 0.16°. The simulation errors of traditional cable model and PSCAD are as follows: amplitude error percentage 0.22%, Angle error 0.19°. At the same time, the error under different voltage frequency excitation is verified as shown in Table 3.

The results in Table 3 can show the rationality of the cable model proposed in this paper, and the error is small, which is suitable for the leakage current analysis of the power supply cable of high-speed maglev train (Fig. 6).

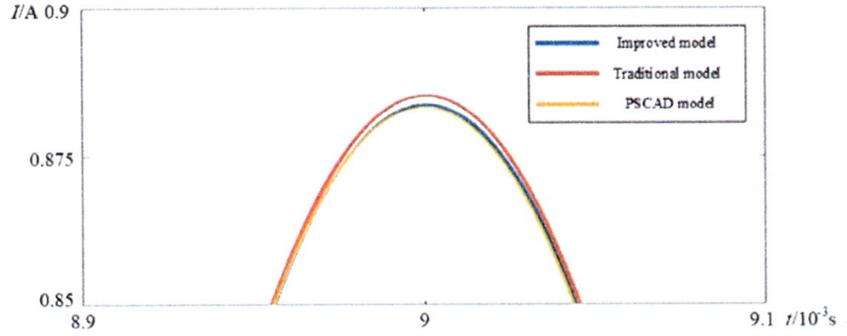

Fig. 6. Leakage current simulation results of traditional model, improved model and PSCAD.

Table 3. Structure parameters of the power supply cable.

Frequency	Amplitude error (traditional/improved)	Angle error (traditional/improved)
10 Hz	−2.78%/−2.47%	2.88°/2.88°
100 Hz	−2.09%/−1.77%	0.07°/0.07°
500 Hz	−1.09%/−0.39%	0.11°/0.08°
1000 Hz	0.22%/0.06%	0.19°/0.16°
1500 Hz	0.38%/0.14%	−0.08°/−0.02°
2000 Hz	0.74%/0.92%	−0.09°/−0.08°

5 Conclusion

Aiming at the problem of variable frequency power supply for the length and distance of high-speed maglev trains, this paper builds the power supply cable model of actual high-speed maglev trains in PSCAD software, and obtains the conclusion that the leakage current is approximately proportional to the voltage source excitation voltage amplitude, the voltage source excitation voltage frequency, and the cable length under different voltage excitation. It can be shown that the leakage current is closely related to the output voltage of the converter under the condition that the cable structure is determined. At the same time, in order to reduce the error between the leakage current in the simulation and the actual situation, this paper proposes a power supply cable model considering the parameters of the grounding wire. By calculating the relevant parameters in the model, the proposed model is simulated in Simulink software, and the simulation results are compared with those in PSCAD software. From the comparison results, it can be found that the error of the cable model proposed in this paper is smaller than that of the traditional model, which verifies the rationality of the model. By analyzing the characteristics of leakage current and the model of power supply cable, this paper provides a theoretical basis for the in-depth extraction and analysis of leakage current of power supply cable of high-speed maglev train.

Acknowledgments. This work is supported by major project in Shanxi Province: Full-scale experimental line of ultra-high-speed low-vacuum pipeline maglev transportation system (Phase I).

References

1. Sun, P., Ge, Q., Wang, X., Zhang, B.: Speed sensorless control of maglev train with double-end power supply. Trans. China Electrotech. Soc. **33**(18), 4249–4256 (2018). (in Chinese)
2. Lu, G., Zhang, Q., Zheng, D., Zhang, P.: A novel online monitoring strategy for the localized grounding insulation defect of converter transformers based on converter switching states control. IEEE Trans. Power Electron. **37**(9), 11124–11134 (2022)
3. Wu, Y., Zhang, P.: High-accuracy condition sensing for power cables in distribution grids based on leakage current measurement approach: An overview. Proc. CSEE **42**(08), 2929–2940 (2022). (in Chinese)
4. Diab, M.S., Yuan, X.: A Quasi-three-level PWM scheme to combat motor overvoltage in SiC-based single-phase drives. IEEE Trans. Power Electron. **35**(12), 12639–12645 (2020)
5. Cao, X., Ge, Q., Zhu, J., Sun, P., Wang, X.: Traction-system research of high-speed maglev train based on integral sliding mode control. Trans. China Electrotech. Soc. **37**(14), 3598–3607 (2022) (in Chinese) doi:https://doi.org/10.19595/j.cnki.1000-6753.tces.210261
6. Chen, X., Zhang, Y., Wang, S., Chen, J., Gong, C.: Impedance-phased dynamic control method for grid-connected inverters in a weak grid. IEEE Trans. Power Electron. **32**(1), 274–283 (2017)
7. Zhao, Y., Zheng, S.: Diagnosis of XLPE cable local tip defects based on analysis of harmonic characteristics of loss current. Trans. China Electrotech. Soc., 1–14 (in Chinese)
8. Qi, M.: Research on circulation characteristics and suppression measures of metal sheath of XLPE cable. Northeast. Electr. Power Univ. (2023) (in Chinese)
9. Wang, X., Blaabjerg, F., Liserre, M.: "An active damper to suppress multiple resonances with unknown frequencies." 2014 IEEE Applied Power Electronics Conference and Exposition—APEC 2014, pp. 2184–2191. Fort Worth, TX, USA (2014)
10. Phinney, J.W., Perreault, D.J., Lang, J.H.: Synthesis of lumped transmission-line analogs. IEEE Trans. Power Electron. **22**(4), 1531–1542 (2007)
11. Yu, S.: Modeling and analysis of conducted EMI of underground high power frequency control system. Beijing Jiaotong Univ. (2022) (in Chinese)

Superconducting Electrodynamic Suspension Traction Numerical Simulation

Xu Ji(✉), Jiansheng Hao, Ying Kang, and Kekang Wei

China Aerospace Science and Industry Co., Ltd., Beijing 100143, China
446818564@qq.co

Abstract. Superconducting electrodynamic suspension technology, characterized by its large levitation gap, passive levitation stability, and energy efficiency, is a key research direction in the field of future rail transportation. Unlike traditional wheel-rail traction systems and electromagnetic levitation systems, the Superconducting electrodynamic suspension traction drive system exhibits features such as a seamless transition between running and levitation states, stable superconducting excitation magnetic fields, and stepless speed control. In this study, we have developed a traction drive system calculation model based on the characteristics of the Superconducting electrodynamic suspension drive system. The calculation and simulation system enables train operation calculations, step simulation, electrical parameter calculations for power supply equipment, and system energy consumption calculations. Through simulation examples, the functionality of the Computational simulation system has been tested, confirming its effectiveness and providing valuable insights for optimizing the configuration of future Superconducting electrodynamic suspension traction systems.

Keywords: Superconducting electrodynamic suspension · Traction drive system · Traction computational numerical simulation

1 Introduction

Modern rail transit has the characteristics of timing, speed, large volume, green energy saving and so on, and has become an important way of travel in human life. High-speed maglev can break through the wheel-rail relationship and the bow network relationship, and the running speed is higher than that of traditional wheel-rail vehicles, reducing the maintenance cost of wheelset maintenance and pantograph (boots) caused by mechanical friction. In recent years, the proposed low-vacuum tube maglev system has further improved the operating speed range, and has outstanding advantages in vibration and noise reduction, green energy saving, all-weather operation, etc., and has become a key research and development direction in the field of rail transit in the future [1–3].

High-speed maglev is divided into two technical routes: electromagnetic suspension and electrodynamic suspension, while electrodynamic suspension is further divided into permanent magnet electrodynamic suspension and superconducting electrodynamic suspension [4–8]. The superconducting electrodynamic suspension has a larger air gap than

L. Jia et al. (Eds.): EITRT 2023, LNEE 1135, pp. 613–627, 2024.
https://doi.org/10.1007/978-981-99-9307-9_67

the electromagnetic suspension, which reduces the requirement of track irregularity. The passive self-stabilizing characteristic does not require active control, which reduces the technical difficulty of the control system. There is no need for a more expensive permanent magnet track, and it has a larger buoyancy to weight ratio than permanent magnet electrodynamic suspension [9, 10]. In 2015, the Yamanashi Superconducting magnetic levitation test line in Japan set the world's highest operating record in the field of rail transit at 603 km/h [11], further demonstrating the system advantages of superconducting electrodynamic suspension in the field of high-speed.

Traction transmission system is used to control the train to realize the acceleration—uniform speed—deceleration movement process, and the numerical simulation of traction transmission system can be used to analyze the train operation process, operation characteristics, fault simulation and other aspects in the early stage of design, and play an important role in measuring the suitability and system economy of each system. In the field of wheel-rail vehicles, the numerical calculation simulation systems represented by the North American RAILSIM system, the European TrainStar system and the Japanese UTRS system have been relatively mature [11–13], and the traction calculation software developed by the Chinese Academy of Railway Sciences, Beijing Jiaotong University, Southwest Jiaotong University and other units has gradually achieved certain results. However, in the field of high-speed maglev, because the traction drive system is located on the ground, there is a big difference from the wheel-rail vehicle traction drive system on the vehicle [14–16], Zhejiang University, Southwest Jiaotong University and other universities have carried out certain research on the traction calculation of electromagnetic levitation [17–20]. Compared with electromagnetic suspension, superconducting electrodynamic suspension has its own characteristics, such as sliding/suspension state transition, superconducting excitation field stability, and stepless system speed regulation. In this paper, by establishing the calculation model of superconducting electrodynamic suspension traction system, the numerical simulation of superconducting electrodynamic suspension traction system is studied, which can provide some reference for the optimization configuration of superconducting electrodynamic suspension traction system in the future.

2 Working Principle of Superconducting Electrodynamic Suspension Traction Drive System

Superconducting electrodynamic suspension is different from the suction suspension actively controlled by electromagnetic suspension, which cannot achieve static suspension. Its traction transmission system supplies power to the long-stator linear motor where the superconducting maglev train is located by adjusting the amplitude, frequency and phase of the output voltage and current of the traction converter in real time. After controlling the motion of the train, the induced electromotive force generated by the superconducting magnet in the zero-flux coil generates an induced current in the closed zero-flux coil, and the induced current generates an induced magnetic field, which interacts with the direct current of the superconducting coil to produce magnetic resistance, suspension force and guiding force [21] (Fig. 1).

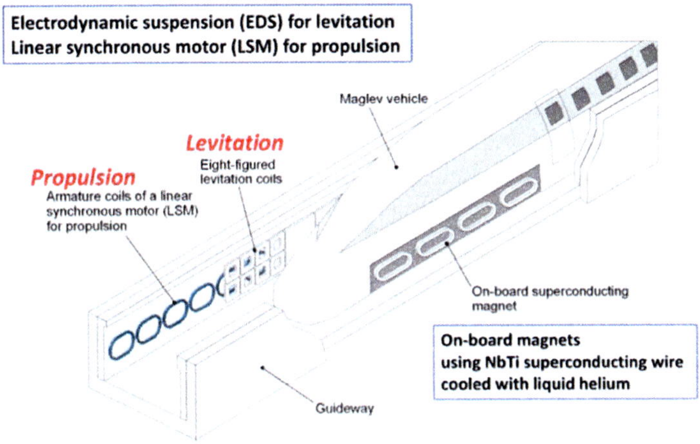

Fig. 1. Superconducting electrodynamic suspension topology

The key to the transition of the Superconducting electrodynamic suspension train is to control the retracting and retracting of the support wheels. When the support wheel is supporting, the center line of the superconducting magnet is flush with the center line of the zero-flux coil, and the total magnetic flux in the zero-flux coil is zero, no induced current is generated, and no magnetic resistance, suspension force, and guiding force are generated. By retracting the support wheel, the center line of the superconducting magnet is lower than the center line of the zero-flux coil, and then induced current is generated, resulting in magnetic resistance, suspension force and guiding force (Fig. 2).

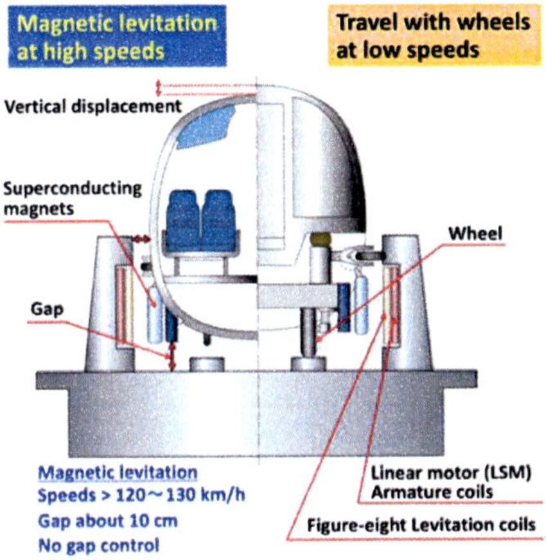

Fig. 2. Schematic diagram of superconducting electrodynamic suspension state transition

3 Superconducting Electrodynamic Suspension Traction Drive System Structure

The Superconducting electrodynamic suspension traction drive system is composed of traction control system, traction converter, feed cable, stator switch station, superconducting linear motor and positioning speed measurement system. The power frequency power supply provided by the external power grid outputs the variable voltage and variable frequency current to the feed cable after the traction converter depressurizes, rectifies, inverts and boosts the voltage. The feed cable sends the current output of the traction converter to the stator switch station distributed along the line. According to the location of the train, the stator switching station determines which sub-section to supply power to drive the train [22] (Fig. 3).

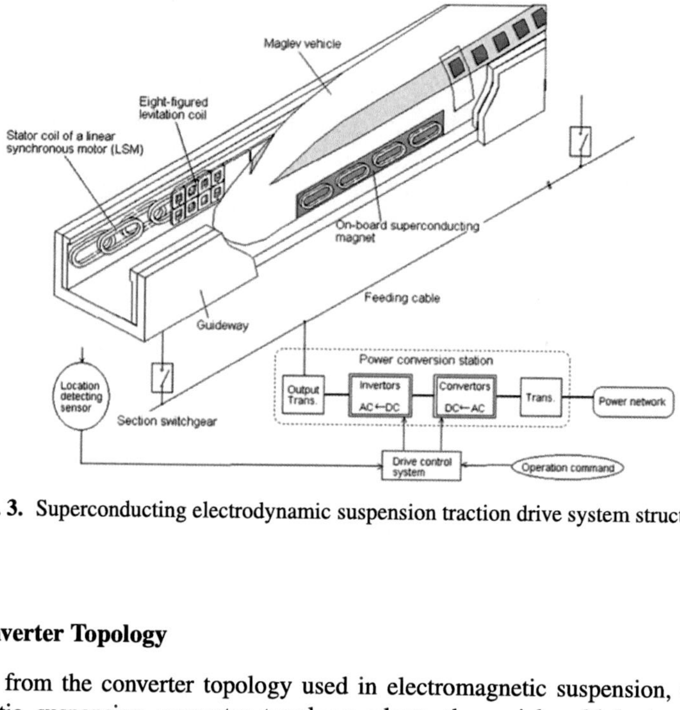

Fig. 3. Superconducting electrodynamic suspension traction drive system structure

3.1 Converter Topology

Different from the converter topology used in electromagnetic suspension, the electromagnetic suspension converter topology adopts the serial-multiple topology of output transformers, which has several advantages over the converter topology in electromagnetic suspension (Fig. 4):

1. It can flexibly adjust the proportional relationship between the output voltage and other heavy transformer modes in a heavy inductance mode, so as to achieve stepless switching between inductance mode and transformer mode, and improve the voltage utilization rate in the inductance mode;
2. Multi-mode single-load inverter unit fault can be cut off at any time to reduce the weight operation, higher reliability;

3. Multiplicity can adopt carrier phase-shift modulation technology, the equivalent switching frequency is higher, the output current harmonic content is lower.

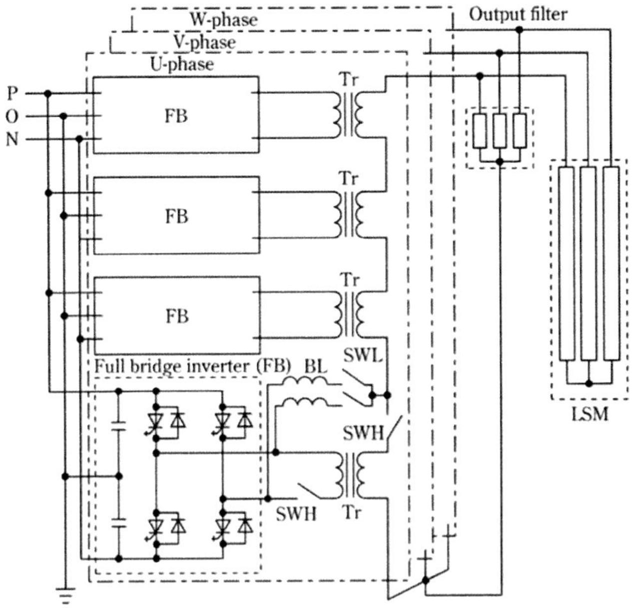

Fig. 4. Superconducting electrodynamic suspension converter topology

3.2 Power Supply Mode

Similar to electromagnetic levitation, the traction transmission system can be divided into single-ended power supply and double-ended power supply according to the different power supply modes of the stator winding in the traction section.

1. Single-end power supply

The stator segment winding is connected to a feeder cable system through a railside stator switch station, and is supplied by a traction converter through the corresponding feeder switch. The single-end power supply mode is suitable for the traction section with short power supply distance, in which the acceleration required by the train can be satisfied by a traction converter.

2. Double-end power supply mode

The stator segment winding is connected to a set of feed cable system through the stator switching station, and is supplied by two traction converters through the corresponding feed switch. Since the traction modules at both ends supply power to the traction section through the feed cable in parallel, the current is distributed between the relevant two traction modules. Therefore, compared with the single-ended power supply mode, the double-ended power supply mode can reduce the current output requirement of a single traction converter when the train obtains the same traction force. In addition, in the case of a long power supply interval and limited power supply capacity, it is more appropriate to use the double-end power supply mode to achieve higher speeds.

3.3 Step-Change Strategy

The step-change strategy of the stator segment of the Superconducting electrodynamic suspension refers to the method of current switching control in the stator segment of the long-stator linear synchronous motor when it passes through the adjacent stator segment. The purpose of using the step change power supply mode is mainly used to reduce the loss and heating of the stator motor, improve the power factor, and reduce the capacity demand of the power supply system. According to the different power supply switching modes of the stator segment, the step changing strategies of the high-speed maglev stator segment are mainly divided into two steps and three steps.

4 Traction Calculation Model of Superconducting Electrodynamic Suspension Traction Drive System

Superconducting electrodynamic suspension traction calculation model is a steady-state mathematical model of train operation built by combining train kinematics model, superconducting synchronous linear motor model and power supply system model. It is based on parameters such as line station information, vehicle information, power supply network information and traction control strategy. The running state of the train and the output voltage, current, capacity, power factor and efficiency of each stator section and traction converter are calculated. Different from the traction calculation of wheel-rail trains, the traction transmission system of maglev trains is located on the ground, so the above data has a certain time variation with the running position.

4.1 Kinematic Model of Train Operation Process

According to the basic laws of classical physics, the basic kinematics formula of the train operation process is as follows:

$$F_t - F_z = M \cdot a \tag{1}$$

where: F_t is the traction force, which depends on the relationship between the stator current and the position of the train, which is determined by the mathematical model of the superconducting linear synchronous motor. F_z is the running resistance, which is specifically divided into aerodynamic resistance and magnetic resistance related to speed, friction resistance of support wheel guide wheel, slope resistance and mechanical braking force independent of speed. M is the total mass of the train, and a is the running acceleration of the train.

1. Aerodynamic resistance

Aerodynamic resistance is mainly related to variables such as front profile section, operating pressure, number of formations and running speed, etc. Aerodynamic resistance coefficient is used to design, and the data obtained by wind tunnel test is modified, and finally converted into a calculation and fitting formula for numerical simulation.

2. Magnetic resistance

The magnetic resistance of superconducting electrokinetic suspension mainly includes eddy current magnetic resistance caused by passive metal induction related to orbit and suspension magnetic resistance caused by suspension sinking height.

3. Support guide wheel friction

The friction force of the supporting guide wheel is estimated by the friction coefficient method. The guiding friction force is positively correlated with the pressure of the supporting guide wheel, while the supporting friction force is positively correlated with the friction coefficient and train weight, etc. When the train changes from the sliding state to the suspension state, there is a retracting and releasing action process of the supporting guide wheel, which is related to the execution speed and time of the retracting and releasing action mechanism of the supporting guide wheel.

4. Slope resistance

It is positively correlated with specific line slope and train weight.

5. Mechanical braking force

Positive correlation with mechanical braking pressure.

According to the train acceleration, the recursive formula of the train kinematics equation can be obtained as follows:

$$V_{i+1} = V_i + a\Delta t \tag{2}$$

$$S_{i+1} = S_i + V_i\Delta t + \frac{1}{2}a\Delta t^2 \tag{3}$$

where: V_i is the speed of the train with the time step i, is the speed of the train with the $i + 1$ time step, and t is the calculated time step; S_i is the position of the train at time step i; S_{i+1} is the position of the $i + 1$ time step train.

4.2 Superconducting Synchronous Linear Motor Model

According to the basic formula of linear motor, the synchronous frequency of long-stator linear motor is $f = \frac{V}{2\sigma}$; The synchronous angular frequency is $\omega_r = \frac{\pi V}{\sigma}$; Where: V is the train speed and the σ is the pole distance of the superconducting magnet.

The equations of voltage, flux and thrust can be obtained according to the coordinate transformation of long-stator linear synchronous motor.

The stator voltage equation is [23]:

$$U_d = Ri_d + \frac{d}{dt}\psi_d - \omega_r\psi_q \tag{4}$$

$$U_q = Ri_q + \frac{d}{dt}\psi_q + \omega_r(L_d i_d + \psi_{sm}) \tag{5}$$

The stator flux equation is:

$$\psi_d = L_d i_d + \psi_{\text{sm}} \tag{6}$$

$$\psi_q = L_q i_q \tag{7}$$

Since the superconducting linear motor can be considered as a surface mounted motor, there is no obvious convex pole effect, that is $L_d = L_q$, the motor thrust equation is:

$$F_t = \frac{3\pi}{2\sigma} p_n \psi_{sm} i_q \tag{8}$$

where: U_d is the voltage of d axis and U_q is the voltage of q axis, i_d is the current of d axis and i_q is the current of q axis, ψ_d is the flux of d axis and ψ_q is the flux of q axis, ψ_{sm} is the flux generated by the superconducting magnet, L_d is the inductance of d axis and L_q is the inductance of q axis, R is the segmented length stator resistance, and p_n is the pole number of the superconducting magnet.

4.3 Power Supply Network Model

The power supply network includes traction converter, feed cable, stator switching station and stator winding, which can be divided into four different power supply networks according to different power supply methods and step change methods, two-step single-ended power supply, two-step double-ended power supply, three-step single-ended power supply, three-step double-ended power supply.

1. Reverse potential calculation under step-change power supply

Different from rotary motors, superconducting linear motors have dynamic changes in stator side reverse potential in excess segments, which depend on the relative position and speed of superconducting magnets and stator side. Without considering reverse potential harmonics, the equivalent calculation of effective value of reverse potential under step-change power supply is as follows:

$$E = \psi_{sm0} L_{\text{eff}} \frac{\pi v}{\sqrt{2}\sigma} \tag{9}$$

where: ψ_{sm0} is a superconducting linear motor flux per unit length, L_{eff} is the effective section length of the coupling between the superconducting magnet and the stator side.

2. Stator segment end voltage calculation

When the train runs on the energized stator section, the electric energy not only provides traction for the train, but also is partly consumed by the heating of the stator section. In the rotor field oriented control mode where the magnetomotive force of the superconducting magnet is constant and the stator direct axis current is constant zero, the back electromotive force of the linear motor and the stator cross axis current vector Iq are in the same line.

Under traction conditions, the back electromotive force and the stator quadrature axis current vector are in the same direction, so the stator segment end voltage is:

$$U_s = \sqrt{(E + R \cdot I_s)^2 + (\omega_s L \cdot I_s)^2} \tag{10}$$

Under braking conditions, the back electromotive force and the stator quadrative axis current vector are reversed, so the stator segment end voltage is:

$$U_s = \sqrt{(E - R \cdot I_s)^2 + (\omega_s L \cdot I_s)^2} \tag{11}$$

where, R is the total resistance of the stator segment, and L is the sum of the main inductance, leakage inductance, and invalid inductance of the stator segment.

3. Converter output voltage calculation

The voltage drop at the output end of the converter includes the voltage of the stator segment and the voltage drop of the long feeder cable, which is related to the power supply mode.

Under single-end power supply, the output current of the converter I_c is the current I_s in the stator section of the linear motor, and the output voltage of the converter under traction condition is:

$$U_s = \sqrt{(E + R \cdot I_s + R_L \cdot I_s)^2 + (\omega_s L \cdot I_s + \omega_s L_L \cdot I_s)^2} \tag{12}$$

The output voltage of the converter under braking condition is:

$$U_s = \sqrt{(E - R \cdot I_s - R_L \cdot I_s)^2 + (\omega_s L \cdot I_s + \omega_s L_L \cdot I_s)^2} \tag{13}$$

Under double-ended power supply, the output current of the converter I_c is the current aI_s in the stator section of the linear motor, where a is the distribution coefficient of the double-ended power supply. The output voltage of the converter under traction condition is:

$$U_s = \sqrt{(E + R \cdot I_s + R_L \cdot aI_s)^2 + (\omega_s L \cdot I_s + \omega_s L_L \cdot aI_s)^2} \tag{14}$$

The output voltage of the converter under braking condition is:

$$U_s = \sqrt{(E - R \cdot I_s - R_L \cdot aI_s)^2 + (\omega_s L \cdot I_s + \omega_s L_L \cdot aI_s)^2} \tag{15}$$

4. Power factor and efficiency calculation of converter output side

The basic formula for the output capacity of the converter is: $S = 3UI = \sqrt{P^2 + Q^2}$, active power P includes system loss P_{loss} and mechanical active power P_m. Because the superconducting linear motor adopts hollow winding and has no core loss, its system loss only includes the copper loss of the feeder cable and the stator winding.

5 Numerical Simulation of Superconducting Electrodynamic Suspension Traction Drive System

5.1 Input Parameters

The input parameters for the numerical simulation of Superconducting electrodynamic suspension traction system mainly include the information of line station, vehicle, power supply network and traction control.

Station Information

Line station information mainly includes: station mileage, number of stations, line section slope, line section speed limit and other parameters.

Vehicle Information

Vehicle information mainly includes: vehicle formation number, vehicle total weight, suspension frame distance, support wheel rolling friction coefficient, guide wheel rolling friction coefficient, lifting control speed, landing control speed, support wheel retracting actuation time, mechanical brake input speed, vehicle-ground communication delay time and other parameters.

Power Supply Network Information

Power supply network information mainly includes: superconducting linear motor information, traction section power supply cable information, stator switch station information, traction converter information.

Traction Control Strategy

Traction control strategy mainly includes:

ATO operation mode, motor control mode, stator step change mode, maximum acceleration limit, maximum deceleration limit, inductor-transformer mode switching frequency, transformer-inductor mode switching frequency.

5.2 Step Calculation

The operation of the Superconducting electrodynamic suspension train itself is the step of the linear motor, and its step is limited by the slope of the line, the speed limit, the power supply capacity of the traction system and the acceleration and deceleration speed. The step calculation process is shown (see Fig. 5).

5.3 Numerical Simulation Results of Traction Drive Under Typical Working Conditions

Taking the superconducting maglev working condition of a line as an example, the input parameters are set as follows:

Table 1 describes the station parameters.

Table 2 shows the configuration of vehicle information parameters.

Table 3 describes the power supply network parameters.

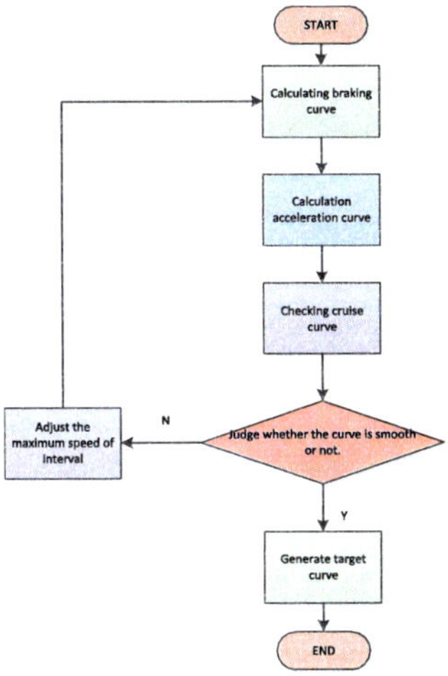

Fig. 5. Superconducting electrokinetic levitation target curve step calculation flow chart

Table 1. Line station parameter configuration table

Item	Value	Unit
Mileage	100	km
Number of stations	2	
Track grade	0.4%	
Line zone speed limit	660	km/h

Table 2. Vehicle parameter configuration table

Item	Value	Unit
Number of vehicle formations	8	
Gross combination	240	t
Suspension frame spacing	21.6	m
Support wheel rolling friction coefficient	0.02	
Lifting control speed	130	km/h
Landing control speed	180	km/h
Mechanical brake entry speed	20	km/h
Vehicle-ground communication delay time	500	ms

Traction control parameters are set as follows Table 4.

According to the typical parameters, the main calculation results are as follows:

Table 3. Power supply network information parameter configuration table

Item	Value	Unit
Single suspension superconducting magnet pole number	4	/
Pole distance of superconducting magnet	1.35	m
Superconducting magnet flux linkage	1.74	Wb
Stator average sectional length	1.2	km
Total resistance within the length of stator segments	0.8	Ω
Total inductance in stator segment length	0.024	H
Number of feeder cables in parallel	2	/
Cable resistance per unit length (km)	0.08	Ω
Cable inductance per unit length (km)	0.000275	H
Upper limit of output current of stator switch station	2000	A
Upper limit of output voltage of stator switch station	24	kV
Upper limit of output current of traction converter power unit	1000	A
Traction converter output transformer ratio	0.8	/
Power supply mode	Double-ended power supply	

Table 4. Traction control parameter configuration table

Item	Value	Unit
ATO operation	Minimum time	
Motor control mode	Rotor field-oriented	
Stator step change mode	Three step change	
Maximum acceleration limit	1.5	m/s^2
Maximum deceleration limit	1.5	m/s^2

When the line speed limit is 660 km/h, the maximum speed of the vehicle is 655 km/h, which does not exceed the speed limit. The maximum acceleration in the starting stage is located in the inductive mode, and the maximum acceleration is 1.3 m/s^2, which does not exceed the limit value of 1.5 m/s^2. The maximum deceleration in the braking stage is -1.38 m/s^2 at the initial braking stage, which does not exceed the limit value of 1.5 m/s^2.

According to the system simulation, the total running time of the train in the interval is 798 s, and the train enters the suspension state after 130 km/h. According to the traction curve, the maximum output capacity of the traction system converter is 26 MVA, the maximum output voltage is 13.8 kV, and the total energy consumption of the traction transmission system during the whole train operation is 10,000 kWh (Fig. 6).

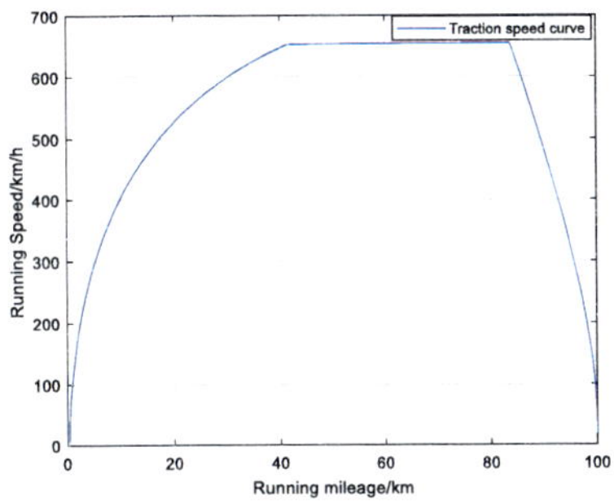

No.1 Mileage - speed curve

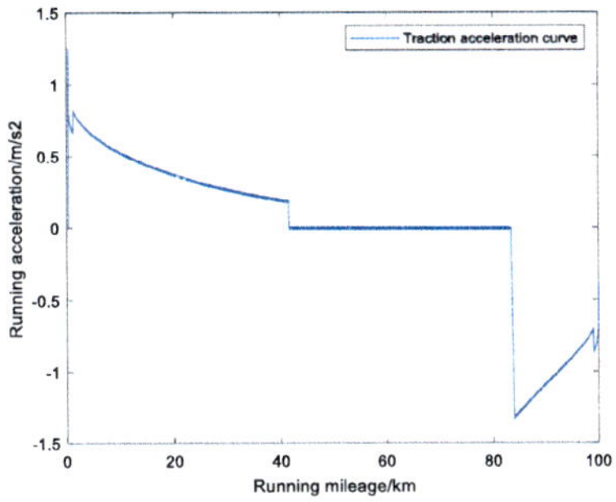

No.2 Mileage - acceleration curve

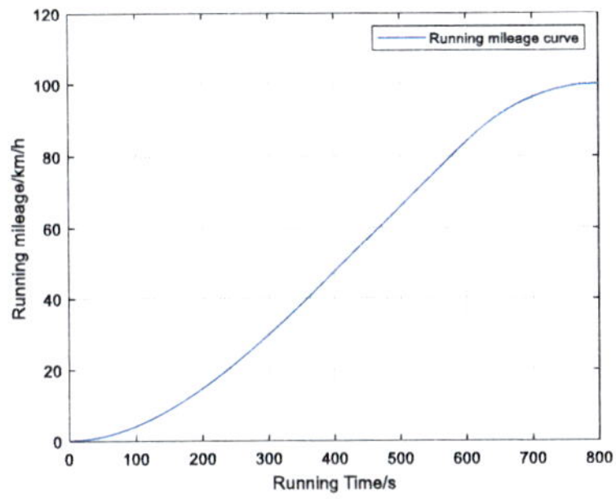

No.3 Run time - mileage curve

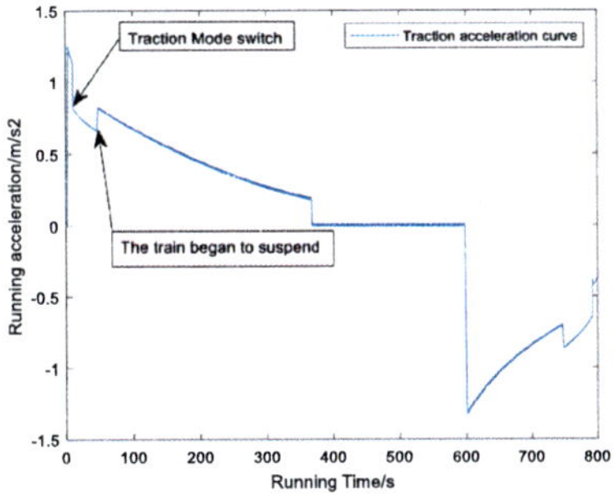

No.4 Run time - acceleration curve

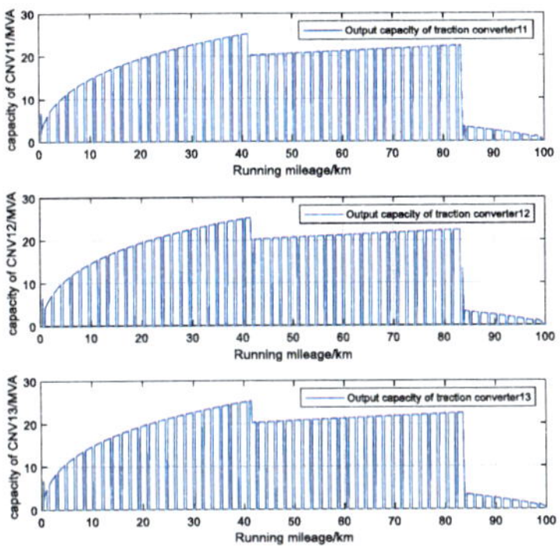

No.5 Mileage - Converter (Station A) output capacity curve

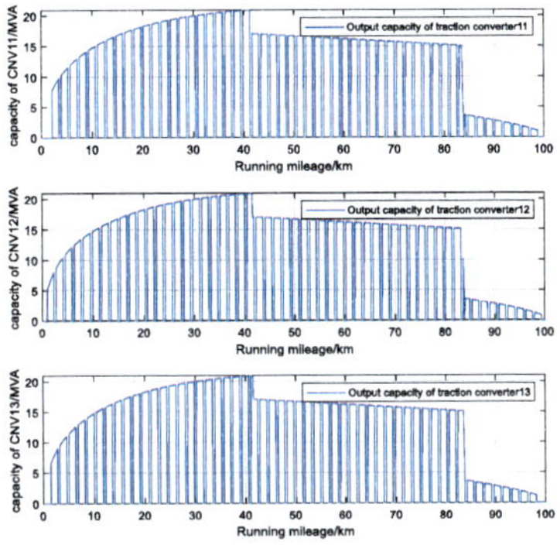

No.6 Mileage - Converter (Station B) output capacity curve

Fig. 6. Figure Numerical simulation results of traction drive under typical working conditions

6 Conclusions

In this paper, based on the characteristics of Superconducting electrodynamic suspension traction transmission system, a mathematical model of Superconducting electrodynamic suspension traction calculation is established, and a numerical simulation and simulation

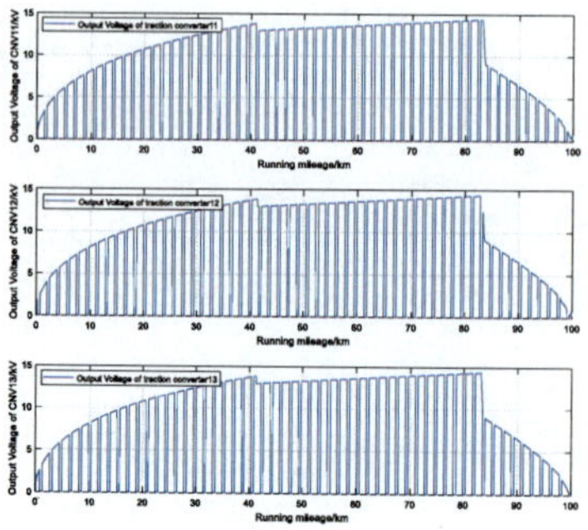

No.7 Mileage - Converter (Station A) output voltage curve

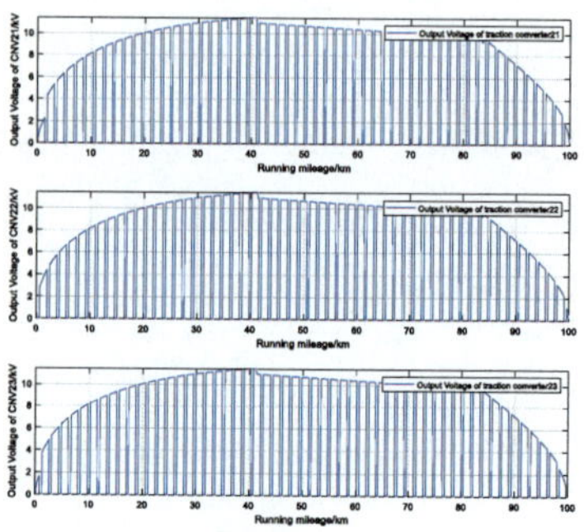

No.8 Mileage - Converter (Station B) output voltage curve

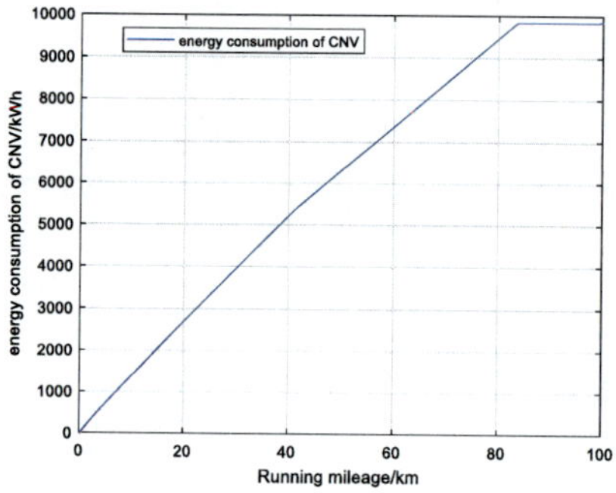

No.9 Mileage - total system energy consumption curve

Fig. 6. (*continued*)

system of Superconducting electrodynamic suspension traction is built by using simulation software, so that the calculation and simulation system can perform functions such as train operation calculation, step change simulation, electrical parameter calculation of power supply equipment and system energy consumption calculation. Finally, the function of the system is tested by a simulation example, and the simulation results are analyzed to verify that the system has certain effectiveness.

Acknowledgments. This work is supported by major project in Shanxi Province: Full-scale experimental line of ultra-high-speed low-vacuum pipeline maglev transportation system (Phase I).

References

1. Yan, L., Xu, S., Sun, G., et al.: Strategic progress of high-speed maglev and the development strategy in China[J]. Adv. Technol. Electr. Eng. Energy **20**(1), 1–7 (2003) (in Chinese)
2. Xu, F., Luo, S., Deng, Z.: Study on key technologies and whole speed range application of maglev rail transport[J]. J. China Railw. Soc. **41**(3), 40–49 (2019) (in Chinese)
3. Lyu, G.: Review of the application and key technology in the linear motor for the rail transit[J]. Proc. CSEE **40**(17), 5665–5675 (2020) (in Chinese)

4. Chen, Y., Zhang, K.: Calculation of electromagnetic force of plate type null double side permanent magnet electrodynamic suspension[J]. Trans. China Electrotech. Soc. **31**(24), 150–156 (2016) (in Chinese)
5. Bruce Montgomery, D.: Overview of the 2004 Magplane design[C]. Maglev 2004 Conference (2004)
6. Lee, J., Bae, D.K., Sim, K., et al.: Characteristics on electrodynamic suspension simulator with HTS levitation magnet[J]. Physica C **469**, 1820–1824 (2009)
7. Su, Y.C., Chang, Y.L., Jung M., et al.: Sub-Sonic linear synchronous motors using superconducting magnets for the Hyperloop[J]. Energies **12**, 4611 (2019)
8. Zigang, D., Weihua, Z., Jun, Z., et al.: A high-temperature superconducting maglev ring test line developed in Chengdu China[J]. IEEE Trans. Appl. Supercond. **26**(6), 3602408 (2016)
9. Hu, D., Feng X., Ma X., Shen S., Zhang Z.: Vertical vibration suppression methods of superconducting electrodynamic suspension system with a field-circuit-motion model[J]. Proc CSEE 43(10), 3972–3985 (2023)
10. Hu, D., Feng, X., Zhang, Z., Powell, J.R., Danby, G.T.: Study on the damping characteristics of superconducting electrodynamic suspension system high speed transport by magnetically suspended trains[J]. Proc. CSEE **41**(13), 4679–4687 (2021). (in Chinese)
11. Shi, H., Peng, Q., Guo, H.: Traction calculation model of urban mass transit[J]. J. Traffic Transp. Eng. **2005**(04), 20–26 (in Chinese)
12. Hans True, J.C.J.: Parameters study of hunting and chaos in railway vehicle dynamic [J]. Veh. Syst. Dyn., 23 (1993)
13. Zhang, D.: Research and development of traction calculation simulation system[D]. Central South University (2010)
14. Hirao, Y.B.L.: Development of a Universal train simulator (UTRAS) and evaluation of signaling system[J]. TRRI **336**(4), 180–185 (1995)
15. Yang, F., Ping, L.I., Zhang, L.: Study on train group running simulation system[J]. Railw. Comput. Appl. **08**, 8–10 (2005) (in Chinese)
16. Yang, Y.: Realization of interactive train calculation software for power supply calculation[D]. Beijing Jiaotong University (2017) (in Chinese)
17. Li, L.: Numerical simulation of maglev propulsion system[D]. Zhejiang University (2009) (in Chinese)
18. Xiong, X.: Research on traction calculation and energy-saving application of high-speed maglev train[D]. Southwest Jiaotong University (2017) (in Chinese)
19. Pan, R., Shaoke, L.: Traction calculation algorithm research for low-speed maglev train[J]. Electr. Drive Locomot. **5**, 66–69 (2012)
20. Gao, X., Zhang, Z., Zheng, J.: Energy consumption and efficiency analysis of high-speed maglev system[J]. Technol. Innov. Appl. (2023)
21. Hiroyuki, O.: Review and update on MAGLEV.: EUROPEAN CRYOGENICS DAYS 2017. Karlsruhe, Germany (2017) (in Chinese)
22. Haruo, I., Shigeo, K., Yutaka, O., Jun-ichi, K., et al.: Development of power supply system for yamanashi maglev test line[J]. Proc. Power Convers. Conf. (1997)
23. Jia, G., Sun, Z., Huang, C., Zhou, W., Mao, Y., Lei, Z.: Magnetic field analysis and optimization of thrust fluctuation of permanent magnet synchronous linear motor for electromagnetic emission launch[J]. Acta Armamentarii (2023)

A Low Switching Frequency Model Predictive Control for Surface-Mounted Permanent Magnet Synchronous Motor Without Weight Factor

Jiangming Deng[1], Han Xiao[2(✉)], Wei Xu[2], and Yirong Tang[2]

[1] CRRC Zhuzhou Locomotive Co., Ltd, Zhuzhou 412005, China
[2] State Key Laboratory of Advanced Electromagnetic Technology, Huazhong University of Science and Technology, Wuhan 430074, China
hanxiao2023@foxmail.com, weixu@hust.edu.com, yirtang@hust.edu.cn

Abstract. An improved low switching frequency model predictive control (MPC) strategy is proposed to improve the efficiency of the surface-mounted permanent magnet synchronous motor (SPMSM) drive system without weight factor. By evaluating the number of switch actions in the previous cycle, it is determined whether to add a constraint on the number of switch actions to the optimal vector selected in the next cycle to reduce the switching frequency of the inverter and improve SPMSM drive system efficiency. Compared to conventional multi-objective optimisation MPC strategies, the proposed strategy avoids complex weight factor design. Finally, the proposed strategy is implemented on the MATLAB simulation platform. The results show that the proposed strategy can effectively reduce the switching frequency of the inverter while ensuring the dynamic and static performance of SPMSM.

Keywords: Surface-mounted permanent magnet synchronous motor (SPMSM) · Model predictive control (MPC) · Multi-objective control · Low switching frequency · Weight factor

1 Introduction

Surface-mounted permanent magnet synchronous motor (SPMSM) has the advantages of high reliability and simple control, so it has been widely applied in fields such as electric vehicles and new energy generation [1, 2]. Many control strategies have been proposed to achieve smooth torque control of SPMSM, such as vector control, direct torque control, and model predictive control (MPC). Among them, MPC has been widely studied due to its unique advantage of low switching frequency, especially in high-power applications [3, 4].

L. Jia et al. (Eds.): EITRT 2023, LNEE 1135, pp. 628–637, 2024.
https://doi.org/10.1007/978-981-99-9307-9_68

The basic idea of MPC is to evaluate the error of all voltage vectors through a predictive model and cost function, then select the voltage vector with the smallest control error. Due to the evaluation of all voltage vectors, the switching frequency of conventional MPC fluctuates, and the current and adjacent switching states may be the same or completely opposite [5]. By utilizing the characteristic, the adjacent switch states of MPC may be the same, and the low energy consumption control of SPMSM in high-power fields can be achieved [6]. The switching frequency and inverter loss are considered to be positively proportional, so reducing the switching frequency can reduce the inverter loss, thereby improving the efficiency of the SPMSM drive system [7, 8]. Therefore, low switching frequency MPC has been widely studied [9, 10]. Usually, multi-objective MPC requires weight factors to balance various control objectives, so the design of weight factors needs to be balanced, but often the weight factors are too dependent on the experience of engineers [11]. Then, an algorithm combining hysteresis control and MPC is proposed, which achieves the two objectives of low switching frequency and traction control of PMSM without weight factor [12]. But the hysteresis width introduced by this algorithm is also time-consuming work.

Therefore, a low switching frequency MPC is proposed to remove the weight factor of multi-objective control. The proposed method determines whether to continue using the current switch state in the next cycle based on the number of switch actions at the previous time. If the number of switch actions at the previous moment is large, the optimal vector selected for the next cycle must be the vector with a small number of switch actions; If the number of switch actions at the previous moment is small, the optimal vector selected for the next cycle does not need to consider the number of switch actions. Finally, the simulation results indicate that the proposed strategy can reduce the switching frequency while maintaining similar control performance.

2 Improved Low Switching Frequency MPC

2.1 Conventional MPC Strategy

The topology of a conventional voltage source inverter driven SPMSM system is shown in Fig. 1a. The three-phase bridge arms of the inverter each have 2 switch states, forming a total of 8 switch states for the inverter drive system. And the corresponding voltage vectors are shown in Fig. 1b, where "1" represents the upper bridge arm conducting and the lower bridge arm closing; "0" represents the upper bridge arm closing and the lower bridge arm conducting.

Typically, the state space equation of SPMSM can be expressed as:

$$\boldsymbol{u}_{\alpha\beta}(k) - R_s \boldsymbol{i}_{\alpha\beta}(k) - L_s \frac{di_{\alpha\beta}}{dt} - \boldsymbol{e}_{\alpha\beta}(k) = 0 \tag{1}$$

where $\boldsymbol{u}_{\alpha\beta}(k) = [u_{\alpha}(k), u_{\beta}(k)]^{\mathrm{T}}$ is the stator terminal voltage of SPMSM; $\boldsymbol{i}_{\alpha\beta}(k) = [i_{\alpha}(k), i_{\beta}(k)]^{\mathrm{T}}$ is current; L_{s} and R_{s} are the stator inductance and resistance, respectively; $\boldsymbol{e}_{\alpha\beta} = [e_{\alpha}(k), e_{\beta}(k)]^{\mathrm{T}} = w_r y_f\ [-\sin q, \cos q]^{\mathrm{T}}$ is the back electromotive force (EMF). q is the rotor angle; w_r is synchronous rotation speed; y_f is the permanent magnet flux linkage.

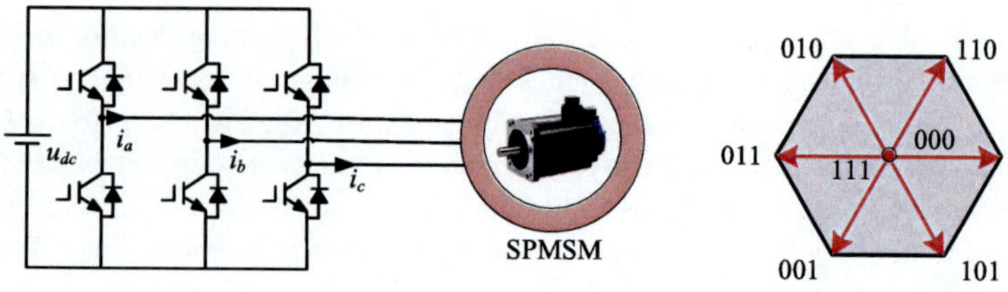

(a) Topological structures (b) Voltage vectors

Fig. 1. Topological structures and voltage vectors of SPMSM drive system. **a** Topological structures **b** Voltage vectors

After forward Euler discretization, the predicted current can be derived:

$$\boldsymbol{i}_{\alpha\beta}(k+1) = \left(1 - \frac{R_s T_s}{L_s}\right)\boldsymbol{i}_{\alpha\beta}(k) + \frac{T_s}{L_s}[\boldsymbol{u}_{\alpha\beta}(k) - \boldsymbol{e}_{\alpha\beta}(k)] \tag{2}$$

where T_s is the sampling period.

To obtain the optimal voltage vector, it is necessary to establish a cost function (3) to evaluate its control error. Then, traversing all 8 voltage vectors, the voltage vector that minimizes the cost function is obtained as the optimal voltage vector for the next cycle.

$$g = \left(i_{\alpha ref} - i_{\alpha}(k+1)\right)^2 + \left(i_{\beta ref} - i_{\beta}(k+1)\right)^2 \tag{3}$$

where i_{aref} and $i_{\mathrm{b}ref}$ is the reference current.

According to (3), it can be seen that the optimal voltage vector selected by (3) may have a significant difference from the current applied voltage vector, resulting in a higher switching frequency. Therefore, further improvements are needed.

2.2 Design of Low Switching Frequency MPC

To suppress the switching frequency of MPC, the switching frequency term is usually added to the cost function, as shown in (4).

$$g = \left(i_{\alpha ref} - i_{\alpha}(k+1)\right)^2 + \left(i_{\beta ref} - i_{\beta}(k+1)\right)^2 + k * \|\boldsymbol{S}_{\mathrm{abc}}(k+1) - \boldsymbol{S}_{\mathrm{abc}}(k)\|_1 \tag{4}$$

where k is the weight factor.

To suppress the switching frequency, a cost function considering the switching states of the previous cycles is established as (5):

$$g_0(k) = \|\Delta \boldsymbol{S}_{abc}(k)\|_1 = \|\boldsymbol{S}_{abc}(k) - \boldsymbol{S}_{abc}(k-1)\|_1 \tag{5}$$

where $\boldsymbol{S}_{\mathrm{abc}} = [S_a, S_b, S_c]^T$ is switch states, $\| \, \|_1$ represents the 1-norm of the vector, and $\Delta \boldsymbol{S}_{\mathrm{abc}}$ represents the difference in switch states between adjacent periods.

According to (5), the values of g_0 are 0, 1, 2, 3. $g_0 < 2$ indicates that the number of switch actions is lower at the previous time, and vice versa indicates that the number of switch actions is higher. Therefore, 0, 1, and 2, 3 can be divided into two groups. If the number of switch actions at the previous moment is 2 and 3, the number of switch actions at the next moment is forced to 0. Therefore, the cost function can be defined as follows:

$$g = \begin{cases} \left(i_{\alpha ref} - i_\alpha(k+1)\right)^2 + \left(i_{\beta ref} - i_\beta(k+1)\right)^2, \; if \; g_0(k) \leq 1 \\ \infty, \; others \end{cases} \tag{6}$$

According to (6), the improved cost function can achieve low switching frequency MPC of SPMSM without weight factor.

3 Simulation Verification and Analysis

To verify the effectiveness of the proposed low switching frequency MPC strategy, a simulation model is built based on MATLAB for verification. The simulation parameters are shown in Table 1.

Table 1. Simulation parameters

Parameters	Symbol	Values
Stator resistance	R_s	1.15 Ω
Stator inductance	L_s	0.029 H
Permanent magnet flux linkage	y_f	0.458 Wb
Pole pairs	p	4
DC voltage	u_{dc}	600 V
Sampling period	T_s	400 μs (2.5 kHz)
Load torque	T_e	10 N·m

3.1 Verification of Static Simulation

To verify the static simulation effect of the proposed strategy, the motor operating conditions are set as follows: the load is 0N m and 10N m, respectively; The motor speeds are 1000 r/min and 600 r/min. Under these operating conditions, the superiority of the proposed method and conventional methods can be judged by switching frequency, speed, current, and load.

Figure 2 shows the comparison of control performance between the proposed low switching frequency MPC and conventional MPC under no-load conditions. Where mean(f_s) represents the average value within the window time in this state, and f_s is the switching frequency, defined as follows:

$$g_p = \frac{\omega_r}{6\pi T_s} \sum \|\Delta S_{abc}\|_1 \tag{7}$$

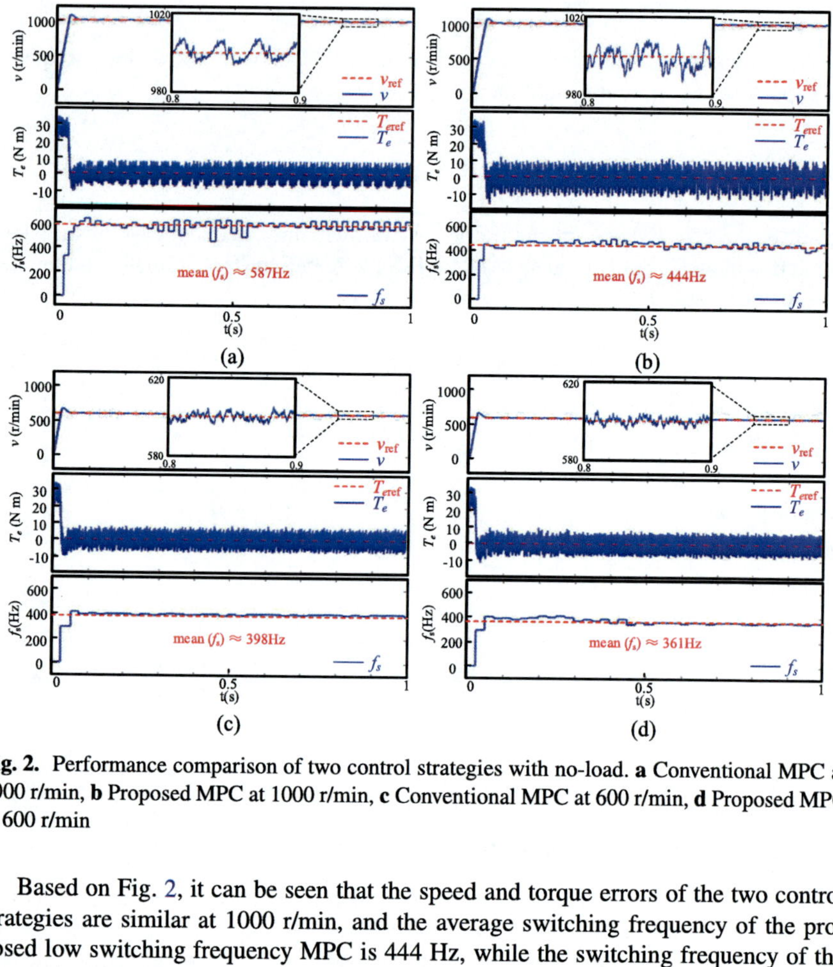

Fig. 2. Performance comparison of two control strategies with no-load. **a** Conventional MPC at 1000 r/min, **b** Proposed MPC at 1000 r/min, **c** Conventional MPC at 600 r/min, **d** Proposed MPC at 600 r/min

Based on Fig. 2, it can be seen that the speed and torque errors of the two control strategies are similar at 1000 r/min, and the average switching frequency of the proposed low switching frequency MPC is 444 Hz, while the switching frequency of the conventional MPC is 587 Hz. At 600 r/min, the speed and torque errors of the two control strategies are similar, and the average switching frequency of the proposed low switching frequency MPC is 361 Hz, while the switching frequency of the conventional MPC is 398 Hz. Therefore, it can be seen that under no-load conditions, the proposed MPC can achieve similar control performance to conventional MPC at a lower switching frequency.

Furthermore, by increasing the load to 10N m and repeating the previous test, Fig. 3 can be obtained. Based on Fig. 3, it can be seen that under a load condition of 10N m, the proposed MPC has switching frequencies of 416 Hz and 405 Hz at 1000 r/min and 600 r/min, respectively, while the switching frequency of conventional MPC are 578 Hz and 433 Hz. Compared to conventional MPC, the switching frequency of the proposed MPC is significantly reduced. However, under the condition of 1000 r/min, the speed error of the proposed strategy is slightly greater than that of conventional MPC, and the performance slightly decreases, but it can still operate normally. Table 2 shows a detailed comparison of the switching frequency in Figs. 2 and 3.

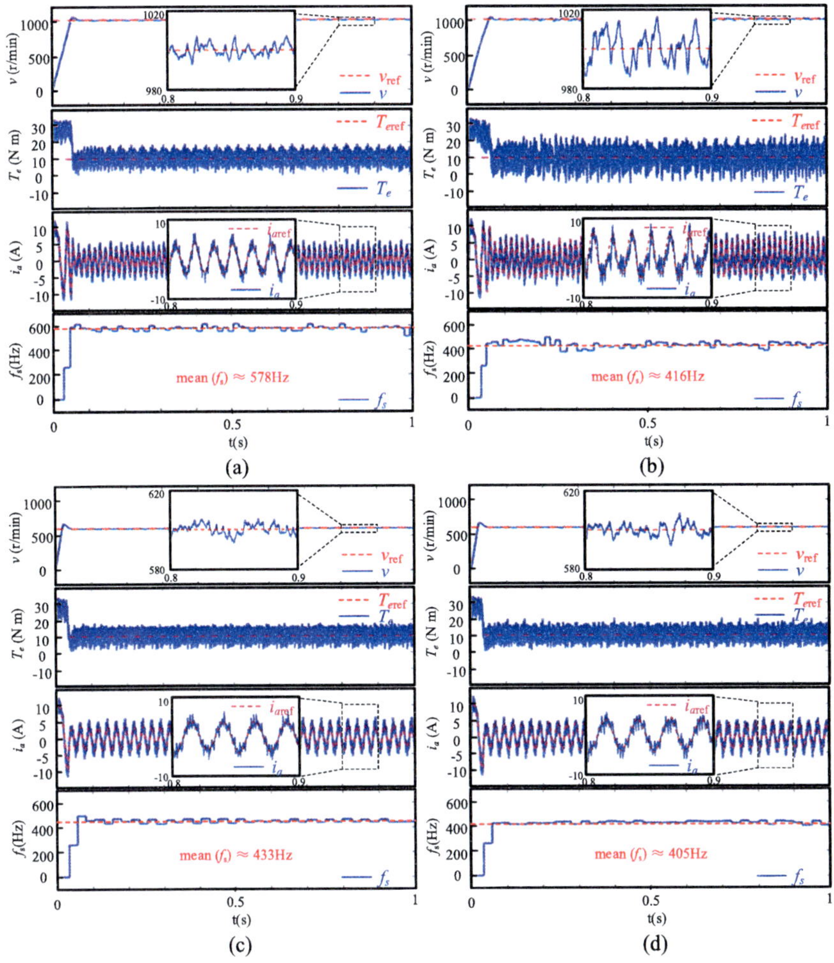

Fig. 3. Performance comparison of two control strategies with 10N m load. **a** Conventional MPC at 1000 r/min, **b** Proposed MPC at 1000 r/min. **c** Conventional MPC at 600 r/min, (d) Proposed MPC at 600 r/min

Table 2. Comparison of control performance between two strategies

Load	Speed	Switching frequency (Hz)		Control performance	
		Conventional MPC	Proposed MPC	Conventional MPC	Proposed MPC
No-load	1000 r/min	587	444 (−24.4%)	High	High
	600 r/min	398	361 (−9.3%)	High	High
10 N m	1000 r/min	578	416 (−28.0%)	High	Medium
	600 r/min	433	405 (−6.5%)	High	Medium

Based on Table 2, it can be seen that the proposed MPC can achieve similar control performance to conventional MPC at a lower switching frequency. Although the performance slightly decreases with 10N m load, the normal operation of the motor can still be maintained, indicating the effectiveness of the proposed MPC.

3.2 Verification of Dynamic Simulation

Two operating conditions, speed and load sudden changes, are set to test the motor. Figure 4 shows the test results of two control strategies when the speed suddenly drops from 1000 r/min to 600 r/min at 0.5 s. And Fig. 5 shows the test results of two control strategies when the load suddenly increases from no-load to 10N m at 0.5 s.

Based on Fig. 4, it can be seen that the dynamic response speed of the two strategies is similar. Under no-load conditions, the switching frequency of the proposed strategy is 468 Hz and 387 Hz, respectively, while the switching frequency of conventional MPC is 578 Hz and 389 Hz. Under 10N m load conditions, the switching frequency of the proposed strategy is 413 Hz and 415 Hz, respectively, while the switching frequency of conventional MPC is 576 Hz and 388 Hz. The switching frequency has significantly increased. Although there are slight burrs in current and torque, this is because the optimization of switching frequency is achieved at the cost of reducing current control performance, which is a common problem in multi-objective optimization control.

Based on Fig. 5, it can be seen that when the speed is constant and the load suddenly changes, the control performance of the proposed MPC is similar to that of conventional MPC, but the switching frequency has decreased from 589 Hz and 580 Hz to 457 Hz and 421 Hz. The switching frequency decreased by 22.4% and 27.4%, respectively, further validating the proposed MPC's superiority.

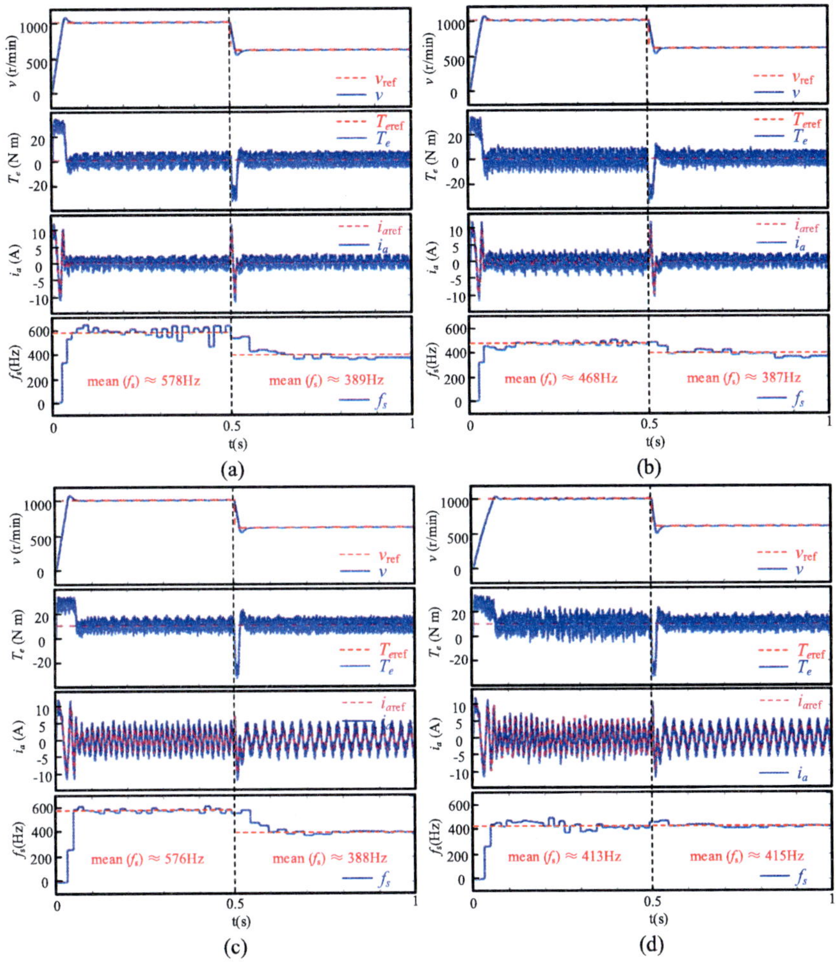

Fig. 4. Comparison of performance between two control strategies when the speed suddenly drops from 1000 r/min to 600 r/min. **a** Conventional MPC with no-load **b** Proposed MPC with no-load **c** Conventional MPC with 10N m load **d** Proposed MPC with 10N m load

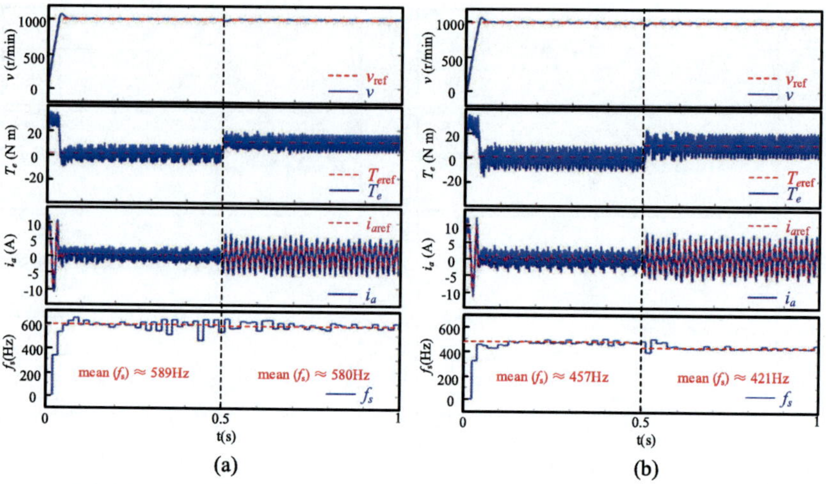

Fig. 5. Comparison of performance between two control strategies when the load suddenly rises from no-load to 10N m. **a** Conventional MPC with 10N m load **b** Proposed MPC with 10N m load

4 Conclusion

An improved low switching frequency MPC without weight factor is proposed. Based on conventional MPC, a constraint condition that the number of adjacent switching actions cannot be too high is introduced. The feasibility and superiority of the proposed low switching frequency MPC has been verified through simulation comparison with conventional MPC. The proposed strategy can ensure the performance of the permanent magnet synchronous traction system while reducing the inverter switching frequency and improving system efficiency.

Acknowledgments. This work is supported by the science and technology innovation Program of Hunan Province (2021RC3136).

References

1. Tang, Y., Xu, W., Liu, Y., Dong, D.: Dynamic performance enhancement method based on improved model reference adaptive system for SPMSM sensorless drives. IEEE Access 9, 135012–135023 (2021)
2. Xu, W., Liao, K., Ge, J., Qu, G., Cheng, S., Wang, A., Boldea, I.: Improved position sensorless control for PMLSM via an active disturbance rejection controller and an adaptive full-order observer. IEEE Trans. Ind. Appl. 59(2), 1742–1753 (2023)
3. Jin, T., et al.: Low complexity model predictive flux control based on discrete space vector modulation and optimal switching sequence for induction motors. IEEE Trans. Industr. Electron. **71**(1), 305–315 (2024)
4. Y. Tang, W. Xu, D. Dong, Y. Liu, and M. Ismail.: Low-complexity multistep sequential model predictive current control for three-level inverter-fed linear induction machines. IEEE Trans. Ind. Electron. 70(6), 5537–5548

5. Liu, C., Xing, X., Du, C., Zhang, B. Zhang and F. Blaabjerg.: An improved model predictive control method using optimized voltage vectors for Vienna rectifier with fixed switching frequency. IEEE Trans. Power Electron. 38, no. 1, pp 358–371 (2023)
6. Wang, F., Wei, Y., Young, H., Ke, D., Xie, H., Rodríguez, J.: Continuous-control-set model-free predictive fundamental current control for PMSM system. IEEE Trans. Power Electron. 38(5), 5928–5938 (2023)
7. Gu, M., Wang, Z., Yu, K., Wang, X., Cheng, M.: Interleaved model predictive control for three-level neutral-point-clamped dual three-phase PMSM drives with low switching frequency. IEEE Trans. Power Electron. 36(10), 11618–11630 (2021)
8. Nian, H., Hu, W.: A sensorless drive strategy for open-end winding PMSM with common DC voltage based on lower switching frequency. IEEE Trans. Energy Conversion 34(3), 1553–1562 (2019)
9. Wu, M., Tian, H., Li, Y. W., Wang, K.: Hybrid voltage balancing control for four-level hybrid-clamped converters with low switching frequency. IEEE Trans. Ind. Electron. 69(11), 11165–11176
10. Zhang, Y., Bai, Y., Yang, H., Zhang, B.: Low switching frequency model predictive control of three-level inverter-fed IM drives with speed-sensorless and field-weakening operations. IEEE Trans. Ind. Electron. 66(6), 4262–4272 (2019)
11. Feng Ling, F., Jianguo, L.L., et al.: An improved low switching frequency model predictive direct torque control strategy for traction permanent magnet synchronous motor. Proc. CSEE 41(21), 7507–7517 (2021)
12. Karamanakos, P., Geyer, T., Kennel, R.: On the choice of norm in finite control set model predictive control. IEEE Trans. Power Electron. 33(8), 7105–7117 (2018)

Optimization Control of Energy Management for Diesel-Electric Hybrid Trains Based on MPC-PMP

Zhe Chen[1(✉)], Xiaoxu Wang[2], Xiongwen Zhu[2], Chi Zhang[2], and Xinjie Zheng[2]

[1] CRRC Zhuzhou Locomotive Co., Ltd, Zhuzhou 412001, Hunan, China
Chenzhe715@126.com

[2] National Active Distribution Network Technology Research Center, Beijing Jiaotong University, Beijing 100044, China
{22121515,21126229,18117027,23111457}@bjtu.edu.cn

Abstract. Appropriate energy management strategies can enhance the efficiency of train hybrid power systems and reduce fuel consumption costs. This paper addresses the problem of requiring global operating conditions for the Pontryagin Minimum Principle (PMP) strategy and proposes an energy management strategy based on model predictive control (MPC), with the goal of minimizing fuel consumption. This strategy can be implemented online. Firstly, a speed prediction model is established using Long Short-Term Memory (LSTM) neural network. Then, the PMP algorithm is applied for rolling optimization within the time domain of speed prediction, solving the optimal power control sequence at each moment. Finally, based on actual operating conditions, the energy management results of the MPC-PMP strategy, offline PMP strategy, and conventional threshold method are compared. Simulation results demonstrate that although the fuel economy of the MPC-PMP strategy is slightly lower than that of the offline PMP strategy, it improves fuel economy by 14.45% compared to the traditional threshold method, while meeting real-time requirements.

Keywords: Energy management strategy · Hybrid power system · Pontryagin minimum principle · Model predictive control · Long short-term memory neural network

1 Introduction

As rail transportation technology continues to evolve and mature, as of 2022, China's railway electrification rate has reached 73.8% [1]. Nevertheless, due to constraints such as economic and natural conditions in certain regions, non-electrified railways remain relatively prevalent. In the context of the "dual carbon" development, in response to the national call for energy conservation and emissions reduction, researchers have developed a novel diesel-electric hybrid power train to address problems posed by traditional pure internal combustion locomotives, including high fuel consumption, low efficiency, and high emissions [2, 3].

L. Jia et al. (Eds.): EITRT 2023, LNEE 1135, pp. 638–651, 2024.
https://doi.org/10.1007/978-981-99-9307-9_69

In comparison with conventional pure internal combustion locomotives, the diesel-electric hybrid power train retains diesel generators as the primary power source while incorporating power batteries as auxiliary power sources. This configuration enables the hybrid power train to recover braking energy and effectively enhance the efficiency of the diesel generators, thereby improving fuel efficiency [4].

The energy management system serves as the control center for the train's hybrid power system [5], significantly impacting the overall fuel economy through optimized power allocation. Currently, the field of automotive research has achieved maturity in this regard. Studies on energy management primarily fall into two categories: rule-based energy management and optimization-based energy management [6]. The former relies on expert experience to design threshold rules, offering simplicity and real-time control, yet struggling to guarantee optimal fuel economy [7]. In contrast, the latter establishes optimization problems with the goal of minimizing fuel consumption, utilizing optimization algorithms to derive optimal control sequences. This approach not only yields superior fuel economy but also allows for coupling with other goals, such as battery life considerations, making it a favored research methodology.

Presently, optimization-based energy management strategies predominantly include dynamic programming (DP), Pontryagin's Minimum Principle (PMP), and equivalent fuel consumption strategy (ECMS) [8]. While the DP strategy seeks global optimization and achieves favorable energy management results, its computational demands restrict its online applicability. Conversely, extensive research indicates that the PMP strategy significantly reduces computational complexity while yielding fuel consumption results similar to those of DP. Therefore, the application of the PMP algorithm to online energy management holds substantial promise [9–11].

The optimization control solved by the PMP strategy relies on a priori global operating conditions, which may not be known during the actual online operation of the train. Thus, while this strategy obtains theoretically optimal offline solutions, it cannot be applied directly to real-time online energy management control. To address this problem, this paper focuses on a diesel-electric hybrid power train and introduces model predictive control (MPC) to enhance the PMP strategy. A novel energy management strategy based on MPC-PMP is proposed. This strategy employs rolling optimization through the PMP algorithm within a prediction time domain, enabling online adaptive adjustment of control parameters for the PMP strategy. This, in turn, achieves optimized online energy management control.

The structure of this paper is as follows: Chap. 1 establishes the topology of the diesel-electric hybrid power system and constructs the hybrid power system model. Chapter 2 utilizes Long Short-Term Memory (LSTM) neural networks to develop a short-term speed prediction model for the train. Chapter 3 analyzes and formulates the energy management optimization problem for the hybrid power train, utilizing actual operating conditions as inputs, and employs the MPC-PMP method for online energy management optimization. Chapter 4 presents simulation analyses of the aforementioned methods. Chapter 5 summarizes the primary research findings and draws conclusions.

2 Hybrid Power System Model

2.1 Hybrid Power System Structure

The subject of this paper is a hybrid diesel-electric train, and its hybrid power system structure is depicted in Fig. 1.

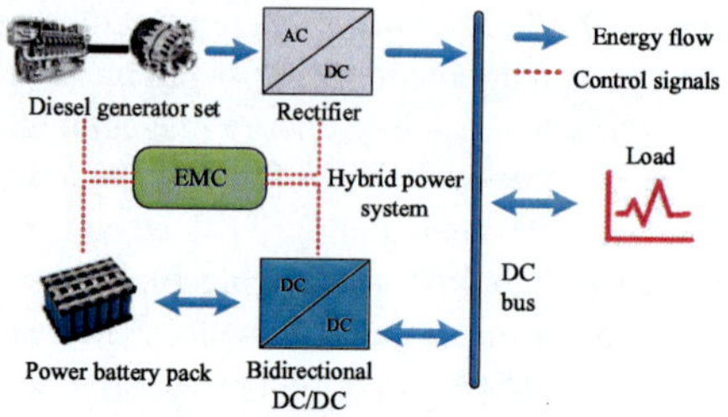

Fig. 1. Hybrid system structure

The diesel generator set is connected to the busbar through a rectifier, while the power battery pack is connected to the busbar through a bidirectional DC/DC converter in parallel. These two components collectively constitute the entire diesel-electric hybrid power system, facilitating energy exchange with the traction and auxiliary systems. The energy management controller dynamically computes and issues control commands through an energy management strategy, facilitating the allocation of energy between the diesel generator set and the power battery pack. To ensure the train's power requirements, a power balance constraint must be satisfied at the DC bus:

$$P_{\text{bus}} = P_{\text{ice}}\eta_{\text{acdc}} + P_{\text{bat}}\eta_{\text{dcdc}}^{k} \tag{1}$$

where, P_{bus}, P_{ice}, and P_{bat} represent the bus demand power, the output power of the diesel generator set, and the output power of the battery pack, respectively. η_{acdc} and η_{dcdc} correspond to the rectifier efficiency and the DC/DC converter efficiency, respectively. The coefficient k is associated with the battery power and its specific expression is given by:

$$k = \begin{cases} 1 & P_{\text{bat}} \geq 0 \\ -1 & P_{\text{bat}} < 0 \end{cases} \tag{2}$$

2.2 Equivalent Circuit Model of Battery Pack

This paper disregards the inconsistency among individual battery cells and adopts a first-order RC equivalent circuit model as illustrated in Fig. 2.

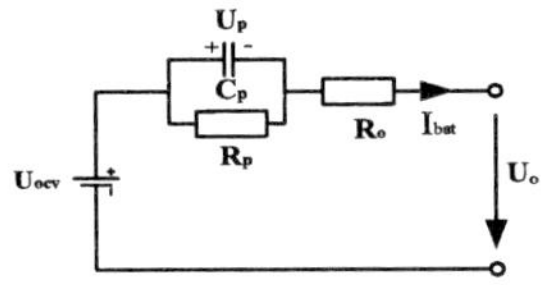

Fig. 2. First-order RC model of battery

In the diagram, U_{ocv} is the open-circuit voltage of the battery, U_p is the polarization voltage of the battery, C_p is the polarization capacitance of the battery, R_p is the polarization resistance of the battery, R_o is the Ohmic internal resistance of the battery, I_{bat} is the output current of the battery, and U_o is the output voltage of the battery.

The mathematical representation of the first-order RC model for the battery is as follows:

$$\begin{cases} U_o = U_{ocv} - I_{bat}R_o - U_p \\ I_{bat} = \dfrac{U_p}{R_p} + C_p\dfrac{dU_p}{dt} \\ P_{bat} = U_o I_{bat} \\ SOC_{bat} = SOC_{bat,0} - \dfrac{1}{3600Q_{bat}}\displaystyle\int I_{bat}dt \end{cases} \tag{3}$$

where, P_{bat} represents the output power of the battery pack, SOC_{bat} represents the state of charge of the battery pack, $SOC_{bat,0}$ represents the initial state of charge of the battery pack at time, and Q_{bat} represents the rated capacity of the battery pack.

2.3 Diesel Generator Fuel Consumption Model

In order to accurately assess the fuel consumption of the diesel generator set, this paper models it using an approach based on experimental data. By utilizing a substantial amount of experimental data, the fuel consumption characteristics of the diesel generator set under various operating conditions are obtained. When coupled with the generator's efficiency, the fuel consumption efficiency data of the diesel generator pack can be derived and presented using a MAP diagram, as shown in Fig. 3.

The fuel consumption of the diesel generator set $\dot{m}$ (g/s) can be expressed as follows:

$$\dot{m}(T_{ice}, w_{ice}) = \frac{\dot{m}_g(T_{ice}, w_{ice})P_{ice}}{3600} \tag{4}$$

where, $\dot{m}_g$ (g/kWh) represents the fuel consumption rate of the diesel generator set, T_{ice} and w_{ice} represent the torque and speed of the diesel generator, respectively.

For the hybrid train studied in this paper, there exists a physical decoupling relationship between the wheel axle speed and the diesel engine speed. Based on this, it can be assumed that the train can achieve optimal control, implying that the diesel generator

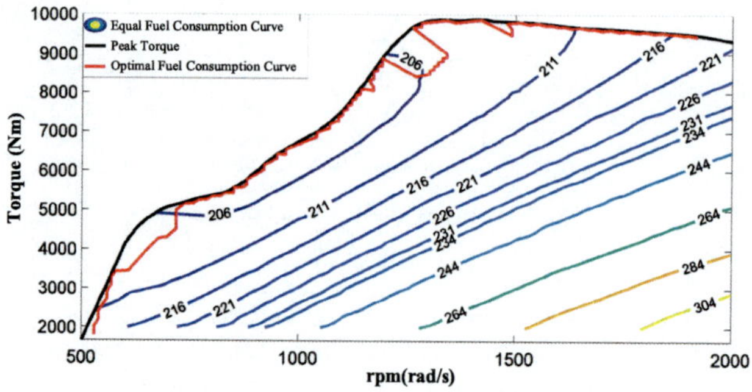

Fig. 3. Brake specific fuel consumption of 1864kW diesel generator set

pack operates at the optimal fuel consumption point for each output power level. Consequently, the fuel consumption of the diesel generator pack can be further represented as follows:

$$\dot{m}(P_{\text{ice}}) = \frac{\dot{m}_{\text{g}}(P_{\text{ice}})P_{\text{ice}}}{3600} \tag{5}$$

3 Speed Prediction Model Based on LSTM Neural Network

3.1 Overview of Speed Prediction

Speed prediction plays a crucial role in energy management, enabling the formulation of effective energy management strategies by forecasting the speed for a future time horizon. With the continuous advancement of artificial intelligence, neural network-based speed prediction has gained widespread adoption. In comparison to conventional approaches, neural networks exhibit enhanced nonlinear modeling capabilities and the ability to capture long-term dependencies, thereby achieving more accurate speed sequence predictions.

In neural network speed prediction, typically, the past m historical time instances of speed are used as inputs to forecast the speed for the upcoming n time instances. This can be expressed as follows (assuming the current time instance is k):

$$\left[v_{\text{k+1}}, v_{\text{k+2}}, \ldots v_{\text{k+n}}\right] = f\left(\left[v_{\text{k - m + 1}}, v_{\text{k - m + 2}}, \ldots v_{\text{k}}\right]\right) \tag{6}$$

3.2 Speed Prediction Model Based on LSTM

Comparative to conventional neural networks, the LSTM neural network introduces memory units, endowing it with distinct advantages in time series prediction tasks such as speed sequence forecasting. The structure of an LSTM neural network unit, as depicted in Fig. 4, comprises a forget gate, an input gate, an output gate, and a memory cell.

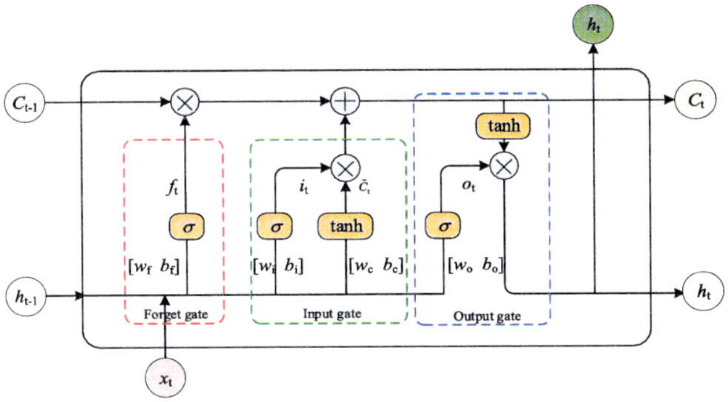

Fig. 4. LSTM neural network unit structure diagram

The computational formulas for LSTM are as follows:

$$\begin{cases} f_t = \sigma(b_f + w_f \cdot [h_{t-1}, x_t]) \\ i_t = \sigma(b_i + w_i \cdot [h_{t-1}, x_t]) \\ \hat{C}_t = \tanh(b_c + w_c \cdot [h_{t-1}, x_t]) \\ o_t = \sigma(b_o + w_o \cdot [h_{t-1}, x_t]) \\ h_t = \tanh(C_t) \cdot o_t \\ C_t = C_{t-1} f_t + i_t \cdot \hat{C}_t \end{cases} \quad (7)$$

where, f_t, i_t, o_t represent the forget gate, input gate, and output gate at time t, respectively, $\widehat{C_t}$ represents the memory cell, C_{t-1} and C_t are the states at the previous time step and the current time step, respectively, w_f, w_i, w_c, w_o are the corresponding weights, b_f, b_i, b_c, b_o are the respective biases, x_t represents the current input at the given time, h_{t-1} and h_t represent the hidden layer vectors at the previous and current time steps, respectively, σ and *tanh* are the activation functions associated with each gate.

The sampling frequency of the speed in this study is 1Hz, implying that the historical time domain for input in the speed prediction model spans m seconds, while the predicted time domain for output spans n seconds. The speed prediction model established in this paper is illustrated in Fig. 5.

To comprehensively evaluate the impact of historical and predicted time domains on the model accuracy, simulations are performed with varying values of $m = 5, 10, 15$ and $n = 5, 10, 15$. The root mean square error (RMSE) is adopted in this paper to assess the model's accuracy.

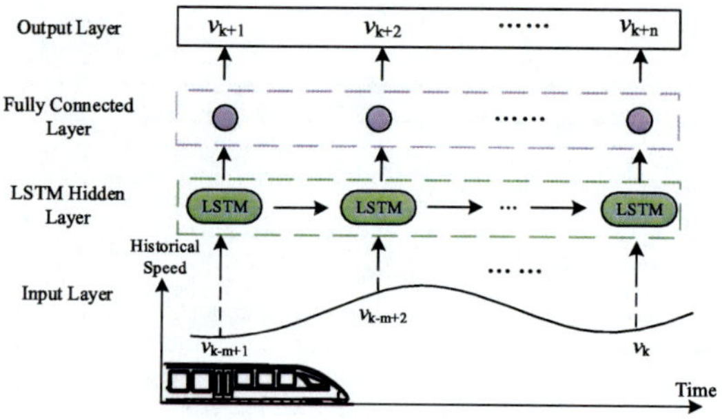

Fig. 5. Speed prediction model structure diagram

4 Online Energy Management Based on MPC-PMP

4.1 Energy Management Optimization Problem Formulation

Train energy management involves the allocation of energy between the diesel generator set and the power battery pack, with the optimization goal being the minimization of total fuel consumption:

$$J = \min \int_{t_0}^{t_f} \dot{m}(P_{ice})dt \tag{8}$$

where, t_0 represents the initial time, and t_f represents the final time, $\dot{m}$ represents the train's fuel consumption rate. Assuming a prediction time domain of n seconds, the optimization goal for each time step can be expressed as follows:

$$J = \min \int_{t_0}^{t_0+n} \dot{m}(P_{ice})dt \tag{9}$$

Choose the battery State of Charge (SOC) as the state variable, and the battery current (I_{bat}) as the control variable. The battery SOC must satisfy boundary conditions:

$$SOC(t_0) = SOC_0, SOC(t_f) = SOC_f \tag{10}$$

Additionally, due to certain physical constraints, the hybrid power system should adhere to the following constraints:

$$\begin{cases} I_{bat\_min} \le I_{bat} \le I_{bat\_max} \\ SOC_{min} \le SOC \le SOC_{max} \\ 0 \le P_{ice} \le P_{ice\_max} \\ P_{aux} + P_{tra} = P_{ice}\eta_{acdc} + P_{bat}\eta_{dcdc}^{k} \end{cases} \tag{11}$$

where, I_{bat_min} and I_{bat_max} represent the maximum charging current and the maximum discharging current of the battery, respectively, SOC_{min} and SOC_{max} represent to the minimum allowable State of Charge (SOC) value and the maximum SOC value for the train's battery, P_{ice_max} represents the maximum discharge power of the diesel generator set, P_{aux} and P_{tra} represent the total power of the auxiliary system and the total power of the traction transmission system, respectively.

4.2 Energy Management Based on PMP

PMP is a classic algorithm in control theory, evolved from the calculus of variations. According to the PMP principle, the Hamiltonian function H of this problem can be expressed as follows:

$$H = \dot{m}(P_{ice}) + \lambda^T(t)S\dot{O}C \tag{12}$$

where λ represents the co-state variable.

Assuming the existence of optimal control, the optimal control sequence $I^*{}_{bat}(t)$ as well as the optimal co-state variable sequence $\lambda^*(t)$ must satisfy the necessary conditions:

$$\begin{cases} S\dot{O}C^*(t) = \frac{\partial H}{\partial \lambda} \\ \dot{\lambda}^*(t) = -\frac{\partial H}{\partial SOC} \end{cases} \tag{13}$$

This optimization problem can be further formulated as follows:

$$I_{bat}^*(t) = \arg \min_{U(t)} H(SOC^*(t), I_{bat}(t), \lambda^*(t), t) \tag{14}$$

where $U(t)$ is the control domain.

From the above procedure, it is evident that when the initial co-state variable $\lambda(0)$ is known, the entire control sequence $I_{bat}(t)$ for the train's entire operation process can be determined. Thus, the entire problem can be transformed into the task of solving the optimal initial co-state variable. Once the optimal initial co-state variable $\lambda^*(0)$ is obtained, the optimal control sequence $I^*{}_{bat}(t)$ can be derived.

4.3 Online Energy Management Optimization Based on MPC-PMP

During the train's online operation, the MPC-PMP method collects and stores historical data. Based on the speed prediction model and the SOC reference curve, the energy management is subjected to rolling optimization using the PMP algorithm within the prediction time domain. In each step, the first control input is applied to the energy management controller, thus resolving the problem of the PMP algorithm requiring knowledge of the global operating conditions. Simultaneously, this approach enables flexible adaptation to changes in operating conditions, achieving real-time online energy management control.

As illustrated in Fig. 6, the schematic diagram depicts the process of online energy management implementation based on MPC-PMP.

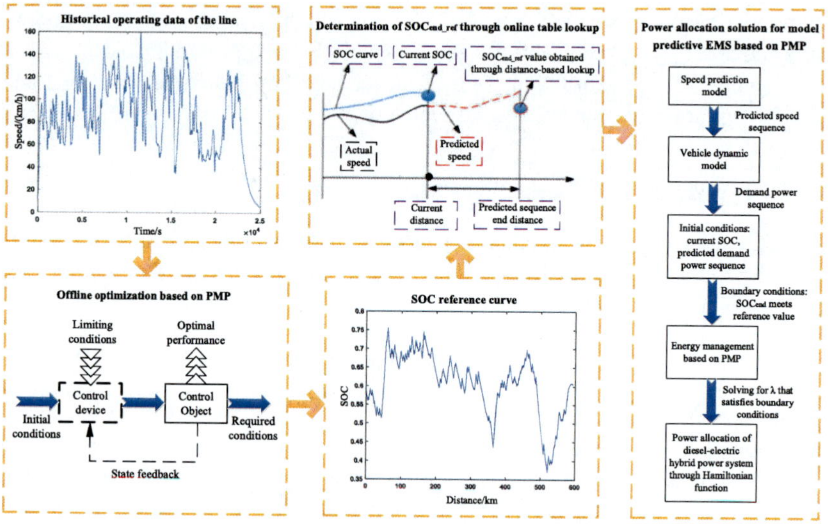

Fig. 6. Energy management strategy based on MPC-PMP

5 Simulation Analysis

5.1 Analysis of LSTM Speed Prediction Model Results

In this study, a set of historical operating data from a railway route in Australia was utilized as the training foundation for the LSTM neural network. Utilizing the speed prediction model, the accuracy of speed prediction results was evaluated using the root mean square error (RMSE) of the trained models across different historical and prediction time domains. Tab. 1 presents the RMSE values for the LSTM speed prediction model across varying historical and prediction time domains.

Table 1. RMSE of LSTM speed prediction model

Historical time domain (s)	Prediction time domain (s)	RMSE
5	5	0.0231
	10	0.0248
	15	0.0281
10	5	0.0176
	10	0.0195
	15	0.0236
15	5	0.0114
	10	0.0163
	15	0.0205

From Table 1, it can be observed that the speed prediction model follows the following patterns: with a fixed historical time domain, increasing the prediction time domain leads to a reduction in model accuracy; with a fixed prediction time domain, extending the historical time domain tends to enhance the model accuracy, albeit with a marginal improvement. The highest accuracy is achieved when the prediction time domain is $p = 5$ s.

In conclusion, this study will further compare results from energy management simulations when the prediction time domain is $p = 5$ s, and historical time domains are $r = 5$ s, 10 s, and 15 s. Thus facilitating the selection of a more suitable online model.

5.2 Analysis of Online Energy Management Results

In this chapter, the influence of different speed prediction models on online energy management is analyzed, including fuel consumption and simulated speed. To facilitate a more convenient fuel consumption comparison, the battery SOC is maintained at a consistent level for each scenario upon reaching the destination. This approach avoids an unreasonable comparison of fuel consumption alone under different terminal SOC conditions, neglecting battery energy consumption. It also ensures that the train does not experience insufficient charging time from the diesel generator to the battery due to short station stops, which could result in excessively low battery SOC during the subsequent phase of operation. The input conditions for simulation are represented by the total power demand curve of the train as shown in Fig. 7.

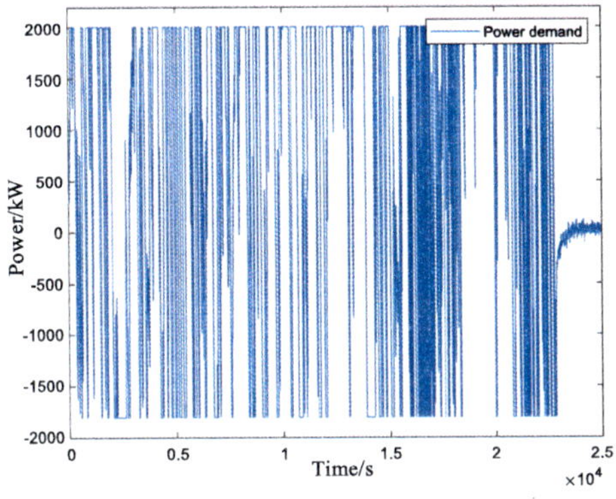

Fig. 7. Power curve of diesel-electric hybrid power system

Basic parameters of diesel-electric hybrid train are presented in Table 2.

Figure 8 and 9 illustrate the power allocation results and SOC curves under three different speed prediction models.

Due to varying prediction accuracy among these models, energy management decisions lead to trajectory discrepancies. Table 3 provides a comparison of fuel consumption and simulated speed under the three speed prediction models. From the table, it can be observed that the fuel economy is optimal when $r = 5$s and $p = 5$s, resulting in a faster

Table 2. Basic parameters of diesel-electric hybrid train

Parameters	Value	Parameters	Value
Vehicle weight	400	Battery type	Energy-type LTO
Rated power of diesel generator set/kW	1864	Battery cell capacity	40
Auxiliary system power/kW	100	Battery series count	480
Traction system efficiency	0.87	Battery parallel count	30
Rectifier efficiency	0.98	Total battery capacity/kWh	1324.8
Bidirectional DC/DC efficiency	0.98	Maximum charge/discharge rate	1.5C

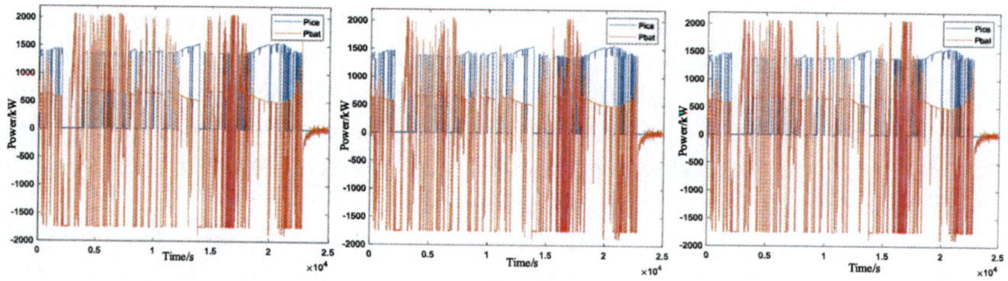

(a) r=5s, p=5s. (b) r=10s, p=5s. (c) r=15s, p=5s.

Fig. 8. Online energy management power allocation under three speed prediction models

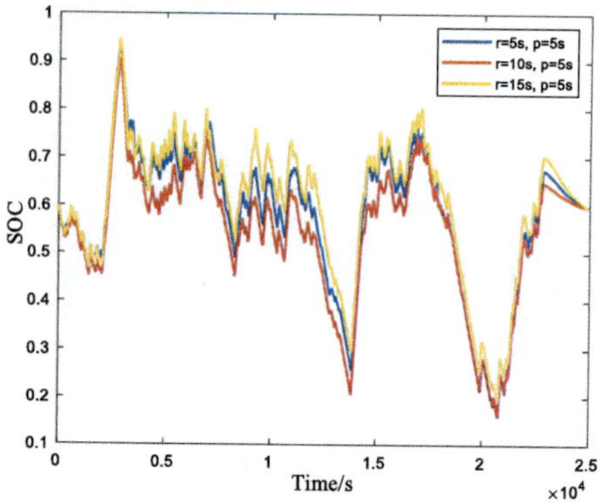

Fig. 9. Online energy management SOC curve under three speed prediction models

simulation speed. In order to ensure both fuel economy and real-time performance, the LSTM neural network model with $r = 5$s and $p = 5$s is selected as the online application's speed prediction model.

To provide a more objective assessment of the online energy management results, a comparison was conducted between the proposed online strategy, the offline PMP strategy, and the rule-based strategy in terms of fuel economy. The input conditions for

Table 3. Comparison of total fuel consumption and simulation time for online energy management under three speed prediction models

Type	Fuel consumption	End SOC	Simulation time
$r = 5$ s, $p = 5$ s	899.69 kg	0.5995	101.17 s
$r = 10$ s, $p = 5$ s	968.98 kg	0.5995	109.93 s
$r = 15$ s, $p = 5$ s	916.68 kg	0.5995	98.952 s

simulation remained consistent, as represented by the total power demand curve of the train shown in Fig. 7. Figures 10 and 11 depict the power allocation results and SOC curves obtained under the three different strategies for this operating condition.

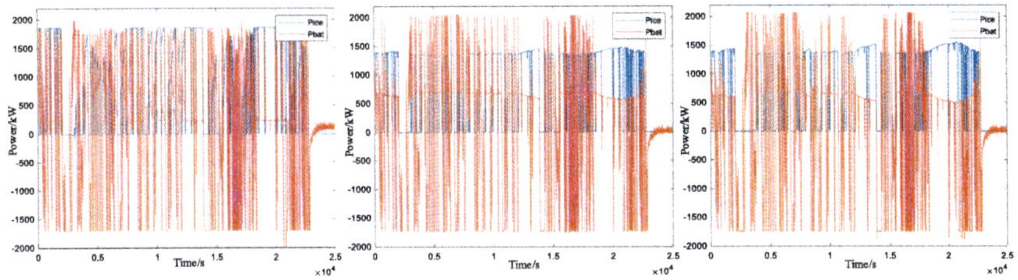

(a) Rule-based strategy. (b) Offline PMP strategy. (c) MPC-PMP strategy.

Fig. 10. Power allocation result curves for three different strategies

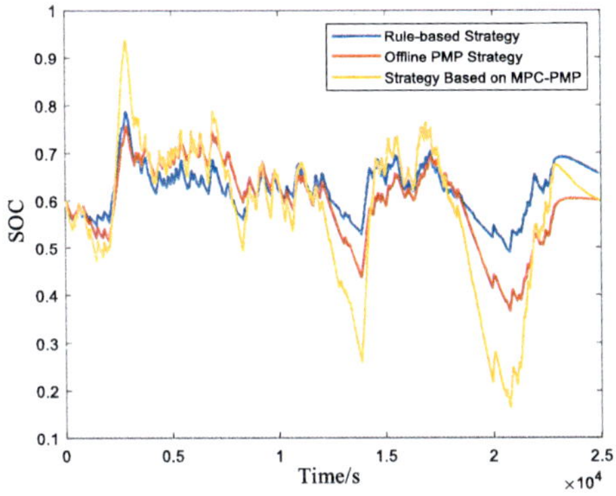

Fig. 11. Power allocation result curves under three different strategies

The terminal SOC may not always be guaranteed to reach the designated reference value using rule-based strategy. To mitigate the impact of terminal SOC, the concept of equivalent fuel consumption is employed here, whereby the fuel consumption of the rule-based strategy is equated to the fuel consumption of the other two methods at the same terminal SOC. Based on this approach, a comparative analysis of fuel consumption results among three strategies is presented in Table 4.

Table 4. Comparison of fuel consumption results among three strategies.

Control strategy	Fuel consumption	End SOC	Fuel saving efficiency
Rule-based strategy	1051.6kg	0.6554	/
Offline PMP strategy	895.54kg	0.5995	14.84%
Strategy based on MPC-PMP	899.69kg	0.5995	14.45%

As the offline PMP strategy optimizes globally, ensuring the diesel generator operates at its optimal efficiency point when starting, it achieves the highest fuel economy. The rule-based energy management strategy, though computationally simple and user-friendly, pre-determines threshold parameters, which cannot guarantee operation within the optimal efficiency range of the diesel generator at every point. As a result, its fuel economy is relatively lower. On the other hand, the proposed online strategy, similar to the offline process, achieves optimization through short-term planning. However, due to the limited scope of short-term optimization, it cannot perform global searches, resulting in slightly lower fuel economy compared to the offline PMP strategy. Nonetheless, the fuel economy of the online strategy is still 14.45% higher than that of the rule-based strategy. This observation underscores the favorable fuel economy of the proposed energy management strategy based on MPC-PMP.

6 Conclusion

In the context of train hybrid power system, this paper addresses the problem of the inapplicability of the PMP strategy for online energy management of actual trains. To overcome this challenge, a novel online energy management strategy based on MPC-PMP is proposed.

Utilizing the adjustment of co-state variables based on the speed prediction model and SOC reference curve, this approach demonstrates favorable performance in maintaining battery SOC. Meanwhile, although the MPC-PMP strategy yields slightly inferior results compared to the offline PMP strategy, it notably outperforms rule-based strategies, indicating its favorable fuel economy characteristics.

Acknowledgments. This research is funded by the Open Project Fund of the State Key Laboratory of Heavy-duty and Express High-power Electric Locomotive.

References

1. Ministry of Development and Reform. Statistical Bulletin of China National Railway Group Co., Ltd. for the Year 2022. People's Railway, 2023–03–17(002) (in Chinese)
2. Cipek, M., Pavković, D., Krznar, M., Kljaić, Z., Mlinarić, T.J.: Comparative analysis of conventional diesel-electric and hypothetical battery-electric heavy haul locomotive operation in terms of fuel savings and emissions reduction potentials. Energy **232**, 121097 (2021)
3. Kapetanović, M., Núñez, A., Oort, N., Goverde, R.: Reducing fuel consumption and related emissions through optimal sizing of energy storage systems for diesel-electric trains. Appl. Energy **294**, 117018 (2021)
4. Wu, J., Zhang, C., Zhang, W.G., Li, T., et al.: Comparison of different simulation models and simulation methods for energy storage trams. Proc. Chinese Soc. Electr. Eng. **40**(16), 5148–5162 (2020). (in Chinese)
5. Therrera, V., Milo, A., et al.: Adaptive energy management strategy and optimal sizing applied on a battery-supercapacitor based tramway. Appl. Energy **169**, 831–845 (2016)
6. Nci, M., Büyük, M., Demir, H., et al.: A review and research on fuel cell electric vehicles: topologies, power electronic converters, energy management methods, technical challenges, marketing and future aspects. Renew. Sustain. Energy Rev. 137, 110648 (2021)
7. Lü, X., Wu, Y., Lian, J., Zhang, Y., Chen, C., Wang, P., et al.: Energy management of hybrid electric vehicles: a review of energy optimization of fuel cell hybrid power system based on genetic algorithm. Energy Convers. Manage. **205**, 112474 (2020)
8. Santucci, A., Sorniotti, A., Lekakou, C.: Power split strategies for hybrid energy storage systems for vehicular applications. J. Power Sources **258**, 395–407 (2014)
9. Zhang, C.N., Zhou, W., Li, J.Q., et al.: An Online Energy management strategy for extended range electric vehicles based on Pontryagin's minimum principle. J. Beijing Inst. Technol. **35**(09), 931–935 (2015). (in Chinese)
10. Meng, X., Li, Q., Chen, W.R., Zhang, G.R.: An energy management method based on Pontryagin minimum principle satisfactory optimization for fuel cell hybrid systems. Proc. Chinese Soc. Electr. Eng. 39(03), 782–792+957 (2019) (in Chinese)
11. Yuan, Z., Teng, L., Fengchun, S., et al.: Comparative study of dynamic programming and Pontryagin's minimum principle on energy management for a parallel hybrid electric vehicle. Energies **6**(4), 2305–2318 (2013)

A High Pulse Power and Efficiency Integrated Supercapacitor Energy Package

Chuanwei Guo[1], Tongzhao Li[2,3], Xinchen Cui[4], Wenli Lin[5], Weiyao Mei[2,3], and Lijun Diao[2,3](✉)

[1] Beijing Spacecrafts Co., Ltd, Beijing 100080, China
[2] School of Electrical Engineering, Beijing Jiaotong University, Beijing 100044, China
ljdiao@bjtu.edu.cn
[3] Key Lab. of Vehicular Multi-Energy Drive Systems, Ministry of Education, Beijing 100044, China
[4] Dongfang Electronics Co, Ltd, Yantai 264000, China
[5] China Academy of Space Technology, Beijing 100081, China

Abstract. In order to improve the power density of the power system and provide power to various loads in a timely and efficient manner, it is particularly necessary to use supercapacitors as energy storage units. This article focuses on the research and integrated development of supercapacitors and bidirectional DC/DC converters under special operating conditions such as load changes in distributed power systems. Firstly, the capacitance of the supercapacitor was calculated and its monomer combination method was designed. Secondly, the bidirectional Buck/Boost circuit was analyzed, passive components such as inductors and capacitors were calculated and selected, and the driving circuit was analyzed. Thirdly, based on the charging and discharging characteristics of supercapacitors, an experimental prototype with a rated power of 2 kW and a pulse power of 10kW was built. The maximum efficiency under rated operating conditions was 96.53%, and the maximum efficiency under pulse operating conditions was 89.56%.

Keywords: Supercapacitor · Bidirectional DC/DC converter · Pulse discharge

1 Introduction

Energy storage is a key support point for the energy revolution. As an effective means to fully leverage the advantages of energy storage, the energy storage structure of distributed power systems has been basically formed after long-term development. With the increasing range of applications, the power demand for electrical equipment is increasing, and the pulse characteristics of the load are becoming increasingly prominent. For example, electromagnetic pulse loads, wind power and photovoltaic grid connection, new energy vehicle startup, etc. all require pulse high power or short-term high power output for the power system [1, 2]. The typical characteristic is that the average power is low and the peak power is so high that high power density and high energy density should be considered more in its energy storage system. Combining supercapacitors

L. Jia et al. (Eds.): EITRT 2023, LNEE 1135, pp. 652–659, 2024.
https://doi.org/10.1007/978-981-99-9307-9_70

with bidirectional DC/DC converters is an effective means to solve the problem of pulse load power supply [3, 4], and can play a role in regulating the stability of the primary bus voltage. The power converter is an important hub for completing energy conversion [5–7]. As the most common energy conversion device in power systems, the topology of DC/DC converters affects the efficiency and power to weight ratio of the system to a certain extent [2].

This article focuses on the research and experimental verification of integrated supercapacitor energy packs. Firstly, the calculation and design of the supercapacitor were completed. Secondly, the design of a bidirectional Buck/Boost was completed. Finally, an experimental prototype was built and the experimental process was designed to verify the system performance.

2 Design of Supercapacitor

Traditional supercapacitors have the advantage of large capacity, but they also have disadvantages with low voltage and current withstand values. Therefore, in high-power applications, it is often necessary to perform series and parallel connections to improve their voltage and current levels.

2.1 Method of Selecting and Combining Monomers

Due to the maximum power of the pulse load being 10 kW and the duration of the current peak being 1 s, the supercapacitor design is carried out according to this standard, and the calculation formula is as follows.

Energy released by supercapacitor discharge

$$W_{sc} = \frac{1}{2}CV_1^2 - \frac{1}{2}CV_2^2 \geq 10\text{kW} \times 1\text{s} = 10\text{kJ} \tag{1}$$

where V_1 and V_2 are the voltages before and after the discharge of the supercapacitor respectively, while C is the capacitance value of the supercapacitor and W_{sc} is the energy released during the discharge process.

System current stress constraint

$$I_{\max} = \frac{P_{sc}}{V_2} \tag{2}$$

where P_{sc} is the charging and discharging power of the supercapacitor, and $I_{\max}$ is the maximum current during the charging and discharging process of the supercapacitor.

System efficiency constraints

$$P_{mos} + P_L + P_{Drv} \leq (1 - \eta)P \tag{3}$$

where P_{mos} is the loss of the switches, P_L is the loss of the inductor, P_{Drv} is the loss of the driving circuit, and P is the input power, η is system efficiency.

Selecting the Hezhong Huineng series supercapacitor monomer (monomer voltage of 2.7 V, capacitance value of 220 F, weight of 40 g) as a feasible solution, using a 32 series and 2 parallel grouping form, the maximum voltage of the formed supercapacitor module is 86.4 V, and the capacitance value is 13.75 F.

For the sake of work safety, the working voltage range of the supercapacitor module is limited to 49 V ~ 75 V. According to the energy storage expression of the supercapacitor in the above formula, when the supercapacitor operates within this range, it can release 22165J of energy per cycle to the load.

2.2 Design of Series Voltage Equalization

In actual batch production, it is difficult to have good consistency in factors such as material selection and manufacturing process of supercapacitors, and there are also varying degrees of differences in performance parameters, such as equivalent series parallel resistance, actual capacitance, leakage current, etc. Taking the capacitance value of supercapacitors as an example, the accuracy is usually -10% to 30%.

As shown in Fig. 1, in order to maximize the balance between reliability and response speed, the design uses a TL431 device based equalization circuit as the voltage sharing strategy for supercapacitors.

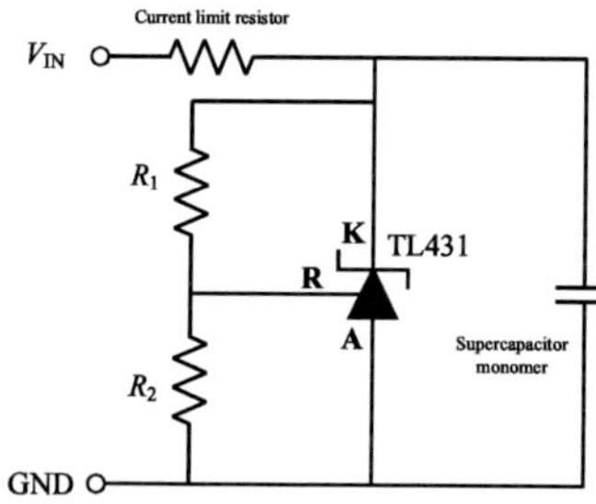

Fig. 1. Equalizing circuit of supercapacitor

The reference voltage will be dynamically maintained at 2.5V, so that the output voltage remains unchanged. When charging supercapacitors, the first fully charged unit will maintain the same voltage, while the unit that is not fully charged will continue to charge, and ultimately all units will reach a voltage equilibrium state.

3 Principle and Design of Bidirectional Buck/Boost Circuit

The system circuit diagram is shown in Fig. 2, where V_{bus} is the DC bus voltage, V_{sc} is the supercapacitor voltage, Q_1 and Q_2 are the top and bottom MOSFETs of the converter, L is the filter inductance of the converter, C_1 is the high-voltage side filter capacitor, C_2 is the low-voltage side filter capacitor, SC is the supercapacitor, R_s is the equivalent series internal resistance of the supercapacitor, R_p is the equivalent parallel resistance of the supercapacitor, and R is the system variable load (rated load 2 kW, pulse load 10 kW).

The bidirectional Buck/Boost converter operates in Buck mode when the top tube is used as the switch and the bottom tube is normally closed. While the top tube is normally closed and the bottom tube is used as the switch tube, the system operates in Boost mode.

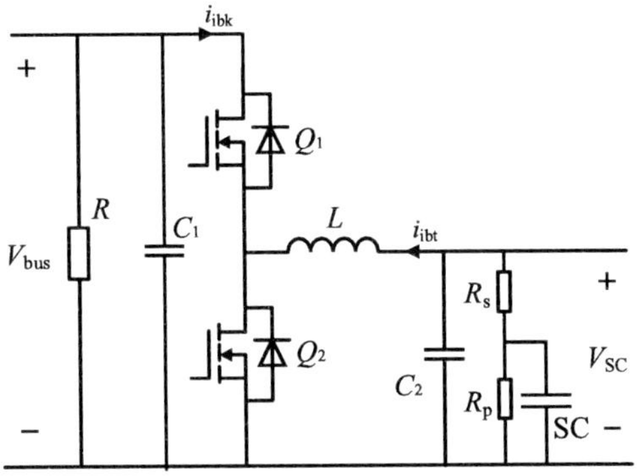

Fig. 2. System topology

3.1 Design of Passive Components

In Buck and Boost modes, the inductance can be expressed respectively as

$$L_{\text{Buck}} = \frac{V_{\text{bus}} D_{\text{Q1}} (1 - D_{\text{Q1}})}{\Delta i_L} T_s \tag{4}$$

$$L_{\text{Boost}} = \frac{V_{\text{sc}} D_{\text{Q2}}}{\Delta i_L} T_s \tag{5}$$

where L_{Buck} and L_{Boost} are the inductance values in Buck and Boost modes, respectively. D_{Q1} and D_{Q2} are the duty cycles of the top and bottom MOSFETs, respectively. T_{s} represent the switching period and $\Delta i_L$is the current ripple of the inductance.

Taking the ripple coefficient of the inductance current as 15% and the switching frequency as 50kHz, the inductance can be obtained as 160μH.

In Buck mode, the filtering capacitor C_2 can be expressed as

$$C_2 = \frac{T_{\text{s}}^2 V_{\text{bus}} D_{\text{Q1}} (1 - D_{\text{Q1}})}{8 L \Delta V_{\text{C2}}} \tag{6}$$

In Boost mode, the filtering capacitor C_1 can be expressed as

$$C_1 = \frac{I_{\text{o}} D_{\text{Q2}} T_{\text{s}}}{\Delta V_{\text{C1}}} \tag{7}$$

where, ΔV_{C1} and ΔV_{C2} are the ripple voltages of two filter capacitors, respectively. And I_{o} is the load current.

According to calculations, C_2 is 40.82μF, and C_1 is 1275.51μF. C_2 can select 68 μF/250V series aluminum electrolytic capacitors. C_1 can select four 470 μF/250V capacitors which are connected in parallel.

3.2 Design of Drive Circuit

The commonly used MOSFETs driver circuit structure is shown in Fig. 3 L_{k} is the parasitic inductance of the driving circuit, and R_{g} is the opening resistance of the driving

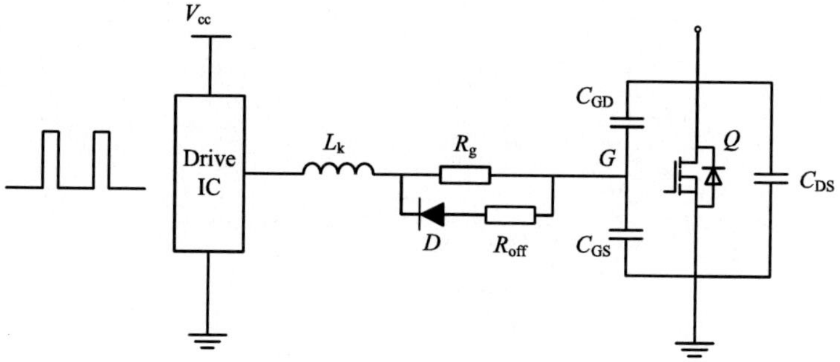

Fig. 3. Circuit diagram of drive circuit

circuit. R_{off} is the turning off resistance of the drive circuit. C_{GD}, C_{GS}, and C_{DS} are parasitic capacitors of MOSFETs.

To prevent oscillation of the driving current, it is necessary to meet the following requirements

$$R_g \geq 2\sqrt{\frac{L_k}{C_{GS}}} \tag{8}$$

In order to meet the current stress requirements under pulse load mode, two IRF300P226 MOSFETs are used in parallel in this design.

According to the datasheet, C_{gs} is 10025pF. Using Ansys/Q3D simulation software to extract stray inductance from the PCB, the results show that L_k is approximately 37.2nH, so the lower limit value of the R_g is 3.85 Ω. Secondly, calculate the upper limit value of the opening resistance to ensure that the rise time during the opening process of the pipe does not exceed 3% of the switching cycle, so take the opening resistance as 20 Ω. MOSFETs usually adopt a slow on and fast off driving method, and in this design, the R_{off} value is 10 Ω.

4 Experimental Verification

4.1 Experimental Platform

In order to minimize system volume, improve power density, and facilitate the installation of the converter and improve compressive strength, the shell structure of the converter was designed. Based on the size of the main circuit board and control board, a structural box was designed. The system adopts PI closed-loop control. The experimental platform is shown in Fig. 4.

4.2 Experimental Waveform

Due to the initial voltage of the supercapacitor being zero, a small current is required to charge first. The experimental flowchart is shown in Fig. 5. The experiment was

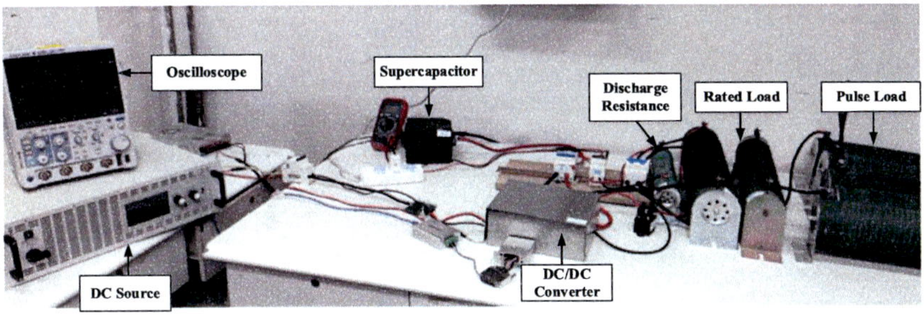

Fig. 4. Experimental platform

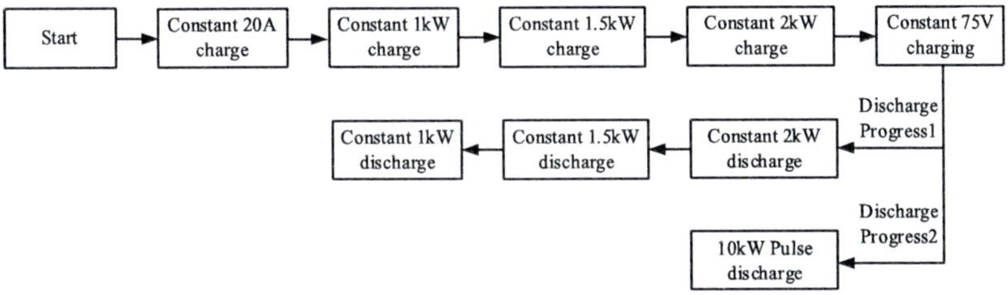

Fig. 5. Experimental flowchart

conducted twice, and after each charge, the system entered a constant voltage charging mode. The rated power discharge and pulse discharge were selected for each discharge progress to verify the performance of the system.

The waveforms of different working modes are shown in Fig. 6. In Fig. 6a is the 20A constant current charging mode, while (b), (c), and (d) are the 1 kW, 1.5 kW, and 2 kW constant power charging modes. (e) is a constant voltage charging mode, during which the supercapacitor voltage remains constant at 75V. (f), (g) and (h) are 2 kW, 1.5 kW, and 1 kW constant power discharge modes, respectively. The waveforms of switching from constant voltage charging mode to pulse load discharge mode are shown in Fig. 6i. It can be concluded that when converted to pulse load, the output power of the supercapacitor is basically stable at around 10 kW.

The efficiency under different modes is shown in Table 1, and it can be concluded that the system efficiency under the charging and discharging modes is greater than 95%. In the pulse load discharge mode, the system efficiency is 89.56%. Overall, the system efficiency meets the design requirements.

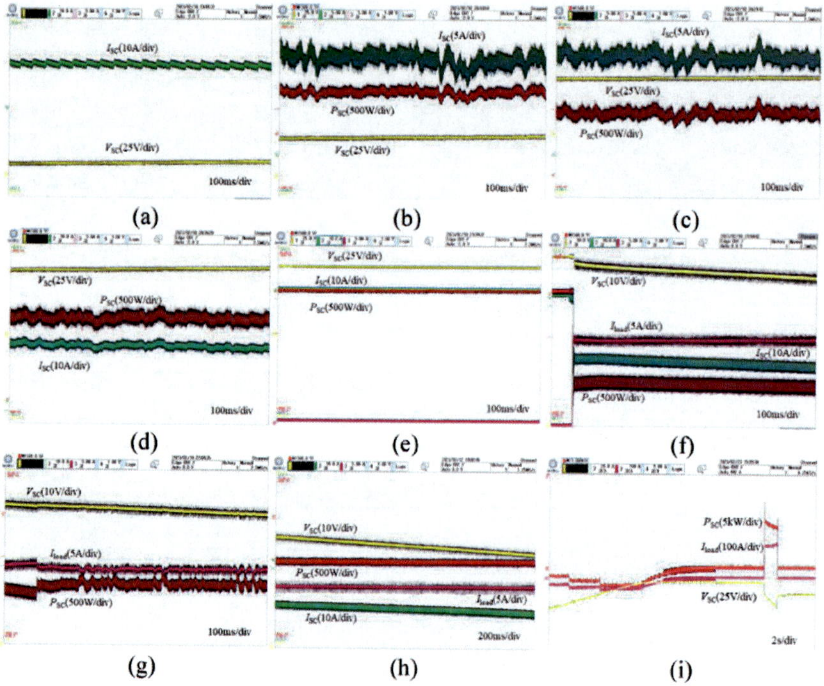

Fig. 6. Waveforms of different working modes

Table 1. System efficiency test results under different modes

Mode	Vbus(V)	I_{in}(A)	Vsc(V)	Isc (A)	η(%)
Constant 20A charge	100.00	12.50	60.00	20.00	95.97
Constant 1 kW charge	102.90	10.08	60.00	16.67	96.37
Constant 1.5 kW charge	103.15	15.16	60.00	25.00	95.93
Constant 2 kW charge	103.40	20.24	60.00	33.33	95.58
Constant 2 kW discharge	96.10	19.99	60.00	33.33	96.08
Constant 1.5 kW discharge	96.35	14.94	60.00	25.00	95.97
Constant 1 kW discharge	96.60	9.99	60.00	16.67	96.53
10 kW Pulse discharge	100.00	89.56	60.00	166.67	89.56

5 Conclusion

The experimental prototype for supercapacitors and bidirectional DC/DC converters has been built, and the feasibility of the proposed scheme has been verified through experiments. Charging and discharging with the rated power of 2 kW and pulse discharging with the power of 10 kW have been achieved. The full load efficiency of the entire system at rated operating point is 95.58%, and the efficiency under pulse operating conditions was 89.56%. Under natural cooling conditions, the power density including the casing is as high as 9.51W/in^3.

References

1. Lai, C.M., Cheng, Y.H., Hsieh, M.H., et al.: Development of a bidirectional DC/DC converter with dual-battery energy storage for hybrid electric vehicle system. IEEE Trans. Veh. Technol. **67**(2), 1036–1052 (2018)
2. Dorn-Gomba, L., Ramoul, J., Reimers, J., et al.: Power electronic converters in electric aircraft: current status, challenges, and emerging technologies. IEEE Trans. Transport. Electrif. **6**(4), 1648–1664 (2020)
3. Spyker, R.L., Nelms, R.M.: Classical equivalent circuit parameters for a double-layer capacitor. IEEE Trans. Aerosp. Electron. Syst. **36**(3), 829–836 (2000)
4. Dougal, R.A., Gao, L., Liu, S.: Ultracapacitor model with automatic order selection and capacity scaling for dynamic system simulation. J. Power Sources **126**(1–2), 250–257 (2004)
5. Tanaka, T., Takahashi, Y., Natori, K., et al. High-efficiency floating bidirectional power flow controller for next-generation DC power network. IEEJ J. Industry 7(1), 29–34 (2018)
6. Hasan, M.K., Mahmud, M., Habib, A., et al.: Review of electric vehicle energy storage and management system: standards, issues, and challenges. J. Energy Storage **41**(4), 1–10 (2021)
7. Chen, H.: Research on battery pack dynamic equalization technology with improved flying capacitor. Int. J. Low-Carbon Technol. **16**(1), 199–204 (2020)

Research on Improving the Elman Neural Network for the INS/GNSS Integrated Velocity and Positioning Algorithm

Junteng Wu[1,2], Jie Yang[1,2](✉), Chuanshu Meng[3], Jusong Jiang[1,2], and Yueyun Zhu[1,2]

[1] Jiangxi University of Science and Technology, Ganzhou 341000, China
yangjie@jxust.edu.cn
[2] Key Laboratory of Maglev Technology of Jiangxi Province, Ganzhou 341000, China
[3] China Academy of Railway Sciences Co, Ltd, Beijing 100081, China

Abstract. Speed and positioning technology is a crucial foundation and key technology for the operation control and safety assurance of maglev trains. In this study, an improved bee algorithm (IABC) optimized Elman neural network-assisted INS/GNSS combination positioning method is proposed to address the issues of satellite signal loss and INS/GNSS error accumulation, which can affect the accuracy of the system. When the satellite signals are reliable, the system adopts the INS/GNSS combination speed and positioning mode. The IABC is used to optimize the weights of the Elman neural network, thereby improving its prediction accuracy. The optimized Elman neural network is trained accordingly. When satellite signals are lost, the trained neural network is utilized to predict the measurement input of the Kalman filter, providing error correction values for the output of the inertial navigation system to enhance speed and positioning accuracy. Simulation experiments have demonstrated that the IABC optimized Elman neural network-assisted combination navigation system significantly improves the speed and positioning accuracy when satellite signals are lost.

Keyword: Speed and positioning · Combination navigation · Bee algorithm · Neural network

1 Introduction

1.1 A Subsection Sample

The maglev train has many advantages, such as high acceleration, unrestricted operating speed due to the absence of wheel-rail adhesion, strong climbing ability, safety and environmental friendliness, low noise, and easy maintenance. It is a very promising mode of rail transportation [1]. The speed and positioning system is an essential component of the train control system, and its accuracy and reliability are directly related to the safety of train operation. Unlike conventional wheel-rail trains, maglev trains operate in a non-contact manner with the track, making it impossible to use traditional axle counting devices for speed and positioning measurements [2]. Due to the unique operating

L. Jia et al. (Eds.): EITRT 2023, LNEE 1135, pp. 660–668, 2024.
https://doi.org/10.1007/978-981-99-9307-9_71

characteristics of maglev trains, this paper adopts an INS/GNSS-based combined speed and positioning system [3]. However, satellite positioning has limitations in dynamic performance and is susceptible to external environmental factors. When encountering obstacles or adverse weather conditions, satellite signals can weaken or even become unattainable. During satellite signal loss, the combined navigation system operates in pure inertial mode. Since the inertial sensors involve multiple integration processes in the speed and position calculations, measurement errors continuously accumulate, leading to a significant degradation in system accuracy [5].

To address the problem of reduced accuracy in speed and positioning due to satellite signal loss and rapid accumulation of inertial sensor errors without increasing hardware costs, various methods have been proposed. For example, Li et al. [6] proposed a time-series-based autoregressive integrated moving average (ARIMA) assisted model that utilizes data prior to interruption to predict GNSS measurements during interruptions and bridges the data gaps. Malleswaran et al. [7] proposed using evolutionary algorithm optimization of a recurrent neural network to overcome the slow convergence and susceptibility to local minima of backpropagation algorithms. Abdolkarimi et al. [8] proposed a single hidden layer feedforward neural network for estimating accuracy and speed by minimizing errors. Zhang et al. [9] used an Elman neural network to predict GNSS observations, storing and utilizing past moment outputs to ensure the continuity of the combined navigation system. Chuang et al. [10] proposed a combination positioning algorithm using a long short-term memory (LSTM) neural network, making full use of LSTM's advantages in long-distance time series applications.

In conclusion, this paper proposes an Elman neural network model prediction algorithm based on the improved artificial bee colony (IABC) optimization. In conditions of good satellite signal quality, the combined INS/GNSS positioning process is performed to correct the measurement error of INS using a Kalman filter output. Simultaneously, the weights of the Elman neural network are optimized using the IABC to improve the prediction accuracy of the network and address the drawbacks of slow iteration and susceptibility to local minima. The output of the inertial sensor and navigation solution is then used as input for the Elman neural network, with the measurement values of the Kalman filter as the training target, to train the neural network. During satellite signal loss, the trained neural network is used to predict the measurement information of the Kalman filter, ensuring that the Kalman filter can correct the velocity and position errors of the INS. Simulation experiments demonstrate the effectiveness of the IABC-assisted Elman combined navigation positioning method.

2 System Model

2.1 Elman-Assisted Combined Navigation System Structure Design

In order for the Kalman filter to remain effective when satellite signals are lost, this paper selects to predicting the measurement input of the Kalman filter to design the system structure. The system structure is shown in Fig. 1.

The INS/GNSS combined speed and positioning system, which incorporates Elman neural network, can be divided into training mode and prediction mode. Both modes use the output of the Kalman filter to correct the output errors of the INS. When satellite

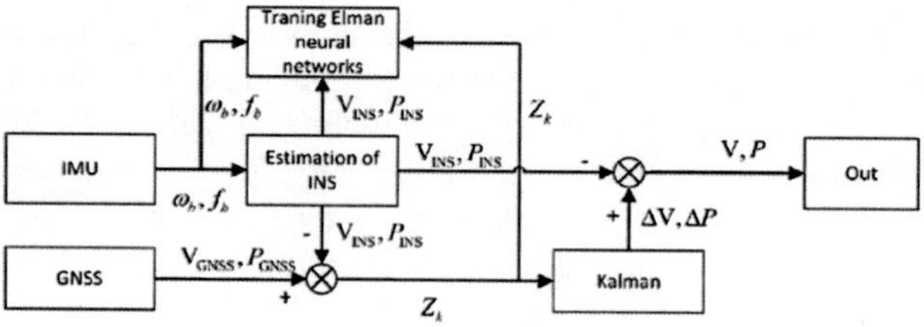

Fig. 1. Training mode

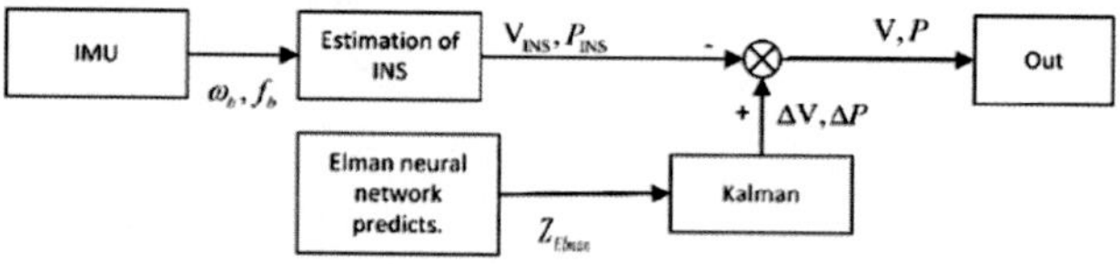

Fig. 2. Predictive mode

signals are strong, the system operates in the training mode as shown in Fig. 1. In this mode, the difference between the velocity and position obtained from GNSS (V_{GNSS}, P_{GNSS}) and the velocity and position obtained from inertial sensors (V_{INS}, P_{INS}) is used as the measurement input of the Kalman filter. The output errors of the INS are corrected using the optimal estimation values (ΔV, ΔP) from the Kalman filter. Additionally, the neural network is trained using the backpropagation algorithm to continuously adjust its weights. When satellite signals are lost or weak, the system operates in the prediction mode as shown in Fig. 2. In this mode, the trained neural network's predictions (Z_{Elman}) are used as the measurement input of the Kalman filter to enhance positioning accuracy in the absence of satellite signals [11].

2.2 The INS/GNSS Combined Positioning System

Considering the complexity and efficiency of the maglev train positioning system, the INS/GNSS combined positioning system in this paper adopts a loosely coupled approach for position and velocity integration [12].

The mathematical model of the INS/GNSS combined positioning system is as follows:

$$\begin{cases} X_k = F_{k,k-1}X_{k-1} + \Gamma_{k,k-1}W_{k-1} \\ Z_k = H_kX_k + V_k \end{cases} \infty \tag{1}$$

X_k is the system state vector, $X_k = \left[\Delta\varphi, \Delta v, \Delta p, \varepsilon_g, \varepsilon_a\right]^T$, $\Delta\varphi, \Delta v, \Delta p, \varepsilon_g, \varepsilon_a$ represents the errors in three-axis attitude angles, three-axis velocity errors, three-axis position errors, gyroscope bias, and accelerometer bias. $Z_k = [\Delta V_E, \Delta V_N, \Delta V_U, \Delta P_E, \Delta P_N, \Delta P_U]$ is the measurement vector of the system, which is the velocity and position differences in the north-east-up directions between the GNSS output and the INS output. $F_{k,k-1}$ is the one-step state transition matrix of the system,

H_k is the measurement matrix of the system, and $\Gamma_{k,k-1}$ is the noise covariance matrix of the system. W_k and V_k are the system noise vector and measurement noise vector, respectively:

Kalman Filter Design:

$$\begin{cases} \hat{X}_{k,k-1} = F_{k,k-1}X_{k-1} \\ P_{k,k-1} = F_{k,k-1}P_{k-1}F_{k,k-1}^T + \Gamma_{k,k-1}Q_{k-1}\Gamma_{k,k-1}^T \\ K_k = P_{k,k-1}H_k^{\mathrm{T}}\left[H_kP_{k,k-1}H_k^{\mathrm{T}} + R_k\right]^{-1} \\ \hat{X}_k = X_{k,k-1} + K_k\left(Z_k - H_k\hat{X}_{k,k-1}\right) \\ P_k = P_{k,k-1}(1 - K_kH_k^T) \end{cases} \tag{2}$$

In the equation: $\hat{X}_{k,k-1}$ is the one-step predicted value of the state; $P_{k,k-1}$ is the predicted covariance matrix; K_k is the Kalman filter gain; $\hat{X}_k$, P_k is the updated state and covariance matrix.

3 Methodology

3.1 Overview of Elman Neural Network

The Elman neural network is a typical type of local regression neural network with a network structure as shown in Fig. 3. It adds a recurrent layer on top of the traditional feedforward neural network, which records and feeds back the hidden layer information from the previous time step. This enables the neural network to be sensitive to historical data and enhances its ability to handle dynamic information. In this paper, the inertial sensor angular velocity (ω_b), acceleration (f_b), and calculated velocity (V_{INS}) and position (P_{INS})are used as input features for the Elman neural network. The measurement inputs from the Kalman filter are used as training targets to train the Elman neural network.

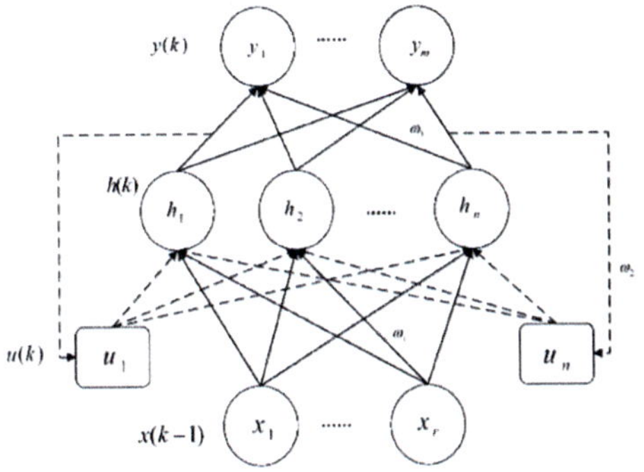

Fig. 3. The neural architecture of Elman network

The specific formula is shown as Eq. (3).

$$\begin{cases} y(k) = g(w_3h(k)) \\ h(k) = f(w_1x(k-1) + w_2u(k)) \\ u(k) = h(k-1) \end{cases} \tag{3}$$

In the formula, y represents an m-dimensional output node vector, h represents an n-dimensional hidden layer node unit vector, b represents an n-dimensional recurrent layer vector. w_3, w_2 and w_1 represent the connection weights from the hidden layer to the output layer, from the input layer to the hidden layer, and from the recurrent layer to the hidden layer, respectively. $f(x)$ represents the transfer function for the hidden layer, and $g(x)$ represents the transfer function for the output layer.

The transfer function for the hidden layer of the Elman neural network is set as "tansig", and the activation function for the output layer is set as "purelin". The training function used is "traingdm". According to empirical formulas, different combinations of the number of hidden layer neurons and hidden layers are explored to obtain the optimal structure of the Elman neural network. In this case, the optimal number of hidden layers is 1, and the number of neurons in the hidden layer is 11.

Although Elman neural networks can capture historical information and have better predictive performance compared to traditional feedforward neural networks, they still suffer from slow convergence and a tendency to get stuck in local optima when trained using gradient descent. Therefore, in this study, the IABC (Improved Artificial Bee Colony) algorithm is proposed to be combined with the Elman neural network to improve its global optimization capability and convergence speed.

3.2 Improved Artificial Bee Colony (IABC) Algorithm

In the Artificial Bee Colony (ABC) algorithm [12], the bee colony is divided into scout bees, follower bees, and onlooker bees, each with different roles. The position of a food source represents a feasible solution in the solution space, and the nectar amount of a food source corresponds to the fitness of the feasible solution.

During the algorithm exploration process, N initial positions of food sources are generated according to Eq. (4):

$$x_{id} = lb_d + \text{rand}()(ub_d - lb_d) \tag{4}$$

In the equation, x_{id} represents the position of the ith food source in the dth dimensionof the search space. ub_j and lb_j are the upper and lower boundaries of the search domain, respectively. $d = 1, 2, 3 \ldots D$ represents the dimension being updated in the current position update.

Since both the leader bee and follower bees search for food sources within a certain neighborhood, their search range is limited. In order to improve the performance of the artificial bee colony algorithm, an adaptive inertia factor and is introduced, which incorporates the global best solution. The improved formula for updating the food sources in the artificial bee colony algorithm is as follows [13]:

$$\begin{cases} V_{ij}^{\text{new}} = \omega_1 x_{id} + \omega_1 r_1 (x_{id} - x_{kd}) + \omega_2 r_2 \left(y_{j\,\text{best}} - x_{id}\right) \\ \omega_1 = \omega_{\min} - (\omega_{\min} - \omega_{\max})\left(\frac{t}{t_{\max}}\right)^2; \\ \omega_2 = \omega_{\max} + (\omega_{\min} - \omega_{\max})\left(\frac{t}{t_{\max}}\right)^2 \end{cases} \tag{5}$$

In Eq. (5): $t_{\max}$ represents the current iteration count; $\omega_{\max}$ and $\omega_{\min}$ are the maximum and minimum values of the inertia factor, $\omega_{\max} = 0.8$ and $\omega_{\min} = 0.5$; ω_1 represents the velocity for the new food source to move towards the original food source and neighborhood food sources, while ω_2 represents the velocity for the new food source to move towards the best food source in the population. From Eq. (10), we can see that at the beginning of the iteration, ω_2 has a larger value while ω_1 has a smaller value, indicating a higher priority in searching for the optimal solution. As the iteration count increases, the value of ω_1 increases, and the influence of the random food sources gradually increases, indicating a higher preference for the global solution.

4 Simulation Experiments and Result Analysis

To evaluate the proposed method of optimizing Elman neural networks assisted INS/GNSS integrated positioning using the improved artificial bee colony algorithm (IABC), INS/GNSS integrated positioning experiments were conducted using MATLAB 2020B and the PSINS Toolbox. In order to simulate realistic conditions, relative errors of the gyroscopes and accelerometers were taken into account. The specific parameters are listed in Table 1.

Table 1. The initial values and parameters of the experiment

Simulation parameters	Type	Metric
Initial values for simulation	Initial position	E115.273°
		N26.323°
	Initial velocity	0m/s
Accelerometer	Constant bias	100 μg
	Random walk for angular velocity	10 $\mu g \sqrt{HZ}$
Gyroscope	Constant bias	0.01(°)/h
	Random walk for velocity	0.001(°)/h
INS	Data update rate	10 Hz
GNSS	Data update rate	1 Hz
Simulation run time	Total duration of simulation	2089 s
	Loss of signal	120 s
Elman neural network	Training set	0–800 s
	Test set	1201–1320 s

To validate the superior suppression effect of IABCELman + kf proposed in this article in the event of INS/GNSS signal loss, a comparison was made between BP + KF, Elman + KF, and IABCElman + KF in the speed and positioning accuracy of the system when satellite signals were lost for 120s. The comparison results are shown in Fig. 4.

From Fig. 4, it can be observed that both Bp and Elman have some effect in suppressing errors, but the IABCELman mode exhibits a more pronounced suppression effect

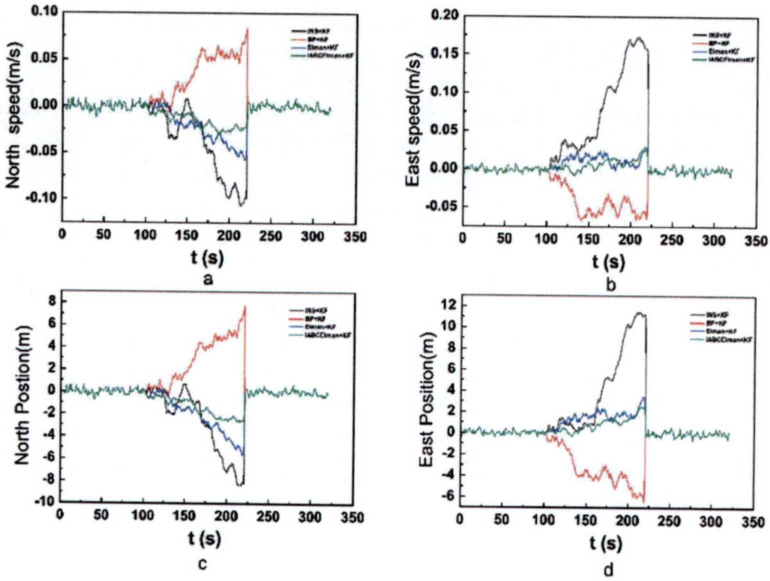

Fig. 4. Different patterns of speed and position errors comparison

on errors. This is mainly attributed to the Elman neural network's ability to remember the previous hidden state, and, with the optimization of the IABC algorithm, there is an improvement in the precision and speed of prediction by the Elman neural network.

Table 2. Comparative results of position and velocity errors in different modes

	INS		Bp		Elman		IABCELman	
	RMSE	SD	RMSE	SD	RMSE	SD	RMSE	SD
East speed(m/s)	0.0988	0.0581	0.0459	0.0184	0.0142	0.0075	0.0106	0.0067
North speed(m/s)	0.0545	0.0358	0.0429	0.0233	0.0293	0.0166	0.0176	0.0080
East Position(m)	5.8698	4.1189	3.7527	1.7206	1.6981	0.7384	1.0960	0.6791
North Postion(m)	3.9882	2.7855	3.6555	2.1634	2.7153	1.6432	1.6298	0.8217

Table 2 displays the comparative performance of the system in the Bp, Elman, and IABCElman modes. From the table, it can be observed that within the 120-s period of satellite signal loss, the IABCElman model exhibits a significant improvement in the root mean square error (RMSE) for both eastward velocity and position, with values of 0.0106 and 1.0960 respectively. This represents an enhancement of 76.87% and

70.79% over the Bp and Elman models for eastward velocity, and 25.68% and 35.46% for position. Similarly, the RMSE for northward velocity and position, at 0.0176and 1.6298respectively, display notable improvements of 58.99% and 55.42% over the Bp and Elman models for velocity, and 39.93% and 39.98% for position. Comparatively, when compared to the pure INS mode, the IABCElman mode showcases a substantial increase in accuracy, with an 89.26% improvement in eastward velocity and a 81.33% improvement in position. Furthermore, the northward velocity and position precision are enhanced by 67.73% and 59.14% respectively.

5 Conclusion

This paper proposes a method for enhancing INS/GNSS integration navigation positioning using an IABC-optimized Elman neural network to address the issue of declining positioning accuracy when satellite signals are unavailable. When satellite signals are lost, the IABC-Elman neural network predicts the measurement input of the Kalman filter, enabling the filter to continue operating and providing error correction for the INS system. As a result, the proposed method improves the speed and positioning accuracy of the system during periods of signal unavailability. Simulation experiments validate the feasibility and effectiveness of the proposed method.

Acknowledgment. This research is supported by Research and Development Fund Key Project of China Academy of Railway Sciences Co., Ltd. (2022YJ077).

References

1. Cheng, S., Liu, C., Song L.: Combination speed positioning method for maglev trains based on multi-source information fusion. Urban Rapid Rail Transit Res 25(08), 136–140 (2022) (in Chinese)
2. Gu, C., Wang, S., Shen, S.: Theoretical research on combined speed positioning of mag lev trains base-d on GNSS and INS. J Railway Sci Eng 17(11), 2756–2766(2020) (in Chinese)
3. Zhou, X., Zhang, H., He, T.: Research on adaptive Kalman filter algorithm for GPS/INS loosely coupled integrated navigation. J. Time-Frequency Wavelet Anal. **43**(03), 222–230 (2020). (in Chinese)
4. Wang, F., Han, B., Hu, L.: Research on GNSS/INS integrated navigation algorithm under complex urban environment. Geomat. Nat. Haz. Risk **42**(01), 15–20 (2022). (in Chinese)
5. Li, Q., Zhang, L., Wang, X.: Loosely coupled GNSS/INS integration based on factor graph and aided by ARIMA model. IEEE Sens. J. **21**(21), 24379–24387 (2021)
6. Malleswaran, M., Vaidehi, V., Sivasankari, N.: A novel approach to the integration of GPS and INS using recurrent neural networks with evolutionary optimization techniques. Aerosp. Sci. Technol. **32**, 169–179 (2014)
7. Abdolkarimi ES., Abaei G., Mosavi MR.: A wavelet-extreme learning machine for low-cost INS/GPS navigation system in high-speed applications. GPS Solut 22(1), 15 (2018)
8. Zhang, K., Cheng Y., Wu Yg.: Research on Elman network-assisted SINS/GNSS integrated navigation method. J. Missile Guidance 42(05), 1–5 (2022) (in Chinese)
9. Iang, C., Zhang, S., Zhang Q.: A new adaptive h-infinity filtering algorithm for the GPS/INS integrated navigation. Sensors (2016)

10. Ning, Y., Wang, J., Han, H.: An optimal radial basis function neural network enhanced adaptive robust Kalman filter for GNSS/INS integrated systems in complex urban areas. Sensors 8(9), 3091–3112 (2018)
11. Cui, X., Zhang, D., Ma, Y.: Short-term forecasting of photovoltaic power generation based on CGABC-elman mode. Power Technol **46**(09), 1043–1047 (2022). (in Chinese)
12. Guo, P.: Path planning of library service robot based on improved ABC algorithm. Inform. Technol. **46**(12), 19–23 (2022). (in Chinese)
13. Ma, S. L., Dai, Y.,. Xu, J. C.: Short-term load forecasting using improved artificial bee colony optimization neural network. Mech. Design Manuf. (07), 50–53+57 (2021) (in Chinese)

Calculation of Air Gap Electromagnetic Field of High-Speed Magnetic Levitation Linear Motor Based on Equivalent Current Layer

Jiahui Zhang, Guodong Wang, and Dan Zhang(✉)

School of Electronic and Information Engineering, Beijing Jiaotong University, Beijing 100044, China
{gdwang,zhang.dan}@bjtu.edu.cn

Abstract. The linear synchronous motor (LSM) realizes energy conversion based on the air gap magnetic field between the stator and rotor, and the air gap magnetic field distribution directly affects the performance of the motor, which in turn has an impact on the operation of the electromagnetic suspension (EMS) maglev system. In order to study the air gap magnetic field distribution of the linear motor, the analytical expression of air gap magnetic density of LSM considering the open structure, cogging effect and harmonic influence was deduced based on electromagnetic field theory and the equivalent current layer model. To reshape the stator and rotor excitation source, the equivalent current sheets are calculated by Fourier series expansion. And the equivalent current sheets are placed in the motor cogging layers when the equivalent current layer model is constructed. The accuracy of the proposed method is verified by comparing the results with the finite element solution.

Keywords: Analytic method · Linear synchronous motor · Maglev

1 Introduction

Long stator linear synchronous motor is the driving force supply and electrical power supply for electromagnetic suspension (EMS) maglev system, which directly affects the operation performance of the magnetic levitation train, so it is very important to carry out the research of analyzing the magnetic field distribution characteristics of linear synchronous motor (LSM) [1, 2]. However, due to the influence of motor core reflection and diffraction, it makes the magnetic field calculation a difficult task. Moreover, the slotted structure of stator/rotor leads to the uneven magnetic field distribution of the linear motor air gap, which also increases the difficulty of direct calculation [3].

At present, the main research methods for the magnetic field distribution of linear synchronous motors are numerical method [4], but calculation time is long when the structure of the motor is complex, and the model needs to be re-modeled when the parameters of the model are changed [5, 6]. The analytical method has clear physical concepts and can intuitively reflect the relationship between the field and the motor

L. Jia et al. (Eds.): EITRT 2023, LNEE 1135, pp. 669–677, 2024.
https://doi.org/10.1007/978-981-99-9307-9_72

parameters [7]. At present, the analytical methods for electromagnetic field calculation of permanent magnet rotary synchronous motors and permanent magnet linear synchronous motors are proposed. However, since the excitation windings are powered by DC and the LSM is quite different from the permanent magnet motor in terms of structure and control methods, this will affect the air gap magnetic field distribution characteristics [8]. It is necessary to propose a solution method for linear synchronous motors. Although the structure of motors is different, they are all based on the magnetic field of the air gap between the stator and the rotor to realize the energy conversion. Therefore, the solution methods of other motors can be referenced. Zesheng Deng derived an analytical solution for the air gap magnetic field strength of the permanent magnet synchronous motor, based on equivalent current layer method, and the results are in the same order of magnitude as the magnetic field measurements [9]. Zhihao Liu proposed an analytical method for calculating the induced electromotive force of LSM based on Faraday's law, but it requires the derivation of the permeability coefficients of different regions, which makes the solution more complicated [10].

The aim of this paper is to derive analytical expressions for the magnetic field and motor parameters, and to obtain the motor magnetic field distribution characteristics quickly by using the layered current model and solving Maxwell's equations. Cogging effects and harmonic influences of the motor are considered. In Sect. 2, we illustrate the structural characteristics and parameters of LSM and simplify the motor with the equivalent current layer model. In Sect. 3, the analytical equations relating the electromagnetic field in each region of the equivalent model to the motor parameters are derived. In Sect. 4, we calculate the air gap magnetic field distribution and the accuracy of the proposed method is verified by comparing the results with the finite element solution. Section 5, gives our conclusions.

2 Equivalent Current Layer Model of Motor

2.1 Structure of Motor

Routine conducting high-speed maglev train use long-stator linear synchronous motor. The primary winding on the stator is excited by three-phase alternating current. Stator produces moving magnetic field that interacts with the rotor poles to produce thrust. The speed of the maglev is adjusted by frequency and amplitude modulation of the three-phase alternating current. The stator/rotor surface has teeth and grooves arranged alternately at a certain pole pitch. Windings mounted in the grooves. The model of long stator linear is shown in Fig. 1.

The spatial coordinate established in the linear motor model is shown in Fig. 1. Take the position at the moment when the axis of the excitation pole coincides with the axis of the armature pole. The stator is stationary relative to the coordinate system, and the rotor moves uniformly to the right and the positive direction of the x-axis is set to the right.

2.2 Equivalent Current Layer Method

In order to calculate the air gap magnetic field distribution of the motor, the following assumptions are firstly made:

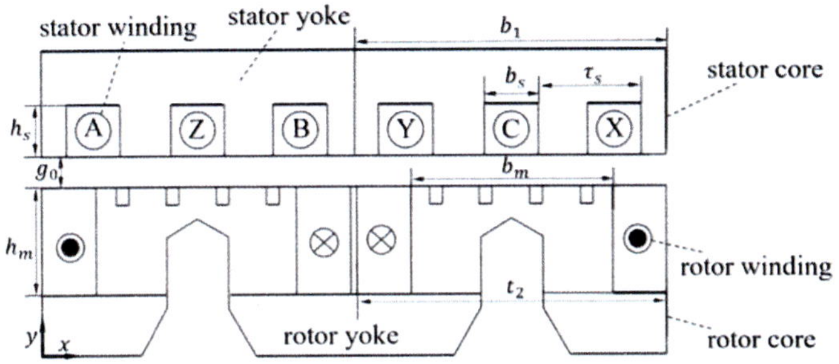

Fig. 1. Model of long stator linear (b_s is the slot width of stator, τ_s is the slot pitch of stator, b_1 is the pole pitch of stator, b_m is the tooth pitch of rotor, and t_2 is the pole pitch of rotor, h_s is the height of stator tooth groove, and h_m is the height of rotor tooth groove).

(1) The iron core is always considered not saturated and μ_r is constant.
(2) Neglect eddy current and hysteresis effects.
(3) Consider the electromagnetic field generated by the motor as a linear superposition of the electromagnetic field generated by the armature winding and the excitation winding.
(4) The generating winding slot size and current are relatively small and have little effect on the air gap magnetic field, so the generating winding slot effect is neglected.

Because of the effect of the cogging of the stator/rotor cores, the air gap magnetic field of the long stator linear motor is not uniform. Therefore, the notched part is layered separately to build the 2D motor equivalent model as shown in Fig. 2. Based on the actual currents in the armature windings and excitation windings in the cogging, the stator equivalent current sheet current density A_{Is} and the rotor equivalent current sheet current density A_{Im} are respectively solved by Fourier transform.

The magnetic permeability of the stator/rotor yoke is much larger than the air gap and the slotted region. Therefore, the yoke part is set as an ideal magnet boundary condition. The magnetic permeability of the air gap layer is μ_0. The slotted structure of stator/rotor is replaced by an anisotropic region where the permeabilities along O_x and O_y are solved by (1) and (2), respectively.

$$\mu_x = \mu_0\mu_r/(1 + b_s(\mu_r - 1)/\tau_s) \tag{1}$$

$$\mu_y = \mu_0[b_s/\tau_s + \mu_r(1 - b_s/\tau_s)] \tag{2}$$

where b_s is the slot width, τ_s is the slot pitch, μ_0 is the permeability of free space, and μ_r is the relative permeability of stator iron. As shown in Fig. 2, the equivalent model has three regions: the stator cogging region I, the air gap region II and the moving cogging region III. Both regions I and III are divided into two parts bounded by the assumed positions of the current sheets A_{Is} and A_{Im}.

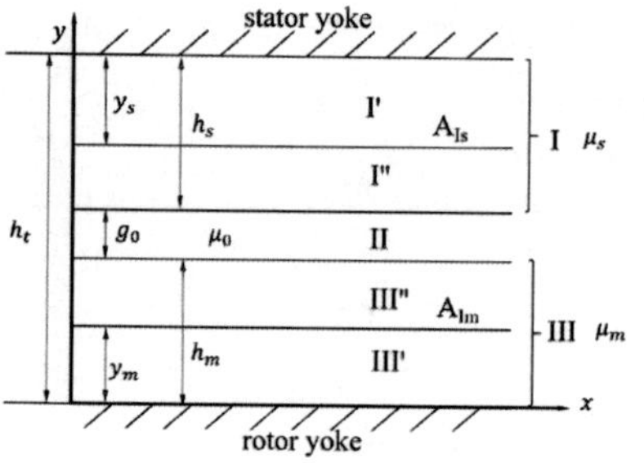

Fig. 2. Equivalent current layer model (y_s is the distance of stator equivalent current sheet from the stator yoke, and y_m is the distance of rotor equivalent current sheet from the rotor yoke).

3 Analytical Model

This section focuses on deriving analytical equations for the air gap magnetic field and motor parameters based on Maxwell's equations. First, an infinitely thin current sheet is solved, which produces a magnetic kinetic potential consistent with the original one. And then the equivalent current sheet is substituted to solve for the magnetic field strength in each region.

3.1 Solving for Equivalent Current Layer Current Density

The three-phase alternating current passed through the armature windings. The infinitely thin stator current sheet is uniformly distributed along the stator slot width, and the assumed current sheet is distributed in the manner shown in the Fig. 3, taking into account the effects of the phase sequences of A, B, and C. X, Y and Z correspond to phases A, B and C with opposite excitation directions, respectively.

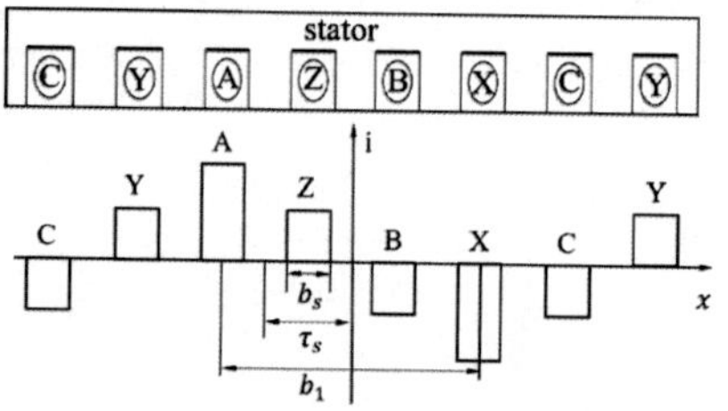

Fig. 3. Equivalent current sheet distribution of stator current.

where the slot width b_s is 43 mm, the slot pitch τ_s is 86 mm, the stator pole pitch b_1 is 258 mm, the air gap height g_0 is 10 mm. Taking the phase A current as an example, the expression containing each harmonic component of the current is obtained by Fourier series expansion as shown in the following equation:

$$j_a(x) = \frac{4}{\pi}\frac{n_s i_a}{2pb_s}\sum_{n=1}^{\infty}\left[\frac{1}{2n-1}\sin\left((2n-1)\pi\left(\frac{n\pi b_s}{6n}\right)\right)\cdot\sin\left((2n-1)\left(\frac{\pi x}{b_1}\right)\right)\right] \quad (3)$$

where n_s is the number of stator winding turns, i_a is the A-phase current, the number of pole pairs p is 1, and the stator current sheet density is distributed with a period of $2b_1$. Phase B and C currents lag and overshoot by 120° respectively. Similarly, they are calculated as j_b and j_c through the Fourier series expansion.The stator equivalent current sheet distribution A_{Is} is obtained by superposition.

Long stator linear motor uses DC excitation. Referring to the equivalent method of stator winding current, rotor equivalent current sheet distribution $\mathrm{A_{Im}}$ can be obtained.

3.2 Derivation of Analytic Expressions

Considering the fact that the windings are not infinitely thin, it is necessary to set the current sheet at different positions and then calculate the average value. Using the A_{Is} and A_{Im} derived above, the control equations for each region are as follows:

(1) Region III' ($0 \leq y \leq y_m$): Solving Maxwell's equations under the A_{Im} distribution condition yields:

$$\frac{\partial^2 H_x}{\partial y} - \frac{\mu_{xm}\pi^2}{b_1^2 \omega \mu_{ym}} H_x = 0 \tag{4}$$

By solving (4) can get:

$$H_{x\mathrm{III}'}(y) = H_{x3'}\left[\sinh K_m y + A\cosh K_m y\right] \tag{5}$$

$$H_{y\mathrm{III}'}(y) = jK_m(b_1/\pi)H_{x3'}\left[A\sinh K_m y + \cosh K_m y\right] \tag{6}$$

where, $K_m = (\pi/b_1)\sqrt{\mu_{xm}/\mu_{ym}}$, and $H_{x3'}$ is the tangential field at the position of rotor in zone III$'$.

Using similar methods, the magnitude of the x and y components of the magnetic field in other regions can be found.

(2) Region III$''$ ($y_m \leq y \leq h_m$):

$$H_{x\mathrm{III}''}(y) = H_{x3''}\left[\sinh K_m(y - y_m) + B\cosh K_m(y - y_m)\right] \tag{7}$$

$$H_{y\mathrm{III}'}(y) = jK_m(b_1/\pi)H_{x3''}\left[B\sinh K_m(y - y_m) + \cosh K_m(y - y_m)\right] \tag{8}$$

(3) Region II ($h_m \leq y \leq h_m + g_0$):

$$H_{x\mathrm{II}}(y) = H_{x2}[\sinh \pi(y - h_m)/b_1 + C\cosh \pi(y - h_m)/b_1)] \tag{9}$$

$$H_{y\mathrm{II}}(y) = jH_{x2}[\cosh \pi(y - h_m)/b_1 + C\sinh \pi(y - h_m)/b_1)] \tag{10}$$

(4) Region I$'$ ($h_m + g_0 \leq y \leq h_t - y_s$):

$$H_{x\mathrm{I}'}(y) = H_{x1'}\left[\sinh K_s(y - (h_t - y_s)) + D\cosh K_s(y - (h_t - y_s))\right] \tag{11}$$

$$H_{y\mathrm{I}'}(y) = jK_m(b_1/\pi)H_{x1'}\left[\cosh K_s(y - (h_t - y_s)) + D\sinh K_s(y - (h_t - y_s))\right] \tag{12}$$

where: $K_s = (\pi/b_1)\sqrt{\mu_{xs}/\mu_{ys}}$.

(5) Region I'' ($h_t - y_s \leq y \leq h_t$):

$$H_{xI''}(y) = H_{x1''}[\sinh K_s(y - h_t)) + E\cosh K_s(y - h_t))] \tag{13}$$

$$H_{yI''}(y) = jK_s(b_1/\pi)H_{x1''}[\cosh K_s(y - h_t)) + E\sinh K_s(y - h_t))] \tag{14}$$

And $H_{x3''}, H_{x2}, H_{x1'}, H_{x1''}$ are the tangential field at the corresponding position of the motor. Collation shows that $H_{x3''}, H_{x3'}, H_{x2}, H_{x1'}/, H_{x1''}, A, B, C, D$ and E are unknown. The solution is performed by substituting the boundary conditions satisfied by each layer. So far, the magnetic field distribution taking into account motor slotting and harmonic influence has been given as analytical expressions.

4 Calculation of Air Gap Magnetic Field

4.1 Air Gap Magnetic Field

Following the program developed in the previous section, we calculate the motor air gap magnetic field distribution. The magnetic field distribution at the center of the air gap is shown in Fig. 4. A complete cycle of air gap magnetic field distribution can be observed at a pair of pole lengths. From Fig. 4, it can be seen that the component y of the magnetic field strength in the air gap is the main part, in other words, the magnetic field component in the vertical direction is much larger than that in the horizontal direction. When the motor parameters are changed, the proposed method can quickly solve the magnetic density of the motor air gap under the new parameters.

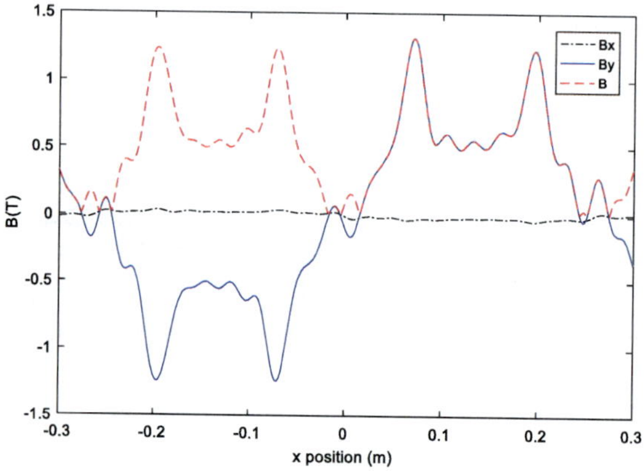

Fig. 4. Distribution of magnetic density along the *x*-direction at the center of the air gap.

4.2 Effect of Equivalent Current Sheets

This section discusses whether the position of the equivalent current sheet has an effect on the solution of the magnetic field distribution. When calculating the armature field distribution, it is only necessary to set to the excitation winding equivalent current surface as 0. The same applies for the excitation winding. The effect of the position of the stator/rotor equivalent current sheets on the solution of the magnetic field is discussed separately below, and the calculation results are shown in Fig. 5. The analysis of the results shows that the position y_s and y_m of the equivalent current sheets have almost no effect on the solution of the magnetic field.

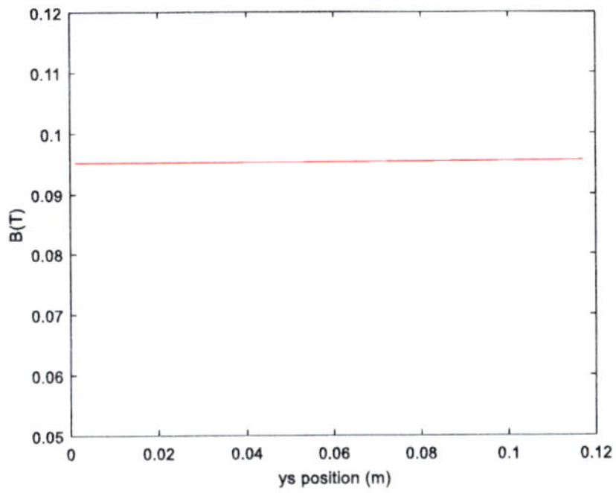

(a) Effect of y_s

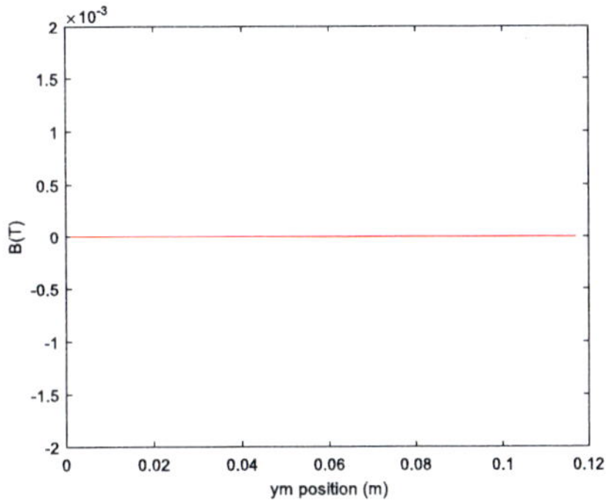

(b) Effect of y_m

Fig. 5. Effect of the position y_s and y_m of equivalent current sheets.

4.3 Validation of Results

A comparison of the air gap magnetic density obtained by the proposed method with the finite element results is shown in Fig. 6. It can be found that the distribution of the air gap magnetic density obtained by the proposed algorithm is consistent with that of the finite element results. However, due to the neglect of eddy current effect and magnetic saturation effect, the peak value of the result of proposed method is higher and the peak is narrower. The mean square error MSE with finite elements is 0.073. However, the proposed method can visualize the relationship between motor parameters and magnetic field strength. And the finite element method takes more time in both remodeling and solving.

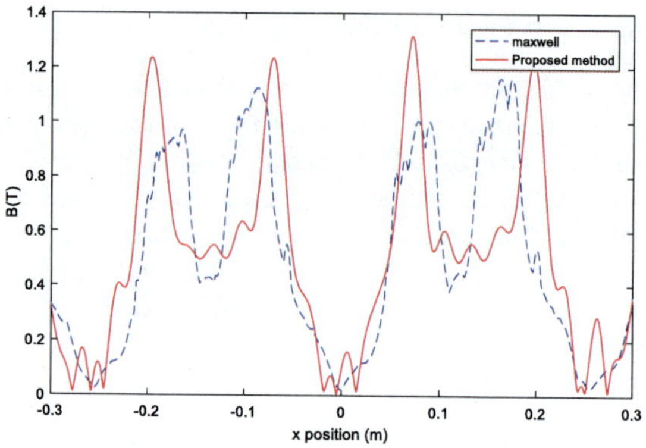

Fig. 6. Comparison of the results of the proposed method with finite element simulation results.

5 Conclusion

As expected, the analytical method satisfactorily predicts the air gap flux density distribution produced by a linear motor with a long stator. A certain amount of error is introduced by neglecting eddy current losses in the primary and secondary, and magnetic saturation. However, the errors are small and the calculations are done in a relatively short period of time. The relationship between the air gap magnetic field and the motor parameters can be visualized through the analytical equation. The conclusions and formulas derived are largely suitable for the analysis and application of linear motors, thus laying a foundation for the motor design and operation control of the LSM.

Acknowledgments. This work is supported by the Fundamental Research Funds for the Central Universities, (2023JBZY015).

References

1. Liao, Z., Yue, J., Lin, G.: Application research of HTS linear motor based on Halbach array in high speed maglev system. IEEE Trans. Appl. Supercond. **31**(5), 1–7 (2021)
2. Chi, S., Yan, J., Guo, J., Zhang, Y.: Analysis of mechanical stress and vibration reduction of high-speed linear motors for electromagnetic launch system. IEEE Trans. Ind. Appl. **58**(6), 7226–7240 (2022)
3. Ogam, E., Fellah, Z. E. A., Ogam, G., Ongwen, N. O., and Oduor, A. O. (2021) Investigation of long acoustic waveguides for the very low frequency characterization of monolayer and stratified air-saturated poroelastic materials. Appl. Acoustics, pp. 182–108200
4. Wang, Y., and Wen, Y.: Electromagnetic emission of maglev vehicle by linear synchronous motor. In: 2019 IEEE 3rd international electrical and energy conference (CIEEC), pp. 1507–1512 (2019)
5. Zhang, H., Zhang, J., Deng, Z., Wang, Q., Li, X., Tang, X., and Zhang, W. (2023) Study on electromagnetic radiation characteristics based on HTS maglev levitation test line. Electronics 12(8), p. 1776

6. Barzegaran, M.R., Sarikhani, A., Mohammed, O.A.: An equivalent source model for the study of radiated electromagnetic fields in multi-machine electric drive systems. In: 2011 IEEE international symposium on electromagnetic compatibility, 442–447 (2011)
7. Dreishing, F, and Kreischer, C.: Experimental and analytical analysis of a compact ironless permanent-magnet tubular linear motor (PM-TLSM). In: Electromechanical drive systems 2021 ETG Symposium, pp. 1–6 (2021)
8. Lv, G., Liu, Y., Zhang, Z., Zhou, T.: Characteristic analysis of coreless-type linear synchronous motor with the racetrack coils as secondary for EDS maglev train. IEEE/ASME Trans. Mechatron. **27**(6), 4654–4664 (2022)
9. Deng, Z., Boldea, I., Nasar, S.: Fields in permanent magnet linear synchronous machines. IEEE Trans. Magn. **22**(2), 107–112 (1986)
10. Liu, Z..: Characterization of linear harmonic generators for electromagnetic high-speed magnetic levitation systems. J. Beijing Jiaotong Univ. 49–52 (2022) (in Chinese)

Author Index

B
Bi, Xin I-572

C
Cao, Gaotao I-1
Chen, Fuwang I-36, I-518
Chen, Guo I-396
Chen, Huixing I-448
Chen, Jiale I-149
Chen, Jie I-80
Chen, Jing I-117, I-125, I-134, I-141
Chen, Peng I-413
Chen, Ruichao I-379
Chen, Tianmu I-108
Chen, Xiaoqiang I-170, I-255
Chen, Xinghu I-556, I-587
Chen, Yong I-371
Chen, Zhe I-638
Cheng, Long I-364
Cheng, Zhiping I-448
Cui, Xinchen I-652

D
Dai, Xuhui I-20
Deng, Jiangming I-628
Diao, Lijun I-198, I-217, I-225, I-287, I-305, I-331, I-339, I-356, I-364, I-652
Dong, Kan I-364
Dong, Xueyan I-313
Du, Huiqing I-188
Du, Pengcheng I-605
Du, Xiaogong I-28

F
Fan, Chenguang I-572
Fang, Zhankai I-396

G
Gao, Jiawei I-141
Gao, Shibin I-11
Gao, Yukang I-422
Gao, Zichen I-297, I-305, I-331
Ge, Qiongxuan I-422
Gou, Jiaxi I-396
Guo, Chen I-439
Guo, Chuanwei I-652
Guo, Xiaoran I-11
Guo, Yongyong I-530, I-548, I-596, I-605

H
Han, Wei I-225
Han, Yajing I-157
Hao, Jiansheng I-613
He, Shan I-28
He, Shuiyuan I-198
He, Xinyu I-322
He, Xizheng I-371
Hou, Chunhua I-371
Hou, Lintao I-371
Hu, Haining I-322
Hu, Jingtai I-506
Huang, Jingyu I-56
Huang, Tingli I-46
Huang, Xianjin I-247
Huang, Yuanfeng I-404

I
Igarshi, Seiki I-247

J
Ji, Xu I-613
Jiang, Jusong I-538, I-660
Jiang, Mingjie I-572
Jiang, Shuai I-80
Jiang, Xin I-498, I-530
Jiao, Deyu I-264
Jiaxi, Gou I-439
Jin, Jie I-157
Jin, Yong I-247
Jin, Zheming I-198, I-339, I-356, I-364

K
Kan, Liu I-339
Kang, Mingming I-572

L. Jia et al. (Eds.): EITRT 2023, LNEE 1135, pp. 679–681, 2024.
https://doi.org/10.1007/978-981-99-9307-9

Kang, Ying I-613

L
Lai, Qiang I-396
Li, Bailin I-448
Li, Dawei I-157
Li, Guosheng I-379, I-387
Li, Jie I-489
Li, Kuanxin I-556, I-587
Li, Liran I-339
Li, Qiang I-322
Li, Qinglai I-605
Li, Tongzhao I-287, I-652
Li, Xiaolong I-431
Li, Xiaona I-28
Li, Xuefei I-225
Li, Xuekun I-564
Li, Yanming I-477
Li, Yaoheng I-331
Li, Yuanzhi I-264
Li, Zhixin I-538
Liang, Haiquan I-506
Liang, Shuai I-178, I-188
Lin, Feng I-157
Lin, Wenli I-652
Liu, Cunguo I-20
Liu, Donghui I-99, I-339
Liu, Jianqiang I-89, I-99
Liu, Jingdou I-149
Liu, Kan I-89, I-99
Liu, Shuang I-396
Liu, Suyao I-217, I-225
Liu, Xinbo I-198
Liu, Yuefeng I-556
Liu, Yuming I-178, I-188
Liu, Yuqi I-404
Liu, Zhigang I-80
Long, Zhiqiang I-431
Lu, Shaofeng I-36, I-518
Lu, Xiaoxi I-207
Luo, Jiaming I-11
Luo, Kangli I-287
Lv, Qing I-68

M
Ma, Chi I-339
Ma, Ruiqi I-198
Ma, Ruiqing I-413
Ma, Yichao I-347
Ma, Yingtao I-198, I-356, I-364
Mai, Ruikun I-463
Mei, Lei I-456
Mei, Weiyao I-287, I-652
Meng, Chuanshu I-538, I-660
Meng, Fanyang I-89, I-99
Miao, Chunhui I-178
Miao, Yifei I-313
Mu, Xiuqing I-255

P
Peng, Xinni I-506
Peng, Yang I-36, I-518
Peng, Yuhao I-28
Pu, Weijie I-247

Q
Qiao, Fengpu I-477

R
Ren, Wanrui I-20

S
Shi, Liming I-587
Shi, Xiaolei I-264
Shuang, Liu I-439
Song, Chenyu I-170
Song, Keling I-68
Su, Di I-379, I-387
Su, Xuan I-322
Sun, Ke I-178
Sun, Qingwen I-322
Sun, Xiran I-188

T
Tang, Fen I-108
Tang, Yirong I-628
Tian, Di I-234

W
Wang, Baomin I-46
Wang, Chao I-356, I-364
Wang, Chengtao I-225
Wang, Guodong I-669
Wang, Haifeng I-404
Wang, Haiting I-331
Wang, Haoyun I-463
Wang, Jing I-149
Wang, Lei I-234, I-305, I-347

Wang, Qingyong I-46
Wang, Xiaonong I-56
Wang, Xiaoxu I-572, I-638
Wang, Yanchao I-264
Wang, Yilin I-207, I-276
Wang, Ying I-170, I-255
Wang, Yonggang I-556, I-587
Wang, Yongxiang I-89, I-99
Wang, Zhiyuan I-477
Wang, Zhonghai I-564
Wang, Zhuo I-117, I-125
Wei, Kekang I-498, I-530, I-548, I-596, I-605, I-613
Wei, Rong I-422
Wei, Xiaoguang I-11
Wei, Yonggang I-1
Wen, Tao I-371
Wu, Jianfan I-157
Wu, Junteng I-660
Wu, Luoqi I-108
Wu, Qiya I-297, I-305, I-331
Wu, Wei I-117, I-125

X
Xi, He I-264
Xiao, Han I-628
Xie, Qubo I-347
Xie, Zheng I-322
Xiong, Bin I-404
Xu, Chengbo I-134
Xu, Chunmei I-217, I-297, I-305
Xu, Lu I-477
Xu, Wei I-628
Xue, Zhigang I-422

Y
Yang, Congding I-134, I-141
Yang, Haotian I-68
Yang, Huchuan I-117, I-125
Yang, Jianyu I-207
Yang, Jie I-538, I-660
Yang, Mu I-141
Yang, Qiaoli I-255
Yang, Yanbiao I-489
Yang, Zhihong I-548, I-596
Yang, Zhongping I-247
Yao, Bo I-572
Yao, Shuncai I-448
Yi, Pinglang I-431
Yin, Hang I-564
Yin, Shaobo I-498, I-530
Yin, Zhenggang I-587
Yu, Yichen I-489
Yu, Yong I-456
Yuan, Jiangjun I-287
Yue, Xuelei I-234
Yue, Zheng I-339
Yunbiao, Huang I-439

Z
Zeng, Guohong I-28, I-108, I-149
Zhai, Mingda I-431
Zhan, Yanwen I-489
Zhang, Bo I-322
Zhang, Chi I-638
Zhang, Dan I-669
Zhang, Jia I-297
Zhang, Jiahui I-669
Zhang, Jiepin I-198
Zhang, Junjie I-80
Zhang, Meng I-117, I-125
Zhang, Shanhui I-305
Zhang, Shuo I-20
Zhang, Weige I-28
Zhang, Yan I-1
Zhang, Yifei I-217
Zhang, Yunpeng I-170, I-255
Zhankai, Fang I-439
Zhao, Jiahao I-46
Zhao, Junbo I-322
Zhao, Junjie I-134, I-141
Zhao, Leiting I-339
Zhao, Zhiyong I-548
Zheng, Caihui I-564
Zheng, Xinjie I-638
Zheng, Yuhao I-56
Zhong, Tianbao I-11
Zhou, Lujie I-46
Zhu, Benchao I-347
Zhu, Hao I-297, I-305
Zhu, Qinyue I-489
Zhu, Xiongwen I-638
Zhu, Yueyun I-660
Ziqing, Yang I-439